Advances in Chitin/Chitosan Characterization and Applications

Advances in Chitin/Chitosan Characterization and Applications

Special Issue Editors

Marguerite Rinaudo
Francisco M. Goycoolea

MDPI • Basel • Beijing • Wuhan • Barcelona • Belgrade

MDPI

Special Issue Editors

Marguerite Rinaudo
University of Grenoble Alpes
France

Francisco M. Goycoolea
University of Leeds
UK

Editorial Office
MDPI
St. Alban-Anlage 66
4052 Basel, Switzerland

This is a reprint of articles from the Special Issue published online in the open access journal *Polymers* (ISSN 2073-4360) from 2017 to 2018 (available at: https://www.mdpi.com/journal/polymers/special_issues/chitin_chitosan)

For citation purposes, cite each article independently as indicated on the article page online and as indicated below:

LastName, A.A.; LastName, B.B.; LastName, C.C. Article Title. *Journal Name* **Year**, *Article Number*, Page Range.

ISBN 978-3-03897-802-2 (Pbk)
ISBN 978-3-03897-803-9 (PDF)

Contents

About the Special Issue Editors

Marguerite Rinaudo is a Professor at the University of Grenoble Alpes, France. She has been researching polysaccharides since 1960. In 1966, Prof. Rinaudo was awarded her PhD for her work on the fundamental properties of polyelectrolytes. For 12 years, she was the director of the Research Centre on "Macromolecules Végétales" (CERMAV-CNRS, Grenoble, France). Prof. Rinaudo is the author and co-author of over 500 journal papers and book chapters on the following topics: properties and applications of polysaccharides and water-soluble polymers; specific chemical modifications of polysaccharides and production of adaptative materials; hydration of polysaccharides in relation to their chemical structure and their environment; polyelectrolyte complexes; rheology in solution and gel states; polysaccharide–surfactant interactions; decoration and stabilization of liposomes with polyelectrolytes; biomaterials from polysaccharides; applications of polysaccharides in cosmetics, food, and biomedicine (drug release).

Francisco M. Goycoolea received his PhD from Cranfield University, UK. He has worked as a research scientist and scholar at CIAD, Mexico, University of Santiago de Compostela, Spain, and University of Münster, Germany, and has been a Professor in Biopolymers at University of Leeds, UK, since 2016. Prof. Goycoolea has published over 117 research articles, reviews, and book chapters, and serves on the editorial board of several journals of repute. His research interests and expertise focus on physicochemical and biological properties of water-soluble biopolymers (mainly polysaccharides) derived from biomass (e.g., plants, seaweeds, and bacteria) and food waste; development of soft nanostructured bioinspired materials obtained by physical, chemical, or enzymatic methods (e.g., hydrogels, polyelectrolyte complexes, colloidal micro- and nanoparticles, and emulsions); mechanistic aspects of the interaction of these biomaterials with biological systems (e.g., epithelia, mucosa, bacteria, and biofilms); colloidal nanoparticles for drug and gene delivery and for antibacterial targeting; association and release of bioactive compounds relevant to food, pharmacy, and biomedical applications.

Preface to "Advances in Chitin/Chitosan Characterization and Applications"

Chitin is an aminopolysaccharide that is widely abundant in the biosphere. Chitin is obtained at an industrial scale from different sources of biomass and food waste (e.g., crustacean and insect exoskeletons, fungi cell walls, and squid pen). Chitosan derived from chitin is water-soluble in acidic conditions and has been used in numerous applications, especially in the biomedical, pharmaceutical, agricultural, and food industries. One of the major remaining issues that has precluded the wider use of chitosan is the difficulty in obtaining a complete and accurate characterization of samples, including the quantities and distribution of the acetyl groups along the chains, the molecular weight distribution, and the choice of a suitable solvent for dissolution. This knowledge allows for assigning a suitable chitosan to fit a given function. Another important issue is the preparation of well-defined derivatives by chemical methods, allowing for a random distribution of the substituents (or grafted chains) along the chains to afford polymers with reproducible properties for specific applications.

Many products developed from chitin/chitosan research are innovative biomaterials that are comprised of well-defined chitin and chitosan samples, including films, fibers, nanoparticles, composite materials (involving fibers, solid particles, etc.), hydrogels, polymeric complexes, nanoporous scaffolds, and membranes. The wide range of applications encompass scaffold biomaterials (e.g., in tissue engineering or regenerative medicine), non-viral gene delivery systems, nano- and microparticles for biological transmucosal delivery, fruit and vegetable coatings, biosensors, matrices for water purification, and novel antimicrobials. This Special Issue presents an updated series of papers addressing some of these applications, including the chemical and enzymatic modifications of oligos and polymers. A better understanding of the properties that underpin the use of this large family of biopolymers in different fields is key to stimulating more extensive industrial utilization, as well as to aid regulatory agencies in establishing specifications, guidelines, and standards for different types of products and applications.

We are extremely grateful to Polymers and to all the contributing authors for giving us the opportunity to materialize this project.

Marguerite Rinaudo, Francisco M. Goycoolea

Special Issue Editors

Review

Chitosan Derivatives: Introducing New Functionalities with a Controlled Molecular Architecture for Innovative Materials †

Waldo M. Argüelles-Monal [1,*] , Jaime Lizardi-Mendoza [1,*] , Daniel Fernández-Quiroz [2] ,
Maricarmen T. Recillas-Mota [1] and Marcelino Montiel-Herrera [3]

[1] Centro de Investigación en Alimentación y Desarrollo, Hermosillo 83304, Sonora, Mexico; mrecillas@ciad.mx
[2] Departamento de Investigación en Física, Universidad de Sonora, Hermosillo 83000, Sonora, Mexico;
    daniel.fernandez@unison.mx
[3] Departamento de Medicina y Ciencias de la Salud, Universidad de Sonora, Hermosillo 83000, Sonora,
    Mexico; marcelino.montiel@unison.mx
* Correspondence: waldo@ciad.mx (W.M.A.-M.); jalim@ciad.mx (J.L.-M.);
    Tel.: +52-662-289-2400 (W.M.A.-M. & J.L.-M.)
† Dedicated to the memory of the late Prof. Ruth D. Henríques, who introduced me to the research on
    chitin [W.M.A.-M.].

Received: 28 February 2018; Accepted: 17 March 2018; Published: 20 March 2018

Abstract: The functionalization of polymeric substances is of great interest for the development of innovative materials for advanced applications. For many decades, the functionalization of chitosan has been a convenient way to improve its properties with the aim of preparing new materials with specialized characteristics. In the present review, we summarize the latest methods for the modification and derivatization of chitin and chitosan under experimental conditions, which allow a control over the macromolecular architecture. This is because an understanding of the interdependence between chemical structure and properties is an important condition for proposing innovative materials. New advances in methods and strategies of functionalization such as the click chemistry approach, *grafting onto* copolymerization, coupling with cyclodextrins, and reactions in ionic liquids are discussed.

Keywords: chitin; chitosan; derivatization; controlled functionalization; click chemistry; graft copolymer; cyclodextrin; dendrimer; ionic liquids

---

1. Introduction

Polysaccharides are widely found in the biosphere, fulfilling various important functions in living organisms, such as energy storage and structural materials, among others. Cellulose and chitin are the most abundant natural polymers in nature. However, chitin has very few applications compared to cellulose. This is for several reasons, including the scarce natural structures of chitin that are available to be used with low processing and the poor solubility properties of this polysaccharide. Therefore, most of the obtained chitin is processed by extensive alkaline deacetylation to obtain chitosan. This amino-polysaccharide is composed of β(1→4) linked units of *N*-acetyl-D-glucosamine and D-glucosamine (Figure 1). Due to its key properties such as its biodegradability and biocompatibility, and its being mucoadhesive and non-toxic, chitosan is of great interest in many applications such as biomedicine, pharmacy, biotechnology, food industry, nanotechnology, etc. [1–7].

One constant topic in materials research is the modification of natural polymers, which results in the development of new derivatives with unique properties. There is a great variety of methods to modify polysaccharides. Chitosan is prone to chemical modification at free amino groups from the deacetylated units at C-2, and hydroxyl groups at C-3 and C-6 positions [2]. Commonly, the chemical derivatization of chitosan is carried out to improve some specific characteristics, such as solubility, hydrophilic character, gelling properties, and affinity toward bioactive molecules, among others [8].

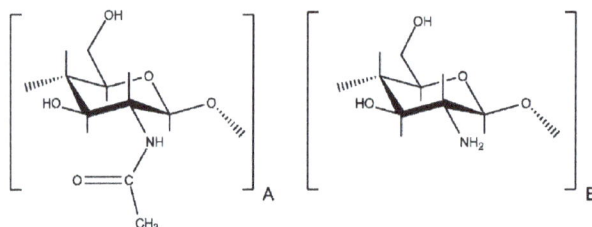

Figure 1. Chemical structure of chitosan composed of β(1→4) linked units of (**A**) *N*-acetyl-D-glucosamine and (**B**) D-glucosamine.

Chemical modification of chitosan is usually done in bulk, giving rise to randomly reacted units. When specific new functionalities are pursued, other approaches are preferred in which the reactions can be controlled stoichiometrically. In this scenario, polymer science is taking advantage of diverse strategies to design new polymer-based hydrogels, drug and gene delivery systems, scaffolds for tissue engineering, toxic substance, mineral chelation, and materials for the electronics and aerospace industry, among others. For example, introducing lipophilic or hydrophilic molecules to chitosan may result in altering or improving its properties, such as solubility in acidic solution or organic solvents, improving thermal and mechanical properties and others [9].

There are previous reviews covering important and specific aspects of the chemical modification of chitin and chitosan [10–16]. In the present paper, we aim to review and analyze recent developments found in literature dealing with the chemical modification of chitin and chitosan, with emphasis on proposed methods to obtain chitosan derivatives with a controlled macromolecular architecture. An understanding of the interdependence between chemical structure and properties is an important condition for proposing innovative materials.

Some aspects of the chemistry of these polysaccharides (and their modification conditions) could have an impact on the properties of the products and should be taken into account:

1. Chitin and chitosan are, in fact, a family of polymers, differing in terms of not only the molecular weight and extent of acetylation but also in the dispersion of the degree of polymerization and the distribution of the acetylated and deacetylated units along the polymer chain. All these parameters will depend mainly on the natural source and isolation processes. Therefore, it is very important to know these intrinsic characteristics, as far as they shall affect the properties of the derivatives.

2. Due to their insolubility in certain solvents (particularly chitin), some chemical reactions are carried out under heterogeneous conditions. This will have a determinant influence on the structure of the obtained derivatives. Using words from Kurita, "reactions under heterogeneous conditions are usually accompanied by problems including poor extents of reaction, difficulty in regioselective substitution, structurally ununiformly products, and partial degradation due to severe reaction conditions" [1]. Nowadays, these drawbacks could be circumvented using some novel solvent systems like ionic liquids. This topic will be revised herein as well.

3. Usually, non-selective chemical derivatization could lead to the development of products with an irregular distribution and uncontrolled growth of the substituent groups in the main chain, or undesired depolymerization of the polysaccharide.

4. Although chitosan presents valuable functional groups for derivatization reactions, often it is necessary to obtain some precursors to facilitate subsequent reactions or, in other cases, to protect the reactive amine in order to favor the chemoselectivity of the modification. Due to their frequent

use in chitosan functionalization processes, we will first refer to those reactions whose use is more or less recurrent under diverse experimental conditions.

1.1. Some Frequent Reactions in Chitosan Chemistry

There is a group of organic reactions, whose use in chitin chemistry is recurrent, due to their experimental simplicity, and because their products can be used as a kind of wildcard during other chemical modification strategies. These reactions provide the researcher with valuable tools for specificity control and the regioselectivity of the functionalization, with minimal possibilities of side reactions or chain degradation. Herein, we will only make a brief summary of them, and the reader can get more details in other excellent reviews and compilations [1,10,17–20].

Among these reactions, the following will be frequently used: (i) *formation of Schiff bases (and reductive amination)*. It refers to the formation of imine products between amine and carbonylic (aldehydes and ketones) groups. This reaction is very simple and takes place under mild conditions. The imine could be easily reduced with a suitable reducing agent like sodium borohydride (or, preferably, sodium cyanoborohydride), producing selectively *N*-alkyl derivatives; (ii) *carbodiimide-mediated amidation*. There is a group of activators for the amidation of amines with carboxylic groups. When using these agents, the amidation is straightforward and usually takes place under mild conditions at room temperature. Most frequently used activators are *N*,*N'*-dicyclohexylcarbodiimide (DCC), 1-ethyl-3-(3-(dimethylamino)propyl) carbodiimide (EDC), and 4-(4,6-dimethoxy-1,3,5-triazin-2-yl)-4-methylmorpholinium chloride (DMTMM). In order to increase yields and decrease side reactions, *N*-hydroxysuccinimide (NHS) is often added.

1.2. Chitosan Precursors with Protected Amino Groups

Due to the high reactivity of the amino groups, these should be protected in order to encourage the functionalization reaction to take place through the hydroxyl groups. Several methods have been proposed, but until now the most frequent has been the *N*-phthaloylation of chitosan [21–23].

With this purpose, typically, amino groups of chitosan could be protected from unwanted reactions by means of phthalic anhydride, whose derivative, *N*-phthaloyl chitosan, protects the amine moieties for further chemical modifications. *N*-phthaloylation should be carried out in DMF/water (95/5) in order to avoid *O*-phthaloyl substitution [22]. At the end of the chemical modification, the phthaloyl protection must be removed from the polymer by reaction with hydrazine monohydrate to regenerate the free amino groups. The *N*-phthaloylation is considered one of the best ways to protect the amine moeties of chitosan; however, this strategy has two drawbacks:

- the *N*-phthaloylation of chitosan affects the solubility of chitosan in aqueous solutions. It is only soluble in aprotic polar organic solvents, which have been attributed to an increase in the crystallinity of the derivative as compared with the pristine chitosan [22,23]. Obviously, the solubility of the precursor in some organic solvents could be advantageous when the *O*-substitution reaction needs to be carried out in the latter.
- the unblocking reaction with hydrazine monohydrate severely depolymerizes chitosan chain, with the consequent weakening of its mechanical properties [24–26].

Another approach for protecting amines is the formation of Schiff bases with aldehydes. It has been found that *N*-aryl aldehydes are the best option, because their Schiff bases seem to disrupt interchain packing, making hydroxyl groups more accessible for reactions [17,27]. Among them, *N*-salicylidene, *N*-*m*-toluylidene, and *N*-benzylidene chitosan usually give better results [27–32]. The deprotection is easily carried out at acidic pH values, giving *O*-substituted chitosan derivatives with almost no chain degradation. Nevertheless, under certain conditions, the protection of the amine is far from complete, which is the main disadvantage of this strategy [33].

Dissolving chitosan in methanesulfonic acid has also been used for the protection of amino groups [34–37]. As it should be expected, there is an important degradation of chitosan polymer chain due to the strong acidic conditions used to dissolve chitosan [34,35].

2. Click Chemistry Reactions

Among the different approaches developed to produce new chitin and chitosan derivatives, the polymer scientist should take into account the way to reduce cost, time consumption during the experiment, undesired byproducts of reactions, and reduce the possible pollution to a minimum. Click chemistry is an expression coined by Sharpless for "a set of near-perfect" reactions [38], in which two functional groups exclusively react with each other. They are quick reactions and exhibit high yields; they are carried out under mild temperature conditions (25–37 °C), in a wide range of hydrogen potential (pH 4–12); they are insensitive to water and oxygen. They must generate highly regioselective products that need no complicated purification processes. An important characteristic is that they are modular reactions resembling biochemical processes in nature. Among these reactions are the following [38,39]:

- cycloaddition reactions, including those from the 1,3-dipolar family (like Huisgen reaction), and hetero-Diels-Alder reactions,
- nucleophilic ring-opening reactions in strained heterocyclic electrophiles,
- carbonyl chemistry of the non-aldol type, and
- additions to carbon-carbon multiple bonds.

Among the reactions that are considered as click chemistry, cycloadditions are the most used reactions in chitosan derivatization. On the one hand, the Huisgen's reaction is a cycloaddition between alkynes and azides yielding two regioisomer triazoles [40]. It could be carried out with or without Cu(I) as a catalyst, as could be appreciated in Figure 2. This is one of the most investigated chemoselective "click" reactions, which takes place in aqueous medium at room temperature, and is almost instantaneous. On the other hand, the Diels-Alder is a [4 + 2] cycloaddition between a diene and a dienophile, producing products with an unsaturated six-membered ring (Figure 3). This is also an important click reaction, with the peculiarity that it can be thermodynamically reversible, depending on the reactants, but its reaction product is absolutely stable [41–43].

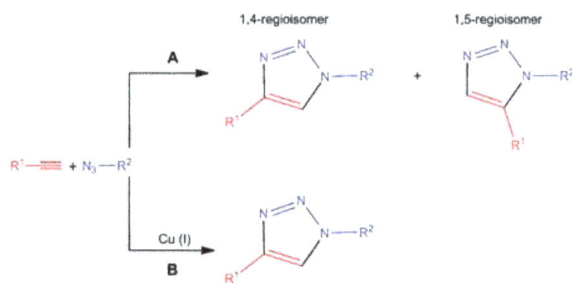

Figure 2. Huisgen cycloaddition reactions in the absence (**A**) or presence (**B**) of Cu(I) catalyst.

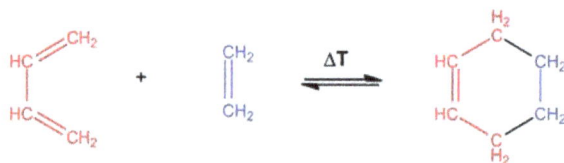

Figure 3. Diels-Alder reaction between a diene and a dienophile.

The use of "click chemistry" has expanded the possibilities of producing new materials with outstanding properties [44–48]. Chitosan derivatives synthesized by click chemistry have shown tunable thermosensitive characteristics, photochromic behaviors, pH-sensitivity macromolecular networks, and highly soluble chemoselective properties [49–52].

The main application of click chemistry on chitosan derivatization seems to be in the preparation of grafting copolymers. Due to the chemoselectivity of these reactions, it was possible to obtain *N*- [53,54] or *O*-grafted [55,56] chitosan-*g*-poly(ethylene glycol). Other homopolymers grafted onto the chitosan backbone are poly(*N*-isopropyl acrylamide) [57,58], β-cyclodextrin (on O-6 position [31] or in the amine [59]), poly(caprolactone) [60,61], and others [62,63].

An interesting example of what could be prepared with this powerful tool is the study of Jung et al. in which chitosan-poly(ethylene glycol) hybrid hydrogel microparticles were prepared and then conjugated with single-stranded DNAs via Cu-free click chemistry [64]. The authors consider this strategy to be an example of a robust biomolecular assembly platform that could be replicated as a biomolecular target and in therapeutic applications.

Furthermore, there are other chitosan derivatives developed via click chemistry reactions [32,50,65–74], some of which exhibit diverse properties like antimicrobial, antifungal, enhanced solubility in acidic and basic conditions, and an antigen detection system initiated by click chemistry, etc. Other materials synthesized are a cellulose-click-chitosan material [75], click-coupled graphene sheet with chitosan [76], and chitosan functionalized multiwalled carbon nanotubes [77].

The use of the Diels-Alder cycloaddition gives an advantage to design materials that can vary their physicochemical properties with temperature, a characteristic that Huisgen's cycloaddition does not possess. In this sense, it could be very relevant to combine the properties of chitosan with the potential capacity of furans to carry out Diels-Alder reactions. The structure of the furan gives it a markedly dienic character, which is very suitable for the development of this type of reaction [78]. This feature opens up opportunities to investigate the potentialities of the Diels-Alder cycloaddition between furan-chitosan derivatives and dienophiles such as maleimides. This approach has been used to obtain a novel chitosan hydrogel with interesting drug-carrier characteristics that are suitable for the development of novel biotechnological and biomedical materials (Figure 4) [79].

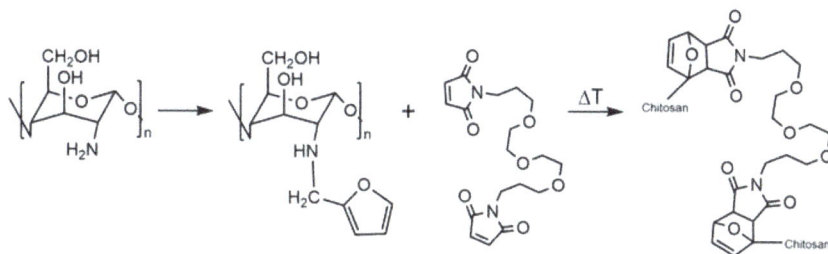

Figure 4. Synthetic scheme for the preparation of *N*-(furfural) chitosan by Schiff base formation/reductive amination process and Diels–Alder cycloaddition with a bismaleimide giving chitosan hydrogel.

The number of reports about the use of click chemistry to modify chitosan is growing fast. However, the application of click chemistry to the generation of regioselectively controlled chitosan derivatives is only in its early stages, and we should expect its use to have an even greater momentum in the coming years. This is undoubtedly due to the simplicity of the reactions and especially to the excellent opportunities offered by these tools to introduce chemical modifications with a high control of the molecular architecture.

3. Graft Copolymerization

Among the strategies available to produce chitin and chitosan derivatives, grafting procedures are a strong chemical tool that is used to develop innovative materials [11]. The structure of a typical *graft*-copolymer consists of a long sequence of the backbone polymer chain (chitin or chitosan in this case), containing one or more side polymer chain of distinct chemical nature [80]. The properties of this kind of copolymer are widely dependent on the molecular characteristics of the grafted side chains, such as molecular structure, length of the chain, and the degree of grafting [81].

There are three main techniques for the grafting copolymerization: *grafting from*, *grafting onto*, and *grafting through*. As far as in this case the polysaccharide backbone chain is already formed, only the two former methods are of interest. On the one hand, the *grafting from* method involves the in situ polymerization of the grafting monomer (Figure 5a). This reaction is initiated directly from the main chain, but its free homopolymerization could not be discarded as well. This procedure is usually accomplished by one-step, but no control over the macromolecular structure is possible. On the other hand, the *grafting onto* method is carried through the reaction between pendant functional groups of the backbone chain and end-functional groups of previously synthesized polymer chains (Figure 5b) [80]. This procedure allows the elaboration of polymer systems with a well-defined structure. This technique affords the preparation of versatile macromolecular materials from chitin and chitosan, allowing the development of novel hybrid materials with specific properties for advanced applications in several fields as food processing, biotechnology, water treatment, biomedicine, among others.

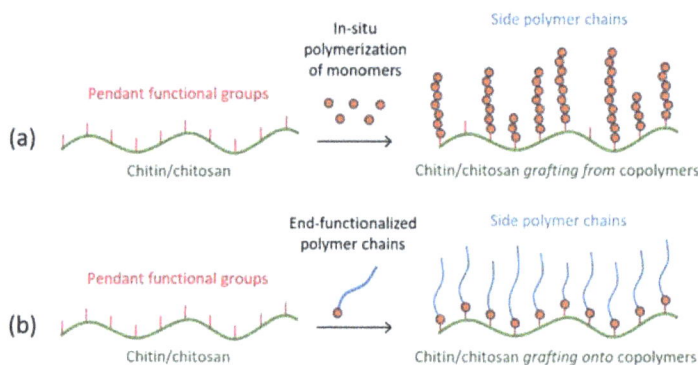

Figure 5. Schematic representation of the (**a**) *grafting from* and (**b**) *grafting onto* methods for graft copolymerization.

3.1. Chitin "Grafting from" Copolymers

The type of polymerization to be selected depends on the type of monomer to be grafted; in most cases, radical polymerization has been used [82–89], although there is also a report of anionic ring-opening polymerization [90]. Acrylic monomers (especially acrylic acid) are among the most frequently grafted into chitin [82–85,88,91–93]. For the development of experimental procedures, it must be taken into account that chitin is not soluble in aqueous media and, therefore, the reaction must be carried out mostly under heterogeneous conditions. Hence, almost no-control over the macromolecular structure is attained, giving rise to a heterogeneous distribution of the grafting chains along the chitin backbone, and, in some cases, only low degrees of grafting could be reached.

The *grafting from* copolymerization of acrylic monomers onto chitin using cerium (IV) as redox initiator has been the subject of some studies [82–84,93]. In the pioneering work by Kurita et al., the influence of several conditions of the copolymerization reaction of acrylamide and acrylic acid onto chitin was investigated [82]. These authors reported a procedure that allows to reach the percentages of grafts above 200%. The obtained copolymers showed enhanced solubility and hygroscopic behavior [82]. Methyl acrylate [83] and methyl methacrylate [84] are other acrylic monomers that have been grafted on chitin in the past under similar conditions.

The other free radical initiator that has been successfully used for the grafting of acrylic monomers onto chitin is potassium persulfate [85,88,91,92]. Hydrogels prepared with chitin-*g*-poly(acrylic acid) by the *grafting from* method have been proposed as a wound dressing material [85,91,92]. The highly water-absorbable film showed a good capacity of absorbing exudates from wounds, thus keeping a moist wound environment [85]. Subsequently, it was shown that the inclusion of glycidyltrimethylammonium chloride improves the wound healing properties of this hydrogel [91,92].

Acrylic acid was also grafted on chitin nanofibers by the *grafting from* method using potassium persulfate. This material showed a stable dispersion in aqueous media at alkaline pH due to the stabilizing effect of electrostatic repulsions between nanofibers [88].

Chitin-*g*-polystyrene copolymer has also been prepared by the *grafting from* method using ammonium persulfate. The effect of some experimental parameters was evaluated, and the resulted material was a copolymer grafted at the C-6 position of chitin backbone [89].

Recently, polypyrrole, an electrically conducting polymer, was grafted on chitin to enhance its mechanical properties. The copolymerization reaction was carried out by the *grafting from* method using ammonium persulfate. The crystallinity of the graft copolymers decreased as a function of the increment of grafting percentage [94]. Itaconic acid, indole, and ε-caprolactone are also other examples of monomers that have been grafted into chitin by the *grafting from* procedure [86,87,90].

3.2. Chitosan "Grafting from" Copolymers

Unlike chitin, the copolymerization reaction of chitosan could be accomplished by the *grafting from* procedure under homogeneous conditions in aqueous media. To some degree, it allows having more control over the macromolecular structure of the obtained copolymer as compared with chitin.

In this case, a greater variety of monomers have been grafted to chitosan via the *grafted from* procedure, for example, acrylic monomers [95–113], styrene [105], oligoethylene glycol methacrylate [114], *N*-vinyl-2-pyrrolidone [115,116], ε-caprolactone [36,37,116–118], lactide [119], urethane [120], indole [87], and aniline [121], among others. The types of polymerization and initiator to be employed depend on the selected monomer.

One of the problems of the *grafting from* procedure is the difficulty of effectively controlling the chemoselectivity of the reaction. To overcome this drawback, the *protection-graft-deprotection* method has been employed [98,116–119]. With this purpose, chitosan amino groups are initially protected by *N*-phthaloylation [22]. Then, the copolymerization reaction is conducted with *N*-phthaloyl chitosan, and, finally, amino groups are regenerated with hydrazine monohydrate. Thus, the side chains are anchored at the C-3 and C-6 hydroxyl groups of chitosan backbone, while amino groups remain free. The main disadvantage of this technique is that the copolymerization reaction should be carried out in organic solvents.

N-isopropyl acrylamide (NIPAm) is one of the most frequently grafted acrylic monomers on chitosan backbone, perhaps due to its thermosensitive properties and its promising applications for the preparation of advanced materials, especially on the biomedical field including drug delivery systems and tissue engineering [96–102,106–108,113]. Ammonium and potassium persulfate is the preferred radical initiator [97,99,100,102,106–108], but also cerium ammonium nitrate [96,101] and azo compounds [113] have been utilized. In general, the *grafting from* synthesis of poly(NIPAm) copolymers are simple and could be accomplished in one step. A strategy proposed by Chen et al. involves the synthesis via atom-transfer radical polymerization from the bromo isobutyryl-terminated chitosan at the C-6 hydroxyl group [98].

The thermosensitive properties of PNIPAm are governed by the variation of hydrophilic and hydrophobic interactions by increasing the temperature. At low temperatures, water molecules form regular ice-like structures around hydrophobic methyl groups. When the temperature increases, that hydrophobic hydration collapses. As a result, hydrophobic interactions are generated between the methyl groups of different segments of PNIPAm chains, giving rise to a polymer network. From a thermodynamic point of view, this phase transition should generate a loss of conformational entropy, due to the ordering of the polymer in the network, which must be compensated for by the translational entropy gain of the ejected water molecules. Therefore, as a result of the phase transition, there is an increase in the total entropy, which is greater than the enthalpy gain (the transition is endothermic), all of which results in a decrease in Gibbs free energy [13]. That is, the phase transition depends on the size and closeness of the grafted PNIPAm chain segments that are involved in the transition. Therefore, an adequate control of the molecular architecture allows one to effectively modulate the properties of the materials and their response to changes in temperature [13,96,122].

The rheological response of the solutions of chitosan-*g*-PNIPAm to changes in temperature is completely reversible. It has been postulated that the increase in the elastic response is due to the formation of hydrophobic crosslinked points at the expense of the amount of sol fraction; for that reason, "the connectivity in the gel network is governed by the net number of formed enthalpic-hydrophobic driven-junctions" [96]. The fast thermoreversible response exhibited by these copolymers could be associated with this phenomenon.

Polyelectrolyte complex membranes formed between chitosan-*g*-PNIPAm and pectin exhibit temperature and pH dual-stimuli responses. Figure 6 shows the release of a model substance as a function of pH and temperature [123], and it can be appreciated how this type of material can respond simultaneously to both parameters.

More information about the structure, properties, and potential applications of chitosan-*g*-PNIPAm copolymers could be found in other specific reviews [13,124].

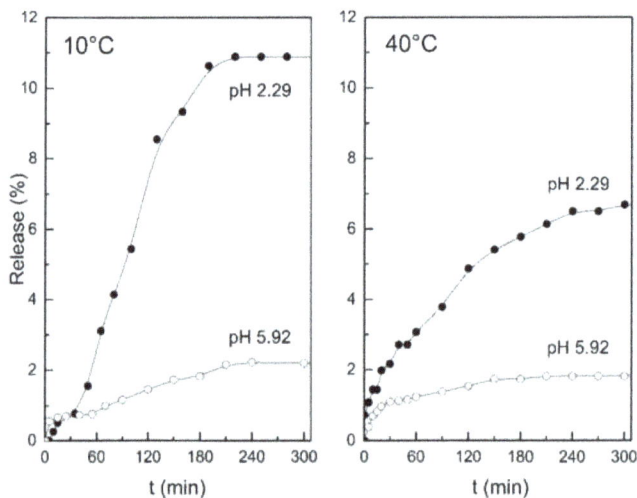

Figure 6. Release profile of Coomassie Blue dye from polyelectrolyte complex membranes formed between this chitosan-*g*-PNIPAm and pectin as a function of pH and temperature. Reprinted from [123], Copyright 2011, with permission from Elsevier.

ε-caprolactone is the other monomer also often grafted onto chitosan. Poly(ε-caprolactone) is a hydrophobic, biodegradable, and biocompatible polymer with excellent mechanical properties. Therefore, the search for hybrid, chitosan-based materials exhibiting the properties of both polymers is an advantageous strategy. Because one of the properties of chitosan that is important to take advantage of is its hydrophilicity and solubility in acidic aqueous solutions, the *protect-graft-deprotect* strategy has been the preferred method [116–118]. In this case, different synthetic approaches have been tested. The typical amino group protection by *N*-phthaloylation has been followed in most of the cases [116–118]. In these cases, tin octanoate was selected as a catalyst [116,117], but it has been also shown that *N*-phthaloyl chitosan is also by itself a catalyst for the ring-opening polymerization of caprolactone monomers and hydroxyl groups acting as initiators [118]. Moreover, there are other reports in which methanesulfonic acid was used as a solvent for chitosan and at the same time served to protect the amino groups, and as a catalyst for the ring-opening reaction [36,37]. This copolymer could be used as an efficient stabilizer of gold nanoparticles [116] and could form amphiphilic copolymer micelles suitable for the drug delivery of hydrophobic anticancer molecules [37].

In conclusion, it can be highlighted that the *grafting from* method has the advantage that chitosan can be functionalized in a fairly easy manner, generally in a single reaction step. If it is required to control the chemoselectivity of the copolymerization and to direct the reaction towards the hydroxyl groups at C-6 and C-3 positions, it is necessary to follow the strategy of protecting amino groups before copolymerization. The main drawback of the *grafting from* method is that there is poor control over the structure of the copolymer, both in terms of the dispersion of the grafted chain length and its distribution throughout the chitosan backbone (Figure 5a).

3.3. Chitosan "Grafting onto" Copolymers

As it is discussed above, the *grafting onto* is the other technique of preparing graft copolymers. Its main advantage is that it is possible to obtain derivatives with better control of the macromolecular architecture, and, therefore, it should be possible to have a greater possibility of modulating the properties and applications of these materials. There are several types of homopolymers that have been grafted onto chitosan, including poly(ε-caprolactone) [125–127], poly(ethylene glycol) and Pluronic [24,53–56,128–141], poly(*N*-isopropyl acrylamide) [57,58,98,142–146], and poly(*N*-vinylcaprolactam) (PVCL) [147–154], among others.

The grafting of end-functionalized poly(ε-caprolactone) has been conducted by the *protect-graft-deprotect* procedure, using EDC condensing agent for carboxylic-terminated poly(caprolactone) (Figure 7) [126,127], or the reaction of isocyanate groups with chitosan hydroxyl groups [125]. It has been reported that the resultant material could be self-assembled into micelles and used as stabilizers to prepare silver nanoparticles with good antimicrobial activity [127].

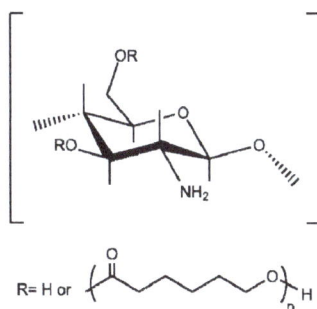

Figure 7. Chemical structure of chitosan-g-poly(ε-caprolactone).

Chitosan grafted with Pluronic, poly(ethylene oxide)-*b*-poly(propylene oxide)-*b*-poly(ethylene oxide), copolymer have also been synthesized by the *grafting onto* method. For this purpose, Pluronic was "activated" with succinic anhydride, and the resulted carboxylated Pluronic was grafted onto chitosan in the presence of EDC/NHS system (Figure 8) [139–141]. This water-soluble thermosensitive copolymer has been evaluated as a potential injectable cell delivery carrier with the aim of using it as a scaffold for cartilage regeneration [140]. Its suitability in the preparation of nanocapsules for drug delivery was also verified [141].

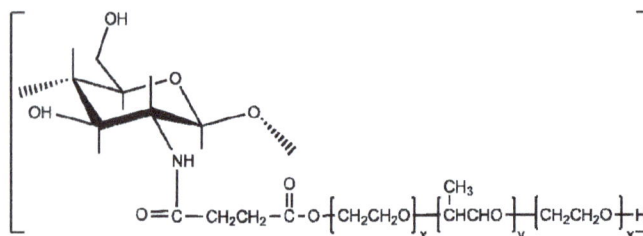

Figure 8. Chemical structure of chitosan-*g*-Pluronic.

The grafting of poly(ethylene glycol) (PEG) onto chitosan backbone has been accomplished via PEGylation of amino groups throughout conjugation with methoxy PEG-nitrophenyl carbonate [131], methoxy PEG-succinimidyl carbonate [133], amidation with carboxylated PEG [134], or reductive amination (Figure 9) [129,130,132,135,136]. The use of "click chemistry" tools has also been reported for the *N*- [53,54] or *O*-PEGylation of chitosan [55,56]. However, the grafting onto the -OH groups at C-6 of chitosan structure is an alternative option of chitosan modification, because it allows the total availability of free amino groups. In this sense, some studies related to *O*-substitution graft copolymers have been developed by etherification reaction [24,128,137,138]. For this purpose, the amino groups of chitosan were protected with phthalic anhydride by the above-mentioned procedure. The resultant material (degree of substitution about 15%) shows solubility in a wide range of pH [128]. PEGylated chitosan has been considered as a bioactive delivery carrier for insulin [138], DNA [131], heparin [133], and albumin [135], among others. A detailed review of the methods of synthesis, characterization, and pharmaceutical applications of PEGylated chitosan derivatives could be consulted [155].

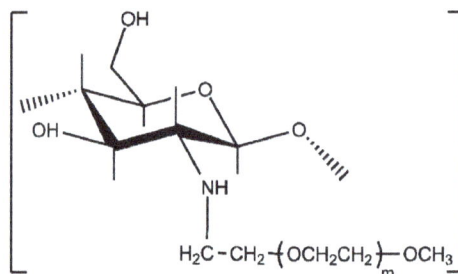

Figure 9. Chemical structure of chitosan-*g*-poly(ethylene glycol).

Chitosan-*g*-PNIPAm copolymer has also been synthesized by the *graft onto* method via the amidation between carboxylic-end PNIPAm chains and chitosan amino groups using carbodiimide compounds like DCC [142], or EDC (Figure 10) [143–145]. Similarly, the same reaction, but between *O*-carboxymethyl chitosan and amino-end PNIPAm chains, has also been proposed [146], having the advantage of leaving the amino groups free. Bao et al. have also made use of "click chemistry" reactions to anchor PNIPAm chains onto chitosan backbone [57,58]. Due to its thermoresponsive behavior, this copolymer forms hydrogels in situ, which favors some properties as enhancement of drug residential time, ocular absorption, pharmacokinetics, and bioavailability of hydrophobic drugs [13,142,145].

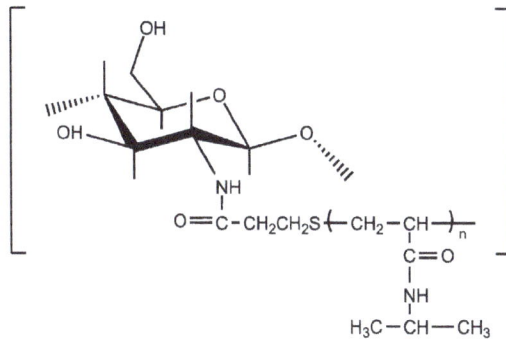

Figure 10. Chemical structure of chitosan-*g*-poly(*N*-isopropyl acrylamide).

The *grafting onto* approach to synthesize chitosan-*graft*-PVCL has been conducted by the amidation between PVCL-COOH and chitosan amino groups using EDC/NHS system [147–150,152,153] or DMTMM (Figure 11) [151,154].

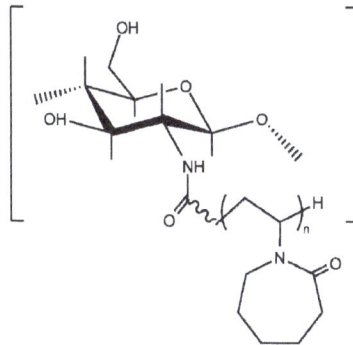

Figure 11. Chemical structure of chitosan-*g*-poly(*N*-vinyl caprolactam).

It has been established that the molecular architecture of this copolymer plays a prominent role in their thermoresponsive properties (LCST within 34–45 °C) [151,154]. Figure 11 shows the dependence of the phase transition on the length of the grafted chain or the closeness between them along the chitosan backbone. The increment of the length of the grafted chains implies that longer hydrophobic segments appear, which favors polymer-polymer long-range interactions and as a result, a lowering of the phase transition temperature takes place (Figure 12a). The spacing between PVCL chains along the chitosan backbone also affects the transition: as they are closer, the lower the cloud point temperature and the greater the enthalpic change (Figure 12b). As the spacing between grafted chains is more reduced, the hydrophobic intercatenary interactions between PVCL segments are favored, giving rise to the above-mentioned behavior [151]. Indulekha et al. reported the study of the chitosan-*g*-PVCL gel as a transdermal drug delivery system for pain management, which showed biocompatibility and drug permeation through in vitro skin test [153]. Jayakumar et al. have studied chitosan-*g*-PVCL-based nanoparticles as a promising candidate for cancer drug delivery [147–150,152].

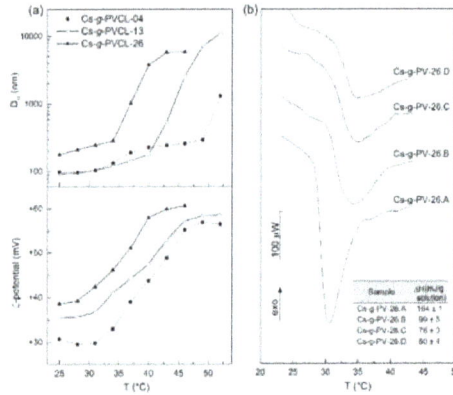

Figure 12. (**a**) Dependence of the hydrodynamic diameter, D$_H$, on temperature of chitosan-*g*-PVCL aqueous solutions (pH 6) for different number-average molecular weights of PVCL-grafted chains (4, 13, and 26 kDa). Reprinted by permission from Springer Nature: [154], Copyright 2015. (**b**) Micro-DSC curve of 10 wt % aqueous solutions (pH 6) of chitosan-*g*-PVCL, varying the spacing between grafted side chains. Reprinted from [151], Copyright 2015, with permission from Elsevier.

Table 1 presents a compendium of the most representative monomers that have been grafted into chitosan, as well as the main applications proposed.

Table 1. Main monomers used for derivatization of chitin and chitosan by grafting copolymerization.

Monomers	Applications	References
Chitin "grafting from" copolymers		
Acrylamide	Water absorbents, chelating agents	[82]
Acrylic acid	Water absorbents, chelating agents. Wound dressing. Nanofibers	[82,85,88]
Methyl methacrylate	Gel-like mass for biomedicine	[84]
Itaconic acid	Waste-water treatment	[86]
Indole	Antimicrobial activity	[87]
ε-caprolactone	Biomedical field	[90]
Glycidytrimethylammonium chloride	Wound healing	[91]
Pyrrole	Electrically-conducting material	[94]
Chitosan "grafting from" copolymers		
Acrylic acid	Controlled release devices, ion-exchange bioseparation, antibacterial activity, removal of heavy metal ions	[95,103,104,111]
N-butyl acrylate	Biodegradable packaging materials, recovery of heavy metals from waste waters	[105,109]
Iodine	Cervical antibacterial biomembrane	[110]
acrylamide-co-acrylic acid	Drug release hydrogels	[112]
Styrene	Recovery of heavy metals from waste waters	[105]
Aniline	Antibacterial activity	[121]
PNIPAm	Biomedical field: tissue engineering, drug delivery systems.	[96–102,106–108,113,123,145]
Lactide	Gene delivery, complex with DNA	[119]
ε-caprolactone	Nanoparticle stabilizer, drug delivery systems	[37,116–118]
N-vinyl-2-pyrrolidone	Antimicrobial activity, nanoparticle stabilizer	[115,116]
Carbamate (urethane)	Drug delivery systems	[120]
Indole	Antimicrobial activity	[87]
Chitosan "grafting onto" copolymers		
Pluronic	Injectable cell delivery carrier, gene expression, controlled release	[140,141]
ε-caprolactone	Drug carriers, antimicrobial activity	[123–127]
Ethylene glycol	Bioactive molecules delivery, polymeric surfactants, gene delivery, apoptosis-inducing activity.	[128,131,133,135,137,138]
PNIPAm	Drug/gene delivery,	[57,58,98,142–146]
PVCL	Controlled drug delivery systems	[147–154]

3.4. Chitosan Network Systems Prepared by Radiation

Ionizing radiation constitutes an environmentally friendly tool for preparing graft copolymers from chitin and chitosan. Fundamentally, UV- and γ-radiation have been used for the preparation of chitosan derivatives. UV-initiated polymerization has some benefits, such as lower reaction temperature, fewer amounts of initiator, higher reaction rate, and shorter polymerization times, among others. The principal disadvantage of this method of modification is the absence of specificity. Usually, the resultant radiation-based graft copolymers tend to exhibit a crosslinked network structure.

On the one hand, chitosan-based graft copolymers have received special attention for applications as flocculants due to their biodegradability, absorption, and charge neutralization ability, among others. Some of those materials are based on acrylic monomers [156–160] and display important flocculation properties. There are also reports of radiation-induced chitosan grafted with poly(maleic acid) showing high sorption capacity of Co(II) [161,162]. On the other hand, Burillo et al. have developed thermosensitive graft copolymers based on chitosan derivatives by gamma radiation [163–166].

4. Chitosan-*grafted*-Cyclodextrin Derivatives

Supramolecular polymer chemistry has gained interest in macromolecular research. A number of molecular architectures have been introduced to develop new materials, in which cyclodextrins (CDs) have been extensively used. CDs are non-toxic cyclic oligosaccharides, formed by 6 to 9 α-D-glucose units linked by (1→4) glycosidic bonds (Figure 13). They possess a truncated cone-shape geometry, with a hydrophilic external surface and a relatively more hydrophobic internal cavity [167]. This arrangement favors host-guest interactions through the inclusion of a wide variety of small organic molecules (which aremainly hydrophobic) such as adamantane, eugenol, doxorubicin, etc. (Figure 14) [168–170]. This important property makes CDs an effective molecular carrier during the design of advanced drug delivery systems. According to Rekharsky and Inoue, the general tendencies of the dependence on thermodynamic quantities can be understood in terms of hydrophobicity, steric effects during the guest-host interaction, the involved guest-host hydrogen bonding, and the flexibility of the guest molecule [171]. Thermodynamic studies about the stability of the inclusion complex demonstrated that enthalpy gain due to the guest inclusion is compensated for by the loss of entropy that results from the considerable conformational changes that take place in the CDs during the complexation and the entropy gain due to the desolvation of both host and guest [171,172].

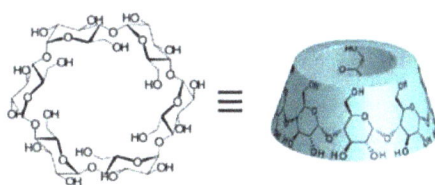

Figure 13. Structure of α-cyclodextrin (formed by six glucosidic units). The arrangement of the external hydrophilic surface and the relatively hydrophobic internal cavity is evident. Reproduced from [173] with permission of The Royal Society of Chemistry.

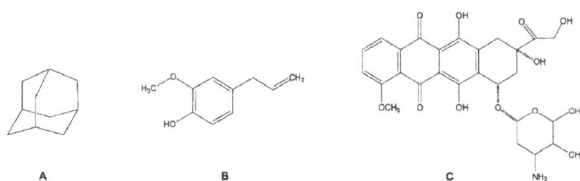

Figure 14. Chemical structure of (**A**) adamantane, (**B**) eugenol, and (**C**) doxorubicin.

Many investigations have been carried out with the aim of proposing methods to prepare chitosan grafted CDs derivatives in order to take advantage of both the mucoadhesive properties and reactive functional groups of chitosan and the ability of CDs to interact with hydrophobic guest substances [174–176]. In the presence of a guest molecule, chitosan-*g*-CDs solutions could form intramolecular and intermolecular complexes, which can lead to a large increase in the viscosity or to the formation of temporary and reversible supramolecular network systems. Consequently, an adequate control of the grafting reaction is of utmost importance for the regulation of molecular architecture, and therefore the behavior and properties of the polymer materials. For this purpose, the methods available for chemical modification of chitosan can also be used to graft cyclodextrin. So far, the following main procedures have been proposed:

1. Reductive amination reaction. Usually, the CD is modified in order to attach an aldehyde group. The inclusion of the CD moieties into the chitosan backbone is carried out by the formation of a Schiff base, followed by the reduction with a proper agent. The reductive amination procedure is one of the most used, because it is a simple, easy, and slightly degradative method [169,177–179].

2. The second most important method is via amidation of CDs modified with a carboxylic group with the amino groups of chitosan. In this case, two strategies have been applied: (i) the classic condensation reaction [180,181], and (ii) by amidation using coupling activators of the carboxylic acid group, like EDC/NHS [175,182–186]. The former reaction requires high temperatures due to the high activation energies involved, while the use of condensation agents in the later selectively promotes the formation of the amide bond in aqueous solution under mild and controllable conditions.

3. The nucleophilic substitution of halides or tosyl groups by chitosan amino groups is another recurrent way to attach CDs into the chitosan backbone [168,170,187–192].

4. A method so far little used but which, in the future, can provide derivatives with a high regioselectivity, is anchoring β-cyclodextrin onto chitosan by click chemistry. In this way, using the Huisgen cycloaddition reaction, β-CD chains have been grafted onto the chitosan backbone through the amino group (position 2) [59] or to the O-6 [31].

5. Other methods, among which (i) the preparation of epoxy-activated chitosan and its reaction with hydroxyl groups of CD [193] or (ii) the anchoring of CD into chitosan using 1,6-hexamethylene diisocyanate [194–196], among others, can be mentioned.

In this sense, Auzély-Velty and Rinaudo have reported a procedure in which a monosubstituted β-CD, possessing a D-galacturonic acid group on the primary face of CD, was grafted onto the chitosan backbone by reductive amination reaction [169,178]. The characterization of the graft copolymer confirmed a successful inclusion of CD on the chitosan chain with almost no degradation of the polymer. These authors also observed a slight reduction in the solubility of the derivative (at grafting degrees 10–12%) as compared with that of the pristine chitosan [178]. At a given concentration, the viscosity of the copolymer solution is higher than that the original chitosan, confirming the presence of interchain interactions induced by the presence of grafted CD [169].

The host properties of CD and chitosan-*g*-CD were comparatively studied toward a low or high molecular weight guest. In the former case, 4-tert-butylbenzoic acid and (+)-catechin low molecular weight guests were chosen, and the inclusion complex was analyzed by means of NMR [178]. Experimental data corroborated that the complexation of 4-tert-butylbenzoic acid is a dynamic process, in the sense that the guest molecule is constantly switching between the free and bound states. Moreover, it was possible to conclude that chitosan-*g*-CD exhibits the same host-guest properties as the native CD toward the low molecular weight hydrophobic guest, suggesting that the grafting process does not have a significant influence over the binding capacity of CD [178]. In the second case, the interaction of chitosan-*g*-CD with two macromolecular guests (adamantane attached to chitosan or poly(ethylene glycol)) was evaluated [169]. On the one hand, NMR analyses demonstrated that the hydrophobic sites of the macromolecular guest interact with the grafted CD moieties in

Polymers **2018**, *10*, 342

the same way as with the non-grafted one. On the other hand, rheological experiments showed that PEG end-capped with adamantane mixed with CD-chitosan solutions promote a significant increase in the viscosity due to cross-linking of CD-chitosan chains through host-guest inclusion complexation with PEG-di-adamantane guest. Nevertheless, when the complexation takes place with the chitosan-di-adamantane derivative, a gel-like behavior is appreciated [169]. These characteristics of the inclusion complex with di-adamantane macromolecular derivatives open interesting possibilities to produce advanced materials with controlled sol-gel properties.

One of the drawbacks of chitosan-*g*-CDs as a drug delivery system is the poor solubility of chitosan at neutral pH values. In this context, Sajomsang et al. have proposed the quaternization of chitosan amino groups in order to obtain a water-soluble grafted biopolymer [189,190]. Synthesis strategy involves the quaternization of previously prepared chitosan-*g*-CDs, carried out by the nucleophilic substitution of the remained free amino groups, yielding a water-soluble quaternized chitosan-*g*-CD. The degree of quaternization (DQ) reached values between 60 and 85%. The mucoadhesive properties of the grafted polymer were dependent on the DQ being stronger as the DQ increases, while its cytotoxicity does not show any dependence with the DQ [190]. The formation of an inclusion complex between the quaternized chitosan-*g*-CDs and eugenol as model guest molecule has also been studied. In this case, it was confirmed that eugenol was included in the hydrophobic cavity of CDs, but a self-aggregated micelle-type structure was formed, within which extra eugenol molecules were entrapped as illustrated in Figure 15. The greatest mucoadhesion was attained with the complex having 11% CD substitution, which suggests that in this case, electrostatic interaction has a key role in governing the adhesion between mucin and the chitosan derivative [168]. Moreover, an enhanced mucoadhesion was reported for this system when CDs were attached to the chitosan backbone throughout a citric acid molecule. This effect is possibly due to additional intermolecular hydrogen bonding between the carboxyl and hydroxyl groups from the citric acid spacer and mucus glycoprotein [181,197].

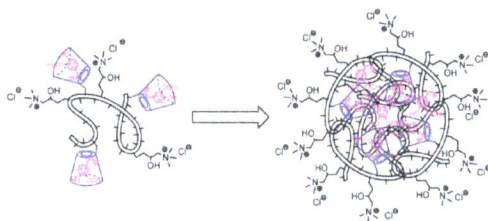

Figure 15. Schematic structure of inclusion complex between eugenol (pink) and quaternized chitosan (black) grafted with β-cyclodextrin (blue) forming self-aggregated micellar structures. Reprinted from reference [168], Copyright 2012, with permission from Elsevier.

Another extensive coupling method used to graft CD into chitosan chain is based on amidation reaction. This reaction occurs among a component containing a free amino group, like chitosan, with a substituted carboxylic acid-cyclodextrin to generate the amide bond. This reaction is mediated by diimide derivatives; among them, EDC is the most used due to its water solubility. Daimon et al. described the preparation of a chitosan-*g*-CDs by the condensation reaction of chitosan and β-CD-carboxylate [175,176]. The interaction between chitosan-*g*-CDs and insulin was evaluated. Insulin was strongly bound to β-CD residues due to the specific host-guest inclusion complex with insulin. The electrostatic interactions between chitosan-*g*-CDs and insulin allowed a strong binding in a wide range of pH [175]. The conclusion of several studies is that chitosan-*g*-CDs have the remarkable potential to be applied in the delivery of peptides and proteins as an efficient delivery carrier [175,176,186].

Kono et al. described the preparation of a hydrogel based on carboxymethyl chitosan and carboxymethyl CD. A reductive amidation reaction was conducted employing EDC-NHS as the coupling agent. It allowed simultaneous grafting of CD onto chitosan and crosslinking. Acetylsalicylic acid was chosen as a model drug to explore its properties as a carrier for drug delivery system. According to their results, the observed drug release profile could be attributed to the formation of an inclusion complex of aspirin inside CD cavity [183].

Apart from the aforementioned applications for controlled release systems, other studies that aimed at the use of chitosan/cyclodextrin materials for the removal of metals or organic micropollutants from wastewaters have been described as well [198,199]. For example, Zhao et al. prepared chitosan- β-cyclodextrin absorbent material using EDTA as cross-linker. According to these authors, "chitosan chain is considered as the backbone, and the immobilized cyclodextrin cavities capture the organic compounds via host-guest inclusion complexation, while EDTA-groups complex metals" [199]. A β-cyclodextrin-chitosan-graphene oxide composite material has also been proposed. It is claimed that this material is appropriate for the removal of manganese ions [198].

Finally, it should be noted that there is an increasing number of publications in which chitosan and cyclodextrin are used as important components in the preparation of nano-vehicles or stimuli-sensitive carriers [170,185,191,200–203].

5. Dendronized Chitosan

Dendrimers are commonly represented as highly symmetrical molecules, displayed in tiers with an algorithmic growth. They are characterized by high-end functional groups located on the surface of a spherical conformation, which lead a molecule that owes a large number of functional sites that are easily accessible. This typical architecture influences the physical properties, like solution behavior, especially at high molecular weights. In dendrimer construction, two synthetic approaches have been employed: divergent and convergent. In the former, stepwise growing occurs from the center by means of a series of highly selective reactions over a single molecule, whereas in the latter, the synthesis begins in the periphery and ends in the core. Despite the important biomedical applications of dendrimers as viral and pathogenic cell adhesion inhibitors, references about dendronized chitosan derivatives are still scarce [204]. Here is a general brief description of these novel chitosan derivatives.

Some of the first reports of the preparation and characterization of chitosan dendrimers are those presented by Sashiwa et al. [205–208]. They reported the preparation of sialic acid-bound dendronized chitosan using gallic acid as the focal point and tri(ethylene glycol) as spacer arm. It was suggested to be a non-toxic alternative and an inhibitor of hemagglutination of influenza viruses. Sashiwa and Aiba have proposed two main strategies for the synthesis of chitosan dendrimers (Figure 16): in *Method A*, dendrimers bearing an aldehyde group are previously synthesized, and then reacted with chitosan amino groups by reductive *N*-alkylation; while *Method B* is based on the binding of chitosan to the dendrimer surface, allowing the use of available amino-dendrimers [12].

A bioabsorbent for heavy metals, composed of different generations of poly(amido amine) (PAMAM) dendrimers, was achieved by divergent approach synthesis. The addition of methyl acrylate over amino groups of chitosan powder surface was driven by the Michael addition reaction followed by the amidation of terminal groups with ethylenediamine. Different generations of PAMAM were obtained by the subsequent propagation of PAMAM. Results indicate that materials with higher generations of dendrimer exhibit greater Pb^{2+} adsorption capacity [209].

The preparation of a water-soluble *O*-carboxymethyl *N*-[(2-hydroxy-3-trimethylammonium) propyl] chitosan (CM-HTCC)/PAMAM dendrimer has also been described [210,211]. These nanoparticles are composed of a PAMAM core dendrimer and an outer CM-HTCC shell attached to it (core-shell nanoparticles), as can be appreciated in Figure 17. The synthesis of this dendronized chitosan involves a two-step reaction: the activation of carboxylic groups in CM-HTCC and the subsequent condensation reaction. The obtained chitosan dendrimer could self-aggregate into core-shell nanoparticles due to the combination of hydrophobic and electrostatic interactions and hydrogen bonding. These dendrimer nanoparticles exhibited antibacterial activity against Gramm negative bacteria as *E. coli.*

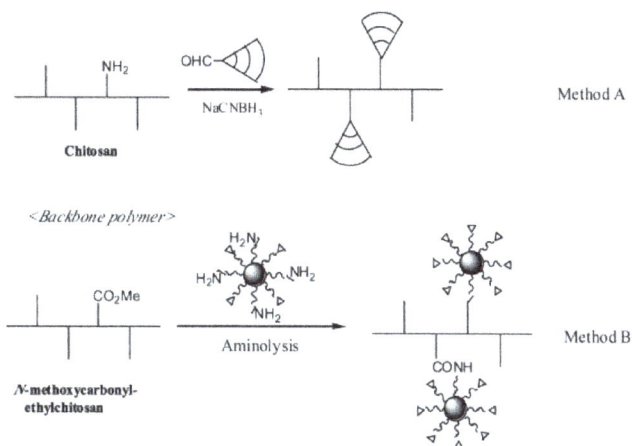

Figure 16. Synthetic strategy on chitosan–dendrimer hybrid. Reprinted from [12], Copyright 2004, with permission from Elsevier.

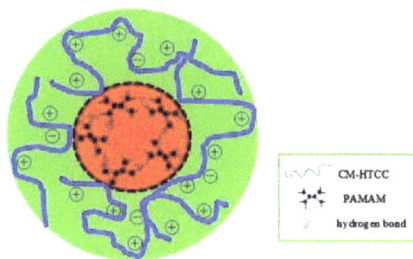

Figure 17. Putative schematic structure of CM-HTCC/PAMAM dendrimer core-shell nanoparticles. Reprinted from [210], Copyright 2012, with permission from Elsevier.

Similar nanostructures were also prepared with magnetite nanoparticles and dendritic branches with carboxymethyl chitosan terminal groups [212]. These dendrimers exhibit selective adsorption for anionic and cationic compounds at specific pH, and their potential use to remove dyes was successfully proved.

6. Chitosan Modification Using Ionic Liquids

The ionic liquids (IL) have become a versatile media in which to perform chitin and chitosan derivatization that was not available few of decades ago. Ionic liquids are salts that remain liquid below 100 °C, in a practical sense, are those salts that should be handled as liquids at room temperature. Most of them are formed by uneven ionic moieties, usually large cations paired with anions of relatively smaller size. The combination and modification of cations and anions make it possible to obtain ionic liquids with diverse chemical characteristics and functional properties. Thus, IL have been praised as customizable solvents, some of them with remarkable properties that have found their way into industrial scale applications. Many IL have been also classified as "green" solvents due to their reduced vapor pressure, conventional non-flammability, and exceptional solvation potential [213,214].

IL's capacity to dissolve polysaccharides was first reported in 1934. However, this does has not received considerable scientific attention, until recently. One of the main focuses of interest has been the capacity of some IL to dissolve typically intractable polysaccharides as cellulose or chitin [215–218]. Imidazolium-based IL, particularly 1-ethyl-3-methylimidazolium (Emim) and 1-butyl-3-methylimidazolium (Bmim) in chloride or acetate form (Figure 18), is commonly used to prepare chitin and chitosan solutions that could reach relatively high concentration (over 10 w%). Other types of IL have been reported to dissolve chitosan to different extents, for example, pyridinium-based IL functionalized with sulfonic acid [219] or amino acid-based IL [220]. The chitosan-IL solutions provide alternative media to get homogeneous reaction conditions and also enable derivatizations that are not favored in aqueous environments. The availability of this type of chitin-chitosan solvent system began to gain relevance in scientific research and applications development.

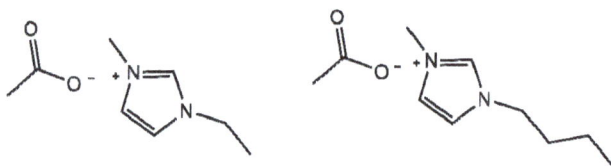

Figure 18. Chemical structure of the acetate salts of 1-ethyl-3-methylimidazolium, Emim, and 1-butyl-3-methylimidazolium, Bmim.

Actually, several types of chemical modifications of chitin-chitosan in IL have been reported. Some of them have been compiled in focused reviews [221,222]. Chitosan has several functional chemical groups that are susceptible to react, which allow the production of a range of derivatives and grafting. Below is a succinct summary of the most relevant chitosan derivation procedures in IL reported in the literature, and some examples of the obtained products are included in Figure 19.

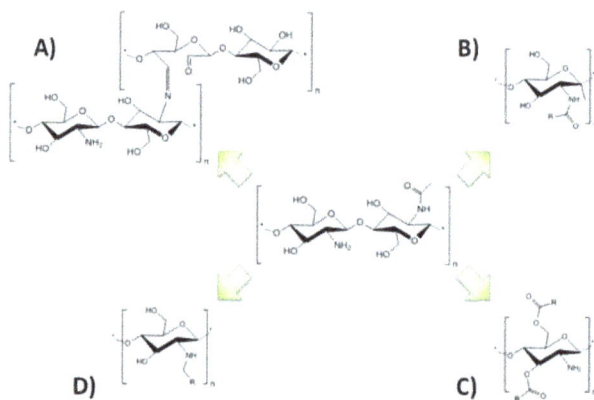

Figure 19. Some examples of chitosan derivatization made in IL. (**A**) Chitosan-graft-oxicellulose, (**B**) *N*-acylation, (**C**) *O*-acylation, and (**D**) Alkylation.

6.1. Acylation

Acetylation was one of the first chemical modification procedures performed on chitin-chitosan dissolved in ionic liquids. Homogeneous acetylation of chitin and chitosan in halide imidazolium-based IL has been reported [223,224]. Based on the degrees of substitution and spectroscopic evidence reported, both *N*-acetylation and *O*-acetylation was achieved indistinctly. With IL, the acetylation of chitosan proceeds in mild and homogeneous conditions, making this methodology more straightforward compared to usual procedures [222]. Other acylation procedures have been reported. The IL Bmim acetate (BmimAc) was used as the reaction solvent to obtain *N*-linoleyl chitosan oligomers. Narrow-distribution low molecular chitosan was used as starting material that was acylated with linoleic acid using EDC and 4-(dimethylamino) pyridine (DMAP) as catalysts under mild reaction conditions. The nanomicelles of the obtained amphiphilic molecules are proposed as drug vector [225]. Similarly, the use of glycine chloride ([Gly]Cl) aqueous solution as media to synthesize *N*-acyl chitosan derivatives (i.e., *N*-maleyl, *N*-succinyl chitosan, and *N*-acetylated) was reported as a procedure for obtaining fibers with improved mechanical properties [226]. Another acylation type modification was achieved by reacting chitosan with monomethyl fumaric acid mediated by EDC. The reaction media was an aqueous solvent system that included 4 w% of the IL, 1-sulfobutyl-3-methylimidazolium trifluoromethanesulfonate (BSmimCF$_3$SO$_3$). The product, monomethyl fumaric-chitosan amide, has improved water solubility and antioxidant activity [227]. Chitosan has been also reacted with a carboxyl group-bearing IL (1-carboxypropyl-3-methyl imidazolium chloride) to obtain an acyl conjugate. Spectroscopic techniques (NMR and FTIR) were used to elucidate the structure of the chitosan-ionic liquid conjugate. This compound shows good anion adsorption performance and was proposed for wastewater treatment [228].

6.2. Alkylation

Several alkylation-type modifications of chitosan have been done using IL as media and catalyst. The nucleophilic substitution of 2,3-epoxypropyltrimethyl ammonium chloride (EPTAC) onto chitosan, using ionic liquid of 1-allyl-3-methylimidazole chloride (AmimCl) as a homogeneous reaction media, produced *N*-[(2-hydroxyl)-propyl-3-trimethyl ammonium] chitosan chloride (HTCC). In this system, the attack of the amino groups of chitosan on the C atom with less steric hindrance in EPTAC is thermodynamically favored, according to quantum chemistry calculations [229]. Chitosan was reacted with four alkyl halides in a basic form of the Bmim IL to prepare a series of alkylated chitosans with different carbon chain substituents (i.e., ethyl-, butyl-, dodecyl-, and cetyl-chitosan). The analysis of FTIR spectra indicates the occurrence of *O*-alkylation; however, the *N*-alkylation prevails at the reaction conditions used. The antibacterial activity of alkylated chitosans decreased with the growth of the DS or the growth of the carbon chain [230]. Another report of *N*-alkylation of chitosan in IL is the production of HTCC in AmimCl [231]. In contrast, there are few examples of *O*-alkylation of chitosan achieved in IL. Dodecanol was selectively linked to hydroxyl groups of chitosan using *N*,*N*'-carbonyldiimidazole as a bonding agent and BmimCl as homogeneous media. The authors attribute the selective alkylation of hydroxyl groups of chitosan, without protecting amino groups, to the particular properties of the ionic liquid solvent [232].

6.3. Grafting

The solvent capacity of several IL has been used to achieve grafting on chitin or chitosan. Chitin graft polystyrene was obtained by atom-transfer radical polymerization (ATRP) in AmimBr [233]. Methacryloyloxyethyl trimethylammonium brushes were formed on chitosan by single electron transfer living radical polymerization in BmimCl [234]. The synthesis of chitosan graft polyethylenimine copolymers was developed in BmimAc [235]. Two different research groups have reported the chitosan grafting with polycaprolactone using IL as a solvent. Wang and collaborators used EmimCl as solvent and stannous octoate as catalyst [236], whereas Yang and co-workers used a ring-opening graft polymerization route with *N*-protected chitosan dissolved in BmimAc [229]

Ionic liquids allow the homogeneous mixture of polysaccharides in solution. This has been used to produce several composite materials. Furthermore, these solvent systems have enabled the possibility of carrying out inter-polysaccharide reactions that have been proven to be difficult to do in other media. Thus, it was possible to produce chitosan graft oxycellulose using a mixture of two

IL, AmimCl as the solvent, and 1-sulfobutyl-3-methylimidazolium hydrogen sulfate (SmimHSO$_4$) IL as the catalyst of the reaction [237]. Another example is the covalent linking of chitosan and xylan through the Maillard reaction in BmimCl [238].

6.4. Other Derivatizations

The crosslinking of chitosan in IL has been explored. Chemical ionogels were obtained crosslinking chitosan with glutaraldehyde in EmimAc [239]. Recently, the design of a dicationic IL (1,10-(butane-1,4-diyl)bis(3-(4-bromobutyl)-1H-imidazole-3-ium)bromide) used as crosslinking agent for chitosan was reported. The composite materials of chitosan crosslinked with IL were tested as catalysts of the cycloaddition reaction of CO$_2$ with various epoxides [240].

Other derivatization reactions of chitosan performed in ionic liquids solutions include the formation of a Schiff base conjugate using BmimCl as solvent [241] and the sulfonation of chitosan in an aqueous solvent system containing [Gly]Cl [242].

6.5. Degradation

A homogeneous reaction media like that obtained using IL represents an opportunity window to test diverse modifications in the chemical structure of chitin and chitosan. One of the basic modifications of these polysaccharides is the deacetylation. This has been achieved by hydrothermal treatment using aqueous BmimAc as reaction medium and catalyst [243]. However, there are more scientific reports on the hydrolysis of chitin and chitosan in IL.

A mixture of BmimCl, BmimBr, and hydrochloric acid was effectively used to depolymerize chitin [244]. Improved reaction rates were reported when chitosan dissolved in AmimCl was treated with sulfonic acid-functionalized ionic liquids based on propylpyridinium and microwave irradiation [219]. An aqueous solution-ionic liquid biphasic catalytic system was proposed for the oxidative degradation of chitosan. Chitosan was dissolved in diluted HCl and the hydrophobic ionic liquid 1-*N*-butyl-3-methylimidazolium bis((trifluoromethyl)sulfonyl) imide ([bmim][Tf$_2$N]) containing with iron(II) phthalocyanine (FePc) complete the oxidative catalytic system [245]. Furthermore, a nitrogen-containing furan derivative has been obtained directly from chitin dissolved in a range of imidazolium-based IL, containing HCl or HBr as additives, after a thermal treatment [246].

6.6. Biocatalyzed Reactions

Ionic liquids have been also used as effective media for biocatalyzed reactions. It is considered that many enzymes, particularly those that tolerate conventional organic solvents, can achieve comparable activities in ionic liquids. Moreover, ionic liquid solvent systems could overcome some limitations that are observed in the biotransformation of highly polar substrates, such as polysaccharides [213]. Consequently, several research groups have studied enzymatic modifications of chitin-chitosan using IL as reaction media or additive. Bacterial and fungal chitinases dispersed in an aqueous solvent system containing EmimAc were applied to produce monomers and oligosaccharides from chitin. A notorious enzymatic activity reduction was observed when IL concentration was over 20 v% [247]. Chitosan oligomers were produced with amylose in a [Gly]BF$_4$ aqueous medium. Similarly, an enzymatic activity reduction was observed when the IL concentration went over 8 v% [248]. On the other hand, commercial lipase was used for the synthesis of chitosan esters via transesterification with methyl palmitate. The reaction media contain a mixture of a hydrophilic IL, EmimAc, and a hydrophobic IL, Bmim tetrafluoroborate [249].

The ionic liquids have become a promising solvent platform for controlled chemical modification of chitosan. There are examples of controlled reactions, even regioselective, derivatization of chitosan using IL as media, additives, or catalysts. The "customization" of IL could provide tunable homogeneous phase media to circumvent the common drawbacks of heterogeneous conditions (i.e., require harsh reaction settings, high variability, low product yields, extended reaction times, etc.) [221,222]. Most of the cited authors in this section remark the "green" solvent condition of IL referred to their low vapor pressure, non-flammability, and thermal and chemical stability. Furthermore, the reuse and recycling of IL has received particular attention. After reaction, the chitosan derivatives are usually recovered by precipitation using miscible non-solvents, e.g., alcohol-water mixtures or organic solvents, depending on the utilized IL and the obtained chitosan derivative. The residual solvents and washing liquids could be distilled to recover the IL. However, the main concerns about the use of IL focus on their biocompatibility and their cost, as they are not readily available yet. The application of IL for polysaccharide processing is relatively recent subject; the possibilities enabled are numerous; thus, considerable research effort is ongoing worldwide.

7. Conclusions

In this review, we summarize and discuss the latest advances in methods and strategies of chitosan functionalization, such as the click chemistry approach, *grafting onto* copolymerization, coupling with cyclodextrins, as well as reactions in ionic liquids. A better understanding of the close relationship between chemical structure and properties is an imperative condition for designing innovative materials. This involves synthesizing substances whose macromolecular architecture is controllable so that the required properties can be optimized and maximized. Throughout this article, several possibilities for combining the distinctive characteristics of chitosan with other interesting molecules, within a context that allows the governance of the architecture of the resulting derivatives, have been demonstrated. The materials produced could be used in exciting technological applications such as sensors, actuators, as a controllable membrane for separations, in the food industry, nanotechnology, and in biomedical and biotechnological fields including drug delivery and tissue engineering. The application of novel techniques of polymer modification with an adequate control of the regioselectivity has allowed important advances in recent years. We should expect that in the near future, new experimental techniques will appear, and materials with properties for specific applications will surely arise.

Conflicts of Interest: The authors declare no conflict of interest.

References

1. Kurita, K. Chitin and Chitosan: Functional Biopolymers from Marine Crustaceans. *Mar. Biotechnol.* **2006**, *8*, 203–226. [CrossRef] [PubMed]
2. Peniche, C.; Argüelles-Monal, W.; Goycoolea, F.M. Chitin and Chitosan: Major Sources, Properties and Applications. In *Monomers, Polymers and Composites from Renewable Resources*; Belgacem, M.N., Gandini, A., Eds.; Elsevier: Amsterdam, The Netherlands, 2008; pp. 517–542. ISBN 978-0-08-045316-3.
3. Younes, I.; Rinaudo, M. Chitin and Chitosan Preparation from Marine Sources. Structure, Properties and Applications. *Mar. Drugs* **2015**, *13*, 1133–1174. [CrossRef] [PubMed]
4. Zargar, V.; Asghari, M.; Dashti, A. A Review on Chitin and Chitosan Polymers: Structure, Chemistry, Solubility, Derivatives, and Applications. *ChemBioEng Rev.* **2015**, *2*, 204–226. [CrossRef]
5. Lizardi-Mendoza, J.; Argüelles Monal, W.M.; Goycoolea Valencia, F.M. Chemical Characteristics and Functional Properties of Chitosan. In *Chitosan in the Preservation of Agricultural Commodities*; Bautista-Baños, S., Romanazzi, G., Jiménez-Aparicio, A., Eds.; Academic Press: San Diego, CA, USA, 2016; pp. 3–31, ISBN 978-0-12-802735-6.
6. Verlee, A.; Mincke, S.; Stevens, C.V. Recent developments in antibacterial and antifungal chitosan and its derivatives. *Carbohydr. Polym.* **2017**, *164*, 268–283. [CrossRef] [PubMed]
7. Rocha, M.A.M.; Coimbra, M.A.; Nunes, C. Applications of chitosan and their derivatives in beverages: A critical review. *Curr. Opin. Food Sci.* **2017**, *15*, 61–69. [CrossRef]
8. Sabaa, M.W. Chitosan-g-Copolymers: Synthesis, Properties, and Applications. In *Polysaccharide Based Graft Copolymers*; Kalia, S., Sabaa, M.W., Eds.; Springer: Berlin/Heidelberg, Germany, 2013; pp. 111–147, ISBN 978-3-642-36565-2.

9. Amato, A.; Migneco, L.M.; Martinelli, A.; Pietrelli, L.; Piozzi, A.; Francolini, I. Antimicrobial activity of catechol functionalized-chitosan versus Staphylococcus epidermidis. *Carbohydr. Polym.* **2018**, *179*, 273–281. [CrossRef] [PubMed]

10. Kurita, K. Controlled functionalization of the polysaccharide chitin. *Prog. Polym. Sci.* **2001**, *26*, 1921–1971. [CrossRef]

11. Jenkins, D.W.; Hudson, S.M. Review of Vinyl Graft Copolymerization Featuring Recent Advances toward Controlled Radical-Based Reactions and Illustrated with Chitin/Chitosan Trunk Polymers. *Chem. Rev.* **2001**, *101*, 3245–3274. [CrossRef] [PubMed]

12. Sashiwa, H.; Aiba, S. Chemically modified chitin and chitosan as biomaterials. *Prog. Polym. Sci.* **2004**, *29*, 887–908. [CrossRef]

13. Argüelles-Monal, W.; Recillas-Mota, M.; Fernández-Quiroz, D. Chitosan-Based Thermosensitive Materials. In *Biological Activities and Application of Marine Polysaccharides*; Shalaby, E.A., Ed.; InTech: Rijeka, Croatia, 2017; pp. 279–302, ISBN 978-953-51-2859-5.

14. Zhang, Y.; Chan, J.W.; Moretti, A.; Uhrich, K.E. Designing polymers with sugar-based advantages for bioactive delivery applications. *J. Control. Release* **2015**, *219*, 355–368. [CrossRef] [PubMed]

15. Sahariah, P.; Másson, M. Antimicrobial Chitosan and Chitosan Derivatives: A Review of the Structure–Activity Relationship. *Biomacromolecules* **2017**, *18*, 3846–3868. [CrossRef] [PubMed]

16. Yu, C.; Kecen, X.; Xiaosai, Q. Grafting Modification of Chitosan. In *Biopolymer Grafting*; Thakur, V.K., Ed.; Elsevier: Amsterdam, The Netherlands, 2018; pp. 295–364. ISBN 978-0-323-48104-5.

17. Roberts, G.A.F. *Chitin Chemistry*; Macmillan: London, UK, 1992; ISBN 978-0-333-52417-6.

18. Uragami, T.; Tokura, S. (Eds.) *Material Science of Chitin and Chitosan*, 2006 ed.; Springer: Tokyo, Japan; Berlin, Germany; New York, NY, USA, 2006; ISBN 978-3-540-32813-1.

19. Kim, S.-K. (Ed.) *Chitin and Chitosan Derivatives: Advances in Drug Discovery and Developments*, 1st ed.; CRC Press: Boca Raton, FL, USA, 2013; ISBN 978-1-4665-6628-6.

20. Carvalho, L.C.R.; Queda, F.; Santos, C.V.A.; Marques, M.M.B. Selective Modification of Chitin and Chitosan: En Route to Tailored Oligosaccharides. *Chem. Asian J.* **2016**, *11*, 3468–3481. [CrossRef] [PubMed]

21. Nishimura, S.; Kohgo, O.; Kurita, K.; Kuzuhara, H. Chemospecific manipulations of a rigid polysaccharide: Syntheses of novel chitosan derivatives with excellent solubility in common organic solvents by regioselective chemical modifications. *Macromolecules* **1991**, *24*, 4745–4748. [CrossRef]

22. Kurita, K.; Ikeda, H.; Yoshida, Y.; Shimojoh, M.; Harata, M. Chemoselective Protection of the Amino Groups of Chitosan by Controlled Phthaloylation: Facile Preparation of a Precursor Useful for Chemical Modifications. *Biomacromolecules* **2002**, *3*, 1–4. [CrossRef] [PubMed]

23. Kurita, K.; Ikeda, H.; Shimojoh, M.; Yang, J. *N*-Phthaloylated Chitosan as an Essential Precursor for Controlled Chemical Modifications of Chitosan: Synthesis and Evaluation. *Polym. J.* **2007**, *39*, 945–952. [CrossRef]

24. Makuška, R.; Gorochovceva, N. Regioselective grafting of poly(ethylene glycol) onto chitosan through C-6 position of glucosamine units. *Carbohydr. Polym.* **2006**, *64*, 319–327. [CrossRef]

25. Hiroyuki, I.; Yoshie, T.; Jin, Y.; Keisuke, K. Phthaloylated Chitosan: Protection-Deprotection and the Influence on the Molecular Weight. *Chitin Chitosan Res.* **2009**, *15*, 7–12.

26. Montiel-Herrera, M.; Gandini, A.; Goycoolea, F.M.; Jacobsen, N.E.; Lizardi-Mendoza, J.; Recillas-Mota, M.T.; Argüelles-Monal, W.M. Furan–chitosan hydrogels based on click chemistry. *Iran. Polym. J.* **2015**, *24*, 349–357. [CrossRef]

27. Plisko, E.A.; Nud'ga, L.A.; Danilov, S.N. Chitin and Its Chemical Transformations. *Russ. Chem. Rev.* **1977**, *46*, 764. [CrossRef]

28. Moore, G.K.; Roberts, G.A.F. Reactions of chitosan: 3. Preparation and reactivity of Schiff's base derivatives of chitosan. *Int. J. Biol. Macromol.* **1981**, *3*, 337–340. [CrossRef]

29. Yang, Z.; Wang, Y.; Tang, Y. Synthesis and adsorption properties for metal ions of mesocyclic diamine-grafted chitosan-crown ether. *J. Appl. Polym. Sci.* **2000**, *75*, 1255–1260. [CrossRef]

30. Yang, Z.; Yuan, Y. Studies on the synthesis and properties of hydroxyl azacrown ether-grafted chitosan. *J. Appl. Polym. Sci.* **2001**, *82*, 1838–1843. [CrossRef]

31. Chen, Y.; Ye, Y.; Li, R.; Guo, Y.; Tan, H. Synthesis of chitosan 6-OH immobilized cyclodextrin derivates via click chemistry. *Fibers Polym.* **2013**, *14*, 1058–1065. [CrossRef]

32. Chen, Y.; Wang, F.; Yun, D.; Guo, Y.; Ye, Y.; Wang, Y.; Tan, H. Preparation of a C6 quaternary ammonium chitosan derivative through a chitosan schiff base with click chemistry. *J. Appl. Polym. Sci.* **2013**, *129*, 3185–3191. [CrossRef]

33. Guinesi, L.S.; Cavalheiro, É.T.G. Influence of some reactional parameters on the substitution degree of biopolymeric Schiff bases prepared from chitosan and salicylaldehyde. *Carbohydr. Polym.* **2006**, *65*, 557–561. [CrossRef]

34. Wang, X.; Ma, J.; Wang, Y.; He, B. Structural characterization of phosphorylated chitosan and their applications as effective additives of calcium phosphate cements. *Biomaterials* **2001**, *22*, 2247–2255. [CrossRef]

35. Sashiwa, H.; Kawasaki, N.; Nakayama, A.; Muraki, E.; Yamamoto, N.; Aiba, S. Chemical Modification of Chitosan. 14: Synthesis of Water-Soluble Chitosan Derivatives by Simple Acetylation. *Biomacromolecules* **2002**, *3*, 1126–1128. [CrossRef] [PubMed]

36. Duan, K.; Chen, H.; Huang, J.; Yu, J.; Liu, S.; Wang, D.; Li, Y. One-step synthesis of amino-reserved chitosan-graft-polycaprolactone as a promising substance of biomaterial. *Carbohydr. Polym.* **2010**, *80*, 498–503. [CrossRef]

37. Gu, C.; Le, V.; Lang, M.; Liu, J. Preparation of polysaccharide derivates chitosan-graft-poly(ε-caprolactone) amphiphilic copolymer micelles for 5-fluorouracil drug delivery. *Colloids Surf. B Biointerfaces* **2014**, *116*, 745–750. [CrossRef] [PubMed]

38. Kolb, H.C.; Finn, M.G.; Sharpless, K.B. Click Chemistry: Diverse Chemical Function from a Few Good Reactions. *Angew. Chem. Int. Ed.* **2001**, *40*, 2004–2021. [CrossRef]

39. Kolb, H.C.; Sharpless, K.B. The growing impact of click chemistry on drug discovery. *Drug Discov. Today* **2003**, *8*, 1128–1137. [CrossRef]

40. Huisgen, R. 1,3-Dipolar cycloaddition–Introduction, survey, mechanism. In *1,3-Dipolar Cycloaddition Chemistry*; Padwa, A., Ed.; John Wiley and Sons: New York, NY, USA, 1984; Volume 1, pp. 1–176, ISBN 978-0-471-08364-1.

41. Wang, Z. Diels-Alder Reaction. In *Comprehensive Organic Name Reactions and Reagents*; John Wiley & Sons, Inc.: Hoboken, NJ, USA, 2010; pp. 886–891. ISBN 978-0-470-63885-9.

42. Wang, Z. Retro-Diels-Alder Reaction. In *Comprehensive Organic Name Reactions and Reagents*; John Wiley & Sons, Inc.: Hoboken, NJ, USA, 2010, 2010; pp. 2367–2372, ISBN 978-0-470-63885-9.

43. Gandini, A. The furan/maleimide Diels–Alder reaction: A versatile click–unclick tool in macromolecular synthesis. *Prog. Polym. Sci.* **2013**, *38*, 1–29. [CrossRef]

44. Elchinger, P.-H.; Faugeras, P.-A.; Boëns, B.; Brouillette, F.; Montplaisir, D.; Zerrouki, R.; Lucas, R. Polysaccharides: The "Click" Chemistry Impact. *Polymers* **2011**, *3*, 1607–1651. [CrossRef]

45. Uliniuc, A.; Popa, M.; Hamaide, T.; Dobromir, M. New approaches in hydrogel synthesis—Click chemistry: A review. *Cellul. Chem. Technol.* **2012**, *46*, 1–11.

46. Barbosa, M.; Martins, C.; Gomes, P. "Click" chemistry as a tool to create novel biomaterials: A short review. *UPorto J. Eng.* **2015**, *1*, 22–34.

47. Meng, X.; Edgar, K.J. "Click" reactions in polysaccharide modification. *Prog. Polym. Sci.* **2016**, *53*, 52–85. [CrossRef]

48. Kritchenkov, A.S.; Skorik, Y.A. Click reactions in chitosan chemistry. *Russ. Chem. Bull.* **2017**, *66*, 769–781. [CrossRef]

49. Bertoldo, M.; Nazzi, S.; Zampano, G.; Ciardelli, F. Synthesis and photochromic response of a new precisely functionalized chitosan with "clicked" spiropyran. *Carbohydr. Polym.* **2011**, *85*, 401–407. [CrossRef]

50. Ifuku, S.; Wada, M.; Morimoto, M.; Saimoto, H. A short synthesis of highly soluble chemoselective chitosan derivatives via "click chemistry". *Carbohydr. Polym.* **2012**, *90*, 1182–1186. [CrossRef] [PubMed]

51. Li, X.; Yuan, W.; Gu, S.; Ren, J. Synthesis and self-assembly of tunable thermosensitive chitosan amphiphilic copolymers by click chemistry. *Mater. Lett.* **2010**, *64*, 2663–2666. [CrossRef]

52. Zampano, G.; Bertoldo, M.; Ciardelli, F. Defined Chitosan-based networks by C-6-Azide–alkyne "click" reaction. *React. Funct. Polym.* **2010**, *70*, 272–281. [CrossRef]

53. Kulbokaite, R.; Ciuta, G.; Netopilik, M.; Makuska, R. N-PEG'ylation of chitosan via "click chemistry" reactions. *React. Funct. Polym.* **2009**, *69*, 771–778. [CrossRef]

54. Truong, V.X.; Ablett, M.P.; Gilbert, H.T.J.; Bowen, J.; Richardson, S.M.; Hoyland, J.A.; Dove, A.P. In situ-forming robust chitosan-poly(ethylene glycol) hydrogels prepared by copper-free azide–alkyne click reaction for tissue engineering. *Biomater. Sci.* **2013**, *2*, 167–175. [CrossRef]

55. Oliveira, J.R.; Martins, M.C.L.; Mafra, L.; Gomes, P. Synthesis of an O-alkynyl-chitosan and its chemoselective conjugation with a PEG-like amino-azide through click chemistry. *Carbohydr. Polym.* **2012**, *87*, 240–249. [CrossRef]

56. Tirino, P.; Laurino, R.; Maglio, G.; Malinconico, M.; d'Ayala, G.G.; Laurienzo, P. Synthesis of chitosan–PEO hydrogels via mesylation and regioselective Cu(I)-catalyzed cycloaddition. *Carbohydr. Polym.* **2014**, *112*, 736–745. [CrossRef] [PubMed]

57. Bao, H.; Li, L.; Leong, W.C.; Gan, L.H. Thermo-Responsive Association of Chitosan-graft-Poly(*N*-isopropylacrylamide) in Aqueous Solutions. *J. Phys. Chem. B* **2010**, *114*, 10666–10673. [CrossRef] [PubMed]

58. Bao, H.; Li, L.; Gan, L.H.; Ping, Y.; Li, J.; Ravi, P. Thermo-and pH-Responsive Association Behavior of Dual Hydrophilic Graft Chitosan Terpolymer Synthesized via ATRP and Click Chemistry. *Macromolecules* **2010**, *43*, 5679–5687. [CrossRef]

59. Lu, L.; Shao, X.; Jiao, Y.; Zhou, C. Synthesis of chitosan-graft-β-cyclodextrin for improving the loading and release of doxorubicin in the nanopaticles. *J. Appl. Polym. Sci.* **2014**, *131*, 41034. [CrossRef]

60. Guerry, A.; Bernard, J.; Samain, E.; Fleury, E.; Cottaz, S.; Halila, S. Aniline-Catalyzed Reductive Amination as a Powerful Method for the Preparation of Reducing End-"Clickable" Chitooligosaccharides. *Bioconjug. Chem.* **2013**, *24*, 544–549. [CrossRef] [PubMed]

61. Guerry, A.; Cottaz, S.; Fleury, E.; Bernard, J.; Halila, S. Redox-stimuli responsive micelles from DOX-encapsulating polycaprolactone-g-chitosan oligosaccharide. *Carbohydr. Polym.* **2014**, *112*, 746–752. [CrossRef] [PubMed]

62. Yuan, W.; Li, X.; Gu, S.; Cao, A.; Ren, J. Amphiphilic chitosan graft copolymer via combination of ROP, ATRP and click chemistry: Synthesis, self-assembly, thermosensitivity, fluorescence, and controlled drug release. *Polymer* **2011**, *52*, 658–666. [CrossRef]

63. Zhang, K.; Zhuang, P.; Wang, Z.; Li, Y.; Jiang, Z.; Hu, Q.; Liu, M.; Zhao, Q. One-pot synthesis of chitosan-g-(PEO-PLLA-PEO) via "click" chemistry and "SET-NRC" reaction. *Carbohydr. Polym.* **2012**, *90*, 1515–1521. [CrossRef] [PubMed]

64. Jung, S.; Yi, H. Fabrication of Chitosan-Poly(ethylene glycol) Hybrid Hydrogel Microparticles via Replica Molding and Its Application toward Facile Conjugation of Biomolecules. *Langmuir* **2012**, *28*, 17061–17070. [CrossRef] [PubMed]

65. Gao, Y.; Zhang, Z.; Chen, L.; Gu, W.; Li, Y. Synthesis of 6-*N,N,N*-Trimethyltriazole Chitosan via "Click Chemistry" and Evaluation for Gene Delivery. *Biomacromolecules* **2009**, *10*, 2175–2182. [CrossRef] [PubMed]

66. Ifuku, S.; Wada, M.; Morimoto, M.; Saimoto, H. Preparation of highly regioselective chitosan derivatives via "click chemistry". *Carbohydr. Polym.* **2011**, *85*, 653–657. [CrossRef]

67. Zhao, Q.; Zhang, J.; Song, L.; Ji, Q.; Yao, Y.; Cui, Y.; Shen, J.; Wang, P.G.; Kong, D. Polysaccharide-based biomaterials with on-demand nitric oxide releasing property regulated by enzyme catalysis. *Biomaterials* **2013**, *34*, 8450–8458. [CrossRef] [PubMed]

68. Ifuku, S.; Matsumoto, C.; Wada, M.; Morimoto, M.; Saimoto, H. Preparation of highly regioselective amphiprotic chitosan derivative via "click chemistry". *Int. J. Biol. Macromol.* **2013**, *52*, 72–76. [CrossRef] [PubMed]

69. Jirawutthiwongchai, J.; Krause, A.; Draeger, G.; Chirachanchai, S. Chitosan-Oxanorbornadiene: A Convenient Chitosan Derivative for Click Chemistry without Metal Catalyst Problem. *ACS Macro Lett.* **2013**, *2*, 177–180. [CrossRef]

70. Jirawutthiwongchai, J.; Draeger, G.; Chirachanchai, S. Rapid Hybridization of Chitosan-Gold-Antibodies via Metal-free Click in Water-based Systems: A Model Approach for Naked-eye Detectable Antigen Sensors. *Macromol. Rapid Commun.* **2014**, *35*, 1204–1210. [CrossRef] [PubMed]

71. Hu, L.; Zhao, P.; Deng, H.; Xiao, L.; Qin, C.; Du, Y.; Shi, X. Electrical signal guided click coating of chitosan hydrogel on conductive surface. *RSC Adv.* **2014**, *4*, 13477–13480. [CrossRef]

72. Li, Q.; Tan, W.; Zhang, C.; Gu, G.; Guo, Z. Novel triazolyl-functionalized chitosan derivatives with different chain lengths of aliphatic alcohol substituent: Design, synthesis, and antifungal activity. *Carbohydr. Res.* **2015**, *418*, 44–49. [CrossRef] [PubMed]

73. Sarwar, A.; Katas, H.; Samsudin, S.N.; Zin, N.M. Regioselective Sequential Modification of Chitosan via Azide-Alkyne Click Reaction: Synthesis, Characterization, and Antimicrobial Activity of Chitosan Derivatives and Nanoparticles. *PLOS ONE* **2015**, *10*, e0123084. [CrossRef] [PubMed]

74. Koshiji, K.; Nonaka, Y.; Iwamura, M.; Dai, F.; Matsuoka, R.; Hasegawa, T. C6-Modifications on chitosan to develop chitosan-based glycopolymers and their lectin-affinities with sigmoidal binding profiles. *Carbohydr. Polym.* **2016**, *137*, 277–286. [CrossRef] [PubMed]

75. Peng, P.; Cao, X.; Peng, F.; Bian, J.; Xu, F.; Sun, R. Binding cellulose and chitosan via click chemistry: Synthesis, characterization, and formation of some hollow tubes. *J. Polym. Sci. Part A Polym. Chem.* **2012**, *50*, 5201–5210. [CrossRef]

76. Ryu, H.J.; Mahapatra, S.S.; Yadav, S.K.; Cho, J.W. Synthesis of click-coupled graphene sheet with chitosan: Effective exfoliation and enhanced properties of their nanocomposites. *Eur. Polym. J.* **2013**, *49*, 2627–2634. [CrossRef]

77. Yadav, S.K.; Mahapatra, S.S.; Yadav, M.K.; Dutta, P.K. Mechanically robust biocomposite films of chitosan grafted carbon nanotubes via the [2 + 1] cycloaddition of nitrenes. *RSC Adv.* **2013**, *3*, 23631–23637. [CrossRef]

78. Gandini, A. Polymers from Renewable Resources: A Challenge for the Future of Macromolecular Materials. *Macromolecules* **2008**, *41*, 9491–9504. [CrossRef]

79. Montiel-Herrera, M.; Gandini, A.; Goycoolea, F.M.; Jacobsen, N.E.; Lizardi-Mendoza, J.; Recillas-Mota, M.; Argüelles-Monal, W.M. N-(furfural) chitosan hydrogels based on Diels–Alder cycloadditions and application as microspheres for controlled drug release. *Carbohydr. Polym.* **2015**, *128*, 220–227. [CrossRef] [PubMed]

80. Odian, G. *Principles of Polymerization*, 4th ed.; John Wiley & Sons: Hoboken, NJ, USA, 2004; ISBN 978-0-471-27400-1.

81. Mourya, V.K.; Inamdar, N.N. Chitosan-modifications and applications: Opportunities galore. *React. Funct. Polym.* **2008**, *68*, 1013–1051. [CrossRef]

82. Kurita, K.; Kawata, M.; Koyama, Y.; Nishimura, S.-I. Graft copolymerization of vinyl monomers onto chitin with cerium (IV) ion. *J. Appl. Polym. Sci.* **1991**, *42*, 2885–2891. [CrossRef]

83. Lagos, A.; Yazdani-Pedram, M.; Reyes, J.; Campos, N. Ceric Ion-Initiated Grafting of Poly(Methyl Acrylate) onto Chitin. *J. Macromol. Sci. Part A* **1992**, *29*, 1007–1015. [CrossRef]

84. Ren, L.; Miura, Y.; Nishi, N.; Tokura, S. Modification of chitin by ceric salt-initiated graft polymerisation—Preparation of poly(methyl methacrylate)-grafted chitin derivatives that swell in organic solvents. *Carbohydr. Polym.* **1993**, *21*, 23–27. [CrossRef]

85. Tanodekaew, S.; Prasitsilp, M.; Swasdison, S.; Thavornyutikarn, B.; Pothsree, T.; Pateepasen, R. Preparation of acrylic grafted chitin for wound dressing application. *Biomaterials* **2004**, *25*, 1453–1460. [CrossRef] [PubMed]

86. Mostafa, T.; Naguib, H.; Sabaa, M.; Mokhtar, S. Graft copolymerization of itaconic acid onto chitin and its properties. *Polym. Int.* **2005**, *54*, 221–225. [CrossRef]

87. Mokhtar, S.M.; Mostafa, T.B.; Hewedy, M.A. Chemical induced grafting of indole onto chitin & chitosan-and their antimicrobial activity. *Aust. J. Basic Appl. Sci.* **2010**, *4*, 3268–3279.

88. Ifuku, S.; Iwasaki, M.; Morimoto, M.; Saimoto, H. Graft polymerization of acrylic acid onto chitin nanofiber to improve dispersibility in basic water. *Carbohydr. Polym.* **2012**, *90*, 623–627. [CrossRef] [PubMed]

89. Abu Naim, A.; Umar, A.; Sanagi, M.M.; Basaruddin, N. Chemical modification of chitin by grafting with polystyrene using ammonium persulfate initiator. *Carbohydr. Polym.* **2013**, *98*, 1618–1623. [CrossRef] [PubMed]

90. Jayakumar, R.; Tamura, H. Synthesis, characterization and thermal properties of chitin-g-poly(ε-caprolactone) copolymers by using chitin gel. *Int. J. Biol. Macromol.* **2008**, *43*, 32–36. [CrossRef] [PubMed]

91. Pilakasiri, K.; Molee, P.; Sringernyuang, D.; Sangjun, N.; Channasanon, S.; Tanodekaew, S. Efficacy of chitin-PAA-GTMAC gel in promoting wound healing: Animal study. *J. Mater. Sci. Mater. Med.* **2011**, *22*, 2497–2504. [CrossRef] [PubMed]

92. Uppanan, P.; Channasanon, S.; Veeranondh, S.; Tanodekaew, S. Synthesis of GTMAC modified chitin–PAA gel and evaluation of its biological properties. *J. Biomed. Mater. Res. Part A* **2011**, *98*, 185–191. [CrossRef] [PubMed]

93. Huang, C.-M.; Chen, L.-C.; Yang, H.-C.; Li, M.-H.; Pan, T.-C. Preparation of acrylic acid-modified chitin improved by an experimental design and its application in absorbing toxic organic compounds. *J. Hazard. Mater.* **2012**, *241–242*, 190–196. [CrossRef] [PubMed]

94. Rao, V.; Praveen, P.; Latha, D. A novel method for synthesis of polypyrrole grafted chitin. *J. Polym. Res.* **2016**, *9*, 1–6. [CrossRef]

95. Peniche, C.; Argüelles-Monal, W.; Davidenko, N.; Sastre, R.; Gallardo, A.; San Román, J. Self-curing membranes of chitosan/PAA IPNs obtained by radical polymerization: Preparation, characterization and interpolymer complexation. *Biomaterials* **1999**, *20*, 1869–1878. [CrossRef]

96. Recillas, M.; Silva, L.L.; Peniche, C.; Goycoolea, F.M.; Rinaudo, M.; Argüelles-Monal, W.M. Thermoresponsive behavior of chitosan-*g*-N-isopropylacrylamide copolymer solutions. *Biomacromolecules* **2009**, *10*, 1633–1641. [CrossRef] [PubMed]

97. Lavrič, P.K.; Warmoeskerken, M.M.C.G.; Jocic, D. Functionalization of cotton with poly-NiPAAm/chitosan microgel. Part I. Stimuli-responsive moisture management properties. *Cellulose* **2012**, *19*, 257–271. [CrossRef]

98. Chen, C.; Liu, M.; Gao, C.; Lü, S.; Chen, J.; Yu, X.; Ding, E.; Yu, C.; Guo, J.; Cui, G. A convenient way to synthesize comb-shaped chitosan-graft-poly(N-isopropylacrylamide) copolymer. *Carbohydr. Polym.* **2013**, *92*, 621–628. [CrossRef] [PubMed]

99. Spizzirri, U.G.; Iemma, F.; Cirillo, G.; Altimari, I.; Puoci, F.; Picci, N. Temperature-sensitive hydrogels by graft polymerization of chitosan and N-isopropylacrylamide for drug release. *Pharm. Dev. Technol.* **2013**, *18*, 1026–1034. [CrossRef] [PubMed]

100. Huang, C.-H.; Wang, C.-F.; Don, T.-M.; Chiu, W.-Y. Preparation of pH- and thermo-sensitive chitosan-PNIPAAm core–shell nanoparticles and evaluation as drug carriers. *Cellulose* **2013**, *20*, 1791–1805. [CrossRef]

101. Raskin, M.M.; Schlachet, I.; Sosnik, A. Mucoadhesive nanogels by ionotropic crosslinking of chitosan-g-oligo(NiPAam) polymeric micelles as novel drug nanocarriers. *Nanomedicine* **2016**, *11*, 217–233. [CrossRef] [PubMed]

102. Liu, S.; Zhang, J.; Cui, X.; Guo, Y.; Zhang, X.; Hongyan, W. Synthesis of chitosan-based nanohydrogels for loading and release of 5-fluorouracil. *Colloids Surf. A Physicochem. Eng. Asp.* **2016**, *490*, 91–97. [CrossRef]

103. Cao, L.; Wang, X.; Wang, G.; Wang, J. A pH-sensitive porous chitosan membrane prepared via surface grafting copolymerization in supercritical carbon dioxide. *Polym. Int.* **2015**, *64*, 383–388. [CrossRef]

104. Metzler, M.; Chylińska, M.; Kaczmarek, H. Preparation and characteristics of nanosilver composite based on chitosan-*graft*-acrylic acid copolymer. *J. Polym. Res.* **2015**, *22*, 146. [CrossRef]

105. García-Valdez, O.; Champagne-Hartley, R.; Saldívar-Guerra, E.; Champagne, P.; Cunningham, M.F. Modification of chitosan with polystyrene and poly(N-butyl acrylate) via nitroxide-mediated polymerization and grafting from approach in homogeneous media. *Polym. Chem.* **2015**, *6*, 2827–2836. [CrossRef]

106. Khan, A.; Othman, M.B.H.; Chang, B.P.; Akil, H.M. Preparation, physicochemical and stability studies of chitosan-PNIPAM based responsive microgels under various pH and temperature conditions. *Iran. Polym. J.* **2015**, *24*, 317–328. [CrossRef]

107. Echeverria, C.; Soares, P.; Robalo, A.; Pereira, L.; Novo, C.M.M.; Ferreira, I.; Borges, J.P. One-pot synthesis of dual-stimuli responsive hybrid PNIPAAm-chitosan microgels. *Mater. Des.* **2015**, *86*, 745–751. [CrossRef]

108. Fernández-Gutiérrez, M.; Fusco, S.; Mayol, L.; Román, J.S.; Borzacchiello, A.; Ambrosio, L. Stimuli-responsive chitosan/poly(N-isopropylacrylamide) semi-interpenetrating polymer networks: Effect of pH and temperature on their rheological and swelling properties. *J. Mater. Sci. Mater. Med.* **2016**, *27*, 109. [CrossRef] [PubMed]

109. Anbinder, P.; Macchi, C.; Amalvy, J.; Somoza, A. Chitosan-graft-poly(N-butyl acrylate) copolymer: Synthesis and characterization of a natural/synthetic hybrid material. *Carbohydr. Polym.* **2016**, *145*, 86–94. [CrossRef] [PubMed]

110. Chen, Y.; Yang, Y.; Liao, Q.; Yang, W.; Ma, W.; Zhao, J.; Zheng, X.; Yang, Y.; Chen, R. Preparation, property of the complex of carboxymethyl chitosan grafted copolymer with iodine and application of it in cervical antibacterial biomembrane. *Mater. Sci. Eng. C* **2016**, *67*, 247–258. [CrossRef] [PubMed]

111. Lin, Y.; Hong, Y.; Song, Q.; Zhang, Z.; Gao, J.; Tao, T. Highly efficient removal of copper ions from water using poly(acrylic acid)-grafted chitosan adsorbent. *Colloid Polym. Sci.* **2017**, *295*, 627–635. [CrossRef]

112. Bashir, S.; Teo, Y.Y.; Ramesh, S.; Ramesh, K. Physico-chemical characterization of pH-sensitive N-Succinyl chitosan-g-poly(acrylamide-co-acrylic acid) hydrogels and in vitro drug release studies. *Polym. Degrad. Stab.* **2017**, *139*, 38–54. [CrossRef]

113. Khalaj Moazen, M.; Ahmad Panahi, H. Magnetic iron oxide nanoparticles grafted N-isopropylacrylamide/chitosan copolymer for the extraction and determination of letrozole in human biological samples. *J. Sep. Sci.* **2017**, *40*, 1125–1132. [CrossRef] [PubMed]

114. Munro, N.H.; Hanton, L.R.; Moratti, S.C.; Robinson, B.H. Synthesis and characterisation of chitosan-graft-poly(OEGMA) copolymers prepared by ATRP. *Carbohydr. Polym.* **2009**, *77*, 496–505. [CrossRef]
115. Aziz, M.S.A.; Naguib, H.F.; Saad, G.R. Nanocomposites Based on Chitosan-Graft-Poly(*N*-Vinyl-2-Pyrrolidone): Synthesis, Characterization, and Biological Activity. *Int. J. Polym. Mater. Polym. Biomater.* **2015**, *64*, 578–586. [CrossRef]
116. Leiva, A.; Bonardd, S.; Pino, M.; Saldías, C.; Kortaberria, G.; Radić, D. Improving the performance of chitosan in the synthesis and stabilization of gold nanoparticles. *Eur. Polym. J.* **2015**, *68*, 419–431. [CrossRef]
117. Liu, L.; Wang, Y.; Shen, X.; Fang, Y. Preparation of chitosan-g-polycaprolactone copolymers through ring-opening polymerization of ε-caprolactone onto phthaloyl-protected chitosan. *Biopolymers* **2005**, *78*, 163–170. [CrossRef] [PubMed]
118. Liu, L.; Chen, L.; Fang, Y. Self-Catalysis of Phthaloylchitosan for Graft Copolymerization of ε-Caprolactone with Chitosan. *Macromol. Rapid Commun.* **2006**, *27*, 1988–1994. [CrossRef]
119. Liu, L.; Shi, A.; Guo, S.; Fang, Y.; Chen, S.; Li, J. Preparation of chitosan-g-polylactide graft copolymers via self-catalysis of phthaloylchitosan and their complexation with DNA. *React. Funct. Polym.* **2010**, *70*, 301–305. [CrossRef]
120. Mirabbasi, F.; Dorkoosh, F.A.; Moghimi, A.; Shahsavari, S.; Babanejad, N.; Seifirad, S. Preparation of Mesalamine Nanoparticles Using a Novel Polyurethane-Chitosan Graft Copolymer. *Pharm. Nanotechnol.* **2017**, *5*, 230–239. [CrossRef] [PubMed]
121. Cabuk, M.; Yavuz, M.; Unal, H.I. Electrokinetic properties of biodegradable conducting polyaniline-graft-chitosan copolymer in aqueous and non-aqueous media. *Colloids Surf. Physicochem. Eng. Asp.* **2014**, *460*, 494–501. [CrossRef]
122. Feil, H.; Bae, Y.H.; Feijen, J.; Kim, S.W. Effect of comonomer hydrophilicity and ionization on the lower critical solution temperature of *N*-isopropylacrylamide copolymers. *Macromolecules* **1993**, *26*, 2496–2500. [CrossRef]
123. Recillas, M.; Silva, L.L.; Peniche, C.; Goycoolea, F.M.; Rinaudo, M.; Román, J.S.; Argüelles-Monal, W.M. Thermo- and pH-responsive polyelectrolyte complex membranes from chitosan-g-*N*-isopropylacrylamide and pectin. *Carbohydr. Polym.* **2011**, *86*, 1336–1343. [CrossRef]
124. Marques, N.D.N.; Maia, A.M.D.S.; Balaban, R.D.C. Development of dual-sensitive smart polymers by grafting chitosan with poly (*N*-isopropylacrylamide): An overview. *Polímeros* **2015**, *25*, 237–246. [CrossRef]
125. Liu, L.; Li, Y.; Liu, H.; Fang, Y. Synthesis and characterization of chitosan-graft-polycaprolactone copolymers. *Eur. Polym. J.* **2004**, *40*, 2739–2744. [CrossRef]
126. Liu, L.; Xu, X.; Guo, S.; Han, W. Synthesis and self-assembly of chitosan-based copolymer with a pair of hydrophobic/hydrophilic grafts of polycaprolactone and poly(ethylene glycol). *Carbohydr. Polym.* **2009**, *75*, 401–407. [CrossRef]
127. Gu, C.; Zhang, H.; Lang, M. Preparation of mono-dispersed silver nanoparticles assisted by chitosan-g-poly(ε-caprolactone) micelles and their antimicrobial application. *Appl. Surf. Sci.* **2014**, *301*, 273–279. [CrossRef]
128. Gorochovceva, N.; Makuška, R. Synthesis and study of water-soluble chitosan-O-poly(ethylene glycol) graft copolymers. *Eur. Polym. J.* **2004**, *40*, 685–691. [CrossRef]
129. Bhattarai, N.; Matsen, F.A.; Zhang, M. PEG-Grafted Chitosan as an Injectable Thermoreversible Hydrogel. *Macromol. Biosci.* **2005**, *5*, 107–111. [CrossRef] [PubMed]
130. Bhattarai, N.; Ramay, H.R.; Gunn, J.; Matsen, F.A.; Zhang, M. PEG-grafted chitosan as an injectable thermosensitive hydrogel for sustained protein release. *J. Control. Release* **2005**, *103*, 609–624. [CrossRef] [PubMed]
131. Jiang, X.; Dai, H.; Leong, K.W.; Goh, S.-H.; Mao, H.-Q.; Yang, Y.-Y. Chitosan-g-PEG/DNA complexes deliver gene to the rat liver via intrabiliary and intraportal infusions. *J. Gene Med.* **2006**, *8*, 477–487. [CrossRef] [PubMed]
132. Du, J.; Hsieh, Y.-L. PEGylation of chitosan for improved solubility and fiber formation via electrospinning. *Cellulose* **2007**, *14*, 543–552. [CrossRef]
133. Bae, K.H.; Moon, C.W.; Lee, Y.; Park, T.G. Intracellular Delivery of Heparin Complexed with Chitosan-g-Poly(Ethylene Glycol) for Inducing Apoptosis. *Pharm. Res.* **2009**, *26*, 93–100. [CrossRef] [PubMed]

134. Liang, Y.; Deng, L.; Chen, C.; Zhang, J.; Zhou, R.; Li, X.; Hu, R.; Dong, A. Preparation and properties of thermoreversible hydrogels based on methoxy poly(ethylene glycol)-grafted chitosan nanoparticles for drug delivery systems. *Carbohydr. Polym.* **2011**, *83*, 1828–1833. [CrossRef]

135. Papadimitriou, S.A.; Achilias, D.S.; Bikiaris, D.N. Chitosan-*g*-PEG nanoparticles ionically crosslinked with poly(glutamic acid) and tripolyphosphate as protein delivery systems. *Int. J. Pharm.* **2012**, *430*, 318–327. [CrossRef] [PubMed]

136. Jiang, G.; Sun, J.; Ding, F. PEG-g-chitosan thermosensitive hydrogel for implant drug delivery: Cytotoxicity, in vivo degradation and drug release. *J. Biomater. Sci. Polym. Ed.* **2014**, *25*, 241–256. [CrossRef] [PubMed]

137. Vasquez, D.; Milusheva, R.; Baumann, P.; Constantin, D.; Chami, M.; Palivan, C.G. The Amine Content of PEGylated Chitosan Bombyx mori Nanoparticles Acts as a Trigger for Protein Delivery. *Langmuir* **2014**, *30*, 965–975. [CrossRef] [PubMed]

138. Ho, T.H.; Le, T.N.T.; Nguyen, T.A.; Dang, M.C. Poly(ethylene glycol) grafted chitosan as new copolymer material for oral delivery of insulin. *Adv. Nat. Sci. Nanosci. Nanotechnol.* **2015**, *6*, 035004. [CrossRef]

139. Chung, H.J.; Go, D.H.; Bae, J.W.; Jung, I.K.; Lee, J.W.; Park, K.D. Synthesis and characterization of Pluronic® grafted chitosan copolymer as a novel injectable biomaterial. *Curr. Appl. Phys.* **2005**, *5*, 485–488. [CrossRef]

140. Park, K.M.; Lee, S.Y.; Joung, Y.K.; Na, J.S.; Lee, M.C.; Park, K.D. Thermosensitive chitosan–Pluronic hydrogel as an injectable cell delivery carrier for cartilage regeneration. *Acta Biomater.* **2009**, *5*, 1956–1965. [CrossRef] [PubMed]

141. Zhang, W.; Gilstrap, K.; Wu, L.; Bahadur, R.K.C.; Moss, M.A.; Wang, Q.; Lu, X.; He, X. Synthesis and Characterization of Thermally Responsive Pluronic F127−Chitosan Nanocapsules for Controlled Release and Intracellular Delivery of Small Molecules. *ACS Nano* **2010**, *4*, 6747–6759. [CrossRef] [PubMed]

142. Cao, Y.; Zhang, C.; Shen, W.; Cheng, Z.; Yu, L. (Lucy); Ping, Q. Poly(*N*-isopropylacrylamide)–chitosan as thermosensitive in situ gel-forming system for ocular drug delivery. *J. Control. Release* **2007**, *120*, 186–194. [CrossRef] [PubMed]

143. Mao, Z.; Ma, L.; Yan, J.; Yan, M.; Gao, C.; Shen, J. The gene transfection efficiency of thermoresponsive *N,N,N*-trimethyl chitosan chloride-g-poly(*N*-isopropylacrylamide) copolymer. *Biomaterials* **2007**, *28*, 4488–4500. [CrossRef] [PubMed]

144. Lee, E.J.; Kim, Y.H. Synthesis and thermo-responsive properties of chitosan-g-poly (*N*-isopropylacrylamide) and HTCC-*g*-poly(*N*-isopropylacrylamide) copolymers. *Fibers Polym.* **2010**, *11*, 164–169. [CrossRef]

145. Sanoj Rejinold, N.; Sreerekha, P.R.; Chennazhi, K.P.; Nair, S.V.; Jayakumar, R. Biocompatible, biodegradable and thermo-sensitive chitosan-*g*-poly (*N*-isopropylacrylamide) nanocarrier for curcumin drug delivery. *Int. J. Biol. Macromol.* **2011**, *49*, 161–172. [CrossRef] [PubMed]

146. Antoniraj, M.G.; Kumar, C.S.; Kandasamy, R. Synthesis and characterization of poly (*N*-isopropylacrylamide)-*g*-carboxymethyl chitosan copolymer-based doxorubicin-loaded polymeric nanoparticles for thermoresponsive drug release. *Colloid Polym. Sci.* **2016**, *294*, 527–535. [CrossRef]

147. Rejinold, N.S.; Chennazhi, K.P.; Nair, S.V.; Tamura, H.; Jayakumar, R. Biodegradable and thermo-sensitive chitosan-*g*-poly(*N*-vinylcaprolactam) nanoparticles as a 5-fluorouracil carrier. *Carbohydr. Polym.* **2011**, *83*, 776–786. [CrossRef]

148. Sanoj Rejinold, N.; Muthunarayanan, M.; Divyarani, V.V.; Sreerekha, P.R.; Chennazhi, K.P.; Nair, S.V.; Tamura, H.; Jayakumar, R. Curcumin-loaded biocompatible thermoresponsive polymeric nanoparticles for cancer drug delivery. *J. Colloid Interface Sci.* **2011**, *360*, 39–51. [CrossRef] [PubMed]

149. Rejinold, N.S.; Thomas, R.G.; Muthiah, M.; Chennazhi, K.P.; Park, I.-K.; Jeong, Y.Y.; Manzoor, K.; Jayakumar, R. Radio frequency triggered curcumin delivery from thermo and pH responsive nanoparticles containing gold nanoparticles and its in vivo localization studies in an orthotopic breast tumor model. *RSC Adv.* **2014**, *4*, 39408–39427. [CrossRef]

150. Sanoj Rejinold, N.; Thomas, R.G.; Muthiah, M.; Chennazhi, K.P.; Manzoor, K.; Park, I.-K.; Jeong, Y.Y.; Jayakumar, R. Anti-cancer, pharmacokinetics and tumor localization studies of pH-, RF- and thermo-responsive nanoparticles. *Int. J. Biol. Macromol.* **2015**, *74*, 249–262. [CrossRef] [PubMed]

151. Fernández-Quiroz, D.; González-Gómez, Á.; Lizardi-Mendoza, J.; Vázquez-Lasa, B.; Goycoolea, F.M.; San Román, J.; Argüelles-Monal, W.M. Effect of the molecular architecture on the thermosensitive properties of chitosan-g-poly(*N*-vinylcaprolactam). *Carbohydr. Polym.* **2015**, *134*, 92–101. [CrossRef] [PubMed]

152. Rejinold, N.S.; Thomas, R.G.; Muthiah, M.; Lee, H.J.; Jeong, Y.Y.; Park, I.-K.; Jayakumar, R. Breast Tumor Targetable Fe$_3$O$_4$ Embedded Thermo-Responsive Nanoparticles for Radiofrequency Assisted Drug Delivery. *J. Biomed. Nanotechnol.* **2016**, *12*, 43–55. [CrossRef] [PubMed]

153. Indulekha, S.; Arunkumar, P.; Bahadur, D.; Srivastava, R. Thermoresponsive polymeric gel as an on-demand transdermal drug delivery system for pain management. *Mater. Sci. Eng. C* **2016**, *62*, 113–122. [CrossRef] [PubMed]

154. Fernández-Quiroz, D.; González-Gómez, Á.; Lizardi-Mendoza, J.; Vázquez-Lasa, B.; Goycoolea, F.M.; Román, J.S.; Argüelles-Monal, W.M. Conformational study on the thermal transition of chitosan-*g*-poly(*N*-vinylcaprolactam) in aqueous solution. *Colloid Polym. Sci.* **2016**, *294*, 555–563. [CrossRef]

155. Casettari, L.; Vllasaliu, D.; Castagnino, E.; Stolnik, S.; Howdle, S.; Illum, L. PEGylated chitosan derivatives: Synthesis, characterizations and pharmaceutical applications. *Prog. Polym. Sci.* **2012**, *37*, 659–685. [CrossRef]

156. Wang, J.-P.; Chen, Y.-Z.; Ge, X.-W.; Yu, H.-Q. Gamma radiation-induced grafting of a cationic monomer onto chitosan as a flocculant. *Chemosphere* **2007**, *66*, 1752–1757. [CrossRef] [PubMed]

157. Wang, J.-P.; Chen, Y.-Z.; Zhang, S.-J.; Yu, H.-Q. A chitosan-based flocculant prepared with gamma-irradiation-induced grafting. *Bioresour. Technol.* **2008**, *99*, 3397–3402. [CrossRef] [PubMed]

158. Zheng, H.; Sun, Y.; Zhu, C.; Guo, J.; Zhao, C.; Liao, Y.; Guan, Q. UV-initiated polymerization of hydrophobically associating cationic flocculants: Synthesis, characterization, and dewatering properties. *Chem. Eng. J.* **2013**, *234*, 318–326. [CrossRef]

159. Sun, Y.; Ren, M.; Zhu, C.; Xu, Y.; Zheng, H.; Xiao, X.; Wu, H.; Xia, T.; You, Z. UV-Initiated Graft Copolymerization of Cationic Chitosan-Based Flocculants for Treatment of Zinc Phosphate-Contaminated Wastewater. *Ind. Eng. Chem. Res.* **2016**, *55*, 10025–10035. [CrossRef]

160. Sun, Y.; Zhu, C.; Sun, W.; Xu, Y.; Xiao, X.; Zheng, H.; Wu, H.; Liu, C. Plasma-initiated polymerization of chitosan-based CS-g-P(AM-DMDAAC) flocculant for the enhanced flocculation of low-algal-turbidity water. *Carbohydr. Polym.* **2017**, *164*, 222–232. [CrossRef] [PubMed]

161. Zhuang, S.; Yi, Y.; Wang, J. Removal of Cobalt Ions from Aqueous Solution Using Chitosan Grafted with Maleic Acid by Gamma Radiation. *Nucl. Eng. Technol.* **2018**, *50*, 211–215. [CrossRef]

162. Saleh, A.S.; Ibrahim, A.G.; Elsharma, E.M.; Metwally, E.; Siyam, T. Radiation grafting of acrylamide and maleic acid on chitosan and effective application for removal of Co.(II) from aqueous solutions. *Radiat. Phys. Chem.* **2018**, *144*, 116–124. [CrossRef]

163. Sosnik, A.; Imperiale, J.C.; Vázquez-González, B.; Raskin, M.M.; Muñoz-Muñoz, F.; Burillo, G.; Cedillo, G.; Bucio, E. Mucoadhesive thermo-responsive chitosan-*g*-poly(*N*-isopropylacrylamide) polymeric micelles via a one-pot gamma-radiation-assisted pathway. *Colloids Surf. B Biointerfaces* **2015**, *136*, 900–907. [CrossRef] [PubMed]

164. Cruz, A.; García-Uriostegui, L.; Ortega, A.; Isoshima, T.; Burillo, G. Radiation grafting of *N*-vinylcaprolactam onto nano and macrogels of chitosan: Synthesis and characterization. *Carbohydr. Polym.* **2017**, *155*, 303–312. [CrossRef] [PubMed]

165. Montes, J.Á.; Ortega, A.; Burillo, G. Dual-stimuli responsive copolymers based on *N*-vinylcaprolactam/chitosan. *J. Radioanal. Nucl. Chem.* **2014**, *303*, 2143–2150. [CrossRef]

166. Pérez-Calixto, M.P.; Ortega, A.; Garcia-Uriostegui, L.; Burillo, G. Synthesis and characterization of *N*-vinylcaprolactam/*N,N*-dimethylacrylamide grafted onto chitosan networks by gamma radiation. *Radiat. Phys. Chem.* **2016**, *119*, 228–235. [CrossRef]

167. Harada, A.; Hashidzume, A.; Takashima, Y. Cyclodextrin-Based Supramolecular Polymers. In *Supramolecular Polymers Polymeric Betains Oligomers*; Advances in Polymer Science; Springer: Berlin/Heidelberg, Germany, 2001; Volume 201.

168. Sajomsang, W.; Nuchuchua, O.; Gonil, P.; Saesoo, S.; Sramala, I.; Soottitantawat, A.; Puttipipatkhachorn, S.; Ruktanonchai, U.R. Water-soluble β-cyclodextrin grafted with chitosan and its inclusion complex as a mucoadhesive eugenol carrier. *Carbohydr. Polym.* **2012**, *89*, 623–631. [CrossRef] [PubMed]

169. Auzély-Velty, R.; Rinaudo, M. New Supramolecular Assemblies of a Cyclodextrin-Grafted Chitosan through Specific Complexation. *Macromolecules* **2002**, *35*, 7955–7962. [CrossRef]

170. Yu, N.; Li, G.; Gao, Y.; Liu, X.; Ma, S. Stimuli-sensitive hollow spheres from chitosan-graft-β-cyclodextrin for controlled drug release. *Int. J. Biol. Macromol.* **2016**, *93*, 971–977. [CrossRef] [PubMed]

171. Rekharsky, M.V.; Inoue, Y. Complexation Thermodynamics of Cyclodextrins. *Chem. Rev.* **1998**, *98*, 1875–1918. [CrossRef] [PubMed]

172. Liu, Y.; Han, B.-H.; Li, B.; Zhang, Y.-M.; Zhao, P.; Chen, Y.-T.; Wada, T.; Inoue, Y. Molecular Recognition Study on Supramolecular System. 14.1 Synthesis of Modified Cyclodextrins and Their Inclusion Complexation Thermodynamics with l-Tryptophan and Some Naphthalene Derivatives. *J. Org. Chem.* **1998**, *63*, 1444–1454. [CrossRef]

173. Harada, A.; Takashima, Y.; Yamaguchi, H. Cyclodextrin-based supramolecular polymers. *Chem. Soc. Rev.* **2009**, *38*, 875–882. [CrossRef] [PubMed]

174. Auzély, R.; Rinaudo, M. Controlled Chemical Modifications of Chitosan. Characterization and Investigation of Original Properties. *Macromol. Biosci.* **2003**, *3*, 562–565. [CrossRef]

175. Daimon, Y.; Izawa, H.; Kawakami, K.; Żywicki, P.; Sakai, H.; Abe, M.; Hill, J.P.; Ariga, K. Media-dependent morphology of supramolecular aggregates of β-cyclodextrin-grafted chitosan and insulin through multivalent interactions. *J. Mater. Chem. B* **2014**, *2*, 1802–1812. [CrossRef]

176. Daimon, Y.; Kamei, N.; Kawakami, K.; Takeda-Morishita, M.; Izawa, H.; Takechi-Haraya, Y.; Saito, H.; Sakai, H.; Abe, M.; Ariga, K. Dependence of Intestinal Absorption Profile of Insulin on Carrier Morphology Composed of β-Cyclodextrin-Grafted Chitosan. *Mol. Pharm.* **2016**, *13*, 4034–4042. [CrossRef] [PubMed]

177. Tojima, T.; Katsura, H.; Han, S.-M.; Tanida, F.; Nishi, N.; Tokura, S.; Sakairi, N. Preparation of an α-cyclodextrin–linked chitosan derivative via reductive amination strategy. *J. Polym. Sci. Part A Polym. Chem.* **1998**, *36*, 1965–1968. [CrossRef]

178. Auzély-Velty, R.; Rinaudo, M. Chitosan Derivatives Bearing Pendant Cyclodextrin Cavities: Synthesis and Inclusion Performance. *Macromolecules* **2001**, *34*, 3574–3580. [CrossRef]

179. Venter, J.P.; Kotzé, A.F.; Auzély-Velty, R.; Rinaudo, M. Synthesis and evaluation of the mucoadhesivity of a CD-chitosan derivative. *Int. J. Pharm.* **2006**, *313*, 36–42. [CrossRef] [PubMed]

180. El-Tahlawy, K.; Gaffar, M.A.; El-Rafie, S. Novel method for preparation of β-cyclodextrin/grafted chitosan and it's application. *Carbohydr. Polym.* **2006**, *63*, 385–392. [CrossRef]

181. Chaleawlert-umpon, S.; Nuchuchua, O.; Saesoo, S.; Gonil, P.; Ruktanonchai, U.R.; Sajomsang, W.; Pimpha, N. Effect of citrate spacer on mucoadhesive properties of a novel water-soluble cationic β-cyclodextrin-conjugated chitosan. *Carbohydr. Polym.* **2011**, *84*, 186–194. [CrossRef]

182. Furusaki, E.; Ueno, Y.; Sakairi, N.; Nishi, N.; Tokura, S. Facile preparation and inclusion ability of a chitosan derivative bearing carboxymethyl-β-cyclodextrin. *Carbohydr. Polym.* **1996**, *29*, 29–34. [CrossRef]

183. Kono, H.; Teshirogi, T. Cyclodextrin-grafted chitosan hydrogels for controlled drug delivery. *Int. J. Biol. Macromol.* **2015**, *72*, 299–308. [CrossRef] [PubMed]

184. Takechi-Haraya, Y.; Tanaka, K.; Tsuji, K.; Asami, Y.; Izawa, H.; Shigenaga, A.; Otaka, A.; Saito, H.; Kawakami, K. Molecular Complex Composed of β-Cyclodextrin-Grafted Chitosan and pH-Sensitive Amphipathic Peptide for Enhancing Cellular Cholesterol Efflux under Acidic pH. *Bioconjug. Chem.* **2015**, *26*, 572–581. [CrossRef] [PubMed]

185. Song, M.; Li, L.; Zhang, Y.; Chen, K.; Wang, H.; Gong, R. Carboxymethyl-β-cyclodextrin grafted chitosan nanoparticles as oral delivery carrier of protein drugs. *React. Funct. Polym.* **2017**, *117*, 10–15. [CrossRef]

186. Zhang, X.; Wu, Z.; Gao, X.; Shu, S.; Zhang, H.; Wang, Z.; Li, C. Chitosan bearing pendant cyclodextrin as a carrier for controlled protein release. *Carbohydr. Polym.* **2009**, *77*, 394–401. [CrossRef]

187. Chen, S.; Wang, Y. Study on β-cyclodextrin grafting with chitosan and slow release of its inclusion complex with radioactive iodine. *J. Appl. Polym. Sci.* **2001**, *82*, 2414–2421. [CrossRef]

188. Martel, B.; Devassine, M.; Crini, G.; Weltrowski, M.; Bourdonneau, M.; Morcellet, M. Preparation and sorption properties of a β-cyclodextrin-linked chitosan derivative. *J. Polym. Sci. Part A Polym. Chem.* **2001**, *39*, 169–176. [CrossRef]

189. Gonil, P.; Sajomsang, W.; Ruktanonchai, U.R.; Pimpha, N.; Sramala, I.; Nuchuchua, O.; Saesoo, S.; Chaleawlert-umpon, S.; Puttipipatkhachorn, S. Novel quaternized chitosan containing β-cyclodextrin moiety: Synthesis, characterization and antimicrobial activity. *Carbohydr. Polym.* **2011**, *83*, 905–913. [CrossRef]

190. Sajomsang, W.; Gonil, P.; Ruktanonchai, U.R.; Pimpha, N.; Sramala, I.; Nuchuchua, O.; Saesoo, S.; Chaleawlert-umpon, S.; Puttipipatkhachorn, S. Self-aggregates formation and mucoadhesive property of water-soluble β-cyclodextrin grafted with chitosan. *Int. J. Biol. Macromol.* **2011**, *48*, 589–595. [CrossRef] [PubMed]

191. Yuan, Z.; Ye, Y.; Gao, F.; Yuan, H.; Lan, M.; Lou, K.; Wang, W. Chitosan-graft-β-cyclodextrin nanoparticles as a carrier for controlled drug release. *Int. J. Pharm.* **2013**, *446*, 191–198. [CrossRef] [PubMed]

192. Phunpee, S.; Suktham, K.; Surassmo, S.; Jarussophon, S.; Rungnim, C.; Soottitantawat, A.; Puttipipatkhachorn, S.; Ruktanonchai, U.R. Controllable encapsulation of α-mangostin with quaternized β-cyclodextrin grafted chitosan using high shear mixing. *Int. J. Pharm.* **2018**, *538*, 21–29. [CrossRef] [PubMed]

193. Zhang, X.; Wang, Y.; Yi, Y. Synthesis and characterization of grafting β-cyclodextrin with chitosan. *J. Appl. Polym. Sci.* **2004**, *94*, 860–864. [CrossRef]

194. Chiu, S.-H.; Chung, T.-W.; Giridhar, R.; Wu, W.-T. Immobilization of β-cyclodextrin in chitosan beads for separation of cholesterol from egg yolk. *Food Res. Int.* **2004**, *37*, 217–223. [CrossRef]

195. Zha, F.; Li, S.; Chang, Y. Preparation and adsorption property of chitosan beads bearing β-cyclodextrin cross-linked by 1,6-hexamethylene diisocyanate. *Carbohydr. Polym.* **2008**, *72*, 456–461. [CrossRef]

196. Sreenivasan, K. Synthesis and preliminary studies on a β-cyclodextrin-coupled chitosan as a novel adsorbent matrix. *J. Appl. Polym. Sci.* **1998**, *69*, 1051–1055. [CrossRef]

197. Sajomsang, W.; Nuchuchua, O.; Saesoo, S.; Gonil, P.; Chaleawlert-umpon, S.; Pimpha, N.; Sramala, I.; Soottitantawat, A.; Puttipipatkhachorn, S.; Ruktanonchai, U.R. A comparison of spacer on water-soluble cyclodextrin grafted chitosan inclusion complex as carrier of eugenol to mucosae. *Carbohydr. Polym.* **2013**, *92*, 321–327. [CrossRef] [PubMed]

198. Yan, J.; Li, X.; Qiu, F.; Zhao, H.; Yang, D.; Wang, J.; Wen, W. Synthesis of β-cyclodextrin–chitosan–graphene oxide composite and its application for adsorption of manganese ion (II). *Mater. Technol.* **2016**, *31*, 406–415. [CrossRef]

199. Zhao, F.; Repo, E.; Yin, D.; Chen, L.; Kalliola, S.; Tang, J.; Iakovleva, E.; Tam, K.C.; Sillanpää, M. One-pot synthesis of trifunctional chitosan-EDTA-β-cyclodextrin polymer for simultaneous removal of metals and organic micropollutants. *Sci. Rep.* **2017**, *7*, 15811. [CrossRef] [PubMed]

200. Krauland, A.; Alonso, M. Chitosan/cyclodextrin nanoparticles as macromolecular drug delivery system. *Int. J. Pharm.* **2007**, *340*, 134–142. [CrossRef] [PubMed]

201. Thanh Nguyen, H.; Goycoolea, F.M. Chitosan/Cyclodextrin/TPP Nanoparticles Loaded with Quercetin as Novel Bacterial Quorum Sensing Inhibitors. *Molecules* **2017**, *22*, 1975. [CrossRef] [PubMed]

202. Zhou, H.Y.; Wang, Z.Y.; Duan, X.Y.; Jiang, L.J.; Cao, P.P.; Li, J.X.; Li, J.B. Design and evaluation of chitosan-β-cyclodextrin based thermosensitive hydrogel. *Biochem. Eng. J.* **2016**, *111*, 100–107. [CrossRef]

203. Anirudhan, T.S.; Divya, P.L.; Nima, J. Synthesis and characterization of novel drug delivery system using modified chitosan based hydrogel grafted with cyclodextrin. *Chem. Eng. J.* **2016**, *284*, 1259–1269. [CrossRef]

204. Sashiwa, H. Chemical Aspects of Chitin and Chitosan Derivatives. In *Chitin and Chitosan Derivatives: Advances in Drug Discovery and Developments*; Kim, S.-K., Ed.; CRC Press: Boca Raton, FL, USA, 2013; pp. 93–111, ISBN 978-1-4665-6628-6.

205. Sashiwa, H.; Shigemasa, Y.; Roy, R. Chemical Modification of Chitosan. 3. Hyperbranched Chitosan−Sialic Acid Dendrimer Hybrid with Tetraethylene Glycol Spacer. *Macromolecules* **2000**, *33*, 6913–6915. [CrossRef]

206. Sashiwa, H.; Shigemasa, Y.; Roy, R. Chemical Modification of Chitosan. 10. Synthesis of Dendronized Chitosan−Sialic Acid Hybrid Using Convergent Grafting of Preassembled Dendrons Built on Gallic Acid and Tri(ethylene glycol) Backbone. *Macromolecules* **2001**, *34*, 3905–3909. [CrossRef]

207. Sashiwa, H.; Shigemasa, Y.; Roy, R. Chemical Modification of Chitosan 11: Chitosan–Dendrimer Hybrid as a Tree Like Molecule. *Carbohydr. Polym.* **2002**, *49*, 195–205. [CrossRef]

208. Sashiwa, H.; Yajima, H.; Aiba, S. Synthesis of a Chitosan−Dendrimer Hybrid and Its Biodegradation. *Biomacromolecules* **2003**, *4*, 1244–1249. [CrossRef] [PubMed]

209. Zarghami, Z.; Akbari, A.; Latifi, A.M.; Amani, M.A. Design of a new integrated chitosan-PAMAM dendrimer biosorbent for heavy metals removing and study of its adsorption kinetics and thermodynamics. *Bioresour. Technol.* **2016**, *205*, 230–238. [CrossRef] [PubMed]

210. Wen, Y.; Tan, Z.; Sun, F.; Sheng, L.; Zhang, X.; Yao, F. Synthesis and characterization of quaternized carboxymethyl chitosan/poly(amidoamine) dendrimer core–shell nanoparticles. *Mater. Sci. Eng. C* **2012**, *32*, 2026–2036. [CrossRef]

211. Wen, Y.; Yao, F.; Sun, F.; Tan, Z.; Tian, L.; Xie, L.; Song, Q. Antibacterial action mode of quaternized carboxymethyl chitosan/poly(amidoamine) dendrimer core–shell nanoparticles against Escherichia coli correlated with molecular chain conformation. *Mater. Sci. Eng. C* **2015**, *48*, 220–227. [CrossRef] [PubMed]

212. Kim, H.-R.; Jang, J.-W.; Park, J.-W. Carboxymethyl chitosan-modified magnetic-cored dendrimer as an amphoteric adsorbent. *J. Hazard. Mater.* **2016**, *317*, 608–616. [CrossRef] [PubMed]

213. Van Rantwijk, F.; Madeira Lau, R.; Sheldon, R.A. Biocatalytic transformations in ionic liquids. *Trends Biotechnol.* **2003**, *21*, 131–138. [CrossRef]

214. Petkovic, M.; Seddon, K.R.; Rebelo, L.P.N.; Silva Pereira, C. Ionic liquids: A pathway to environmental acceptability. *Chem. Soc. Rev.* **2011**, *40*, 1383–1403. [CrossRef] [PubMed]

215. Xie, H.; Zhang, S.; Li, S. Chitin and chitosan dissolved in ionic liquids as reversible sorbents of CO_2. *Green Chem.* **2006**, *8*, 630–633. [CrossRef]

216. Liebert, T.; Heinze, T. Interaction of Ionic Liquids with Polysaccharides 5. Solvents and Reaction Media for the Modification of Cellulose. *Bioresources* **2008**, *3*, 576–601.

217. Zakrzewska, M.E.; Bogel-Łukasik, E.; Bogel-Łukasik, R. Solubility of Carbohydrates in Ionic Liquids. *Energy Fuels* **2010**, *24*, 737–745. [CrossRef]

218. Isik, M.; Sardon, H.; Mecerreyes, D. Ionic Liquids and Cellulose: Dissolution, Chemical Modification and Preparation of New Cellulosic Materials. *Int. J. Mol. Sci.* **2014**, *15*, 11922–11940. [CrossRef] [PubMed]

219. Chen, Q.; Xiao, W.; Zhou, L.; Wu, T.; Wu, Y. Hydrolysis of chitosan under microwave irradiation in ionic liquids promoted by sulfonic acid-functionalized ionic liquids. *Polym. Degrad. Stab.* **2012**, *97*, 49–53. [CrossRef]

220. Tian, T.-C.; Xie, C.-X.; Li, L.; Wei, Q.-L.; Yu, S.-T.; Zhang, T. Research on the structure of amino acid ILs and its solubility for chitosan with chemical software. *Polym. Degrad. Stab.* **2016**, *126*, 17–21. [CrossRef]

221. Yang, X.; Qiao, C.; Li, Y.; Li, T. Dissolution and resourcfulization of biopolymers in ionic liquids. *React. Funct. Polym.* **2016**, *100*, 181–190. [CrossRef]

222. Silva, S.S.; Mano, J.F.; Reis, R.L. Ionic liquids in the processing and chemical modification of chitin and chitosan for biomedical applications. *Green Chem.* **2017**, *19*, 1208–1220. [CrossRef]

223. Mine, S.; Izawa, H.; Kaneko, Y.; Kadokawa, J. Acetylation of α-chitin in ionic liquids. *Carbohydr. Res.* **2009**, *344*, 2263–2265. [CrossRef] [PubMed]

224. Liu, L.; Zhou, S.; Wang, B.; Xu, F.; Sun, R. Homogeneous acetylation of chitosan in ionic liquids. *J. Appl. Polym. Sci.* **2013**, *129*, 28–35. [CrossRef]

225. Liu, Y.; Huang, Y.; Boamah, P.-O.; Cao, L.; Zhang, Q.; Lu, Z.; Li, H. Homogeneous synthesis of linoleic acid-grafted chitosan oligosaccharide in ionic liquid and its self-assembly performance in aqueous solution. *J. Appl. Polym. Sci.* **2015**, *132*, 41727. [CrossRef]

226. Li, L.; Yuan, B.; Liu, S.; Yu, S.; Xie, C.; Liu, F.; Zhang, C. N-acyl chitosan and its fiber with excellent moisture absorbability and retentivity: Preparation in a novel [Gly]Cl/water homogeneous system. *J. Appl. Polym. Sci.* **2013**, *129*, 3282–3289. [CrossRef]

227. Wang, Z.; Zheng, L.; Li, C.; Zhang, D.; Xiao, Y.; Guan, G.; Zhu, W. Modification of chitosan with monomethyl fumaric acid in an ionic liquid solution. *Carbohydr. Polym.* **2015**, *117*, 973–979. [CrossRef] [PubMed]

228. Wei, Y.; Huang, W.; Zhou, Y.; Zhang, S.; Hua, D.; Zhu, X. Modification of chitosan with carboxyl-functionalized ionic liquid for anion adsorption. *Int. J. Biol. Macromol.* **2013**, *62*, 365–369. [CrossRef] [PubMed]

229. Yang, L.; Zhang, J.; He, J.; Zhang, J.; Gan, Z. Homogeneous synthesis of amino-reserved chitosan-graft-polycaprolactone in an ionic liquid and the application in cell cultivation. *Polym. Int.* **2015**, *64*, 1045–1052. [CrossRef]

230. Pei, L.; Cai, Z.; Shang, S.; Song, Z. Synthesis and antibacterial activity of alkylated chitosan under basic ionic liquid conditions. *J. Appl. Polym. Sci.* **2014**, *131*, 40052. [CrossRef]

231. Yang, X.; Zhang, C.; Qiao, C.; Mu, X.; Li, T.; Xu, J.; Shi, L.; Zhang, D. A simple and convenient method to synthesize N-[(2-hydroxyl)-propyl-3-trimethylammonium] chitosan chloride in an ionic liquid. *Carbohydr. Polym.* **2015**, *130*, 325–332. [CrossRef] [PubMed]

232. Chen, H.; Cui, S.; Zhao, Y.; Zhang, S.; Chen, H.; Peng, X.; Wang, B. O-Alkylation of Chitosan for Gene Delivery by Using Ionic Liquid in an in-situ Reactor. *Engineering* **2012**, *4*, 114–117. [CrossRef]

233. Yamamoto, K.; Yoshida, S.; Mine, S.; Kadokawa, J. Synthesis of chitin-graft-polystyrene via atom transfer radical polymerization initiated from a chitin macroinitiator. *Polym. Chem.* **2013**, *4*, 3384–3389. [CrossRef]

234. Lin, C.; Liu, D.; Luo, W.; Liu, Y.; Zhu, M.; Li, X.; Liu, M. Functionalization of chitosan via single electron transfer living radical polymerization in an ionic liquid and its antimicrobial activity. *J. Appl. Polym. Sci.* **2015**, *132*, 42754. [CrossRef]

235. Chen, H.; Cui, S.; Zhao, Y.; Zhang, C.; Zhang, S.; Peng, X. Grafting Chitosan with Polyethylenimine in an Ionic Liquid for Efficient Gene Delivery. *PLoS ONE* **2015**, *10*, e0121817. [CrossRef] [PubMed]

236. Wang, Z.; Zheng, L.; Li, C.; Zhang, D.; Xiao, Y.; Guan, G.; Zhu, W. A novel and simple procedure to synthesize chitosan-graft-polycaprolactone in an ionic liquid. *Carbohydr. Polym.* **2013**, *94*, 505–510. [CrossRef] [PubMed]

237. Zhou, Y.; Fan, M.; Luo, X.; Huang, L.; Chen, L. Acidic ionic liquid catalyzed crosslinking of oxycellulose with chitosan for advanced biocomposites. *Carbohydr. Polym.* **2014**, *113*, 108–114. [CrossRef] [PubMed]

238. Luo, Y.; Ling, Y.; Wang, X.; Han, Y.; Zeng, X.; Sun, R. Maillard reaction products from chitosan–xylan ionic liquid solution. *Carbohydr. Polym.* **2013**, *98*, 835–841. [CrossRef] [PubMed]

239. Guyomard-Lack, A.; Buchtová, N.; Humbert, B.; Bideau, J.L. Ion segregation in an ionic liquid confined within chitosan based chemical ionogels. *Phys. Chem. Chem. Phys.* **2015**, *17*, 23947–23951. [CrossRef] [PubMed]

240. Taheri, M.; Ghiaci, M.; Shchukarev, A. Cross-linked chitosan with a dicationic ionic liquid as a recyclable biopolymer-supported catalyst for cycloaddition of carbon dioxide with epoxides into cyclic carbonates. *New J. Chem.* **2018**, *42*, 587–597. [CrossRef]

241. Hua, D.; Jiang, J.; Kuang, L.; Jiang, J.; Zheng, W.; Liang, H. Smart Chitosan-Based Stimuli-Responsive Nanocarriers for the Controlled Delivery of Hydrophobic Pharmaceuticals. *Macromolecules* **2011**, *44*, 1298–1302. [CrossRef]

242. Wang, Z.; Zheng, L.; Li, C.; Wu, S.; Xiao, Y. Preparation and antimicrobial activity of sulfopropyl chitosan in an ionic liquid aqueous solution. *J. Appl. Polym. Sci.* **2017**, *134*, 44989. [CrossRef]

243. Ishii, D.; Ohashi, C.; Hayashi, H. Facile enhancement of the deacetylation degree of chitosan by hydrothermal treatment in an imidazolium-based ionic liquid. *Green Chem.* **2014**, *16*, 1764–1767. [CrossRef]

244. Zhang, Z.; Li, C.; Wang, Q.; Zhao, Z.K. Efficient hydrolysis of chitosan in ionic liquids. *Carbohydr. Polym.* **2009**, *78*, 685–689. [CrossRef]

245. Zhao, X.; Kong, A.; Hou, Y.; Shan, C.; Ding, H.; Shan, Y. An innovative method for oxidative degradation of chitosan with molecular oxygen catalyzed by metal phthalocyanine in neutral ionic liquid. *Carbohydr. Res.* **2009**, *344*, 2010–2013. [CrossRef] [PubMed]

246. Chen, X.; Liu, Y.; Kerton, F.M.; Yan, N. Conversion of chitin and *N*-acetyl-D-glucosamine into a *N*-containing furan derivative in ionic liquids. *RSC Adv.* **2015**, *5*, 20073–20080. [CrossRef]

247. Husson, E.; Hadad, C.; Huet, G.; Laclef, S.; Lesur, D.; Lambertyn, V.; Jamali, A.; Gottis, S.; Sarazin, C.; Nhien, A.N.V. The effect of room temperature ionic liquids on the selective biocatalytic hydrolysis of chitin via sequential or simultaneous strategies. *Green Chem.* **2017**, *19*, 4122–4131. [CrossRef]

248. Yuan, B.; Li, L.; Xie, C.; Liu, K.; Yu, S. Preparation of oligochitosan via In situ enzymatic hydrolysis of chitosan by amylase in [Gly]BF4 ionic liquid/water homogeneous system. *J. Appl. Polym. Sci.* **2014**, *131*, 41152. [CrossRef]

249. Zhao, G.; Lang, X.; Wang, F.; Li, J.; Li, X. A one-pot method for lipase-catalyzed synthesis of chitosan palmitate in mixed ionic liquids and its characterization. *Biochem. Eng. J.* **2017**, *126*, 24–29. [CrossRef]

polymers

MDPI

Review

Chitin Deacetylases: Structures, Specificities, and Biotech Applications

Laia Grifoll-Romero, Sergi Pascual, Hugo Aragunde, Xevi Biarnés and Antoni Planas *

Laboratory of Biochemistry, Institut Químic de Sarrià, Universitat Ramon Llull, 08017 Barcelona, Spain; laiagrifollr@iqs.edu (L.G.-R.); sergipascualt@iqs.edu (S.P.); hugoaragunde@gmail.com (H.A.); xavier.biarnes@iqs.edu (X.B.)
* Correspondence: antoni.planas@iqs.edu

Received: 19 February 2018; Accepted: 19 March 2018; Published: 22 March 2018

Abstract: Depolymerization and de-*N*-acetylation of chitin by chitinases and deacetylases generates a series of derivatives including chitosans and chitooligosaccharides (COS), which are involved in molecular recognition events such as modulation of cell signaling and morphogenesis, immune responses, and host-pathogen interactions. Chitosans and COS are also attractive scaffolds for the development of bionanomaterials for drug/gene delivery and tissue engineering applications. Most of the biological activities associated with COS seem to be largely dependent not only on the degree of polymerization but also on the acetylation pattern, which defines the charge density and distribution of GlcNAc and GlcNH$_2$ moieties in chitosans and COS. Chitin de-*N*-acetylases (CDAs) catalyze the hydrolysis of the acetamido group in GlcNAc residues of chitin, chitosan, and COS. The deacetylation patterns are diverse, some CDAs being specific for single positions, others showing multiple attack, processivity or random actions. This review summarizes the current knowledge on substrate specificity of bacterial and fungal CDAs, focusing on the structural and molecular aspects of their modes of action. Understanding the structural determinants of specificity will not only contribute to unravelling structure-function relationships, but also to use and engineer CDAs as biocatalysts for the production of tailor-made chitosans and COS for a growing number of applications.

Keywords: chitin deacetylases; chitosan; chitooligosaccharides; carbohydrate esterases; structure; substrate specificity; deacetylation pattern

1. Introduction

Chitin is a linear polysaccharide of β(1→4)-linked *N*-acetylglucosamine monomers. It was first isolated from fungi in 1811 [1] and its structure was determined in 1929 [2]. Chitin is a major structural component of the exoskeletons of arthropods (insects and crustaceans), of the endoskeletons of mollusks, and it is also found in the cell walls of fungi and diatoms [1,3,4]. It is one of the most abundant organic molecules after cellulose, and the most abundant natural amino polysaccharide. Chitin is present as ordered macrofibrils, mainly in two allomorphs: α-chitin, with antiparallel chains [5], is the most abundant and it is isolated from the exoskeleton of crustaceans, particularly from shrimps and crabs; and β-chitin, with parallel chains [6], is present in the cell walls of diatoms and in the skeletal structures of cephalopods, and commonly extracted from squid pens. β-Chitin is easily converted to α-chitin by alkaline treatment followed by flushing in water [5]. Chitin is also found as γ-chitin in fungi and yeast, which is a combination of the α and β allomorphs [7,8].

Depolymerization (hydrolysis) of chitin by chitinases results in chitooligosaccharides (COS), and de-*N*-acetylation of chitin and COS yields chitosan and partially acetylated COS (paCOS) or fully deacetylated glucosamine oligomers (Figure 1). In nature, the deacetylation of chitin is almost never complete, and chitosan refers to a family of heteropolysaccharides composed of *N*-acetylglucosamine and glucosamine, characterized by their degree of polymerization (DP), degree of acetylation (DA), and pattern of acetylation (PA). Only some fungi of the *Zygomycota*, *Basidiomycota* and *Ascomycota* phyla have been reported to be capable of naturally producing chitosans [9]. The free amino groups of the deacetylated units in the polymer are protonated at slightly acidic pH, thus making chitosans the only known natural polycationic polysaccharides [1,3,9]. They interact with polyanionic biomolecules such as polyanionic phospholipidic membranes, glycosaminoglycans at cell surfaces, proteins, and nucleic acids. Depolymerization by hydrolysis of the β-1,4-linkages of chitin and chitosan yields their respective oligosaccharides (COS, paCOS) [10]. Whereas chitin and chitosans act as structural polymers, their oligomers are involved in molecular recognition events, such as cell signaling and morphogenesis, and act as immune response elicitors and host-pathogen mediators [11–15]. The catabolism of chitin and chitosan is summarized in Figure 1. Chitin and chitosan oligosaccharides are further degraded following different organism-dependent pathways to end up in the central energy metabolism.

The de-*N*-acetylation of chitin and chitooligosaccharides is catalyzed by chitin deacetylases (CDAs), which exhibit different substrate specificities leading to fully or partially deacetylated products with diverse degrees of acetylation (DA) and patterns of acetylation (PA). In addition to the role CDAs play in the biology of their natural organisms, there is a growing interest in the biochemical characterization of CDAs in order to use them as biocatalysts for the production of partially deacetylated chitooligosaccharides (paCOS) as bioactive molecules in different application fields, or to inhibit them since they are potential targets against pathogenic microorganisms. The aim of this review is to analyze the current knowledge on the biochemistry of chitin deacetylases with regard to their substrate specificity. Although a large number of enzymes have been experimentally identified as chitin deacetylases able to deacetylate chitin either in vivo or in vitro, few of them have been analyzed at the protein level. Here we will focus on those CDAs with characterized activity on chitooligosaccharides (COS) to address the issue of substrate specificity and the deacetylation pattern.

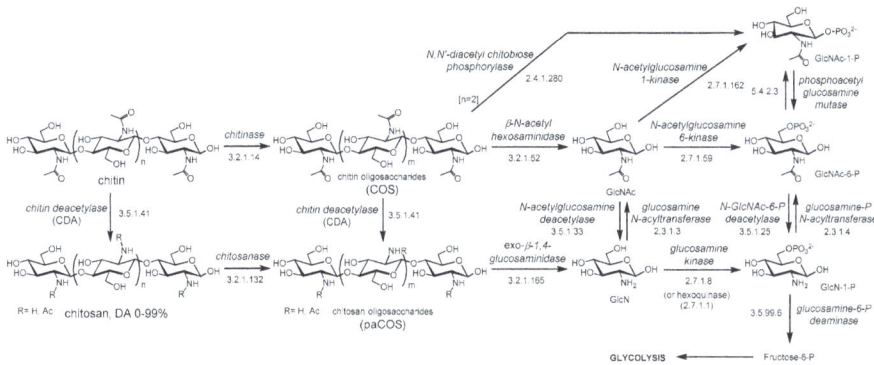

Figure 1. Chitin catabolism. GlcNAc: *N*-acetylglucosamine; GlcN: glucosamine; DA: degree of acetylation. COS: chitooligosaccharides (or chitin oligosaccharides), paCOS: partially acetylated chitooligosaccharides (or chitosan oligosaccharides).

2. Chitin Deacetylases and the Carbohydrate Esterase Family 4 (CE4)

Chitin deacetylases (CDAs, EC 3.5.1.41) and chitooligosaccharides deacetylases (CODs, EC 3.5.1.105) are classified in the carbohydrate esterase family 4 (CE4) in the CAZY database (Carbohydrate Active Enzymes, www.cazy.org) [16]. The CE4 family also contains peptidoglycan *N*-acetylglucosamine deacetylases (EC 3.5.1.104), peptidoglycan *N*-acetylmuramic acid deacetylases (EC 3.5.1.-), poly-β-1,6-*N*-acetylglucosamine deacetylase (EC 3.5.1.-), and some acetyl xylan esterases (EC 3.1.1.72) [17]. These enzymes share a conserved region known as the NodB homology domain due to its similarity to the NodB oligosaccharide deacetylase, one of the first CE4 enzymes to be characterized [18]. Most currently reported and characterized CDAs and CODs are CE4 enzymes, with the exception of diacetylchitobiose deacetylases (Dacs) from archaea and a COD from *Bacillus cereus* (BcZBP) that belong to the CE14 family [19,20]. Few other enzymes, such as insect CDAs and a COD from *E. coli* (ChbG) [21] are in the group of "non-classified" in the CAZY database since they do not share sequence similarities to the other CDA families.

The deacetylase activity from extracts of the fungus *Mucor rouxii* was the first active CDA identified and partially purified in the mid 1970s [22,23]. Later on, the NodB from a rhizobium species was the first biochemically characterized COD in 1993 [18]. Many other CDAs and CODs were later identified and purified from very diverse organisms, including archaea, marine bacteria, fungi, and insects [24]. These enzymes are diverse in their biochemical properties: molecular masses in the range from 12 to 150 kDa, acidic isoelectric points (pI from 2.7 to 4.8), optimum pH for activity from 4.5 to 12, and significant thermal stability, with optimum temperatures for activity in the range from 30 to 60 °C. Most CDAs are highly inactive on crystalline chitin due to the inaccessibility of the acetyl groups in the tightly packed chitin structure, and have a preference for soluble forms of chitins such as glycol-chitin or chitin oligomers, as well as partially deacetylated chitin (chitosans). It has recently been shown that CDA activity on crystalline chitin is greatly enhanced by oxidative cleavage of the surface polymer chains by lytic polysaccharide monooxygenases (LPMO) [25]. Some CDAs contain carbohydrate binding modules (CBM) fused to the catalytic domain that seem to enhance the deacetylase activity by increasing the accessibility of the substrate to the catalytic domain [26].

CDAs are localized in different cellular compartments, in the periplasm, in the cytosol, or secreted as extracellular enzymes. Periplasmic fungal CDAs are generally tightly coupled to a chitin synthase to rapidly deacetylate newly synthesized chitin before their maturation and crystallization. Extracellular fungal CDAs are secreted to alter the physicochemical properties of the cell wall, which results in protection against exogenous chitinases, or initiates sporulation or autolysis. In bacteria, CDAs are either intracellular, as those involved in Nod factors biosynthesis in *Rhizobium* species, or extracellular, as those involved in the catabolism of chitin in marine bacteria [24,27,28].

Some CE4 enzymes classified in a specific subfamily also show activity on typical substrates from other subfamilies. Peptidoglycan GlcNAc deacetylases, involved in the de-*N*-acetylation of the bacterial cell wall peptidoglycan with critical functions in the maturation and turnover of peptidoglycan and in bacterial pathogenicity, are also active on COS. Some CDAs have activity on acetylxylan, as well as some acetylxylan esterases are active on COS, which makes their classification doubtful in some cases.

3. Function and Specificity of CE4 Chitin Deacetylases

3.1. Deacetylation Patterns

Chitin deacetylases exhibit diverse deacetylation patterns, reflecting different substrate specificities and pattern recognition on their linear substrates. The mechanisms of action of enzymes that modify in-chain units on a linear polysaccharide are commonly classified as multiple-attack, multiple-chain, and single-chain mechanisms [29]. In the multiple-attack mechanism, binding of the enzyme to the polysaccharide chain is followed by a number of sequential deacetylations, after which the enzyme binds to another region of the polymeric chain. (i.e., *M. rouxii* [30,31]). On a polymeric substrate, this mechanism will result in a block-copolymer structure with blocks of GlcNH$_2$ units within the GlcNAc chain. On COS, it will usually result in full deacetylation of the oligomer. In the multiple-chain mechanism, the enzyme forms an active enzyme-polymer complex and catalyzes the hydrolysis of only one acetyl group before it dissociates and forms a new active complex (i.e., *C. lindemuthianum* CDA [32,33]). It will result in a random distribution of the GlcNH$_2$ and GlcNAc units along the polymeric chain or, in the case of COS substrates, it will render a number of partially deacetylated oligosaccharide intermediates ending in a specific deacetylation pattern or in full deacetylation, depending on the enzyme and the substrate. Finally, a single-chain mechanism includes processive enzymes in which a number of catalytic events occur on a single substrate molecule leading to sequential deacetylation. Some bacterial chitooligosaccharides deacetylases (CODs), which are specific for a single position leading to mono-deacetylated products (i.e., *Rhizobium* NodB or *Vibrio* CDA or COD) are also included in the last group.

A fundamental question is how these enzymes define their action pattern. This is relevant not only to understand their biological functions but also to use CDAs (native and engineered variants) as biocatalysts for the production of chitosans with non-random deacetylation patterns, and partially deacetylated COS with tailored patterns of acetylation (see Section 6). A structural model on the determinants of substrate specificity is currently emerging from studies on substrate specificity, determination of 3D structures of enzyme-substrate complexes, and multiple sequence alignments. Many CDAs, particularly from fungal origin, have been identified as involved in chitin deacetylation in vivo, but only few of them have been characterized with regard to substrate specificity and mode of action. To the best of our knowledge, Table 1 compiles the CDA enzymes in family CE4 that have been biochemically characterized and have reported activity on COS substrates, some of them with solved three-dimensional structure by X-ray crystallography. Relevant information on their biological function and substrate specificity is summarized below.

CDAs participate in diverse biological processes, which include cell wall morphogenesis and host-pathogen interaction in fungi, generation of signaling molecules in bacteria, and participation in the catabolism of chitin as carbon, nitrogen, and energy sources in marine bacteria and fungi. CDAs were thought to be restricted to fungi and bacteria until a first report in 1986 on their presence in arthropods [34]. CDAs seem to be widely present in insects, in cuticles and the perithrophic midgut matrix, but little is known on the function and properties of insect CDAs [35,36], and they are not included here because scarce information on substrate specificities has been reported.

3.2. Fungal CDAs

Fungal CDAs are involved in fungal nutrition, morphogenesis and development [27,29], participating in cell wall formation and integrity [37], in spore formation [38], germling adhesion [39], fungal autolysis [40], and in defense mechanisms for host infection [41].

Fungi that have chitosan (in addition to chitin) as a structural component of the cell wall, secrete CDAs to the periplasmic space that contribute to chitosan biosynthesis from nascent chitin synthesized by chitin synthases. It occurs during exclusive periods corresponding to their particular biological role in the cell cycle of the fungal species: during cell wall formation (i.e., *M. rouxi* [42] and *A. coerulea* [43]), during sporulation (i.e., *S. cerevisiae* [38,44]), or during vegetative growth (i.e., *C. neoformas* [37]).

Pathogenic fungi secrete CDAs during fungal hyphae penetration to evade plant defense mechanisms and gain access to host tissues. Plants secrete chitinases to break the fungal cell-wall chitin down to chitooligosaccharides (COS), and the released COS are recognized by plant chitin-specific receptors, triggering resistance responses [41]. COS elicitation of resistance mechanisms involve activation of host defense genes [45,46]. There is cumulative evidence that fungi evade plant defense mechanisms by partially deacetylating either their exposed cell wall chitin or the chitooligosaccharides produced by the action of plant chitinases. In both cases, the resulting partially deacetylated oligomers are not well recognized by the specific plant receptors reducing or preventing the elicitation of the defense responses [41,47–49].

Filamentous fungi undergo autolysis by self-digestion of aged hyphal cultures due to carbon starvation [40,50]. During this event there is an increased presence of hydrolytic enzymes, especially those involved in cell wall degradation, and CDAs are secreted to the extracellular medium to deacetylate the chitin oligomers produced by chitinases (i.e., *A. nidulans* [51,52]).

Table 1. CDAs and CODs with characterized activity on COS.

Enzyme [1]	Organism	ID [2]	PDB [3] [Ref.]	COS Substrates [4]	Ref. [5] COS	Metal [6]	PA [7] (on A_n)
MrCDA	*Mucor rouxii*	P50325		≥DP3	[31]	Zn^{2+}	D_n
ClCDA	*Colletotrichum lindemuthianum*	Q6DWK3	2IW0 [53]	DP6 > DP5 > DP4 > DP3 > DP2	[32]	Co^{2+}, Zn^{2+}	D_n
AnCDA	*Aspergillus nidulans*	Q5AQQ0	2Y8U [57]	DP2 > DP3 > DP4 > DP5	[23]	Co^{2+}	D_n
PaCDA	*Podospora anserine*	XP_001912680.1		≥DP2	[51]	Zn^{2+}	D_n
PgtCDA	*Puccinia graminis*	XP_003323413.1		DP6 > DP5 > DP4	[51]	n.r. [7]	AAD_{n-2}
PesCDA	*Pestalotiopsis sp.*	APH81274.1		DP6-DP5-DP4	[49]	n.r.	$AAD_{n-3}A$
PcCDA	*Pochonia chlamydosporia*			DP5 > DP4	[51]	n.r.	$ADDA_{n-3}$
ScCDA1	*Saccharomyces cerevisiae*	Q06702		DP4, DP6	[56]	n.r.	n.r.
ScCDA2	*Saccharomyces cerevisiae*	Q06703		DP6 > DP5 > DP4 > DP3 > DP2	[57]	Co^{2+}	n.r.
MoCDA *	*Mortierella sp.*			DP7 > DP6 > DP5 > DP4 > DP3 > DP2	[8]	(Co^{2+})	n.r.
AcoCDA *	*Absidia corymbifera*			DP5 > DP4 > DP3	[13]	n.r.	n.r.
AcorCDA *	*Absidia corymbifera*			DP7 > DP6 > DP5 > DP4 > DP3 > DP2	[59]	$(Co^{2+}, Ca^{2+}, Mg^{2+})$	n.r.
FvCDA	*Flammulina velutipes*	BAE92728.1		DP5 > DP4 > DP3 > DP2	[60]	$(Co^{2+}, Ca^{2+}, Zn^{2+})$	n.r.
PoCDA *	*Penicillium oxalicum*			DP5 >> DP3 > DP2	[61]	$(Co^{2+}\ Cu^{2+})$	n.r.
AfCDA *	*Aspergillus flavus*			DP4	[62]	(Zn^{2+}, Mn^{2+})	n.r.
SbCDA *	*Scopulariopsis brevicaulis*			DP6 > DP5 > DP4 > DP3 > DP2	[63]	n.r.	n.r.
RcCDA	*Rhizopus circinans*	A7UMZ0		DP6	[64]	(Mn^{2+}, Mg^{2+})	n.r.
RsCDA	*Rhizopus stolonifer (nigricans)*	Q32XH4		n.r.	[64]		
GbCDA	*Gongronella butleri*	Q8J2N6		n.r.	[65]		
PbCDA	*Phycomyces blakesleeanus*	Q9P4U2		n.r.	[66]		
SchCDA	*Schizophyllum commune*	Q9P453		n.r.	[67]		
CnCDA1,2,3	*Cryptococcus neoformans*	Q5KFC8, Q5KIC2, P0CP76		n.r.	[7]		
EhCDA	*Entamoeba histolytica*	XP_656356.1		DP5, DP6	[68]	n.r.	n.r.
NodB	*Sinorhizobium meliloti*	P02963		DP5 > DP2 (DP4, DP3)	[15]	$Mn^{2+}\ Mg^{2+}$	DA_{n-1}
VcCOD (VcCDA)	*Vibrio cholera*	Q9KSH6	4NY2 [69]	DP2 > DP3 > DP4 > DP5 > DP6	[70]	Zn^{2+}	ADA_{n-2}
VpCOD	*Vibrio parahaemolyticus*	Q9KSH6	3WX7 [71]	DP2 > DP3	[72]	Zn^{2+}	n.r.
VaCOD	*Vibrio alginolyticus*	Q9KSH6		DP2	[73]	Zn^{2+}	AD
SwCOD	*Shewanella woodyi*	ACA84860.1		DP2 > DP3 > DP4	[74]	n.r.	AD; $[ADA_{n-2}]$
SbCOD	*Shewanella baltica*	ABN60929.1		DP2 > DP4 > DP3	[75]	n.r.	AD; $[ADA_{n-2}]$
ArCE4A	*Arthrobacter sp.*	A0A2C8C1I7	5LFZ [76]	DP5 > DP6 ≈ DP4 > DP3 >> DP2	[9]	Ni^{2+} 8	$D_{n-1}A$

[1] Characterized recombinant enzymes, except those with an asterisk (*) that have been characterized from the native organism and are not included in the sequence alignment, Figure 4. [2] Uniprot or GenBank accession code, [3] PDB accession code and publication of the 3D structure. [4] Activity on chitooligosaccharides (COS) as a function of the degree of polymerization (DP), [5] Selected publication on substrate specificity. [6] Native metal cation or, in parenthesis, metals that enhanced the enzyme activity when added in the reaction buffer. [6] Pattern of acetylation (PA): structure of the main final deacetylated product (A: GlcNAc; D: GlcNH₂). Other patterns of acetylation with specific substrates are given in the text. [7] n.r.: not reported. [8] Native metal unknown, Ni^{2+} probably from purification/crystallization.

In general, fungal CDAs deacetylate soluble forms of chitin such as glycol-chitin and chitosans of variable DA, but they are inactive, or show low activity, towards insoluble chitins such as crystalline α- and β-chitin and colloidal chitin. Pretreatment of chitin to make surface fibrils more accessible may result in increased deacetylase activity. It has been recently shown that the activity of an *A. nidulans* CDA on crystalline chitin was enhanced by a lytic polysaccharide monooxygenase (LPMO) that increases substrate accessibility by oxidative cleavage of the chitin chains [25]. Some CDAs also appear to be active on acetylxylan (i.e., *An*CDA), but any of them act on peptidoglycans, typical substrates of other CE4 family members. In addition to polymeric substrate, not many CDA have been analyzed with COS as substrates (Table 1). The analysis of the products from enzymatic deacetylation with regards to the extent and pattern of deacetylated provides information about the specificity and mode of action of these enzymes. The first seven entries in Table 1 correspond to CDAs for which the deacetylation pattern on COS has been reported, whereas the rest of the entries are CDAs active on COS but, to the best of our knowledge, the structure of the deacetylated products has not been analyzed.

Mucor rouxii (*Amylomyces rouxii*) (*Mr*CDA). The dimorphic fungus *M. rouxii* has a cell wall mainly composed of chitin, chitosan, and mucoric acid. While chitin accounts for 10% of the total dry weight of the cell wall, chitosan reaches 30% [77]. The *M. rouxii* CDA enzyme was initially found in the cytosol [23], but it is also secreted into the periplasm where it participates in a tandem synthetic mechanism that involves a chitin synthase and a chitin deacetylase working consecutively and synchronously for synthesis and deposition of chitosan polymers at the outer membrane [78]. Decoupling this mechanism prevents the formation of chitosan [42,79]. *M. rouxii*, like other fungi, has been identified as a suitable microorganism for chitosan production by means of biofermentation processes [80,81]. *Mr*CDA is a monomeric high-mannose-type glycosylated protein with an apparent molecular mass of 75–80 kDa [82]. Kinetic studies indicate that the preferred catalytic metal is Zn^{2+} like many other CDAs. In terms of activity, its optimal pH and temperature values are between 4.5–5.5 and 50 °C, respectively [23,33]. *Mr*CDA deacetylates chitinous polymers such as glycol-chitin, colloidal chitin, chitosan, and chitin, but also deacetylates acetylxylan [17,23]. On chitooligosaccharides, triacetylchitotriose is the smallest substrate and the activity increases with the degree of polymerization (DP) [23,31,42,78]. The enzyme follows a multiple-attack mechanism [30] but the resulting pattern of acetylation (PA) depends on the DP of the substrate: whereas DP3, DP6 and DP7 substrates are not fully deacetylated, leaving the reducing GlcNAc unmodified [$D_{n-1}A$], DP4 and DP5 substrates are fully deacetylated [D_n]. In all cases, deacetylation starts at the non-reducing end residue and then proceeds to the neighboring monomer towards the reducing end [31].

Colletotrichum lindemuthianum (*Cl*CDA). The deuteromycete *C. lindemuthianum* is a plant pathogen that causes anthracnose, a disease which affects economically important crop species [83]. *Cl*CDA is a heavily glycosylated secreted enzyme allegedly playing a role in the host-pathogen interaction, deacetylating the chitin oligomers resulting from the activity of plant chitinases on the fungal cell wall [83,84]. Less likely is its function in deacetylating the fungal cell wall chitin to evade degradation by plant chitinases, since no chitosan has been observed in the cell wall ultra-structure [85]. Since its discovery in the 1980s it has been purified from its natural host [84,86] as well as expressed in several eukaryotic and bacterial hosts such as *Pichia pastoris* [87,88] and *E. coli* [89,90]. The enzyme has a preference for Co^{2+} and Zn^{2+} as the catalytic metal cation and its activity is substantially inhibited by Cu^{2+} or Ni^{2+}, but not inhibited by EDTA or acetate [53,86]. It is a quite thermostable enzyme with an optimum temperature of 60 °C, and a pH optimum of 8.0. *Cl*CDA is active on both chitin polymers (glycol-chitin) and COS. It fully deacetylates COS with a DP equal to or greater than 3, while it only deacetylates the non-reducing GlcNAc of diacetylchitobiose [32,91]. *Cl*CDA acts by a multiple-chain mechanism following a pathway in which the first residue to be deacetylated is the second from the reducing end [32,33]. The initial mono-deacetylation reaction shows no dependency of k_{cat} on DP and a decrease of K_M with increasing DP [33,53]. However, kinetics of fully deacetylated products formation show an increase in k_{cat} and reduction in K_M that correlate with the increase of DP [86]. It has been reported that this enzyme is reversible, as it is also able to catalyze the acetylation of chitosan oligomers [92–94].

Aspergillus nidulans (*An*CDA). During cell autolysis, *An*CDA is secreted into the extracellular medium to deacetylate the chitin oligomers produced by chitinases [40,50,52,95]. The enzyme has been purified from A. nidulans cultures as a glycosylated enzyme [51]. The recombinant protein has been expressed in E. coli and purified by refolding from inclusion bodies [96] and recently it has been obtained in soluble form [25]. Like ClCDA, *An*CDA is a thermostable protein with an optimal temperature of 50 °C and retaining 68% activity after 1 h at 80 °C. Its optimum pH is 7–8 [51,96]. The enzyme is active on soluble chitins (CM-chitin, glycol-chitin), colloidal chitin, chitosan, acetylxylan, and acetylated glucuronoxylan, but not on peptidoglycan [25,51]. *An*CDA is active on COS with a DP from 2 to 6 [25]. The enzyme catalyzes mono-deacetylation of $(GlcNAc)_2$ and it is inactive on GlcNAc monosaccharide. Longer substrates than DP2 are fully deacetylated. However, the deacetylation rate exhibits a counter-intuitive relationship with the DP of the substrate: odd-numbered COS (DP5, DP3) have higher apparent rate constants than even-numbered oligomers (DP4, DP2). For the DP6 substrate, time-course monitoring of products formation reveals that the first deacetylation event occurs at random positions except for the reducing end, which reacts much slower to yield the fully deacetylated end product $[D_n]$.

Podospora anserina (*Pa*CDA). The filamentous ascomycete *Podospora anserina* lives as a saprophyte on herbivore dung [97]. It has a limited lifespan and it is a model organism in cell aging studies [98]. *Pa*CDA was identified in a search for CDAs containing chitin binding domains. The enzyme has been recombinantly expressed in *Hansenula polymorpha* as a full length protein composed of the CE4 domain flanked by two CBM18 domains [26]. The low activity of the enzyme on colloidal chitin is significantly reduced by deletion of the CBM domains, which supports the hypothesis that the presence of the CBMs helps the enzyme to act on insoluble substrates. *Pa*CDA is active on soluble glycol-chitin, chitosans with a high DA, and COS, with optimum pH and temperature values of 8.0 and 55 °C, respectively. It fully deacetylates COS with a DP $\geq$ 2 and follows a multiple-chain mechanism. With the DP3 substrate, the first deacetylation event has a clear preference for the reducing end, but all possible isomers are found for both mono- and di-deacetylated intermediate products. With DP4 and DP5 substrates, the residue next to the reducing end is preferentially deacetylated first, with the second deacetylation occurring mainly next to the existing $GlcNH_2$ unit on either side. Deacetylation is faster for longer substrates, with deacetylation of the reducing end occurring as a late event [26].

Puccinia graminis f. sp. *Tritici* (*Pgt*CDA). The biotrophic basidiomycete *Puccinia graminis f.* sp. *Tritici* is the causative agent of the stem rust [99]. The appearance of resistant races of *P. graminis* affecting wheat cultivars has been recognized as a serious threat to food security [100,101], boosting the interest in understanding the virulence and defense mechanism of this fungal pathogen. Rust fungi promote the formation of complex structures in order to invade the plant cells but at the same time they must prevent the triggering of immune responses [102]. A main transition during infection is from the ectophytically growing appressorium to the endophytically growing substomatal vesicle; while the former exposes chitin on its surface, the latter exposes chitosan [47,103]. *Pgt*CDA may not only participate in the chitin to chitosan transition, making the cell wall less susceptible to host chitinases [104], but also could deacetylate the chitooligosaccharide products, reducing its elicitor properties [105]. *Pgt*CDA has been recombinantly expressed in *E. coli* as a fusion protein with the maltose binding protein (MBP) [54]. Its optimal pH for activity is between 8 and 9 and its optimal temperature is 50 °C. It is not active on insoluble polymers such as α- or β-chitin, but efficiently deacetylates colloidal chitin, glycol-chitin and chitosans, on which activity increases with the degree of acetylation. With COS substrates, the minimal substrate is tetraacetylchitotetraose (DP4). The structure of the products from enzymatic deacetylation of DP4 to DP6 substrates reveals that the enzyme acts by a multiple-chain mechanism and specifically deacetylates all but the last two GlcNAc units on the non-reducing end $[AA(D)_{n-2}]$ [54].

Pestalotiopis sp. (*Pes*CDA). The endophytic fungus *Pestalotiopis* sp. lives inside the tissues of its plant hosts in tropical areas [106]. To successfully survive in their hosts, endophytes also need to avoid being detected by the plant immune system. A secreted *Pestalotiopis* CDA has been identified when chitosan was present in the culture medium [49]. The recombinantly expressed *Pes*CDA is active on colloidal chitin as substrate, chitosans with a DA of 10–60% (higher activity with a higher DA), and COS, but inactive on crystalline α- or β-chitin. When analyzing the activity on COS, tetraacetylchitotetraose is the minimal substrate. With a DP5 substrate, the optimum pH and temperature values are 8.0 and 55 °C, respectively. Through a multiple-chain mechanism, the enzyme deacetylates all residues of the substrates except the reducing end and the last two GlcNAc residues from the non-reducing end, with a pattern of deacetylation [AA(D)$_{n-3}$A] [49]. The chitosan oligomers obtained from deacetylation of a DP6 substrate by *Pes*CDA have shown that, as opposed to the fully acetylated oligomer, they are no longer elicitors of the plant immune system in rice cells [49].

Pochonia chlamydosporia (*Pc*CDA). The ascomycete *Pochonia chlamydosporia* infects females and eggs of cyst or root-knot nematodes. It is used as a biocontrol agent against a number of plant parasitic nematodes in food-security crops [107–109]. *P. chlamydosporia* expresses chitosanases and chitin deacetylases during egg infection. Since chitosan is associated with the sites of fungal penetration, it has been suggested that secreted CDAs are involved in nematode infection [110]. A *Pc*CDA has been recently characterized [55]. The full-length protein contains the CE4 catalytic domain flanked by two CBM18 chitin binding domains. The recombinantly expressed *Pc*CDA catalytic domain deacetylates COS with a DP $\geq$ 4, with preference for longer substrates. It starts deacetylating the penultimate residue from the non-reducing end and continues deacetylating the next residue towards the reducing end, with a pattern of acetylation [ADDA$_{n-3}$] [55].

The above described CDAs are well characterized in terms of their deacetylation mode of action on COS and the structure of their deacetylated products. A number of other fungal CDAs (Table 1, and below) have also been assayed on COS substrates but, to the best of our knowledge, the deacetylation pattern of the products has not been reported.

Saccharomyces cerevisiae (*Sc*CDA1 and 2). The *S. cerevisiae* ascospore walls are well ordered structures with two outer layers that confer spore resistance, one made of 95% chitosan and the outermost proteinaceous layer rich in dityrosine [44]. Two CDAs are expressed exclusively during sporulation and are required for spore wall rigidity [38]. Both CDAs have been cloned and expressed in yeast as glycosylated proteins active on glycol-chitin [38,56], and in *E. coli* [57] as soluble proteins with deacetylase activity on glycol-chitin, chitosan (DA 50%) and COS. More detailed characterization of *Sc*CDA2 expressed in *E. coli* revealed that at least two GlcNAc residues are required for activity on COS, with maximum activity on DP6 [57]. When glycol chitin is used as substrate the optimum temperature for enzyme activity is 50 °C and the pH optimum is 8.0. It has also been shown that the *Sc*CDAs may act on nascent chitin chains in an in vitro assay system with chitin synthase [56].

Mortierella sp. (*Mo*CDA). Some *Mortierella* species live as saprotrophs in soil and other organic materials such as decaying plant leafs, fecal pellets or on the exoskeleton of arthropods, whereas other species are endophytes [111]. An extracellular CDA was identified [58] and purified from a *Mortierella* sp. as a highly glycosylated protein with maximum activity at pH 5.5–6 and 60 °C [112]. *Mo*CDA is active on soluble substrates as chitosans and glycol-chitin but with no detectable activity on β-chitin, colloidal chitin, and CM-chitin. It is active on COS with a DP $\geq$ 2, with higher activity with increasing DP of the substrate. With diacetylchitobiose, only monodeacetylation was observed. The structure of the deacetylated products from larger oligomers has not been reported.

Absidia sp. (*Acoe*CDA, *Acory*CDA). *Absidia* strains of *Zygomycetes* produce chitosan in their cell wall through the tandem action of chitin synthases and deacetylases. In *A. coerulea*, chitosan accounts for 10% of the vegetative cells and the DA reaches 95%. *Acoe*CDA was purified and proven to be active on glycol-chitin with a pH optimum of 5 at 50 °C. When the purified enzyme was incubated with a chitin synthase, it converted 90% of the nascent chitin from UDP-GlcNAc into chitosan. It deacetylates COS with more than two GlcNAc units, with increasing activity with longer substrates [43]. Similarly, *Absidia corymbifera* secretes a CDA active on glycol-chitin and chitosans with optimum pH and temperature of 6.5 and 55 °C, respectively, and active on COS with DP $\geq$ 2 [59].

Flammulina velutipes (*Fv*CDA). The basidiomycete *Flammulina velutipes* (called Enokitake in Japan) is commercially cultivated and fruited to produce foods with high nutritional value. A CDA that is expressed at the early stages of fruity body development was recombinantly expressed in *Pichia pastoris* [60]. *Fv*CDA, active on glycol-chitin and colloidal chitin, deacetylates COS from dimer to pentamer, with activity increasing with the DP of the substrate. The enzyme exhibits the maximum activity at 60 °C and pH 7.

Penicilium oxilicum (PoCDA). An extracellular CDA from *Penicilium oxilicum*, purified from culture supernatants, exhibits deacetylase activity on glycol-chitin at pH 9, a common value for extracellularly secreted CDAs as opposed to intracellular CDAs with typical pH optima in the range of 5 to 7. *Po*CDA is active on COS with activity increasing from DP2 to DP5 [61].

Aspergillus flavus (AfCDA). In the search for extracellularly secreted CDAs for industrial applications, optimization of solid substrate fermentation and submerged fermentation of *Aspergillus flavus* has been reported [62,113]. The *Af*CDA enzyme purified from the extracellular medium has optimal activity on glycol-chitin and colloidal chitin at pH 8 and 50 °C. When assayed with COS as substrates, *Af*CDA is active on DP4 but has no activity on shorter substrates [62].

Scopulariopsis brevicaulis (SbCDA). *Scopulariopsis* spp. are common soil saprophytes. Few species have been associated with human diseases, including *S. brevicaulis*. They are dermatomycotic molds and mainly have been associated with onychomycosis [114,115]. *Sb*CDA is an extracellular enzyme that is active on chitin and chitosans. The purified native enzyme is also active on COS with at least two GlcNAc units, and the activity increases with the DP of the substrate. With DP6, optimum conditions for deacetylation are pH 7.5 and 55 °C [63].

Rhizopus sp. (*RcCDA, RsCDA*). *Rhizopus* species have been screened as CDA producers. A *R. circicans* CDA has been cloned and recombinantly expressed in *Pichia pastoris* [64]. *Rc*CDA has maximum activity on glycol-chitin at pH 5–6 and 37 °C. On COS, only activity on a DP6 substrate has been reported. A CDA from *Rhizopus stolonifer* (or *nigricans*) has also been isolated as an active enzyme on glycol-chitin but no activity on COS has been reported [64,116]. Fermentation conditions of other *Rhizopus* species as CDA producers are being studied for the bioconversion of chitin to chitosan [117].

Other chitosan producers have been identified and studied as a source of chitosans, with many reports on screening and fermentation optimization, but the corresponding chitin deacetylases have not been characterized yet. Some examples include *Gongronella butleri* [65,118], *Phycomyces blakesleeanus* [66,119], and *Schizophyllum commune* [67].

Cryptococcus neoformans (CnCDA). *Cryptococcus neoformans* is a dimorphic basidiomycetous human fungal pathogen that causes cryptococcal meningoencephalitis, particularly in immunocompromised patients [37]. *C. neoformans* has substantial chitosan in its cell wall during vegetative growth that is necessary for virulence and persistence in the mammalian host [120,121]. Three CDAs are predicted to be GPI-anchored to the cell wall, suggesting that they transverse the plasma membrane or attach to the cell wall to deacetylate the chitin generated by a chitin synthase as it is extruded through the plasma membrane [37]. The GPI-anchor of *Cn*CDA2 has proven to be required for membrane association but dispensable for cell wall association [122]. Activity of *C. neoformans* CDAs on COS substrates has not been reported. Interestingly, screening studies to identify cryptococcal antigens that stimulate an immune response on murine T cell hybridomas reactive with cryptococcal proteins, have shown that two of the CDAs are immunogenic [123,124].

3.3. Protozoan CDAs

Entamoeba histolytica (EhCDA). *Entamoeba histolytica* is an anaerobic parasitic amoebozoan that predominantly infects humans and other primates causing amoebiasis [125]. The genome contains two putative CDAs, one of which has been cloned and recombinantly expressed in *Saccharomyces cerevisiae* [68]. *Eh*CDA deacetylates COS, being active on DP5 and DP6, but with no detected activity on DP4 [68].

3.4. Bacterial CDAs

The predominant CE4 deacetylases in bacteria are chitin oligosaccharide deacetylases (CODs), active on low molecular mass COS and essentially inactive on polymeric chitin and chitosans. These include rhizobial NodB deacetylases and CODs from marine bacteria. But bacterial CDAs other than CODs are being discovered from screening programs and data mining of sequenced genomes and metagenomes, as in the recent case of an *Arthobacter* CDA.

Sinorhizobium meliloti (NodB). Rhizobial NodB is part of the Nod operon involved in the biosynthesis of Nod factors, the morphogenic signal molecules produced by rhizobia, which initiate the development of root nodules in leguminous plants [126]. NodB is active on chitooligosaccharides from DP2 to DP5 with no differences in k_{cat}, but K_M decreases with increasing DP [18,127–129]. Specifically, k_{cat}/K_M is 5-fold higher for DP5 than for DP2 substrates. DP4 or DP5 substrates are the natural substrates depending on the Rhizobial strain. *Sm*NodB optimum activity between pH 7 and 8 at 30 °C [18]. NodB is highly specific deacetylating only the non-reducing end residue [DA_{n-1}] although traces of a second deacetylation event have been observed upon long incubations [18,130,131].

Vibrio species (*Vc*CDA, *Vp*CDA, *Va*CDA). Chitin oligosaccharide deacetylases (COD) from the *Vibrionaceae* family are involved in the chitin degradation cascades occurring in sea water [132–135]. They have been identified in many *Vibrio* species, such as *V. algynolyticus* [73,136], *V. parahaemolyticus* [72,137], *V. cholera* [70], *V. harveyi* [138] and others. The *V. parahaemolyticus* and *Vibrio* sp. SN184 CDAs only deacetylate DP2 and DP3 substrates, whereas the *Vibrio cholera* chitin deacetylase (*Vc*CDA) has a broader specificity, accepting substrate from DP2 to DP6 [69,70]. *Vc*CDA has a 10-fold higher activity on DP2 than on DP4 [69], and specifically deacetylates the penultimate residue from the non-reducing end, generating monodeacetylated products with the pattern [ADA_{n-2}] [69,70,130].

Shewanella species (*Sw*COD, *Sb*COD). In addition to the *Vibrio* genus, CODs have been recently identified and characterized from the *Shewanella* genus, marine bacteria found in extreme aquatic habitats (low temperature and high pressure). *Shewanella* sp. CODs share high sequence identity (50–60%) with *Vibrio* CODs, and have essentially the same biochemical properties. The *S. woodyi* enzyme (*Sw*COD) contains two CBM12 chitin binding domains at the C-terminus, deacetylates the reducing end on diacetylchitobiose [AD], and the activity drastically decreases from DP2 to DP4 substrates, with no activity detected on a DP5 substrate [74]. The *S. baltica* enzyme (*Sb*COD) contains a single CBM12 at the C-terminus, it is active of diacetylchitobiose with the same deacetylation pattern [AD] but it is less active on a DP3 than on a DP4 substrate [75].

Arthrobacter sp. (*Ar*CE4). A bioinformatics search for monodomain and extracellular CDAs in annotated genomes and metagenomes identified *Ar*CE4 as a CDA from an *Arthrobacter* species [76], a Gram-positive bacteria known to grow on chitin and secrete chitinases [139–141]. *Ar*CE4 is active on chitosan (DA 64%), acetylxylan, and insoluble chitin. It also deacetylates COS substrates with DP ≥ 2. The activity increases with increasing DP, with higher activity against DP5 compared to DP6. As shown with the DP5 substrate, the enzyme follows a multiple-chain mechanism where different mono- and di-deacetylated products are obtained. Whereas the first deacetylation occurs at all three internal positions, di-deacetylation mainly takes place at the GlcNAc unit next to the reducing end and at either of the two other internal units (ADDAA and ADADA). The final products have a pattern of acetylation [$D_{n-1}A$], where the reducing end unit is not deacetylated [76].

4. Structural Determinants of Activity and Specificity

Structural analysis of CE4 enzymes with solved 3D structure have been recently reviewed [142], comparing and highlighting the differences between the different subfamilies based on substrate preferences. Here we focus and summarize the current knowledge on the structure and specificity of CDAs as a subfamily of CE4 enzymes. The closer similarity and activity on the same substrates provides a framework to analyze the structural determinants responsible for the different modes of action that lead to different patterns of deacetylation in their products.

4.1. 3D Structures

Some CDAs are mono-domain proteins and some others have a multi-domain architecture composed of the CE4 catalytic domain (or NodB homology domain), and several other domains, such as carbohydrate binding modules (CBMs [143]) and domains with unknown function. The function of the CBMs is not clear and might be diverse depending on the biological role of each enzyme in its organism. In extracellular CDAs acting on the cell wall chitin, they may facilitate solubilization and access to the substrate (i.e., *Pa*CDA with two CBMs, where deletion of one or both confirmed their proposed function in supporting the enzymatic conversion of insoluble chitin [26]). In CDAs acting on low molecular weight COS, the CBMs may be involved in enzyme localization. This is the case of COD enzymes from marine bacteria (*Vibrio* and *Shewanella* species), where the small substrate does not span out of the active site, and the CBMs might bind to chitinous material in order to keep the COD activity close to the site where COS are generated by the action of chitinases.

The first CE4 enzymes with 3D structure determined by X-ray crystallography were the peptidoglycan deacetylases *Bs*PdaA [144] and *Sp*PgdA [145], and the first CDA was that from *Colletotrichum lindemuthianum* (*Cl*CDA) [53]. Currently, only five CDAs in the CE4 family have known 3D structure (Figure 2). The CE4 catalytic domain is characterized by a distorted $(\beta/\alpha)_8$ barrel fold. The distorted barrel, which often lacks one of the $\alpha\beta$ repeats of regular TIM barrels, creates a groove into which the extended polymer substrate binds [144,146,147]. Seven or eight parallel β-strands form the β-barrel surrounded by α-helices. In addition, a series of loops decorate the β-barrel and make up the majority of the carbohydrate binding pocket as discussed below.

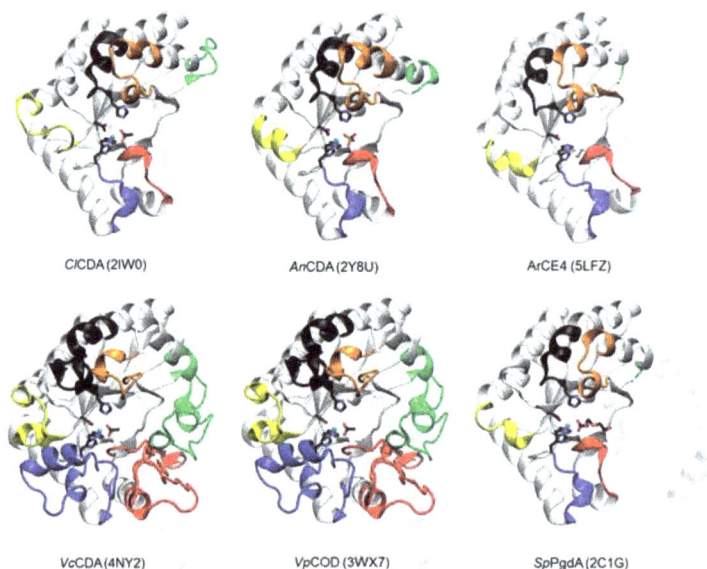

*Cl*CDA (2IW0)   *An*CDA (2Y8U)   ArCE4 (5LFZ)

*Vc*CDA (4NY2)   *Vp*COD (3WX7)   *Sp*PgdA (2C1G)

Figure 2. 3D structures of CDAs determined by X-ray crystallography. Loops 1 to 6 colored as in Figure 3. The peptidoglycan deacetylase *Sp*PgdA is also included for comparison (see text). In parenthesis, PDB accession codes.

4.2. The NodB Homology Domain and Conserved Active Site Motifs

The multiple sequence alignment of the CE4 domain for the CDAs listed in Table 1 was guided by the structural superimposition of the available X-ray structures (Figure 2) and is presented in Figure 3. Compared to most of the CDA members, the *Vc*CDA enzyme has substantially longer insertions, and it was key to defining the loops that differentiate CDAs and shape the binding site cleft of these enzymes. Sequences of enzymes without structural date were incorporated into the alignment by means of Hidden Markov Model comparisons. As seen in Figure 3, the conserved motifs and non-conserved insertions are evenly distributed along the sequences of CDAs.

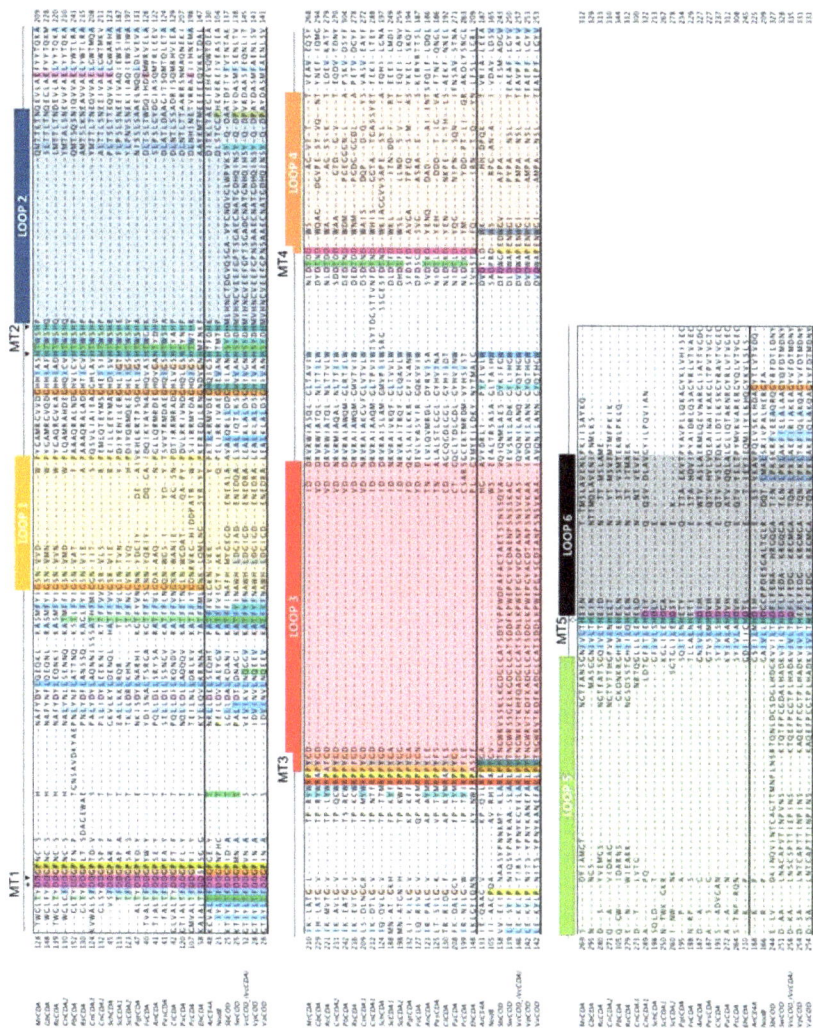

Figure 3. Multiple sequence alignment of the CDA enzymes listed in Table 1. Loops are highlighted with colored boxes according to [69]. Conserved catalytic motifs are labelled MT1-5. The 'His-His-Asp' metal binding triad (▼), catalytic base (*), and catalytic acid (◇) are labelled.

The conserved motifs are related to enzymatic activity (Motifs 1 to 5) and are typically located at the center of the active site structure. The non-conserved insertions correspond to both un-structured and structured loops of variable length, sequence, and geometry that surround the active site. These loops are numbered from Loop 1 to Loop 6 in the sequence alignment (Figure 3). As discussed in Section 4.5, they are key elements in determining the substrate specificity of different CDAs.

As members of the CE4 family, CDAs share the ≈150 aa-long NodB homology domain (CE4 domain). This region is defined by five conserved motifs that, according to the order they appear in the sequence, are named Motif 1 to Motif 5. These consensus motifs were first proposed in 2005 by sequence alignment of representative enzyme members of the CE4 family when the 3D structure of the peptidoglycan deacetylase *Sp*PgdA was solved [144]. Motif 1 (TFDD) is highly conserved in CDAs and contains the general base aspartate (first D) and the metal-binding aspartate (second D). Motif 2 (H(S/T)xxH) is a zinc-binding motif, where the two His residues bind the metal cation and the Ser or Thr residue forms a hydrogen bond with the second His, stabilizing the local conformation of the loop-shaped motif. These two His from Motif 2 plus the metal-binding Asp from Motif 1 are often designated the His-His-Asp metal-binding triad of CE4 enzymes. Motif 3 (RxPY) forms one of the sides of the active site groove and establishes stabilizing interactions with other active site residues. Motif 4 (DxxD(W/Y)) forms the other side of the active site groove, including a hydrophobic residue exposed to the solvent and a buried Asp. Motif 5 (I(V/I)LxHD) contains the catalytic general acid His residues and a Leu, which is part of a hydrophobic pocket that accommodates the acetate methyl group of the substrate.

4.3. Phylogeny of CE4 Chitin Deacetylases

Based on the above multiple sequence alignment, a clustering of the CE4 domain sequences of characterized CDAs based on phylogenetic analysis is presented in Figure 4. This is a reduced phylogenic analysis limited to the CDAs with reported activity on COS as listed in Table 1. Fungal and bacterial CDAs are clearly segregated in two clades, with a protist CDA (*Eh*CDA) located between both groups.

Fungal enzymes from organisms belonging to different phyla (*Zygomycota*, *Basidiomycota*, and *Ascomycota*) are distributed throughout the fungal clade. Within the clade, CDAs appear grouped in two clusters related with their biological function. The first cluster contains orthologous CDAs of different phyla known to have a role in cell wall chitosan biosynthesis at different stages of the fungal cell cycle, such as *Mr*CDA during cell wall formation [42], *Cn*CDAs during vegetative growth [37], or *Sc*CDAs during sporulation [38,44]. Although there is no experimental proof of their biological function, the CDAs from *Gonglonella*, *Phycomyces*, as well as those from *Rhizopus* are likely to be also involved in cell wall formation due to their location in the same cluster of the phylogenetic tree and their taxonomic classification (mucorales inside the *Zygomycota* phylum). Regarding the cellular location of the enzymes in this cluster, most of them are secreted to the periplasm or are GPI-anchored to the cell wall, where they are coupled with chitin synthases for chitosan biosynthesis. The second cluster is mainly composed of extracellular CDAs that participate in host infection, either as a defense mechanism to prevent the elicitation of host defense mechanisms (*Pgt*CDA, *Pes*CDA), or involved in the interaction with the host or as a virulent factor (*Cl*CDA, *Pc*CDA). Extracellular CDAs involved in cell autolysis (*An*CDA) also fall in this group.

Most of the bacterial enzymes included in the alignment are chitin oligosaccharide deacetylases (COD) and are more distantly related to the fungal CDAs. These form a different clade in the phylogenetic tree (Figure 4). The enzymes from marine bacteria (*Vibrio* and *Shewanella* species) are clustered together with high sequence similarity and have similar biological functions and biochemical properties. NodB has a more distant relationship with the other CODs.

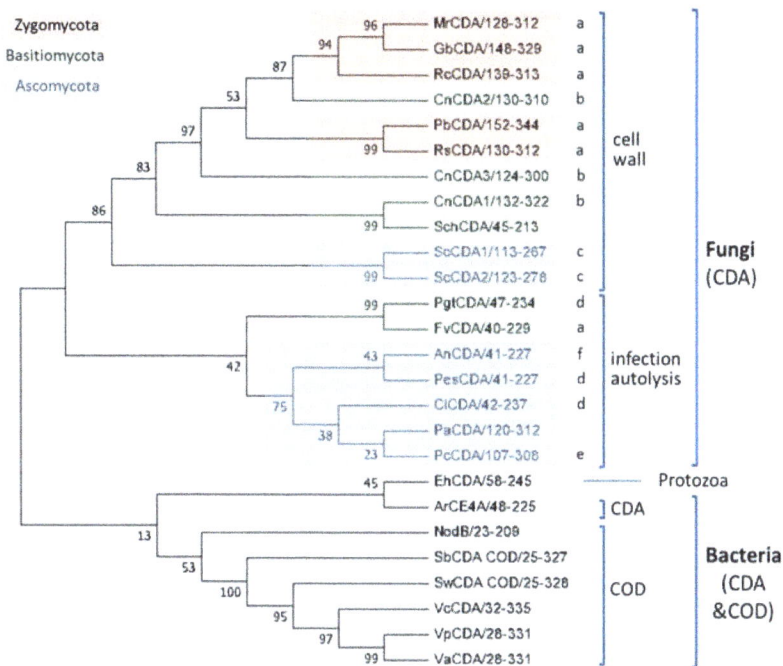

Figure 4. Phylogenetic analysis of CDAs from the multiple sequence alignment presented in Figure 3. A bootstrap analysis with 500 replicates was carried out on the trees inferred from the neighbor joining method. The consensus tree is shown with bootstrap values at each node of the tree. Biological functions: cell wall biosynthesis: (a) cell wall, (b) vegetative growth, (c) sporulation; host infection, (d) defense, (e) interaction/infection, (f) autolysis (see text).

4.4. Catalytic Mechanism

CDA enzymes operate by metal-assisted acid/base catalysis. The general mechanism was first proposed for the peptidoglycan GlcNAC deacetylase *Sp*PgdA when solving its X-ray structure [144] and short after supported by the 3D structures of the acetylxylan esterases *Sl*AxeA and *Ct*AxeA [148]. The catalytic machinery involves the conserved active site motifs containing the metal-binding triad and the general acid and base residues. Only the structures of four different CDA have been solved up to date (*Colletotrichum*, *Aspergillus*, *Vibrio*, and *Arthrobacter* CDAs, Table 1), all consistent with the proposed metal-assisted mechanism. *Vc*CDA was the first CE4 enzyme for which the 3D structure of enzyme-substrate complexes were solved by X-ray crystallography [69]. The structure of complexes of an inactive mutant (at the general base Asp residue) with diacetylchitobiose (DP2) and triacetylchitotriose (DP3) in productive binding for catalysis showed that a sugar hydroxyl group of the substrate also participates in metal coordination. Specifically (Figure 5), the Zn^{2+} cation is coordinated by the imidazole nitrogens of His97 and His101, the carboxylate group of Asp40, and the O7 atom of the *N*-acetyl group and O3 hydroxyl of the GlcNAc ring. The distorted octahedral coordination is completed by a water molecule. Upon activation, this water molecule is proposed to be the nucleophile responsible for removal of the *N*-acetyl group. Just recently, a second structure of an enzyme-substrate complex has been reported for the *Arthrobacter* sp. CDA (*Ar*CE4) [76]. The diacetylchitobiose ligand bound into the active site also shows the same type of interactions with the conserved active site residues.

The proposed mechanism of CDAs and related CE4 enzymes is shown in Figure 5. In the first step, metal coordination polarizes the carbonyl amide of the substrate which reacts with the nucleophilic water molecule activated by the general base (Asp), leading to a tetrahedral oxyanion intermediate. Next, protonation of the nitrogen group of the intermediate by the general acid (His) facilitates C-N bond breaking with release of acetate and the generation of a free amine in product. Kinetic evidence for an oxyanion tetrahedral intermediate and significant charge development at the first transition state was provided by Hammett linear free energy correlations using the *Cl*CDA enzyme with α-haloacetamido substrate analogues [53]. In most of the enzymes, the catalytic acid and base residues are part of two conserved "charge relay" side chain pairs that may contribute to modulate the pK_a of the catalytic residues [53,144]: the catalytic base (Asp) is tethered by a conserved Arg from MT3 (RxxPY) and the catalytic acid (His) is tethered by a conserved Asp from MT4, DxxD(W/Y).

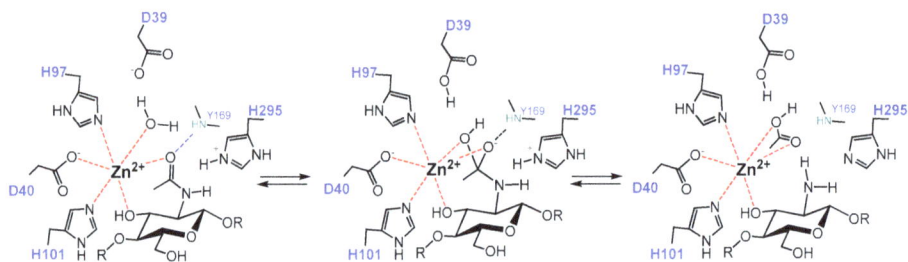

Figure 5. Metal-assisted general acid/base mechanism proposed for CE4 deacetylases. Scheme based on the 3D structure of the enzyme·substrate complex VcCDA$_{D39S}$·DP2 [61]. D39 is the general base and His295 is the general acid.

4.5. Determinants of Substrate Specificity

The series of crystal structures of the *Vibrio cholerae* chitin oligosaccharide deacetylase (*Vc*CDA or *Vc*COD) reported in 2014 [69] were the first 3D structures of a CE4 enzyme in complex with substrates. These data provided a first insight into structure-function relationships for this family of enzymes and highlighted the role of the loops that shape the binding site cleft in substrate binding. Recently by the end of 2017, the 3D structure of an *Arthrobacter* sp. CDA (ArCE4) in complex with substrate has also been reported [76]. Albeit sharing the same molecular function, both enzymes represent two different scenarios regarding the binding site topology and, hence, substrate specificity. *Vc*CDA has a rather closed binding cleft and is highly specific for monodeacetylation of COS, whereas *Ar*CE4 has a more open binding cleft and is able to fully deacetylate their COS substrates (Figure 6). A comparative structural analysis of both enzyme structures has been recently reviewed [142].

4.5.1. *Vc*CDA. Long Loops and High Specificity

Currently available structures of *Vc*CDA include the unliganded form of the enzyme and the binary complexes with *N*-acetylglucosamine (DP1), diacetylchitobiose (DP2), and triacetylchitotriose (DP3). These structures revealed two significant observations: a series of non-conserved loops (labeled Loop 1 to 6 in Figures 3 and 6) that shape the binding cleft, and the dynamics of the loops that assemble the active site for catalysis [69].

In all structures, the substrate is confined in a small binding cleft that is shaped by a series of long loops surrounding the active site (see Figure 6A for the *Vc*CDA·DP3 complex). Given the topology of *Vc*CDA protein surface, the binding of longer COS is prevented because these loops cap both the reducing and non-reducing ends of the substrate. Indeed, the catalytic efficiency of *Vc*CDA drops substantially on oligomers longer than DP2 [69] and attempts to solve the structure of *Vc*CDA bound to substrates longer than DP3 in a catalytically competent mode have been unsuccessful. Another consequence of this constricting topology is the high specificity of *Vc*CDA to exclusively deacetylate the penultimate residue from the non-reducing end of the substrates. There is no room for the ligand to slide along the binding cleft, thus it can only accommodate one GlcNAc unit of the oligomeric chain at the catalytic site where deacetylation takes place. The binding of substrates induces a conformational change of Loop 4 from an open conformation in the unliganded enzyme to a closed conformation in the enzyme·DP2 complex or a semi-closed conformation in the enzyme·DP3 complex. It is triggered by a staking interaction between a Trp residue located in apical site of Loop 4 and the GlcNAc unit at the catalytic center, locking the substrate in the active site in the proper orientation for catalysis.

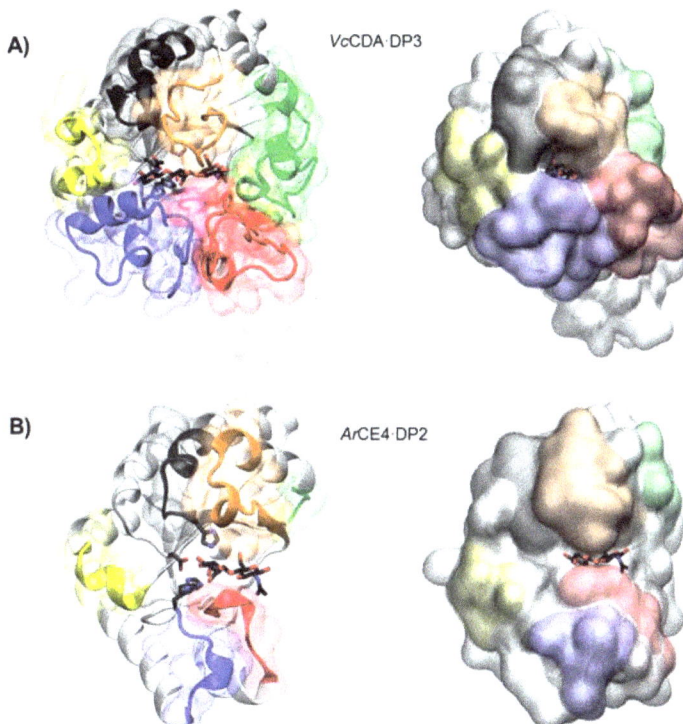

Figure 6. 3D structures of enzyme·substrate complex. (**A**) *Vc*CDA with DP3 substrate and (**B**) *Ar*CE with DP2 substrate. Loops 1 to 6 are colored as in Figure 3.

4.5.2. *Ar*CE4. Short Loops and Broad Specificity

In contrast to *Vc*CDA, the crystal structure of *Ar*CE4 in complex with diacetylchitobiose [76] reveals a flatter protein surface with the substrate bound to a more open binding cleft (Figure 6B). Even though the enzyme was co-crystalized with tetraacetylchitotetraose (DP4), only two GlcNAc units are observed in the structure. This indicates a weak binding of part of the COS substrate on this flat topology of the protein surface. The catalytic center in both enzymes is in the same position with respect to the protein core, being the main difference, the size and shape of the loops surrounding the active site. Since the binding cleft is more open, the enzyme can accommodate longer COS. Indeed, the enzymatic activity of *Ar*CE4 increases as the length of the chitin oligomer chain increases. The lack of protein caps at either the reducing and non-reducing ends of the substrate can also explain the multiple-chain mechanism proposed for this enzyme. Deacetylation takes place at all GlcNAc units of the substrate (except the reducing end) because it can freely bind to *Ar*CE4 in different binding modes exposing different GlcNAc units of the oligomeric chain at the catalytic site.

4.5.3. The Subsite Capping Model

The diversity of deacetylation patterns exhibited by chitin deacetylases and related CE4 enzymes can be attributed to the differential accessibility of the linear chitin oligosaccharide chain to the separate subsites along the substrate binding cleft of their structures. Considering all CE4 enzymes with reported activity on polymeric chitin or COS, these can be classified into two groups. One group is represented by general chitin deacetylases (CDA), and a second group is formed by chitin oligosaccharide deacetylases (COD). The two structures of the enzymes-substrate complexes described above are reference models for the protein surface topologies and substrate binding mechanisms of these two groups: CDAs (*Ar*CE4) and CODs (*Vc*CDA). These two structures provide a unified view of the determinants of substrate specificity in chitin deacetylases in terms of the "subsite capping model" proposed in [69]. According to this model, substrate accessibility is affected by the length, shape, and dynamics of a series of loops surrounding the active site of CE4 enzymes. These loops are numbered from 1 to 6 and their location in the sequences and structures of CDAs and CODs is highlighted in Figures 2 and 3.

The group of CDAs bears short loops, and their structures exhibit a flat and open binding cleft. The substrate binding mechanism in this group of enzymes may be similar to that described for the reference structure of *Ar*CE4. According to the model, the substrate may be able to slide along the binding cleft or to bind in different modes resulting in processive or multiple-chain attack mechanisms of deacetylation. This can already be anticipated for CDA enzymes of known structure (Figure 2) because the flat protein surface is already evident, but also for CDA enzymes of unknown structure given the similar sequence lengths of the loops evidenced in the alignment (Figure 3). This could be an explanation of why CDA enzymes in general are not specific for the deacetylation at a single *N*-acetylglucosamine unit. However, the patterns of deacetylation differ among the different CDAs. The surface charge distribution along the binding cleft and other structural features yet to be disclosed may also participate in defining the mode of action and deacetylation pattern by each particular enzyme.

On the contrary, the group of COD enzymes bears longer loops and their structures have narrower binding pockets and buried active sites. According to the subsite capping model, the substrate is constrained to bind in very specific binding modes resulting in single-site deacetylations. This is the case for the reference structure of *Vc*CDA in complex with substrates, but it can also be anticipated for other COD enzymes for which the 3D structure is still unknown. For instance, Loop 6 in *Rm*NodB is longer than in other CDAs. This loop is located on the non-reducing end site of the binding cleft and may cap the accessibility of the substrate after subsite 0 (the catalytic site) thus defining the deacetylation specificity for the non-reducing end of the substrate. Likewise, the *Shewanella* CODs have a Loop 6 with the same length than the *Vibrio* CODs, but shorter than NodB, and both exhibit the same mono-deacetylation specificity for the penultimate GlcNAc residue from the reducing end of the substrate.

For most CDAs, the reducing end of the substrate is not deacetylated, or it is the least reactive GlcNAc unit. As seen in the *Ar*CE4·DP2 complex 3D structure [64], binding to the +1 subsite seems to be dominated by the stacking interaction of the GlcNAc unit of the substrate with a Trp in Motif 4 at the beginning of Loop 4. This aromatic residue is highly conserved (MT4, DxxD(W/Y), Figure 3). CDA enzymes having this aromatic residue prefer a sugar bound in the +1 subsite; they do not deacetylate the reducing end of their substrates, as it is the case for *Ar*CE4, *Pes*CDA [89] and *Pc*CDA [95], or the reducing end is the slowest position to be deacetylated, as shown for *Cl*CDA [49] and *An*CDA [12]. On the contrary, *Pgt*CDA, which deacetylates the reducing end GlcNAc unit of all substrates from DP4 to DP6, lacks the +1 aromatic residue [89]. Different is the case of *Vibrio* and *Sewanella* CODs that have the equivalent aromatic residue in a slightly different position after a two-amino acid insertion in the MT4 motif, and it is located farther in Loop 4 (Figure 3). In the *Vc*CDA enzyme this loop moves from an open to a closed conformation upon substrate binding, and the same is expected for the other closely related CODs that have the same Loop 4 size. As a consequence of the induced fit, the Trp residue now establishes a stacking interaction with the GlcNAc unit in subsite 0. DP2 is the preferred substrate for this group of COD enzymes, and it is deacetylated at the reducing end [62].

Gaining further structural information of protein-ligand complexes of CDA and COD enzymes, and other CE4 in general, will contribute further to decipher the structural and sequential determinants of substrate specificity in this family of enzymes. This will pave the way to the rational design or discovery of novel CDA with controlled specificities on the deacetylation of oligomeric and polymeric chitin for the biotechnological production of chitosans and paCOS with defined patterns of acetylation.

5. Application of Chitin Deacetylases

5.1. Targets for Antifungals

Fungal infections have an enormous impact on human health. Fungi are generally opportunistic pathogens affecting immunocompromised individuals including those with AIDS, receiving immunosuppressive drugs or undergoing cancer treatments. The cell wall is a meaningful target for antifungal therapies. Current major classes of antifungal drugs target cell membrane ergosterol biosynthesis (azoles), ergosterol function by disrupting membrane integrity (polyenes), or 1,3-β-glucan synthase preventing the formation of the cell-wall structural polysaccharide 1,3-β-glucan (echinocandins) [149]. New targets to overcome the emerging drug resistance by pathogenic fungi are becoming critical to treat life-threatening fungal infections. Other promising targets are the so called cell wall proteins (CWP) which mediate important cellular processes, including adhesion, invasion, biofilm formation and flocculation [122]. In fungal chitosan producers, chitin deacetylases are a class of CWP and potential targets for drug design. *Cryptococcus neoformans*, one of the most deadly pathogens, requires chitosan for virulence. Lack of chitosan in the cell wall has detrimental consequences in fungal growth and results in the complete loss of sporulation [120,121]. Thus, CDAs represent a promising target for anticryptococcal therapeutics [37,120], but no CDA inhibitors have been reported yet.

In pathogenic plants, major strategies to prevent fungal pathogenesis are related to the inhibition of fungal chitinases, which are required for chitin remodeling in the cell wall [150–152]. Different types of chitinase inhibitors have been reported, including potent natural inhibitors such as allosamidin [153] and the cyclic pentapeptides argifin and argadin [154]. However, inhibition must be selective so as not to interfere with the plant chitinases involved in triggering the plant defense mechanisms. Another potential and promising strategy is the inhibition of extracellularly secreted fungal CDAs since they constitute a defense mechanism to evade the plant immune system, as discussed in Section 3.2. As in the case of human fungal pathogens, no inhibitors have been yet reported against CDAs from plant pathogenic fungi.

5.2. Biocatalysts for the Enzymatic Production of Chitosans and paCOS

Chitosans can be found in a large number of applications in such distant areas as agriculture, cosmetics, water treatment, medicine and the food industry [155–159]. In addition to chitosan polymers, their oligomers (paCOS) have also proven to have relevant potential applications in agriculture and pharmaceutical industries [160]. The physicochemical and biological properties of chitosans and paCOS have been shown to be strongly dependent on their degree of polymerization and their degree of acetylation [161,162]. Many of the identified CDAs arose from screening programs addressed to find efficient biocatalysts to overcome the current industrial chitosan production by highly concentrated alkali treatment of chitin. Some examples include CDAs from *Mortiriella* sp. [58], *Rhizopus* sp. [117], or *Gongronella* sp. [118].

Chemical methods for the production of COS and paCOS are based on chemical depolymerization of chitosan [163,164], total synthesis of chitosan oligomers [165,166], partial chemical deacetylation of fully acetylated COS, or chemical re-*N*-acetylation of glucosamine oligomers based in two-step procedures or one-pot synthesis [167–169]. The drawbacks of chemical strategies are the unwanted side reactions and the randomness of the chemical reactions. Current efforts are addressed to develop enzymatic routes for COS and paCOS production with defined DP, DA, and PA.

Enzymatic approaches include depolymerization of chitin or chitosan polymers using hydrolytic enzymes, chitinases and chitosanases, enzymatic polymerization by transglycosidation using transglycosylating hydrolases, and enzymatic de-acetylation and re-acetylation of chitin oligomers using chitin deacetylases, strategies recently reviewed in [170]. Based on the current knowledge on the specificity of a number of fungal and bacterial CDAs, recent reports have combined enzymes with different specificities to have access to a large family of paCOS with defined structures. The first proof of concept was to show that two specific CODs, NodB and VcCDA, each accept the monodeacetylated product from the other, leading to specific di-deacetylation, and that both enzymes can work in a one-pot process [130]. Recently, the use of different recombinant CDAs from bacterial and fungal origin to produce all of fourteen possible partially acetylated chitosan tetramers combining different enzymatic deacetylations and enzymatic *N*-acetylations has been reported [171] (Figure 7).

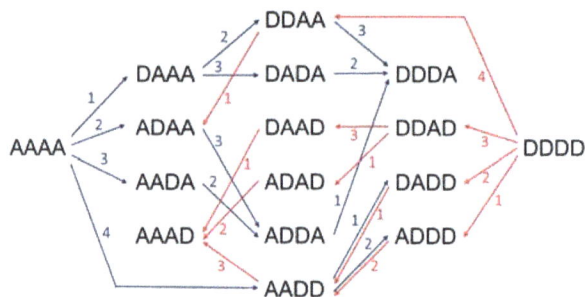

Figure 7. Production routes of all possible chitin and chitosan tetramers using 4 different CDAs to specifically deacetylate or *N*-acetylate paCOS. A: GlcNAc, D: GlcNH$_2$. Blue arrows, deacetylation reactions, red arrows, *N*-acetylation reactions in the presence of excess acetate.

The production of paCOS using in vivo strategies is an alternative to increase the scalability of the process. The first example towards a more general cell factory approach for the in vivo synthesis of paCOS was based on NodB deacetylase. By in vivo studies with *Escherichia coli* expressing different combinations of the nodABCS genes of *Azorhizobium caulinodans*, Nod factor intermediates were identified, as well as the sequence of the biosynthetic steps [172]. The nod gene cluster encodes a series of enzymes, which include the NodC chitin oligosaccharide synthase that produces fully acetylated chitin oligomers, the NodB chitin oligosaccharide deacetylase that deacetylates the non-reducing end unit, the NodA *N*-acyl transferase that transfer a fatty acid chain to the free amine group, and the NodS *N*-methyl transferase. Further transformations by other nod proteins elaborate the final Nod signaling factors [173]. In a first cell factory approach, high density cells of *E. coli* expressing nodC or nodBC genes produced in high yield (up to 2.5 g/L) penta-*N*-acetyl-chitopentaose and its deacetylated derivative tetra-*N*-acetyl-chitopentaose, which were easily purified by charcoal adsorption and ion-exchange chromatography [174]. The strategy was further extended to the production of sulfated and *O*-acetylated derivatives of these two compounds by coexpressing nodC or nodBC with nodH and/or nodL that encode chitooligosaccharide sulfotransferase and chitooligosaccharide *O*-acetyltransferase, respectively [175]. Other Nod analogues have also been generated with further modifications [176–178]. The cell factory approach, currently limited to one deacelylation pattern based on the use of NodB, is a promising technology to be developed by incorporating the diversity of CDAs with different deacetylation patterns in order to access a large family of paCOS and derivatives.

6. Conclusions

We have here summarized the current knowledge on substrate specificity of fungal and bacterial chitin deacetylases, their modes of action, and their use as biocatalysts for the production of chitosans and chitosan oligosaccharides with defined pattern of acetylation. By combining multiple sequence alignments and 3D structures of enzyme·substrate complexes of representative enzymes, a unified view of the determinants of substrate specificity is proposed in terms of the "subsite capping model." According to this model, substrate accessibility is affected by the length, shape, and dynamics of a series of loops surrounding the active site of CE4 enzymes. The group of CDAs active on polymeric substrates and COS bear short loops, and their structures exhibit a flat and open binding cleft. The substrate may be able to slide along the binding cleft or to bind in different modes, resulting in processive or multiple-chain attack mechanisms of deacetylation. Other structural features not yet disclosed, such as the charge distribution along the binding cleft may also participate in defining the mode of action and deacetylation pattern by each particular enzyme. The group of COD enzymes active on low molecular mass COS bear longer loops and their structures have narrower binding pockets and buried active sites. The substrate is constrained to bind in very specific binding modes resulting in single-site deacetylations.

But a deeper knowledge on substrate specificity requires further structural information of protein-ligand complexes of CDA and COD enzymes in order to decipher the structural and sequential determinants of substrate specificity in this family of enzymes aimed at the rational design or discovery of novel CDAs with controlled specificities on the deacetylation of oligomeric and polymeric chitin for biotechnological applications.

Although CDAs have been proposed as targets for antifungal drugs, no specific inhibitors have been yet reported. This is an open field that deserves attention not only for drug design but also to probe the signaling function of CDAs and CODs through their specific deacetylation of COS substrates.

Applications of CDAs and CODs as biocatalyst are currently being developed as a novel methodology to produce partially acetylated COS with tailored patterns of acetylation. Since not all patterns for COS of different sizes are yet available, enzyme discovery and protein engineering offer new opportunities for the biotechnological production of chitosans and paCOS with defined patterns of acetylation.

Acknowledgments: Work supported by the European Union's Seventh Framework Programme for research, technological development and demonstration under grant agreement n°613931, and grant BFU2016-77427-C2-1-R from MINECO, Spain. Laia Grifoll and Sergi Pascual acknowledge a predoctoral contract from the NANO3BIO project. Hugo Aragunde acknowledges a predoctoral fellowship from Generalitat de Catalunya.

Author Contributions: Antoni Planas conceived and designed the review; Laia Grifoll, Sergi Pascual, Hugo Aragunde, Xevi Biarnés, and Antoni Planas compiled the references and prepared the figures. Xevi Biarnés and Antoni Planas wrote the paper.

Conflicts of Interest: The authors declare no conflict of interest.

Abbreviations

AXE	Acetylxylan esterase
A_n	$(GlcNAc)_n$
CDA	Chitin deacetylase
COS	Chitooligosaccharides
D_n	$(GlcNH_2)_n$
DA	Degree of acetylation
DP	Degree of polymerization
GlcNAc	*N*-acetylglucosamine
PA	Pattern of acetylation
paCOS	Partially acetylated chiton oligosaccharides
PDB	Protein data bank

References

1. Peniche Covas, C.A.; Argüelles-Monal, W.; Goycoolea, F.M. Chitin and chitosan: Major sources, properties and applications. In *Monomers, Polymers and Composites from Renewable Resources*; Belgacem, M.N., Gandini, A., Eds.; Elsevier: Amsterdam, the Netherlands, 2008; Volume 1, pp. 517–542. ISBN 9780080453163.
2. Karrer, P.; Hofmann, A. Über den enzymatischen Abbau von Chitin und Chitosan I. *Helv. Chim. Acta* **1929**, *12*, 616–637. [CrossRef]
3. Rinaudo, M. Chitin and chitosan: Properties and applications. *Prog. Polym. Sci.* **2006**, *31*, 603–632. [CrossRef]
4. Dutta, P.K.; Duta, J.; Tripathi, V.S. Chitin and Chitosan: Chemistry, properties and applications. *J. Sci. Ind. Res. (India)* **2004**, *63*, 20–31. [CrossRef]
5. Noishiki, Y.; Takami, H.; Nishiyama, Y.; Wada, M.; Okada, S.; Kuga, S. Alkali-induced conversion of β-chitin to α-chitin. *Biomacromolecules* **2003**, *4*, 896–899. [CrossRef] [PubMed]
6. Jang, M.K.; Kong, B.G.; Jeong, Y.I.; Lee, C.H.; Nah, J.W. Physicochemical characterization of α-chitin, β-chitin, and γ-chitin separated from natural resources. *J. Polym. Sci. Part A Polym. Chem.* **2004**, *42*, 3423–3432. [CrossRef]
7. Kumirska, J.; Czerwicka, M.; Kaczyński, Z.; Bychowska, A.; Brzozowski, K.; Thöming, J.; Stepnowski, P. Application of spectroscopic methods for structural analysis of chitin and chitosan. *Mar. Drugs* **2010**, *8*, 1567–1636. [CrossRef] [PubMed]
8. Kaya, M.; Mujtaba, M.; Ehrlich, H.; Salaberria, A.M.; Baran, T.; Amemiya, C.T.; Galli, R.; Akyuz, L.; Sargin, I.; Labidi, J. On chemistry of γ-chitin. *Carbohydr. Polym.* **2017**, *176*, 177–186. [CrossRef] [PubMed]
9. Dhillon, G.S.; Kaur, S.; Brar, S.K.; Verma, M. Green synthesis approach: Extraction of chitosan from fungus mycelia. *Crit. Rev. Biotechnol.* **2013**, *33*, 379–403. [CrossRef] [PubMed]
10. Hoell, I.A.; Vaaje-Kolstad, G.; Eijsink, V.G.H. Structure and function of enzymes acting on chitin and chitosan. *Biotechnol. Genet. Eng. Rev.* **2010**, *27*, 331–366. [CrossRef]
11. Xia, W.; Liu, P.; Zhang, J.; Chen, J. Biological activities of chitosan and chitooligosaccharides. *Food Hydrocoll.* **2011**, *25*, 170–179. [CrossRef]
12. Yu, R.; Liu, W.; Li, D.; Zhao, X.; Ding, G.; Zhang, M.; Ma, E.; Zhu, K.Y.; Li, S.; Moussian, B.; et al. Helicoidal organization of chitin in the cuticle of the migratory locust requires the function of the chitin deacetylase2 enzyme (LmCDA2). *J. Biol. Chem.* **2016**, *291*, 24352–24363. [CrossRef] [PubMed]
13. Winkler, A.J.; Dominguez-Nuñez, J.A.; Aranaz, I.; Poza-Carrión, C.; Ramonell, K.; Somerville, S.; Berrocal-Lobo, M. Short-chain chitin oligomers: Promoters of plant growth. *Mar. Drugs* **2017**, *15*, 40. [CrossRef] [PubMed]
14. Li, X.; Min, M.; Du, N.; Gu, Y.; Hode, T.; Naylor, M.; Chen, D.; Nordquist, R.E.; Chen, W.R. Chitin, chitosan, and glycated chitosan regulate immune responses: The novel adjuvants for cancer vaccine. *Clin. Dev. Immunol.* **2013**, *2013*. [CrossRef] [PubMed]
15. Varki, A.; Sharon, N. *Essentials of Glycobiology*, 2nd ed.; Varki, A., Cummings, R., Esko, J., Freeze, H., Stanley, P., Bertozzi, C.R., Hart, G., Etzler, M.E., Eds.; Cold Spring Harbor Laboratory Press: Cold Spring Harbor, NY, USA, 2009.
16. Lombard, V.; Golaconda Ramulu, H.; Drula, E.; Coutinho, P.M.; Henrissat, B. The carbohydrate-active enzymes database (CAZy) in 2013. *Nucleic Acids Res.* **2014**, *42*, 490–495. [CrossRef] [PubMed]
17. Caufrier, F.; Martinou, A.; Dupont, C.; Bouriotis, V. Carbohydrate esterase family 4 enzymes: Substrate specificity. *Carbohydr. Res.* **2003**, *338*, 687–692. [CrossRef]
18. John, M.; Rohrig, H.; Schmidt, J.; Wieneke, U.; Schell, J. Rhizobium NodB protein involved in nodulation signal synthesis is a chitooligosaccharide deacetylase. *Proc. Natl. Acad. Sci. USA* **1993**, *90*, 625–629. [CrossRef] [PubMed]
19. Mine, S.; Niiyama, M.; Hashimoto, W.; Ikegami, T.; Koma, D.; Ohmoto, T.; Fukuda, Y.; Inoue, T.; Abe, Y.; Ueda, T.; et al. Expression from engineered *Escherichia coli* chromosome and crystallographic study of archaeal *N,N*-diacetylchitobiose deacetylase. *FEBS J.* **2014**, *281*, 2584–2596. [CrossRef] [PubMed]
20. Fadouloglou, V.E.; Deli, A.; Glykos, N.M.; Psylinakis, E.; Bouriotis, V.; Kokkinidis, M. Crystal structure of the BcZBP, a zinc-binding protein from *Bacillus cereus*. *FEBS J.* **2007**, *274*, 3044–3054. [CrossRef] [PubMed]
21. Verma, S.C.; Mahadevan, S. The ChbG gene of the chitobiose (chb) operon of *Escherichia coli* encodes a chitooligosaccharide deacetylase. *J. Bacteriol.* **2012**, *194*, 4959–4971. [CrossRef] [PubMed]
22. Araki, Y.; Ito, E. A pathway of chitosan formation in *Mucor rouxii*: Enzymatic deacetlation of chitin. *Biochem. Biophys. Res. Commun.* **1974**, *56*, 669–675. [CrossRef]

23. Araki, Y.; Ito, E. A Pathway of Chitosan Formation in *Mucor rouxii* Enzymatic Deacetylation of Chitin. *Eur. J. Biochem.* **1975**, *55*, 71–78. [CrossRef] [PubMed]

24. Zhao, Y.; Ju, W.; Jo, G.; Jung, W.; Park, R. Perspectives of Chitin Deacetylase Research. In *Biotechnology of Biopolymers*; InTech: London, UK, 2011; pp. 131–145. [CrossRef]

25. Liu, Z.; Gay, L.M.; Tuveng, T.R.; Agger, J.W.; Westereng, B.; Mathiesen, G.; Horn, S.J.; Vaaje-Kolstad, G.; van Aalten, D.M.F.; Eijsink, V.G.H. Structure and function of a broad-specificity chitin deacetylase from *Aspergillus nidulans* FGSC A4. *Sci. Rep.* **2017**, *7*, 1746. [CrossRef] [PubMed]

26. Hoßbach, J.; Bußwinkel, F.; Kranz, A.; Wattjes, J.; Cord-Landwehr, S.; Moerschbacher, B.M. A chitin deacetylase of *Podospora anserina* has two functional chitin binding domains and a unique mode of action. *Carbohydr. Polym.* **2018**, *183*, 1–10. [CrossRef] [PubMed]

27. Ghormade, V.; Kulkarni, S.; Doiphode, N.; Rajamohanan, P.R.; Deshpande, M.V. Chitin deacetylase: A comprehensive account on its role in nature and its biotechnological applications. In *Current Research, Technology and Education Topics in Applied Microbiology and Microbial Biotechnology*; Méndez-Vilas, A., Ed.; Formatex Research Center: Badajoz, Spain, 2010; pp. 1054–1066.

28. Tsigos, I.; Martinou, A.; Kafetzopoulos, D.; Bouriotis, V. Chitin deacetylases: New, versatile tools in biotechnology. *Trends Biotechnol.* **2000**, *18*, 305–312. [CrossRef]

29. Zhao, Y.; Park, R.D.; Muzzarelli, R.A.A. Chitin deacetylases: Properties and applications. *Mar. Drugs* **2010**, *8*, 24–46. [CrossRef] [PubMed]

30. Martinou, A.; Bouriotis, V.; Stokke, B.T.; Vårum, K.M. Mode of action of chitin deacetylase from *Mucor rouxii* on partially N-acetylated chitosans. *Carbohydr. Res.* **1998**, *311*, 71–78. [CrossRef]

31. Tsigos, I.; Zydowicz, N.; Martinou, A.; Domard, A.; Bouriotis, V. Mode of action of chitin deacetylase from *Mucor rouxii* on N-acetylchitooligosaccharides. *Eur. J. Biochem.* **1999**, *261*, 698–705. [CrossRef] [PubMed]

32. Tokuyasu, K.; Mitsutomi, M.; Yamaguchi, I.; Hayashi, K.; Mori, Y. Recognition of chitooligosaccharides and their N-acetyl groups by putative subsites of chitin deacetylase from a Deuteromycete, *Colletotrichum lindemuthianum*. *Biochemistry* **2000**, *39*, 8837–8843. [CrossRef] [PubMed]

33. Hekmat, O.; Tokuyasu, K.; Withers, S.G. Subsite structure of the endo-type chitin deacetylase from a deuteromycete, *Colletotrichum lindemuthianum*: An investigation using steady-state kinetic analysis and MS. *Biochem. J.* **2003**, *374*, 369–380. [CrossRef] [PubMed]

34. Gooday, G.W. Chitin deacetylases in invertebrates. In *Chitin in Nature and Technology*; Springer: Boston, MA, USA, 1986; pp. 263–267.

35. Muthukrishnan, S.; Merzendorfer, H.; Arakane, Y.; Yang, Q. Chitin Metabolic Pathways in Insects and Their Regulation. In *Extracellular Composite Matrices in Arthropods*; Cohen, E., Moussian, B., Eds.; Springer: Berlin, Germany, 2016; pp. 31–65. ISBN 9783319407401.

36. Dixit, R.; Arakane, Y.; Specht, C.A.; Richard, C.; Kramer, K.J.; Beeman, R.W.; Muthukrishnan, S. Domain organization and phylogenetic analysis of proteins from the chitin deacetylase gene family of *Tribolium castaneum* and three other species of insects. *Insect Biochem. Mol. Biol.* **2008**, *38*, 440–451. [CrossRef] [PubMed]

37. Baker, L.G.; Specht, C.A.; Donlin, M.J.; Lodge, J.K. Chitosan, the deacetylated form of chitin, is necessary for cell wall integrity in *Cryptococcus neoformans*. *Eukaryot. Cell* **2007**, *6*, 855–867. [CrossRef] [PubMed]

38. Christodoulidou, A.; Bouriotis, V.; Thireos, G. Two sporulation-specific chitin deacetylase-encoding genes are required for the ascospore wall rigidity of *Saccharomyces cerevisiae*. *J. Biol. Chem.* **1996**, *271*, 31420–31425. [CrossRef] [PubMed]

39. Geoghegan, I.A.; Gurr, S.J. Chitosan Mediates Germling Adhesion in Magnaporthe oryzae and Is Required for Surface Sensing and Germling Morphogenesis. *PLoS Pathog.* **2016**, *12*, 1–34. [CrossRef] [PubMed]

40. White, S.; McIntyre, M.; Berry, D.R.; McNeil, B. The autolysis of industrial filamentous fungi. *Crit. Rev. Biotechnol.* **2002**, *22*, 1–14. [CrossRef] [PubMed]

41. Sánchez-Vallet, A.; Mesters, J.R.; Thomma, B.P. The battle for chitin recognition in plant-microbe interactions. *FEMS Microbiol. Rev.* **2015**, *39*, 171–183. [CrossRef] [PubMed]

42. Davis, L.L.; Bartnicki-Garcia, S. Chitosan Synthesis by the Tandem Action of Chitin Synthetase and Chitin Deacetylase from *Mucor rouxii*. *Biochemistry* **1984**, *23*, 1065–1073. [CrossRef]

43. Gao, X.-D.; Katsumoto, T.; Onodera, K. Purification and characterization of chitin deacetylase from *Absidia coerulea*. *J. Biochem.* **1995**, *117*, 257–263. [CrossRef] [PubMed]

44. Christodoulidou, A.; Briza, P.; Ellinger, A.; Bouriotis, V. Yeast ascospore wall assembly requires two chitin deacetylase isozymes. *FEBS Lett.* **1999**, *460*, 275–279. [CrossRef]

45. Hadwiger, L.A. Anatomy of a nonhost disease resistance response of pea to Fusarium solani: PR gene elicitation via DNase, chitosan and chromatin alterations. *Front. Plant Sci.* **2015**, *6*, 373. [CrossRef] [PubMed]

46. Hadwiger, L.A. Pea-Fusarium solani interactions contributions of a system toward understanding disease resistance. *Phytopathology* **2008**, *98*, 372–379. [CrossRef] [PubMed]

47. El Gueddari, N.E.; Rauchhaus, U.; Moerschbacher, B.M.; Deising, H.B. Developmentally regulated conversion of surface-exposed chitin to chi-tosan in cell walls of plant pathogenic fungi. *New Phytol.* **2002**, *156*, 103–112. [CrossRef]

48. Liu, T.; Liu, Z.; Song, C.; Hu, Y.; Han, Z.; She, J.; Fan, F.; Wang, J.; Jin, C.; Chang, J.; et al. Chitin-Induced Dimerization Activates a Plant Immune Receptor. *Science* **2012**, *336*, 1160–1164. [CrossRef] [PubMed]

49. Cord-Landwehr, S.; Melcher, R.L.J.; Kolkenbrock, S.; Moerschbacher, B.M. A chitin deacetylase from the endophytic fungus *Pestalotiopsis* sp. efficiently inactivates the elicitor activity of chitin oligomers in rice cells. *Sci. Rep.* **2016**, *6*, 38018. [CrossRef] [PubMed]

50. Emri, T.; Molnár, Z.; Szilágyi, M.; Pócsi, I. Regulation of autolysis in *Aspergillus nidulans*. *Appl. Biochem. Biotechnol.* **2008**, *151*, 211–220. [CrossRef] [PubMed]

51. Alfonso, C.; Nuero, O.M.; Santamaría, F.; Reyes, F. Purification of a heat-stable chitin deacetylase from *Aspergillus nidulans* and its role in cell wall degradation. *Curr. Microbiol.* **1995**, *30*, 49–54. [CrossRef] [PubMed]

52. Reyes, F.; Calatayud, J.; Martinez, M.J. Endochitinase from *Aspergillus nidulans* implicated in the autolysis of its cell wall. *FEMS Microbiol. Lett.* **1989**, *51*, 119–124. [CrossRef] [PubMed]

53. Blair, D.E.; Hekmat, O.; Schüttelkopf, A.W.; Shrestha, B.; Tokuyasu, K.; Withers, S.G.; van Aalten, D.M. Structure and Mechanism of Chitin Deacetylase from the Fungal Pathogen Colletotrichium lindemuthianum. *Biochemistry* **2006**, *45*, 9416–9426. [CrossRef] [PubMed]

54. Naqvi, S.; Cord-Landwehr, S.; Singh, R.; Bernard, F.; Kolkenbrock, S.; El Gueddari, N.E.; Moerschbacher, B.M. A recombinant fungal chitin deacetylase produces fully defined chitosan oligomers with novel patterns of acetylation. *Appl. Environ. Microbiol.* **2016**, *82*, 6645–6655. [CrossRef] [PubMed]

55. Aranda-Martinez, A.; Grifoll-Romero, L.; Aragunde Pazos, H.; Enea Sancho-Vaello; Biarnés, X.; Lopez-Llorca, L.V.; Planas, A. Expression and specificity of a chitin deacetylase catalytic doain from the nematophagous fungus *Pochonia chlamydosporia* potentially involved in pathogenicity. *Sci. Rep.* **2018**, *8*, 2170. [CrossRef] [PubMed]

56. Mishra, C.; Mishra, C.; Semino, C.; Semino, C.; Mccreath, K.J.; Mccreath, K.J.; de la Vega, H.; de la Vega, H.; Jones, B.J.; Jones, B.J.; et al. Cloning and expression of two chitin deacetylase gens of *Saccharomyces cerevisiae*. *Yeast* **1997**, *13*, 327–336. [CrossRef]

57. Martinou, A.; Koutsioulis, D.; Bouriotis, V. Cloning and expression of a chitin deacetylase gene (CDA2) from *Saccharomyces cerevisiae* in *Escherichia coli*: Purification and characterization of the cobalt-dependent recombinant enzyme. *Enzyme Microb. Technol.* **2003**, *32*, 757–763. [CrossRef]

58. Kim, Y.J.; Zhao, Y.; Oh, K.T.; Nguyen, V.N.; Park, R.D. Enzymatic deacetylation of chitin by extracellular chitin deacetylase from a newly screened *Mortierella* sp. DY-52. *J. Microbiol. Biotechnol.* **2008**, *18*, 759–766. [PubMed]

59. Zhao, Y.; Kim, Y.J.; Oh, K.T.; Nguyen, V.N.; Park, R.D. Production and characterization of extracellular chitin deacetylase from *Absidia corymbifera* DY-9. *J. Appl. Biol. Chem.* **2010**, *53*, 119–126. [CrossRef]

60. Yamada, M.; Kurano, M.; Inatomi, S.; Taguchi, G.; Okazaki, M.; Shimosaka, M. Isolation and characterization of a gene coding for chitin deacetylase specifically expressed during fruiting body development in the basidiomycete *Flammulina velutipes* and its expression in the yeast *Pichia pastoris*. *FEMS Microbiol. Lett.* **2008**, *289*, 130–137. [CrossRef] [PubMed]

61. Pareek, N.; Vivekanand, V.; Saroj, S.; Sharma, A.K.; Singh, R.P. Purification and characterization of chitin deacetylase from *Penicillium oxalicum* SAEM-51. *Carbohydr. Polym.* **2012**, *87*, 1091–1097. [CrossRef]

62. Karthik, N.; Binod, P.; Pandey, A. SSF production, purification and characterization of chitin deacetylase from *Aspergillus flavus*. *Biocatal. Biotransform.* **2017**. [CrossRef]

63. Cai, J.; Yang, J.; Du, Y.; Fan, L.; Qiu, Y.; Li, J.; Kennedy, J.F. Purification and characterization of chitin deacetylase from *Scopulariopsis brevicaulis*. *Carbohydr. Polym.* **2006**, *65*, 211–217. [CrossRef]

64. Gauthier, C.; Clerisse, F.; Dommes, J.; Jaspar-Versali, M.F. Characterization and cloning of chitin deacetylases from *Rhizopus circinans*. *Protein Expr. Purif.* **2008**, *59*, 127–137. [CrossRef] [PubMed]

65. Maw, T.; Tan, T.K.; Khor, E.; Wong, S.M. Complete cDNA sequence of chitin deacetylase from *Gongronella butleri* and its phylogenetic analysis revealed clusters corresponding to taxonomic classification of fungi. *J. Biosci. Bioeng.* **2002**, *93*, 376–381. [CrossRef]

66. Mélida, H.; Sain, D.; Stajich, J.E.; Bulone, V. Deciphering the uniqueness of Mucoromycotina cell walls by combining biochemical and phylogenomic approaches. *Environ. Microbiol.* **2015**, *17*, 1649–1662. [CrossRef] [PubMed]

67. Smirnou, D.; Krcmar, M.; Prochazkova, E.V.A. Chitin-Glucan complex production by *Schizophyllum commune* submerged cultivation. *Pol. J. Microbiol.* **2011**, *60*, 223–228. [PubMed]

68. Das, S.; Van Dellen, K.; Bulik, D.; Magnelli, P.; Cui, J.; Head, J.; Robbins, P.W.; Samuelson, J. The cyst wall of *Entamoeba invadens* contains chitosan (deacetylated chitin). *Mol. Biochem. Parasitol.* **2006**, *148*, 86–92. [CrossRef] [PubMed]

69. Andrés, E.; Albesa-Jové, D.; Biarnés, X.; Moerschbacher, B.M.; Guerin, M.E.; Planas, A. Structural basis of chitin oligosaccharide deacetylation. *Angew. Chem. Int. Ed.* **2014**, *53*, 6882–6887. [CrossRef] [PubMed]

70. Li, X.; Wang, L.X.; Wang, X.; Roseman, S. The chitin catabolic cascade in the marine bacterium *Vibrio cholerae*: Characterization of a unique chitin oligosaccharide deacetylase. *Glycobiology* **2007**, *17*, 1377–1387. [CrossRef] [PubMed]

71. Hirano, T.; Sugiyama, K.; Sakaki, Y.; Hakamata, W.; Park, S.Y.; Nishio, T. Structure-based analysis of domain function of chitin oligosaccharide deacetylase from *Vibrio parahaemolyticus*. *FEBS Lett.* **2015**, *589*, 145–151. [CrossRef] [PubMed]

72. Kadokura, K.; Rokutani, A.; Yamamoto, M.; Ikegami, T.; Sugita, H.; Itoi, S.; Hakamata, W.; Oku, T.; Nishio, T. Purification and characterization of *Vibrio parahaemolyticus* extracellular chitinase and chitin oligosaccharide deacetylase involved in the production of heterodisaccharide from chitin. *Appl. Microbiol. Biotechnol.* **2007**, *75*, 357–365. [CrossRef] [PubMed]

73. Ohishi, K.; Yamagishi, M.; Ohta, T.; Motosugi, M.; Izumida, H.; Sano, H.; Adachi, K.; Miwa, T. Purification and Properties of Two Deacetylases Produced by *Vibrio alginolyticus* H-8. *Biosci. Biotechnol. Biochem.* **1997**, *61*, 1113–1117. [CrossRef]

74. Hirano, T.; Uehara, R.; Shiraishi, H.; Hakamata, W.; Nishio1, T. Chitin Oligosaccharide Deacetylase from *Shewanella woodyi* ATCC51908. *J. Appl. Glycosci.* **2015**, *62*, 153–157. [CrossRef]

75. Hirano, T.; Shiraishi, H.; Ikejima, M.; Uehara, R.; Hakamata, W.; Nishio, T. Chitin oligosaccharide deacetylase from *Shewanella baltica* ATCC BAA-1091. *Biosci. Biotechnol. Biochem.* **2017**, *81*, 547–550. [CrossRef] [PubMed]

76. Tuveng, T.R.; Rothweiler, U.; Udatha, G.; Vaaje-Kolstad, G.; Smalås, A.; Eijsink, V.G.H. Structure and function of a CE4 deacetylase isolated from a marine environment. *PLoS ONE* **2017**, *12*, e0187544. [CrossRef] [PubMed]

77. Bartnicki-Garcia, S.; Nickerson, W.J. Isolation, composition, and structure of cell walls of filamentous and yeast-like forms of *Mucor rouxii*. *Biochim. Biophys. Acta* **1962**, *58*, 102–119. [CrossRef]

78. Kafetzopoulos, D.; Martinou, A.; Bouriotis, V. Bioconversion of chitin to chitosan: Purification and characterization of chitin deacetylase from *Mucor rouxii*. *Proc. Natl. Acad. Sci. USA* **1993**, *90*, 2564–2568. [CrossRef] [PubMed]

79. Davis, L.L.; Bartnicki-Garcia, S. The co-ordination of chitosan and chitin synthesis in *Mucor rouxii*. *J. Gen. Microbiol.* **1984**, *130*, 2095–2102. [CrossRef] [PubMed]

80. Chatterjee, S.; Adhya, M.; Guha, A.K.; Chatterjee, B.P. Chitosan from *Mucor rouxii*: Production and physico-chemical characterization. *Process Biochem.* **2005**, *40*, 395–400. [CrossRef]

81. Synowiecki, J.; Al-Khateeb, N.A.A.Q. Mycelia of *Mucor rouxii* as a source of chitin and chitosan. *Food Chem.* **1997**, *60*, 605–610. [CrossRef]

82. Martinou, A.; Kafetzopoulos, D.; Bouriotis, V. Isolation of chitin deacetylase from *Mucor rouxii* by immunoaffinity chromatography. *J. Chromatogr. A* **1993**, *644*, 35–41. [CrossRef]

83. Kauss, H.; Bauch, B. Chitin deacetylase from *Colletotrichum lindemuthianum*. *Methods Enzymol.* **1988**, *161*, 518–523. [CrossRef]

84. Tsigos, I.; Bouriotis, V. Purification and characterization of chitin deacetylase from *Colletotrichum lindemuthianum*. *J. Biol. Chem.* **1995**, *270*, 26286–26291. [CrossRef] [PubMed]

85. O'Connell, R.J.; Ride, J.P. Chemical detection and ultrastructural localization of chitin in cell walls of *Colletotrichum lindemuthianum*. *Physiol. Mol. Plant Pathol.* **1990**, *37*, 39–53. [CrossRef]
86. Tokuyasu, K.; Ohnishi-Kameyama, M.; Hayashi, K. Purification and characterization of extracellular chitin deacetylase from *Colletotrichum lindemuthianum*. *Biosci. Biotechnol. Biochem.* **1996**, *60*, 1598–1603. [CrossRef] [PubMed]
87. Shrestha, B.; Blondeau, K.; Stevens, W.F.; Hegarat, F.L. Expression of chitin deacetylase from *Colletotrichum lindemuthianum* in *Pichia pastoris*: Purification and characterization. *Protein Expr. Purif.* **2004**, *38*, 196–204. [CrossRef] [PubMed]
88. Kang, L.; Chen, X.; Zhai, C.; Ma, L. Synthesis and high expression of chitin deacetylase from *Colletotrichum lindemuthianum* in *Pichia pastoris* GS115. *J. Microbiol. Biotechnol.* **2012**, *22*, 1202–1207. [CrossRef] [PubMed]
89. Tokuyasu, K.; Kaneko, S.; Hayashi, K.; Mori, Y. Production of a recombinant chitin deacetylase in the culture medium of *Escherichia coli* cells. *FEBS Lett.* **1999**, *458*, 23–26. [CrossRef]
90. Tokuyasu, K.; Ohnishi-Kameyama, M.; Hayashi, K.; Mori, Y. Cloning and expression of chitin deacetylase gene from a deuteromycete, *Colletotrichum lindemuthianum*. *J. Biosci. Bioeng.* **1999**, *87*, 418–423. [CrossRef]
91. Tokuyasu, K.; Ono, H.; Ohnishi-Kameyama, M.; Hayashi, K.; Mori, Y. Deacetylation of chitin oligosaccharides of dp 2-4 by chitin deacetylase from *Colletotrichum lindemuthianum*. *Carbohydr. Res.* **1997**, *303*, 353–358. [CrossRef]
92. Tokuyasu, K.; Ono, H.; Hayashi, K.; Mori, Y. Reverse hydrolysis reaction of chitin deacetylase and enzymatic synthesis of β-D-GlcNAc-(1→4)-GlcN from chitobiose. *Carbohydr. Res.* **1999**, *322*, 26–31. [CrossRef]
93. Tokuyasu, K.; Ono, H.; Mitsutomi, M.; Hayashi, K.; Mori, Y. Synthesis of a chitosan tetramer derivative, β-D-GlcNAc-(1→4)-β-D-GlcNAc-(1→4)-β-D-GlcNAc-(1→4)-D-GlcN through a partial *N*-acetylation reaction by chitin deacetylase. *Carbohydr. Res.* **2000**, *325*, 211–215. [CrossRef]
94. Kang, L.X.; Liang, Y.X.; Ma, L.X. Novel characteristics of chitin deacetylase from *Colletotrichum lindemuthianum*: Production of fully acetylated chitooligomers, and hydrolysis of deacetylated chitooligomers. *Process Biochem.* **2014**, *49*, 1936–1940. [CrossRef]
95. Reyes, F.; Calatayud, J.; Vazquez, C.; Martínez, M.J. β-*N*-Acetylglucosaminidase from *Aspergillus nidulans* which degrades chitin oligomers during autolysis. *FEMS Microbiol. Lett.* **1990**, *65*, 83–87. [CrossRef]
96. Wang, Y.; Song, J.Z.; Yang, Q.; Liu, Z.H.; Huang, X.M.; Chen, Y. Cloning of a heat-stable chitin deacetylase gene from *Aspergillus nidulans* and its functional expression in *Escherichia coli*. *Appl. Biochem. Biotechnol.* **2010**, *162*, 843–854. [CrossRef] [PubMed]
97. Espagne, E.; Lespinet, O.; Malagnac, F.; Da Silva, C.; Jaillon, O.; Porcel, B.M.; Couloux, A.; Aury, J.-M.; Ségurens, B.; Poulain, J.; Anthouard, V.; et al. The genome sequence of the model ascomycete fungus *Podospora anserina*. *Genome Biol.* **2008**, *9*, R77. [CrossRef] [PubMed]
98. Lorin, S.; Dufour, E.; Sainsard-Chanet, A. Mitochondrial metabolism and aging in the filamentous fungus *Podospora anserina*. *Biochim. Biophys. Acta Bioenerg.* **2006**, *1757*, 604–610. [CrossRef] [PubMed]
99. Figueroa, M.; Upadhyaya, N.M.; Sperschneider, J.; Park, R.F.; Szabo, L.J.; Steffenson, B.; Ellis, J.G.; Dodds, P.N. Changing the Game: Using Integrative Genomics to Probe Virulence Mechanisms of the Stem Rust Pathogen *Puccinia graminis* f. sp. tritici. *Front. Plant Sci.* **2016**, *7*, 1–10. [CrossRef] [PubMed]
100. Singh, R.P.; Hodson, D.P.; Jin, Y.; Lagudah, E.S.; Ayliffe, M.A.; Bhavani, S.; Rouse, M.N.; Pretorius, Z.A.; Szabo, L.J.; Huerta-Espino, J.; et al. Emergence and Spread of New Races of Wheat Stem Rust Fungus: Continued Threat to Food Security and Prospects of Genetic Control. *Phytopathology* **2015**, *105*, 872–884. [CrossRef] [PubMed]
101. Jin, Y.; Singh, R.P. Resistance in U.S. Wheat to Recent Eastern African Isolates of *Puccinia graminis* f. sp. tritici with Virulence to Resistance Gene Sr31. *Plant Dis.* **2006**, *90*, 476–480. [CrossRef]
102. Mendgen, K.; Hahn, M. Plant infection and the establishment of fungal biotrophy. *Trends Plant Sci.* **2002**, *7*, 352–356. [CrossRef]
103. Broeker, K.; Fehser, S.; Tenberge, K.B.; Moerschbacher, B.M. Two class III chitin synthases specifically localized in appressoria and haustoria of *Puccinia graminis* f. sp. tritici. *Physiol. Mol. Plant Pathol.* **2011**, *76*, 27–33. [CrossRef]
104. Ride, J.P.; Barber, M.S. Purification and characterization of multiple forms of endochitinase from wheat leaves. *Plant Sci.* **1990**, *71*, 185–197. [CrossRef]

105. Vander, P.; Vårum, K.M.; Domard, A.; El Gueddari, N.E.; Moerschbacher, B.M. Comparison of the Ability of Partially *N*-Acetylated Chitosans and Chitooligosaccharides to Elicit Resistance Reactions in Wheat Leaves. *Plant Physiol.* **1998**, *118*, 1353–1359. [CrossRef] [PubMed]

106. Maharachchikumbura, S.S.N.; Hyde, K.D.; Groenewald, J.Z.; Xu, J.; Crous, P.W. *Pestalotiopsis* revisited. *Stud. Mycol.* **2014**, *79*, 121–186. [CrossRef] [PubMed]

107. Lopez-Llorca, L.V.; Olivares-Bernabeu, C.; Salinas, J.; Jansson, H.-B.; Kolattukudy, P.E. Pre-penetration events in fungal parasitism of nematode eggs. *Mycol. Res.* **2002**, *106*, 499–506. [CrossRef]

108. Manzanilla-Lopez, R.H.; Esteves, I.; Finetti-Sialer, M.M.; Hirsch, P.R.; Ward, E.; Devonshire, J.; Hidalgo-Diaz, L. *Pochonia chlamydosporia*: Advances and Challenges to Improve Its Performance as a Biological Control Agent of Sedentary Endo-parasitic Nematodes. *J. Nematol.* **2013**, *45*, 1–7. [PubMed]

109. Larriba, E.; Jaime, M.D.L.A.; Carbonell-Caballero, J.; Conesa, A.; Dopazo, J.; Nislow, C.; Martín-Nieto, J.; Lopez-Llorca, L.V. Sequencing and functional analysis of the genome of a nematode egg-parasitic fungus, *Pochonia chlamydosporia*. *Fungal Genet. Biol.* **2014**, *65*, 69–80. [CrossRef] [PubMed]

110. Aranda-Martinez, A.; Lenfant, N.; Escudero, N.; Zavala-Gonzalez, E.A.; Henrissat, B.; Lopez-Llorca, L.V. CAZyme content of *Pochonia chlamydosporia* reflects that chitin and chitosan modification are involved in nematode parasitism. *Environ. Microbiol.* **2016**, *18*, 4200–4215. [CrossRef] [PubMed]

111. Wani, Z.A.; Kumar, A.; Sultan, P.; Bindu, K.; Riyaz-Ul-Hassan, S.; Ashraf, N. *Mortierella* alpina CS10E4, an oleaginous fungal endophyte of *Crocus sativus* L. enhances apocarotenoid biosynthesis and stress tolerance in the host plant. *Sci. Rep.* **2017**, *7*, 8598. [CrossRef] [PubMed]

112. Zhao, Y.; Jo, G.-H.; Ju, W.-T.; Jung, W.-J.; Park, R.-D. A Highly *N*-Glycosylated Chitin Deacetylase Derived from a Novel Strain of *Mortierella* sp. DY-52. *Biosci. Biotechnol. Biochem.* **2011**, *75*, 960–965. [CrossRef] [PubMed]

113. Narayanan, K.; Parameswaran, B.; Pandey, A. Production of chitin deacetylase by *Aspergillus flavus* in submerged conditions. *Prep. Biochem. Biotechnol.* **2016**, *46*, 501–508. [CrossRef] [PubMed]

114. Cuenca-Estrella, M.; Gomez-Lopez, A.; Mellado, E.; Buitrago, M.J.; Monzón, A.; Rodriguez-Tudela, J.L. *Scopulariopsis brevicaulis*, a fungal pathogen resistant to broad-spectrum antifungal agents. *Antimicrob. Agents Chemother.* **2003**, *47*, 2339–2341. [CrossRef] [PubMed]

115. Tosti, A.; PiracciniI, B.M.; Stinchi, C.; Lorenzi, S. Onychomycosis due to *Scopulariopsis brevicaulis*: Clinical features and response to systemic antifungals. *Br. J. Dermatol.* **1996**, *135*, 799–802. [CrossRef] [PubMed]

116. El Ghaouth, A.; Arul, J.; Grenier, J.; Asselin, A. Effect of chitosan and other polyions on chitin deacetylase in *Rhizopus stolonifer*. *Exp. Mycol.* **1992**, *16*, 173–177. [CrossRef]

117. Zhang, H.; Yang, S.; Fang, J.; Deng, Y.; Wang, D.; Zhao, Y. Optimization of the fermentation conditions of *Rhizopus japonicus* M193 for the production of chitin deacetylase and chitosan. *Carbohydr. Polym.* **2014**, *101*, 57–67. [CrossRef] [PubMed]

118. Maw, T.; Tan, T.K.; Khor, E.; Wong, S.M. Selection of *Gongronella butleri* strains for enhanced chitosan yield with UV mutagenesis. *J. Biotechnol.* **2002**, *95*, 189–193. [CrossRef]

119. Yonemura, A.; Nagashima, T.; Murayama, T. *Expression of Chitin Deacetylase Gene from Phycomyces blakesleeanus in Aspergillus oryzae and Neurospora crassa*; The Society for Bioscience and Bioengineering: Osaka, Japan, 2007; p. 129. Available online: http://dl.ndl.go.jp/view/download/digidepo_10529404_po_ART0009175183.pdf?contentNo=1&alternativeNo= (accessed on 19 February 2018).

120. Baker, L.G.; Specht, C.A.; Lodge, J.K. Cell wall chitosan is necessary for virulence in the opportunistic pathogen *Cryptococcus neoformans*. *Eukaryot. Cell* **2011**, *10*, 1264–1268. [CrossRef] [PubMed]

121. Doering, T.L. How Sweet it is! Cell Wall Biogenesis and Polysaccharide Capsule Formation in *Cryptococcus neoformans*. *Annu. Rev. Microbiol.* **2009**, *63*, 223–247. [CrossRef] [PubMed]

122. Gilbert, N.M.; Baker, L.G.; Specht, C.A.; Lodge, J.K. A glycosylphosphatidylinositol anchor is required for membrane localization but dispensable for cell wall association of chitin deacetylase 2 in *Cryptococcus neoformans*. *mBio* **2012**, *3*, e00007–e00012. [CrossRef] [PubMed]

123. Levitz, S.M.; Nong, S.-H.; Mansour, M.K.; Huang, C.; Specht, C.A. Molecular characterization of a mannoprotein with homology to chitin deacetylases that stimulates T cell responses to *Cryptococcus neoformans*. *Proc. Natl. Acad. Sci. USA* **2001**, *98*, 10422–10427. [CrossRef] [PubMed]

124. Biondo, C.; Beninati, C.; Delfino, D.; Oggioni, M.; Mancuso, G.; Midiri, A.; Tomaselli, G.; Teti, G.; Biondo, C.; Beninati, C.; et al. Identification and Cloning of a Cryptococcal Deacetylase That Produces Protective Immune Responses. *Infect. Immun.* **2002**, *70*, 2383–2391. [CrossRef] [PubMed]

125. Loftus, B.; Anderson, I.; Davies, R.; Alsmark, U.C.M.; Samuelson, J.; Amedeo, P.; Roncaglia, P.; Berriman, M.; Hirt, R.P.; Mann, B.J.; et al. The genome of the protist parasite *Entamoeba histolytica*. *Nature* **2005**, *433*, 865–868. [CrossRef] [PubMed]

126. Roche, P.; Maillet, F.; Plazanet, C.; Debellé, F.; Ferro, M.; Truchet, G.; Promé, J.C.; Dénarié, J. The common nodABC genes of *Rhizobium meliloti* are host-range determinants. *Proc. Natl. Acad. Sci. USA* **1996**, *93*, 15305–15310. [CrossRef] [PubMed]

127. Egelhoff, T.T.; Long, S.R. *Rhizobium meliloti* nodulation genes: Identification of nodDABC gene products, purification of nodA protein, and expression of nodA in *Rhizobium meliloti*. *J. Bacteriol.* **1985**, *164*, 591–599. [PubMed]

128. Spaink, H.P.; Wijfjes, A.H.M.; der van Drift, K.M.G.M.; Haverkamp, J.; Thomas-Oates, J.E.; Lugtenberg, B.J.J. Structural identification of metabolites produced by the NodB and NodC proteins of *Rhizobium leguminosarum*. *Mol. Microbiol.* **1994**, *13*, 821–831. [CrossRef] [PubMed]

129. Chambon, R.; Pradeau, S.; Fort, S.; Cottaz, S.; Armand, S. High yield production of *Rhizobium* NodB chitin deacetylase and its use for in vitro synthesis of lipo-chitinoligosaccharide precursors. *Carbohydr. Res.* **2017**, *442*, 25–30. [CrossRef] [PubMed]

130. Hamer, S.N.; Cord-Landwehr, S.; Biarnés, X.; Planas, A.; Waegeman, H.; Moerschbacher, B.M.; Kolkenbrock, S. Enzymatic production of defined chitosan oligomers with a specific pattern of acetylation using a combination of chitin oligosaccharide deacetylases. *Sci. Rep.* **2015**, *5*, 8716. [CrossRef] [PubMed]

131. Röhrig, H.; Schmidt, J.; Wieneke, U.; Kondorosi, E.; Barlier, I.; Schell, J.; John, M. Biosynthesis of lipooligosaccharide nodulation factors: *Rhizobium* NodA protein is involved in N-acylation of the chitooligosaccharide backbone. *Proc. Natl. Acad. Sci. USA* **1994**, *91*, 3122–3126. [CrossRef] [PubMed]

132. Keyhani, N.O.; Roseman, S. Physiological aspects of chitin catabolism in marine bacteria. *Biochim. Biophys. Acta Gen. Subj.* **1999**, *1473*, 108–122. [CrossRef]

133. Zobell, C.; Rittenberg, S. The occurrence and characteristics of chitinoclastic bacteria in the sea. *J. Bacteriol.* **1937**, *35*, 275–287.

134. Meibom, K.L.; Li, X.B.; Nielsen, A.T.; Wu, C.-Y.; Roseman, S.; Schoolnik, G.K. The *Vibrio cholerae* chitin utilization program. *Proc. Natl. Acad. Sci. USA* **2004**, *101*, 2524–2529. [CrossRef] [PubMed]

135. Li, X.; Roseman, S. The chitinolytic cascade in Vibrios is regulated by chitin oligosaccharides and a two-component chitin catabolic sensor/kinase. *Proc. Natl. Acad. Sci. USA* **2004**, *101*, 627–631. [CrossRef] [PubMed]

136. Ohishi, K.; Murase, K.; Ohta, T.; Etoh, H. Cloning and sequencing of the deacetylase gene from *Vibrio alginolyticus* H-8. *J. Biosci. Bioeng.* **2000**, *90*, 561–563. [CrossRef]

137. Kadokura, K.; Sakamoto, Y.; Saito, K.; Ikegami, T.; Hirano, T.; Hakamata, W.; Oku, T.; Nishio, T. Production of a recombinant chitin oligosaccharide deacetylase from *Vibrio parahaemolyticus* in the culture medium of *Escherichia coli* cells. *Biotechnol. Lett.* **2007**, *29*, 1209–1215. [CrossRef] [PubMed]

138. Hirano, T.; Maebara, Y.; Uehara, R.; Sakaki, Y.; Shiraishi, H.; Ichimura, S.; Hakamata, W.; Nishio, T. Chitin oligosaccharide deacetylase from Vibrio harveyi ATCC BAA-1116: Gene cloning, overexpression, purification, and characterization. *Chitin Chitosan Res.* **2012**, *19*, 321–324.

139. Jacquiod, S.; Franqueville, L.; Cécillon, S.; Vogel, T.M.; Simonet, P. Soil bacterial community shifts after Chitin enrichment: An integrative metagenomic approach. *PLoS ONE* **2013**, *8*. [CrossRef] [PubMed]

140. Dsouza, M.; Taylor, M.W.; Turner, S.J.; Aislabie, J. Genomic and phenotypic insights into the ecology of *Arthrobacter* from Antarctic soils. *BMC Genom.* **2015**, *16*, 36. [CrossRef] [PubMed]

141. Lonhienne, T.; Mavromatis, K.; Vorgias, C.E.; Buchon, L.; Gerday, C.; Bouriotis, V. Cloning, sequences, and characterization of two chitinase genes from the Antarctic *Arthrobacter* sp. strain TAD20: Isolation and partial characterization of the enzymes. *J. Bacteriol.* **2001**, *183*, 1773–1779. [CrossRef] [PubMed]

142. Aragunde-pazos, H.; Biarnés, X.; Planas, A. Substrate recognition and specificity of chitin deacetylases and related family 4 carbohydrate esterases. *Int. J. Mol. Sci.* **2018**, *19*, 412. [CrossRef] [PubMed]

143. Boraston, A.B.; Bolam, D.N.; Gilbert, H.J.; Davies, G.J. Carbohydrate-Binding modules: Fine-Tuning polysaccharide recognition. *Biochem. J.* **2004**, *382*, 769–781. [CrossRef] [PubMed]

144. Blair, D.E.; Schuttelkopf, A.W.; MacRae, J.I.; van Aalten, D.M.F. Structure and metal-dependent mechanism of peptidoglycan deacetylase, a streptococcal virulence factor. *Proc. Natl. Acad. Sci. USA* **2005**, *102*, 15429–15434. [CrossRef] [PubMed]

145. Blair, D.E.; Van Aalten, D.M.F. Structures of *Bacillus subtilis* PdaA, a family 4 carbohydrate esterase, and a complex with *N*-acetyl-glucosamine. *FEBS Lett.* **2004**, *570*, 13–19. [CrossRef] [PubMed]

146. Nakamura, A.M.; Nascimento, A.S.; Polikarpov, I. Structural diversity of carbohydrate esterases. *Biotechnol. Res. Innov.* **2017**, *1*, 35–51. [CrossRef]

147. Nishiyama, T.; Noguchi, H.; Yoshida, H.; Park, S.Y.; Tame, J.R.H. The structure of the deacetylase domain of *Escherichia coli* PgaB, an enzyme required for biofilm formation: A circularly permuted member of the carbohydrate esterase 4 family. *Acta Crystallogr. Sect. D Biol. Crystallogr.* **2013**, *69*, 44–51. [CrossRef] [PubMed]

148. Hernick, M.; Fierke, C.A. Zinc hydrolases: The mechanisms of zinc-dependent deacetylases. *Arch. Biochem. Biophys.* **2005**, *433*, 71–84. [CrossRef] [PubMed]

149. Xie, J.L.; Polvi, E.J.; Shekhar-Guturja, T.; Cowen, L.E. Elucidating drug resistance in human fungal pathogens. *Future Microbiol.* **2014**, *9*, 523–542. [CrossRef] [PubMed]

150. Takaya, N.; Yamazaki, D.; Horiuchi, H.; Ohta, A.; Takagi, M. Cloning and characterization of a chitinase-encoding gene (chiA) from *Aspergillus nidulans*, disruption of which decreases germination frequency and hyphal growth. *Biosci. Biotechnol. Biochem.* **1998**, *62*, 60–65. [CrossRef] [PubMed]

151. Hartl, L.; Zach, S.; Seidl-Seiboth, V. Fungal chitinases: Diversity, mechanistic properties and biotechnological potential. *Appl. Microbiol. Biotechnol.* **2012**, *93*, 533–543. [CrossRef] [PubMed]

152. Aoun, M. Host defense mechanisms during fungal pathogenesis and how these are overcome in susceptible plants: A review. *Int. J. Bot.* **2017**, *13*, 82–102. [CrossRef]

153. Huang, G. Chitinase Inhibitor Allosamidin and Its Analogues: An Update. *Curr. Org. Chem.* **2012**, *16*, 115–120. [CrossRef]

154. Rao, F.V.; Houston, D.R.; Boot, R.G.; Aerts, J.M.F.G.; Hodkinson, M.; Adams, D.J.; Shiomi, K.; Omura, S.; Van Aalten, D.M.F. Specificity and affinity of natural product cyclopentapeptide inhibitors against *A. fumigatus*, human, and bacterial chitinases. *Chem. Biol.* **2005**, *12*, 65–76. [CrossRef] [PubMed]

155. Younes, I.; Rinaudo, M. Chitin and chitosan preparation from marine sources. Structure, properties and applications. *Mar. Drugs* **2015**, *13*, 1133–1174. [CrossRef] [PubMed]

156. Cheung, R.C.F.; Ng, T.B.; Wong, J.H.; Chan, W.Y. Chitosan: An update on potential biomedical and pharmaceutical applications. *Mar. Drugs* **2015**, *13*, 5156–5186. [CrossRef] [PubMed]

157. Anitha, A.; Sowmya, S.; Kumar, P.T.T.S.; Deepthi, S.; Chennazhi, K.P.; Ehrlich, H.; Tsurkan, M.; Jayakumar, R. Chitin and chitosan in selected biomedical applications. *Prog. Polym. Sci.* **2014**, *39*, 1644–1667. [CrossRef]

158. Abdul Khalil, H.P.S.; Saurabh, C.K.; Adnan, A.S.; Nurul Fazita, M.R.; Syakir, M.I.; Davoudpour, Y.; Rafatullah, M.; Abdullah, C.K.; Haafiz, M.K.M.; Dungani, R. A review on chitosan-cellulose blends and nanocellulose reinforced chitosan biocomposites: Properties and their applications. *Carbohydr. Polym.* **2016**, *150*, 216–226.

159. Pestov, A.; Bratskaya, S. Chitosan and Its Derivatives as Highly Efficient Polymer Ligands. *Molecules* **2016**, *21*, 330. [CrossRef] [PubMed]

160. Das, S.N.; Madhuprakash, J.; Sarma, P.V.S.R.N.; Purushotham, P.; Suma, K.; Manjeet, K.; Rambabu, S.; Gueddari, N.E.E.; Moerschbacher, B.M.; Podile, A.R. Biotechnological approaches for field applications of chitooligosaccharides (COS) to induce innate immunity in plants. *Crit. Rev. Biotechnol.* **2015**, *35*, 29–43. [CrossRef] [PubMed]

161. Sorlier, P.; Denuzière, A.; Viton, C.; Domard, A. Relation between the degree of acetylation and the electrostatic properties of chitin and chitosan. *Biomacromolecules* **2001**, *2*, 765–772. [CrossRef] [PubMed]

162. Omura, Y.; Shigemoto, M.; Akiyama, T.; Saimoto, H.; Shigemasa, Y.; Nakamura, I.; Tsuchido, T. Antimicrobial Activity of Chitosan with Different Degrees of Acetylation and Molecular Weights. *Biocontrol Sci.* **2003**, *8*, 25–30. [CrossRef]

163. Domard, A.; Cartier, N. Glucosamine oligomers: 1. Preparation and characterization. *Int. J. Biol. Macromol.* **1989**, *11*, 297–302. [CrossRef]

164. Einbu, A.; Varum, K.M. Depolymerization and de-*N*-acetylation of chitin oligomers in hydrochloric acid. *Biomacromolecules* **2007**, *8*, 309–314. [CrossRef] [PubMed]

165. Kuyama, H.; Nakahara, Y.; Nukada, T.; Ito, Y.; Nakahara, Y.; Ogawa, T. Stereocontrolled synthesis of chitosan dodecamer. *Carbohydr. Res.* **1993**, *243*, C1–C7. [CrossRef]

166. Barroca-Aubry, N.; Pernet-Poil-Chevrier, A.; Domard, A.; Trombotto, S. Towards a modular synthesis of well-defined chitooligosaccharides: Synthesis of the four chitodisaccharides. *Carbohydr. Res.* **2010**, *345*, 1685–1697. [CrossRef] [PubMed]

167. Weinhold, M.X.; Sauvageau, J.C.M.; Kumirska, J.; Thöming, J. Studies on acetylation patterns of different chitosan preparations. *Carbohydr. Polym.* **2009**, *78*, 678–684. [CrossRef]

168. Abla, M.; Marmuse, L.; Delolme, F.; Vors, J.P.; Ladavière, C.; Trombotto, S. Access to tetra-*N*-acetyl-chitopentaose by chemical *N*-acetylation of glucosamine pentamer. *Carbohydr. Polym.* **2013**, *98*, 770–777. [CrossRef] [PubMed]

169. Trombotto, S.; Ladavière, C.; Delolme, F.; Domard, A. Chemical preparation and structural characterization of a homogeneous series of chitin/chitosan oligomers. *Biomacromolecules* **2008**, *9*, 1731–1738. [CrossRef] [PubMed]

170. Naqvi, S.; Moerschbacher, B.M. The cell factory approach toward biotechnological production of high-value chitosan oligomers and their derivatives: An update. *Crit. Rev. Biotechnol.* **2017**, *37*, 11–25. [CrossRef] [PubMed]

171. Hembach, L.; Cord-Landwehr, S.; Moerschbacher, B.M. Enzymatic production of all fourteen partially acetylated chitosan tetramers using different chitin deacetylases acting in forward or reverse mode. *Sci. Rep.* **2017**, *7*, 17692. [CrossRef] [PubMed]

172. Mergaert, P.; D'Haeze, W.; Geelen, D.; Promé, D.; Van Montagu, M.; Geremia, R.; Promé, J.C.; Holsters, M. Biosynthesis of *Azorhizobium caulinodans* Nod factors: Study of the activity of the nodABCS proteins by expression of the genes in *Escherichia coli*. *J. Biol. Chem.* **1995**, *270*, 29217–29223. [CrossRef] [PubMed]

173. Poinsot, V.; Crook, M.B.; Erdn, S.; Maillet, F.; Bascaules, A.; Ané, J.M. New insights into Nod factor biosynthesis: Analyses of chitooligomers and lipo-chitooligomers of *Rhizobium* sp. IRBG74 mutants. *Carbohydr. Res.* **2016**, *434*, 83–93. [CrossRef] [PubMed]

174. Samain, E.; Drouillard, S.; Heyraud, A.; Driguez, H.; Geremia, R.A. Gram-scale synthesis of recombinant chitooligosaccharides in *Escherichia coli*. *Carbohydr. Res.* **1997**, *302*, 35–42. [CrossRef]

175. Samain, E.; Chazalet, V.; Geremia, R.A. Production of O-acetylated and sulfated chitooligosaccharides by recombinant *Escherichia coli* strains harboring different combinations of nod genes. *J. Biotechnol.* **1999**, *72*, 33–47. [CrossRef]

176. Cottaz, S.; Samain, E. Genetic engineering of *Escherichia coli* for the production of N^I, N^{II}-diacetylchitobiose (chitinbiose) and its utilization as a primer for the synthesis of complex carbohydrates. *Metab. Eng.* **2005**, *7*, 311–317. [CrossRef] [PubMed]

177. Bettler, E.; Samain, E.; Chazalet, V.; Bosso, C.; Heyraud, A.; Joziasse, D.H.; Wakarchuk, W.W.; Imberty, A.; Geremia, R.A. The living factory: In Vivo production of *N*-acetyllactosamine containing carbohydrates in *E. coli*. *Glycoconj. J.* **1999**, *16*, 205–212. [CrossRef] [PubMed]

178. Southwick, A.M.; Wang, L.X.; Long, S.R.; Lee, Y.C. Activity of *Sinorhizobium meliloti* NodAB and nodH enzymes on thiochitooligosaccharides. *J. Bacteriol.* **2002**, *184*, 4039–4043. [CrossRef] [PubMed]

polymers

MDPI

Article

Biosynthetic Pathway and Genes of Chitin/Chitosan-Like Bioflocculant in the Genus *Citrobacter*

Masahiro Takeo [1,*], Kazuyuki Kimura [2], Shanmugam Mayilraj [3], Takuya Inoue [1], Shohei Tada [1], Kouki Miyamoto [1], Masami Kashiwa [1], Keishi Ikemoto [1], Priyanka Baranwal [1], Daiichiro Kato [4] and Seiji Negoro [1]

[1]   Department of Applied Chemistry, Graduate School of Engineering, University of Hyogo, 2167 Shosha, Himeji, Hyogo 671-2280, Japan; takuya.i@eco-feed.org (T.I.); tada@bell-c.co.jp (S.T.); miyamoto-kouki@nikke.co.jp (K.M.); mashia@occn.zaq.ne.jp (M.K.); bambootail2016@gmail.com (K.I.); barnpriyanka@gmail.com (P.B.); negoro@eng.u-hyogo.ac.jp (S.N.)

[2]   Hyogo Analysis Center Co., Ltd., 4-10-8 Seimondori, Hirohata, Himeji, Hyogo 671-1116, Japan; kimura@hyobun.co.jp

[3]   Microbial Type Culture Collection & Gene Bank (MTCC), CSIR-Institute of Microbial Technology, Sector 39-A, Chandigarh-160 036, India; mayil@imtech.res.in

[4]   Department of Chemistry and Bioscience, Graduate School of Science and Engineering, Kagoshima 890-8580, Japan; kato@sci.kagoshima-u.ac.jp

*   Correspondence: takeo@eng.u-hyogo.ac.jp; Tel.: +81-79-267-4893

Received: 23 January 2018; Accepted: 22 February 2018; Published: 27 February 2018

Abstract: Chitin/chitosan, one of the most abundant polysaccharides in nature, is industrially produced as a powder or flake form from the exoskeletons of crustaceans such as crabs and shrimps. Intriguingly, many bacterial strains in the genus *Citrobacter* secrete a soluble chitin/chitosan-like polysaccharide into the culture medium during growth in acetate. Because this polysaccharide shows strong flocculation activity for suspended solids in water, it can be used as a bioflocculant (BF). The BF synthetic pathway of *C. freundii* IFO 13545 is expected from known bacterial metabolic pathways to be as follows: acetate is metabolized in the TCA cycle and the glyoxylate shunt via acetyl-CoA. Next, fructose 6-phosphate is generated from the intermediates of the TCA cycle through gluconeogenesis and enters into the hexosamine synthetic pathway to form UDP-*N*-acetylglucosamine, which is used as a direct precursor to extend the BF polysaccharide chain. We conducted the draft genome sequencing of IFO 13545 and identified all of the candidate genes corresponding to the enzymes in this pathway in the 5420-kb genome sequence. Disruption of the genes encoding acetyl-CoA synthetase and isocitrate lyase by homologous recombination resulted in little or no growth on acetate, indicating that the cell growth depends on acetate assimilation via the glyoxylate shunt. Disruption of the gene encoding glucosamine 6-phosphate synthase, a key enzyme for the hexosamine synthetic pathway, caused a significant decrease in flocculation activity, demonstrating that this pathway is primarily used for the BF biosynthesis. A gene cluster necessary for the polymerization and secretion of BF, named *bfpABCD*, was also identified for the first time. In addition, quantitative RT-PCR analysis of several key genes in the expected pathway was conducted to know their expression in acetate assimilation and BF biosynthesis. Based on the data obtained in this study, an overview of the BF synthetic pathway is discussed.

Keywords: *Citrobacter*; biosynthesis; bioflocculant; chitosan; metabolic pathway

Polymers **2018**, *10*, 237

1. Introduction

Chitin/chitosan, a polysaccharide consisting of *N*-acetylglucosamine (GlcNAc) and/or glucosamine (GlcN) linked through β-1,4-glycosidic linkages, is one of the most abundant polysaccharides in nature [1,2]. Although chitin/chitosan is distributed in a wide range of living organisms [2,3], it is industrially produced from the exoskeletons of crustaceans such as crabs, shrimps, prawns, lobsters, and krill because waste from food-processing industry provides an abundant source of this material [2,3]. Due to its useful biological properties (e.g., biocompatibility, biodegradability, and antimicrobial activity) and chemical modification potentials through its reactive functional groups (–OH, –NH$_2$, and –COOH) [2,4,5], chitin/chitosan has been used in a wide range of fields including biomedical, pharmaceutical, food production, and wastewater treatment fields [3,5,6]. However, time-consuming and costly extraction/purification processes (for the solubilization of chitin and the removal of proteins, minerals, and colors) are required for the production [3]. In addition, the supply of the waste depends on seasonal yields of the crustaceans and geographic conditions. Therefore, as an alternative, the production of chitin/chitosan from microbial sources has been considered and attempted [7]. In the case of microbial sources, the production can be conducted throughout the year using biotechnological processes, and the quality of the product is generally stable. The major source of such chitin/chitosan is fungal mycelia, because the cell walls of fungi contain large amounts of chitin, and fungal mycelia can be obtained as waste from mushroom production and from the fermentation industry [3,7]. However, the extraction/purification processes cannot be omitted, although they may be somewhat simpler than that from the waste of the crustaceans. If chitin/chitosan were secreted by microbial cultures in a soluble form, its production and the downstream processes for its production could be greatly simplified.

In 2000, Fujita et al. reported that the enterobacterial strain *Citrobacter* sp. TKF04, which produces a bioflocculant (BF) from acetate and propionate [8]. By the chemical analysis, the BF was found to be a high-molecular-weight (around 320 kDa) polysaccharide consisting of GlcNAc and GlcN, similar to chitin/chitosan [8]. Infrared spectroscopic analysis showed that it has a very similar structure to those of commercial chitin/chitosan products [8] and the deacetylation degree was estimated as 50–60%. Chitinase and chitosanase preferentially degraded it, resulting in significantly-reduced flocculation activities [8]. Later, Son et al. and Kim et al. also reported similar chitin/chitosan-like polysaccharide-producing enterobacterial strains, *Enterobacter* sp. BL-2 and *Citrobacter* sp. BL-4 [9,10]. Hence, we collected 36 *Citrobacter* strains from various microbial culture collection centers and measured their flocculation activities. We found that 21 strains belonging to four species (*C. freundii*, *C. braakii*, *C. youngae*, and *C. werkmanii*) showed flocculation activity when grown on acetate [11]. We confirmed that five selected strains with high flocculation activity secreted the chitin/chitosan-like BF with wide molecular weight distributions (with the peak tops of >1660 kDa in the gel filtration chromatography) into the culture medium [11]. These results demonstrate that many strains belonging to the genus *Citrobacter* have metabolic potential to produce the chitin/chitosan-like BF as a soluble form from acetate. We are very interested in the chitin/chitosan-like BF biosynthetic pathway.

Based on known bacterial acetate metabolic and peptidoglycan synthetic pathways [12,13], we expected the BF synthetic pathway of *Citrobacter* strains to be as follows (Figure 1): acetate is first converted into acetyl-CoA, which is further transformed into fructose 6-phosphate (Fru 6-P) through the TCA cycle, the glyoxylate shunt, and gluconeogenesis. Next, Fru 6-P enters into the hexosamine synthetic pathway (hereafter referred to as the hexosamine pathway) where it is transformed into UDP-*N*-acetylglucosamine (UDP-GlcNAc), which is finally used as a direct precursor for the extension of the BF polysaccharide chain. Unfortunately, little information is available on the polymerization and secretion of bacterial chitin/chitosan. Recently, we performed the draft genome sequencing of four *Citrobacter* strains, including *C. freundii* IFO 13545. We identified all the candidate genes corresponding to all the enzymes in the chitin/chitosan-like BF biosynthetic pathway in the 5420-kb genome sequence of IFO 13545 (Figure 1 and Table 1). To confirm the involvement of the products of these genes in acetate assimilation and BF biosynthesis, we conducted a gene disruption study involving several key genes of the pathway (Figure 1) in which we investigated the effects of the gene disruption of these genes on cell growth on acetate and on flocculation activity. In addition, through this gene disruption study, we identified a candidate gene cluster for the polymerization and secretion of the BF and report it here.

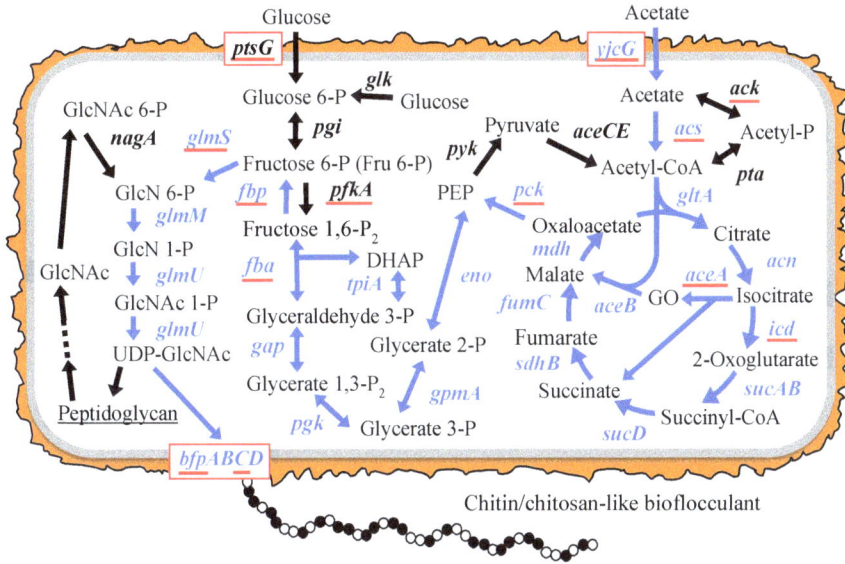

Figure 1. Proposed bioflocculant (BF) synthetic pathway and the genes of *Citrobacter freundii* IFO 13545 involved in the pathway. Blue arrows and gene names indicate the proposed BF synthetic pathway and the genes encoding the enzymes involved in this pathway. The genes underlined in red were used in a qRT-PCR study to determine the level of gene expression in glucose medium (GM) and in acetate medium (AM). Abbreviations: BF, bioflocculant; GO, glyoxylate; P, phosphate; P_2, diphosphate; DHAP, dihyroxyacetone-P; PEP, phosphoenolpyruvate; GlcN, glucosamine; GlcNAc, *N*-acetylglucosamine.

Table 1. Accession numbers of the registered genes in *Citrobacter freundii* IFO 13545.

Accession No.	Gene	Enzyme	Strain with Homologous Gene	Homologous Gene Accession No.	Identity at nt Sequence Level (%)	Identity at aa Sequence Level (%)
AB823554	*aceA*	isocitrate lyase	*E. coli* MG1655	U00096	86	96
AB823555	*aceB*	malate synthase	*E. coli* MG1655	U00096	84	92
AB823558	*acn*	aconitate hydratase I	*E. coli* MG1655	U00096	87	96
AB823559	*acs*	acetyl-CoA synthetase	*E. coli* MG1655	U00096	85	94
LC020545	*ack*	acetate kinase	*E. coli* MG1655	U00096	91	96
AB823580	*bfpA*	PGA export porin	*C. werkmanii*	KF057877	91	98
AB823581	*bfpB*	PGA *N*-deacetylase	*C. freundii* CFNIH1	CP007557	90	97
AB823582	*bfpC*	PGA synthase	*C. werkmanii* strain BF-6	KF057878	94	99
AB823583	*bfpD*	PGA biosynthesis protein	*C. freundii* CFNIH1	CP007557	88	95

Table 1. *Cont.*

Accession No.	Gene	Enzyme	Strain with Homologous Gene	Homologous Gene Accession No.	Identity at nt Sequence Level (%)	Identity at aa Sequence Level (%)
LC020058	*cpsA*	undecaprenyl-phosphate glucose phosphotransferase	*C. freundii* MTCC1658	CP007557	86	100
AB823561	*eno*	enolase	*E. coli* MG1655	U00096	93	97
AB823562	*fba*	fructose bisphosphate aldolase	*E. coli* MG1655	U00096	84	97
LC027370	*fbp*	fructose 1,6-bisphosphatase	*E. coli* MG1655	U00096	96	97
AB823563	*fumC*	fumarate hydratase	*Salmonella enterica*	CP007530	81	93
AB823564	*g6pd*	glucose 6-phosphate dehydrogenase	*E. coli* MG1655	U00096	86	97
AB823565	*gap*	glyceraldehyde 3-phosphate dehydrogenase	*E. albertii* KF-1	CP007025	79	91
AB823566	*glk*	glucokinase	*E. coli* MG1655	U00096	80	93
AB823568	*glmM*	phosphoglucosamine mutase	*E. coli* MG1655	U00096	85	96
AB823569	*glmS*	glucosamine-6-phosphate synthase	*E. coli* MG1655	U00096	87	95
AB823570	*glmU*	uridyltransferase/glucosamine-1-phosphate acetyltransferase	*E. coli* MG1655	U00096	83	91
AB823571	*gltA*	type II citrate synthase	*E. coli* MG1655	U00096	86	96
AB823572	*gpmA*	phosphoglyceromutase	*E. coli* MG1655	U00096	88	96
AB823573	*icd*	isocitrate dehydrogenase	*E. coli* MG1655	U00096	88	97
AB823574	*mdh*	malate dehydrogenase	*E. coli* MG1655	U00096	87	97
LC363529	*nagA*	*N*-acetylglucosamine 6-phosphate deacetylase	*E. coli* MG1655	U00096	84	91
AB823577	*pck*	phosphoenolpyruvate carboxykinase	*E. coli* MG1655	U00096	86	93
AB823579	*pfkA*	phosphofructokinase I	*E. coli* MG1655	U00096	87	95
AB823585	*pgk*	phosphoglycerate kinase	*E. coli* MG1655	CP007025	79	90
LC020430	*pta*	phosphotransacetylase	*C. freundii* MTCC1658	EKS56947	98	100
AB823586	*ptsG*	PTS system glucose-specific transporter subunits IIBC	*E. coli* MG1655	U00096	89	97
AB823587	*pyk*	pyruvate kinase	*E. albertii* KF1	CP007025	85	95
AB823588	*sdhB*	succinate dehydrogenase iron-sulfur subunit	*E. coli* MG1655	U00096	87	96
AB823589	*sucA*	2-oxoglutarate dehydrogenase E1 component	*E. coli* MG1655	U00096	89	94
AB823590	*sucB*	dihydrolipoamide succinyltransferase	*E. coli* MG1655	U00096	87	94
AB823591	*sucD*	succinyl-CoA synthetase subunit alpha	*E. coli* MG1655	U00096	88	95
AB823592	*tpiA*	triosephosphate isomerase	*E. coli* MG1655	U00096	90	95
LC018665	*yjcG*	acetate permease	*E. coli* strain ST2747	CP007394	86	96

2. Materials and Methods

2.1. Bacterial Strains, Media, and Culture Conditions

Two BF-producing strains, *C. freundii* IFO 13545 (Figure S1) and *C. freundii* GTC 09479 [11], were used in this study. *Escherichia coli* JM109 (Takara Bio, Kyoto, Japan) was also used as a host for the construction of recombinant DNA molecules. Acetate medium (AM) (10 g CH_3COONa, 0.1 g yeast extract, 1.0 g $(NH_4)_2SO_4$, 1.0 g K_2HPO_4, 0.05 g NaCl, 0.2 g $MgSO_4 \cdot 7H_2O$, 0.05 g $CaCl_2$, and 0.01 g $FeCl_3$ in 1 L, pH 7.2) [11] and glucose medium (GM, to which glucose was added at 7.32 $g \cdot L^{-1}$ instead of acetate in AM to yield the same carbon content in both media) were used for the growth and BF production by *Citrobacter* strains, whereas LB medium [14] was used for the cell growth of pre-cultures and for plasmid preparation. Kanamycin (Nacalai Tesque, Kyoto, Japan) and tetracycline (Nacalai Tesque) were added to the growth medium at 25 $mg \cdot L^{-1}$ when necessary. The cultivation conditions were the same as those used in a previous study [11]. Briefly, twenty-five milliliters of each medium was added to 100-mL flasks that had been sterilized by autoclaving, and the medium was inoculated with each bacterial strain. Then, the flasks were shaken on a reciprocal shaker at 30 °C and 125 rpm. For the selection and growth of the *glmS* disruptant, LB agar plates containing kanamycin were used after 50 µL of 10 mM glucosamine 6-phosphate (GlcN 6-P) was spread on the agar surface. GlcN 6-P was added to the liquid medium to a final concentration of 1 mM.

2.2. Measurement of Flocculation Activity and Determination of Flocculation Titer

Flocculation activity was measured using kaolin suspensions as described previously [11]. A two-fold dilution series of the culture supernatant was prepared with distilled water and the flocculation activity of each sample was measured. The flocculation titer (Ft) was defined as the fold dilution that resulted in 50% flocculation activity according to plots of fold dilution vs flocculation activity; it was calculated using the two data points of the flocculation activities that were closest to the 50% flocculation activity.

2.3. Preparation of DNA Fragments by PCR for Gene Disruption

Total DNA was extracted from the cells of *Citrobacter* strains as described previously [11]. The primers used in this study for PCR are listed in Table S1. To disrupt genes in the *Citrobacter* strains, we employed a two-step PCR method based on the method of Yamamoto et al. [15]. As shown in Figure 2a, in the first step of this method, a kanamycin resistance gene cassette flanked by FLP recognition target (FRT) sequences was amplified from plasmid pKD4 [16] by PCR using the paired primers pKD4-Km-F and pKD4-Km-R. The 0.5-kb upstream and downstream flanking regions of the target gene to be disrupted were then amplified using primer pairs corresponding to both regions (Figure 2b). In the second step, a larger DNA fragment for homologous recombination was amplified by PCR using the outer primers (Figure 2c, H3 and H4) and the three PCR-amplified fragments as the templates (Figure 2c). The PCR mixture contained template DNA (50 ng), each primer (10 pmol), 4 µL of dNTP mixture (2.5 mM each), 5 µL of 10 × ExTaq buffer (Takara Bio), and 1.25 U of ExTaq polymerase (Takara Bio), and dH_2O in 50 µL. The thermal program employed was as follows: initial denaturation at 94 °C for 5 min; 30 cycles of denaturation at 94 °C for 30 s, annealing at 55 °C for 30 s, and extension at 72 °C for 1.5 min for the first-step PCR (Figure 2a,b) or for 2.5 min for the second-step PCR (Figure 2c); and final extension at 72 °C for 5 min. In these reactions, the following primer pairs were used: acs-up-F and acs-up-R for the upstream region of *acs*; acs-down-F and acs-down-R for the downstream region of *acs*; aceA-up-F and aceA-up-R for the upstream region of *aceA*; aceA-down-F and aceA-down-R for the downstream region of *aceA*; glmS-up-F and glmS-up-R for the upstream region of *glmS*; glmS-down-F and glmS-down-R for the downstream region of *glmS*; cpsA-up-F and cpsA-up-R for the upstream region of *cpsA*; cpsA-down-F and cpsA-down-R for the downstream region of *cpsA*; bfpC-up-F and bfpC-up-R for the upstream region of *bfpC*; and bfpC-down-F and bfpC-down-R for the downstream region of *bfpC* (Table S1). For *nagA* disruption, a one-step PCR method was employed, in which nagA-Km-F with the 50-bp upstream flanking region of *nagA* at the 5′ end and nagA-Km-F with the 50-bp downstream flanking region of *nagA* at the 5′ end (Table S1) were used to amplify the Km^r gene cassette from pKD4. The resulting PCR fragment was used for homologous recombination after gel purification. The PCR products were analyzed by electrophoresis using 0.8% agarose gels (L03, Takara Bio).

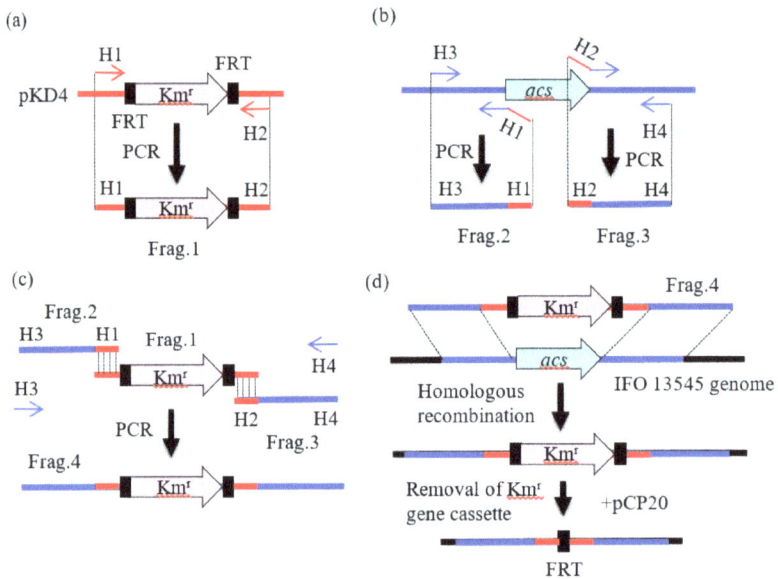

Figure 2. Construction of an acetyl-CoA synthetase gene-disruptant (Δ*acs*) from *C. freundii* IFO 13545: (**a**) preparation of a Km^r gene cassette from pKD4 (Frag. 1); (**b**) amplification of the upstream and downstream regions of the target gene (*acs*) from the IFO 13545 genome (Frag. 2 and Frag. 3); (**c**) preparation of a DNA fragment for gene disruption (Frag. 4) from the amplified three fragments; and (**d**) disruption of the target gene (*acs*) by homologous recombination using Frag. 4. H1, H2, H3, and H4 indicate specific primers that were used or their sequence regions.

2.4. Gene Disruption by Homologous Recombination

Citrobacter strains were first transformed with the Red helper plasmid pKD119 by electroporation (the conditions are described below). This plasmid promotes homologous recombination using the functions of the γ, β, and *exo* gene products of the λ phage [15–17]. Then, a selected transformant was grown at 30 °C in 100 mL of SOB medium [14] supplemented with tetracycline and L-arabinose (10 mM) (Nacalai Tesque). The cultured cells were harvested when the culture reached optical density at 600 nm (O.D.$_{600}$) of 0.6, washed three times with ice-cold 10% (*v/v*) glycerol, and resuspended in 10% glycerol solution at O.D.$_{600}$ = 12. The PCR products described in the previous section were purified by electrophoresis, and they were dissolved in a small amount of 10 mM Tris buffer (pH 8.0). The purified DNA fragments (10–100 ng) were electroporated into the cells (150 μL of the cell suspension) in a 0.1-cm gap cuvette using a Gene Pulser II (Bio-Rad Japan, Tokyo) under the following conditions: voltage, 2.25 kV; capacitance, 25 μF; resistance, 400 Ω. After electroporation, 1 mL of SOC medium (SOB medium containing 20 mM glucose) was added to the recovered cell suspension and the culture was incubated at 30 °C with agitation at 150 rpm for 1 h. Aliquots of the culture were spread on LB agar plates containing tetracycline and kanamycin, and the plates were incubated at 30 °C. The addition of kanamycin provided the basis for the first selection of the gene-disrupted strains (Figure 2d), and the addition of tetracycline ensured the maintenance of pKD119 in the cell. The kanamycin resistance gene cassette was removed from the genome of each strain using the FLP helper plasmid pCP20 (Figure 2d) [16] when necessary.

2.5. Southern Hybridization

To confirm that the target gene was successfully replaced with the kanamycin resistance gene cassette (or successfully deleted), Southern hybridization was performed using a G-Capillary Blotter (TAITEC, Saitama, Japan), a Hybond-N+ membrane (GE Healthcare, Buckinghamshire, UK), and a DIG-High Prime DNA Labeling and Detection Kit (Roche Diagnostics, Mannheim, Germany) according to the manufacturer's instructions [18]. The DNA regions used for the preparation of the probe are shown in Figure 3c for *acs* and *aceA*, in Figure S2c for *glmS*, and in Figure S3c for *bfpC*.

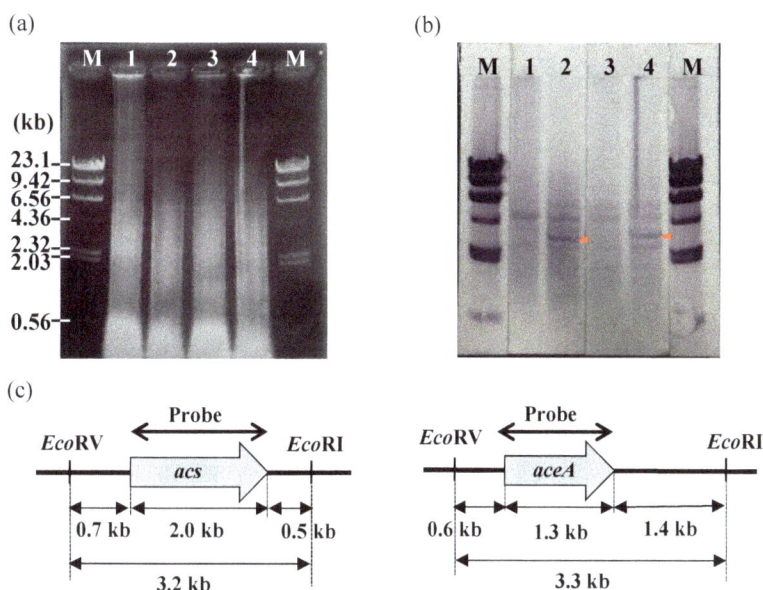

Figure 3. Electrophoresis of *Eco*RI-*Eco*RV-digested total DNA of *C. freundii* IFO 13545 and its gene-disrupted mutants, Δ*acs* and Δ*aceA* (**a**); corresponding Southern blots (**b**); and restriction maps of the regions around *acs* and *aceA* in the IFO 13545 genome (**c**). Lane M: λDNA *Hind*III digest (size marker); Lane 1: Δ*acs*; Lanes 2 and 4: IFO 13545 (wild type); Lane 3: Δ*aceA*. A probe based on the *acs* sequence was used in Lanes 1 and 2, whereas a different probe corresponding to the *aceA* sequence was used in Lanes 3 and 4. The arrow heads indicate the original gene fragments in the wild-type strain.

2.6. Batch Cultivation in a Mini-Jar Fermentor

Batch cultivation of *C. freundii* IFO 13545 was conducted in 500 ml of AM or GM using a mini-jar fermenter, NBS BioFlo115 (1.4-L vessel) (New Brunswick Scientific, Edison, NJ, USA) at 30 °C. During the cultivation, the pH value was automatically maintained below 8.5 in AM by addition of 0.5 M H_2SO_4 and above 7.2 in GM by addition of 1 M NaOH. The amount of dissolved oxygen was set at 20% of the saturated concentration and was automatically controlled by a combination of agitation (<150 rpm) and air supply (<2 vvm). The glucose and acetate in the culture were measured using enzyme assay kits (R-BIOPHARM AG, Darmstadt, Germany) according to the instructions provided with the kits.

2.7. Quantitative Reverse-Transcription-PCR (qRT-PCR)

During the abovementioned batch cultivation, cells were harvested by centrifugation (3000×
g, 4 °C, 10 min) from 30 mL of each culture at three sampling points during the logarithmic growth
phases. Total RNA was extracted from the cells using a PureLink™ RNA Mini Kit (Life Technologies,
Carlsbad, CA, USA) according to the manufacturer's instructions. The RNA samples were digested
with *Dpn*I (Takara Bio) under the conditions recommended by the supplier to remove any remaining
host DNA (methylated DNA), followed by qRT-PCR analysis using a One Step SYBR® PrimeScript™
RT-PCR Kit II (Takara Bio) and a Bio-Rad Real-Time PCR system MiniOpticon (Bio-Rad Japan, Tokyo,
Japan). The following primers were used for cDNA synthesis and amplification of the target genes:
aceA-RT-F and aceA-RT-R for *aceA*, acs-RT-F and acs-RT-R for *acs*, ack-RT-F and ack-RT-R for *ack*,
bfpC-RT-F and bfpC-RT-R for *bfpC*, fba-RT-F and fba-RT-R for *fba*, fbp-RT-F and fbp-RT-R for *fbp*,
glmS-RT-F and glmS-RT-R for *glmS*, icd-RT-F and icd-RT-R for *icd*, pck-RT-F and pck-RT-R for *pck*,
pfkA-RT-F and pfkA-RT-R for *pfkA*, ptsG-RT-F and ptsG-RT-R for *ptsG*, yjcG-RT-F and yjcG-RT-R for
yjcG, and 16S-RT-F and 16S-RT-R for the 16S rRNA gene (Table S1). These paired primers were designed
to amplify an approximately 200-bp internal region of each target gene. To check the suitability of the
paired primers for this analysis, amplification of the DNA fragments of the expected size from the total
DNA of IFO 13545 was confirmed in advance by PCR using these paired primers (data not shown). A
typical reaction mixture for qRT-PCR contained template RNA (50 ng), 2 × One Step SYBR® RT-PCR
Buffer 4 (Takara Bio) (12.5 μL), PrimeScript One Step Enzyme Mix 2 (Takara Bio) (1.0 μL), each primer
(0.4 μM), and RNase-free dH$_2$O (up to 25 μL). The thermal program employed was as follows: initial
denaturation at 42 °C for 5 s, and at 95 °C for 10 s; 40 cycles of denaturation at 95 °C for 5 s, annealing
and extension at 60 °C for 30 s; and a melting-curve step. Gene expression level was evaluated as the
relative amount of specific RNA calculated from the qRT-PCR data using the $2^{-\Delta\Delta CT}$ method [19].

2.8. Nucleotide Sequences Registered in the Databases

The genes involved in the acetate assimilation and BF synthesis in *C. freundii* IFO 13545 are listed
in Table 1. The sequences of these genes have been registered in DDBJ/EMBL/GenBank under the
accession numbers that are shown in the same table.

3. Results

3.1. Putative BF Synthetic Pathway of C. freundii IFO 13545 and Its Related Genes

As described in the Introduction, based on known bacterial carbon metabolic pathways and on
the peptidoglycan synthetic pathway [12,20], we expected the BF synthetic pathway of *C. freundii* IFO
13545 to be as follows (Figure 1): acetate is metabolized in the TCA cycle and the glyoxylate shunt
via acetyl-CoA; Fru 6-P is generated from TCA cycle intermediates through gluconeogenesis; Fru 6-P
enters into the hexosamine pathway and used to form UDP-GlcNAc, which serves as a direct precursor
for extension of the BF polysaccharide chain. From the genome sequence of IFO 13545, we identified
all the candidate genes corresponding to the enzymes that act during this metabolic pathway (Figure 1
and Table 1). These genes showed 79–98% nt sequence identity (90–100% aa sequence identities
for the encoded proteins) with the corresponding genes of *Escherichia coli* K-12 MG1655 and with
those of some other enteric bacterial strains (Table 1); the genome sequences of these strains were
previously registered in the DNA databases, and the putative functions of the corresponding genes
had been annotated.

3.2. Effects of the Disruption of Acs or AceA on the Growth of IFO 13545 on Acetate

In *Escherichia coli*, there are two well-known major routes for the initial conversion of acetate into
acetyl-CoA (Figure 1) [12,21]. In the first route, two enzymes, acetate kinase and phosphotransacetylase,
catalyze two successive reactions that result in the conversion of acetate to acetyl-CoA via acetyl
phosphate. The second route involves direct conversion of acetate into acetyl-CoA by acetyl-CoA

synthetase. The reactions that occur in the first route are reversible, whereas the latter reaction is irreversible. To understand the first step in the acetate assimilation of IFO 13545, based on its genome sequence, a putative acetyl-CoA synthetase gene (*acs*, 1956 bp) was first disrupted by homologous recombination, and the gene disruption was confirmed by Southern hybridization. The Southern blotting revealed that the *acs* disruptant (Δ*acs*) lacked a 3.2-kb *Eco*RV-*Eco*RI-digested DNA fragment, to which the *acs* probe hybridized in the wild-type strain (cf. Lanes 1 and 2 in Figure 3b). Although the disruptant displayed almost the same magnitude of growth on glucose as that of the wild-type strain, it showed little growth on acetate (Figure 4). This result suggests that IFO 13545 mainly uses the second route for acetate assimilation. Because Δ*acs* was still able to grow very slowly on acetate, the first route may exist in this strain and may contribute slightly to acetate assimilation. In fact, two candidate genes (*ack* and *pta*) for acetate kinase and phosphotransacetylase were identified in the IFO 13545 genome sequence (Table 1) and the gene expression of *ack* was detected by qRT-PCR analysis as described later.

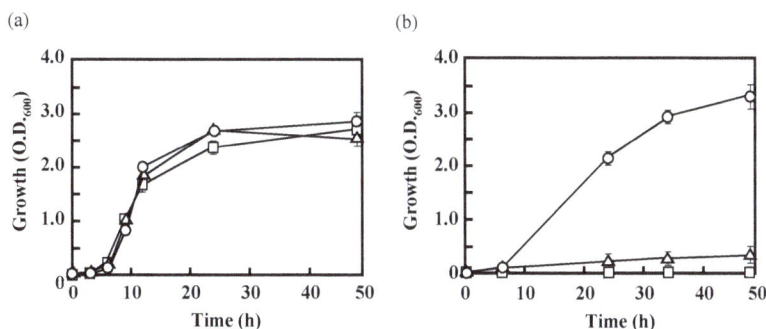

Figure 4. Growth of *C. freundii* IFO 13545 and its *acs* and *aceA* disruptants (Δ*acs* and Δ*aceA*) on: glucose (**a**); and acetate (**b**). Symbols: circles, IFO 13545 (wild type); triangles, Δ*acs*; squares, Δ*aceA*. This experiment was conducted in triplicate; the averages ± standard deviations of the values obtained are shown.

Next, if the acetyl-CoA that was formed enters into the TCA cycle and is used to make cell components and energy, the glyoxylate shunt should be used. The glyoxylate shunt includes two successive enzymatic reactions that catalyze the conversion of isocitrate to glyoxylate and succinate and then of glyoxylate and acetyl-CoA to L-malate (Figure 1). Passage through the shunt can shortcut two successive decarboxylation steps in the TCA cycle (from isocitrate to succinyl-CoA via 2–oxoglutarate), which release two carbon atoms as CO_2 [12]. Therefore, using the glyoxylate shunt, the carbon atoms of acetate can be fixed to the cell. To confirm this assumption, a putative gene encoding isocitrate lyase (*aceA*, 1317 bp), which catalyzes the first reaction of the glyoxylate shunt, was disrupted in a manner similar to that described above. Southern hybridization demonstrated that the *aceA* disruptant (Δ*aceA*) lacked a 3.3-kb *Eco*RV-*Eco*RI-digested DNA fragment, to which the *aceA* probe hybridized in the wild-type strain (cf. Lanes 3 and 4 in Figure 3b). This disruptant, Δ*aceA*, no longer grew on acetate, although like the wild-type strain it grew well on glucose (Figure 4). This result suggests that the glyoxylate shunt mainly contributes to the acetate assimilation in IFO 13545.

3.3. Effects of the Disruption of GlmS and/or NagA on the Growth of IFO 13545 on Acetate and on Its Flocculation Activity

Because the chitin/chitosan-like BF of IFO 13545 is composed of GlcNAc and/or GlcN [11], the precursor of the BF polymer was expected to be UDP-GlcNAc, which can be synthesized through the hexosamine pathway in the early stage of cell wall (peptidoglycan) biosynthesis (Figure 1) [13]. The starting compound of this pathway is Fru 6-P in glycolysis/gluconeogenesis, which can be formed from the intermediates of the TCA cycle through gluconeogenesis. In the first reaction of the hexosamine pathway, Fru 6-P is converted into GlcN 6-P by GlcN 6-P synthase (Figure 1). A putative gene encoding GlcN 6-P synthase (*glmS*, 1827 bp) was also identified in the IFO 13545 genome sequence and was found to be coupled with another hexosamine pathway gene, *glmU* (1368 bp), which encodes a bifunctional enzyme, GlcN 1-phosphate (GlcN 1-P) transacetylase/GlcNAc 1-phosphate (GlcNAc 1-P) uridyltransferase, in an operon (*glmUS*). To examine the effects of the disruption of *glmS* on cell growth on acetate and on flocculation activity, *glmS* was disrupted as described above. After homologous recombination, we attempted to isolate the *glmS* disruptant (Δ*glmS*) on LB agar plates containing kanamycin and tetracycline. However, no colonies were obtained in repeated experiments. We surmised that this disruption could cause a shortage of GlcN 6-P in the cell, resulting in severely decreased levels of UDP-GlcNAc, the common precursor for peptidoglycan and BF biosynthesis [13]. Thus, GlcN 6-P was added to the surfaces of the LB selection plates. This resulted in the appearance of several colonies on the plates used for selection of the transformants. Southern hybridization revealed that *glmS* had been successfully replaced with the kanamycin resistance gene cassette in these colonies (Figure S2). The growth of Δ*glmS* on acetate (in AM) was almost identical to that of the wild-type strain (Figure 5a). However, its flocculation activity was only 11.9% of that of the wild-type strain after 48 h-incubation (Figure 5b), indicating that *glmS* contributes significantly to BF biosynthesis. The remaining activity indicates the presence of alternative routes for supplying amino sugars for the BF biosynthesis. To identify the alternative routes, we decided to disrupt one possible route in which GlcNAc taken up from the outside of the cell and recycled from peptidoglycan is assimilated (Figure 6) [21,22]. The key step in the peptidoglycan recycling and the usage of external GlcNAc for BF synthesis is conversion of GlcNAc 6-P into GlcN 6-P by GlcNAc 6-P deacetylase, which is encoded by *nagA* (Figures 1 and 6). Its candidate gene (1146 bp) was found in the IFO 13545 genome sequence (Table 1) and disrupted by homologous recombination. The *nagA* disruption was confirmed by PCR (data not shown). As shown in Figure 5a, the growth of the *nagA*-disruptant (Δ*nagA*) on acetate was almost the same as that of the wild-type strain, whereas its flocculation activity was reduced to 36.3% of that of the wild-type strain (Figure 5b). Thus, *nagA* disruption affects BF biosynthesis to a certain extent. We also constructed a double disruptant (Δ*glmS*Δ*nagA*) by the same procedure to determine the simultaneous effect of the disruption of both genes on flocculation activity and on growth on acetate. The double disruptant completely lost flocculation activity (Figure 5b). This result indicates that the disruption of both genes almost completely blocked the supply of amino sugars for BF biosynthesis. However, the double disruptant was still able to grow on acetate (Figure 5a), suggesting the possibility of the existence of other amino sugar supply resources for the growth.

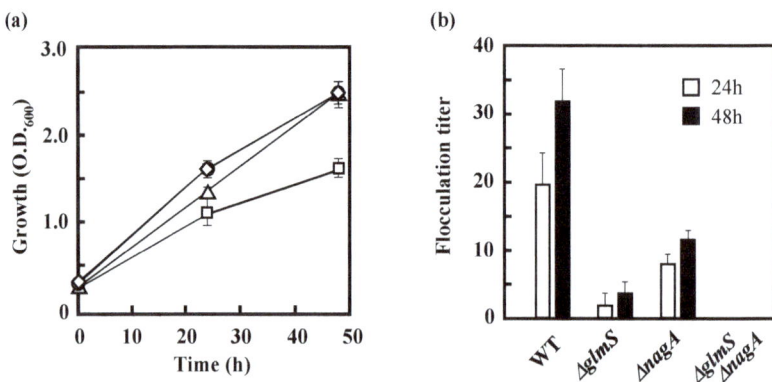

Figure 5. Growth of *C. freundii* IFO 13545 and its *glmS*, *nagA*, and *glmSnagA* disruptants (Δ*glmS*, Δ*nagA*, and Δ*glmS*Δ*nagA*) on acetate (**a**); and their flocculation activities (**b**). Symbols: circles, IFO 13545 (wild type, WT); triangles, Δ*glmS*; diamonds, Δ*nagA*; squares, Δ*glmS*Δ*nagA*. This experiment was conducted in triplicate; the averages ± standard deviations of the values obtained are shown.

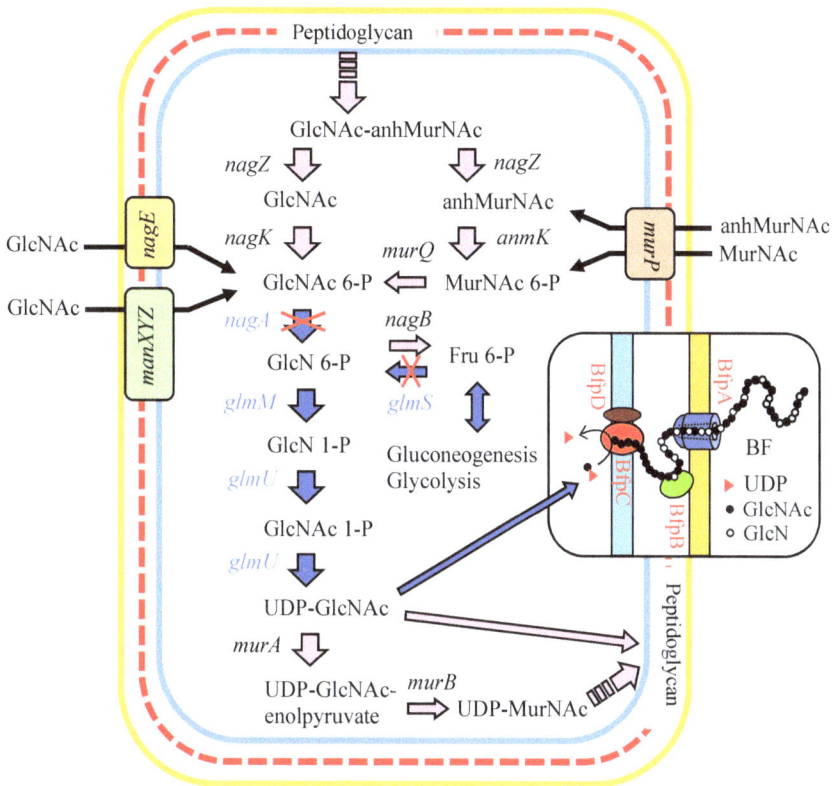

Figure 6. Amino sugar metabolism, peptidoglycan synthesis and recycling, and the putative polymerization and secretion mechanisms involved in production of the chitin/chitosan-like bioflocculant. These pathways and mechanisms are depicted based on information obtained from published reports [13,21–23]. Blue arrows and gene names indicate the proposed BF synthetic pathway and genes, respectively. Abbreviations: BF, bioflocculant; GlcN, glucosamine; GlcNAc, *N*-acetylglucosamine; MurNAc, *N*-acetylmuramic acid; P, phosphate; anh, anhydro.

To investigate the effects of the yeast extract used in AM as a supply of amino sugars from outside the cell, these *glmS* and/or *nagA* disruptants were cultivated in AM lacking yeast extract. Under these conditions, the wild-type strain showed growth almost identical to its growth in AM (Figure 7a), whereas the single disruptants, Δ*glmS* and Δ*nagA*, showed considerably slower growth than they displayed in the presence of yeast extract (Figure 7b,c). The double disruptant failed to grow in the absence of yeast extract (Figure 7d), suggesting that yeast extract, despite its presence at a low concentration (0.1 g·L^{-1} in AM), provides important factors necessary for the growth of the disruptants. Addition of GlcNAc at 0.2 g·L^{-1} to AM lacking yeast extract greatly improved the growth of Δ*glmS* and Δ*nagA* (Figure 7b,c), demonstrating that one of the necessary factors was GlcNAc. However, the double disruptant showed little growth on acetate even after the addition of GlcNAc (Figure 7d). The fact that the double disruptant was able to grow in the presence of yeast extract indicates that yeast extract contains other compounds that support the cell growth (probably other sugars that can be used in peptidoglycan synthesis).

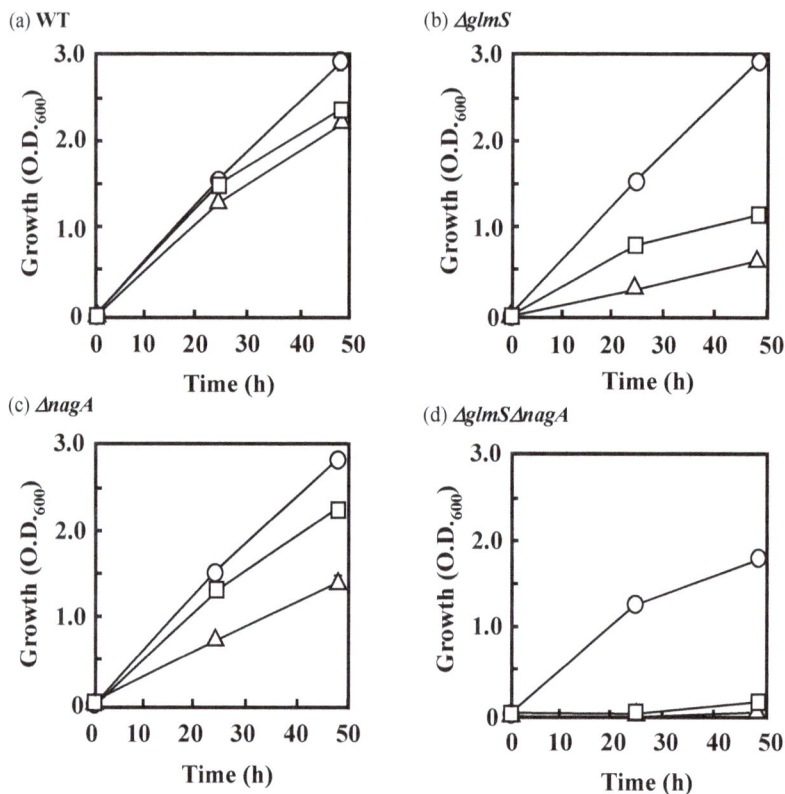

Figure 7. Growth of *C. freundii* IFO 13545 (wild type, WT) (**a**) and of its *glmS*, *nagA*, and *glmSnagA* disruptants (Δ*glmS* (**b**); Δ*nagA* (**c**); and Δ*glmS*Δ*nagA* (**d**)) in acetate medium containing 0.1 g·L^{-1} yeast extract (circles), in acetate medium lacking yeast extract (triangles), and in acetate medium lacking yeast extract and supplemented with 0.2 g·L^{-1} GlcNAc (squares). This experiment was performed in triplicate; the averages ± standard deviations of the values obtained are shown. Almost all the standard deviations were within the symbol sizes.

3.4. Identification of Genes that Are Involved in the Polymerization and Secretion of the Chitin/Chitosan-Like BF

In this gene disruption study, we attempted to identify genes that are involved in the polymerization and secretion of the chitin/chitosan-like BF, indispensable processes for BF production. However, limited information about these genes is currently available. From the IFO 13545 genome sequence, several genes that potentially encode enzymes involved in polysaccharide polymerization or membrane channel proteins involved in polysaccharide secretion were selected and independently disrupted by homologous recombination. In this way, we identified two intriguing genes, *cpsA* (1413 bp) and *bfpC* (1332 bp), whose disruption significantly affected the flocculating activity of IFO 13545. The aa sequence of the *cpsA* gene product exhibited 99–100% aa sequence identity with the undecaprenyl phosphate glucose phosphotransferases of several *Citrobacter* and *E. coli* strains (e.g., accession Nos. EKS56046, EOQ22059, and KEL78484), which catalyze the addition of the first hexose to undecaprenyl phosphate (a lipid carrier) that is anchored at the cell membrane for the biosynthesis of lipopolysaccharides (LPS) and capsular polysaccharides (CPS) [24,25]. When *cpsA* was disrupted by homologous recombination, the flocculation activity of the *cpsA* disruptant (Δ*cpsA*) was reduced to 62% of that of the wild-type strain (data not shown), although it grew considerably better on acetate than did the wild-type strain.

The *bfpC* gene product exhibited 96–100% aa sequence identity with the putative glycosyltransferases (polysaccharide *N*-glycosyltransferases) of several *Citrobacter* and *E. coli* strains that are involved in the biosynthesis of poly-β-1,6-*N*-acetyl-D-glucosamine (PGA or PNAG) (e.g., accession nos. EEH94348, EHL86151, and KEL79567). PGA is required for the structural development and integrity of biofilms in a wide variety of Gram-positive and Gram-negative bacteria, and it has profound effects on host–microbe interactions [26,27]. A *bfpC*-disruptant (Δ*bfpC*) was also constructed from IFO 13545 by homologous recombination (Figure S3). It showed no flocculation activity. This result strongly suggests that *bfpC* encodes a glycosyltransferase that is required for the BF polymerization. To confirm the importance of *bfpC* in the BF biosynthesis, a *bfpC* homolog in a different BF-producing strain, *C. freundii* GTC 09479 (99.4% identity with the *bfpC* of IFO 13545, nt355104-nt356435 in accession No. AOMS01000025) was also disrupted by homologous recombination. The GTC 09479 disruptant also showed no flocculation activity. Accordingly, we concluded that this gene is indispensable for the BF biosynthesis in IFO 13545. Based on the sequence analysis, the *bfpC* gene product (BfpC) was determined to belong to the nucleotide-diphospho-sugar glycosyltransferase family 2 [28]; five aa residues, Asp^{163}, Asp^{256}, Gln^{292}, Arg^{295}, and Trp^{296}, which are thought to be the catalytic residues [29], are conserved in this protein.

3.5. Expression of Key Genes Involved in Acetate Assimilation and BF Production in IFO 13545

To determine the expression levels of the abovementioned genes in IFO 13545 cells grown on acetate or glucose, qRT-PCR analysis was conducted using RNA samples obtained from cells grown in AM or GM. The RNA samples were prepared from IFO 13545 cells sampled at the early, middle, and late stages of logarithmic growth during batch cultivation in a mini-jar fermenter (Figure 8). The genes studied in this analysis were as follows: *yjcG*, *acs*, and *ack*, which encode proteins involved in the uptake of acetate and its initial conversion into acetyl-CoA; *aceA* and *icd*, which encode enzymes that act at the branching point in the TCA cycle for the glyoxylate shunt; *pck*, *fba*, and *fbp*, which encode proteins involved in gluconeogenesis; *ptsG*, *pfkA*, and *fba*, which are necessary for glucose uptake and glycolysis; and *glmS* and *bfpC*, which are required for the hexosamine pathway and BF polymerization, respectively (Table 1). Figure 9 shows the profile of gene expression normalized to the level of expression of the 16S rRNA gene. The expression levels of *yjcG*, *acs*, and *aceA* were higher in cells grown in acetate than in cells grown in glucose at the early- and mid-log phases. The expression of *aceA* was especially prominent, indicating its strong induction by acetate. This result is in good agreement with the result from the growth experiments using Δ*acs* and Δ*aceA* (Figure 4): the disruption of these genes negatively affected the growth of the cells on acetate, resulting in no or little growth. The gene expression level of *icd* was also high in the acetate-grown cells as well as in the glucose-grown cells. The contribution of *aceA* and *icd* to carbon metabolism at the branching point for the glyoxylate shunt is discussed later in this work. It is worth noting that an increased level of *pck* expression for gluconeogenesis was detected in the acetate-grown cells. In addition, the expression of *fba*, and *fbp* was also higher during the early- and mid-log phases of growth, indicating that gluconeogenesis was promoted under conditions of acetate assimilation. The expression of *bfpC* was somewhat higher in acetate-grown cells than that in glucose-grown cells, whereas the expression of *glmS* was similar under the two growth conditions. At the late-log phase, the expression of all of the studied genes in acetate-grown cells declined, likely due to exhaustion of the carbon source (acetate) (Figure 8a).

In contrast to the results for cells grown in acetate, in the glucose-grown cells, the gene expression levels of *ptsG* and *ack*, which encode a glucose-specific transporter and acetate kinase, respectively, were very high. Acetate kinase, the *ack* gene product, is known to produce acetate from acetyl-CoA during the overflow metabolism (fast unbalanced metabolism) of glucose in *E. coli*, whereas acetyl-CoA synthetase scavenges the accumulated acetate [30–32]. Notably, at the late-log phase, a low level of acetate formed from glucose had accumulated (Figure 8b); at that point, a metabolic adaptation from glucose to acetate (as carbon source) could have occurred because the expression levels of *yjcG*, *acs*, *aceA*, *pck*, *fba*, *fbp* and *bfpC*, which showed higher expression in the acetate-grown cells, increased at the late-log phase in the glucose-grown cells (Figure 9).

(a) **In AM**   (b) **In GM**

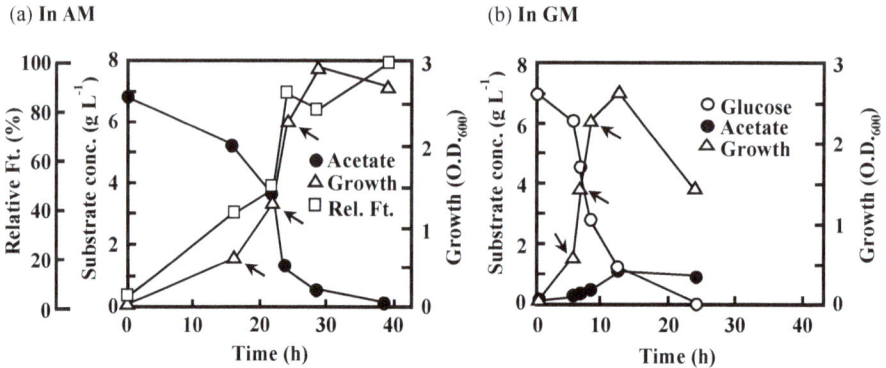

Figure 8. Changes in substrate concentration, growth, and relative flocculation activity (flocculation titer, Ft) during the batch cultivation of *C. freundii* IFO 13545 in: AM (**a**); and GM (**b**). Arrows indicate the sampling points of the cultures for qRT-PCR analysis (early-, mid-, and late-log phases in Figure 9).

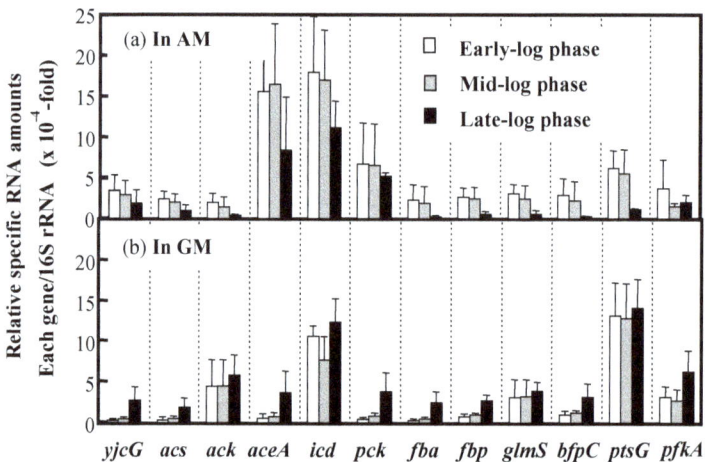

Figure 9. Relative gene expressions in the cells of *C. freundii* IFO 13545 grown in AM (on acetate) (**a**) or GM (on glucose) (**b**) compared to that of the 16S rRNA gene. The cells were grown in AM or GM and harvested at $O.D._{600}$ = 0.52, 1.15, and 2.00 (early-, mid- or late-log phase, respectively). RNA samples were extracted from the cells and qRT-PCR was conducted using specific paired primers for each gene (Table 1). The average ± the standard deviation of the values obtained in three repeated experiments is shown. The relative amounts of RNA from the 16S rRNA gene in AM were 3.27×10^{-2} (early), 2.63×10^{-2} (mid), and 3.88×10^{-2} (late), whereas the amounts in GM were 3.39×10^{-2} (early), 2.11×10^{-2} (mid), and 3.35×10^{-2} (late).

4. Discussion

In this study, to identify the BF synthetic pathway and the genes associated with this pathway in *C. freundii* IFO 13545, several key genes from the expected pathway were independently disrupted by homologous recombination. The poor growth of Δ*acs* on acetate (Figure 4b) suggested that acetate is mainly converted into acetyl-CoA by acetyl-CoA synthetase (Acs). The higher gene expression level of *acs* observed in the acetate-grown cells compared to the glucose-grown cells (Figure 9) also supports this hypothesis. The acetyl-CoA that is formed from acetate can be easily metabolized in the TCA cycle. At the branching point in the TCA cycle for the glyoxylate shunt, isocitrate lyase (ICL, the *aceA* gene product) and isocitrate dehydrogenase (IDH, the *icd* gene product) compete for the same substrate, isocitrate (Figure 1). The level of expression of *aceA* in the acetate-grown cells was very high, and expression of this gene appeared to be strongly induced by acetate, as indicated by the gene expression study (Figure 9). Unexpectedly, the expression level of *icd* was also high in the acetate-grown cells. Thus, in the presence of acetate, both enzymes may be present in large amounts in the same cell. In *E. coli*, ICL has a much lower affinity for isocitrate (K_m = 604 μM) than IDH (K_m = 8 μM) [12]. If this is the case, major carbon flux tends to flow into the TCA cycle side (to 2-oxoglutarate) at the branching point. However, in *E. coli*, the activity of IDH is regulated by its reversible phosphorylation by IDH kinase/phosphatase, which is encoded by *aceK* in the *aceBAK* operon [12]. This operon was also detected in IFO 13545 (data not shown). Concomitant-up-regulation of *aceA* and *aceK* by acetate (>10-fold compared to by glucose) has been reported in the gene expression profiling studies of acetate-grown *E. coli* using DNA microarrays [33]. Therefore, in cells grown on acetate, IDH is always inactivated by phosphorylation even if a large amount of IDH protein exists in the cell, and most carbon flux flows into the glyoxylate shunt. Based on the data, we obtained on the growth of Δ*aceA* on acetate (Figure 4) and on the gene expression of *aceA* (Figure 9), this may occur in the IFO 13545 strain.

Our qRT-PCR analysis of *pck*, *fbp*, and *fba* expression revealed that gluconeogenesis is greatly promoted in the acetate-grown cells. In addition, the disruption of *glmS* caused a noticeable decrease in flocculation activity, suggesting that the major BF synthetic pathway involves gluconeogenesis and the hexosamine pathway. Kim et al. conducted proteomic analysis of acetate metabolism in the polyglucosamine (PGB-2)-producing strain *Citrobacter* sp. BL-4 and found that 124 proteins were produced only in the acetate metabolism; these proteins included Acs, which acts during acetyl-CoA biosynthesis, and ICL, which acts in the glyoxylate shunt [34]. This result is reasonable and in good agreement with our results. Those authors also detected GlcN 6-P synthase (the *glmS* gene product, GlmS), which functions in the hexosamine pathway. However, in IFO 13545, the gene expression level of *glmS* in acetate-grown cells was relatively low compared to that in glucose-grown cells. One possible explanation for this is that amino sugar formation by GlmS is always necessary for peptidoglycan synthesis irrespective of the carbon source. However, the reaction catalyzed by GlmS may be a limiting step for BF biosynthesis by IFO 13545. Son et al. hypothesized that the genes associated with the hexosamine pathway (*glmU*, *glmS*, and *glmM*) are essential for the production of microbial polyglucosamine PGB-1 in *Enterobacter* sp. BL-2, and they introduced additional *glmS* copies (as a pBBR1MCS-2-based plasmid) into the BL-2 strain by conjugation to strengthen the pathway [9,35]. The resulting BL-2S strain produced a 1.5-fold more PGB-1 compared than the wild-type BL-2 strain. This strategy has been used to enhance the production of GlcNAc/GlcN by recombinant *E. coli* strains [36,37]. Therefore, this strategy would be a promising way to improve BF production by IFO 13545. In addition to *glmS*, we showed that *nagA* is also very important for growth and BF biosynthesis in IFO 13545 because it enables the cells to utilize amino sugars from outside the cell, and probably from the peptidoglycan recycling. By disruption of *nagA* and *glmS*, we showed that yeast extract can act as an amino sugar supply source in these disruption mutants. Based on the above results, the overview of the BF synthetic pathway from acetate (acetate assimilation, gluconeogenesis, and hexosamine pathway) was confirmed, although it will be important to complete the biochemical characterization of each enzyme step according to this pathway.

In this study, we searched for the genes involved in the polymerization and secretion of BF in the genome of IFO 13545. Currently, two major bacterial CPS/EPS biosynthesis systems, the Wzy- and ABC transporter-dependent CPS/EPS biosynthesis systems, are recognized [25]. We identified many gene clusters for ABC transporters in the IFO 13545 genome; however, based on sequence homology, those genes appeared to encode proteins that function in the transportation of amino acids and low-molecular weight ions from the sequence homologies (data not shown). We also found a gene cluster encoding a typical Wzy-dependent polysaccharide synthesis system that occupies over 20 kb in the genome; this cluster is very similar to the gene cluster that encode the enzymes involved in colanic acid biosynthesis in *E. coli* [38]. Some of the genes in the gene cluster encoding a Wzx homolog (flippase) and a Wza homolog (outer membrane channel protein) were independently disrupted, but this did not affect the flocculation activity of the disruptants; it was similar to that of the wild-type strain. We then identified *bfpC*, which is clustered with three other genes, named *bfpA* (2460 bp), *bfpB* (2019 bp), and *bfpD* (504 bp) as *bfpABCD*. These genes are transcribed in the same direction in that order. *bfpA* and *bfpB* are separated by a 12-bp gap and the stop codon of *bfpB* overlaps with the start codon of *bfpC*. Although *bfpD* lacks a typical start codon ATG coding for methionine, its tentative start codon, GTG, also overlaps with the stop codon of *bfpC*. These observations indicate that *bfpABCD* forms a single transcriptional unit (an operon). The *bfpABCD* gene cluster is similar to the *pgaABCD* gene cluster of *Escherichia coli* K-12 MG1655, which is involved in the biosynthesis and secretion of poly-β-1,6-*N*-acetyl-D-glucosamine (PGA) [23,39], although the sequence identities of the gene products encoded by *bfpABCD* and *pgaABCD* are relatively low. The *bfpABCD* gene products showed 44.5%, 52.3%, 67.2%, and 26.2% aa sequence identities with the corresponding *pgaABCD* gene products. The predicted functions of the *pgaABCD* gene products are as follows: PgaA is an 807-aa outer membrane protein that functions in the translocation of PGA; PgaB is a 672-aa polysaccharide *N*-deacetylase that is required for PGA deacetylation; PgaC is a 441-aa *N*-glycosyltransferase that is required for PGA polymerization; and PgaD is a 137-aa inner membrane protein that likely functions as a helper protein for PgaC [23]. Based on the available information on the functions of the *pgaABCD* gene products, we expected that the mechanisms of the polymerization and secretion of the chitin/chitosan-like BF are as shown in Figure 6. We performed the disruption of *bfpA*, *bfpB*, and *bfpD* in the same manner as for *bfpC*. All of these disruptions resulted in complete loss of the flocculation activity in IFO 13545 (data not shown). In addition, a trial, in which *bfpABCD* was introduced into the *bfpC* disruptant via an *E. coli* expression vector, resulted in the recovery of the flocculation activity (data not shown). Therefore, the *bfpABCD* genes are likely to encode proteins necessary for the polymerization and secretion of BF. As described above, Kim et al. conducted a proteomic analysis of the PGB-2-producing strain *Citrobacter* sp. BL-4 [34]. However, those authors did not identify *pgaABCD*- and *bfpABCD*-related gene products as acetate-induced proteins. Our qRT-PCR analysis indicated that the gene expression level of *bfpC* is not so high (Figure 9). Therefore, this gene cluster would not be found out without this gene disruption study.

Supplementary Materials: The following material is available online at www.mdpi.com/2073-4360/10/3/237/s1: Figure S1. SEM images of *Citrobacter freundii* IFO 13545, which is a Gram-negative rod-shaped bacterium that is phylogenetically very close to *Escherichia coli*; Figure S2. Electrophoresis of *Eco*RV-digested total DNA of *C. freundii* IFO 13545 and its *glmS*-disrupted mutant ($\Delta glmS$) (a); corresponding Southern blots (b); and restriction maps of the regions around *glmS* in the genomes of IFO 13545 and the disruptant (c). Lane M: 1-kb DNA ladder (size marker); Lane 1: IFO 13545 (wild type); Lane 2: $\Delta glmS$; Figure S3. Electrophoresis of *Afl*II-*Eco*RI-*Sph*I-digested total DNA of *C. freundii* IFO 13545 and its *bfpC*-disruptant ($\Delta bfpC$) (a); corresponding Southern blots (b); and restriction maps of the regions around *bfpC* in the genomes of IFO 13545 and the disruptant (c). Lane M: λDNA *Hind*III digest (size marker); Lane 1: IFO 13545 (wild type); Lane 2: $\Delta bfpC$; Table S1: Oligonucleotide primers used in this study.

Acknowledgments: This work was supported in part by the Hyogo COE Program Promotion Project (2010, Hyogo prefecture, Japan); the Regional Innovation Strategy Support Program (2012–2017), Ministry of Education, Culture, Sports, Science and Technology, Japan; a Grant-in Aid for Scientific Research (C) from Japan Society for the Promotion of Science (JSPS) (2015–2017); and the JSPS bilateral collaboration program with the Curtin University (Perth, Australia) research team (2016–2017).

Author Contributions: M.T. and S.N. conceived and designed the experiments; K.K., T.I., S.T., K.M., and M.K. performed the gene disruption experiments; K.I. conducted the qRT-PCR experiment; D.K., P.B., and S.M. performed the genome sequence analysis of *Citrobacter* strains; and M.T. wrote the paper with the assistance of P.B.

Polymers **2018**, *10*, 237

Conflicts of Interest: The authors declare no conflict of interest. The founding sponsors had no role in the design of the study; in the collection, analyses, or interpretation of data; in the writing of the manuscript, and in the decision to publish the results.

References

1. Li, Q.; Dunn, E.; Grandmaison, E.W.; Goosen, M.F.A. Applications and properties of chitosan. *J. Bioact. Compat. Polym.* **1992**, *7*, 370–397. [CrossRef]
2. Struszczyk, M.H. Chitin and chitosan, Part I properties and production. *Polimery* **2002**, *47*, 316–325.
3. Hamed, I.; Özogul, F.; Regenstein, J.M. Industrial application of crustacean by-products (chitin, chitosan, and chitooligosaccharides): A review. *Trends Food Sci. Technol.* **2016**, *48*, 40–50. [CrossRef]
4. Struszczyk, M.H. Chitin and chitosan, Part III some aspects of biodegradation and bioactivity. *Polimery* **2002**, *47*, 619–629.
5. Prashanth, K.V.H.; Tharanathan, R.N. Chitin/chitosan: Modifications and their unlimited application potential-an overview. *Trends Food Sci. Technol.* **2007**, *18*, 117–131. [CrossRef]
6. Struszczyk, M.H. Chitin and chitosan, Part II applications of chitosan. *Polimery* **2002**, *47*, 396–403.
7. Knezevic-Jugovic, Z.; Petronijevic, Z.; Smelcerovic, A. *Chitin, Chitosan, Oligosaccharides and Their Derivatives: Biological Activities and Applications*; Kim, S.-K., Ed.; CRC Press: New York, NY, USA, 2010; Chapter 3; pp. 24–36. ISBN 978-1-4398-1604-2.
8. Fujita, M.; Ike, M.; Tachibana, S.; Kitada, G.; Kim, S.M.; Inoue, Z. Characterization of a bioflocculant produced by *Citrobacter* sp. TKF04 from acetic and propionic acids. *J. Biosci. Bioeng.* **2000**, *89*, 40–46. [CrossRef]
9. Son, M.K.; Shin, H.D.; Huh, T.L.; Jang, J.H.; Lee, Y.H. Novel cationic microbial polyglucosamine biopolymer from new *Enterobacter* sp. BL-2 and its bioflocculation efficacy. *J. Microbiol. Biotechnol.* **2005**, *15*, 626–632.
10. Kim, L.S.; Hong, S.J.; Son, M.K.; Lee, Y.H. Polymeric and compositional properties of novel extracellular microbial polyglucosamine biopolymer from new strain of *Citrobacter* sp. BL-4. *Biotechnol. Lett.* **2006**, *28*, 241–245. [CrossRef] [PubMed]
11. Kimura, K.; Inoue, T.; Kato, D.; Negoro, S.; Ike, M.; Takeo, M. Distribution of chitin/chitosan-like bioflocculant-producing potential in the genus *Citrobacter*. *Appl. Microbiol. Biotechnol.* **2013**, *97*, 9569–9577. [CrossRef] [PubMed]
12. Cozzone, A.J. Regulation of acetate metabolism by protein phosphorylation in enteric bacteria. *Annu. Rev. Microbiol.* **1998**, *52*, 127–164. [CrossRef] [PubMed]
13. Barreteau, H.; Kovac, A.; Boniface, A.; Sova, M.; Gobec, S.; Blanot, D. Cytoplasmic steps of peptidoglycan biosynthesis. *FEMS Microbiol. Rev.* **2008**, *32*, 168–207. [CrossRef] [PubMed]
14. Atlas, R.M. *Handbook of Microbiological Media*, 4th ed.; CRC Press: New York, NY, USA, 2010; pp. 807, 934. ISBN 978-1-4398-0406-3.
15. Yamamoto, S.; Izumiya, H.; Morita, M.; Arakawa, E.; Watanabe, H. Application of lambda Red recombination system to *Vibrio cholerae* genetics: Simple methods for inactivation and modification of chromosomal genes. *Gene* **2009**, *438*, 57–64. [CrossRef] [PubMed]
16. Datsenko, K.A.; Wanner, B.L. One-step inactivation of chromosomal genes in *Escherichia coli* K-12 using PCR products. *Proc. Natl. Acad. Sci. USA* **2000**, *97*, 6640–6645. [CrossRef] [PubMed]
17. Murphy, K.C. Use of bacteriophage lambda recombination functions to promote gene replacement in *Escherichia coli*. *J. Bacteriol.* **1998**, *180*, 2063–2071. [PubMed]
18. DIG Application Manual for Filter Hybridization, in ROCHE Homepage. Available online: http://lifescience. roche.com/wcsstore/RASCatalogAssetStore/Articles/05353149001_08.08.pdf (accessed on 6 December 2017).
19. Livak, K.J.; Schmittgen, T.D. Analysis of relative gene expression data using real-time quantitative PCR and the $2^{-\Delta\Delta CT}$ Method. *Methods* **2001**, *25*, 402–408. [CrossRef] [PubMed]
20. Kumari, S.; Beatty, C.M.; Browning, D.F.; Busby, S.J.; Simel, E.J.; Hovel-Miner, G.; Wolfe, A.J. Regulation of acetyl coenzyme A synthetase in *Escherichia coli*. *J. Bacteriol.* **2000**, *182*, 4173–4179. [CrossRef] [PubMed]
21. Park, J.T.; Uehara, T. How bacteria consume their own exoskeletons (turnover and recycling of cell wall peptidoglycan). *Microbiol. Mol. Biol. Rev.* **2008**, *72*, 211–227. [CrossRef] [PubMed]
22. Plumbridge, J. An alternative route for recycling of *N*-acetylglucosamine from peptidoglycan involves the *N*-acetylglucosamine phosphotransferase system in *Escherichia coli*. *J. Bacteriol.* **2009**, *191*, 5641–5647. [CrossRef] [PubMed]

23. Itoh, Y.; Rice, J.D.; Goller, C.; Pannuri, A.; Taylor, J.; Meisner, J.; Beveridge, T.J.; Preston, J.F., 3rd; Romeo, T. Roles of *pgaABCD* genes in synthesis, modification, and export of the *Escherichia coli* biofilm adhesin poly-β-1,6-*N*-acetyl-D-glucosamine. *J. Bacteriol.* **2008**, *190*, 3670–3680. [CrossRef] [PubMed]

24. Reeves, P.R.; Cunneen, M.M. Biosynthesis of *O*-antigen chains and assembly. In *Microbial Glycobiology*; Moran, A.P., Holst, O., Brennan, P.J., von Itzstein, M., Eds.; Academic Press: Amsterdam, The Netherland, 2009; Chapter 18; pp. 319–335. ISBN 978-0-1237-4546-0.

25. Reid, A.N.; Szymanski, C.M. Biosynthesis and assembly of capsular polysaccharides. In *Microbial Glycobiology*; Moran, A.P., Holst, O., Brennan, P.J., von Itzstein, M., Eds.; Academic Press: Amsterdam, The Netherland, 2009; Chapter 20; pp. 351–373. ISBN 978-0-1237-4546-0.

26. Izano, E.A.; Sadovskaya, I.; Vinogradov, E.; Mulks, M.H.; Velliyagounder, K.; Ragunath, C.; Kher, W.B.; Ramasubbu, N.; Jabbouri, S.; Perry, M.B.; et al. Poly-*N*-acetylglucosamine mediates biofilm formation and antibiotic resistance in *Actinobacillus pleuropneumoniae*. *Microb. Pathog.* **2007**, *43*, 1–9. [CrossRef] [PubMed]

27. Itoh, Y.; Wang, X.; Hinnebusch, B.J.; Preston, J.F., 3rd; Romeo, T. Depolymerization of β-1,6-*N*-acetyl-D-glucosamine disrupts the integrity of diverse bacterial biofilms. *J. Bacteriol.* **2005**, *187*, 382–387. [CrossRef] [PubMed]

28. Campbell, J.A.; Davies, G.J.; Bulone, V.; Henrissat, B. A classification of nucleotide-diphospho-sugar glycosyltransferases based on amino acid sequence similarities. *Biochem. J.* **1997**, *326*, 929–939. [CrossRef] [PubMed]

29. Gerke, C.; Kraft, A.; Sussmuth, R.; Schweitzer, O.; Gotz, F. Characterization of the *N*-acetylglucosaminyltransferase activity involved in the biosynthesis of the *Staphylococcus epidermidis* polysaccharide intercellular adhesin. *J. Biol. Chem.* **1998**, *273*, 18586–18593. [CrossRef] [PubMed]

30. Vemuri, G.N.; Altman, E.; Sangurdekar, D.P.; Khodursky, A.B.; Eiteman, M.A. Overflow metabolism in *Escherichia coli* during steady-state growth: Transcriptional regulation and effect of the redox ratio. *Appl. Environ. Microbiol.* **2006**, *72*, 3653–3661. [CrossRef] [PubMed]

31. Brown, T.D.K.; Jones-Mortimer, M.C.; Kornbero, H.L. The enzyme interconversion of acetate and acetyl-coenzyme A in *Escherichia coli*. *J. Gen. Microbiol.* **1977**, *102*, 327–336. [CrossRef] [PubMed]

32. Luli, G.W.; Strohl, W.R. Comparison of growth, acetate production, and acetate inhibition of *Escherichia coli* strains in batch and fed-batch fermentations. *Appl. Environ. Microbiol.* **1990**, *56*, 1004–1011.

33. Oh, M.K.; Rohlin, L.; Kao, K.C.; Liao, J.C. Global expression profiling of acetate-grown *Escherichia coli*. *J. Biol. Chem.* **2002**, *277*, 13175–13183. [CrossRef] [PubMed]

34. Kim, Y.M.; Lee, S.E.; Park, B.S.; Son, M.K.; Jung, Y.M.; Yang, S.O.; Choi, H.K.; Hur, S.H.; Yum, J.H. Proteomic analysis on acetate metabolism in *Citrobacter* sp. BL-4. *Int. J. Sci.* **2012**, *8*, 66–78. [CrossRef]

35. Son, M.K.; Hong, S.J.; Lee, Y.H. Acetate-mediated pH-stat fed-batch cultivation of transconjugant *Enterobacter* sp. BL-2S over-expressing *glmS* gene for excretive production of microbial polyglucosamine PGB-1. *J. Ind. Microbiol. Biotechnol.* **2007**, *34*, 799–805. [CrossRef] [PubMed]

36. Deng, M.D.; Severson, D.K.; Grund, A.D.; Wassink, S.L.; Burlingame, R.P.; Berry, A.; Running, J.A.; Kunesh, C.A.; Song, L.; Jerrell, T.A.; et al. Metabolic engineering of *Escherichia coli* for industrial production of glucosamine and *N*-acetylglucosamine. *Metab. Eng.* **2005**, *7*, 201–214. [CrossRef] [PubMed]

37. Deng, M.D.; Grund, A.D.; Wassink, S.L.; Peng, S.S.; Nielsen, K.L.; Huckins, B.D.; Burlingame, R.P. Directed evolution and characterization of *Escherichia coli* glucosamine synthase. *Biochimie* **2006**, *88*, 419–429. [CrossRef] [PubMed]

38. Stevenson, G.; Andrianopoulos, K.; Hobbs, M.; Reeves, P.R. Organization of the *Escherichia coli* K-12 gene cluster responsible for production of the extracellular polysaccharide colanic acid. *J. Bacteriol.* **1996**, *178*, 4885–4893. [CrossRef] [PubMed]

39. Wang, X.; Preston, J.F., 3rd; Romeo, T. The *pgaABCD* locus of *Escherichia coli* promotes the synthesis of a polysaccharide adhesin required for biofilm formation. *J. Bacteriol.* **2004**, *186*, 2724–2734. [CrossRef] [PubMed]

polymers

MDPI

Review

Cosmetics and Cosmeceutical Applications of Chitin, Chitosan and Their Derivatives

Inmaculada Aranaz [1,2,*], Niuris Acosta [1,2], Concepción Civera [1], Begoña Elorza [1], Javier Mingo [1,2], Carolina Castro [1,2], María de los Llanos Gandía [1,2] and Angeles Heras Caballero [1,2,*]

[1] Department of Chemistry in Pharmaceutical Sciences, Pharmacy Faculty, Complutense University, Plaza de Ramón y Cajal, s/n, 28040 Madrid, Spain; facosta@ucm.es (N.A.); mccivera@ucm.es (C.C.); belorza@ucm.es (B.E.); javier.mingo@ucm.es (J.M.); caroca03@ucm.es (C.C.); gandia.mllanos@gmail.com (M.d.l.G.)

[2] Bifunctional Studies Institute, Complutense University, Paseo Juan XXIII, 1, 28040 Madrid, Spain

[*] Correspondence: iaranaz@ucm.es (I.A.); aheras@farm.ucm.es (A.H.C.); Tel.: +34-913-943-284 (I.A. & A.H.C.)

Received: 4 February 2018; Accepted: 20 February 2018; Published: 22 February 2018

Abstract: Marine resources are well recognized for their biologically active substances with great potential applications in the cosmeceutical industry. Among the different compounds with a marine origin, chitin and its deacetylated derivative—chitosan—are of great interest to the cosmeceutical industry due to their unique biological and technological properties. In this review, we explore the different functional roles of chitosan as a skin care and hair care ingredient, as an oral hygiene agent and as a carrier for active compounds, among others. The importance of the physico-chemical properties of the polymer in its use in cosmetics are particularly highlighted. Moreover, we analyse the market perspectives of this polymer and the presence in the market of chitosan-based products.

Keywords: chitin; chitosan; chitosan derivative; chitin derivative; oral care; skin care; hear care; marine resources; over-the counter-drug; polymer carrier

---

1. Introduction

A cosmetic is defined as any substance or preparation intended to be placed in contact with the various external parts of the human body (epidermis, hair system, nails, lips and external genital organs) or with the teeth and the mucous membranes of the oral cavity with a view exclusively or mainly to cleaning them, perfuming them, changing their appearance and/or correcting body odours and/or protecting them or keeping them in good condition [1]. That means that cosmetics are intended to be applied outside the body and no treatment against any specific disease can be claimed. These types of products are strictly regulated by different governmental agencies.

The term cosmeceutical was first coined by R.E Reed in 1962 [2]—Reed´s definition included four statements related to the idea of an increase in the quality of cosmetic products, which are included in the cosmetic industry nowadays (Table 1).

Table 1. Statements included in Reed´s cosmeceutical definition.

(i) A Cosmeceutical is a scientifically designed product intended for external application to the human body
(ii) A Cosmeceutical produces a useful and desired result
(iii) A Cosmeceutical has desirable aesthetic properties
(iv) A Cosmeceutical meets rigid chemical, physical and medical standards

Nowadays, the use of the term cosmeceutical is widespread [3–6] with a different meaning that was established by Klingman in 1993 [7]. Cosmeceuticals are intended to carry out their functions as protection, whitening, tanning, anti-wrinkling, deodorants, antiaging and nail and hair care as a cosmetic product but cosmeceuticals claim to have biologically active ingredients with medicinal or drug-like benefits. In this new vision, the term cosmeceutical suggests that drugs and some cosmetics could share some possible borders which generates some controversial [8]. Cosmeceuticals include not only high-quality cosmetics as defined by Reed but also some prescription drugs such as topical retinoids, topical minoxidil, eflornithine and over-the-counter-drug (OTC) included in sun creams and antiperspirants.

Although it is claimed that a new regulation is needed to differentiate cosmeceuticals from cosmetics or drugs, these products have not been legally recognized yet. In fact, a consensus definition does not exit.

Since the antique, natural resources are well known as a source of biologically active substance to be used as cosmetics or cosmeceutical products. In recent years, much attention has been paid to marine resources as a new source of inexpensive and safe substances. Moreover, environmental issues are one of the main driven factors in the growing of this type of ingredients.

Table 2. Some potential cosmeceutical ingredients from marine resources and their use [9,10].

Ingredient	Source	Activity/Use
Alginate	Seaweed (brown algae)	Texture and emulsion stabilizer Vehicle for controlled delivery Thickening agent
Fucoidans		Wound-healing
Phlorotannin		Sunscreen and antioxidant activities
Fucoxanthin		UV protective and antioxidant activities
Carrageenan	Seaweed (red algae)	Viscosity altering Thickening agent
MAAs		Antioxidant
Ulvans	Seaweed (green algae)	Antioxidant
Glycogen	Mussel	UVB protection Moisturizing
Aluminium silicate	Sea mud	Absorbent
Squalene	Shark	Skin lubrication
Chitin	Crustaceans shells	Vehicle for controlled delivery Antiaging Skin protecting
Chitosan	Crustaceans shells	Vehicle for controlled delivery Antimicrobial Antioxidant Emulsifying Skin protecting

MAAs: Mycosporine-like amino acids.

Form the active ingredients shown in Table 2, this review is going to focus on chitin, chitosan and their derivatives. Chitin is one of the most abundant polysaccharides found in nature which is widely distributed in the animal and vegetal kingdom. It appears in the exoskeleton of invertebrates such as crustaceans, mollusc or insects. Chemically chitin is a copolymer of *N*-acetylglucosamide and glucosamide. Chitosan is the *N*-deacetylated derivative of chitin which is only found in some fungi in nature. Chitosan is industrially produced from chemical *N*-deacetylation of chitin isolated from crustaceans. The main difference between chitin and chitosan is its solubility. Chitin is insoluble in aqueous media while chitosan is soluble in acidic aqueous systems being this property the criteria to differentiate chitin from chitosan [11]. Chitin and chitosan can be chemically modified using simple chemical tools producing a large number of derivatives with different properties that will be described in this review.

2. Target Organs for Cosmetics Products

Chitin and chitosan can be used in different body sites such as skin, hair, gums and teeth. In this section, a brief anatomical description and the main problems/illness related to these body sites are described to better understand how chitin and chitosan can be used. In Figure 1, a brief description of these body structures is shown.

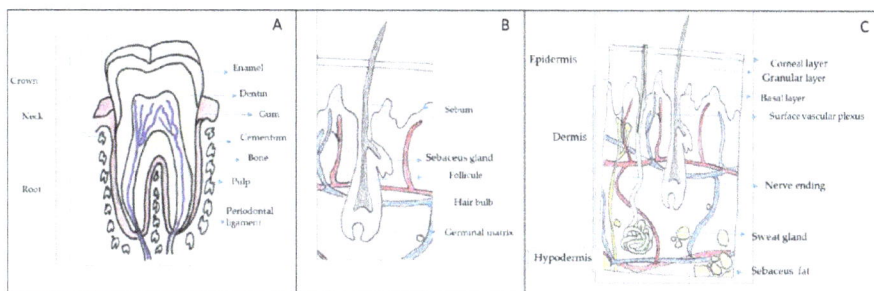

Figure 1. Target organs for cosmetic and cosmeceutical products. (**A**) Gums and tooth, (**B**) hair and (**C**) skin. Adapted from Wikipedia Commons (authors: KDS4444, Human tooth diagram-en.svg CC-BY-SA 4.0, Madhero88, Skinlayers.svg CC-BY-SA 3.0).

2.1. Gums

The periodontium is made up of four parts, namely tooth sockets (dental alveoli), periodontal ligament, cementum and gums (gingivae).

Gums, a soft connective tissue covered with mucous membrane, are found in the oral cavity that surrounds the necks of the teeth and adjacent alveolar bone (Figure 1A). The teeth are connected with the walls of the tooth sockets and anchored in the jaws by the periodontal ligament and the cementum. Each tooth consists of a hard shell that surrounds a cavity of soft tissue, known as pulp. The crown (the exposed part of the shell) is coated with a tough layer of enamel, beneath which is a layer of a yellowish substance similar to ivory, called dentin. The dentin and pulp form long, pointed roots that extend into the jawbone.

The gum tissue inflammation, known as gingivitis, causes the gum slightly detaches from the neck of the tooth. The space created between the tooth and the gum can then deepen to form a pocket where bacteria can rapidly build up. Gingivitis can be visible as bleeding gums but often does not cause any symptoms. Gingivitis can lead to periodontitis, an inflammation of the periodontium due to specific microorganisms that attacks the gums as well as the bone of the jaw. The presence of any of the 6–12 bacteria species responsible for the majority of the bacterial periodontitis can start periodontitis [12]. Advanced periodontitis can cause teeth to become loose and fall out or require them to be removed. It is very remarkable that illnesses related to the oral cavity are not only an aesthetic problem but also a serious health problem. The scientific evidence has shown a direct link between periodontal disease and cardiovascular disease [13]. The relationships between periodontal disease and other non-communicable diseases (hypertension, diabetes mellitus, osteoporosis, cerebral infarction, angina pectoris, myocardial infarction and obesity) have also been evaluated. This study demonstrated that the presence of periodontal disease is associated with a significantly elevated risk of non-communicable diseases in the Korean adult population, especially obesity, osteoporosis and angina pectoris [14].

2.2. Teeth

Dental caries and tooth wear are the main factors related to mineral lost and enamel demineralization. Dental caries is defined as a localized chemical dissolution of the tooth surface caused by metabolic events that occur in the dental plaque (biofilm that grows on surfaces within the mouth) [15]. The role of *Streptococcus mutans* is well recognized in the initiation of dental caries being attributed at least, in part, to its ability to colonise the tooth surface. Therefore, factors which prevent oral bacterial attachment to tooth are of considerable interest for the prophylaxis of this infectious disease.

Tooth wear is every mineral lost non-related to dental caries or dental trauma. Tooth wear can be produced by physical tooth-to-tooth contact (attrition), other physical contacts non-related to teeth (erosion) or chemical dissolution of tooth substance caused by acids, unrelated to the acid produced by bacteria in dental plaque [16]. Dental abrasion is a particularly worrying process in primary teeth, because the enamel layer in primary teeth is quite thin and, therefore, abrasions are frequently observed. With the age, the permanent teeth also accumulate dental abrasion. Dental erosion produces bulk mineral loss with a partly demineralized surface of reduced micro-hardness. Moreover, in early stages, coronal dentin often is exposed. The primary objective goal of active ingredients against dental erosion is to increase the resistance of tooth surfaces or pellicles to acids.

Several approaches including physical dental plaque remonition, use of antimicrobials and the use of fluorides are used to manage dental care. Caries-prophylactic agents may be delivered to the oral cavity by various delivery formulations (vehicles) such as mouth rinses, sprays, dentifrices, gels, chewing guns, lozenges and sustained-release formulations or devices (including nanoparticles/microparticles and functionalized films) which different characteristics are described in Table 3.

Table 3. Delivery formulations used in oral care.

Vehicle	Characteristics
Mouth rinses	Simplest vehicle formulation Compatible with most antimicrobial agents
Sprays	Relatively small doses to achieve efficacy Good compliance Easy usage
Dentifrices	Complex formulation Possible interaction among components Tooth brushing with a dentifrice is a well-adopted habit
Gels	Thickened aqueous system Non-abrasive No foaming agents Compatible with relevant antimicrobials Specific devices are needed for applications
Chewing gum/lozenges	Larger contact time Stimulated salivary secretion Useful for patients with low tooth-brushing compliance
Sustained-release formulations/devices	Long-term effect The efficacy is independent of patient compliance

2.3. Hair

Hair is mainly composed of a protein called keratin. The second most abundant component are lipids, mainly ceramides. These lipids avoid the interfibre friction which is related to the sensory perception of hair. Hair shaft structure is very complex and consists of several concentric layers known as medulla, cortex and cuticle (Figure 1B). The medulla is the innermost layer, it is soft and fragile. Its function is unclear not appearing in some types of hair. The cortex is formed of elongated cells aligned along the axis of the fibre and it is filled with keratins that are arranged in a coiled-coil configuration. The cortex is responsible for the mechanical strength of hair and it controls water uptake. Moreover, the cortex contains melanin which accounts for hair colour.

Finally, the cuticle, the outer covering layer, is a thin laminar-like structure composed of layers of overlapping, flat, scale-like cells acting as a protective hair sheath. The cuticle is responsible for hair hydrophobicity since it is covered with a single molecular layer of lipids that makes the hair repels water.

Several factors account for hair protein denaturalization such as UV-exposition, heat stress from curling irons and blow dryers, chlorine, harsh chemicals in colouring, straightening and perming. Hair care products need to fulfil several requisites as summarized in Table 4.

Table 4. Hair care products requisites.

Low stickiness
Lack of powdering or flaking
Preferably being clear
Preferably transparent
Preferably glossy
Good film formation
Good holding power
High level of style retention
Prolonged curl retention
Improved combability
Easily removed upon washing the hair

2.4. Skin

The human´s largest body organ is the skin. The surface area of adults is about 1.6 m^2 (female) to 1.8 m^2 (male) [17]. The skin is composed of three layers: epidermis, dermis and hypodermis (Figure 1C) [18]. In the epidermis, there are five sublayers called *stratum corneum, stratum lucidum stratum granulosum, stratum spinosum and stratum germinativum* or *stratum basale*. The *stratum basale*, the deepest layer, is primarily made up of basal keratinocytes. These cells divide to form the keratinocytes of the *stratum spinosum*, which migrate superficially in a journey that takes approximately fourteen days and become flat, keratenised, water resistant dead cells of the corneum. At the transition between the *stratum granulosum* and the *stratum corneum*, cells secrete lamellar bodies (containing lipids and proteins) into the extracellular space. This results in the formation of the hydrophobic lipid envelope responsible for the skin's barrier properties. The *stratum corneum* forms a barrier to protect underlying tissue from infection, dehydration, chemicals and mechanical stress. Desquamation, the process of cell shedding from the surface of the *stratum corneum*, balances proliferating keratinocytes that form in the *stratum basale*.

The dermis is the structure beneath the epidermis, it is approximately 15 to 40 times as thick as the epidermis and it is composed of three layers (papillary layer, subpapillary layer and reticular layer). The components of the dermis comprise the fibrous tissue and the dermal matrix formed by cells in the interstitial components. Interstitial components are a dense network of structural proteins. Collagen fibres account for 70% of the weight of dry dermis; elastic fibres are composed of elastin. In between fibres and between cells a gelatinous amorphous substance called ground substance is observed. This matrix is mainly composed of proteoglycans and glycoproteins. The dermis is responsible for the skin´s elasticity and the senses of touch and heat.

The hypodermis or subcutaneous tissue is the layer between the dermis and the fascia and consist of a fat-rich tissue. The fat tissue acts to preserve neutral fat, cushion against external physical pressure, retain moisture and generate heat.

Three functions have been described for the skin: protection, regulation and sensation. The skin offers a protective barrier against harmful environmental factors, such as heat, solar ultraviolet (UV) irradiation, infection, injury and water loss. However, lifestyle and environmental factors cause both cosmetics and dermatological problems.

There are two primary skin ageing processes, intrinsic and extrinsic. Intrinsic skin ageing depends on genetic background, therefore seems to be inevitable and not subject to modification through changes in human behaviour. On the contrary, extrinsic ageing is caused by environmental factors such as light, heat, cold, etc. and therefore can be altered by modifying our style life.

Skin cosmeceuticals have been developed after research on common skin problems like hyperpigmentation, skin cancer, skin microbial infections, wound healing and wrinkles associated with sun damage and ageing (Figure 2) [19].

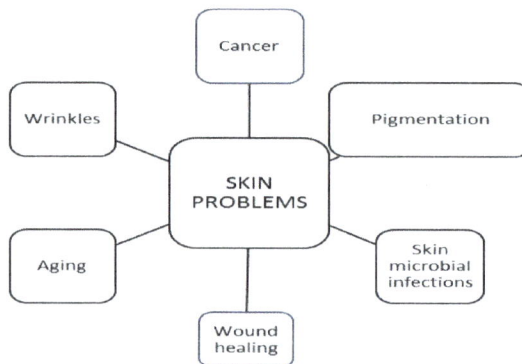

Figure 2. Major worldwide skin problems.

3. Chitin, Chitosan and their Derivatives in Cosmetics and Cosmeceutical Industry

Chitin exhibits low solubility and it is quite difficult to handle. On the contrary, chitosan is soluble in acid aqueous solutions and can easily produce different conformations such as micro, nano and milli particles, films, scaffolds and fibres among others [20]. Oligosaccharides and some chitin and chitosan derivatives are water soluble at physiological pH and exhibit improved or even new properties. When taking into consideration the use of chitin, chitosan and derivatives in cosmetics one should keep in mind that these polymers can act as ingredients due to their specific properties or as a carrier of other active ingredients due to their technological properties.

The cosmetic ingredients are on the EU market regulated over Cosmetic Product Regulation adopted in 2009 [21]. CosIng is the European Commission database for information on cosmetic substances and ingredients (http://ec.europa.eu/growth/tools-databases/cosing/, last accessed 23 January 2018). A search using the term chitosan on this database retrieved 44 results while 8 results were retrieved with the term chitin.

According to CosIng database, the functions assigned to chitin are abrasive and bulking whereas the functions assigned to chitosan are film forming and hair fixing. A large number of functions, which will be described in this review, are identified for different chitosan derivatives. It is remarkable that not all activities and derivatives discussed in this review are included in CosIng database. Besides CosIng, this paper includes data from research journals and patents. Moreover, beyond the properties described in CosIng, other relevant biological properties of interest for the cosmeceutical can be found in chitin, chitosan and derivatives such as antifungic and wound healing activities, bio-adhesivity, non-toxicity and biodegradability [22].

3.1. Chitosan and Derivatives in Oral Healthcare

The use of chitosan has been proposed in all fields of dentistry including preventive dentistry, conservative dentistry, endodontics, surgery, periodontology, prosthodontics and orthodontics [23].

In this review, we are going to focus on the use of chitosan in preventive oral care. Oral healthcare is focused on preventive dentistry which aim is to avoid dental and gums illness.

In Figure 3, a summary of the activity of the polymers in oral healthcare and the type of product tested is shown. In this section, a description of each activity and its relationship to the physicochemical properties of the polymers, when possible, is given.

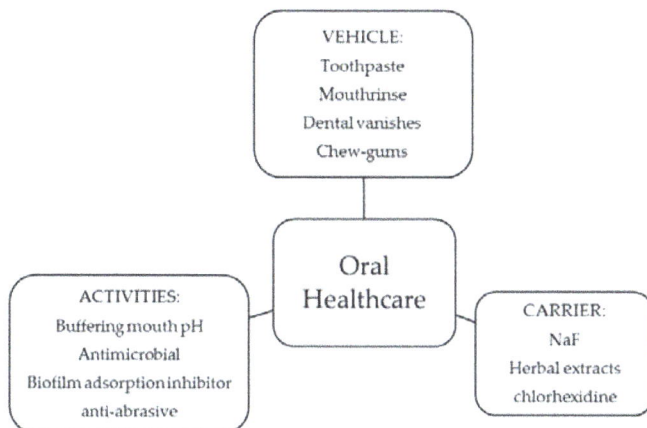

Figure 3. Use of chitosan and derivatives in preventive oral healthcare.

3.1.1. Reduction of Dental Plaque

Dental caries started via a carbohydrate fermentation process due to bacteria metabolism, in which the production of strong organic acids such as lactate, formate and pyruvate cause demineralization of the tooth surface. It has been reported that the optimum pKa value for buffering substances against plaque pH fall in vitro maybe around 6.3 [24]. This value falls in the pKa range of chitosan which is found around 5.1 and 6.5 depending on the chitosan M_w and degree of acetylation (DA). It has been described that pKa decreases with chitosan M_w while decreased with acetylation degree [25]. The effect of six low molecular weight chitosans (500–3000 Da) with acetylation degrees ranging from 0.05–0.5 were tested in vitro and in vivo for dental plaque pH reduction [26]. Evaluations using *S. mutans* cell suspensions in vitro revealed no differences among chitosan samples regarding their ability to reduce pH fall in dental plaque. Moreover, chitosan samples did not have any influence on the glycolytic activity of *S. mutans*. The clinical evaluation showed differences between the different chitosan samples. The most effective sample was a chitosan with a molecular weight of 3000 kDa and DA of 0.5. The second ranked had a M_w of 500 Da and DA of 0.05.

Another strategy to control dental plaque is the use of molecules with antimicrobial activity. The gold standard to avoid dental plaque and gingivitis is chlorhexidine gluconate (0.2%) due to its strong antimicrobial activity against *S. mutans*. However, this product has several side effects such as tooth staining, alteration in taste perception, changes in tongue sensitivity and pain due to the alcoholic content of the oral rinse. Moreover, chlorhexidine gluconate has a strong effect against *S. mutans* but little effect against other bacteria presented in dental plaque. Therefore, there is a great interest in new products with wide antimicrobial activity against oral pathogens with fewer side effects than chlorhexidine gluconate.

Chitosan has a wide antibacterial activity and therefore its activity against other bacterial strains, besides *S. mutans*, related to dental caries has been evaluated by several authors [27–30]. The evaluation of chitosan's antimicrobial action mechanism testing two chitosan samples (624 kDa, DA < 0.25 and 107 kDa, DA 0.15–0.25) showed that both MWs acted upon the bacterial cell wall and were not capable of interacting with the intracellular substances, showing little to non eflocculation capability [29]. In Table 5, the reported minimal inhibitory concentration (MIC) values from different chitosan samples against different oral pathogen are summarized. In general, the higher the Mw of the polymers the lower was the MIC value. The exceptions were *P. buccae* and *P. intermedia*. Against *P. Buccae* the lower was chitosan´s M_w the lower was the MIC; while against *P. intermedia* the lowest MIC was observed with medium M_w chitosan samples.

The antimicrobial activity of chitosan (low M_w, DA < 0.15) and chitosan nanoparticles *against S mutans, S sobrinus, S sanguis and S salivarius* was tested. The MIC of chitosan nanoparticles was lower in some strains than the MIC of chitosan [31]. This result is in good agreement with previous results that showed a better antimicrobial activity of chitosan nanoparticles when comparing to chitosan due to their small size [32]. The anti-inflammatory activity of chitosan nanoparticles was also studied [33]. This study showed that chitosan nanoparticles exerted a predominantly anti-inflammatory activity by modulating PGE2 levels through the c-Jun N-terminal kinases (JNK) pathway.

No antimicrobial activity was detected against *S. sanguis, S. gordonii, S. constellatus, S. anginosus, S. intermedius, S. oralis, S. salivarius and S. vestibularis* when low-molecular-weight produced by enzymatic chitosan depolymerisation was tested [34]. This result is contradictory to the previous one in which low Mw is claimed to exhibit antimicrobial activity against *S. sanguis*. In this case, authors did not describe properly the chitosan characteristics. Other authors have found that chitosan gels did not exhibit antimicrobial activity per se against *A. actinomycetemcomitans* and *S. mutans* but no data regarding chitosan properties were included in this paper [35].

Table 5. Effect of chitosan M_w and deacetylation degree on minimum inhibitory concentrations (MIC) against oral pathogens [27–30].

Bacterial Strain	Chitosan Properties		MIC mg/mL
	M_w, kDa	DA	
S. mutants	1400	0.2	0.08
	1080	0.14	2.5
	624	<0.25	3
	107	0.15–0.25	5
P. intermedia	1080	0.14	2.5
	624	<0.25	1
	107	0.15–0.25	3
P. buccae	624	<0.25	3
	107	0.15–0.25	1
T. forythensis	624	<0.25	1
	107	0.15–0.25	3
A. actinomycetemcomitans	1080	0.14	2.5
	624	<0.25	5
	107	0.15–0.25	3
P. gingivalis	1080	0.14	0.5
	624	<0.25	1
	272	0.05	3.8
	272	0.16	3.8
	272	0.27	3.6
	107	0.15–0.25	1

MIC: minimum inhibitory concentration. DA: Acetylation degree.

Besides antimicrobial activity, other chitosan effects such as cell stimulation and biofilm adsorption inhibition have been reported. Chitosan (1080 kDa, DA 0.14) stimulated the proliferation of human periodontal ligament cells [27]. Chitosan micro and nanoparticles were well tolerated by gingival fibroblasts and were able to stimulate cell proliferation through the Ras-Erk-ETS 1/2 (ERK1/2) signalling pathway. Furthermore, it was hypothesized that a synergistic response between chitosan particles and growth factors (such as PDGF-BB) may stimulate cell proliferation in gingival fibroblasts [36].

The inhibition of *S. mutant's* attachment to hydroxyapatite (HA) and saliva-treated hydroxyapatite (S-HA) by low M_w chitosan produced by enzymatic depolymerisation (membrane cut-off 1500) has been described. When S-HA and HA beads were treated with low molecular weight chitosan, a reduction in *S. mutans* adsorption ranging from 47 to 66% was observed [37]. The effects of chitosan Mw upon *S mutant's* adherence and biofilm formation was studied testing two chitosan samples (High Mw sample: 624 kDa, DA < 0.25 and medium M_w sample 107 kDa, DA 0.15–0.25) [38]. Both samples showed an evident strong effect on *S. mutans* adherence and biofilm formation, inhibiting both adherence and biofilm formation in the initial stage of dental plaque formation as well as disrupting mature biofilms. The high Mw sample exhibited better performance regarding inhibiting dual-species biofilm formation. The effect of chitosan M_w (0.8–6 kDa) and DA (0.05–0.90) on the adsorption of *S. sobrinus* to S-HA was tested. The inhibition of *S. sobrinus* adsorption on S-HA showed a positive correlation with the Mw of chitosan and the optimal degree of acetylation was determined to be 0.4–0.5 [39].

Chitosan has been added to tooth pastes, rinses and other vehicles to test its activity against dental plaque formation in real formulations. Chitosan included in tooth paste and mouthwashes to avoid biofilm formation due to the presence of *S mutans* in the mouth demonstrated a reduction of *S mutans* colonies in early childhood caries [40]. Moreover, daily use of chitosan rinse (6 kDa, DA 0.4, 0.5% *w/w*) was effective to reduce dental plaque formation and count of salivary *S mutants* [41]. A chitosan mouthwash, composed of a mixture of two chitosans (DA < 0.25; M_w 624 kDa and DA 0.15–0.22; M_w 107 kDa, the final concentration of either chitosan being 0.4% *v/v*), was compared with two commercial mouth rinsers [42]. The chitosan-based product exhibited a wide range of antibacterial activity (against *S. mutans*, *E. faecium*, *P. intermedia* and *L. acidophilus*) and antifungal activity against *C. albanis*. Moreover, its activity was superior to both commercial mouthwashes examined. In a further study, chitosan mouthwash safety was evaluated and the antimicrobial activity was corroborated through in vivo assays [43]. Chlorhexidine gluconate showed minor efficacy on other cariogenic bacteria than *S. mutans* such as *S. sanguinis* or *lactobacilli* even when applied at high concentrations [44]. The effect of two chitosan samples, at a concentration 0.2% which corresponds to the chitosan MIC for oral *streptococci* according to previous studies, alone or in combination with chlorhexidine gluconate on dental plaque was studied. The mixtures containing chitosan and chlorhexidine exhibit a better antibacterial activity than chlorhexidine alone. Differences were observed regarding the chitosan sample used but unfortunately, the polymer characteristics were not described [45]. In another study, chlorhexidine gluconate mucoadhesive gels were produced with different chitosan samples (High M_w (1400 kDa, DA 0.2) and medium Mw (272 kDa, DA ranging from 0.05 to 0.27). In this case, the combined gels exhibited lower MIC than pure chitosan or chlorhexidine gluconate against *P. gingivalis*. Interestingly, the best combined formulation did not correspond to the chitosan that exerted more remarkable antimicrobial activity alone [28]. A chitosan tooth paste containing herbal extracts with antimicrobial activity against dental pathogens was formulated. The pharmaceutical evaluation of toothpaste was carried out as per the US Government Tooth Paste Specifications. After 4-weeks experiment the chitosan-containing polyherbal toothpaste significantly reduced the plaque index and bacterial count during in vivo tests [46]. Chewing gums containing chitosan demonstrated its ability to suppress bacterial growth and to increase salivary secretion in vivo which may improve the quality of life of dry-mouth patients. These findings suggest that a supplementation of chitosan to gum is an effective method for controlling the number of cariogenic bacteria in situations where it is difficult to brush one's teeth [47,48].

Besides chitin or chitosan, several derivatives of both polymers have been synthetized and tested against different bacterial strains. As seen in Table 6, most of these derivatives have antimicrobial activity against *S. mutants* while only some derivatives have exhibited activity against other *streptococci* strains [27,34,37,49–52].

Table 6. Use of chitin and chitosan derivatives in oral healthcare [27,34,37,49–52].

Polymer Derivative	Bacterial Strain	Effect
Ethylenglycol chitin	*S. mutants* *S. sanguis* *S mitis*	Reduce bacterial adsorption on S-HA in vitro Dose dependent effect Better activity on *S. mutants*
Carboxymethyl chitin	*S. mutants* *S. sanguis* *S mitis*	Reduce bacterial adsorption on S-HA in vitro Dose dependent effect Better activity on *S. mutants*
N-Carboxymethyl chitosan	*S mutants*	Prevent bacterial adsorption to *HA* in vitro
	S. sanguis, S. gordonii, S. constellatus, S. anginosus, S. intermedius, S. oralis, S. salivarius, S. vestibularis	Adsorption reduction on HA and S-HA (60%–98%) in vitro
Imidazolyl chitosan	*S mutants*	Prevent bacterial adsorption to HA in vitro
	S. sanguis, S. gordonii, S. constellatus, S. anginosus, S. intermedius, S. oralis, S. salivarius, S. vestibularis	No effect on bacterial adhesion to HA or S-HA in vitro
Sulphated chitosan	*S. muntants* *S. sanguis* *S. mitis*	Reduce bacterial adsorption on S-*HA* in vitroDose dependent effect. Better activity on *S. mutants*
Phosphorylated chitosan	*S. muntants* *S. sanguis* *S. mitis*	Reduce bacterial adsorption on S-HA in vitro. Dose dependent effect. Better activity on *S. mutants* Plaque reduction Slight plaque buffering capacity
N-1hydroxy 3 trimethyl ammonium chitosan HCl	*P. gingivalis* *P. intermedia* *A. actinomycetemcomitans* *S. mutans*	Antibacterial activity in vitro MIC: 0.5–1 mg/mL
Glucosamine Maillard chitosan derivative	*S. mutants* *L. brevis*	CBM *S. mutants* 0.4 mg/mL CBM *L. brevis* 0.5 mg/mL No cytotoxicity in vivo
Water-soluble reduced chitosan	*S. mutans* *S sanguinis*	MIC *S mutans* 1.25 mg/mL MIC *S sanguinis* 10 mg/mL Reduction plaque index Reduction vital fluorescence

MIC: minimum inhibitory concentration. MBC: minimum bactericidal concentration; HA: Hydroxyapatite S-HA Saliva coated hydroxyapatite.

3.1.2. Reduction of Dental Abrasion

Another use of chitosan in dentistry is related to avoid dental abrasion. Dental tooth, which otherwise is beneficial to avoid dental plaque, contain abrasive components that may cause brush scratches or micro wear on sound enamel surfaces. It is thought that the presence of fluoride may cause less brushing abrasion on teeth with artificial decay lesions than toothpaste without fluoride since fluoride has a remineralization effect. Different ingredients, such as polyvalent cations (Sn^{2+}, nano-phosphate salts (Ca/P), proteins and biopolymers such as chitosan, have been proposed to design more effective formulations as fluoride substitutes or included in fluoride tooth pastes [53].

A tooth paste containing chitosan (Chitodent®) was compared with a propolis-containing tooth paste (Aagaard®), a fluoridated (500 ppm) tooth paste (Elmex®) and a control group without treatment. The brushing abrasion depths formed in primary tooth enamel caused using the different tooth paste were compared. In this study, both propolis tooth paste and chitosan tooth paste showed an average brushing abrasion value on healthy surfaces lower than the observed using the fluoride tooth paste [54]. Divalent cations such as Sn^{2+} in combination with fluoride are more effective than fluoride alone to prevent dental erosion. The use of Sn^{2+} combined with chitosan and fluoride was proposed not only in terms of anti-erosive activity but also as anti-abrasive. The F/Sn/chitosan toothpaste (1.400 ppm F^-, 3.500 ppm Sn^{2+}, 0.5% chitosan) reduced the erosive/abrasive tissue loss significantly compared to placebo (tooth paste without fluoride, Sn^{2+}, chitosan) and showed an efficacy in the order of the positive control (GelKam = 3.000 ppm Sn^{2+},1.000ppm F^-) or experimental toothpaste (1.400 ppm F^-,

3.500 ppm Sn^{2+}). The effect of chitosan was ascribed to its ability to bind to enamel creating a protective shell [55].

3.1.3. Chitosan as Vehicle in Oral Healthcare

Chitosan has also been used as a vehicle for other therapeutic products with different activities. For instance, chitosan gels containing herbal extracts showed a dental plaque reduction of 70% and a reduction of bacterial counting of 85% [46]. Chitosan microparticles containing NaF produced by spray drying exhibited bioadhesive properties in the oral cavity acting as a fluoride reservoir both in fluoride burst release or fluoride controlled delivery systems [56]. Chitosan nanoparticles (100 nm) loading NaF were produced by ionotropic gelation using tripolyphosphate (TPP) as crosslinker agent [57]. Dental varnishes containing chitosan nanoparticles as NaF carrier was developed with antimicrobial activity against *S. mutants* and the ability to inhibit demineralization was proved [58]. Chitosan (DA 0.15–0.25 viscosity 200–800 cps) microparticles containing chlorhexidine diacetate were produced by spray-drying. Chlorhexidine-chitosan microspheres were dissolved faster in vitro than chlorhexidine powder. The antimicrobial activity was tested against *Escherichia coli*, *Pseudomonas aeruginosa*, *Staphylococcus aureus* and *C. albicans*. Buccal tablets were prepared by direct compression of the microparticles with mannitol alone or with sodium alginate. After their in vivo administration, the determination of chlorhexidine in saliva showed the capacity of these formulations to give a prolonged release of the drug in the buccal cavity [59].

Chitosan- hydroxypropyl methylcellulose 3D hydrogels containing O-toluidine for antimicrobial photodynamic inactivation were produced and tested against *S. aureus*, *A. actinomycetemcomitans* and *P. gingivalis* biofilms. These hydrogels showed promising results regarding their clinical use with an appropriate delivery of o-toluidine [60]

3.2. *Chitin, Chitosan and Derivatives in Haircare*

The usage of polymers in haircare products is gaining attention due to their ability to improve the rheological behaviour of the product or to enhance the adhesion of other ingredients to the hair. As previously mentioned in Section 2.3, the proteinic structure in damaged hair is denaturalized by different processes. It has been reported that the use of cationic polymers can help in the treatment of damaged hair. Polymers need to fulfil some requirements to be considered appropriate for hair care. As seen in Table 7, chitosan fulfils most of the requirements while each chitosan derivative needs to be addressed individually for most of them.

Table 7. Polymer requisites in hair care.

Requisite	Chitosan	Water Soluble Derivatives
Heat stability	Up to 170 °C	To be checked
Very good solubility	Only acidic media, depends on DA and Mw	Yes
Compatibility with cosmetic bases	Yes	To be checked
pH stability in the range of 4 to 9	Yes	To be checked
Processability into a variety of products	Yes	Yes
Compatibility with other ingredients and with the packaging materials	Yes	Yes
Free of colour	White to yellowish	To be checked
Neutral or pleasant odour	Yes	Yes
Low volatility	Non-volatile	Non-volatile

3.2.1. Chitosan and Derivatives as Hair Care Ingredient

Chitosan and its derivatives have been included in a large variety of hair products such as shampoos, rinses, permanent wave agents, hair colorants, styling lotions, hair sprays and hair tonics [61].

Chitosan and its cationic derivatives have the ability to interact with keratin forming transparent, elastic films over hair fibres. These films increase hair softness, hair strength and avoid hair damage. Chitosan was blended with hyaluronic acid and collagen and the produced films on hair were studied. The covering of hair led to an increase in hair thickness and to the improvement hair mechanical properties with an enhancement in the general appearance and conditioning of the hair [62]. Beyond chitosan's filmogenic properties, chitosan gelling ability in hydroalcoholic mixtures was used to formulate chitosan in gel form [61,63]. Chitosan salts in different formulations such as hair lotions, hair conditioners and hair shampoos were tested as haircare products taking advance of the filming properties, emulsifier activity and cationic surface, respectively [64,65]. Chitosan, microcrystalline chitosan and quaternized chitosan were added to shampoos and hair sprays due to its film former activity and moisturizing effect [66–68]. Glycerol chitosan was included as a component into liquid hair strengthens or hair sprays due to their improved solubility and film forming capacity [69] and glyceryl chitosan forms foam and has an emulsifying action, so it could be used directly in shampoos [70].

Alkyl-hydroxypropyl-substituted chitosan derivatives were added as a component of hairsprays in substitution of synthetic resins in order to avoid long drying time, hair sticky feeling and helmet formation [71]. Moreover, the solubility in organic media allowed the use of halogen free propellant gases and their solubility in basic media allowed their use in alkaline permanent wave agents or hair colouring agents. Quaternary chitosan derivatives with solubility in aqueous and basic media were also developed for the same purpose [72,73].

Chitosan hair fixing, hair conditioning and hair filming functions, as well as Carboxymethyl chitosan gel forming functions, are described in CosIng database. Moreover, hair conditioning and film forming functions have also been described for a large number of chitosan derivatives (Table 8).

Table 8. Chitin and chitosan derivatives with hair conditioning and film forming functions according to CosIng Database.

Hair Conditioning	Film Forming
Butoxy CH	Butoxy CH
Carboxybutyl CH	Calcium CH
Carboxymethyl CH Succinamide	Carboxybutyl CH
CHT Propylsulfonate	Carboxymethyl Caprooyl CHT
Hydrolyzed CHT	Carboxymethyl CHT
CHT Hydroxypropyltrimonium Chloride	Carboxymethyl CHT Succinamide
CHT Isostearamide Hydroxypropyltrimonium Chloride	CHT Adipate
CHT Lauramide Succinamide	CHT Ascorbate
CHT Lauroyl Glycinate	CHT Formate
CHT Rice Branamide Hydroxypropyltrimonium Chloride	CHT Glycolate
Sodium Carboxymethyl CHT Lauramide	CHT Lactate
Sodium CHT Caprylamide Hydroxypropylsulfonate	CHT PCA Palmitamide Succinamide
Sodium CHT Cocamide Hydroxypropylsulfate	CHT Propylsulfonate
Sodium CHT Cocamide Hydroxypropylsulfonate	CHT Salicylate
Sodium CHT Isostearamide Hydroxypropylsulfonate	CHT Succinamide
Sodium CHT Lauramide Hydroxypropylsulfate	Hydrolyzed CH
Sodium CHT Lauramide Hydroxypropylsulfonate	Hydroxyethyl CHTT
Sodium CHT Rice Branamide Hydroxypropylsulfonate	Hydroxypropyl CHT
Sodium CHT Stearamide Hydroxypropylsulfonate	Polyquaternium-29
Carboxymethyl CH	

CHT: Chitosan; CH: Chitin.

3.2.2. Chitosan as a Vehicle in Hair Care

Androgenetic (or pattern) alopecia is a very common disease affecting both male and female. It is a genetically determined disorder characterized by the gradual conversion of terminal hairs into indeterminate and finally into vellus hairs. The standard treatment for male and female androgenic alopecia is Minoxidil sulphate in topic treatment. Minoxidil solubility in water is low and the formulations in the market included large amounts of ethanol or propylene glycol which are highly irritating in continuous use. Moreover, due to its potent antihypertensive activity dermal exposure and a consequent systemic effect should be avoided to minimize adverse side effects. Therefore, formulations with capacity to target minoxidil to hair follicles in a sustained manner are very desirable. Minoxidil sulphate was encapsulated in chitosan microparticles and nanoparticles. Chitosan microparticles were produced by spray-drying (Medium Mw chitosan 190–310 kDa, DA: 0.15–0.25). Even after swelling, the microparticles presented an appropriate diameter for drug target to hair follicle with a sustained release of the drug [74]. Minoxidil loaded chitosan nanoparticles (Low Mw chitosan, DA 0.15–0.25) provided a sustained release avoiding dermal exposure with a minoxidil two-fold increase in the follicle when compared to a controlled drug solution [75].

3.3. Chitin, Chitosan and Their Derivatives in Skin Care

The interest of the cosmetic and cosmeceutical industry on chitin, chitosan and their derivatives rely on the unique biological and technological properties of these polymers. Their main functions in skin care are summarized in Figure 4. In this section, these functions are reviewed and a general overview of the use of these polymers as vehicles for active ingredients is given.

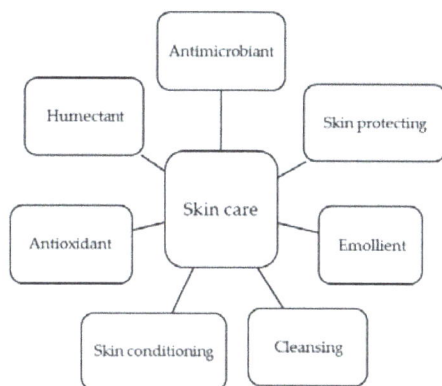

Figure 4. Main functions of chitosan derivatives in skin care.

3.3.1. Application of Chitin, Chitosan and its Derivatives in UV Protection

Skin photoaging is a premature skin-aging damage after repeated exposure to ultraviolet (UV) radiation, mainly characterized by oxidative stress and inflammatory disequilibrium, which makes skin show the typical symptoms of photoaging such as coarse wrinkling, dryness, irregular pigmentation and laxity. Moreover, UV exposition is directly related to skin cancer.

UV spectra of chitosan and chitosan films reveal absorption below 400 nm and therefore they may be used as sunscreens. A chitosan gel with an in vitro Sun Protector Factor (SPF of 0.89) was reported [76]. When comparing two chitosan films, the transmittance of mushroom chitosan film for UVA-UVB (300–250nm) was lower than that of shrimp chitosan film. This result implies the capacity of UV resistance of chitosan extract from mushroom was better than that of shrimp chitosan film. Whether this difference is due to chitosan origin or due to other properties (M_w or DA) is not possible to be determined since no proper polymer characterization was carried out [77].

Chitosan oligosaccharides demonstrated its ability to reduce skin photoaging in hairless mouse dorsal skin after UV radiation for 10 weeks. Results indicated that chitooligosaccharides were able to regulate antioxidant and anti-inflammatory status [78].

Although along this review little examples of the use of chitin in cosmetic has been shown, due to its poor solubility in most aqueous and organic solvents, several examples regarding the use of chitin as protective UV molecule can be found in the literature as nanofibers and nanocrystals.

The potential use of chitin nanofibrils (NF) and chitin nanocrystals (NC) as components of skin-protective formulations was evaluated. The application of nanofibrils and nanocrystals to skin improved the epithelial granular layer and increased granular density. Furthermore, NF and NC application reduced the production of transforming growth factor beta (TGF-β) compared to that of the control group [79].

It is well known that urocanic acid is a major ultraviolet (UV)-absorbing chromophore. Chitin nanofibers (NFs) were produced by acid hydrolysis with urocanic acid and the protective effect of these urocanic acid chitin NFs (UNFs) and acetic acid chitin NFs (ANFs) against UVB radiation was tested [80]. A lower UVB radiation-induced cutaneous erythema than in the control was observed and sunburn cells were rarely detected in the epidermis of UNFs-or ANFs coated mice after UVB irradiation. These results showed that ANFs also exhibited a protective effect against UVB. This could be explained considering the inherent anti-inflammatory and antioxidant activity of chitin nanofibers.

3.3.2. Use as Skin Cleansing, Skin Conditioning and Emollient

Cleansing products are those to help to maintain the body surface clean. Chitosan argininamide is the only chitosan derivative identified as skin cleansing according to CosIng database. Skin conditioning products are those which function is to maintain the skin in good conditions. A large number of chitosan derivatives and some chitin derivatives have been classified as ingredients with skin conditioning function (Table 9). Emollients function is to soften and smooths the skin. Some chitosan derivatives present this function whereas no chitin derivative has been described as emollient.

Table 9. Chitin and Chitosan derivatives with function as a skin conditioning and emollient in CosIng Database.

Skin Conditioning	Emollient
Calcium CHT	
Carboxybutyl CHT	
Carboxymethyl Caprooyl CHT	
Carboxymethyl CHT Succinamide	
CHT Ascorbate	
CHT Caprylamide Hydroxypropyl trimonium Chloride	
Hydrolyzed CHT	
Hydrolyzed CHT Ferulyl Linoleate	
Myristoyl/PCA CH	
Polyquaternium-29	
Sodium Carboxymethyl CHT Lauramide	
Sodium CHT Cocamide Hydroxypropyl sulfate	CHT Rice Branamide Hydroxypropyl
Sodium CH Cocamide Hydroxypropyl sulfonate	trimonium Chloride
Sodium CHT Isostearamide Hydroxypropyl sulfonate	Sodium CHT Caprylamide Hydroxypropyl sulfonate
Sodium CHT Lauramide Hydroxypropyl sulfate	Sodium CHT Cocamide Hydroxypropyl sulfate Sodium CHT
Sodium CHT Lauramide Hydroxypropyl sulfonate	Cocamide Hydroxypropyl sulfonate
Sodium CHT Rice Branamide Hydroxypropyl sulfonate	Sodium CHT Isostearamide Hydroxypropyl sulfonate
Sodium CHT Stearamide Hydroxypropyl sulfonate	Sodium CHT Rice Branamide Hydroxypropyl sulfonate
Sodium Carboxymethyl CH	Sodium CHT Stearamide Hydroxypropyl sulfonate
CHT Glycolate	CHT Caprylamide Hydroxypropyl trimonium Chloride
CHT Isostearamide Hydroxypropyl trimonium Chloride	
CHT Lauramide Hydroxypropyltrimonium Chloride	
CHT PCA Palmitamide Succinamide	
CHT Propylsulfonate	
CHT Rice Branamide Hydroxypropyl trimonium Chloride	
CHT Salicylate	
Carboxymethyl CH	
CH glycolate	
CH sulfate	
Hydrolyzed CH	
Mystoil PCA CH	

CHT: Chitosan; CH: Chitin.

3.3.3. Use as Humectant and Moisturizing Agent

Humectants are cosmetics intended to increase water content on top layers of the skin. Cationic humectants absorb to the negatively changes skin surface. As a humectant, chitosan was combined with pyrrolidone carboxylic acid (PCA) producing a film-forming humectant material [81]. Acyl and alkylated chitosan derivatives were produced to improve chitosan humectant properties [82]. Other chitin and chitosan derivatives with humectant function are summarized in Table 10. Humectant polymers activity depends on the cationic charge, molecular weight and hydrophobicity of the polymer [83]. In the case of chitosan, therefore samples with low DA and Mw may exhibit better humectant properties.

Moisturizing products increase the water content of the skin and helps keep it soft and smooth. Several chitosan samples with different M_w (1.2×10^3 to 30×10^4 kDa) were prepared by oxidative degradation with H_2O_2. The moisture-absorption and moisture-retention capacities of resulting chitosans were dependent on both the molecular weight and the degree of acetylation (DA). The moisture-absorption capacity increased as the M_w decreased. As the M_w was reduced the moisture- retention capacity first increased and then declined with a maximum moisture-retention capacity when the chitosan sample of 0.45×10^4 kDa was used. Moreover, the moisture retention and moisture-absorption increased with chitosan DD [84]. From all tested samples, the best sample was a chitosan of 0.45×10^4 kDa and a DA of 0.1

When a family of carboxymethyl derivatives were tested (6-O-CM-Chitin, 6-O-CM-Chitosan, 6-3-O-CM-Chitin, 6-3-O-CM-Chitosan and N-CM-chitosan all samples showed good moisture-retention ability and the moisture-absorption properties were quite similar to those of hyaluronic acid. Results indicated that substitution in 6-OH position is the main active site for moisture absorption and retention with a lower contribution of 3-OH and *N* position. In 6-O-CM-Chitin and 6-O-CM-Chitosan as higher was the molecular weight, the better moisture-retention ability was observed [85]. Because chitosan has lower cost, this polymer might compete with hyaluronic acid as moisturizing agent in cosmetics.

Besides chitosan derivatives, some chitin derivatives also have moisturizing and humectant properties (Table 10). Moreover, chitin–glucan complexes, that are composed of chitin and beta-glucan units covalently linked, exhibited appropriated properties regarding basic moisturization in creams and after prolonged used no erythema was observed. Furthermore, skin physiology was positively modified and as a consequence some signs of skin ageing were improved [86]. Quaternized carboxymethyl chitosan-montmorillonite nanocomposites also exhibited appropriate properties against skin ageing. Not only due to their moisturizing effect but also due to their good UV-protection ability [87].

Table 10. Chitin and chitosan derivatives with humectant or moisturizing activity included in CosIng Database.

Humectant	Moisturizing
Carboxymethyl Caprooyl CHT	CHT Rice Branamide Hydroxypropyltrimonium Cl
Carboxymethyl CHT Myristamide	Sodium CHT Cocamide Hydroxypropylsulfate
Carboxymethyl CHT Succinamide	Sodium CH Cocamide Hydroxypropylsulfonate
CHT Hydroxypropyltrimonium Cl	Sodium CHT Isostearamide Hydroxypropylsulfonate
CHT Lauroyl Glycinate	Sodium CHT Rice Branamide Hydroxypropylsulfonate
CHT PCA Palmitamide Succinamide	Sodium CHT Stearamide Hydroxypropylsulfonate
CH sulfate	CHT Caprylamide Hydroxypropyltrimonium Cl
Sodium Carboxymethyl CH	Carboxybutyl CHT
	Carboxymethyl CHT Succinamide
	CHT Propylsulfonate
	Hydrolyzed CHT
	CHT Isostearamide Hydroxypropyltrimonium Cl
	Sodium Carboxymethyl CHT Lauramide
	Sodium CHT Lauramide Hydroxypropylsulfate
	Sodium CHT Lauramide Hydroxypropylsulfonate
	Calcium CHT
	Carboxymethyl Caprooyl CHT
	CHT Ascorbate
	CHT Glycolate
	CHT PCA Palmitamide Succinamide
	CHT Salicylate
	Polyquaternium-29
	CHT Lauramide Hydroxypropyltrimonium Cl
	CHT PCA Palmitamide Succinamide
	Hydrolyzed CHT Ferulyl Linoleate
	Myristoyl/PCA CH

CHT: Chitosan; CH: Chitin.

3.3.4. Use as Surfactant, Emulsifier, Stabilizer and Viscosifier

Surfactants are very useful in the cosmetics industry being their function to lower the surface tension of cosmetics as well as to aid the even distribution of the product when used.

Chitosan yields stable water-in-oil-in-water (w/o/w) multiple emulsions without adding any surfactant. Emulsification of sunflower oil by chitosan solutions with the degree of acetylation (DA) between 0.25 and 0.05 was studied. DA only affects droplet size distribution which was unimodal at high DA and at low DA, for all tested concentrations. On the contrary at intermediate DA, distribution was unimodal only when we used the most concentrated solutions. Chitosan DA did not affect emulsion stability or ageing [88].

Chitosan can interact with anionic surfactants to form complexes which exhibit interesting surface-active properties related to surface tension and viscoelastic properties, even at very low surfactant concentrations (much lower than the CMC of pure surfactant). These properties allow their use as emulsion stabilizers in cosmetic formulations [89]. Chitosan derivatives containing fatty acid chains such as Chitosan Lauramide Succinamide, Chitosan Lauroyl Glycinate, Hydrolysed Chitosan Ferulyl Linoleate have been described as stabilizers in the CosIng database (Table 11).

Viscosity controlling ingredients are useful in the cosmetic industry to increase or reduce the viscosity of formulations. Viscosity can be modified by controlling the polymer concentration or the polymer Mw. The effect of chitosan Mw on the viscosity of a vitamin E-containing cream was studied. The apparent viscosity of vitamin E-containing creams increased with increasing chitosan Mw. At 0.5% w/w chitosan concentration, the apparent viscosity was higher than the control (2% glycerol monostearate). Apart from higher viscosity, larger storage life and better skin hydration than the control was observed [90].

N-Carboxybutyl chitosans were soluble in water and water ethanol mixtures giving more viscous solutions than the corresponding chitosans [91].

Table 11. Chitin and chitosan derivatives with surfactant, emulsifier and viscosity controlling function.

Surfactant	Emulsifier	Viscosifier
Carboxymethyl Caprooyl CHT CHT Argininamide CHT Lauramide Hydroxypropyl trimonium Cl CHT Stearamide Hydroxypropyl trimonium Cl Sodium Carboxymethyl CHT Lauramide Sodium CHT Caprylamide Hydroxypropyl sulfonate Sodium CHT Cocamide Hydroxypropyl sulphate Sodium CHT Cocamide Hydroxypropyl sulfonate Sodium CHT Lauramide Hydroxypropyl sulphate Sodium CHT Rice Branamide Hydroxypropyl sulfonate Sodium CHT Stearamide Hydroxypropyl sulfonate	CHT Isostearamide Hydroxypropyl trimonium Cl CHT Lauramide Hydroxypropyl trimonium Cl CHT PCA Palmitamide Succinamide CHT Rice Branamide Hydroxypropyl trimonium Cl CHT Stearamide Hydroxypropyl trimonium Cl Sodium Carboxymethyl CHT Lauramide Sodium CHT Caprylamide Hydroxypropyl sulfonate Sodium CHT Cocamide Hydroxypropyl sulphate Sodium CHT Cocamide Hydroxypropyl sulfonate Sodium CHT Lauramide Hydroxypropyl sulphate Sodium CHT Lauramide Hydroxypropyl sulfonate Sodium CHT Rice Branamide Hydroxypropyl sulfonate	Carboxymethyl CH Butoxy CHT Carboxybutyl CHT Carboxymethyl CHT Hydroxyethyl CHT

CHT: Chitosan; CH: Chitin.

3.3.5. Use as Antioxidant and Antimicrobial Agent

Antioxidant activity of chitin, chitosan and their derivatives can be attributed to in vitro and in vivo free radical-scavenging activities [92]. This activity is very valuable from the point of view of skin protection against oxidative damage. Moreover, their use as additive may prevent the oxidation of other active ingredients such as essential oils or vitamins. Antioxidant activity is particularly remarkable in the case of chitooligosaccharides (COS) [93,94]. COS, enzymatically produced, antioxidant activity depended on the type of enzyme used to hydrolyse chitosan (Chitosanase or lysozyme). COS produced by chitosanase have better antioxidant activity compared to COS produced with lysozyme. This behaviour seems to be related to the different distribution of acetylglucosamine and glucosamine residues in the COS backbone due to the different cleavage site of each enzyme [95]. Chitosan has been widely modified with well-known antioxidant molecules such as ferulic acid, glycolic acid, ascorbic acid and salicylic acid to improve its antioxidant activity (Table 12).

Chitosan and some derivatives (mainly cationic derivatives) exhibit antimicrobial activity against gram positive and gram negative bacteria [96]. Chitosan (low M_w, viscosity below 30 mPa·s. at pH 5 and degree of acetylation most preferable to be at least 0.10) has been added in a fine emulsion for use as preservative in home care or personal care products against microbial spoilage [97]. Moreover, due to antibiotic resistance and chitosan activity against *Propionibacterium acnes* and *Staphylococcus aureus*, this polymer was evaluated as possible antimicrobial molecule in acne vulgaris. To determine the effect of molecular weight on antibacterial activity, chitosan of low Mw (50–190 kDa), medium M_w (190–310 kDa) and high M_w (310–375 kDa) was tested against *P. acnes* and *S. aureus*. The sample with the highest M_w had greater effect against both strains, particularly against *S. aureus* [98].

Table 12. Chitosan derivatives with antioxidant and antimicrobial activity included in CosIng data base.

Antioxidant	Antimicrobial
CHT Ascorbate CHT Glycolate CHT Salicylate Hydrolysed CHT Ferulyl Linoleate	Hydrolysed CHT Ferulyl Linoleate CHT Benzamide

CHT: Chitosan.

3.3.6. Chitosan and Derivatives as Vehicle for Active Ingredients in Skin Care

Chitosan is a good polymer matrix for the delivery of active ingredients. Taking advance of the extraordinary technological properties of this polymer, chitosan has been processed into different formulations such as gels, micro and nanoparticles etc.

Several examples of sunscreen molecules included in chitosan formulations can be found in the literature. Phenylbenzimidazole sulphonic acid (PBSA) is a relatively photostable UV-B filter that was successfully encapsulated into chitosan microparticles with an encapsulation yield higher than 70%. The UV screening effect of a chitosan gel was increased by incorporating the microparticles containing

PBSA into the gel [76]. In another example, benzophenone-3 was loaded into polycaprolactone nanocapsules coated with chitosan and the release was tested. Penetration profiles showed that a higher amount of benzophenone-3 remained at the skin surface and a lower amount was found in the receptor compartment after the application of the formulation containing chitosan-coated nanocapsules compared to a formulation containing its free form [99]. An oil-in-water photoprotective and antioxidant nanoemulsion (NE) containing chitosan was developed with the aim of protecting skin against ultraviolet radiation while toxic sunscreens substances included in the formulation are retained in the skin. Several molecules were tested: benzophenone-3, diethylamino hydroxybenzoyl hexylbenzoate, octocrylene and octylmethoxycinnamate, pomegranate antioxidant extract and chitosan. Formulations containing chitosan were stable for at least 6 months, were photostable when irradiated in a solar simulator and effective. Moreover, chitosan promoted retention of the formulation in the epidermis, thus increasing formulation safety [100]. A multifunctional hydroxyapatite-chitosan gel that works as an antibacterial sunscreen agent for skin care was developed. This gel conferred protection against UV-radiation and antibacterial activity [101]. Chitin Nanofibril-Hyaluronan nanoparticles (CN-HA) has the ability of easily loading active ingredients, facilitating penetration through the skin layers and increasing their effectiveness and safety as an anti-aging agent. CN-HA nanoparticles were evaluated in vitro measuring their antioxidant capacity, anti-collagenase activity and metalloproteinase and pro-inflammatory release. The efficacy was also shown in vivo by a double-blind vehicle-controlled study for 60 days on 60 women affected by photo-aging [102]. Finally, mycosporines and mycosporine-like amino proteins were grafted to chitosan to produce multifunctional materials based only in natural components. These materials were biocompatible, photoresistant and thermoresistant and exhibited a highly efficient absorption of both UV-A and UV-B radiations [103].

Retinols have shown promising results for photoaging and depigmentation at high concentrations which are higher than those found in OTC products. For acne treatment, the evidence is more limited and more studies are needed [104]. Formulations of retinols into appropriate vehicles may improve control delivery, stability, bioavailability and potency. Several chitosan formulations have been proposed with this aim. Retinol-encapsulated chitosan nanoparticles (100 nm) were produced using a water soluble chitosan (18 kDa, DA 0.04), these nanoparticles protect retinol against degradation [105]. Retinol was encapsulated in zein nanoparticles (300 nm) that were coated with chitosan. Chitosan Mw affected the particle size and polydispersity which increased as the Mw increased. On the contrary, a slight reduction of encapsulation efficiency was observed when Mw increased. This system improves retinol photostability [106]. In another example, retinol was encapsulated into succinic-chitosan nanoparticles by means of retinol complexation with the chitosan derivatives through H-bonding. The antioxidant activity of the encapsulated retinol was significantly greater than pure retinol [107].

Chitin Nanofibrils/bio-lignin non-woven tissues were electrospun with different antiaging ingredients. The beauty masks were controlled for their safeness and effectiveness both in vitro on keratinocyte and fibroblast cultures and in vivo on 30 volunteer women showing signs of premature skin aging (photoaging) for a period of 30 days. Results showed that the beauty masks that were not only effective on the aged and sensitive skin but also very safe and stable for a long period of time because they were free of water [108].

Polyphenols with a strong antioxidant activity have also been encapsulated in nanoparticles and mircrospheres produced by ionic gelation and by spray-drying, respectively [109].

Chitosan is also a useful matrix to encapsulate essential oils. For instance, Mentha Piperita essential oil was encapsulated in chitosan beads Two chitosan samples (High Mw chitosan 2.000 kDa, DA < 0.25 and medium M_w chitosan 750 kDa, DA 0.15–0.25) were used as polymeric matrix to encapsulate the oil. Chitosan glycolic solutions (High Mw, Low Mw and mixture of both chitosans) containing the essential oil were dropped to NaOH or TPP solutions. NaOH beads showed much better swelling properties and greater release of the essential oil. Encapsulation efficiency was not affected by chitosan sample and during beads breakage assay, medium Mw chitosan-based particles showed better behaviour [110].

4. Functional Characterization of Chitin and Chitosan in Cosmetics

There is a large variety of chitin and chitosan samples differing in polymer molecular weight and its distribution, deacetylation degree, acetyl distribution and crystallinity among other physical-chemical properties. Moreover, a myriad of chitin and chitosan derivatives have been produced from different chitin and chitosan samples by taking advance of the simplicity of chemical modification of these polymers. The wide range of applications in different fields is sustained by this structural variability. The question that arises at this point is which sample is the most appropriate for a specific application. Therefore, the proper sample characterization is an essential issue that needs to be carefully considered. Ambiguities in research studies regarding different properties and functions of these polymers are frequently found in the literature, these ambiguities coming, at least in part, from inadequate data on the polymer characteristics. During the elaboration of this review, we have frequently found research work in which the polymer samples were not characterized or generic data were given, such as deacetylation degree and molecular weight in a range rather than a specific value. It is also worth mentioning that accurate determination of the polymer characteristics is difficult since most of the proposed techniques are easy and simple but not accurate [111].

Keeping in mind these considerations, in Table 13, some general recommendations are included to assist researchers and industrials to select the most appropriate polymer for a specific purpose.

It is worth mentioning that differences can be observed between in vitro and in vivo assays due to the more complex environment found in in vivo assays. For instance, no effect of chitosan properties in its mouth pH buffering activity was observed in vitro while in vivo a specific chitosan performance a better behaviour as described in Section 3.3.1.

It has been reported that antibacterial activity depends on several parameters such as molecular weight, the degree of acetylation, the type of substituents and the type of bacterium. The exact mechanism of the antimicrobial action of chitin, chitosan and their derivatives is still unknown. Both low and high M_W polymers exhibit antimicrobial activity but different mechanisms have been proposed for each family of chitosan samples. High molecular weight polymers seem to form a film that protects cells against nutrient transport through the microbial cell membrane. Low molecular weight chitosan derivatives are water soluble and can better incorporate the active molecule into the cell. Gram-negative bacteria have an anionic bacterial surface on which cationic chitosan derivatives may interact electrostatically. Gram-positive bacteria seem to be inhibited by the binding of lower molecular weight chitosan derivatives to DNA or RNA. Chitosan nanoparticles exhibit an increase in loading capacity and efficacy [112]. Antibacterial activity of chitosan against oral pathogens seems to be mainly affected by chitosan molecular weight with better activity as the M_W increase. Antimicrobial activity against skin pathogens showed a similar trend although the evidence is lower since little research regarding chitosan characteristics has been carried out.

Regarding chitosan ability to avoid bacterial adsorption to hydroxyapatite, the chitosan samples described in the examined papers were not appropriately characterized so it is not possible to conclude which are the properties of the best chitosan or derivative. Authors described their chitosan samples and derivatives as low Mw polymers; but as no clear definition exits for low, medium and high molecular weight chitosan it is not possible to evaluate the polymer size. Moreover, it cannot be dismissed that medium or high molecular weight samples avoid bacterial adsorption since they had not been tested.

Regarding the use of chitosan as sunscreen, it seems that this activity depends on the chitosan sample but from the reported data it is not possible to determine which chitosan property is involved in this activity.

Table 13. Some general recommendations on chitin, chitosan and derivatives in cosmetics and cosmeceutical applications.

Field	Property	Effect of Physico-Chemical Properties
Oral healthcare	ability to inhibit pH fall in dental plaque	In vitro: No effect of M_w or DA in buffering capacity. In vivo: best sample 3000 Da, DA: 0.05
Oral healthcare	Antimicrobial Activity	Depends on M_w, DA and bacterial strain. Very active against *S. mutants* and other oral *streptococci*.
Oral healthcare	Bacterial adsorption inhibition	Dual-species inhibition (High M_w :624 kDa, DA < 0.25). Inhibition *S. sobrinus* High Mw, optimum DA 0.40–0.50
Product conservation	Antioxidant	Best activity COS. Chitosanase- COS are preferred over lysozyme-COS
Product conservation	Antimicrobial Activity	Depends on M_w, DA, bacterial strain
Skin care	Antioxidant	Best activity COS Chitosanase-COS are preferred over lysozyme-COS
Skin care (Acne treatment)	Antimicrobial activity	High M_w
Skin/hair care	Humectant	Low DA
Skin/hair care	Moisturizing agent	High M_w Carboxylmethyl derivatives
Skin care	Sunscreen	Possible effect of the chitosan source

Chitosanase-COS: COS produced by chitosanase; Lysozyme-COS: COS produced by lysozyme, DA: Acetylation degree.

5. Conclusions

Chitin, chitosan and its derivatives exhibit many different relevant properties as active ingredient in dental, skin, hair and nails care. Moreover, they have optimal properties to vehiculate active ingredients for the cosmetics and cosmeceutical industry.

These valuable properties are strongly related to the polymer physico-chemical characteristics and therefore an accurate polymer characterization is needed to determine which characteristics are more relevant for a specific application.

It is very remarkable that several properties can be found in a single chitosan derivative, for instance, Carboxymethyl Caprooyl Chitosan (Chitosonic® acid) is a water-soluble derivative with high HLB value; that can form a nano-network structures at higher concentration than 0.5% and can self-assemble into a nanosphere structure at lower concentration than 0.2%. Besides, it has potent antimicrobial activity against gram-positive bacteria, gram-negative bacteria and fungus; moderate DPPH radical scavenging activity and exhibits good hydration activity for absorbing and retaining water molecules [113].

This multi-functional behaviour is quite frequent and implies that each polymer must be characterized not only in terms of its physico-chemical properties but also in terms of functional properties.

An accurate polymer characterization will boost our knowledge in these polymers and promote their use in the industry.

Acknowledgments: Authors acknowledge financial support from Spanish Ministerio de Economia y Competividad (MAT-2015 65184–C2-1-R) and InFiQuS S.L.

Conflicts of Interest: The authors declare no conflict of interest. The founding sponsors had no role in the design of the study; in the collection, analyses, or interpretation of data; in the writing of the manuscript and in the decision to publish the results."

References

1. European Commission. *Glossary and Acronyms Related to Cosmetics Legislation*; European Commission: Brussels, Belgium, 2015.
2. Reed, E. The definition of "cosmeceuticals". *J. Soc. Cosm. Chem.* **1962**, *13*, 103–110.

3. Morganti, P. Reflections on cosmetics, cosmeceuticals, and nutraceuticals. *Clin. Dermatol.* **2008**, *26*, 318–320. [CrossRef] [PubMed]

4. Morganti, P.; Sud, M. Cosmeceuticals. *Clin. Dermatol.* **2008**, *26*, 317. [CrossRef] [PubMed]

5. Gao, X.-H.; Zhang, L.; Wei, H.; Chen, H.-D. Efficacy and safety of innovative cosmeceuticals. *Clin. Dermatol.* **2008**, *26*, 367–374. [CrossRef] [PubMed]

6. Vermeer, B.J.; Gilchrest, B.A.; Vermeer; Friedel, S.L. Cosmeceuticals: A proposal for rational definition, evaluation, and regulation. *Arch. Dermatol.* **1996**, *132*, 337–340. [CrossRef] [PubMed]

7. Kligman, A.M. Why cosmeceuticals? *Cosmet. Toilet.* **1993**, *108*, 37–38.

8. Saint-Leger, D. "Cosmeceuticals". Of men, science and laws.. *Int. J. Cosmet. Sci.* **2012**, *34*, 396–401. [CrossRef] [PubMed]

9. Kim, S.K.; Wijesekara, I. Cosmeceuticals from marine resources. Prospects and commercial trends. In *Marine Cosmeceuticals. Trends and Prospects*; Kim, S.K., Ed.; CRC Press, Taylor & Francis: Boca Raton, FL, USA, 2012; pp. 1–9.

10. Senevirathne, W.S.M.; Kim, S.K. 23-Cosmeceuticals from algae. In *Functional Ingredients from Algae for Foods and Nutraceuticals*; Domínguez, H., Ed.; Woodhead Publishing: Cambridge, UK, 2013; pp. 694–713. ISBN 978-0-85709-512-1.

11. Peniche, C.; Argüelles-Monal, W.; Goycoolea, F.M. Chapter 25—Chitin and Chitosan: Major Sources, Properties and Applications A2—Belgacem, Mohamed Naceur. In *Monomers, Polymers and Composites from Renewable Resources*; Gandini, A., Ed.; Elsevier: Amsterdam, The Netherland, 2008; pp. 517–542. ISBN 978-0-08-045316-3.

12. Kinney, J.S.; Morelli, T.; Braun, T.; Ramseie, C.A.; Herr, A.E.; Sugai, J.V.; Shelburne, C.E.; Rayburn, L.A.; Singh, A.K.; Giannobile, W.V. Saliva/Pathogen Biomarker Signatures and Periodontal Disease Progression. *J. Dent. Res.* **2011**, *90*, 752–758. [CrossRef] [PubMed]

13. Dhadse, P.; Gattani, D.; Mishra, R. The link between periodontal disease and cardiovascular disease: How far we have come in last two decades? *J. Indian Soc. Periodontol.* **2010**, *14*, 148–154. [CrossRef] [PubMed]

14. Lee, J.H.; Oh, J.Y.; Youk, T.M.; Jeong, S.N.; Kim, Y.T.; Choi, S.H. Association between periodontal disease and non-communicable diseases: A 12-year longitudinal health-examinee cohort study in South Korea. *Medicine* **2017**, *96*, e7398. [CrossRef] [PubMed]

15. Fejerskov, O.; Kidd, E.A.M.; Nyvad, B.; Baelum, V. Defining the disease: An introduction. In *Dental Caries. The Disease and Its Clinical Management*, 2nd ed.; Fejerskov, O., Kidd, E., Eds.; Blackwell Munksgaard Ltd.: Oxford, UK, 2003.

16. Imfeld, T. Dental erosion Definition, classification and links. *Eur. J. Oral Sci.* **1996**, *104*, 151–155. [CrossRef] [PubMed]

17. Bender, A.E. Body Surface Area. In *A Dictionary of Food and Nutrition*; DA, B., Ed.; Oxford University Press: New York, NY, USA, 1995.

18. Shimizu, H. Chapter 1: Structure and Function of the Skin. In *Shimizu's Textbook of Dermatology*; Nakayama Shoten Publishers; Hokkaido University Press: Hokkaido, Japan, 2007.

19. Agrawal, S.; Adholeya, A.; Barrow, C.J.; Deshmukh, S.K. Marine fungi: An untapped bioresource for future cosmeceuticals. *Phytochem. Lett.* **2018**, *23*, 15–20. [CrossRef]

20. Acosta, N.; Sánchez, E.; Calderón, L.; Cordoba-Diaz, M.; Cordoba-Diaz, D.; Dom, S.; Heras, Á. Physical Stability Studies of Semi-Solid Formulations from Natural Compounds Loaded with Chitosan Microspheres. *Mar. Drugs* **2015**, *13*, 5901–5919. [CrossRef] [PubMed]

21. Commission, E. Regulation (EC) No 1223/2009 of the European Parliament and of the Council of on cosmetic products. *Off. J. Eur. Union* **2009**, *1223*, 342–359.

22. Aranaz, I.; Mengíbar, M.; Harris, R.; Paños, I.; Miralles, B.; Acosta, N.; Galed, G.; Heras, Á. Functional characterization of chitin and chitosan. *Curr. Chem. Biol.* **2009**, *3*. [CrossRef]

23. Wieckiewicz, M.; Boening, K.W.; Grychowska, N.; Paradowska-Stolarz, A. Clinical Application of Chitosan in Dental Specialities. *Mini-Rev. Med. Chem.* **2017**, *17*, 401–409. [CrossRef] [PubMed]

24. Shibasaki, K.; Sano, H.; Matsukubo, T.; Takaesu, Y. The influences of the buffer capacity of various substances on pH changes in dental plaque. *Bull. Tokyo Dent. Coll.* **1994**, *35*, 27–32. [PubMed]

25. Wang, Q.Z.; Chen, X.G.; Liu, N.; Wang, S.X.; Liu, C.S.; Meng, X.H.; Liu, C.G. Protonation constants of chitosan with different molecular weight and degree of deacetylation. *Carbohydr. Polym.* **2006**, *65*, 194–201. [CrossRef]

26. Shibasaki, K.; Sano, H.; Matsukubo, T.; Takaesu, Y. Effects of low molecular chitosan on pH changes in human dental plaque. *Bull. Tokyo Dent. Coll.* **1994**, *35*, 33–39. [PubMed]

27. Ji, Q.X.; Zhong, D.Y.; Lv, R.; Zhang, W.Q.; Deng, J.; Chen, X.G. In vitro evaluation of the biomedical properties of chitosan and quaternized chitosan for dental applications. *Carbohydr. Res.* **2009**. [CrossRef] [PubMed]

28. İkinci, G.; Şenel, S.; Akıncıbay, H.; Kaş, S.; Erciş, S.; Wilson, C.G.; Hıncal, A.A. Effect of chitosan on a periodontal pathogen Porphyromonas gingivalis. *Int. J. Pharm.* **2002**, *235*, 121–127. [CrossRef]

29. Costa, E.M.; Silva, S.; Pina, C.; Tavaria, F.K.; Pintado, M.M. Evaluation and insights into chitosan antimicrobial activity against anaerobic oral pathogens. *Anaerobe* **2012**, *18*, 305–309. [CrossRef] [PubMed]

30. Costa, E.M.; Silva, S.; Veiga, M.; Tavaria, F.K.; Manuela, M. A review of chitosan´s effect on oral bio films: Perspectives from the tube to the mouth. *J. Oral Biosci.* **2017**, *59*, 1–6. [CrossRef]

31. Aliasghari, A.; Rabbani Khorasgani, M.; Vaezifar, S.; Rahimi, F.; Younesi, H.; Khoroushi, M. Evaluation of antibacterial efficiency of chitosan and chitosan nanoparticles on cariogenic streptococci: An in vitro study. *Iran J. Microbiol.* **2016**, *8*, 93–100. [PubMed]

32. Qi, L.; Xu, Z.; Jiang, X.; Hu, C.; Zou, X. Preparation and antibacterial activity of chitosan nanoparticles. *Carbohydr. Res.* **2004**, *339*, 2693–2700. [CrossRef] [PubMed]

33. Arancibia, R.; Maturana, C.; Silva, D.; Tobar, N.; Tapia, C.; Salazar, J.C.; Martínez, J.; Smith, P.C. Effects of Chitosan Particles in Periodontal Pathogens and Gingival Fibroblasts. *J. Dent. Res.* **2013**, *92*, 740–745. [CrossRef] [PubMed]

34. Tarsi, R.; Corbin, B.; Pruzzo, C.; Muzzarelli, R.A. Effect of low-molecular-weight chitosans on the adhesive properties of oral streptococci. *Oral Microbiol. Immunol.* **1998**, 217–224. [CrossRef]

35. Ganguly, A.; Ian, C.K.; Sheshala, R.; Sahu, P.S.; Al-Waeli, H.; Meka, V.S. Application of diverse natural polymers in the design of oral gels for the treatment of periodontal diseases. *J. Mater. Sci. Mater. Med.* **2017**, *28*, 39. [CrossRef] [PubMed]

36. Silva, D.; Arancibia, R.; Tapia, C.; Acuña-Rougier, C.; Diaz-Dosque, M.; Cáceres, M.; Martínez, J.; Smith, P.C. Chitosan and platelet-derived growth factor synergistically stimulate cell proliferation in gingival fibroblasts. *J. Periodontal Res.* **2013**, *48*, 677–686. [CrossRef] [PubMed]

37. Tarsi, R.; Muzzarelli, R.A.; Guzmán, C.A.; Pruzzo, C. Inhibition of Streptococcus mutans adsorption to hydroxyapatite by low-molecular-weight chitosans. *J. Dent. Res.* **1997**, *76*, 665–672. [CrossRef] [PubMed]

38. Costa, E.M.; Silva, S.; Tavaria, F.K.; Pintado, M.M. Study of the effects of chitosan upon Streptococcus mutans adherence and biofilm formation. *Anaerobe* **2013**, *20*, 27–31. [CrossRef] [PubMed]

39. Sano, H.; Shibasaki, K.; Matsukubo, T.; Takaesu, Y. Effect of molecular mass and degree of deacetylation of chitosan on adsorption of Streptococcus sobrinus 6715 to saliva treated hydroxyapatite. *Bull. Tokyo Dent. Coll.* **2002**, *43*, 75–82. [CrossRef] [PubMed]

40. Achmad, H.; Ramadhany, Y.F. Effectiveness of chitosan tooth paste from white shrimp (Litopenaeus vannamei) to reduce number of Streptococcus mutans in the case of early childhood Caries. *J. Int. Dent. Med. Res.* **2017**, *10*, 358–363.

41. Sano, H.; Shibasaki, K.-I.; Matsukubo, T.; Takaesu, Y. Effect of chitosan rinsing or reduction of dental plaque formation. *Bull. Tokyo Dent. Coll. J.* **2003**. [CrossRef]

42. Costa, E.M.; Silva, S.; Madureira, A.R.; Cardelle-Cobas, A.; Tavaria, F.K.; Pintado, M.M. A comprehensive study into the impact of a chitosan mouthwash upon oral microorganism's biofilm formation in vitro. *Carbohydr. Polym.* **2014**, *101*, 1081–1086. [CrossRef] [PubMed]

43. Costa, E.M.; Silva, S.; Costa, M.R.; Pereira, M.; Campos, D.A.; Odila, J.; Madureira, A.R.; Cardelle-Cobas, A.; Tavaria, F.K.; Rodrigues, A.S.; Pintado, M.M. Chitosan mouthwash: Toxicity and in vivo validation. *Carbohydr. Polym.* **2014**, *111*, 385–392. [CrossRef] [PubMed]

44. Emilson, C.G. Susceptibility of various microorganisms to chlorhexidine. *Scand. J. Dent. Res.* **1977**, *85*, 255–265. [CrossRef] [PubMed]

45. Decker, E.M.; von Ohle, C.; Weiger, R.; Wiech, I.; Brecx, M. A synergistic chlorhexidine/chitosan combination for improved antiplaque strategies. *J. Periodontal Res.* **2005**, *40*, 373–377. [CrossRef] [PubMed]

46. Mohire, N.C.; Yadav, A. V Chitosan-based polyherbal toothpaste: As novel oral hygiene product. *Indian J. Dent. Res.* **2010**, *21*, 380–384. [CrossRef] [PubMed]

47. Hayashi, Y.; Ohara, N.; Ganno, T.; Yamaguchi, K.; Ishizaki, T.; Nakamura, T.; Sato, M. Chewing chitosan-containing gum effectively inhibits the growth of cariogenic bacteria. *Arch. Oral Biol.* **2007**, *52*, 290–294. [CrossRef] [PubMed]

48. Hayashi, Y.; Ohara, N.; Ganno, T.; Ishizaki, H.; Yanagiguchi, K. Chitosan-containing gum chewing accelerates antibacterial effect with an increase in salivary secretion. *J. Dent.* **2007**, *35*, 871–874. [CrossRef] [PubMed]

Polymers **2018**, *10*, 213

49. Sano, H.; Matsukubo, T.; Shibasaki, K.; Itoi, H.; Takaesu, Y. Inhibition of adsorption of oral streptococci to saliva treated hydroxyapatite by chitin derivatives. *Bull. Tokyo Dent. Coll.* **1991**, *32*, 9–17. [PubMed]

50. Sano, H.; Shibasaki, K.; Matsukubo, T.; Takaesu, Y. Effect of rinsing with phosphorylated chitosan on four-day plaque regrowth. *Bull. Tokyo Dent. Coll.* **2001**, *42*, 251–256. [CrossRef] [PubMed]

51. Chen, C.-Y.U.; Chung, Y.-C. Antibacterial effect of water-soluble chitosan on representative dental pathogens Streptococcus mutans and Lactobacilli brevis. *J. Appl. Oral Sci.* **2012**, *20*, 620–627. [CrossRef] [PubMed]

52. Bae, K.; Jun, E.J.; Lee, S.M.; Paik, D.I.; Kim, J.B. Effect of water-soluble reduced chitosan on Streptococcus mutans, plaque regrowth and biofilm vitality. *Clin. Oral Invest.* **2006**, *10*, 102–107. [CrossRef] [PubMed]

53. Ganss, C.; Schulze, K.; Schlueter, N. Toothpaste and Erosion. In *Toothpastes. Monogr Oral Sci. Basel*; van Loveren, C., Ed.; Karger: Amsterdam, The Netherland, 2013; Volume 23, pp. 88–99.

54. Ozalp, S.; Tulunoglu, O. SEM–EDX analysis of brushing abrasion of chitosan and propolis based toothpastes on sound and artificial carious primary enamel surfaces. *Int. J. Paediatr. Dent.* **2014**, *24*, 349–357. [CrossRef] [PubMed]

55. Schlueter, N.; Klimek, J.; Ganss, C. Effect of a chitosan additive to a Sn^{2+}-containing toothpaste on its anti-erosive/anti-abrasive efficacy—A controlled randomised in situ trial. *Clin. Oral Investig.* **2014**, *18*, 107–115. [CrossRef] [PubMed]

56. Keegan, G.M.; Smart, J.D.; Ingram, M.J.; Barnes, L.-M.; Burnett, G.R.; Rees, G.D. Chitosan microparticles for the controlled delivery of fluoride. *J. Dent.* **2012**, *40*, 229–240. [CrossRef] [PubMed]

57. Sanko, N.; Escudero, C.; Sediqi, N.; Smistad, G.; Hiorth, M. Fluoride loaded polymeric nanoparticles for dental delivery. *Eur. J. Pharm. Sci.* **2017**, *104*, 326–334. [CrossRef]

58. Wassel, M.O.; Khattab, M.A. Antibacterial activity against Streptococcus mutans and inhibition of bacterial induced enamel demineralization of propolis, miswak, and chitosan nanoparticles based dental varnishes. *J. Adv. Res.* **2017**, *8*, 387–392. [CrossRef] [PubMed]

59. Giunchedi, P.; Juliano, C.; Gavini, E.; Cossu, M.; Sorrenti, M. Formulation and in vivo evaluation of chlorhexidine buccal tablets prepared using drug-loaded chitosan microspheres. *Eur. J. Pharm. Biopharm.* **2002**, *53*, 233–239. [CrossRef]

60. Peng, P.C.; Hsieh, C.M.; Chen, C.P.; Tsai, T.; Chen, C.T. Assessment of photodynamic inactivation against periodontal bacteria mediated by a chitosan hydrogel in a 3D gingival model. *Int. J. Mol. Sci.* **2016**, *17*. [CrossRef] [PubMed]

61. Dutta, P.K.; Dutta, J.; Tripathi, V.S. Chitin and chitosan: Chemistry, properties and applications. *J. Sci. Ind. Res.* **2004**, *63*, 20–31. [CrossRef]

62. Sionkowska, A.; Kaczmarek, B.; Michalska, M.; Lewandowska, K.; Grabska, S. Preparation and characterization of collagen/chitosan/hyaluronic acid thin films for application in hair care cosmetics. *Pure Appl. Chem.* **2017**, *89*, 1829–1839. [CrossRef]

63. Brigham, C.J. Chitin and Chitosan: Sustainable, Medically Relevant Biomaterials. *Int. J. Biotechnol. Wellness Ind.* **2017**, *6*, 41–47. [CrossRef]

64. Gross, P.; Konrad, E.; Mager, H. Hair Shampoo and Conditioning Lotion. U.S. Patent 4202.881, 13 May 1980.

65. Gross, P.; Konrad, E.; Mager, H. Hair Setting Lotion Containing a Chitosan Derivative. US4134412A, 16 January 1979.

66. Xiuzhen, S. Chitosan Hair Spray. CN1082883 A, 2 March 1994.

67. Derks, F.J.M.; Foster, S.; Lochhead, R.; Maini, A.; Weber, D. Shampoo Preparations. US9339449 B2, 17 May 2016.

68. Beumer, R.; Derks, F.; Mendrok, C. Hair Care Compositions. US20140348771 A1, 2014.

69. Lang, G.; Wendel, H.; Konrad, E. Cosmetic Composition Based Upon Chitosan Derivatives, New Chitosan Derivatives as well as Processes for the Production Thereof. US4528283 A, 1985.

70. McClements, D.J.; Gumus, C.E. Natural emulsifiers—Biosurfactants, phospholipids, biopolymers, and colloidal particles: Molecular and physicochemical basis of functional performance. *Adv. Colloid Interface Sci.* **2016**, *234*, 3–26. [CrossRef] [PubMed]

71. Lang, G.; Maresch, G.; Lenz, H.R.; Konrad, E.; Breuer, L.; Hoch, D. Cosmetic Compositions on the Basis of Alkyl-Hydroxypropyl-Substituted Chitosan Derivatives, New Chitosan Derivatives and Processes for the Production Thereof. US4845204 A, 4 July 1989.

72. Lang, G.; Wendel, H.; Konrad, E. Cosmetic Agent on the Basis of Quaternary Chitosan Derivatives, Novel Quaternary Chitosan Derivatives as well as Processes for Making Same. US4822598 A, 18 April 1989.

73. Lang, G.; Wendel, H. Macromolecular, Surface-Active, Quaternary, *N*-Substituted Chitosan Derivatives as well as Cosmetic Composition Based on These New Chitosan Derivatives. US4976952 A, 11 December 1990.

74. Gelfuso, G.M.; Gratieri, T.; Simão, P.S.; de Freitas, L.A.P.; Lopez, R.F.V. Chitosan microparticles for sustaining the topical delivery of minoxidil sulphate. *J. Microencapsul.* **2011**, *28*, 650–658. [CrossRef] [PubMed]

75. Matos, B.N.; Reis, T.A.; Gratieri, T.; Gelfuso, G.M. Chitosan nanoparticles for targeting and sustaining minoxidil sulphate delivery to hair follicles. *Int. J. Biol. Macromol.* **2015**, *75*, 225–229. [CrossRef] [PubMed]

76. Gomaa, Y.A.; El-Khordagui, L.K.; Boraei, N.A.; Darwish, I.A. Chitosan microparticles incorporating a hydrophilic sunscreen agent. *Carbohydr. Polym.* **2010**, *81*, 234–242. [CrossRef]

77. Chou, K.F.; Wang, L.F. Addition of Titanium Dioxide and Sources Effects on UV Transmittance and Hydrophilicity of Chitosan Film. In Proceedings of the 5th International Conference on Biomedical Engineering and Technology; IACSIT Press: Seoul, Korea, 2015; Volume 81, pp. 20–24. [CrossRef]

78. Kong, S.-Z.; Li, D.-D.; Luo, H.; Li, W.-J.; Huang, Y.-M.; Li, J.-C.; Hu, Z.; Huang, N.; Guo, M.-H.; Chen, Y.; Li, S.-D. Anti-photoaging effects of chitosan oligosaccharide in ultraviolet-irradiated hairless mouse skin. *Exp. Gerontol.* **2018**, *103*, 27–34. [CrossRef] [PubMed]

79. Ito, I.; Osaki, T.; Ifuku, S.; Saimoto, H.; Takamori, Y.; Kurozumi, S.; Imagawa, T.; Azuma, K.; Tsuka, T.; Okamoto, Y.; Minami, S. Evaluation of the effects of chitin nanofibrils on skin function using skin models. *Carbohydr. Polym.* **2014**, *101*, 464–470. [CrossRef] [PubMed]

80. Ito, I.; Yoneda, T.; Omura, Y.; Osaki, T.; Ifuku, S.; Saimoto, H.; Azuma, K.; Imagawa, T.; Tsuka, T.; Murahata, Y.; Ito, N.; Okamoto, Y.; Minami, S. Protective Effect of Chitin Urocanate Nanofibers against Ultraviolet Radiation. *Mar. Drugs* **2015**, *13*, 7076. [CrossRef] [PubMed]

81. Zocchi, G. Skin-feel agents. In *Handbook of Cosmetic Science and Technology*; Barel, A.O., Paye, M., Maibach, H., Eds.; Marcel Dekker, Inc.: New York, NY, USA, 2001.

82. Zhang, J.; Qi, Y.; Liu, Y.; Zhang, D.; Hu, X.; Li, Z. Preparation of Chitosan Derivative Cosmetic Humectant. CN1253769 A, 24 May 2000.

83. Zheng, Y.; Lai, X.; Ipsen, H.; Larsen, J.N.; Løwenstein, H.; Søndergaard, I.; Jacobsen, S. Structural changes of protein antigens due to adsorption onto and release from aluminium hydroxide using FTIR-ATR. *Spectroscopy* **2007**, *21*, 211–226. [CrossRef]

84. Qin, C.; Du, Y.; Xiao, L.; Liu, Y.; Yu, H. Moisture retention and antibacterial activity of modified chitosan by hydrogen peroxide. *J. Appl. Polym. Sci.* **2002**, *86*, 1724–1730. [CrossRef]

85. Chen, L.; Du, Y.; Wu, H.; Xiao, L. Relationship between molecular structure and moisture-retention ability of carboxymethyl chitin and chitosan. *J. Appl. Polym. Sci.* **2002**, *83*, 1233–1241. [CrossRef]

86. Gautier, S.; Xhauflaire-Uhoda, E.; Gonry, P.; Piérard, G.E. Chitin-glucan, a natural cell scaffold for skin moisturization and rejuvenation. *Int. J. Cosmet. Sci.* **2008**, *30*, 459–469. [CrossRef] [PubMed]

87. Chen, K.; Guo, B.; Luo, J. Quaternized carboxymethyl chitosan/organic montmorillonite nanocomposite as a novel cosmetic ingredient against skin aging. *Carbohydr. Polym.* **2017**, *173*, 100–106. [CrossRef] [PubMed]

88. Rodríguez, M.S.; Albertengo, L.A.; Agulló, E. Emulsification capacity of chitosan. *Carbohydr. Polym.* **2002**, *48*, 271–276. [CrossRef]

89. Desbrieres, J.; Bousquet, C.; Babak, V. Surfactant-chitosan interactions and application to emulsion stabilization. *Cell. Chem. Technol.* **2010**, *44*, 395–406.

90. Chen, R.H.; Heh, R.S. Skin hydration effects, physico-chemical properties and vitamin E release ratio of vital moisture creams containing water-soluble chitosans. *Int. J. Cosmet. Sci.* **2000**, *22*, 349–360. [CrossRef]

91. Muzzarelli, R.; Weckx, M.; Filippini, O.; Lough, C. Characteristic properties of *N*-Carboxybutyl chitosan. *Carbohydr. Polym.* **1989**, *11*, 307–320. [CrossRef]

92. Ngo, D.-H.; Kim, S.-K. Chapter Two—Antioxidant Effects of Chitin, Chitosan, and Their Derivatives. In *Advances in Food and Nutrition Research*; Kim, S.-K., Ed.; Academic Press: Cambridge, MA, USA, 2014; Volume 73, pp. 15–31. ISBN 1043-4526.

93. Sun, T.; Zhou, D.; Xie, J.; Mao, F. Preparation of chitosan oligomers and their antioxidant activity. *Eur. Food Res. Technol.* **2007**, *225*, 451–456. [CrossRef]

94. Sánchez, Á.; Mengíbar, M.; Rivera-Rodríguez, G.; Moerchbacher, B.; Acosta, N.; Heras, A. The effect of preparation processes on the physicochemical characteristics and antibacterial activity of chitooligosaccharides. *Carbohydr. Polym.* **2017**, *157*, 251–257. [CrossRef] [PubMed]

95. Mengíbar, M.; Mateos-Aparicio, I.; Miralles, B.; Heras, Á. Influence of the physico-chemical characteristics of chito-oligosaccharides (COS) on antioxidant activity. *Carbohydr. Polym.* **2013**, *97*, 776–782. [CrossRef] [PubMed]

96. Jarmila, V.; Eva, V. Chitosan Derivatives with Antimicrobial, Antitumour and Antioxidant Activities—A Review. *Curr. Pharm. Des.* **2011**, *17*, 3596–3607. [CrossRef] [PubMed]

97. Fankhauser, P.; Müller, B.W. Chitosan emulsion formulation. EP1190702 A1, 2002.

98. Champer, J.; Patel, J.; Fernando, N.; Salehi, E.; Wong, V.; Kim, J. Chitosan against cutaneous pathogens. *AMB Express* **2013**, *3*, 37. [CrossRef] [PubMed]

99. Siqueira, N.M.; Contri, R.V.; Paese, K.; Beck, R.C.R.; Pohlmann, A.R.; Guterres, S.S. Innovative Sunscreen Formulation Based on Benzophenone-3-Loaded Chitosan-Coated Polymeric Nanocapsules. *Skin Pharmacol. Physiol.* **2011**, *24*, 166–174. [CrossRef] [PubMed]

100. Cerqueira-Coutinho, C.; Santos-Oliveira, R.; dos Santos, E.; Mansur, C.R. Development of a photoprotective and antioxidant nanoemulsion containing chitosan as an agent for improving skin retention. *Eng. Life Sci.* **2015**, *15*, 593–604. [CrossRef]

101. Morsy, R.; Ali, S.S.; El-Shetehy, M. Development of hydroxyapatite-chitosan gel sunscreen combating clinical multidrug-resistant bacteria. *J. Mol. Struct.* **2017**, *1143*, 251–258. [CrossRef]

102. Morganti, P.; Palombo, M.; Tishchenko, G.; Yudin, E.V.; Guarneri, F.; Cardillo, M.; Del Ciotto, P.; Carezzi, F.; Morganti, G.; Fabrizi, G. Chitin-Hyaluronan Nanoparticles: A Multifunctional Carrier to Deliver Anti-Aging Active Ingredients through the Skin. *Cosmetics* **2014**, *1*. [CrossRef]

103. Fernandes, S.C.M.; Alonso-Varona, A.; Palomares, T.; Zubillaga, V.; Labidi, J.; Bulone, V. Exploiting Mycosporines as Natural Molecular Sunscreens for the Fabrication of UV-Absorbing Green Materials. *ACS Appl. Mater. Interfaces* **2015**, *7*, 16558–16564. [CrossRef] [PubMed]

104. Higgins, S.; Wesley, N.O. Topical Retinoids and Cosmeceuticals: Where Is the Scientific Evidence to Recommend Products to Patients? *Curr Derm. Rep.* **2015**, *4*, 56–62. [CrossRef]

105. Kim, D.-G.; Jeong, Y.-I.; Choi, C.; Roh, S.-H.; Kang, S.-K.; Jang, M.-K.; Nah, J.-W. Retinol-encapsulated low molecular water-soluble chitosan nanoparticles. *Int. J. Pharm.* **2006**, *319*, 130–138. [CrossRef] [PubMed]

106. Park, C.-E.; Park, D.-J.; Kim, B.-K. Effects of a chitosan coating on properties of retinol-encapsulated zein nanoparticles. *Food Sci. Biotechnol.* **2015**, *24*, 1725–1733. [CrossRef]

107. Huang, S.-J.; Sun, S.-L.; Chiu, C.-C.; Wang, L.-F. Retinol-encapsulated water-soluble succinated chitosan nanoparticles for antioxidant applications. *J. Biomater. Sci. Polym. Ed.* **2013**, *24*, 315–329. [CrossRef] [PubMed]

108. Morganti, P.; Palombo, M.; Carezzi, F.; Nunziata, L.M.; Morganti, G.; Cardillo, M.; Chianese, A. Green Nanotechnology Serving the Bioeconomy: Natural Beauty Masks to Save the Environment. *Cosmetics* **2016**, *3*. [CrossRef]

109. Harris, R.; Lecumberri, E.; Mateos-Aparicio, I.; Mengíbar, M.; Heras, A. Chitosan nanoparticles and microspheres for the encapsulation of natural antioxidants extracted from Ilex paraguariensis. *Carbohydr. Polym.* **2011**, *84*, 803–806. [CrossRef]

110. Anchisi, C.; Meloni, M.C.; Maccioni, A.M. Chitosan beads loaded with essential oils in cosmetic formulations. *J. Cosmet. Sci.* **2006**, *57*, 205–214. [CrossRef] [PubMed]

111. Bellich, B.; D'Agostino, I.; Semeraro, S.; Gamini, A.; Cesàro, A. "The Good, the Bad and the Ugly" of Chitosans. *Mar. Drugs* **2016**, *14*, 99. [CrossRef] [PubMed]

112. Xia, W.; Liu, P.; Zhang, J.; Chen, J. Biological activities of chitosan and chitooligosaccharides. *J. Food Hydrocoll* **2011**, *25*, 170–179. [CrossRef]

113. Lee, S.-M.; Liu, K.-H.; Liu, Y.-Y.; Chang, Y.-P.; Lin, C.-C.; Chen, Y.-S. Chitosonic(®) Acid as a Novel Cosmetic Ingredient: Evaluation of its Antimicrobial, Antioxidant and Hydration Activities. *Materials (Basel)* **2013**, *6*, 1391–1402. [CrossRef] [PubMed]

polymers

Review
Recent Advances of Chitosan Applications in Plants

Massimo Malerba [1] and Raffaella Cerana [2],*

[1] Dipartimento di Biotecnologie e Bioscienze, Università degli Studi di Milano-Bicocca, 20126 Milan, Italy; massimo.malerba@unimib.it

[2] Dipartimento di Scienze dell'Ambiente e della Terra, Università degli Studi di Milano-Bicocca, 20126 Milan, Italy

* Correspondence: raffaella.cerana@unimib.it; Tel.: +39-02-64482932

Received: 18 December 2017; Accepted: 19 January 2018; Published: 26 January 2018

Abstract: In recent years, the search for biological methods to avoid the application of chemical products in agriculture has led to investigating the use of biopolymers-based materials. Among the tested biomaterials, the best results were obtained from those based on the biopolymer chitosan (CHT). CHT, available in large quantities from the deacetylation of chitin, has multiple advantages: it is safe, inexpensive and can be easily associated with other compounds to achieve better performance. In this review, we have summarized the latest researches of the application of CHT on plant productivity, plant protection against the attack of pathogens and extension of the commercial life of detached fruits.

Keywords: chitosan; defense responses; fruits; nanoparticles; plant growth; pesticides

1. Introduction

In recent years, the always growing demand for food worldwide, the ongoing climate change, the dangerous consumption of farmlands and the increasing attention of consumers to high quality, safe and environmental-friendly food products have stimulated the search for alternative biological methods that can meet this demand. Among the alternatives that are currently under investigation to avoid the use of chemical products to control plant diseases and increase crop productivity, are the biopolymer-based materials. In several cases, these materials have shown adequate activity against pathogens with low toxic effects on mammals and marginal impact on the environment. In addition, these biomaterials are also able to increase the productivity of many agricultural plants avoiding the use of large amounts of chemical fertilizers and dangerous farming practices. Among the tested biomaterials, the best results were obtained from those based on the biopolymer chitosan (CHT). CHT, chemically a linear unbranched polymer of β-1,4-D-glucosamine, is obtained from chitin, a co-polymer of N-acetyl-D-glucosamine and D-glucosamine constituting the main component of the exoskeleton of arthropods. Chitin is also present in diverse organisms such as fungi, mollusks, diatoms, and marine and fresh water sponges [1]. This natural polymer is convenient and largely available as waste from shell of shrimps and crabs processed by the seafood industry. In fact, chitin is the second largest renewable carbon source in the word after cellulose with a production of over 10^{11} tons per year. This makes its utilization of commercial interest for the production of CHT. Worldwide, industrial preparation produces more than 2000 tons per year of CHT [1]. This preparation is easy to perform and consists in the treatment of solid chitin with 40–50% (w/v) NaOH at 120–150 °C. This treatment removes the majority of the acetyl groups, converting N-acetyl-D-glucosamine in β-1,4-D-glucosamine. While chitin is insoluble in the principal solvents, thus inhibiting its direct utilization, CHT can be easily solubilized in weak organic acids, for example acetic acid, and its limited solubility in water can be overcome by chemical modification such as carboxymethylation [2]. However, the industrial method of preparation implies that the term "chitosan" does not refer to a unique compound, but is ascribed to many polymers of heterogeneous deacetylation and polymerization degrees, viscosity, molecular mass, acid and dissociation constant. It should be noted that this heterogeneity can greatly affect biological properties of CHT [3].

Everything considered CHT has multiple advantages over other biopolymers (cellulose, starch, galactomannans, etc.): it is safe, inexpensive and its chemical structure easily allows the introduction of specific molecules to design polymers for selected applications. These characteristics confer to CHT a role of great importance for a wide range of potential users ranging from medical and biotechnological industries to agricultural applications [4,5]. In particular, in recent years, an increasing number of researchers investigated the effects of CHT-based compounds on plants. In this review, we have summarized the latest research of the application of CHT on plant productivity, plant protection against the attack of pathogens and extension of the commercial life of detached fruits.

2. CHT Effects on Plant Productivity

The increasing demand for food to feed the rising world population has led to the development of agronomic practices able to significantly raise plant productivity. However, this has led to an ever-increasing use of chemical fertilizers and pesticides and high soil consumption. To stop this deleterious trend, many researchers investigated agricultural applications of CHT-based materials and in several cases these materials resulted able to increase plant productivity (Table 1).

For example, CHT (250–500 ppm, from Sigma-Aldrich (St. Louis, MO, USA), two applications at seven-day intervals, from pre-flowering to post-flowering stage) induces 56% higher fruit production in tissue-cultured plants of cv. Strawberry Festival compared to non-treated control [6]. High (124 kDa) and low (66.4 kDa) molecular weight CHT polymers (prepared by the authors with basic deacetylation of chitin and acetylation degree 13.7% and 15.2%, respectively) as well as a hydrolyzed CHT derivative (13.2 kDa) applied at 31, 45 and 59 days after planting enhance the tuber size in two different cultivars of potato (*Solanum tuberosum* L.) [7]. Foliar application of 0.5% CHT (origin and characteristics not specified, applications at seven-day intervals, starting from two weeks after transplanting) increases fruit weight, fruit diameter, and yield of Bell pepper (*Capsicum annuum*) [8]. Interesting results were obtained using CHT to protect plants against abiotic stresses. CHT (1 mg/mL, origin and characteristics not further specified, viscosity 5–30 mPa s, added 2 days before exposition of plants to dehydration stress) improves drought resistance in white clover (*Trifolium repens*) by enhancing the accumulation of stress protective metabolites [9]. Foliar application of CHT (200–400 µL/L, origin and characteristics not further specified, sprayed three times, just prior to flowering stage, at 50% flowering and at full bloom) reduces the negative impact of drought condition on dry matter and oil yield of *Thymus daenensis* Celak [10]. Exogenous application of CHT (0.2–0.4 g/L, origin and characteristics not further specified, sprayed thrice, before flowering and two weeks later) increases plant growth parameters in two species of sweet basil (*Ocimum ciliatum* and *O. basilicum*) under drought stress [11]. In experiments conducted in indoor climate controlled chambers, rice (*Oryza sativa* L.) plants soaked and sprayed with 0.05% CHT (origin and characteristics not further specified, molecular weight 50 kDa, germinated seeds soaked for 14 days, plants transferred in clay sprayed every day for additional three weeks before ozone fumigation) show significant reduction of the harmful effects of ozone compared with the control plants [12]. In hydroponic pot experiments, foliar application of CHT with different molecular weight (10 kDa, 5 kDa and 1 kDa, deacetylation degree of 80%, purchased from Qingdao Yunzhou Biochemistry Co., Ltd. (Qingdao, Shandong Province, China), applied every day for one week) could alleviate toxic effects of cadmium (Cd) on growth and leaf chlorophyll content of edible rape (*Brassica rapa* L.) [13]. In the same experimental material grown under greenhouse conditions, similar protective effect on Cd toxicity is obtained with a chitooligosaccharide, a hydrolysis product of CHT directly produced by the authors (50–200 mg/L, average molecular weights 1.6 kDa, deacetylation degree 82%, applied every day for one week) [14]. In addition to the effect on plant growth, in several medicinal plants, CHT could increase the commercial content of secondary metabolites. In stevia (*Stevia rebaudiana* Bertoni) plants, spraying of leaves with CHT (0.5%, 0.1% and 0.2%) has significant effect on the content of phenols and the glycoside rebaudiosides A [15]. In two hairy root clones of

Gentiana dinarica Beck, CHT (50 mg/L, obtained from Sigma Chemical Company (Saint Louis, MO, USA), added to growth medium after 28 days of cultivation, experiments performed after three or seven additional days) strongly increases the content of xanthone aglycone norswertianin and causes the occurrence of new xanthone compounds not detectable in control samples [16]. In cultured cells of the medicinal plant *Phyllanthus debilis* Klein ex Willd, CHT (50, 100, 150 and 200 mg/L, obtained from Himedia, Mumbai, India, added during the stationary phase of culture growth) significantly increases the content of hydrolysable tannins, the main therapeutically active constituents of the medicinal plant [17]. Treatment of dark-germinated sprouts of two malting barley (*Hordeum vulgare* L.) cultivars with CHT oligosaccharide (1–10 g/L, obtained from Kong Poong Bio, Jeju, South Korea, time of incubation 8 h on a rotary shaker) results in accumulation of antioxidant-linked, anti-hyperglycemic, bioactive high phenolic compounds [18].

Table 1. CHT effects on plant growth.

Plant Species	CHT Formulation and Administration	CHT Effect	References
Strawberry (*Fragaria* × *ananassa* Duch.)	0.025–0.05% (plant spraying)	higher fruit yield	[6]
Potato (*Solanum tuberosum* L.)	200, 325 and 558 mg/ha (foliar spraying)	enhancement of tuber size	[7]
Bell pepper (*Capsicum annuum*)	0.3–0.5% (leaves and fruits spraying)	increase in fruit weight, fruit diameter, and yield	[8]
Basil (*Ocimum ciliatum* and *Ocimum basilicum*)	0.02–0.04% (foliar spraying)	increase in plant growth and total phenol content	[11]
Rice (*Oryza sativa* L.)	0.05% (plants soaking and spraying)	increase in plant growth, higher photosynthesis rate.	[12]
Rape (*Brassica rapa* L.)	0.05–0.1% (foliar spraying)	increase in plant growth and leaf chlorophyll content	[13]
Barley (*Hordeum vulgare* L.)	0.1–1% (germinating seeds)	higher phenolic content	[18]
Maize (*Zea mays* L.)	0.04–0.16% (seeds soaking and plant spraying)	promotion of plant growth and grain weight	[19]

Some of these promoting effects of CHT on plant growth can be ascribed to fertilizing properties of CHT compounds able to supplement plants with essential nutrients. In both pot and field conditions, maize (*Zea mays* L.) plants treated with Cu-CHT nanoparticles (0.04–0.16%, CHT molecular weight 50–190 kDa, deacetylation degree 80%, obtained from Sigma-Aldrich (St. Louis, MO, USA), applied for 4 h to sterilized seeds and after the transfer in standard clay type soil sprayed every day for 35 days) show enhanced plant height, stem diameter, root length and number, chlorophyll content, ear length and weight/plot, grain yield/plot and weight [19]. CHT-polyvinyl alcohol hydrogels with absorbed copper nanoparticles (CHT obtained from Marine Chemicals, Kerala, India, molecular weight 200 kDa, added once as hydrogel on plants) increase stomata width, primary stem length, and root length of grafted "Jubilee" watermelon (*Cucurbita maxima* x *Cucurbita moschata*) [20]. Foliar application of zinc complexed CHT nanoparticles (CHT obtained from India Sea Foods, Cochin, Kerala, molecular weight 60 kDa, deacetylation degree 85%, applied twice a week for five weeks) efficiently supplements the micronutrient to wheat plant cultivated under zinc deficient conditions [21]. CHT combined with waste silica may allow farmers to reduce the use of NKP fertilizers to improve corn production in Indonesia with environmental and economic advantages [22]. In addition, CHT nanoparticles can be also used as carrier system for plant growth hormones. For example, CHT nanoparticles combined with gibberellic acid (CHT obtained from Sigma-Aldrich, molecular weight 27 kDa, deacetylation degree 75–85%, added once seven days before analyses) significantly increase leaf area and the levels of chlorophylls and carotenoids in *Phaseolus vulgaris* [23].

Collectively, these results support the ability of CHT-based materials to increase plant productivity. In particular, these materials seem useful to increase productivity under stress conditions and this is important for the agronomic utilization of marginal lands.

3. CHT Effects on Plant Pathogens

The promoting effect of CHT-based compounds on plant productivity can easily be ascribed to its ability to control plant pathogens such as virus, bacteria, fungi and nematodes. This ability is documented in several studies, the latest presented in the following (Table 2).

Treatment of seeds with CHT (low molecular weight, 5–20 kDa, obtained from Bioinzheneriya center of RAS, CJSC Bioprogress, added once to seeds and sprayed every day on leaves) induces the resistance of tomato plants to *Phytophtora infestans* and *Alternaria solani* [24]. Both in greenhouse trials as well as in vitro, CHT nanoparticles (1000–5000 ppm, CHT obtained from Biobasic, Canada and Sigma-Aldrich, Germany, molecular mass 161–810 kDa, deacetylation degree 75–90%, added once at anthesis) are effective against *Fusarium graminearum*, the causing agent of Fusarium head blight on wheat [25]. Seed soaking and foliar application of CHT (0.25–2 g/L, origin and characteristics not further specified, seeds soaked for 24 h and leaves sprayed three times at seven-day intervals starting from the second true leaf stage of emerged bean seedlings with half concentrations of used rate for seed soaking treatment) are efficient in the control of *Fusarium solani* and *Rhizoctonia solani* in *Phaseolus vulgaris* L. both in vivo and in vitro [26]. Application of 0.05% and 0.1% CHT to the leaves efficiently controls anthracnose, the disease caused by *Colletotrichum* spp. on cucumber (*Cucumis sativus* L.) plants [27]. Foliar application of 0.01% CHT (obtained from MP Biomedicals, LLC. (Santa Ana, CA, USA), deacetylation degree 90%, added at an interval of 15 days for five months) reduces the blister blight disease caused by the biotrophic fungal pathogen *Exobasidium vexans* in *Camellia sinensis* (L.) O. Kuntze plants [28].

CHT (0.5%, origin and characteristics not further specified, directly mixed to the soil before seeds) is also able to shift the abundance of resident and inoculated biocontrol agents in the rhizosphere to suppress growth of the nematode *Heterodera glycines* in soybean [29]. Both in vitro and in vivo under glasshouse conditions as well as in field experiments CHT (0.5–2 g/L, obtained from Roth, Cat. C0108, Lot 133,115, tuber soaked for 30 min, leaves sprayed twice, 30 days and 45 days after planting) efficiently acts against *Ralstonia solanacearum*, the causal agent of potato bacterial wilt disease, thus increasing plant health [30]. *Pochonia chlamydosporia*, a fungal parasite used as biological control agent for the management of *Meloidogyne* spp., the most damaging plant-parasitic nematodes for horticultural crops worldwide, better develops in soil and endophytically colonize roots of tomato plants irrigated with a 0.1 mg/mL CHT solution (CHT obtained from Marine BioProducts GmbH, Bremerhaven, Germany, molecular weight 70 kDa, deacetylation degree 80.5%, added daily for 10–30 days) thereby increasing its efficiency against nematodes [31]. Finally, CHT (supplemented by the addition of *Cunninghamella elegans*, a fungus that contains chitin and chitosan in its cell wall, added once to soil) mixed with a biofertilizer obtained from phosphate and potassium rocks protects green peppers, but not tomato plants, against *Ralstonia solanacearum* infection [32].

Table 2. CHT effects on plant pathogens.

Plant Species	CHT Formulation and Administration	Pathogen	References
Tomato (*Solanum lycopersicon*)	0.4% (seeds soaking, fruits spraying)	*Phytophtora infestans, Alternaria solani*	[24]
Wheat (*Triticum* spp.)	0.1–0.5% (spikelets spraying)	*Fusarium graminearum*	[25]
Green bean (*Phaseolus vulgaris* L.)	0.025–0.2% (seeds soaking, foliar sprayimg)	*Fusarium solani, Rhizoctonia solani*	[26]
Cucumber (*Cucumis sativus* L.)	0.05–0.1% (foliar spraying)	*Colletotrichum* spp.	[27]
Tea (*Camellia sinensis* L.)	0.01% (foliar spraying)	*Exobasidium vexans*	[28]
Soybean (*Glicine max* L.)	0.5% (soil treatment)	*Heterodera glycines*	[29]
Tomato (*Solanum lycopersicon*)	0.01% (plant irrigation)	*Meloidogyne* spp.	[31]

4. CHT Effects on Detached Fruits

For several agronomic commodities, fruits are the part of the plant with the best economic value. However, fruits tend to have a short shelf life even under strict cold chain management. In addition, the site of cultivation is often far from the consumer markets. Postharvest diseases are the main cause of losses for detached fruits and the management of these diseases is one of the major challenges for the farmers. More and more consumers demand products of high quality and free of pesticide residues. This directs research towards integrated alternative strategies to manage postharvest diseases. In this perspective, edible coating materials such as polysaccharides, proteins, lipids and plant extracts are of high interest [33]. The adhesive nature of CHT combined with its biodegradability makes application of CHT edible coatings the best way to prolong the commercial life of fresh agricultural products. In addition, the use of CHT either alone or in combination with other protectants (e.g., minerals, vitamins, or nutraceutical compounds) that increase the beneficial properties of fresh commodities and in some cases the anti-pathogen activity of CHT permits avoiding the use of chemical products. CHT coating can form a semipermeable film on the surface of fruit and vegetables. This affects the rate of respiration, decreases water loss and weight decrease, and permits maintaining the requested quality for the market. Therefore, in recent years, an impressive number of papers dealing with CHT and protection of fresh agricultural commodities has been published (see [33] for a review). In this section, we summarize the latest results (Table 3).

Application of a CHT edible coating combined with an acetonic extract of *Salvia fruticosa* Mill (1% CHT, obtained from Sigma-Aldrich (St. Louis, MO, USA), medium molecular weight, deacetylation degree 75–85%, viscosity 200–800 cP, added by fruit immersion for 1 min) effectively controls *Botrytis cinerea* infection without affecting quality and physico-chemical properties of table grapes (*Vitis vinifera* cv. "Thompson Seedless") [34]. Similar results have been obtained in table grapes (*Vitis vinifera* L. "Yongyou 1") where preharvest treatment with CHT-*g*-salicylic acid (1% *w/v* CHT, obtained from Zhejiang Aoxing Biotechnology Co., Ltd. (Kanmen, Zhejiang, China), food-grade, deacetylation degree ≥95%, viscosity ≤30 mPa s, added by spraying grape clusters five days before harvest) positively influences postharvest table grape quality, shelf life, and resistance to *Botrytis cinerea*-induced spoilage [35]. Postharvest coating of clusters with CHT/polyvinyl alcohol blended with ascorbic acid (2.8–8.2 mM CHT, obtained from Merck, 64271 Darmstadt Germany, molecular weight 71.3 kDa, deacetylation degree 94%, added by fruit immersion for 5 min) significantly slows down the rate of deterioration of *Vitis vinifera* L. cv "Superior Seedless" grapevines [36]. Combined CHT and nano-SiOx coating (1% CHT, obtained from Aoxing Biotechnology Co., Ltd. (Taizhou, Zhejiang, China), added by fruit immersion for 5 min) extends the shelf life of Chinese cherries (*Prunus pseudocerasus* L.) during postharvest storage by inhibiting pectin chain degradation [37]. Coating with an edible film composed by CHT, quinoa protein and sunflower oil (2% CHT, directly obtained by the authors from giant squid, added by fruit immersion for 1.5 min) is able to control the growth of molds and yeasts during storage of Highbush blueberries (*Vaccinium corymbosum* L. cv. O'Neal) [38]. CHT treatment (2.5–15 g/L CHT, obtained from Sigma-Aldrich, Steinheim, Germany, characteristics not further specified, added by arils and fruit immersion for 1 min) protects arils and whole pomegranate (*Punica granatum* L.) fruits against *Botrytis* spp., *Penicillium* spp. and *Pilidiella granati* infection in in vivo and in vitro experiments [39]. CHT coating (1% CHT directly obtained by the authors from *Daphnia longispina* ephippia, viscosity-average molecular weight 4.16 kDa, deacetylation degree 70–75%, added by two fruit immersions of 10 s each for Ref. [40]; and 1% CHT obtained from Sigma Aldrich, Germany, with the CAS number: 9012-76-4, PCodes of low molecular weight chitosan: 1001654976; medium molecular weight chitosan: 1001567692 and high molecular weight chitosan: 101476130, added by fruit immersion for 10 s for Ref. [41]) extends the commercial life of red kiwifruit (*Actinidia melanandra*) and of fruits of *Actinidia kolomikta* (Maxim.), two species of the genus *Actinidia* with valuable properties in terms of content of biologically active substances and areal of cultivation but currently hardly commercialized due to their very short (less than two days) shelf life [40,41]. CHT nanoparticles loaded with a solution of the natural essential oil sanitizing carvacrol (1% CHT, origin and characteristics not further specified, added by slice immersion for 4 min) reduce the microbial growth in fresh sliced carrots (*Daucus carota* L.) during storage, avoiding the carvacrol-related off-flavors [42]. CHT coating (1% CHT, origin and characteristics not further specified, added by fruit immersion for 5 min) significantly delays fruit senescence and preserves the nutrient content and antioxidant abilities of jujube (*Zizyphus jujube* Miller cv. Dongzao) [43]. Coating with enzymatic hydrolyzed low molecular weight CHT (1% CHT, obtained from Sigma-Aldrich (St. Louis, MO, USA), apparent molecular weight of 50 kDa, deacetylation degree 90%, added by fruit immersion for 10 min) significantly preserves wounded and unwounded pear (*Pyrus bretschneideri* cv. "Huangguan") fruits by *Botryosphaeria* spp. Attack, thus inhibiting postharvest decay and browning processes [44]. Coating with a combination of CHT, alginate and pomegranate peel extract (1% CHT, obtained from E-Merck Ltd., Mumbai, India, low molecular weight, deacetylation degree 75%, added by fruit immersion for 1 min), is an effective treatment to maintain the overall fruit quality and total flavonoids and total phenolics contents in guava (*Psidium guajava* L. cv Allahabad *safeda*) [45]. CHT was also used to protect various cultivated species of mushrooms. For example, coating with a protocatechuic acid-grafted-CHT solution (1% CHT, obtained from Sangon Biotechnology Co. Ltd. (Shanghai, China), average molecular weight of 250 kDa, deacetylation degree 71%, added by fungus immersion for 30 s) efficiently protects king oyster mushroom (*Pleurotus eryngii*) during postharvest storage [46].

Table 3. CHT effects on fruits.

Plant Species	CHT Formulation and Administration	CHT Effect	References
Grape (*Vitis vinifera*)	Coating with 1% CHT and *Salvia fruticosa* extract	Inhibition of *Botrytis cinerea* growth	[34]
Pomegranate (*Punica granatum* L.)	Coating with 1.5% CHT	Inhibition of *Botrytis* spp., *Penicillium* spp. and *Pilidiella granati* growth	[39]
Red kiwifruit (*Actinidia melanandra*)	Coating with 1% CHT	Extension of the fruit commercial life	[40]
Pear (*Pyrus bretschneideri*)	Coating with 1% CHT	Inhibition of *Botryosphaeria* spp. growth	[44]
Mango (*Mangifera indica* L.)	Coating with 1% CHT	Delay of fruit ripening, inhibition of *Colletotrichum gloeosporioides* growth	[47]
Mango (*Mangifera indica* L.)	Coating with 1% CHT and 0.1 ppm spermidine	Delay of fruit softening, accumulation of phenolic compounds during storage, induction of defense enzyme activities, inhibition of *Colletotrichum gloeosporioides* growth	[48]
Mango (*Mangifera indica* L.)	Coating with 1% CHT and *Mentha piperita* L. essential oil	Inhibition of *Colletotrichum asianum*, *Colletotrichum dianesei, Colletotrichum fructicola, Colletotrichum tropicale* and *Colletotrichum karstii* growth	[49]

Mango (*Mangifera indica* L.), one of the most popular tropical fruits with high demand and great market value, has been the object of several studies. In fact, due to its large ethylene production, the fruit quickly ripens and softens after harvest. Moreover, mango fruits are subject to the anthracnose severe disease caused by the pathogen *Colletotrichum gloeosporioides* (Penz.) Penz. & Sacc that can attack immature fruits causing large loss in production. Fungicides currently used to face this problem are under investigation in several countries. Thus, the search for alternative methods is very important. High molecular weight CHT solution applied as fruit coating (1% CHT, obtained from A.N. Lab, Thailand, molecular weight 360 kDa, deacetylation degree 85%, added by fruit immersion for 1 min) significantly delays mango (cv. Nam Dok Mai, the most important export fruit of Thailand) fruit ripening thus impacting its postharvest quality. Moreover, these CHT-treated fruits exhibit no incidences of disease symptoms throughout storage [47]. Better results, especially in terms of protection against anthracnose disease, are obtained in the same mango cultivar using CHT coating combined with spermidine (1% CHT, obtained from A.N. Lab (Thailand), molecular weight 360 Da, deacetylation degree 85%, added by fruit immersion for 1 min) [48]. In fruits of mango cultivar

Tommy Atkins artificially contaminated with the pathogens, CHT alone or in combination with *Mentha piperita* L. essential oil (5–7.5 mg/mL CHT, obtained from Sigma-Aldrich Corp. (St. Louis, MO, USA), medium molecular weight, deacetylation degree 75–85%, batch MKBH1108V, added by fruit immersion for 5 min) effectively inhibits mycelial growth of five different *Colletotrichum* species: *C. asianum, C. dianesei, C. fructicola, C. tropicale* and *C. karstii* [49]. Coating with CHT solutions (1–3% CHT, obtained by Sigma-Aldrich, viscosity 20–300 cP, deacetylation degree 95–98%, added by fruit immersion for 1 min) delays climacteric peak, water loss and preserves fruit firmness in *Mangifera indica* L. cv. Palmer by affecting basic mitochondrial respiration and starch degradation rate [50].

Despite this large number of papers published in recent years, the mode of action of CHT in fruit protection is not yet fully clarified. However, the research summarized in Table 3 indubitably shows that the anti-pathogen activity of CHT is the main component of its protective effect on detached fruits, suggesting the possible general use of this compound for pest control. Very recently, by transcriptomic analyses, researchers started to investigate the genes that are either activated or repressed in fruits treated with CHT. Treatment with 1% CHT (obtained from Chito Plant, ChiPro GmbH, Bremen, Germany, added by spraying the plant canopy) of strawberry cultivar "Alba" (*Fragaria* × *ananassa*; $2n = 8x = 56$) plant canopy induces in fruits harvested at 6, 12, and 24 h post-treatment the different expression (fold change ≥ 2) of more than 5000 genes. These genes are associated with biotic and abiotic stresses, plant immune system, hormone metabolism, systemic acquired resistance, photosynthesis, heat-shock proteins, and reprogramming of protein metabolism with an increment of storage proteins [51]. In addition, expression profiles show that avocado (*Persea americana* Mill) fruits inoculated with *Colletotrichum gloeosporioides* in the presence of 1.5% CHT (obtained by Sigma Aldrich, viscosity 35 cP, deacetylation degree 96.1%, added by fruit immersion for 1 min) present a greater number of differentially expressed genes, compared to the fruits inoculated with the pathogen in the absence of CHT. These differentially expressed genes are involved in many metabolic processes [52].

5. Conclusions

The large number of papers published in the last year show that CHT, as a unique product available in large quantities and at a low price, has a bright future in development of sustainable agricultural practices as well as in food production and preservation. In particular, given the always growing demand for food worldwide, the ongoing climate change and the consumption of farmlands, CHT appears to be a promising tool for cultivation under stress conditions and to permit the cultivation of varieties with interesting organoleptic properties but with severe fruit-bearing duration problems.

Author Contributions: Massimo Malerba and Raffaella Cerana equally contributed to the paper.

Conflicts of Interest: The authors declare that they have no conflict of interest.

References

1. Kurita, K. Chitin and chitosan: Functional biopolymers from marine crustaceans. *Mar. Biotechnol.* **2006**, *8*, 203–226. [CrossRef] [PubMed]
2. Rinaudo, M. Chitin and chitosan: Properties and applications. *Prog. Polym. Sci.* **2006**, *31*, 606–632. [CrossRef]
3. Choi, C.; Nam, J.P.; Nah, J.W. Application of chitosan and chitosan derivatives as biomaterials. *J. Ind. Eng. Chem.* **2016**, *33*, 1–10. [CrossRef]
4. Anitha, A.; Sowmya, S.; Sudheesh Kumar, P.T.; Deepthi, S.; Chennazhi, K.P.; Ehrlich, H.; Tsurkan, M.; Jayakumar, R. Chitin and chitosan in selected biomedical applications. *Prog. Polym. Sci.* **2014**, *39*, 1644–1667. [CrossRef]
5. Malerba, M.; Cerana, R. Chitosan effects on plant systems. *Int. J. Mol. Sci.* **2016**, *17*, 996. [CrossRef] [PubMed]
6. Mutka, J.A.; Rahman, M.; Sabir, A.A.; Gupta, D.R.; Surovy, M.Z.; Rahman, M.; Tofazzal Islam, M. Chitosan and plant probiotics application enhance growth and yield of strawberry. *Biocatal. Agric. Biotechnol.* **2017**, *11*, 9–18.

7. Falcón-Rodríguez, A.B.; Costales, D.; Gónzalez-Peña, D.; Morales, D.; Mederos, Y.; Jerez, E.; Cabrera, J.C. Chitosans of different molecular weight enhance potato (*Solanum tuberosum* L.) yield in a field trial. *Span. J. Agric. Res.* **2017**, *15*, e0902. [CrossRef]

8. Mahmood, N.; Abbasi, N.A.; Hafiz, I.A.; Ali, I.; Zakia, S. Effect of biostimulants on growth, yield and quality of bell pepper cv. Yolo wonder. *Pak. J. Agric. Sci.* **2017**, *54*, 311–317.

9. Li, Z.; Zhang, Y.; Zhang, X.; Merewitz, E.; Peng, Y.; Ma, X.; Huang, L.; Yan, Y. Metabolic pathways regulated by chitosan contributing to drought resistance in white clover. *J. Proteome Res.* **2017**, *16*, 3039–3052. [CrossRef] [PubMed]

10. Bistgani, Z.E.; Siadat, S.A.; Bakhshandeh, A.; Pirbalouti, A.G.; Hashemic, M. Interactive effects of drought stress and chitosan application on physiological characteristics and essential oil yield of *Thymus daenensis* Celak. *Crop J.* **2017**, *5*, 407–415. [CrossRef]

11. Pirbalouti, A.G.; Malekpoor, F.; Salimi, A.; Golparvar, A. Exogenous application of chitosan on biochemical and physiological characteristics, phenolic content and antioxidant activity of two species of basil (*Ocimum ciliatum* and *Ocimum basilicum*) under reduced irrigation. *Sci. Hortic.* **2017**, *217*, 114–122. [CrossRef]

12. Phothi, R.; Theerakarunwong, C.D. Effect of chitosan on physiology, photosynthesis and biomass of rice (*Oryza sativa* L.) under elevated ozone. *Aust. J. Crop Sci.* **2017**, *11*, 624–630. [CrossRef]

13. Zong, H.; Liu, S.; Xing, R.; Chen, X.; Li, P. Protective effect of chitosan on photosynthesis and antioxidative defense system in edible rape (*Brassica rapa* L.) in the presence of cadmium. *Ecotoxicol. Environ. Saf.* **2017**, *138*, 271–278. [CrossRef] [PubMed]

14. Zong, H.; Kecheng, L.; Liu, S.; Song, L.; Xing, R.; Chen, X.; Li, P. Improvement in cadmium tolerance of edible rape (*Brassica rapa* L.) with exogenous application of chitooligosaccharide. *Chemosphere* **2017**, *181*, 92–100. [CrossRef] [PubMed]

15. Mehregan, M.; Mehrafarin, A.; Labbafi, M.R.; Naghdi Badi, H. Effect of different concentrations of chitosan biostimulant on biochemical and morphophysiological traits of stevia plant (*Stevia rebaudiana* Bertoni). *J. Med. Plants* **2017**, *16*, 169–181.

16. Krstić-Milošević, D.; Janković, T.; Uzelac, B.; Vinterhalter, D.; Vinterhalter, B. Effect of elicitors on xanthone accumulation and biomass production in hairy root cultures of *Gentiana dinarica*. *Plant Cell Tissue Organ Cult.* **2017**, *130*, 631–640. [CrossRef]

17. Malayamana, V.; Sisubalan, N.; Senthilkumar, R.P.; Sheik Mohamed, S.; Ranjithkumar, R.; Ghouse Basha, M. Chitosan mediated enhancement of hydrolysable tannin in *Phyllanthus debilis* Klein ex Willd via plant cell suspension culture. *Int. J. Biol. Macromol.* **2017**, *104*, 1656–1663.

18. Ramakrishna, R.; Sarkar, D.; Manduri, A.; Iyer, S.G.; Shetty, K. Improving phenolic bioactive-linked anti-hyperglycemic functions of dark germinated barley sprouts (*Hordeum vulgare* L.) using seed elicitation strategy. *J. Food Sci. Technol.* **2017**, *54*, 3666–3678. [CrossRef] [PubMed]

19. Choudhary, R.C.; Kumaraswamy, R.V.; Kumari, S.; Sharma, S.S.; Pal, A.; Raliya, R.; Biswas, P.; Saharan, V. Cu-chitosan nanoparticle boost defense responses and plant growth in maize (*Zea mays* L.). *Sci. Rep.* **2017**, *7*, 9754–9765. [CrossRef] [PubMed]

20. González Gómez, H.; Ramírez Godina, F.; Ortega Ortiz, H.; Benavides Mendoza, A.; Robledo Torres, V.; Cabrera De la Fuente, M. Use of chitosan-PVA hydrogels with copper nanoparticles to improve the growth of grafted watermelon. *Molecules* **2017**, *22*, 1031. [CrossRef] [PubMed]

21. Deshpande, P.; Dapkekar, A.; Oak, M.D.; Paknikar, K.M.; Rajwade, J.M. Zinc complexed chitosan/TPP nanoparticles: Promising micronutrient nanocarrier suited for foliar application. *Carbohydr. Polym.* **2017**, *165*, 394–401. [CrossRef] [PubMed]

22. Gumilar, T.A.; Prihastanti, E.; Haryanti, S.; Subagio, A.; Ngadiwiyana, A. Utilization of waste silica and chitosan as fertilizer nano chisil to improve corn production in Indonesia. *Adv. Sci. Lett.* **2017**, *23*, 2447–2449. [CrossRef]

23. Espirito Santo Pereira, A.; Mayara Silva, P.; Oliveira, J.L.; Oliveira, H.C.; Fernandes Faceto, L. Chitosan nanoparticles as carrier systems for the plant growth hormone gibberellic acid. *Colloids Surf. B Biointerfaces* **2017**, *150*, 141–152. [CrossRef] [PubMed]

24. Kiprushkina, E.I.; Shestopalova, I.A.; Pekhotina, A.M.; Kuprina, E.E.; Nikitina, O.V. Protective-stimulating properties of chitosan in the vegetation and storing tomatoes. *Prog. Chem. Appl. Chitin Deriv.* **2017**, *23*, 77–81.

25. Kheiri, A.; Moosawi Jorf, S.A.; Malihipour, A.; Saremi, H.; Nikkhah, A. Synthesis and characterization of chitosan nanoparticles and their effect on Fusarium head blight and oxidative activity in wheat. *Int. J. Biol. Macromol.* **2017**, *102*, 526–538. [CrossRef] [PubMed]

26. El-Mohamedy, R.S.R.; Shafeek, M.R.; Abd El-Samad, E.E.-D.H.; Salama, D.M.; Rizk, F.A. Field application of plant resistance inducers (PRIs) to control important root rot diseases and improvement growth and yield of green bean (*Phaseolus vulgaris* L.). *Aust. J. Crop Sci.* **2017**, *11*, 496–505. [CrossRef]

27. Dodgson, J.L.A.; Dodgson, W. Comparison of effects of chitin and chitosan for control of *Colletotrichum* sp. on cucumbers. *J. Pure Appl. Microbiol.* **2017**, *11*, 87–93. [CrossRef]

28. Chandra, S.; Chakraborty, N.; Panda, K.; Acharya, K. Chitosan-induced immunity in *Camellia sinensis* (L.) O. Kuntze against blister blight disease is mediated by nitric-oxide. *Plant Physiol. Biochem.* **2017**, *115*, 298–307. [CrossRef] [PubMed]

29. Mwaheb, M.A.M.A.; Hussain, M.; Tian, J.; Xiaoling Zhang, X.; Imran Hamid, M.; Abo El-Kassim, N.; Hassan, G.M.; Xiang, M.; Liu, X. Synergetic suppression of soybean cyst nematodes by chitosan and *Hirsutella minnesotensis* via the assembly of the soybean rhizosphere microbial communities. *Biol. Control* **2017**, *115*, 86–94. [CrossRef]

30. Farag, S.M.A.; Elhalag, K.M.A.; Mohamed, H.; Hagag, M.H.; Khairy, A.S.M.; Ibrahim, H.M.; Saker, M.T.; Messiha, N.A.S. Potato bacterial wilt suppression and plant health improvement after application of different antioxidants. *J. Phytopathol.* **2017**. [CrossRef]

31. Escudero, N.; Lopez-Moya, F.; Ghahremani, Z.; Zavala-Gonzalez, E.A.; Alaguero-Cordovilla, A.; Caridad Ros-Ibañez, C.; Lacasa, A.; Sorribas, F.J.; Lopez-Llorca, L.V. Chitosan increases tomato root colonization by *Pochonia chlamydosporia* and their combination reduces root-knot nematode damage. *Front. Plant Sci.* **2017**, *8*. [CrossRef] [PubMed]

32. Stamford, N.P.; Santos, L.R.C.; dos Santos, A.B.; de Souza, K.R.; da Silva Oliveira, W.; da Silva, E.V.N. Response of horticultural crops to application of bioprotector and biological control of *Ralstonia* wilt in Brazilian Ultisol. *Aust. J. Crop Sci.* **2017**, *11*, 284–289. [CrossRef]

33. Paloua, L.; Ali, A.; Fallik, E.; Romanazzi, G. GRAS, plant- and animal-derived compounds as alternatives to conventional fungicides for the control of postharvest diseases of fresh horticultural produce. *Postharvest Biol. Technol.* **2016**, *122*, 41–52. [CrossRef]

34. Kanetis, L.; Exarchou, V.; Charalambous, Z.; Goulas, V. Edible coating composed of chitosan and *Salvia fruticosa* Mill. extract for the control of grey mould of table grapes. *J. Sci. Food Agric.* **2017**, *97*, 452–460. [CrossRef] [PubMed]

35. Shen, Y.; Yang, H. Effect of preharvest chitosan-*g*-salicylic acid treatment on postharvest table grape quality, shelf life, and resistance to *Botrytis cinerea*-induced spoilage. *Sci. Hortic.* **2017**, *224*, 367–373. [CrossRef]

36. Lo'ay, A.A.; Dawood, H.D. Active chitosan/PVA with ascorbic acid and berry quality of 'Superior seedless' grapes. *Sci. Hortic.* **2017**, *224*, 286–292. [CrossRef]

37. Xin, Y.; Chen, F.; Lai, S.; Yang, H. Influence of chitosan-based coatings on the physicochemical properties and pectin nanostructure of Chinese cherry. *Postharvest Biol. Technol.* **2017**, *133*, 64–71. [CrossRef]

38. Abugoch, L.; Tapia, C.; Plasencia, D.; Pastor, A.; Castro-Mandujano, O.; López, L.; Escalona, V.H. Shelf-life of fresh blueberries coated with quinoa protein/chitosan/sunflower oil edible film. *J. Sci. Food Agric.* **2016**, *96*, 619–626. [CrossRef] [PubMed]

39. Munhuweyi, K.; Lennox, C.L.; Meitz-Hopkins, J.C.; Caleb, O.J.; Sigge, G.O.; Opara, U.L. Investigating the effects of crab shell chitosan on fungal mycelial growth and postharvest quality attributes of pomegranate whole fruit and arils. *Sci. Hortic.* **2017**, *220*, 78–89. [CrossRef]

40. Kaya, M.; Česoniene, L.; Daubaras, R.; Leskauskaite, D.; Zabulione, D. Chitosan coating of red kiwifruit (*Actinidia melanandra*) for extending of the shelf life. *Int. J. Biol. Macromol.* **2016**, *85*, 355–360. [CrossRef] [PubMed]

41. Drevinskas, T.; Naujokaitytè, G.; Maruška, A.; Kaya, M.; Sargin, I.; Daubaras, R.; Česoniene, L. Effect of molecular weight of chitosan on the shelf life and other quality parameters of three different cultivars of *Actinidia kolomikta* (kiwifruit). *Carbohydr. Polym.* **2017**, *173*, 269–275. [CrossRef] [PubMed]

42. Martínez-Hernández, G.B.; Amodio, M.L.; Colelli, G. Carvacrol-loaded chitosan nanoparticles maintain quality of fresh-cut carrots. *Innov. Food Sci. Emerg. Technol.* **2017**, *41*, 56–63. [CrossRef]

43. Kou, X.; Li, Y.; Wu, J.; Chen, Q.; Xue, Z. Effects of edible coatings on quality and antioxidant activity of *Ziziphus Jujuba* Miller cv. Dongzao during storage. *Trans. Tianjin Univ.* **2017**, *23*, 51–61. [CrossRef]

44. Wang, Y.; Li, B.; Zhang, X.; Peng, N.; Mei, Y.; Liang, Y. Low molecular weight chitosan is an effective antifungal agent against *Botryosphaeria* sp. and preservative agent for pear (*Pyrus*) fruits. *Int. J. Biol. Macromol.* **2017**, *95*, 1135–1143. [CrossRef] [PubMed]

45. Nair, M.S.; Saxena, A.; Kaur, C. Effect of chitosan and alginate based coatings enriched with pomegranate peel extract to extend the postharvest quality of guava (*Psidium guajava* L.). *Food Chem.* **2018**, *240*, 245–252. [CrossRef] [PubMed]

46. Liu, J.; Meng, C.; Wang, X.; Chen, Y.; Kan, J.; Jin, C. Effect of protocatechuic acid-grafted-chitosan coating on the postharvest quality of *Pleurotus eryngii*. *J. Agric. Food Chem.* **2016**, *64*, 7225–7233. [CrossRef] [PubMed]

47. Jongsri, P.; Wangsomboondee, T.; Rojsitthisak, P.; Saraypheap, K. Effect of molecular weights of chitosan coating on postharvest quality and physicochemical characteristics of mango fruit. *LWT Food Sci. Technol.* **2016**, *73*, 28–36. [CrossRef]

48. Jongsri, P.; Rojsitthisak, P.; Wangsomboondee, T.; Saraypheap, K. Influence of chitosan coating combined with spermidine on anthracnose disease and qualities of 'Nam Dok Mai' mango after harvest. *Sci. Hortic.* **2017**, *224*, 180–187. [CrossRef]

49. De Oliveira, K.Á.R.; Berger, L.R.R.; de Araújo, S.A.; Câmara, M.P.S.; de Souza, E.L. Synergistic mixtures of chitosan and *Mentha piperita* L. essential oil to inhibit *Colletotrichum* species and anthracnose development in mango cultivar Tommy Atkins. *Food Microbiol.* **2017**, *66*, 96–103.

50. Cosme Silva, G.M.; Batista Silva, W.; Medeiros, D.B.; Rodrigues Salvador, A.; Menezes Cordeiro, M.H.; Martins da Silva, N.; Bortolini Santana, D.; Polete Mizobutsi, G. The chitosan affects severely the carbon metabolism in mango (*Mangifera indica* L. cv. Palmer) fruit during storage. *Food Chem.* **2017**, *237*, 372–378. [CrossRef] [PubMed]

51. Landi, L.; De Miccolis Angelini, R.M.; Pollastro, S.; Feliziani, E.; Faretra, F.; Romanazzi, G. Global transcriptome analysis and identification of differentially expressed genes in Strawberry after preharvest application of benzothiadiazole and chitosan. *Front. Plant Sci.* **2017**, *8*, 235. [CrossRef] [PubMed]

52. Xoca-Orozco, L.-Á.; Cuellar-Torres, E.A.; González-Morales, S.; Gutiérrez-Martínez, P.; López-García, U.; Herrera-Estrella, L.; Vega-Arreguín, J.; Chacón-López, A. Transcriptomic analysis of avocado hass (*Persea americana* Mill) in the interaction system fruit-chitosan-*Colletotrichum*. *Front. Plant Sci.* **2017**, *8*, 956. [CrossRef] [PubMed]

polymers

MDPI

Review

Chitosan Based Self-Assembled Nanoparticles in Drug Delivery

Javier Pérez Quiñones [1], **Hazel Peniche [2]** and **Carlos Peniche [3],***

[1] Institute of Polymer Chemistry, Johannes Kepler University, Altenberger Strasse 69, 4040 Linz, Austria; javenator@gmail.com

[2] Centro de Biomateriales, Universidad de La Habana, Ave. Universidad S/N entre G y Ronda, 10400 La Habana, Cuba; hazel@biomat.uh.cu

[3] Facultad de Química, Universidad de La Habana, Zapata S/N entre G y Carlitos Aguirre, 10400 La Habana, Cuba

* Correspondence: peniche@fq.uh.cu Tel.: +53-7870-0594

Received: 31 January 2018; Accepted: 23 February 2018; Published: 26 February 2018

Abstract: Chitosan is a cationic polysaccharide that is usually obtained by alkaline deacetylation of chitin poly(N-acetylglucosamine). It is biocompatible, biodegradable, mucoadhesive, and non-toxic. These excellent biological properties make chitosan a good candidate for a platform in developing drug delivery systems having improved biodistribution, increased specificity and sensitivity, and reduced pharmacological toxicity. In particular, chitosan nanoparticles are found to be appropriate for non-invasive routes of drug administration: oral, nasal, pulmonary and ocular routes. These applications are facilitated by the absorption-enhancing effect of chitosan. Many procedures for obtaining chitosan nanoparticles have been proposed. Particularly, the introduction of hydrophobic moieties into chitosan molecules by grafting to generate a hydrophobic-hydrophilic balance promoting self-assembly is a current and appealing approach. The grafting agent can be a hydrophobic moiety forming micelles that can entrap lipophilic drugs or it can be the drug itself. Another suitable way to generate self-assembled chitosan nanoparticles is through the formation of polyelectrolyte complexes with polyanions. This paper reviews the main approaches for preparing chitosan nanoparticles by self-assembly through both procedures, and illustrates the state of the art of their application in drug delivery.

Keywords: chitosan; self-assembled; polyelectrolyte complex; nanoparticle; drug delivery

---

1. Introduction

Chitosan (CS) is a family of linear polysaccharides that is composed of glucosamine and N-acetylglucosamine units linked together by β (1 → 4) glycosidic links (Figure 1). CS is obtained by the partial deacetylation of the naturally occurring polysaccharide, chitin, which is essentially poly(N-acetylglucosamine). Depending on the natural source and the conditions used to isolate and deacetylate chitin, the resulting degree of acetylation (DA) and molecular weight of chitosan will depend on the reaction parameters that are involved [1]. Molecular weight, the DA, and even the pattern of acetylation (the distribution of glucosamine and N-acetylglucosamine units along the chitosan chain) will affect its chemical and biological properties [2,3].

The degree of deacetylation (DD = 100 − DA) of chitosan is about 50% or higher. In dilute aqueous acid solutions, the amino groups of chitosan become protonated, allowing for its dissolution. In fact, the solubility of chitosan in 1% or 0.1 M acetic acid is a simple and practical criterion used to differentiate it from chitin. However, chitosan solubility depends on its DD, the ionic concentration, the pH, and the distribution of acetyl groups along the chain, as well as the conditions of isolation and drying. If deacetylation of chitin is performed under homogeneous conditions chitosans with a DD about 50% might dissolve, but if deacetylation is carried out under heterogeneous conditions, DD of 65% or higher is usually needed to achieve dissolution [4].

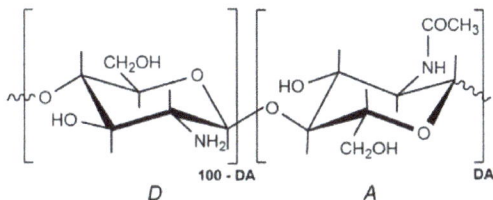

Figure 1. Structural units of chitin and chitosan. (A) *N*-acetylglucosamine unit; (D) Glucosamine unit. In chitosan DA < 50.

Chitosan is a biocompatible, biodegradable, and non-toxic material. It exhibits other significant biological properties, such as wound healing capacity, antimicrobial and hemostatic activities. It is an excellent film former and can be processed into fibers, gels, microspheres-microcapsules, and micro/nanoparticles [5]. Also, because it has free –OH and –NH_2 groups in its structure, it is amenable to chemical modifications that can potentiate some of its properties for certain applications. All of these remarkable physical, chemical, and biological properties have made CS an excellent candidate for applications in cosmetics, food industry, medicine and pharmacy [4]. The preference of chitosan in comparison with other cationic polymers, such as polylysine, polyarginin, or polyethyleneimine for many of these applications relies on its comparatively lower toxicity [6].

Mucoadhesive and absorption-enhancing properties are also found in CS. It opens the tight junctions between cells so that the drug of interest can traverse the mucosal cells. [7,8]. These properties also make CS an ideal candidate for the delivery of drugs and bioactive molecules in general. Numerous reports show the applications of CS in drug delivery, with several reviews on the subject [6,8–10]. Applications include CS as an excipient in tablets, chitosan hydrogels, films, fibers, micro/nanocapsules and micro/nanoparticles.

Nanoparticles of CS are applied in drug delivery, not only by the traditional administration routes (e.g., oral and parenteral routes), but also via mucosal (nasal, pulmonary, vaginal) and ocular routes [11]. Chitosan nanoparticles are as well used in designing non-viral vectors for gene delivery and the delivery of vaccines [12].

Different approaches have been used to produce CS nanoparticles. These include ionotropic gelation [13,14], spray drying [15], water-in-oil emulsion cross-linking [16], reverse micelle formation [17,18], emulsion-droplet coalescence [19,20], nanoprecipitation [21], and by a self-assembling mechanism [22,23].

The self-assembling has been described as the association of certain molecules, macromolecules, or composite materials with themselves to form tridimensional networks or other structures with new distinguishing properties. The self-assembling process can take place at the molecular or supramolecular level [24,25]. It can occur by self-association or by an association with other structures through interactions such as hydrogen bond, van der Waals forces, and ionic or hydrophobic interactions. It can also be caused by an inclusion/complexation mechanism, like the iodine inclusion complex with starch [25].

CS self-assembled (also referred to as self-aggregated) nanoparticles (NPs) are particularly useful for encapsulating hydrophilic as well as lipophilic drugs [26]. Self-assembly can be provoked by the introduction of hydrophobic moieties into the CS molecules by grafting, in order to modify its hydrophobic-hydrophilic balance. The grafting agent can be a hydrophobic moiety, such as cholesterol [27], cholic [28], and deoxycholic acid [29], or 5β-cholanic acid [30], to form micelles that can entrap lipophilic drugs or it can be the drug itself. Frequently, a water soluble CS derivative, such as glycol chitosan [31] or succinyl chitosan [32], is used instead of CS. Another suitable way to generate self-assembled chitosan nanoparticles is through the formation of polyelectrolyte complexes with polyanions [33]. The aim of the present article is to review the main approaches used for preparing chitosan nanoparticles by self-assembly through both procedures, and to illustrate the state of the art in drug delivery.

2. Polyelectrolyte Complexes

Polyelectrolyte complexes (PECs) are formed when the solutions of two polyelectrolytes carrying complementary charges (i.e., a polycation and a polyanion or their corresponding salts) are mixed together. PEC formation is mainly caused by the strong Coulomb interaction between the oppositely charged polyelectrolytes. The formation of complexes brings about at least a partial charge neutralization of polymers [10]. The complexes obtained (also called polysalts) generally precipitate or separate from the solution forming a complex rich liquid phase (coacervate). However, under certain conditions, the polyelectrolytes, with weak ionic groups and significantly different molecular weights at non-stoichiometric mixing ratios, can generate water-soluble PECs on a molecular level [34,35].

The formation of polyelectrolyte complexes is accompanied by the release of small counter-ions into the medium. The increase in entropy produced by the release of these low molecular weight counter-ions to the medium is the main driving force for PEC formation. Although the electrostatic interaction between the complementary ionic groups of polyelectrolytes is responsible for PEC formation, hydrogen bonds, and hydrophobic interactions also contribute to complexing. The arrangement of chains in a PEC can be envisaged as a combination of a disordered scrambled egg-like structure and a highly ordered ladder-like organization (Figure 2). Therefore, the actual structure having hydrophobic and hydrophilic regions makes PECs a particular class of physically cross-linked hydrogels that are sensitive to pH and to other environmental factors such as temperature and ionic strength.

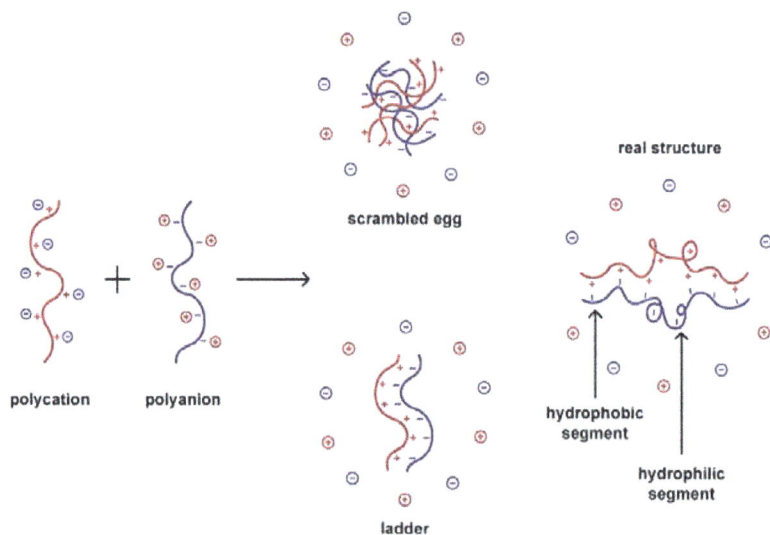

Figure 2. The structure of polyelectrolyte complexes. Scrambled egg and ladder arrangements illustrate extreme situations. The actual structure can be represented as an intermediate one combining hydrophobic ladder-like segments coexisting with disordered hydrophilic regions.

Many factors affect the structure and stability of PECs; these include: the degree of ionization of each one of the polyelectrolytes and their charge density and charge distribution on the polymer chains, polyelectrolytes concentration, mixing ratio (Z), mixing order, the nature of the ionic groups on the polymer chains, molecular weights of the polyelectrolytes, flexibility of the polymer chains, interaction time and temperature and ionic strength, as well as the pH of the medium [36].

As a cationic biopolymer, CS may react with negatively charged polyelectrolytes, giving rise to the formation of PECs [37,38]. Many reports show PECs produced with chitosan and carboxymethyl cellulose (CMC) [39,40], alginate [41–45], poly(acrylic acid) [46,47], pectin [48–51], carrageenans [52,53], heparin [54], and other polyions [55–61].

2.1. Chitosan Based PEC Nanoparticles and Their Application in Drug Delivery

Because of the above mentioned biological properties of CS, many applications of these PECs have been proposed for biomedical purposes, particularly for drug delivery [62]. Hence, researchers have shown special interest in the preparation of chitosan PEC nanoparticles for the delivery of drugs, proteins, genes, and vaccines [36,63,64].

When chitosan PEC particles are formed, they tend to aggregate because of charge neutralization, therefore, at least two conditions are mandatory in order to avoid aggregation and to obtain nanoparticles: the polyelectrolyte solutions must be diluted, and one of the polyions must be in the appropriate excess required so that the charge ratio $(n_+/n_-) \neq 1$ (Figure 3).

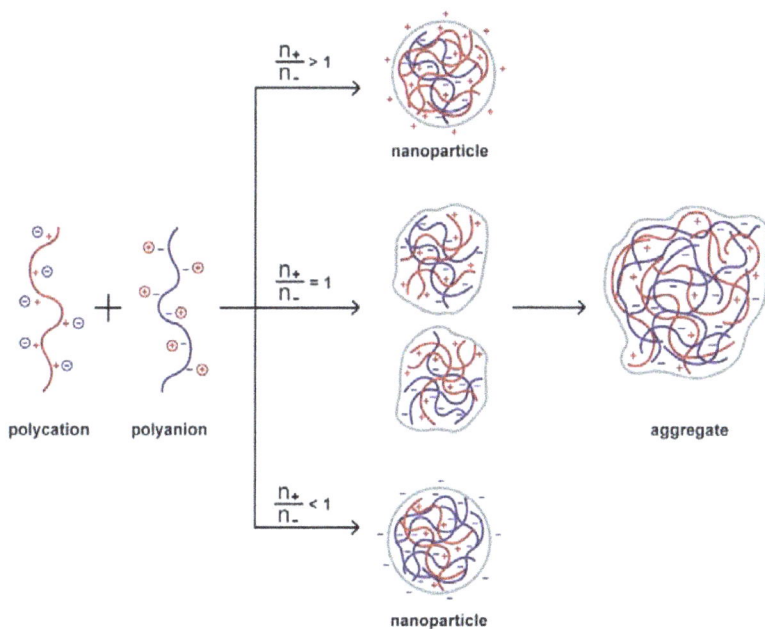

Figure 3. Effect of the polyelectrolytes charge ratio on the size and charge of the polyelectrolyte complexes (PEC) formed. When the charge ratio is different from one, the nanoparticles formed are charged with the same charge as the polyion in excess. If the charge ratio equals one, uncharged particles are formed, thereby producing large aggregates.

Other conditions, such as pH (particularly important in weak polyelectrolyes), ionic strength, and the mixing rate, should be adjusted to the particular chitosan-polyanion pair system selected, since these variables will also influence the size and charge of nanoparticles.

Different preparation methods will result in diverse kinds of nanoparticles, which can be classified as nanoaggregates, nanocapsules or nanospheres. The particular procedure selected can be largely determined by the water solubility of the active agent that will be encapsulated and the polyanion used.

2.1.1. Chitosan-Alginate PEC Nanoparticles

Alginates are a family of anionic polysaccharides extracted from brown algae. They are composed of α-L-guluronic acid (G) and β-D-mannuronic (M) acid units that are linearly linked by 1,4-glycosidic bonds (Figure 4). The M/G ratio and their distribution along the chains (chain microstructure) are strongly dependent on the particular species of algae from which it was extracted [65]. Alginate (ALG) is non-toxic, biocompatible and biodegradable, mucoadhesive, and non-immunogenic. The gelling capacity of ALG in the presence of calcium ions in the so-called "egg-box" model has been extensively employed to prepare gels, capsules, and micro- and nanoparticles for drug delivery [66]. The guluronic units are responsible for the crosslinking reaction; and, the properties of the beads formed, such as strength and porosity, will therefore depend on the alginate source. Other parameters affecting the characteristics of beads are ALG molecular weight, and the concentration of CaCl$_2$ and alginate solutions [65].

Chitosan-alginate PEC nanoparticles are usually prepared by one of the following three procedures.

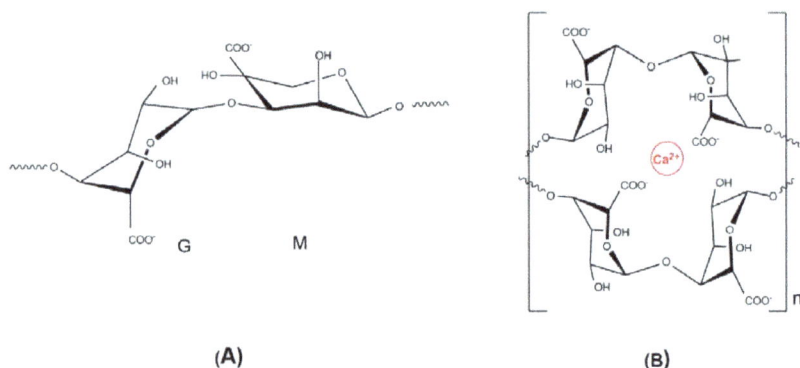

Figure 4. (**A**) Structural units in alginate. (G) Guluronic acid; (M) Mannuronic acid; (**B**) Representation of two G-blocks forming an 'egg box' sequence with a calcium ion.

Plain Complex Coacervation by Mixing Dilute Solutions of CS and ALG

In this procedure, the order of addition of one polysaccharide into the other, the CS/ALG ratio, the molecular weight of both polyelectrolytes and the pH and the ionic strength of the solutions are important factors in determining the relevant parameters of nanoparticles (size, particle charge, stability, encapsulation efficiency).

This procedure was used by Liu and Zhao [67] to prepare negatively or positively charged CS/ALG nanoparticles by dropping a CS solution over the ALG solution. They found that particle sizes varied from 320 to 700 nm, depending on the pH and the ionic strength of the solution. At pH 4.80 in deionized water the sample displayed a narrow unimodal size distribution with an average hydrodynamic diameter (Dh) of 329 ± 9 nm. The Zeta-potential of NPs was also dependent on pH and ranged from +6.34 mV at pH 3.0 to −44.5 mV at pH 10.0. The loading capacities of NPs for ibuprofen and dipyridamole were 14.18% and 13.03%, respectively. Drug release was governed simultaneously by the solubility of the drug and the permeability of the CS/ALG nanoparticles [67].

In a modification of this procedure, a CS solution containing Tween 80 (stirring the chitosan solution with Tween 80 generated chitosan drops) was dropped into a previously prepared solution of an alginate complex with doxorubicin (DOX). The NP suspension was stirred overnight and the doxorubicin loaded CS/ALG NPs were separated by centrifugation. The size of the NPs was 100 ± 35 nm, with a polydispersity index (PDI) of 0.40 ± 0.07, the Zeta-potential was of +35 ± 4 mV, and the encapsulation efficiency (EE) achieved was 95 ± 4% for the optimal formulation (CS/ALG = 2:1) [68].

The reverse procedure was used to encapsulate amoxicillin in CS/ALG nanoparticles. Essentially, a mixture of chitosan, Pluronic F-127 (surfactant), and amoxicillin was prepared in various concentrations of all the components. An aqueous solute on of ALG was sprayed into this mixture with stirring to form NPs. Both of the solutions were at pH 5.0. The process was optimized for variables such as pH and the mixing ratio of polymers, concentrations of polymers, drug and surfactant, using the 33 Box-Behnken design. The resulting particle size, surface charge, drug entrapment percentage, in vitro mucoadhesion, and in vivo mucopenetration of nanoparticles in rat models were analyzed. The optimized formulation with particle size, Zeta-potential and encapsulation efficiencies of 651 nm, +59.76 mV and 91.23%, respectively, showed comparatively low in vitro mucoadhesion as compared to plain chitosan nanoparticles, but excellent mucopenetration and localization [69].

A modified hybrid blending system was developed by Goycoolea et al., which combined the complex coacervation of CS and ALG with the ionotropic gelation of CS with trisodium tripoliphosphate (TTP). The purpose of this combination was to increase the stability in the biological media and for better pharmacological performance than with conventional CS-TPP nanoparticles. In this method, an ALG solution containing TPP was mixed under rapid stirring with the CS solution forming the CS-TPP-ALG nanoparticles. Insulin loaded CS-TPP-ALG nanoparticles were obtained by adding insulin into the ALG-TPP solution before mixing with the CS solution. The average particle size of the insulin-loaded CS-TPP-ALG NPs was 297 ± 4 nm (PDI 0.25), similar to that of the unloaded NPs, which was 307 ± 5 nm (PDI 0.22). High positive Zeta-potential values ~+42 mV were obtained in both cases, providing good stability to the NPs. Insulin loading efficiencies (defined as insulin loaded per weight of nanoparticles) as high as 50.7% were attained [70].

Ionotropic Pregelation of Alginate Followed by Complexation with Chitosan

This is a very common method in which pregelation is usually attained with $CaCl_2$, but other divalent ions may also be used. The active agent can be dissolved or dispersed in the ALG solution or it can be loaded into the resulting CS/ALG nanoparticles. Azevedo et al. used this procedure by setting the initial pH of the ALG and CS solutions to 4.9 and 4.6, respectively. In their formulation, the average size of CS/ALG NPs was 120 ± 50 nm with a Zeta-potential of -30.9 ± 0.5 mV. Vitamin B2 loaded NPs were obtained by dissolving the compound in the ALG solution before the pregelation step. The average size of nanoparticles with vitamin B2 was 104 ± 67 nm (PDI 0.32 ± 0.07) with a Zeta-potential of -29.6 ± 0.1 mV. The nanoparticles showed EE and loading capacity (LC) values of $56 \pm 6\%$ and $2.2 \pm 0.6\%$, respectively [71].

Oil-in-Water (O/W) Microemulsion of Alginate Followed by Ionotropic Gelation and Further Complexation with Chitosan

This method is usually preferred for encapsulating hydrophobic drugs. The preparation of nanocapsules is carried out by emulsifying a solution of the drug (oil phase) into the aqueous sodium alginate solution containing a surfactant, followed by gelation with calcium chloride and CS.

Bhunchu et al. used this method to prepare CS/ALG NPs containing curcumin diethyl disuccinate (CDD). CDD dissolved in acetone (1 mL) was dropped into 20 mL of a dilute ALG solution (0.6 mg/mL) containing a non-ionic surfactant (Pluronic F127, Cremophor RH40™ and Tween 80®). Four mL of a $CaCl_2$ solution (0.67 mg/mL) was added while stirring, followed by sonication. Four mL of the CS solution at various concentrations (0.15–0.45 mg/mL in 1% (v/v) acetic acid) were added with continuous stirring at 1000 rpm for 30 min. After standing overnight for equilibration the CDD loaded CS/ALG NPs were obtained as dispersion in the aqueous solution. Pluronic F127 gave the smallest particle size, 414 ± 16 nm (PDI 0.63 ± 0.05) with the highest Zeta-potential, -22 ± 1 mV. The EE and LC of these NPs were $55 \pm 1\%$ and $3.33 \pm 0.08\%$, respectively. These NPs improved cellular uptake of CDD in Caco-2 cells, as compared to free CDD [72].

A list of some selected examples of CS/ALG PEC nanoparticles based on the different procedures mentioned above is given in Table 1.

Table 1. Chitosan-Alginate PEC nanoparticles. The intervals shown generally indicate extreme values obtained under different preparation conditions.

Procedure	Active agent	Particle size (nm)	Zeta-potential (mV)	Ref.
Complex coacervation				
CS added into ALG CS into ALG-DOX	Ibuprofen Dipyridamole	320 to 700 [b]	+6.34 [b] to −44.5 [b,*]	[67]
	Gatifloxacin [a]	347 [c]	+38.6 [c]	[73]
	Doxorubicin	100 ± 28 [b] 100 ± 35 [c]	+36 ± 3 [b] +35 ± 4 [c]	[68]
ALG added into CS ALG into Thiolated CS	Amoxicillin [a]	264 to >601	+ 35 to + 61.9	[69]
	Fluorescein isothiocyanate	338 ± 16 [b] 266 ± 7 [c]	+34 ± 8 [b] +30 ± 4 [c]	[74]
	Fluorescein isothiocyanate	338 ± 16 [b] 266 ± 7 [c]	+34 ± 8 [b] +30 ± 4 [c]	
ALG + TPP added into CS	Insulin	260–525	+41 to +50	[70]
Ionotropic pregelation of alginate plus PEC coating with CS				
CS into Ca/(ALG + drug)	Insulin	781 ± 61 [b] 748 ± 217	−15 ± 2 [b] −6 ± 2 [c]	[75]
	Vitamin-B2	120 ± 50 [b] 104 ± 67 [c]	−30.9 ± 0.5 [b] −29.6 ± 0.1 [c]	[71]
	Acetamiprid	201.5	−32.1	[76]
CS + EGF into Ca/ALG	EGF-antisense [a]	194–1435	~+30	[77]
CS + plasmid into Ca/ALG	pEGFP plasmid	161	+29.3	[78]
o/w ALG microemulsion followed by ionotropic gelation and further complexation with CS				
	Turmeric oil	522–667	−21.8 to −22.2	[79]
	A.A.	400		[80]
	CDD	410 ± 20	22 ± 1	[72]
LMWAlg + OligoCS	BSA	134–229		[81]

[a] Optimization performed; [b] unloaded particle; [c] loaded particle; A.A., aminoacid derivatives; CDD, curcumin diethyl disuccinate; * pH 3.0.

Inspection of Table 1 reveals that a wide variation in particle size and Zeta-potential is reported for all of the three general procedures devised for preparing CS-ALG PEC nanoparticles. The same happens with the EE and the LC. This is a consequence of the already mentioned dependence of these parameters on multiple variables.

In plain complex coacervation a surfactant is sometimes added to improve the entrapment efficiency and the solubility of the drug [69,73], but it might increase the size of the particles and decrease the Zeta-potential [69]. EE values that are reported in these methods vary from around 50% [69,70] to 95% [67]. The LC is not always declared, but the values of 14.18% and 13.03%, depending on the drug [66], have been reported.

In the method based on the pregelation of alginate encapsulation efficiencies reported are in general higher than 50%. For instance, Azevedo et al. [71] reported an EE of 56 ± 6% for vitamin B2, while other authors declared 73 ± 2% for insulin [75], 62% for acetamiprid [76], and 95.6% for EGF-antisense [77]. However, loading capacities reported were only 2.2 ± 0.6% [71] and 10 ± 2% [75]. For the NPs that are loaded with vitamin B2 the PDI was 0.32 ± 0.07. The other reports did not declare the PDI obtained.

The method based *o/w* microemulsion of ALG followed by ionotropic gelation and complexation with CS produced in general nanocapsules with sizes of about 400 to 660 nm [72,79,80]. The PDI was reported only in reference [74] and was 0.63 ± 0.05. When using low molecular weight polysaccharides, particles sizes ranging from 134 to 229 nm were reported [81]. Encapsulation efficiencies informed were 55 ± 1% for curcumin diethyl disuccinate (LC, 3.33 ± 0.08%) [72] and 88.4% for BSA [81].

2.1.2. Chitosan-Pectin PEC Nanoparticles

Pectin is an anionic hetero-polysaccharide derived from plant cell walls, consisting primarily of 1,4 linked α-D-galactopyranosyl uronic acid residues with 1,2-linked α-L-rhamnopyranose residues interspersed with varying frequencies (Figure 5). Pectin structure also presents a certain amount of neutral sugars (arabinose, galactose, rhamnose, xylose, and glucose). A number of galacturonic acid residues in the pectin are methyl or acetyl esterified. The percentage of galacturonic acid residues that are esterified is known as the degree of esterification (DE).

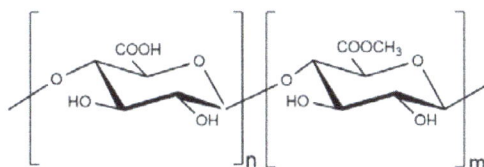

Figure 5. Chemical structure of partially acetylated polygalacturonic acid in pectin.

Pectin is hydrophilic, biocompatible, and biodegradable, and it has low toxicity. As in ALG, pectin with a low methoxyl content (DE < 50%), has the ability to gel in the presence of Ca^{2+} ions generating junction zones between chains with an egg-box structure. Pectins with higher DE can also form gels, provided that there are a sufficient number of blocks of non-esterified uronic acid residues per molecule to allow the formation of a sufficient number of junction zones to form a network. These properties of pectin have been employed to prepare diverse formulations for drug delivery applications.

Galacturonic acid provides pectin with a negative charge in solutions with pH higher than 3.5, permitting the formation of polyelectrolyte complexes with chitosan. The strength of the interaction depended on the degree of esterification of the pectin, with pectins of a relatively low DE (36%) readily forming PECs with CS [82]. PEC formation is also affected by the ratio of pectin to CS and the pH of the solutions [83].

CS-pectin PEC nanoparticles can be prepared by the same methods previously described for CS-ALG PEC nanoparticles. Birch and Schiffman prepared nanoparticles by the complex coacervation technique adding pectin at the appropriate CS-to-pectin ratio to the CS solution. They thereby obtained particle sizes ranging from 560 ± 10 nm to 1000 ± 40 nm. The Zeta-potential varied from +20 ± 1 mV to +26 ± 1 mV. When the addition order was reversed the particle size increased from 460 ± 20 nm to 1110 ± 30 nm and the Zeta-potential changed from +19 ± 1 to +28 ± 1 mV [84].

Rampino et al. prepared CS-pectin PEC nanoparticles by two different procedures: a) coating, by adding a dispersion of low molecular weight CS NPs previously prepared by the ionotropic gelation of CS with TPP to a pectin (from apple and citrus fruit) solution; and, b) blending, by adding a CS solution to a solution of pectin and TPP. Nanoparticles were charged with ovalbumin (OVA) and bovine serum albumin (BSA) as the model proteins. They pointed out that the blending technique could be advantageous because, by being a one-step preparation, it is highly desirable for a scale-up process. Additionally, it gives the possibility of tuning the size and Zeta-potential by properly selecting the ratios of CS, pectin, and TPP. However, they found that there was a decrease in the loading of BSA and OVA in the case of the blending technique (loading efficiency, ranging between 16% and 27%) due to the electrostatic interactions of CS with the protein and pectin, both negatively charged. Therefore, they concluded that the selected technique would depend on the physicochemical characteristics of the polymer and the protein involved [85]. Some of the parameters reported in their work are listed in Table 2, together with some selected examples of CS-pectin preparation procedures reported by other authors.

Table 2. Chitosan-Pectin PEC nanoparticles. The intervals shown generally indicate extreme values obtained under different preparation conditions.

Procedure	Active agent	Particle size (nm)	Zeta-potential (mV)	Ref.
Complex coacervation				
Pectin added into CS	Insulin	441 ± 32 [a] 580–896 [b] * 650 ± 86 [b]	+62 ± 3 [b] +33 ± 4 [b]	[86]
	Curcumin	10–59 (dry NPs)		[87]
	Insulin	1175–2618 [a] 964–2510 [b]	−22.5 to +35.0 [a] −22.4 to +33.2 [b]	[23]
	Nisin	301–712 [b]		[88]
	None	560–1000	+20 to +26	[84]
CS added into Pectin	None	460–1110	+19 to +28	[84]
Combined ionotropic gelation and complex coacervation				
Pectin + TPP added into CS	Insulin	375–7239	+10.6 to +32.7	[86]
CS added into Pectin + TPP	OVA	250–750 [a]	−20 to −29 [a]	[85]
CS + TPP added into Pectin	BSA	200–400 [a] 700–1250 [b]	−15 to −45 [a] −38 [b]	[85]
Ionotropic pregelation of pectin plus PEC coating with CS				
CS added into Pectin + CaCl$_2$	OVA	419 [a]	−30.4 [a]	[84]
		302–409 [b]	−21.9 to −26.0 [b]	

[a] Unloaded particle; [b] loaded particle * The CS solution contained Ca^{2+} ions.

Not all of the references in Table 2 report parameters, such as EE, LC, and PDI. In plain complex coacervation, Maciel, et al. [23] reported microparticles with size less than ~2500 nm using charge ratios (n_+/n_- given by the chitosan/pectin mass ratio) of 0.25 and 5.00, with PDIs of 0.25 ± 0.06 and 0.40 ± 0.06, respectively. The highest EE (~62.0%) of the system was observed at a charge ratio (n_+/n_-) 5.00. Andriani et al. [87] added glutaraldehyde to the chitosan-pectin mixed solution. This way, they obtained encapsulation efficiencies varying from 24.0% (LC 6.30%) to 94.7% (LC 21.05%). Combining ionotropic gelation and complex coacervation, Al-Azi, et al. [86] reported a PDI of 0.67–0.71. Insulin association efficiency varied from 2.40 ± 0.33% (LC 0.31 ± 0.04%) at pH 3, to 4.06 ± 0.12% (LC 0.52 ± 0.01%) at pH 5. Using Ca^{2+} ions caused a marked improvement in insulin association efficiency of nanoparticles.

2.1.3. Chitosan-Dextran Sulfate PEC Nanoparticles

Dextran sulfate (DS) is a biodegradable and biocompatible negatively charged branched polyanion that is able to interact with positively charged polymers. It is a high-molecular weight, branched-chain polysaccharide polymer of D-glucose containing 17–20% sulfur. The straight chain consists of approximately 95% α-(1,6) glycosidic linkages. The remaining α-(1,3) linkages account for the branching of dextran (Figure 6).

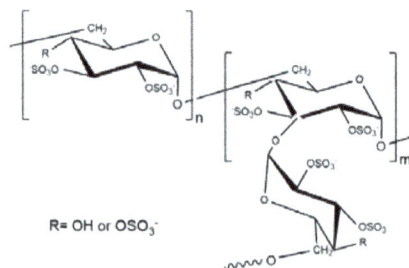

Figure 6. The chemical structure of dextran sulfate.

DS has been used as an anticoagulant and with applications in drug delivery. For instance, it was used to mask the positive charge of doxorubicin (DOX) before its addition to a CS solution for nanoparticle formation by ionotropic gelation with TPP. This modification doubled DOX EE relative to the controls, and made it possible to reach loadings of up to 4.0 wt % DOX [89].

CS-DS PEC nanoparticles are almost invariably prepared by simple coacervation. The factors affecting the mechanism for the formation of these nanoparticles: the mode of addition, charge mixing ratio, pH and ionic strength of the media, and the molar mass of both components have been thoroughly reviewed by Schatz et al. [90,91].

There are numerous reports on the preparation of CS-DS PEC nanoparticles with a potential application for the delivery of proteins (insulin, BSA), growth factors [92–94], immunoglobulin-A [95], and vaccines [96,97]. Recently, fluorescein isothiocyanate loaded CS-DS nanoparticles (FCS-DS NPs; mean size, 400 nm (PDI 0.25 ± 0.01); and, surface charge, +48 mV) were topically applied to the porcine ocular surface where it was retained for more than 4 h. After 6 h under the topical FCS-DS NPs treatment, particles were accumulated in the corneal epithelium but were not found in the corneal stroma. However, when the epithelium was removed, the FCS-DS NPs penetrated the stroma. These results indicate that FCS-DS NPs are potentially useful for drug/gene delivery to the ocular surface and to the stroma when the epithelium is damaged [98].

Most of nanoparticles formulations reported describe processing factors affecting the characteristics of CS-DS nanoparticles, including their physicochemical properties as well as the optimal conditions for their preparation. Some examples are listed in Table 3.

Table 3. Chitosan-Dextran sulfate PEC nanoparticles. The intervals shown generally indicate extreme values obtained under different preparation conditions.

Procedure	Active agent	Particle size (nm)	Zeta-potential (mV)	Ref.
Complex coacervation				
DS added into CS	BSA	>244 [a]	−47.1 to −60 [a]	[99]
		478–1138 [b]	−28.0 to +56.4 [b]	
	Rhodamine 6G	245–3521 [b]	−31.0 to +34.0 [b]	
CS added into DS	Insulin	489–665 [b]	−0.4 to −21.5 [b]	[100]
		527–1577 [b]	−20.6 to +11.5 [b]	[101]
	Amphotericin B	616–891 [a]		[93]
		644–1040 [b]	−27 to −37	
	REPIFERMIN®	239	−18.4	[94]
		306	−15.5	
Mixing with agitation	Hydralazine	290 ± 60 [a]	−7 ± 4 [a]	[102]
		340 ± 50 [b]	−5 ± 1 [b]	

[a] Unloaded particle; [b] loaded particle.

In Table 3, reference [99] illustrates that different results are obtained for the same system when used to encapsulate two different substances, BSA and Rhodamine 6G, by complex coacervation. In this work, the size of the BSA loaded CS-DS NPs varied from 478 nm (PDI 0.64; EE 96.8%; LC 81.6%) to 1138 nm (PDI 0.97; EE 53.2%; LC 29.3%). However, the Rhodamine 6G loaded nanoparticle sizes were higher, varying from 545 nm (PDI 0.60; EE 98%; LC 31%) to 3521 nm (PDI 0.68; EE 42%; LC 18%). In both cases, the bigger NPs were more polydisperse and had lower LC and EE [99].

Sarmento et al. prepared CS-DS PEC nanoparticles containing insulin with association efficiencies varying from 85.4 ± 0.5% to 72 ± 3%, depending on CS/DS mass ratio [100]. In a later article [101], these authors evaluated the pharmacological activity of insulin-loaded CS-DS PEC nanoparticles following oral dosage in diabetic rats. On this occasion, they introduced small changes in the preparation parameters and obtained somewhat lower association efficiencies, ranging from 69 ± 1% (LC, 2.3 ± 0.6%) to 24 ± 2% (LC, 2.0%) [101]. This influence of the preparation parameters on the characteristics of the PEC nanoparticles can be used to modify them to meet the specific requirements of a determined application.

PECs of soluble chitosan derivatives with DS have also been formulated to overcome the insolubility of chitosan in neutral and basic media. Glycol chitosan (GC) and DS solutions were mixed together to prepare GC-DS PEC nanoparticles that were loaded with the antifolic agent methotrexate (MTX), aiming to increase its efficacy for the treatment of brain tumors. EE was as high as 87% for a particle size of 149 ± 41 nm (PDI 0.7 ± 0.1). In vitro experiments indicated their potential for the controlled delivery of the drug to the brain [103].

PEC nanoparticles of water soluble *N,N,N*-Trimethyl chitosan (TMC) and DS were prepared by adding DS solutions to TMC solutions at the desired pH values (5, 8, 10, and 12). The release efficiency and ex vivo nasal toxicity evaluation were assessed after loading a model drug, ropinirole hydrochloride, into an optimized PEC formulation at pH 10 (particle size, 255 ± 10 nm; Zeta-potential, −4 ± 1 mV; LC = 82 ± 2%; EE = 87.9 ± 0.6%). Data indicated that the PECs produced at alkaline pH have a reliable formulation for nasal administration. They are biologically compatible with the mucosal surface, thereby being potentially applicable as carriers for nose to brain drug delivery [104].

2.1.4. Chitosan-Carboxymethyl Chitosan PEC Nanoparticles

O-Carboxymethyl chitosan (CMCS) is a water-soluble amphiphilic derivative of chitosan that conserves the biological properties of native chitosan with increased antibacterial activity [105]. The structural unit of CMCS is shown in Figure 7. CMCS has been applied in biomedicine, especially in drug delivery where CMCS nanoparticles prepared by ionotropic gelation have demonstrated promising results for drug [106,107] and antigen delivery [108].

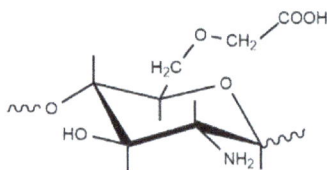

Figure 7. The structural unit of carboxymethyl chitosan.

The pKa of CMCS is 2.0–4.0, so that at pH above 4 it is negatively charged and forms polyelectrolytes complexes with chitosan [109]. CS-CMCS PEC nanoparticles were produced by complex coacervation. Wang et al. developed insulin-loaded nanogels with opposite Zeta-potential by adding a previously prepared insulin-CMCS solution into a CS solution (particle size, 260 ± 5 nm; PDI 0.08 ± 0.02; Zeta-potential, +17.2 ± 0.5 mV for insulin: CMCS/CS-NGs(+)) or inversing the order of addition (particle size, 243 ± 4 nm; PDI 0.03 ± 0.01; Zeta-potential, −15.9 ± 0.5 mV for insulin: CMCS/CS-NGs(−)), respectively. Encapsulation efficiencies of about 75% and loading capacities near 30% were attained in both cases. They observed that negatively charged particles exhibited enhanced mucoadhesion in the small intestine and had better intestinal permeability in the jejunum, indicating there was a better performance in insulin: CMCS/CS-NGs(−) for blood glucose management than in those positively charged [110,111].

CS-CMCS nanoparticles have also been prepared by combining ionotropic gelation and complex coacervation. CMCS and TPP at varying concentrations were blended with a previously prepared mixture of DOX and CS solutions. Nanoparticles sizes between 249 ± 10 nm (Zeta potential, −27.6 ± 0.8 mV) and 362.7 ± 8.4 nm (Zeta potential, −42 ± 1 mV) with encapsulation efficiencies and loading capacities of around 70.5% and 20%, respectively, were obtained depending on the preparation conditions. Results from in vivo experiments indicated that CS/CMCS-NPs were efficient and safe for the oral delivery of DOX [112]. After certain modifications of the preparation procedure, positively charged CS/CMCS-NPs were obtained. This time, the DOX aqueous solution was premixed with CMCS and the CS solution and TPP were subsequently blended with the mixture under agitation. Nanoparticles sizes were of between 197 ± 11 nm (PDI 0.235; Zeta-potential, +37.6 ± 0.8 mV) and 442 ± 7 nm (PDI 0.635; Zeta-potential, +12.2 ± 0.6 mV), depending on the pH of the medium. In vivo studies revealed that CS/CMCS-NGs had a high transport capacity by paracellular and transcellular pathways, which guaranteed the excellent absorption of encapsulated DOX throughout the entire small intestine [113].

2.1.5. Chitosan-Chondroitin Sulfate PEC Nanoparticles

Chitosan-chondroitin sulfate PEC NPs have been prepared by complex coacervation and the influence of the preparation conditions on the properties of nanoparticles was reported [114,115]. Chondroitin sulphate is a linear glycosaminoglycan (GAG) that is composed of alternating D-glucuronate and β(1,3) linked *N*-acetyl-D-galactosamine-4- or 6-sulfate (Figure 8). It is found in cartilage, bone and connective mammalian tissue. Chondroitin sulphate (CHOS) has shown in vivo anti-inflammatory properties in animal models and in vitro regulation of chondrocyte metabolism, such as the stimulation of proteoglycan and collagen synthesis and the inhibition of the production of cytokines that are involved in cartilage degradation [116]. Its biological properties have stimulated the preparation and evaluation of CS-CHOS nanoparticles for drug/gene delivery [117,118] and delivery of platelet lysates [119]. CS-CHOS nanoparticles have been suggested as a novel delivery system for the transport of hydrophilic macromolecules [120].

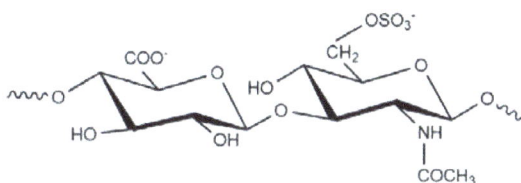

Figure 8. Chemical structure of chondroitin sulfate.

2.1.6. Chitosan-Heparin and Chitosan-Hyaluronan PEC Nanoparticles

CS PECs with other two glycosaminoglycans, hyaluronic acid (hyaluronan, HA) and heparin (HEP), have also been used to prepare nanoparticles. HA is a high molecular weight linear polysaccharide that is composed of β(1,3) linked D-glucuronate and *N*-acetyl-D-glucosamine units. It is present in all soft tissues of higher organisms, and in particularly high concentrations in the synovial fluid and vitreous humor of the eye. It plays a vital role in many biological processes, such as tissue hydration, proteoglycan organization, cell differentiation, and angiogenesis, and acts as a protective coating around the cell membrane. On the other hand, HEP has a more heterogeneous composition, but its main disaccharide unit is composed of D-glucuronate-2-sulfate (or iduronate-2-sulfate) and α(1,3) linked *N*-sulfo-D-glucosamine-6-sulfate, which provides it with the highest negative charge density of any known biological macromolecule (Figure 9). HEP can be primarily found on the cell surface or in the extracellular matrix, attached to a protein core. Heparin is a well-known anticoagulant drug and is extensively used in medical practice [121]. The important bioactivity of both GAGs has stimulated the preparation of CS-HA and CS-HEP PEC nanoparticles due to their high potential in applications as delivery systems for these macromolecules, particularly in tissue engineering [58,122–124].

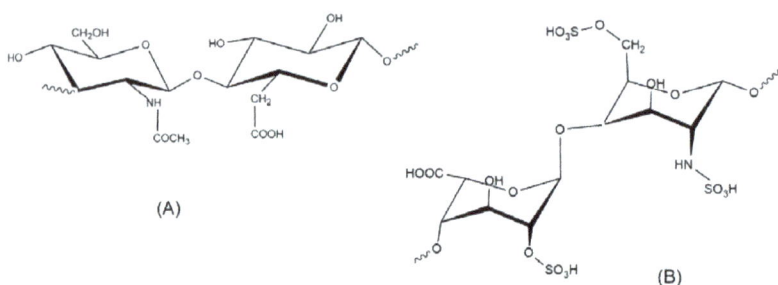

Figure 9. Chemical structures of (**A**) Hyaluronic acid and (**B**) Heparin.

2.1.7. Chitosan and Poly(γ-Glutamic Acid) PEC Nanoparticles

Poly(γ-glutamic acid) (γ-PGA) is an anionic, natural polypeptide that is made of D- and L-glutamic acid units, joined together by amide linkages between the α-amino and γ-carboxylic acid groups (Figure 10). PEC formation between CS and γ-PGA has been evaluated in terms of physical and chemical properties. In experimental trials, it has shown wound-healing efficacy with a potential application as a wound dressing material [125].

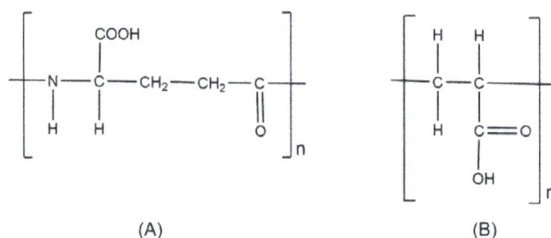

Figure 10. Chemical structures of (**A**) Poly(γ-glutamic acid) and (**B**) Poly(acrylic acid).

PEC nanoparticles of γ–PGA and low molecular weight CS were obtained by complex coacervation by Lin et al. by adding an aqueous γ-PGA solution at pH 7.4 to a low molecular weight CS solution at different pH values. The NPs prepared at pH 6.0 and a CS/γ-PGA ratio of 4.5:1.0 (*w*/*w*) had a Zeta-potential of +32 ± 2 mV with a particle size of 146 ± 2 nm (PDI 0.21 ± 0.02). Insulin loaded NPs were obtained by including insulin in the γ–PGA solution before its addition to the CS solution. Nanoparticles with a mean size of 198 ± 6 nm (PDI 0.30 ± 0.09) and a Zeta-potential of 28 ± 1 mV were obtained when the amount of insulin added was 84 µg/mL (EE 55 ± 3, LC 14.1 ± 0.9. Animal studies indicated that the insulin loaded NPs enhanced insulin adsorption and reduced the blood glucose level in diabetic rats [126]. Hajdu et al. [127] reported the effect of pH, polymer ratios, concentrations, and orders of addition on the physicochemical properties of NPs.

The same procedure was used to prepare exendin-4 loaded NPs, only that in this case, the CS solution contained distinct metal ions (Cu^{2+}, Fe^{2+}, Zn^{2+} or Fe^{3+}) to enhance the drug loading efficiency. Loading efficiency of 61 ± 2% (LC 15 ± 2%) was achieved for exendin-4 loaded NPs formed with Fe^{3+}. Their particle size was 261 ± 26 nm [128].

Nanoparticles of γ-PGA and CS have also been prepared by the combination of ionotropic gelation and complex coacervation. To this end, the insulin and γ-PGA solutions were premixed. Afterwards, TPP and $MgSO_4$ solutions were mixed together and were added to the insulin and γ-PGA mixture. The resulting solution was then added by flush mixing with a pipette tip into the aqueous CS solution and the nanoparticles were then formed. These NPs also resulted in a promising carrier for the improved trans mucosal delivery of insulin in the small intestine [129,130].

More recently Pereira et al. used the pregelation method to prepare CS/γ-PGA PEC nanoparticles to be used as a nanocarrier system for the plant growth regulator gibberellic acid (GA3). To this end, a CaCl$_2$ solution was added to a solution of γ-PGA at pH 4.9. Then, a CS solution at pH 4.5 was added to the γ-PGA/CaCl$_2$ solution while stirring, using a peristaltic pump. To prepare GA3 loaded NPs, the plant hormone was added to the γ-PGA/CaCl$_2$ before the addition of the CS solution. The unloaded γ-PGA/CS nanoparticles presented an average size of 117 ± 9 nm (PDI 0.43 ± 0.07) and Zeta-potential of −29.0 ± 0.5 mV at pH 4.4. The corresponding values for the GA3 loaded γ-PGA/CS nanoparticles were 134 ± 9 nm (PDI 0.35 ± 0.05) and −27.8 ± 0.5 mV at pH 4.4, respectively. The encapsulation efficiency of the GA3 particles was 61%. In laboratory experiments using *Phaseolus vulgaris* seeds, the γ-PGA/CS-GA3 NPs showed high biological activity, with an enhanced rate of germination when compared with the free hormone. The encapsulated GA3 was also more efficient than the free GA3 in the increase of leaf area and the induction of root development, demonstrating the considerable potential of this system for its use in the field [131].

2.1.8. Chitosan-Poly(Acrylic Acid) PEC Nanoparticles

Poly(acrylic acid) (PAA) is a biocompatible linear anionic polyelectrolyte that readily reacts with CS, generating polyelectrolyte complexes by the electrostatic interaction between its COO$^-$ groups and the NH$_3^+$ groups of chitosan [33,38].

Hu et al. prepared CS-PAA PEC nanoparticles by template polymerization of acrylic acid in chitosan solution using chitosan as the template. Positively charged NPs with the mean size and Zeta-potential of 206 ± 22 nm and +25.3 ± 3.2 mV, respectively, were obtained with 70% yield. These NPs were loaded with silk peptide powder (SP) with an encapsulation efficiency of 82%. Release experiments showed a marked pH dependence of the peptide release profile. They also obtained CS-PAA PEC NPs by complex coacervation by dropping the CS solution into the solution of PAA and vice versa, to study the effect of reversing the order of addition on the resulting nanoparticles. When CS was added to PAA, negatively charged particles were obtained with a mean size of 436 ± 78 nm and a Zeta-potential of −22.2 ± 3.6 mV. On the other hand, adding PAA solution into the CS solution produced positively charged NPs with a mean size and Zeta-potential of 358 ± 46 nm and +47 ± 3 mV, respectively. The order of addition also influenced the microstructure of NPs. Transmission electron micrographs of dry nanoparticles showed that NPs that were obtained by adding the CS solution over the solution of PAA had a hollow core, in contrast with the nanoparticles obtained with the reverse addition method, which presented a compact core [132]. In a further study, it was found that nanoparticle size was affected by the molecular weight of CS and PAA, the ratio of the amino group to the carboxyl group (n_a/n_c) and incubation temperature [133].

Davidenko et al. examined the influence of some experimental parameters such as the pH of the polyelectrolyte solutions, their concentrations and the purification procedure on the dimensions of nanoparticles and their size distribution. NPs were formed by the dropwise addition of an aqueous solution of PAA into the corresponding volume of an aqueous solution of CS of a determined concentration with high-speed magnetic stirring (ca. 1300 rpm). The ratio of primary amino groups in CS to carboxylic groups in PAA was fixed at 1.25. They showed that it was possible to obtain nanometric particle suspensions at concentrations of below 0.1%. The most convenient pH values for obtaining CHI-PAA NPs with an optimum yield (nearly 90%) are 4.5–5.5 for CS and 3.2 for PAA. Under these conditions, the size of NPs was 0.477 ± 0.008 nm. Particle sizes of approximately 130–140 nm were obtained at other pH values, but yields were lower than 45%. It was found that purification by dialysis could provoke a drastic change both in the distribution profile and in the particle size of the complex. To avoid this the pH of the NPs dispersion should be as near as possible to the pH of the outer dialysis solution [134]. CS-PAA PEC nanoparticles obtained by this procedure were loaded with 5-fluoruracil (5-Fu) and the release profiles at pH 2 and 7.4 were obtained. At pH 2 almost 100% release was achieved after two hours, whereas at pH 7.4 only 65% of the loaded drug was released after nine hours. At this pH constant release was observed after the first 90 min [135].

The complex coacervation procedure has also been used for preparing CS-PAA PECs nanofiber structures with average fibre diameters of 210 to 910 nm and Zeta-potentials of +39 ± 1 mV to −22 ± 3 mV, respectively. These parameters vary according to the preparation conditions (volume ratio of CS to PAA, final suspension pH, concentration and molecular weight of CS, incubation time and reaction temperature). Nanofibers can bind plasmid DNA very well and show a potential to enhance gene transfer in tissue engineering applications [136,137].

2.1.9. Chitosan PEC Nanoparticles with Other Polyanions

The preparation of CS PEC nanoparticles for the delivery of drug and therapeutic proteins is continuously increasing. They include other polyanions of natural origin, like carrageenan [138,139], carboxymethyl gum kondagogu [140], and gum arabic [141], as well as synthetic ones. Examples of the latter are poly(malic acid) [142], poly(2-acrylamido-2-methylpropanesulfonic acid) [143], and polystyrene-block-poly(acrylic acid) [144]. The methods that were used for the preparation of these nanoparticles are based on the general techniques already described, and will therefore not be discussed here.

3. Hydrophobic Modification of Chitosan and Derivatives for Self-Assembly

The hydrophobic modification of chitosan and chitosan derivatives enables an appropriate hydrophilic/hydrophobic balance to promote self-assembly in an aqueous or polar medium. This modification is usually achieved by grafting hydrophobic moieties to the polysaccharide chains. The hydrophobically modified chitosan chains self-aggregate in the hydrophilic media as illustrated in Figure 11. The following sections are shown to illustrate the state of the art of this method of chitosan and chitosan derivatives NPs preparation.

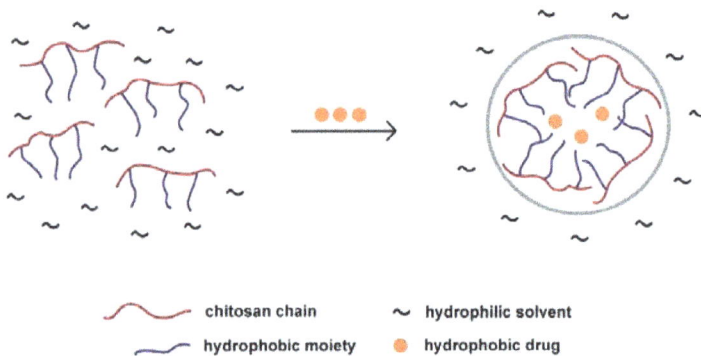

Figure 11. Schematic representation of hydrophobically modified chitosan self-assembly. Aggregates can entrap hydrophobic drugs in their hydrophobic core.

3.1. Hydrophobically Modified Chitosan and Chitosan Oligosaccharides

Deoxycholic acid-modified CS self-aggregates have been proposed as a gene delivery system for DNA transfection in cells [145,146]. This system is based on the complex formation between plasmid DNA and positively charged chitosan self-aggregates, which produces micelle-like nanoparticles having controlled dimensions for the effective gene delivery to cells. The hydrophobic modification of chitosan was accomplished with deoxycholic acid that is mediated by carbodiimide coupling (1-ethyl-3-(3-dimethylaminopropyl) carbodiimide, EDC) for amide bond formation. Self-aggregates were obtained by varying the chitosan/deoxycholic acid ratio (degree of substitution of chitosan, DS from 0.02 to 0.1) and the molecular weight of the reacting CS (molecular weight, MW from 5 to 200 kDa). They exhibited hydrodynamic sizes ranging from 132 to 300 nm. For CS molecular weights higher than 40 kDa, a transition from a bamboo-like cylindrical structure to a poorly organized bird nest-like structure of self-aggregates was proposed. The DNA-CS complex formation had strong dependency on the size and structure of CS self-aggregates and significantly influenced gene transfection efficiency (up to a factor of 10) [146].

Similarly, Wang et al. prepared cholesterol-modified chitosan self-aggregates with succinyl linkages mediated by EDC coupling amidation of CS, thus attaining a DS of 0.073 and hydrodynamic diameters of 417.2 nm. Epirubicin was used as a model anticancer drug. It was physically entrapped into the cholesterol-CS self-aggregates, forming almost spherical nanoparticles of 338.2 to 472.9 nm with the epirubicin loading content increasing from ca. 8 to 14%. The controlled release of epirubicin from the loaded nanoparticles was slow, reaching a total release of 24.9% in 48 h [147].

CS-cholesterol self-aggregates were also synthesized with another approach. Prior phthaloylation of CS enabled the esterification of the primary –OH group at C6 with EDC/N-Hydroxysuccinimide pre-activated cholesterol succinate. Later, CS deprotection afforded 6-O-cholesterol-modified chitosans (DS of 0.017, 0.04, and 0.059) that self-assembled, forming nanoparticles of 100–240 nm sizes. These NPs were capable of physically entrapping the all-trans retinoic acid with different drug loading contents, encapsulation efficiencies, and particle sizes. The sustained release of the all-trans retinoic acid extended over 72 h [148].

Chitosan oligosaccharides (low molecular weight CS produced by depolymerization) are usually preferred over high molecular weight CS for pharmaceutical applications [149]. Chitosans with molecular weights ranging from few hundreds daltons (c.a. trimers, tetramers) to 20 kDa have been referred to as chitosan oligomers [3]. Thus, Hu et al. prepared a CS oligosaccharide (ca. 19 kDa weight average molecular weight) hydrophobically modified with stearic acid and encapsulated paclitaxel or doxorubicin for their controlled delivery [149–152]. CS oligosaccharide (COS) modification was conducted with stearic acid by an EDC mediated amide linkage reaction, achieving COS substitution degrees of 0.035, 0.05, 0.12, 0.255, and 0.42 [150–153]. Further glutaraldehyde cross-linking of COS micelle shells before and after paclitaxel physical entrapping enabled drug loading contents of up to 94% and to control the micelle size and paclitaxel release rate [150]. A reduction of micelle diameters from 322.2 to 272.0 nm was observed after glutaraldehyde cross-linking for the blank COS-stearic acid particles and from 355.0 to 305.3 nm for the doxorubicin-loaded COS-stearic acid particles. The Zeta-potential of particles was reduced from +57.1 to +34.2 mV and from +69.1 to +51.8 mV, respectively [150]. Shell cross-linking of doxorubicin-loaded COS-stearic acid micelles also showed enhanced cytotoxicity to A549, LLC, and SKOV3 cancer cell lines [150].

To reduce the observed initial burst release during the dilution of doxorubicin-loaded COS-stearic acid micelles by body fluid, stearic acid was also physically encapsulated into the micelle core [152]. The hydrodynamic diameter of stearic acid-loaded COS-stearic acid micelles increased significantly from 27.4 nm to ca. 60 nm for a 10 wt % of stearic acid/COS-g-stearic acid micelles, while Zeta-potential decreased from +51.7 mV to ca. +35 mV [151]. The incorporation of stearic acid that was physically entrapped in the core of doxorubicin-loaded COS-g-stearic acid micelle significantly reduced the drug release rate.

Hu et al. also studied the dual functionalization of COS with stearic acid and doxorubicin cis-aconitate [152]. To this end, a previously prepared COS-stearic acid conjugate (DS in stearic acid of ca. 0.06) was further reacted with doxorubicin cis-aconitate by EDC mediated amidation. This produced COS conjugates with doxorubicin contents of 3, 6, and 10%. DOX-g-COS-g-stearic acid self-aggregated in an aqueous medium giving micelle sizes of 40.1, 70.7, and 105.8 nm, respectively, and Zeta-potential values of +43.7, +40.2, and +32.0 mV, respectively [152].

Chitosan has also been hydrophobically modified with different acyl groups mediated by amide linkage formation with different anhydrides and acyl chlorides such as DL-Lactide (PLA unit modifying the CS), propionic and hexanoic anhydrides, nonaoyl chloride, lauroyl chloride, pentadecanoyl chloride, and stearoyl chloride [153,154]. It was observed that the micelle size of blank CS-PLA increased with an increase of the degree of substitution with PLA units or with the increase of side chain length for the different acyl groups (propionate, hexanoate, nonanoate, etc.). Furthermore, the

Zeta-potential changed from +26.0 mV for propionyl chitosan to +10.2 mV for hexanoyl chitosan and remained ca. +13 to +15 mV for the other acyl chitosans. Drug loading content and drug release rate were also influenced by the CS substitution degree or the chain length of the acyl substituents of CS. Rifampin loading content increased and drug release rate decreased with the increase of CS substitution with PLA units [154]. Vitamin C loading content increased and drug release rate decreased with the chain length of the acyl group modifying CS [154].

Water soluble chitosan *N*,*O*6-acetyl chitosan was prepared for future hydrophobic modification with different steroids and DL-α-tocopherol for the sustained release of agrochemicals, testosterone and vitamin E [155]. Drug content achieved values between 11.8 and 56.4 wt %. The CS-steroid and CS-tocopherol micelles formed showed hydrodynamic sizes of ca. 200 to 360 nm in phosphate buffered saline solution (PBS), with Zeta-potential values varying from +7 to +22.7 mV in bi-distilled water. Sustained releases were achieved for the steroids and tocopherol from the CS particles and the biological activity of the released drug appeared unaffected [155].

Amphiphilic block or graft copolymers of phthaloyl chitosan with different materials as poly(ethylene glycol), *N*-vinyl-2-pyrrolidone, and ε-caprolactone have a wide range of pharmaceutical applications [156–162]. For example, *N*-phthaloylchitosan-*g*-mPEG micelles have been physically loaded with camptothecin and all-trans retinoic acid for their controlled release [156–158]. These micelles exerted a protective effect from hydrolysis on the loaded drug (camptothecin, which is sensitive to hydrolysis of the lactone group) or photodegradation (all-trans retinoic acid). Furthermore, a continuous release without an initial burst of prednisone acetate from *N*-phthaloylchitosan-*g*-polyvinylpyrrolidone micelles was achieved [159].

There are also several reports showing that chitosan-*graft*-polycaprolactone nanomicelles have been physically loaded with 7-ethyl-10-hydroxy-camptothecin, BSA, paclitaxel, and 5-fluorouracil [160–163].

Another amphiphilic copolymer of CS was synthesized from *N*-acetyl histidine as the hydrophobic segment and arginine-grafted chitosan by EDC carbodiimide-mediated coupling for the controlled delivery of doxorubicin [164]. The key finding was the effectiveness of doxorubicin loaded *N*-acetyl histidine and arginine-grafted CS for the suppression of both the sensitive and resistant human breast tumor cell line (MCF-7) in a dose- and time-dependent pattern.

More details of prepared hydrophobically modified CS and CS oligosaccharide conjugates can be found in Table 4.

Table 4. Hydrophobically modified chitosan and chitosan oligosaccharides.

Hydrophobic Moiety	Active agent	Particle size (nm)	Zeta-potential (mV)	Ref.
deoxycholic acid	DNA	162 ± 18 [a]		[145]
		~300 [b]		
		130–300 [a]		[146]
cholesterol	Epirubicin	417 ± 18 [a]		[147]
		338–473 [b]		
6-*O*-cholesterol	All-trans retinoic acid	100–240 [a]	+24.5 to +25.9 [a]	[148]
		192–222 [b]		
stearyl	Paclitaxel	28.1–74.6 [a]	+39.0 to +53.2 [a]	[149]
		35.8–175.1 [b]	+44.0 to +58.7 [b]	
	Doxorubicin	272–322 [a]	+34.2 to +57.1 [a]	[150]
		305–355 [b]	+51.8 to +69.1 [b]	
		27.4 ± 2.4 [a]	+52 ± 3 [a]	[151]
		20.4 ± 1.1 [b]	+53.1 ± 14.4 [b]	
stearyl + doxorubicin	Doxorubicin	40.1–105.8 [b]	+32.0 to +43.7 [b]	[152]
Acyl	Rifampin	154–181 [a]		[153]
		163–210 [b]		
	Vitamin C	444–487 [a]	+10.2 to +28.9 [a]	[154]
		216–288 [b]	+5.9 to +18.4 [b]	
N,*O*6-acetyl + steroid	Steroids	197–358 [b]	+7 to +22.7 [b]	[155]
N,*O*6-acetyl + tocopherol	Vitamin E	275 ± 5 [b]	+14.9 ± 0.7 [b]	

Table 4. *Cont.*

Hydrophobic Moiety	Active agent	Particle size (nm)	Zeta-potential (mV)	Ref.
phthaloyl	Camptothecin	~170 [a] ~200–267 [b] ~50–100 [a] ~100–250 [b]		[156] [157]
	All-trans retinoic acid Prednisone acetate	~50–100 [a] ~80–160 [b] 89.8 [a] 143.3 [b]		[158,159]
polycaprolactone, (Chitosan-grafted)	7-Ethyl-10-hydroxy-camptothecin	47–113 [a] 63–152 [b]	+26.7 to +50.8 [a] +25.6 to +48.8 [b]	[160]
	BSA	168.44 [b] 200.7 [b] 435 ± 25 [a]		[161]
	Paclitaxel	408–529 [b] 61.4–108.6 [a]	+27.5 ± 1.1 [a] +30.9 to +33.3 [b]	[162]
	5-Fluorouracil	67.9–96.7 [b]	+18.9 to +43.1 [b]	[163]
N-acetyl histidine	Doxorubicin	218 [a] 185.3–218.3 [b]	+40.1 ± 2.8 [a] +36.3 to +40.1 [b]	[164]

[a] Unloaded particle; [b] loaded particle; * 5 mg/mL.

3.2. Hydrophobically Modified Glycol Chitosan

The limited water solubility of CS and the precipitation of some self-aggregated chitosan conjugates restricts its application in medical practice as a drug delivery system. In contrast, glycol chitosan (GC) exhibits good water solubility at all pHs, good biocompatibility, and is widely applied as a hydrophobic drug and gene carrier [165–170]. The structural units of GC are shown in Figure 12.

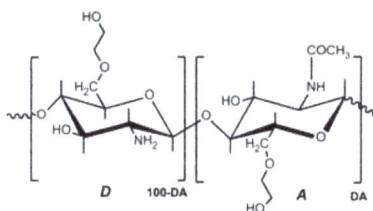

Figure 12. The chemical structure of glycol chitosan.

GC has been functionalized with cholanic acid, cholesterol, deoxycholic acid, vitamins, testosterone, doxorubicin, and other hydrophobic compounds using mostly an EDC-mediated coupling reaction to achieve the amidation of CS amine groups with the desired carboxylic acid or acyl chloride of the hydrophobic substituent. Further physical encapsulation of anticancer drugs or bioactive compounds in the core of self-assembled GC hydrophobically-modified micelles is usually performed.

Hwang et al. introduced cholanic acid in GC. The resulting GC-cholanic acid micelles can be easily loaded with the anticancer drug docetaxel [165]. Docetaxel loaded GC-cholanic acid synthesized spontaneously and was self-assembled as 350 nm aggregates in an aqueous medium. During in vivo experiments in mice, these docetaxel loaded nanoaggregates showed higher anticancer efficacy to A549 lung cancer cells and reduced toxicity when compared to the free drug.

The anticancer drug camptothecin has also been encapsulated into self-aggregates of GC-cholanic acid, with a drug loading efficiency of above 80% [166]. GC-cholanic acid micelles protected the lactone ring of camptothecin from hydrolysis and camptothecin loaded micelles showed significant antitumor activity towards MDA-MB231 human breast cancer cells that were implanted in nude mice. The 5β-cholanic hydrophobic functionalization of both GC and polyethylenimine and later mixing of both modified polymers, made it possible to obtain self-assembled nanoparticles of ca.

350 nm with a Zeta-potential of +23.8 mV, for the delivery of siRNA in tumor-bearing mice [167]. The siRNA-GC-polyethylenimine complex transfected the B16F10 tumor cells, efficiently inhibiting the RFP gene expression of RFP/B16F10-bearing mice. Thus, GC-polyethylenimine self-aggregates were revealed as promising gene carrier for cancer treatment [167]. GC-cholanic acid self-aggregates have also been proposed for the delivery of RGD peptide and indomethacin [168,169].

The hydrophobic modification of GC with deoxycholic acid and the later physical encapsulation of palmityl-acylated exendin-4 peptide in formed self-assembled nanogels for a long-acting anti-diabetic inhalation system was studied by Lee et al. [170]. Results were promising, with the ca. 72 h residence of the administered anti-diabetic drug (palmityl-acylated exendin-4 peptide) in the lungs, good hypoglycemic response, and acceptable toxicity.

In another approach, the hydrophobic modification of GC with the drug to be delivered has been explored. Quiñones et al. synthesized GC hydrophobically-modified with ergocalciferol hemisuccinate, tocopherol hemiesters, and testosterone 17β-hemisuccinate for the controlled release of vitamin D2, vitamin E and testosterone [171–173]. The degrees of substitution of GC with the vitamins and the testosterone reached values of 0.039 for vitamin D2, 0.21 to 0.36 for vitamin E and 0.015 for testosterone. The GC-vitamin and GC-testosterone conjugates formed self-assembled NPs in an aqueous medium with hydrodynamic sizes from 280 to 500 nm andZeta-potential values of +7.7 to +36.5 mV. The sustained release of covalently linked vitamins and testosterone from the GC self-aggregates was observed in an acidic medium for 3 to 4 days.

The hydrophobic modification of GC with an *N,N*-diethylnicotinamide-based oligomer enabled a high paclitaxel loading content with an EE of up to 98% [174]. The hydrodynamic diameter of the blank hydrophobically modified GC was 313 ± 20 nm in PBS. Paclitaxel loaded modified GC particles with a drug loading content of 9.8, 18.9, and 23.9 wt % exhibited hydrodynamic sizes of 331 ± 25 nm, 354 ± 23 nm, and 363 ± 32 nm, respectively. A sustained release of paclitaxel from the GC self-aggregates was observed. Overall, the anticancer assessment of the paclitaxel loaded GC particles appears promising in cancer therapy.

Doxorubicin encapsulation in GC-3-diethylaminopropyl self-aggregates and the hydrophobic functionalization of GC with doxorubicin was also accomplished for the evaluation of doxorubicin delivery systems for cancer therapy [175,176]. The hydrodynamic parameters of GC-based self-aggregates discussed are summarized in the Table 5.

Table 5. Hydrophobically modified glycol chitosan.

Hydrophobic Moiety	Active agent	Particle size (nm)	Zeta-potential (mV)	Ref.
Cholanic acid	Docetaxel	350 [b]		[165]
	Camptothecin	254 [a]		[166]
	siRNA	279–328 [b] 350 [a] 250 [b]	+23.8 ± 0.9 [a] +10.0 ± 0.8 [b]	[167]
	RGD peptide	224 [a] 189–265 [b]		[168]
Cholesterol	Indomethacin	228 [a] 275–384 [b]		[169]
Deoxycholic acid	Palmityl-acylated exendin-4	~52–250 [a]		[170]
Ergocalciferol	Vitamin D2	279 ± 7 (PBS)	+7.7 ± 0.1	[171]
DL-α-tocopherol	Vitamin E	284–496 (PBS)	+11.7 to +36.5	[172]
Testosterone	Testosterone	332 ± 4 (PBS)	+9.7 ± 0.6	[173]
N,N-diethylnicotinamide-based oligomer	Paclitaxel	313 ± 20 [a] 331–363 [b]		[174]
3-Diethylaminopropyl	Doxorubicin	102 [a]	−0.9 [a]	[175]
Doxorubicin	Doxorubicin	238 [a] 342 [b]		[176]

[a] Unloaded particle; [b] loaded particle.

3.3. Hydrophobically Modified Carboxymethyl Chitosan

O-Carboxymethyl chitosan, typically named carboxymethyl chitosan (CMCS), has been hydrophobically modified with oleoyl chloride in pyridine/dichloromethane or with linoleic acid using an EDC-mediated amide linkage reaction [177–179].

Oleoyl-modified CMCS formed self-aggregates in an aqueous medium with average hydrodynamic diameters that depended on the molecular weight of the chitosan used to prepare the CMCS [177,178]. Hydrodynamic diameters of 157.4 nm (CS with molecular weight of 50 kDa), 161.8 nm (CS with molecular weight of 38 kDa), 274.1 nm (CS with molecular weight of 170 kDa), and 396.7 nm (CS with molecular weight of 820 kDa) have been reported for different oleoyl-modified CMCS. The Zeta-potential values observed for blank oleoyl-modified CMCS particles were +16 ± 1 mV, +17.2 ± 0.9 mV, and +20 ± 1 mV. Rifampicin and microbial antigens were physically entrapped in the oleoyl-modified CMCS micelles with drug loading efficiency of 20% for rifampicin and ca. 52 to 62.5% for microbial antigens. The sustained release of encapsulated drugs was extended until 40–48 h [177,178].

Linoleic acid modified CMCS self-aggregated micelles were loaded with the anticancer drug, adriamycin, for a sustained release [179]. The average hydrodynamic diameter of the blank linoleic-modified CMCS was 418 ± 18 nm. Adriamycin was slowly released from the micelles for about three days. Results are summarized in Table 6.

Table 6. Hydrophobically modified carboxymethyl chitosan.

Hydrophobic Moiety	Active agent	Particle size (nm)	Zeta-potential (mV)	Ref.
Oleoyl	Rifampicin	161.8 [a]		[177]
	Microbial	157.4–396.7 [a]	+15.6 to +19.6 [a]	[178]
	antigen	237.6–482.3 [b]	+14.2 to +17.1 [b]	
		331.6–573.9 [b]	+12.8 to +16.3 [b]	
Acyl	Adriamycin	418 ± 18 [a]		[179]

[a] Unloaded particle; [b] loaded particle.

3.4. Hydrophobically Modified Succinyl Chitosan

Water soluble succinyl chitosans have been prepared by amidation (*N*-succinyl chitosan) and esterification (*O*6-succinyl chitosan) of chitosan by its reaction with succinic anhydride (Figure 13).

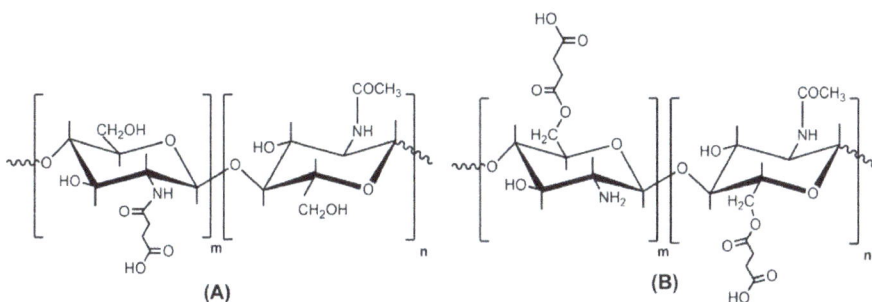

Figure 13. The chemical structure of *N*-succinyl chitosan (**A**) and O6-succinyl chitosan (**B**).

Xiangyang et al. reported the preparation of *N*-succinyl-*N'*-octyl chitosan micelles as doxorubicin carriers for an effective anti-tumor activity [180]. Average hydrodynamic sizes of doxorubicin loaded modified succinyl chitosan (SCS), which depended on the amount of octyl chains and the drug loading content, was between 100 to 200 nm. Doxorubicin loaded SCS particles showed a sustained release and more cytotoxic activity against HepG2, A549, BGC, and K562 cancer cell lines than parent doxorubicin.

In another study on SCS, the interactions between the polymer and BSA in the nanoaggregates are inspected using different techniques [181]. The authors concluded that no significant change on the conformation of BSA occurred during chain entanglements between the protein and the *N*-succinyl chitosan. The hydrodynamic sizes of the micelles formed are reported in Table 7.

Table 7. Hydrophobically modified succinyl chitosan.

Hydrophobic moiety	Active agent	Particle size (nm)	Zeta-potential (mV)	Ref.
Octyl	Doxorubicin	130.4–150.1 [a] 155.4–170.1 [b]		[180]
Acyl	BSA	~50–100 [a] ~100–200 [b]		[181]
DL-α-tocopherol	Vitamin E	254 ± 4	+36.3 ± 0.9	[172]

[a] Unloaded particle; [b] loaded particle.

The synthesis of *O6*-succinyl chitosan involves phthaloyl protection of chitosan, the reaction with succinic anhydride, and deprotection (removal of the phthaloyl groups). Further hydrophobic modification of free amine groups of *O6*-succinyl chitosan with tocopherol succinate mediated by an EDC activated coupling reaction, made it possible to prepare cationic self-assembled SCS nanoparticles with hydrodynamic diameters of 254 ± 4 nm and Zeta-potential values of +36.3 ± 0.9 mV [172]. The sustained release of covalently linked vitamin E (tocopherol) was extended up to 96 h. The results are shown in Table 7.

3.5. Hydrophobically Modified Trimethyl Chitosan

N,N,N-Trimethyl chitosan (TMC) is a water soluble derivative of chitosan that is prepared by exhaustive *N*-methylation of some free amine groups of CS using iodomethane.

TMC has been hydrophobically modified with octyl, decanoyl, lauryl, lactose, and palmitoyl substituents for hydroxycamptothecin and harmine encapsulation in the hydrophobic core [182–185]. *N*-octyl-*N*-trimethyl chitosan and *N*-lauryl-*N*-trimethyl chitosan were self-assembled in an aqueous medium as micelles of 23.5 and 20.8 nm, while *N*-decanyl-*N*-trimethyl chitosan formed micelles with a hydrodynamic diameter of 277.2 nm.

Hydroxycamptothecin loaded *N*-alkyl-*N*-trimethyl chitosan micelles showed a sustained release of the anticancer drug, with improved pharmacokinetic properties and the stability of the camptothecin lactone ring in vivo [182]. On the other hand, the harmine loaded hydrophobically modified TMC released 65.3% of the encapsulated drug in three days at pH 7.4 [183,184].

Mi et al. investigated the preparation of self-assembled NPs by TMC and poly(γ-glutamic acid) for the oral delivery of insulin [185]. The hydrodynamic diameters and Zeta-potential values of blank and insulin loaded TMC/poly(γ-glutamic acid) NPs are presented in Table 8.

Table 8. Hydrophobically modified trimethyl chitosan.

Hydrophobic moiety	Active agent	Particle size (nm)	Zeta-potential (mV)	Ref.
Alkyl	Hydroxy-camptothecin	20.8–277.2 [a] 26.0–273.1 [b]		[182]
Palmitoyl	Harmine	193.4 ± 3.1 [b]	+26.67 [b]	[183,184]
Acyl	Peptide drugs	101.3–106.3 [a] 522 ± 6 [b,*]	+30.6 to +36.2 [a] +14.2 ± 0.6 [b,*]	[185]

[a] Unloaded particle; [b] loaded particle; [*] pH 7.4.

3.6. Other Hydrophobically Modified Chitosan Derivatives

N-octyl-*O*-sulfate chitosan (NOSC) micelles have been prepared from chitosan for the sustained release of physically entrapped paclitaxel for cancer therapy [186–188]. Paclitaxel loaded *N*-octyl-*O*-sulfate chitosan micelles showed hydrodynamic diameters of ca. 200 nm and Zeta-potential values of ca. −30 mV [186,187]. On the other hand, the additional modification of *N*-octyl-*O*-sulfate chitosan with polyethylene glycol monomethyl ether reduced the hydrodynamic sizes of the paclitaxel loaded NOSC to ca. 100 nm [188]. The anticancer drug loaded NPs exhibited reduced toxicity and improved the bioavailability of encapsulated paclitaxel [186–188].

Pedro et al. synthesized *N*-dodecyl-*N'*-glycidyl(chitosan) for the delivery of quercetin [189]. The hydrodynamic parameters of quercetin loaded hydrophobically modified CS micelles were measured by dynamic light scattering showing sizes from 140 to 260 nm and Zeta-potential values from +18.7 to +30.4 mV at pH 7.4. At pH 5.0 the sizes ranged from 150 to 300 nm and the Zeta-potential values varied from +14.1 to +29.9 mV, showing the dependence of both the parameters on sample concentration at both pHs. pH was also found to play a key role on quercetin release from the micelles. The results are summarized in Table 9.

Table 9. Other chitosan derivatives.

Hydrophobic moiety	Active agent	Particle size (nm)	Zeta-potential (mV)	Ref.
Octyl	Paclitaxel	~200 [b]	−31.1 [a] −28.8 [b]	[186]
		200.8 [b]		[187]
		104.3–133.4 [b]		[188]
Acyl	Quercetin	140–300 [a]	+14.1 to +30.4 [a]	[189]

[a] Unloaded particle; [b] loaded particle.

4. Conclusions

A considerable amount of research is going on for the self-assembling preparation of chitosan nanoparticles in drug delivery applications. In particular, the nanoparticle preparations by polyelectrolyte complexation and by the self-assembly of hydrophobically modified chitosans are able to encapsulate the drug under mild conditions without losing their stability and biocompatibility. Therefore, chitosan based self-assembled nanoparticles have great potential, as well as multiple applications for the future in the design of novel drug delivery systems.

Author Contributions: Javier Pérez Quiñones, Hazel Peniche and Carlos Peniche equally contributed to the paper.

Conflicts of Interest: The authors declare no conflict of interest.

References

1. Roberts, G.A.F. (Ed.) Structure of Chitin and Chitosan. In *Chitin Chemistry*; Macmillan: Houndmills, UK, 1992; pp. 1–53, ISBN 978-1-349-11547-1.
2. Dasha, M.; Chiellini, F.; Ottenbrite, R.M.; Chiellini, E. Chitosan—A Versatile Semi-Synthetic Polymer in Biomedical Applications. *Prog. Polym. Sci.* **2011**, *36*, 981–1014. [CrossRef]
3. Aranaz, I.; Mengíbar, M.; Harris, R.; Miralles, B.; Acosta, N.; Calderón, L.; Sánchez, A.; Heras, A. Role of Physicochemical Properties of Chitin and Chitosan on Their Functionality. *Curr. Chem. Biol.* **2014**, *8*, 27–42. [CrossRef]
4. Rinaudo, M. Chitin and Chitosan: Properties and Applications. *Prog. Polym. Sci.* **2006**, *31*, 603–632. [CrossRef]
5. Azuma, K.; Izumi, R.; Osaki, T.; Ifuku, S.; Morimoto, M.; Saimoto, H.; Minami, S.; Okamoto, Y. Chitin, Chitosan, and Its Derivatives for Wound Healing: Old and New Materials. *J. Funct. Biomater.* **2015**, *6*, 104–142. [CrossRef] [PubMed]

6. Bernkop-Schnürch, A.; Dünnhaupt, S. Chitosan-Based Drug Delivery Systems. *Eur. J. Pharm. Biopharm.* **2012**, *81*, 463–469. [CrossRef] [PubMed]

7. Illum, L.; Jabbal-Gill, I.; Hinchcliffe, M.; Fisher, A.N.; Davis, S.S. Chitosan as a Novel Nasal Delivery System for Vaccines. *Adv. Drug Deliv. Rev.* **2001**, *51*, 81–96. [CrossRef]

8. Zhang, Y.; Chan, J.W.; Moretti, A.; Uhrich, K.E. Designing polymers with sugar-based for bioactive delivery applications. *J. Control. Release* **2015**, *219*, 355–368. [CrossRef] [PubMed]

9. Saikia, C.; Gogoi, P.; Maji, T.K. Chitosan: A Promising Biopolymer in Drug Delivery Applications. *J. Mol. Genet. Med.* **2015**, 1–10. [CrossRef]

10. Kumar, M.N.; Muzzarelli, R.A.; Muzzarelli, C.; Sashiwa, H.; Domb, A.J. Chitosan Chemistry and Pharmaceutical Perspectives. *Chem. Rev.* **2004**, *104*, 6017–6084. [CrossRef] [PubMed]

11. Kumar, A.; Vimal, A.; Kumar, A. Why Chitosan? From Properties to Perspective of Mucosal Drug Delivery. *Int. J. Biol. Macromol.* **2016**, *91*, 615–622. [CrossRef] [PubMed]

12. Bravo-Anaya, L.M.; Soltero, J.F.A.; Rinaudo, M. DNA/Chitosan Electrostatic Complex. *Int. J. Biol. Macromol.* **2016**, *88*, 345–353. [CrossRef] [PubMed]

13. Calvo, P.; Remunan-Lopez, C.; Vila-Jata, J.L.; Alonso, M.J. Novel Hydrophilic Chitosan-Polyethylene Oxide Nanoparticles as Protein Carriers. *J. Appl. Polym. Sci.* **1997**, *63*, 125–132. [CrossRef]

14. Hassani, S.; Laouini, A.; Fessi, H.; Charcosset, C. Preparation of Chitosan–Tpp Nanoparticles Using Microengineered Membranes—Effect of Parameters and Encapsulation of Tacrine. *Colloids Surf. A* **2015**, *482*, 34–43. [CrossRef]

15. Ngan, L.T.K.; Wang, S.-L.; Hiep, Đ.M.; Luong, P.M.; Vui, N.T.; Đinh, T.M.; Dzung, N.A. Preparation of Chitosan Nanoparticles by Spray Drying, and Their Antibacterial Activity. *Res. Chem. Intermed.* **2014**, *40*, 2165–2175. [CrossRef]

16. Riegger, B.R.; Bäurer, B.; Mirzayeva, A.; Tovar, G.E.M.; Bach, M. Systematic Approach for Preparation of Chitosan Nanoparticles via Emulsion Crosslinking as Potential Adsorbent in Wastewater Treatment. *Carbohydr. Polym.* **2018**. [CrossRef] [PubMed]

17. Chen, X.G.; Lee, C.M.; Park, H.J. O/W Emulsification for the Self-Aggregation and Nanoparticle Formation of Linoleic Acids Modified Chitosan in the Aqueous System. *J. Agric. Food Chem.* **2003**, *51*, 3135–3139. [CrossRef] [PubMed]

18. Kafshgari, M.H.; Khorram, M.; Mansouri, M.; Samimi, A.; Osfouri, S. Preparation of Alginate and Chitosan Nanoparticles Using a New Reverse Micellar System. *Iran. Polym. J.* **2012**, *21*, 99–107. [CrossRef]

19. Tokumitsu, H.; Ichikawa, H.; Fukumori, Y. Chitosan Gadopentetic Acid Complex for Gadolinium Neutron-Capture Therapy of Cancer Nanoparticles: Preparation by Novel Emulsion-Droplet Coalescent Technique and Characterization. *Pharm. Res.* **1999**, *16*, 1830–1835. [CrossRef] [PubMed]

20. Shering, M.A.; Kannan, C.; Kumar, K.S.; Kumar, V.S.; Suganeshwari, M. Formulation of 5-Fluorouracil Loaded Chitosan Nanoparticles by Emulsion Droplet Coalescence Method for Cancer Therapy. *Int. J. Pharm. Biol. Arch.* **2011**, *2*, 926–931.

21. Luque-Alcaraz, A.G.; Lizardi-Mendoza, J.; Goycoolea, F.M.; Higuera-Ciapara, I.; Arguelles-Monal, W. Preparation of Chitosan Nanoparticles by Nanoprecipitation and Their Ability as a Drug Nanocarrier. *RSC Adv.* **2016**, *6*, 59250–59256. [CrossRef]

22. Liu, L.; Zhou, C.; Xia, X.; Liu, Y. Self-Assembled Lecithin/Chitosan Nanoparticles for Oral Insulin Delivery: Preparation and Functional Evaluation. *Int. J. Nanomed.* **2016**, *11*, 761–769. [CrossRef] [PubMed]

23. Maciel, V.B.V.; Yoshida, C.M.P.; Pereira, S.M.S.S.; Goycoolea, F.M.; Franco, T.T. Electrostatic Self-Assembled Chitosan-Pectin Nano- and Microparticles for Insulin Delivery. *Molecules* **2017**, *22*, 1707. [CrossRef] [PubMed]

24. Lehn, J.-M. Perspectives in Supramolecular Chemistry—From Molecular Recognition towards Molecular Information Processing and Self-Organization. *Angew. Chem. Int. Ed. Engl.* **1990**, *29*, 1304–1319. [CrossRef]

25. Mateescu, M.A.; Ispas-Szabo, P.; Assaad, E. The Concept of Self-Assembling and the Interactions Involved. In *Controlled Drug Delivery. The Role of Self-Assembling Multi-Task Excipients*, 1st ed.; Mateescu, M.A., Ispas-Szabo, P., Assaad, E., Eds.; Elsevier: Cambridge, UK, 2015; pp. 1–20, ISBN 978-1-907568-45-9.

26. Yang, Y.; Wang, S.; Wang, Y.; Wang, X.; Wang, Q.; Chen, M. Advances in Self-Assembled Chitosan Nanomaterials for Drug Delivery. *Biotechnol. Adv.* **2014**, *32*, 1301–1316. [CrossRef] [PubMed]

27. Cheng, L.-C.; Jiang, Y.; Xie, Y.; Qiu, L.-L.; Yang, Q.; Lu, H.-Y. Novel Amphiphilic Folic Acid-Cholesterol-Chitosan Micelles for Paclitaxel Delivery. *Oncotarget* **2017**, *8*, 3315–3326. [CrossRef] [PubMed]

28. You, J.; Li, W.; Yu, C.; Zhao, C.; Jin, L.; Zhou, Y.; Xu, X.; Dong, S.; Lu, X. Amphiphilically Modified Chitosan Cationic Nanoparticles for Drug Delivery. *J. Nanopart. Res.* **2013**, *15*, 21–23. [CrossRef]

29. Pasanphan, W.; Choofong, S.; Rimdusit, P. Deoxycholate-Chitosan Nanospheres Fabricated by γ-Irradiation and Chemical Modification: Nanoscale Synthesis and Controlled Studies. *J. Appl. Polym. Sci.* **2011**, *123*, 3309–3320. [CrossRef]

30. Li, T.; Longobardi, L.; Granero-Molto, F.; Myers, T.J.; Yan, Y.; Spagnoli, A. Use of Glycol Chitosan Modified by 5β-Cholanic Acid Nanoparticles for the Sustained Release of Proteins During Murine Embryonic Limb Skeletogenesis. *J. Control. Release* **2010**, *144*, 101–108. [CrossRef] [PubMed]

31. Yhee, J.Y.; Son, S.; Kim, S.H.; Park, K.; Choi, K.; Kwon, I.C. Self-Assembled Glycol Chitosan Nanoparticles for Disease-Specific Theranostics. *J. Control. Release* **2014**, *193*, 202–213. [CrossRef] [PubMed]

32. Monsalve, Y.; Sierra, L.; López, B.L. Preparation and Characterization of Succinyl-Chitosan Nanoparticles for Drug Delivery. *Macromol. Symp.* **2015**, *354*, 91–98. [CrossRef]

33. Hamman, J.H. Chitosan Based Polyelectrolyte Complexes as Potential Carrier Materials in Drug Delivery Systems. *Mar. Drugs* **2010**, *8*, 1305–1322. [CrossRef] [PubMed]

34. Kabanov, V. Fundamentals of Polyelectrolyte Complexes in Solution and the Bulk. In *Multilayer Thin Films: Sequential Assembly of Nanocomposite Materials*; Decher, G., Schlenoff, J.B., Eds.; Wiley-VCH: Weinheim, Germany, 2002; pp. 47–86.

35. Tsuchida, E.; Osada, Y.; Ohno, H. Formation of Interpolymer Complexes. *J. Macromol. Sci. Part B Phys.* **1980**, *17*, 683–714. [CrossRef]

36. Luo, Y.; Wang, Q. Recent Development of Chitosan-Based Polyelectrolyte Complexes with Natural Polysaccharides for Drug Delivery. *Int. J. Biol. Macromol.* **2014**, *64*, 353–367. [CrossRef] [PubMed]

37. Kubota, N.; Shimoda, K. Macromolecule Complexes of Chitosan. In *Polysaccharides: Structural Diversity and Functional Versatility*, 2nd ed.; Dumitriu, S., Ed.; Marcel Dekker, Inc.: New York, NY, USA, 2005; pp. 679–706.

38. Peniche, C.; Argüelles-Monal, W. Chitosan Based Polyelectrolyte Complexes. *Macromol. Symp.* **2001**, *168*, 103–116. [CrossRef]

39. Chen, H.; Fan, M. Chitosan/Carboxymethyl Cellulose Polyelectrolyte Complex Scaffolds for Pulp Cells Regeneration. *J. Bioact. Compat. Polym.* **2007**, *22*, 475–490. [CrossRef]

40. Fukuda, H.; Kikuchi, Y. Polyelectrolyte Complexes of Sodium Carboxymethylcellulose with Chitosan. *Makromol. Chem.* **1979**, *180*, 1631–1633. [CrossRef]

41. Alsharabasy, A.M.; Moghannem, S.A.; El-Mazny, W.N. Physical Preparation of Alginate/Chitosan Polyelectrolyte Complexes for Biomedical Applications. *J. Biomater. Appl.* **2016**, *30*, 1071–1079. [CrossRef] [PubMed]

42. Caetano, G.F.; Frade, M.A.C.; Andrade, T.A.M.; Leite, M.N.; Bueno, C.Z.; Moraes, A.M.; Ribeiro-Paes, J.T. Chitosan-Alginate Membranes Accelerate Wound Healing. *J. Biomed. Mater. Res. Part B Appl. Biomater.* **2015**, *103*, 1013–1022. [CrossRef] [PubMed]

43. Cárdenas, A.; Argüelles-Monal, W.; Goycoolea, F.M.; Higuera-Ciapara, I.; Peniche, C. Diffusion through Membranes of the Polyelectrolyte Complex of Chitosan and Alginate. *Macromol. Biosci.* **2003**, *3*, 535–539. [CrossRef]

44. Lee, K.Y.; Park, W.H.; Ha, W.S. Polyelectrolyte Complexes of Sodium Alginate with Chitosan or Its Derivatives for Microcapsules. *J. Appl. Polym. Sci.* **1997**, *63*, 425–432. [CrossRef]

45. Sæther, H.V.; Holme, H.K.; Maurstad, G.; Smidsrød, O.; Stokke, B.T. Polyelectrolyte Complex Formation Using Alginate and Chitosan. *Carbohydr. Polym.* **2008**, *74*, 813–821. [CrossRef]

46. Chavasit, V.; Kienzle-Sterzer, C.; Torres, J.A. Formation and Characterization of an Insoluble Polyelectrolyte Complex Chitosan-Polyacrylic Acid. *Polym. Bull.* **1988**, *19*, 223–230. [CrossRef]

47. De Oliveira, H.C.; Fonseca, J.L.; Pereira, M.R. Chitosan-Poly(Acrylic Acid) Polyelectrolyte Complex Membranes: Preparation, Characterization and Permeability Studies. *J. Biomater. Sci. Polym. Ed.* **2008**, *19*, 143–160. [CrossRef] [PubMed]

48. Arguelles-Monal, W.; Cabrera, G.; Peniche, C.; Rinaudo, M. Conductometric Study of the Inter-Polyelectrolyte Reaction between Chitosan and Poly(Galacturonic Acid). *Polymer* **1999**, *41*, 2373–2378. [CrossRef]

49. Bernabe, P.; Peniche, C.; Argüelles-Monal, W. Swelling Behavior of Chitosan/Pectin Polyelectrolyte Complex Membranes. Effect of Thermal Cross-Linking. *Polym. Bull.* **2005**, *55*, 367–375. [CrossRef]

50. Luppi, B.; Bigucci, F.; Abruzzo, A.; Corace, G.; Cerchiara, T.; Zecchi, V. Freeze-Dried Chitosan/Pectin Nasal Inserts for Antipsychotic Drug Delivery. *Eur. J. Pharm. Biopharm.* **2010**, *75*, 381–387. [CrossRef] [PubMed]

51. Yao, K.D.; Tu, H.; Cheng, F.; Zhang, J.W.; Liu, J. pH-Sensitivity of the Swelling of a Chitosan-Pectin Polyelectrolyte Complex. *Angew. Makromol. Chem.* **1997**, *245*, 63–72. [CrossRef]

52. Arguelles-Monal, W.; Goycoolea, F.M.; Lizardi, J.; Peniche, C.; Higuera-Ciapara, I. Chitin and Chitosan in Gel Network Systems. In *Acs Symposium Series*; Bohidar, H., Dubin, P., Osada, Y., Eds.; American Chemical Society: Washington, DC, USA, 2003; pp. 102–121, ISBN13 9780841237612; eISBN 9780841219342. [CrossRef]

53. Carneiro, T.N.; Novaes, D.S.; Rabelo, R.B.; Celebi, P.; Chevallier, D.; Mantovani, M.; Beppu, R.S.; Vieira, R.S. Bsa and Fibrinogen Adsorption on Chitosan/κ-Carrageenan Polyelectrolyte Complexes. *Macromol. Biosci.* **2013**, *13*, 1072–1083. [CrossRef] [PubMed]

54. Martins, A.F.; Piai, J.F.; Schuquel, I.T.A.; Rubira, A.F.; Muniz, E.C. Polyelectrolyte Complexes of Chitosan/Heparin and *N,N,N*-Trimethyl Chitosan/Heparin Obtained at Different pH: I. Preparation, Characterization, and Controlled Release of Heparin. *Colloid Polym. Sci.* **2011**, *289*, 1133–1144. [CrossRef]

55. Pushpa, S.; Srinivasan, R. Polyelectrolyte Complexes of Glycol Chitosan with Some Mucopolysaccharides: Dielectric Properties and Electric Conductivity. *Biopolymers* **1984**, *23*, 59–69. [CrossRef]

56. Stoilova, O.; Koseva, N.; Manolov, N.; Rashkov, I. Polyelectrolyte Complex between Chitosan and Poly(2-Acryloylamido-2-Methylpropanesulfonic Acid). *Polym. Bull.* **1999**, *43*, 67–73. [CrossRef]

57. Berth, G.; Voig, A.; Dautzenberg, H.; Donath, E.; Moehwald, H. Polyelectrolyte Complexes and Layer-by-Layer Capsules from Chitosan/Chitosan Sulfate. *Biomacromolecules* **2002**, *3*, 579–590. [CrossRef] [PubMed]

58. Gamzazade, A.I.; Nasibov, S.M. Formation and Properties of Polyelectrolyte Complexes of Chitosan Hydrochloride and Sodium Dextran sulfate. *Carbohydr. Polym.* **2002**, *50*, 339–343. [CrossRef]

59. Lin, Y.S.; Radzi, R.; Morimoto, M.; Saimoto, H.; Okamoto, Y.; Minami, S. Characterization of Chitosan-Carboxymethyl Dextran Nanoparticles as a Drug Carrier and as a Stimulator of Mouse Splenocytes. *J. Biomater. Sci. Polym. Ed.* **2012**, *23*, 1401–1420. [CrossRef] [PubMed]

60. Wu, D.; Delair, T. Stabilization of Chitosan/Hyaluronan Colloidal Polyelectrolyte Complexes in Physiological Conditions. *Carbohydr. Polym.* **2015**, *119*, 149–158. [CrossRef] [PubMed]

61. Lalevée, G.; Sudre, G.; Montembault, A.; Meadows, J.; Malaise, S.; Crépet, A.; David, L.; Delair, T. Polyelectrolyte Complexes Via Desalting Mixtures of Hyaluronic Acid and Chitosan. Physicochemical Study and Structural Analysis. *Carbohydr. Polym.* **2016**, *154*, 86–95. [CrossRef] [PubMed]

62. Mateescu, M.A.; Ispas-Szabo, P.; Assaad, E. Chitosan-Based Polyelectrolyte Complexes as Pharmaceutical Excipients. In *Controlled Drug Delivery. The Role of Self-Assembling Multi-Task Excipients*, 1st ed.; Mateescu, M.A., Ispas-Szabo, P., Assaad, E., Eds.; Elsevier: Cambridge, UK, 2015; pp. 127–161, ISBN 978-1-907568-45-9.

63. Peniche, H.; Peniche, C. Chitosan Nanoparticles: A Contribution to Nanomedicine. *Polym. Int.* **2011**, *60*, 883–889. [CrossRef]

64. Wang, J.J.; Zen, Z.W.; Xiao, R.Z.; Xie, T.; Zhou, G.L.; Zhan, X.R.; Wang, S.L. Recent Advances of Chitosan Nanoparticles as Drug Carriers. *Int. J. Nanomed.* **2011**, *6*, 765–774. [CrossRef]

65. Thu, B.; Bruheim, O.; Espevik, T.; Smidsrød, O.; Soon-Shiong, P.; Skjåk-Bræk, G. Alginate Polycation Microcapsules I. Interaction between Alginate and Polycation. *Biomaterials* **1996**, *17*, 1031–1040. [CrossRef]

66. Paques, J.P.; van der Linden, E.; van Rijn, C.J.M.; Sagis, L.M.C. Preparation Methods of Alginate Nanoparticles. *Adv. Colloid Interface Sci.* **2014**, *209*, 163–171. [CrossRef] [PubMed]

67. Liu, P.; Zhao, X. Facile Preparation of Well-Defined near-Monodisperse Chitosan/Sodium Alginate Polyelectrolyte Complex Nanoparticles (Cs/Sal Nps) Via Ionotropic Gelification: A Suitable Technique for Drug Delivery Systems. *Biotechnol. J.* **2013**, *8*, 847–854. [CrossRef] [PubMed]

68. Katuwavila, N.P.; Perera, A.D.L.C.; Samarakoon, S.R.; Soysa, P.; Karunaratne, V.; Amaratunga, G.A.J.; Karunaratne, D.N. Chitosan-Alginate Nanoparticle System Efficiently Delivers Doxorubicin to Mcf-7 Cells. *J. Nanomater.* **2016**, *2016*, 1–12. [CrossRef]

69. Arora, S.; Gupta, S.; Narang, R.K.; Budhiraja, R.D. Amoxicillin Loaded Chitosan-Alginate Polyelectrolyte Complex Nanoparticles as Mucopenetrating Delivery System for H. Pylori. *Sci. Pharm.* **2011**, *79*, 673–694. [CrossRef] [PubMed]

70. Goycoolea, F.M.; Lollo, G.; Remuñán-López, C.; Quaglia, F.; Alonso, M.J. Chitosan-Alginate Blended Nanoparticles as Carriers for the Transmucosal Delivery of Macromolecules. *Biomacromolecules* **2009**, *10*, 1736–1743. [CrossRef] [PubMed]

71. Azevedo, M.A.; Bourbon, A.I.; Vicente, A.A.; Cerqueira, M.A. Alginate/Chitosan Nanoparticles for Encapsulation and Controlled Release of Vitamin B2. *Int. J. Biol. Macromol.* **2014**, *71*, 141–146. [CrossRef] [PubMed]

72. Bhunchu, S.; Rojsitthisak, P.; Rojsitthisak, P. Effects of Preparation Parameters on the Characteristics of Chitosanealginate Nanoparticles Containing Curcumin Diethyl Disuccinate. *J. Drug Deliv. Sci. Technol.* **2015**, *28*, 64–72. [CrossRef]

73. Motwani, S.K.; Chopra, S.; Talegaonkar, S.; Kohli, K.; Ahmad, F.J.; Khar, R.K. Chitosan-Sodium Alginate Nanoparticles as Submicroscopic Reservoirs for Ocular Delivery: Formulation, Optimisation and in Vitro Characterisation. *Eur. J. Pharm. Biopharm.* **2008**, *68*, 513–525. [CrossRef] [PubMed]

74. Zhu, X.; Su, M.; Tang, S.; Wang, L.; Liang, X.; Meng, F.; Hong, Y.; Xu, Z. Synthesis of Thiolated Chitosan and Preparation Nanoparticles with Sodium Alginate for Ocular Drug Delivery. *Mol. Vis.* **2012**, *18*, 1973–1982. [PubMed]

75. Sarmento, B.; Ribeiro, A.; Veiga, F.; Sampaio, P.; Neufeld, R.; Ferreira, D. Alginate/Chitosan Nanoparticles Are Effective for Oral Insulin Delivery. *Pharm. Res.* **2007**, *24*, 2198–2206. [CrossRef] [PubMed]

76. Kumar, S.; Chauhan, N.; Gopal, M.; Kumar, R.; Dilbaghi, N. Development and Evaluation of Alginate-Chitosan Nanocapsules for Controlled Release of Acetamiprid. *Int. J. Biol. Macromol.* **2015**, *81*, 631–637. [CrossRef] [PubMed]

77. Gazori, T.; Khoshayan, M.R.; Azizi, E.; Yazdizade, P.; Nomani, A.; Haririan, I. Evaluation of Alginate/Chitosan Nanoparticles as Antisense Delivery Vector: Formulation, Optimization and in Vitro Characterization. *Carbohydr. Polym.* **2009**, *77*, 599–606. [CrossRef]

78. Rafiee, A.; Alimohammadian, M.H.; Gazori, T.; Riazi-rad, F.; Fatemi, S.M.R.; Parizadeh, A.; Haririan, I.; Havaskary, M. Comparison of Chitosan, Alginate and Chitosan/Alginate Nanoparticles with Respect to Their Size, Stability, Toxicity and Transfection. *Asian Pac. J. Trop. Dis.* **2014**, *4*, 372–377. [CrossRef]

79. Lertsutthiwong, P.; Rojsitthisak, P.; Nimmannit, U. Preparation of Turmeric Oil-Loaded Chitosan-Alginate Biopolymeric Nanocapsules. *Mater. Sci. Eng.* **2009**, *29*, 856–860. [CrossRef]

80. Grebinişan, D.; Holban, M.; Şunel, V.; Popa, M.; Desbrieres, J.; Lionte, C. Novel Acyl Derivatives of *N*-(P-Aminobenzoyl)-L-Glutamine Encapsulated in Polymeric Nanocapsules with Potential Antitumoral Activity. *Cellul. Chem. Technol.* **2011**, *45*, 571–577. Available online: http://www.cellulosechemtechnol.ro/pdf/CCT45,9-10%282011%29/p.571-577.pdf (accessed on 25 January 2018).

81. Wang, T.; He, N. Preparation, Characterization and Applications of Low-Molecular-Weight Alginate-Oligochitosan Nanocapsules. *Nanoscale* **2010**, *2*, 230–239. [CrossRef] [PubMed]

82. Marudova, M.; MacDougall, A.J.; Ring, S.G. Pectin-Chitosan Interactions and Gel Formation. *Carbohydr. Res.* **2004**, *339*, 1933–1939. [CrossRef] [PubMed]

83. Morris, G.A.; Kök, S.M.; Harding, S.E.; Adams, G. Polysaccharide Drug Delivery Systems Based on Pectin and Chitosan. *Biotechnol. Genet. Eng. Rev.* **2010**, *27*, 257–284. [CrossRef] [PubMed]

84. Birch, N.P.; Schiffman, J.D. Characterization of Self Assembled Polyelectrolytc Complex Nanoparticles Formed from Chitosan and Pectin. *Langmuir* **2014**, *30*, 3441–3447. [CrossRef] [PubMed]

85. Rampino, A.; Borgogna, M.; Bellich, B.; Blasi, P.; Virgilio, F.; Cesàro, A. Chitosan-Pectin Hybrid Nanoparticles Prepared by Coating and Blending Techniques. *Eur. J. Pharm. Sci.* **2016**, *84*, 37–45. [CrossRef] [PubMed]

86. Al-Azi, O.S.M.; Tan, Y.T.F.; Wong, T.W. Transforming Large Molecular Weight Pectin and Chitosan into Oral Protein Drug Nanoparticulate Carrier. *React. Funct. Polym.* **2014**, *84*, 45–52. [CrossRef]

87. Andriani, Y.; Grasianto; Siswanta; Mudasir. Glutaraldehyde-Crosslinked Chitosan-Pectin Nanoparticles as a Potential Carrier for Curcumin Delivery and Its In Vitro Release Study. *Int. J. Drug Deliv.* **2015**, *7*, 167–173. Available online: http://www.arjournals.org/index.php/ijdd/article/view/1775/pdf (accessed on 25 January 2018).

88. Wan, H.; Yang, B.; Sun, H. Pectin-Chitosan Polyelectrolyte Complex Nanoparticles for Encapsulation and Controlled Release of Nisin. *Am. J. Polym. Sci. Technol.* **2017**, *3*, 82–88. [CrossRef]

89. Janes, K.A.; Fresneau, M.P.; Marazuela, A.; Fabra, A.; Alonso, M.J. Chitosan Nanoparticles as Delivery Systems for Doxorubicin. *J. Control. Release* **2001**, *73*, 255–267. [CrossRef]

90. Schatz, C.; Lucas, J.M.; Viton, C.; Domard, A.; Pichot, C.; Delair, T. Formation and Properties of Positively Charged Colloids Based on Polyelectrolyte Complexes of Biopolymers. *Langmuir* **2004**, *20*, 7766–7778. [CrossRef] [PubMed]

91. Delair, T. Colloidal Polyelectrolyte Complexes of Chitosan and Dextran Sulfate Towards Versatile Nanocarriers of Bioactive Molecules. *Eur. J. Pharm. Biopharm.* **2011**, *78*, 10–18. [CrossRef] [PubMed]

92. Huan, M.; Vitharana, S.N.; Peek, L.J.; Coop, T.; Berkland, C. Polyelectrolyte Complexes Stabilize and Controllably Release Vascular Endothelial Growth Factor. *Biomacromolecules* **2007**, *8*, 1607–1614. [CrossRef] [PubMed]

93. Tiyaboonchai, W.; Limpeanchob, N. Formulation and Characterization of Amphotericin *B*-Chitosan-Dextran Sulfate Nanoparticles. *Int. J. Pharm.* **2007**, *329*, 142–149. [CrossRef] [PubMed]

94. Huang, M.; Berkland, C. Controlled Release of Repifermin® from Polyelectrolyte Complexes Stimulates Endothelial Cell Proliferation. *J. Pharm. Sci.* **2009**, *98*, 268–280. [CrossRef] [PubMed]

95. Sharma, S.; Benson, H.A.E.; Mukkur, T.K.S.; Rigby, P.; Chen, Y. Preliminary Studies on the Development of Iga-Loaded Chitosan-Dextran Sulphate Nanoparticles as a Potential Nasal Delivery System for Protein Antigens. *J. Microencapsul.* **2013**, *30*, 283–294. [CrossRef] [PubMed]

96. Weber, C.; Drogoz, A.; David, L.; Domard, A.; Charles, M.-H.; Verrier, B.; Delair, T. Polysaccharide-Based Vaccine Delivery Systems: Macromolecular Assembly, Interactions with Antigen Presenting Cells, and in Vivo Immunomonitoring. *J. Biomed. Mater. Res. Part A* **2010**, *93*, 1322–1334. [CrossRef]

97. Drogoz, A.; Munier, S.; Verrier, B.; David, L.; Domard, A.; Delair, T. Towards Biocompatible Vaccine Delivery Systems: Interactions of Colloidal Pecs Based on Polysaccharides with Hiv-1 P24 Antigen. *Biomacromolecules* **2008**, *9*, 583–591. [CrossRef] [PubMed]

98. Chaiyasan, W.; Praputbut, S.; Kompella, U.B.; Srinivas, S.P.; Tiyaboonchaia, W. Penetration of Mucoadhesive Chitosan-Dextran Sulfate Nanoparticles into the Porcine Cornea. *Colloids Surf. B* **2017**, *149*, 288–296. [CrossRef] [PubMed]

99. Chen, Y.; Mohanraj, V.J.; Wang, F.; Benson, H.A.E. Designing Chitosan-Dextran Sulfate Nanoparticles Using Charge Ratios. *AAPS PharmSciTech* **2007**, *8*. [CrossRef] [PubMed]

100. Sarmento, B.; Ribeiro, A.; Veiga, F.; Ferreira, D. Development and Characterization of New Insulin Containing Polysaccharide Nanoparticles. *Colloids Surf. B* **2006**, *53*, 193–202. [CrossRef] [PubMed]

101. Sarmento, B.; Ribeiro, A.; Veiga, F.; Ferreira, D.; Neufeld, R. Oral Bioavailability of Insulin Contained in Polysaccharide Nanoparticles. *Biomacromolecules* **2007**, *8*, 3054–3060. [CrossRef] [PubMed]

102. Cho, Y.; Shi, R.; Ben Borgens, R. Chitosan Nanoparticle-Based Neuronal Membrane Sealing and Neuroprotection Following Acrolein Induced Cell Injury. *J. Biol. Eng.* **2010**, *4*, 2. Available online: http://www.jbioleng.org/content/4/1/2 (accessed on 25 January 2018). [CrossRef] [PubMed]

103. Saboktakin, M.R.; Tabatabaie, R.M.; Maharramovb, A.; Ramazanov, M.A. Synthesis and Characterization of pH-Dependent Glycol Chitosan and Dextran Sulfate Nanoparticles for Effective Brain Cancer Treatment. *Int. J. Biol. Macromol.* **2011**, *49*, 747–751. [CrossRef] [PubMed]

104. Kulkarni, A.D.; Vanjari, Y.H.; Sancheti, K.H.; Patel, H.M.; Belgamwar, V.S.; Surana, S.J.; Pardeshi, C.V. New Nasal Nanocomplex Self-Assembled from Charged Biomacromolecules: *N,N,N*-Trimethyl Chitosan and Dextran Sulphate. *Int. J. Biol. Macromol.* **2016**, *88*, 476–490. [CrossRef] [PubMed]

105. Anitha, A.; Rani, V.V.D.; Krishna, R.; Sreeja, V.; Selvamurugan, N.; Nair, S.V.; Tamura, H.; Jayakumar, R. Synthesis, Characterization, Cytotoxicity and Antibacterial Studies of Chitosan, *O*-Carboxymethyl and *N,O*-Carboxymethyl Chitosan Nanoparticles. *Carbohydr. Polym.* **2009**, *78*, 672–677. [CrossRef]

106. Snima, K.S.; Jayakumar, R.; Unnikrishnan, A.G.; Nair, S.V.; Lakshmanan, V.-K. *O*-Carboxymethyl Chitosan Nanoparticles for Metformin Delivery to Pancreatic Cancer Cells. *Carbohydr. Polym.* **2012**, *89*, 1003–1007. [CrossRef] [PubMed]

107. Anitha, A.; Maya, S.; Deepa, N.; Chennazhi, K.P.; Nair, S.V.; Tamurab, H.; Jayakumar, R. Efficient Water Soluble *O*-Carboxymethyl Chitosan Nanocarrier for the Delivery of Curcumin to Cancer Cells. *Carbohydr. Polym.* **2011**, *83*, 452–461. [CrossRef]

108. Gao, P.; Xia, G.; Bao, Z.; Feng, C.; Cheng, X.; Kong, M.; Liu, Y.; Chen, X. Chitosan Based Nanoparticles as Protein Carriers for Efficient Oral Antigen Delivery I. *Int. J. Biol. Macromol.* **2016**, *91*, 716–723. [CrossRef] [PubMed]

109. Mourya, V.K.; Inamdar, N.N.; Tiwari, A. Carboxymethyl Chitosan and Its Applications. *Adv. Mat. Lett.* **2010**, *1*, 11–33. [CrossRef]

110. Wang, J.; Xu, M.; Cheng, X.; Kong, M.; Liu, Y.; Feng, C.; Chen, X. Positive/Negative Surface Charge of Chitosan Based Nanogels and Its Potential Influence on Oral Insulin Delivery. *Carbohydr. Polym.* **2016**, *136*, 867–874. [CrossRef] [PubMed]

111. Wang, J.; Kong, M.; Zhou, Z.; Yan, D.; Yu, X.; Chen, X.; Feng, C.; Liu, Y.; Chen, X. Mechanism of Surface Charge Triggered Intestinal Epithelial Tight Junction Opening Upon Chitosan Nanoparticles for Insulin Oral Delivery. *Carbohydr. Polym.* **2017**, *157*, 596–602. [CrossRef] [PubMed]

112. Feng, C.; Wang, Z.; Jiang, C.; Kong, M.; Zhou, X.; Li, Y.; Cheng, X.; Chen, X. Chitosan/O-Carboxymethyl Chitosan Nanoparticles for Efficient and Safe Oral Anticancer Drug Delivery: In Vitro and in Vivo Evaluation. *Int. J. Pharm.* **2013**, *457*, 158–167. [CrossRef] [PubMed]

113. Feng, C.; Sun, G.; Wang, Z.; Cheng, X.; Park, H.; Cha, D.; Kong, M.; Chen, X. Transport Mechanism of Doxorubicin Loaded Chitosan Based Nanogels across Intestinal Epithelium. *Eur. J. Pharm. Biopharm.* **2014**, *87*, 197–207. [CrossRef] [PubMed]

114. Fajardo, A.R.; Lopes, L.C.; Valente, A.J.M.; Rubira, A.F.; Muniz, E.C. Effect of Stoichiometry and pH on the Structure and Properties of Chitosan/Chondroitin Sulfate Complexes. *Colloid Polym. Sci.* **2011**, *289*, 1739–1748. [CrossRef]

115. Umerska, A.; Corrigan, O.I.; Tajber, L. Design of Chondroitin Sulfate-Based Polyelectrolyte Nanoplexes: Formation of Nanocarriers with Chitosan and a Case Study of Salmon Calcitonin. *Carbohydr. Polym.* **2017**, *156*, 276–284. [CrossRef] [PubMed]

116. Bali, J.P.; Cousse, H.; Neuzil, E. Biochemical Basis of the Pharmacologic Action of Chondroitin Sulfates on the Osteoarticular. *Semin. Arthritis Rheum.* **2001**, *31*, 58–68. [CrossRef] [PubMed]

117. Tsai, H.-Y.; Chiu, C.-C.; Li, P.-C.; Chen, S.-H.; Huang, S.-J.; Wang, L.-F. Antitumor Efficacy of Doxorubicin Released from Crosslinked Nanoparticulate Chondroitin Sulfate/Chitosan Polyelectrolyte Complexes. *Macromol. Biosci.* **2011**, *11*, 680–688. [CrossRef] [PubMed]

118. Zhao, L.; Liu, M.; Wang, J.; Zhai, G. Chondroitin Sulfate-Based Nanocarriers for Drug/Gene Delivery. *Carbohydr. Polym.* **2015**, *133*, 391–399. [CrossRef] [PubMed]

119. Santo, V.E.; Gomes, M.E.; Mano, J.F.; Reis, R.L. Chitosan-Chondroitin Sulfate Nanoparticles for Controlled Delivery of Platelet Lysates in Bone Regenerative Medicine. *J. Tissue Eng. Regener. Med.* **2012**, *6*, s47–s59. [CrossRef] [PubMed]

120. Hu, C.-S.; Chiang, C.-H.; Hong, P.-D.; Yeh, M.-K. Influence of Charge on Fitc-Bsa-Loaded Chondroitin Sulfate-Chitosan Nanoparticles upon Cell Uptake in Human Caco-2 Cell Monolayers. *Int. J. Nanomed.* **2012**, *7*, 4861–4872. [CrossRef]

121. Shriver, Z.; Capila, I.; Venkataraman, G.; Sasisekharan, R. Heparin and Heparan Sulfate: Analyzing Structure and Microheterogeneity. In *Handbook of Experimental Pharmacology*; Springer: Berlin/Heidelberg, Germany, 2012; Volume 207, pp. 159–176. [CrossRef]

122. Lee, H.J.; Park, K.-H.; Park, S.R.; Min, B.-H. Chitosan/Heparin Polyelectrolyte Complex Nanoparticles (100~200 nm) Covalently Bonded with Pei for Enhancement of Chondrogenic Phenotype. In *Key Engineering Materials*; Trans Tech Publications Ltd.: Zürich, Switzerland, 2007; Volume 342, pp. 329–332. [CrossRef]

123. Costalat, M.; Alcouffe, P.; David, L.; Delair, T. Controlling the Complexation of Polysaccharides into Multi-Functional Colloidal Assemblies for Nanomedicine. *J. Colloid Interface Sci.* **2014**, *430*, 147–156. [CrossRef] [PubMed]

124. Peniche, H.; Reyes, F.; Aguilar, R M.R.; Rodríguez, G.; Abradelo, C.; García, L.; Peniche, C.; Román, J.S. Thermosensitive Macroporous Cryogels Functionalized With Bioactive Chitosan/Bemiparin Nanoparticles. *Macromol. Biosci.* **2013**, *13*, 1556–1567. [CrossRef] [PubMed]

125. Tsao, C.T.; Chang, C.H.; Lin, Y.Y.; Wu, M.F.; Wang, J.L.; Young, T.H.; Han, J.L.; Hsieh, K.H. Evaluation of Chitosan/γ-Poly(Glutamic Acid) Polyelectrolyte Complex for Wound Dressing Materials. *Carbohydr. Polym.* **2011**, *84*, 812–819. [CrossRef]

126. Lin, Y.-H.; Chen, C.-T.; Liang, H.-F.; Kulkarni, A.R.; Lee, P.-W.; Chen, C.-H.; Sung, H.-W. Novel Nanoparticles for Oral Insulin Delivery via the Paracellular Pathway. *Nanotechnology* **2007**, *18*, 105102. [CrossRef]

127. Hajdu, I.; Bodnár, M.; Filipcsei, G.; Hartmann, J.F.; Daróczi, L.; Zrínyi, M.; Borbély, J. Nanoparticles Prepared by Self-Assembly of Chitosan and Poly-γ-Glutamic Acid. *Colloid Polym. Sci.* **2008**, *295*, 343–350. [CrossRef]

128. Nguyen, H.-N.; Wey, S.-P.; Juan, J.-H.; Sonaje, K.; Ho, Y.-C.; Chuang, E.-Y.; Hsu, C.-W.; Yen, T.-C.; Lin, K.-J.; Sung, H.-W. The Glucose-Lowering Potential of Exendin-4 Orally Delivered via a pH-Sensitive Nanoparticle Vehicle and Effects on Subsequent Insulin Secretion in Vivo. *Biomaterials* **2011**, *32*, 2673–2682. [CrossRef] [PubMed]

129. Lin, Y.-H.; Sonaje, K.; Li, K.M.; Juang, J.-H.; Mi, F.-L.; Yang, H.-W.; Sung, H.-W. Multi-Ion-Crosslinked Nanoparticles with pH-Responsive Characteristics for Oral Delivery of Protein Drugs. *J. Control. Release* **2008**, *132*, 141–149. [CrossRef] [PubMed]

130. Sonaje, K.; Lin, Y.-H.; Juang, J.-H.; Wey, S.-P.; Chen, C.-T.; Sung, H.-W. In Vivo Evaluation of Safety and Efficacy of Self-Assembled Nanoparticles for Oral Insulin Delivery. *Biomaterials* **2009**, *30*, 2329–3239. [CrossRef] [PubMed]

131. Pereira, A.E.S.; Sandoval-Herrera, I.E.; Zavala-Betancourt, S.A.; Oliveira, H.C.; Ledezma-Pérez, A.S.; Romero, J.; Fraceto, L.F. γ-Polyglutamic Acid/Chitosan Nanoparticles for the Plant Growth Regulator Gibberellic Acid: Characterization and Evaluation of Biological Activity. *Carbohydr. Polym.* **2017**, *157*, 1862–1873. [CrossRef] [PubMed]

132. Hu, Y.; Jiang, X.; Ding, Y.; Ge, H.; Yuan, Y.; Yang, C. Synthesis and Characterization of Chitosan-Poly(Acrylic Acid) Nanoparticles. *Biomaterials* **2002**, *23*, 3193–3201. [CrossRef]

133. Chen, Q.; Hu, Y.; Che, Y.; Jiang, X.; Yang, Y. Microstructure Formation and Property of Chitosan-Poly(Acrylic Acid) Nanoparticles Prepared by Macromolecular Complex. *Macromol. Biosci.* **2005**, *5*, 993–1000. [CrossRef] [PubMed]

134. Davidenko, N.; Blanco, M.D.; Peniche, C.; Becherán, L.; Guerrero, S.; Teijón, J.M. Effects of Different Parameters on the Characteristics of Chitosan-Poly(Acrylic Acid) Nanoparticles Obtained by the Method of Coacervation. *J. Appl. Polym. Sci.* **2009**, *111*, 2362–2371. [CrossRef]

135. Becherán, L.; Bocourt, M.; Pérez, J.; Peniche, C. Chitosan in Biomedicine. From Gels to Nanoparticles. In *Advances in Chitin Science, Proceeding of the 6th Iberoamerican Chitin Symposium and 12th International Conference on Chitin and Chitosan, VI SIAQ/XII ICCC, Fortaleza, Brazil, 2–5 September 2012*; Campana, S.P., Masumi, M.M., Flamingo, A., Eds.; São Carlos-IQSC: São Carlos, Brasil, 2014; pp. 217–224, ISBN 078-85-63191-03-8 (v1.4).

136. Chen, C.-Y.; Wang, J.-W.; Hon, M.-H. Polyion Complex Nanofibrous Structure Formed by Self-Assembly of Chitosan and (Acrylic Acid). *Macromol. Mater. Eng.* **2006**, *291*, 123–127. [CrossRef]

137. Wang, J.-W.; Chen, C.-Y.; Kuo, Y.-M. Effect of Experimental Parameters on the Formation of Chitosan-Poly(Acrylic Acid) Nanofibrous Scaffolds and Evaluation of Their Potential Application as DNA Carrier. *J. Appl. Polym. Sci.* **2010**, *115*, 1769–1780. [CrossRef]

138. Yew, H.-C.; Misran, M. Preparation and Characterization of pH Dependent K-Carrageenan-Chitosan Nanoparticle as Potential Slow Release Delivery Carrier. *Iran. Polym. J.* **2016**, *25*, 1037–1046. [CrossRef]

139. Rodrigues, S.; Rosa da Costa, A.M.; Grenha, A. Chitosan/Carrageenan Nanoparticles: Effect of Cross-Linking with Tripolyphosphate and Charge Ratios. *Carbohydr. Polym.* **2012**, *89*, 282–289. [CrossRef] [PubMed]

140. Kumar, A.; Ahuja, M. Carboxymethyl Gum Kondagogu-Chitosan Polyelectrolyte Complex Nanoparticles: Preparation and Characterization. *Int. J. Biol. Macromol.* **2013**, *62*, 80–84. [CrossRef] [PubMed]

141. Hu, Q.; Wang, T.; Zhou, M.; Xue, J.; Luo, Y. Formation of Redispersible Polyelectrolyte Complex Nanoparticles from Gallic Acid-Chitosan Conjugate and Gum Arabic. *Int. J. Biol. Macromol.* **2016**, *92*, 812–819. [CrossRef] [PubMed]

142. Arif, M.; Raja, M.A.; Zeenat, S.; Chi, Z.; Liu, C. Preparation and Characterization of Polyelectrolyte Complex Nanoparticles Based on Poly (Malic Acid), Chitosan. A pH-Dependent Delivery System. *J. Biomater. Sci. Polym. Ed.* **2017**, *28*, 50–62. [CrossRef] [PubMed]

143. Zhang, L.; Wang, J.; Ni, C.; Zhang, Y.; Shi, G. Preparation of Polyelectrolyte Complex Nanoparticles of Chitosan and Poly(2-Acrylamido-2-Methylpropanesulfonic Acid) for Doxorubicin Release. *Mater. Sci. Eng.* **2016**, *58*, 724–729. [CrossRef] [PubMed]

144. Rolland, J.; Guillet, P.; Schumers, J.-M.; Duhem, N.; Prèat, V.; Gohy, J.-F. Polyelectrolyte Complex Nanoparticles from Chitosan and Poly(Acrylic Acid) and Polystyrene-Block-Poly(Acrylic Acid). *J. Polym. Sci. Part A Polym. Chem.* **2012**, *50*, 4484–4493. [CrossRef]

145. Lee, K.Y.; Kwon, I.C.; Kim, Y.-H.; Jo, W.H.; Jeong, S.Y. Preparation of chitosan self-aggregates as a gene delivery system. *J. Control. Release* **1998**, *51*, 213–220. [CrossRef]

146. Kim, Y.H.; Gihm, S.H.; Park, C.R.; Lee, K.Y.; Kim, T.W.; Kwon, I.C.; Chung, H.; Jeong, S.Y. Structural Characteristics of Size-Controlled Self-Aggregates of Deoxycholic Acid-Modified Chitosan and Their Application as a DNA Delivery Carrier. *Bioconjugate Chem.* **2001**, *12*, 932–938. [CrossRef]

147. Wang, Y.-S.; Liu, L.-R.; Jiang, Q.; Zhang, Q.-Q. Self-aggregated nanoparticles of cholesterol-modified chitosan conjugate as a novel carrier of epirubicin. *Eur. Polym. J.* **2007**, *43*, 43–51. [CrossRef]

148. Chen, M.; Liu, Y.; Yang, W.; Li, X.; Liu, L.; Zhou, Z.; Wang, Y.; Li, R.; Zhang, Q. Preparation and characterization of self-assembled nanoparticles of 6-*O*-cholesterol-modified chitosan for drug delivery. *Carbohydr. Polym.* **2011**, *84*, 1244–1251. [CrossRef]

149. Hu, F.Q.; Ren, G.F.; Yuan, H.; Du, Y.Z.; Zeng, S. Shell cross-linked stearic acid grafted chitosan oligosaccharide self-aggregated micelles for controlled release of paclitaxel. *Colloids Surf. B* **2006**, *50*, 97–103. [CrossRef] [PubMed]

150. Hu, F.-Q.; Wu, X.-L.; Du, Y.-Z.; You, J.; Yuan, H. Cellular uptake and cytotoxicity of shell crosslinked stearic acid-grafted chitosan oligosaccharide micelles encapsulating doxorubicin. *Eur. J. Pharm. Biopharm.* **2008**, *69*, 117–125. [CrossRef] [PubMed]

151. Ye, Y.Q.; Yang, F.L.; Hu, F.Q.; Du, Y.Z.; Yuan, H.; Yu, H.Y. Core-modified chitosan-based polymeric micelles for controlled release of doxorubicin. *Int. J. Pharm.* **2008**, *352*, 294–301. [CrossRef] [PubMed]

152. Hu, F.-Q.; Liu, L.-N.; Du, Y.-Z.; Yuan, H. Synthesis and antitumor activity of doxorubicin conjugated stearic acid-*g*-chitosan oligosaccharide polymeric micelles. *Biomaterials* **2009**, *30*, 6955–6963. [CrossRef] [PubMed]

153. Wu, Y.; Li, M.; Gao, H. Polymeric micelle composed of PLA and chitosan as a drug carrier. *J. Polym. Res.* **2009**, *16*, 11–18. [CrossRef]

154. Cho, Y.; Kim, J.T.; Park, H.J. Size-controlled self-aggregated *N*-acyl chitosan nanoparticles as a vitamin C carrier. *Carbohydr. Polym.* **2012**, *88*, 1087–1092. [CrossRef]

155. Quiñones, J.P.; Gothelf, K.V.; Kjems, J.; Caballero, A.M.H.; Schmidt, C.; Covas, C.P. *N*,*O*6-partially acetylated chitosan nanoparticles hydrophobically-modified for controlled release of steroids and vitamin E. *Carbohydr. Polym.* **2013**, *91*, 143–151. [CrossRef] [PubMed]

156. Opanasopit, P.; Ngawhirunpat, T.; Chaidedgumjorn, A.; Rojanarata, T.; Apirakaramwong, A.; Phongying, S.; Choochottiros, C.; Chirachanchai, S. Incorporation of camptothecin into *N*-phthaloyl chitosan-*g*-mPEG self-assembly micellar system. *Eur. J. Pharm. Biopharm.* **2006**, *64*, 269–276. [CrossRef] [PubMed]

157. Opanasopit, P.; Ngawhirunpat, T.; Rojanarata, T.; Choochottiros, C.; Chirachanchai, S. Camptothecin-incorporating *N*-phthaloylchitosan-*g*-mPEG self-assembly micellar system: Effect of degree of deacetylation. *Colloids Surf. B* **2007**, *60*, 117–124. [CrossRef] [PubMed]

158. Opanasopit, P.; Ngawhirunpat, T.; Rojanarata, T.; Choochottiros, C.; Chirachanchai, S. *N*-Phthaloylchitosan-*g*-mPEG design for all-trans retinoic acid-loaded polymeric micelles. *Eur. J. Pharm. Sci.* **2007**, *30*, 424–431. [CrossRef] [PubMed]

159. Bian, F.; Jia, L.; Yu, W.; Liu, M. Self-assembled micelles of *N*-phthaloychitosan-*g*-polyvinylpyrrolidone for drug delivery. *Carbohydr. Polym.* **2009**, *76*, 454–459. [CrossRef]

160. Duan, K.; Zhang, X.; Tang, X.; Yu, J.; Liu, S.; Wang, D.; Yaping, L.; Huang, J. Fabrication of cationic nanomicelle from chitosan-*graft*-polycaprolactone as the carrier of 7-ethyl-10-hydroxycamptothecin. *Colloids Surf. B* **2010**, *76*, 475–482. [CrossRef] [PubMed]

161. Li, F.; Zhang, X.; Li, H.; Xiang, L.; Chen, Y. Preparation of self-assembled nanoparticles of chitosan oligosaccharide-*graft*-polycaprolactone as a carrier of bovine serum albumin drug. *Bio-Med. Mater. Eng.* **2014**, *24*, 2041–2048. [CrossRef]

162. Almeida, A.; Silva, D.; Goncalves, V.; Sarmento, B. Synthesis and characterization of chitosan-*grafted*-polycaprolactone micelles for modulate intestinal paclitaxel delivery. *Drug Deliv. Transl. Res.* **2017**, 1–11. [CrossRef] [PubMed]

163. Gu, C.; Le, V.; Lang, M.; Liu, J. Preparation of polysaccharide derivates chitosan-*graft*-poly(varepsilon-caprolactone) amphiphilic copolymer micelles for 5-fluorouracil drug delivery. *Colloids Surf. B* **2014**, *116*, 745–750. [CrossRef] [PubMed]

164. Raja, M.A.; Arif, M.; Feng, C.; Zeenat, S.; Liu, C.G. Synthesis and evaluation of pH-sensitive, self-assembled chitosan-based nanoparticles as efficient doxorubicin carriers. *J. Biomater. Appl.* **2017**, *31*, 1182–1195. [CrossRef] [PubMed]

165. Hwang, H.-Y.; Kim, I.-S.; Kwon, I.C.; Kim, Y.-H. Tumor targetability and antitumor effect of docetaxel-loaded hydrophobically modified glycol chitosan nanoparticles. *J. Control. Release* **2008**, *128*, 23–31. [CrossRef] [PubMed]

166. Min, K.H.; Park, K.; Kim, Y.S.; Bae, S.M.; Lee, S.; Jo, H.G.; Park, R.W.; Kim, I.S.; Jeong, S.Y.; Kim, K.; et al. Hydrophobically modified glycol chitosan nanoparticles-encapsulated camptothecin enhance the drug stability and tumor targeting in cancer therapy. *J. Control. Release* **2008**, *127*, 208–218. [CrossRef] [PubMed]

167. Huh, M.S.; Lee, S.Y.; Park, S.; Lee, S.; Chung, H.; Lee, S.; Choi, Y.; Oh, Y.-K.; Park, J.H.; Jeong, S.Y.; et al. Tumor-homing glycol chitosan/polyethylenimine nanoparticles for the systemic delivery of siRNA in tumor-bearing mice. *J. Control. Release* 2010, *144*, 134–143. [CrossRef] [PubMed]

168. Park, J.H.; Kwon, S.; Nam, J.-O.; Park, R.-W.; Chung, H.; Seo, S.B.; Kim, I.-S.; Kwon, I.C.; Jeong, S.Y. Self-assembled nanoparticles based on glycol chitosan bearing 5β-cholanic acid for RGD peptide delivery. *J. Control. Release* 2004, *95*, 579–588. [CrossRef] [PubMed]

169. Yu, J.-M.; Li, Y.-J.; Qiu, L.-Y.; Jin, Y. Self-aggregated nanoparticles of cholesterol-modified glycol chitosan conjugate: Preparation, characterization, and preliminary assessment as a new drug delivery carrier. *Eur. Polym. J.* 2008, *44*, 555–565. [CrossRef]

170. Lee, J.; Lee, C.; Kim, T.H.; Lee, E.S.; Shin, B.S.; Chi, S.C.; Park, E.S.; Lee, K.C.; Youn, Y.S. Self-assembled glycol chitosan nanogels containing palmityl-acylated exendin-4 peptide as a long-acting anti-diabetic inhalation system. *J. Control. Release* 2012, *161*, 728–734. [CrossRef] [PubMed]

171. Quiñones, J.P.; Gothelf, K.V.; Kjems, J.; Caballero, A.M.H.; Schmidt, C.; Covas, C.P. Self-assembled nanoparticles of glycol chitosan—Ergocalciferol succinate conjugate, for controlled release. *Carbohydr. Polym.* 2012, *88*, 1373–1377. [CrossRef]

172. Quiñones, J.P.; Gothelf, K.V.; Kjems, J.; Yang, C.; Caballero, A.M.H.; Schmidt, C.; Covas, C.P. Self-assembled nanoparticles of modified-chitosan conjugates for the sustained release of DL-α-tocopherol. *Carbohydr. Polym.* 2013, *92*, 856–864. [CrossRef] [PubMed]

173. Quiñones, J.P.; Gothelf, K.V.; Kjems, J.; Heras, A.; Schmidt, C.; Peniche, C. Novel Self-assembled Nanoparticles of Testosterone-Modified Glycol Chitosan and Fructose Chitosan for Controlled Release. *J. Biomater. Tissue Eng.* 2013, *3*, 164–172. [CrossRef]

174. Saravanakumar, G.; Min, K.H.; Min, D.S.; Kim, A.Y.; Lee, C.-M.; Cho, Y.W. Hydrotropic oligomer-conjugated glycol chitosan as a carrier of paclitaxel: Synthesis, characterization, and in vivo biodistribution. *J. Control. Release* 2009, *140*, 210–217. [CrossRef] [PubMed]

175. Oh, N.M.; Oh, K.T.; Baik, H.J.; Lee, B.R.; Lee, A.H.; Youn, Y.S.; Lee, E.S. A self-organized 3-diethylaminopropyl-bearing glycol chitosan nanogel for tumor acidic pH targeting: In vitro evaluation. *Colloids Surf. B* 2010, *78*, 120–126. [CrossRef] [PubMed]

176. Son, Y.J.; Jang, J.-S.; Cho, Y.W.; Chung, H.; Park, R.-W.; Kwon, I.C.; Kim, I.-S.; Park, J.Y.; Seo, S.B.; Park, C.R.; Jeong, S.Y. Biodistribution and anti-tumor efficacy of doxorubicin loaded glycol-chitosan nanoaggregates by EPR effect. *J. Control. Release* 2003, *91*, 135–145. [CrossRef]

177. Li, Y.; Zhang, S.; Meng, X.; Chen, X.; Ren, G. The preparation and characterization of a novel amphiphilic oleoyl-carboxymethyl chitosan self-assembled nanoparticles. *Carbohydr. Polym.* 2011, *83*, 130–136. [CrossRef]

178. Liu, Y.; Cheng, X.J.; Dang, Q.F.; Ma, F.K.; Chen, X.G.; Park, H.J.; Kim, B.K. Preparation and evaluation of oleoyl-carboxymethy-chitosan(OCMCS) nanoparticles as oral protein carriers. *J. Mater. Sci. Mater. Med.* 2012, *23*, 375–384. [CrossRef] [PubMed]

179. Liu, C.; Fan, W.; Chen, X.; Liu, C.; Meng, X.; Park, H.J. Self-assembled nanoparticles based on linoleic-acid modified carboxymethyl-chitosan as carrier of adriamycin (ADR). *Curr. Appl. Phys.* 2007, *7*, 25–29. [CrossRef]

180. Xiangyang, X.; Ling, L.; Jianping, Z.; Shiyue, L.; Jir, Y.; Xiaojin, Y.; Jinsheng, R. Preparation and characterization of N-succinyl-N′-octyl chitosan micelles as doxorubicin carriers for effective anti-tumor activity. *Colloids Surf. B* 2007, *55*, 222–228. [CrossRef] [PubMed]

181. Zhu, A.P.; Yuan, L.H.; Chen, T.; Wu, H.; Zhao, F. Interactions between N-succinyl-chitosan and bovine serum albumin. *Carbohydr. Polym.* 2007, *69*, 363–370. [CrossRef]

182. Zhang, C.; Ding, Y.; Yu, L.L.; Ping, Q. Polymeric micelle systems of hydroxycamptothecin based on amphiphilic N-alkyl-N-trimethyl chitosan derivatives. *Colloids Surf. B* 2007, *55*, 192–199. [CrossRef] [PubMed]

183. Bei, Y.Y.; Zhou, X.F.; You, B.G.; Yuan, Z.Q.; Chen, W.L.; Xia, P.; Liu, Y.; Jin, Y.; Hu, X.J.; Zhu, Q.L.; et al. Application of the central composite design to optimize the preparation of novel micelles of harmine. *Int. J. Nanomed.* 2013, *8*, 1795–1808. [CrossRef]

184. Bei, Y.Y.; Zhou, X.F.; You, B.G.; Yuan, Z.Q.; Chen, W.L.; Xia, P.; Liu, Y.; Jin, Y.; Hu, X.J.; Zhu, Q.L.; et al. Novel self-assembled micelles based on palmitoyl-trimethyl-chitosan for efficient delivery of harmine to liver cancer. *Expert Opin. Drug Deliv.* 2014, *11*, 843–854. [CrossRef] [PubMed]

185. Mi, F.L.; Wu, Y.Y.; Lin, Y.H.; Sonaje, K.; Ho, Y.C.; Chen, C.T.; Juang, J.H.; Sung, H.W. Oral delivery of peptide drugs using nanoparticles self-assembled by poly(gamma-glutamic acid) and a chitosan derivative functionalized by trimethylation. *Bioconjugate Chem.* **2008**, *19*, 1248–1255. [CrossRef] [PubMed]

186. Mo, R.; Jin, X.; Li, N.; Ju, C.; Sun, M.; Zhang, C.; Ping, Q. The mechanism of enhancement on oral absorption of paclitaxel by *N*-octyl-*O*-sulfate chitosan micelles. *Biomaterials* **2011**, *32*, 4609–4620. [CrossRef] [PubMed]

187. Zhang, C.; Qu, G.; Sun, Y.; Wu, X.; Yao, Z.; Guo, Q.; Ding, Q.; Yuan, S.; Shen, Z.; Ping, Q.; et al. Pharmacokinetics, biodistribution, efficacy and safety of *N*-octyl-*O*-sulfate chitosan micelles loaded with paclitaxel. *Biomaterials* **2008**, *29*, 1233–1241. [CrossRef] [PubMed]

188. Qu, G.; Yao, Z.; Zhang, C.; Wu, X.; Ping, Q. PEG conjugated *N*-octyl-*O*-sulfate chitosan micelles for delivery of paclitaxel: In vitro characterization and in vivo evaluation. *Eur. J. Pharm. Sci.* **2009**, *37*, 98–105. [CrossRef] [PubMed]

189. Pedro, R.D.O.; Pereira, S.; Goycoolea, F.M.; Schmitt, C.C.; Neumann, M.G. Self-aggregated nanoparticles of *N*-dodecyl,*N'*-glycidyl(chitosan) as pH-responsive drug delivery systems for quercetin. *J. Appl. Polym. Sci.* **2018**, *135*, 1–12. [CrossRef]

polymers

MDPI

Review

The Use of Polymer Chitosan in Intravesical Treatment of Urinary Bladder Cancer and Infections

Andreja Erman and Peter Veranič *

Institute of Cell Biology, Faculty of Medicine, University of Ljubljana, Ljubljana 1000, Slovenia;
andreja.erman@mf.uni-lj.si
* Correspondence: peter.veranic@mf.uni-lj.si; Tel.: +386-1-543-7682; Fax: +386-1-543-7681

Received: 19 January 2018; Accepted: 3 March 2018; Published: 5 March 2018

Abstract: The most frequent diseases of the urinary bladder are bacterial infections and bladder cancers. For both diseases, very high recurrence rates are characteristic: 50–80% for bladder cancer and more than 50% for bladder infections, causing loss of millions of dollars per year for medical treatment and sick leave. Despite years of searching for better treatment, the prevalence of bladder infections and bladder cancer remains unchanged and is even increasing in recent years. Very encouraging results in treatment of both diseases recently culminated from studies combining biopolymer chitosan with immunotherapy, and chitosan with antibiotics for treatment of bladder cancer and cystitis, respectably. In both pathways of research, the discoveries involving chitosan reached a successful long-lasting cure. The property of chitosan that boosted the effectivity of illness-specific drugs is its ability to enhance the accessibility of these drugs to the very sources of both pathologies that individual treatments without chitosan failed to achieve. Chitosan can thus be recognised as a very promising co-player in treatment of bladder cancer and bacterial cystitis.

Keywords: chitosan; biological activity; medical applications

1. Urinary Bladder Function and Malfunctions

The main function of the urinary bladder is to enable controlled micturition by retaining the urine at constant pressure between periodic micturition. To prevent the permeation of potentially toxic and hypertonic urine back into underlying tissues and bloodstream during the retention, the luminal side of the urinary bladder is covered by a three-layered epithelium called the urothelium [1]. The superficial urothelial cells, also called umbrella cells, are mainly responsible for maintaining a blood-urine barrier, which is supposed to be the tightest among the barriers in mammalian tissues [2]. These cells have a highly specialised apical plasma membrane [3] composed of proteins uroplakins, which are arranged in semi-crystalline structures forming rigid-looking urothelial plaques of thickened membrane, interspersed by normally thick membrane areas called hinge regions, altogether give apical plasma membrane a characteristic scalloped appearance (Figure 1) [4]. The permeation of metabolites in the urine is additionally prevented by a thick layer of glycocalix [5], while a paracellular diffusion of the urine is limited by tight junctions [6,7].

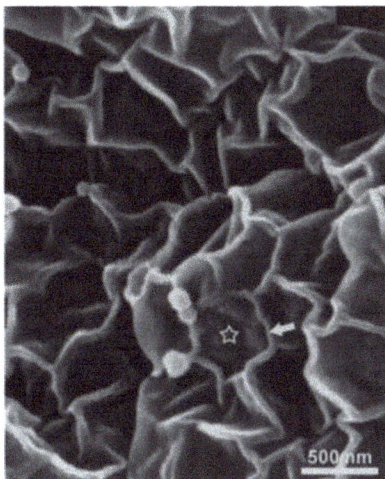

Figure 1. Apical plasma membrane of umbrella cell. The scalloped appearance is due to urothelial plaques (asterisk) of thickened membrane surrounded by hinge regions of normal membrane (arrow), as seen under the scanning electron microscope.

To maintain the barrier function in spite of constant changes of urine volume in the bladder, the umbrella cells have to adjust to extreme changes of the surface between filled and empty bladder [8,9] by reversible adjustment of apical surface and by altering their shape from cuboidal to extremely flat [10]. In order to prevent damage of umbrella cells during the filling of the bladder and distension of the urothelium, the subapical actin cortex, widely present in most epithelial cells, is in umbrella cells replaced by a compact cytokeratin network in the same location [11,12]. This provides cells with a much stronger mechanical support. The cytokeratin network is connected to a dense array of desmosomes concentrated in the subapical area of lateral membranes [13]. Replacement of superficial urothelial cells occurs constantly during the renewal of the urothelium, but the turnover of cells in the normal urothelium is very slow, because the life span of umbrella cells may be 6–7 months in physiological conditions [14]. However, after injury, a proliferation of basal cells increases rapidly, leading to fast renewal of intermediate and umbrella cells and the reestablishment of the permeability barrier function [15,16].

Such a stable epithelium is well adapted for maintaining the barrier function. However, this characteristic of the urothelium is largely responsible also for the limited success in treatment of the most frequent pathologies of the urinary bladder, bladder cancer and bacterial cystitis, because the blood-urine barrier prevents medications from reaching cancer cells in deeper layers of the urothelium and the bacteria hidden inside urothelial cells. To improve the penetration of drugs into urothelial tissue, several inducers of urothelial cell desquamation could be applied. Among them, mainly cyclophosphamide [12,17], 12-o-tetradecanoylphorbol-13-acetate (TPA) [18], sodium saccharin [19], or hyperthermic shock [20] have been used to largely remove urothelial cells. Most procedures that induced cell removal resulted in an inflammatory response, prolonged cell desquamation, and transitional hyperplasia of the urothelium. The most successful way to enhance the permeability of urothelium with minimal inflammatory response to the neighbouring tissue was proven to be the application of biopolymer chitosan that can be used for the controlled removal of the superficial layer of urothelial cells with minimum damage to the rest of the bladder wall [21].

2. Effect of Chitosan to the Urothelium

Chitosan, a cationic polysaccharide composed of glucosamine and *N*-acetyl glucosamine, is obtained by partial deacetylation of chitin. In general, chitosan is regarded as a biodegradable, biocompatible nontoxic polymer [22]. Chitosan has been widely used in biotechnology and reconstructive medicine [23]. Surprisingly, in the urothelium, chitosan (chitosan hydrochloride, 86% degree of deacetylation, 30–400 kDa) causes a very rapid drop in transepithelial resistance, indicating for the effect on tight junctions that was proven in both ex vivo [16] and in vitro experiments [24]. However, the mechanisms responsible for chitosan activity are largely unknown, even though they are essential for the proper use of this biopolymer. It has been proposed that the mechanism for an interruption of epithelial barrier function is mainly due to electrostatic interaction between the positively charged chitosan (85% degree of deacetylation, 80 kDa) and the negatively charged integrin $\alpha(V)\beta(3)$. This electrostatic interaction can lead to the conformational change of integrin $\alpha(V)\beta(3)$ and its clustering along the cell border, F-actin reorganization, and claudin 4 down-regulation, eventually resulting in the disruption of tight junctions and consequent increase of paracellular permeability [25]. Another mechanism by which chitosan (15% degree of deacetylation, M_n = 108,700) increases the permeability of the epithelia is by its interactions with cell membranes. The high density of positive charges is responsible for chitosan to interact strongly with negatively charged [26] and neutral molecules [27,28] in the plasma membrane. The combination of electrostatic and hydrophobic interactions can be crucial for binding of chitosan to phospholipids at the plasma membrane surface [26], and when chitosan is able to enter deeper into the lipid bilayer, the molecules can interact by hydrophobic interactions also with membrane fatty acids. Extraction of cholesterol has been proposed as a major mechanism of chitosan action, however, studies of the interaction between chitosan and membrane models proved that chitosan only binds to cholesterol in membranes but cannot remove this molecule from the membrane [29]. Protonated amino groups of chitosan enable the polymer to interact also with a negatively charged mucus molecules covering the epithelia via electrostatic interactions [30]. A higher charge density is obtained at acidic pH values, since the pK_a value of the D-glucosamine residue of chitosan is about 6.2–7.0 [31]. The charge density of chitosan is therefore considered to be an important factor in the drug absorption enhancement caused by this polymer.

Intravesical application of chitosan (chitosan hydrochloride, 86% degree of deacetylation, 30–400 kDa) to experimental animals resulted in induction of urothelial cell desquamation [15] (Figure 2A). However, the precise mechanism of induction of desquamation is still poorly understood, and further investigations must therefore be performed to explain the mechanism by which chitosan acts on urothelial cells. It has been discovered that chitosan adheres to the apical membrane of umbrella cells and causes necrotic changes and desquamation of superficial cells (Figure 2B). The apical plasma membrane of terminally differentiated superficial cells is composed of rigid urothelial plaques and normally thick membrane areas [32]. The main constituents of urothelial plaques are uroplakins, which are highly glycosylated proteins contributing to the urothelial glycocalyx [33]. Due to cationic nature of chitosan, this polymer can adhere to negatively charged groups of glycosaminoglycan in plaque regions, as well as to hinge regions of the apical plasma membrane of superficial cells. It is therefore possible that breaks in the apical membrane appear because of different viscoelastic properties of the discrete domains, to which the chitosan molecules are attached, which is in agreement with the finding of Pavinatto et al. [26]. When covered by a chitosan layer, the breaks in the apical plasma membrane can appear because of repeated stretching and contracting of the elastically and structurally inhomogeneous apical plasma membrane of umbrella cells. At higher chitosan concentrations (chitosan hydrochloride, 86% degree of deacetylation, 30–400 kDa) and longer exposure times, the release of cellular content and lysosomal enzymes from umbrella cells subsequently triggers cell death of also underlying urothelial cells [21].

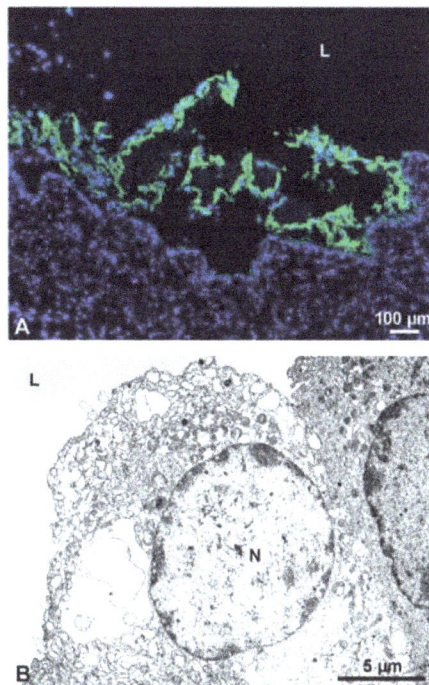

Figure 2. Chitosan induced umbrella cells desquamation (A) and necrosis (B). (**A**) Cytokeratin 20-positive umbrella cells (green fluorescence) detach from the urothelium after intravesical application of chitosan. (**B**) These cells have distinctive signs of necrosis under transmission electron microscope. N-nucleus, L-urinary bladder lumen.

It has been well documented that by choosing the appropriate chitosan concentration and the duration of intravesical application, chitosan can induce a controlled desquamation of urothelial cells in experimental animals. Intravesical application of 0.005% chitosan dispersion (chitosan hydrochloride, 86% degree of deacetylation, 30–400 kDa) caused complete removal of exclusively superficial cells after 20 min of treatment. A complete regeneration of this mild injury was completed within 60 min after an application of chitosan, and no inflammatory cell response was determined in such a short period of time. In addition to the morphological characteristics of the regeneration process after the chitosan-induced disruption of the urothelium, functional regeneration was also shown as a rapid restoration of the transepithelial resistance of urothelial tissue. By in vivo and ex vivo experiments, such intravesical chitosan application was proven to be safe in experimental animals because only weak inflammatory response and no persistent urothelial damage were detected [16].

3. Chitosan in Treatment of Urinary Bladder Cancer

Bladder cancer is the fourth most common cancer in males. Worldwide annual incidence of bladder cancer is estimated to be 14.1 million cancer cases around the world in 2012 [34,35]. The muscle non-invasive bladder cancer, which is the most common cancer of urinary bladder, usually appears as multiple focal tumours distributed throughout the urothelium of the bladder [36]. Urothelial tumours are treated depending on the metastatic potential with transurethral tumour resection (TUR) for non-invasive papillary tumours, and with cystectomy for invasive tumours, followed by radiation treatment or chemotherapy, or immunotherapy. In urothelial tumours, the recurrence rate is 50–80%, being the highest of any major malignancy [37]. Despite advantages of local delivery that overcome systemic adverse effects, intravesical therapy has its limitations, mainly because of the blood-urine barrier limits the penetration of cytostatics to deeper layers of the urothelium. Reduced doses of cytostatic in deeper layers of the urothelium may preserve individual cancer cells as seeds for the growth of new tumours, which make frequent use of invasive cystoscopy necessary. It is thus an urgent need for further development [38] in which mainly investigative work with nanoparticles (NPs) appear to be a promising strategy for improvement [39] in both diagnostics and treatment. As early diagnosis of recurrent tumours is important, a recent development of peptide-targeted glycol chitosan nanoparticles containing ferromagnetic iron oxide for multimodal imaging became a promising tool for non-invasive detection of bladder tumours by magnetic resonance imaging and near infrared fluorescent imaging [40]. NPs are proposed also to be used in the treatment of bladder cancers. Namely, encapsulation of anti-tumour drugs into NPs can protect the drug from degradation, enhance its solubility, and enable controlled release in cancerous tissue [41]. Chitosan functionalisation has been proven to increase transurothelial penetration and tumour cell uptake of commonly used poly(lactic-*co*-glycolic acid) (PLGA) nanoparticles [42]. Chitosan-functionalized nanoparticles demonstrated an increased binding to and uptake in intravesically instilled mouse bladders at 10 times higher level in comparison to PLGA-only nanoparticles. Furthermore, binding a survivin siRNA to chitosan-functionalised nanoparticles significantly decreases the survivin expression and by that decreases the proliferation of bladder cancer cells. Thus, chitosan-functionalised nanoparticles proved to have the capacity to transport large amounts of siRNA across the urothelium and to the tumour site, thus increasing therapeutic response.

In spite of encouraging results of using nanoparticles covered with chitosan, the penetration in deeper layers of the urothelium is still questionable, especially in the case of individual cancer cells migrating from the primary tumours below normal umbrella cells [43]. The most efficient way to temporarily compromise the blood-urine barrier appears to be the controlled removal of umbrella cells with chitosan. In our recent in vitro study, it became evident that chitosan selectively removes highly differentiated urothelial cells and is much less toxic to less differentiated urothelial cells [24]. The higher toxicity of chitosan to umbrella cells versus urothelial cancer cells and less differentiated urothelial cells of lower urothelial layers can be explained by existence of very specific scalloped apical plasma membrane only in umbrella cells but not in less differentiated normal or cancer urothelial cells (Figure 3). Specific toxicity of chitosan to umbrella cells provides an ideal tool to briefly enhance the access of anti-tumour agents to the nests of bladder cancer cells in deeper urothelial layers while not causing major obliteration of the urothelium [15].

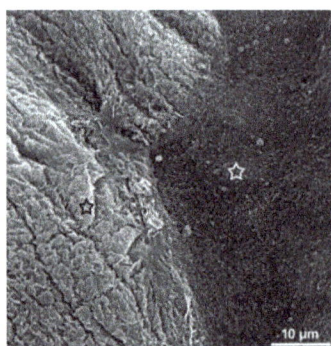

Figure 3. Apical plasma membrane of normal and cancer urothelial cells. Distinctive difference between scalloped apical membrane of normal umbrella cells on the left (black rimmed asterisk) and microvillar apical membrane of cancer cells on the right side (white rimmed asterisk) of the micrograph taken by scanning electron microscope.

A very promising model of treatment of bladder cancer using chitosan as a temporary destructor of the barrier function in the urothelium was developed by D.A. Zaharoff's group combining immune stimulant interleukin 12 (IL-12) with chitosan (chitosan glutamate, 75–90% degree of deacetylation, 200–600 kDa) that allows the penetration of the drug into the urothelium. The four-times-repeated treatments with a combination of IL-12 and chitosan eliminated 90% of bladder tumours in experimental animals and provoked a memory response protecting the animals from bladder tumours for the rest of their life [44] (Figure 4). Further study of the same research group demonstrated the ability of chitosan-enhanced interleukin-12- chitosan (chitosan glutamate, 75–90% degree of deacetylation, 200–600 kDa) based therapies to engage adaptive immunity within the tumour itself as well as throughout the body, and strengthen the case for clinical translation of chitosan-interleukin 12 as an intravesical treatment for bladder cancer [45].

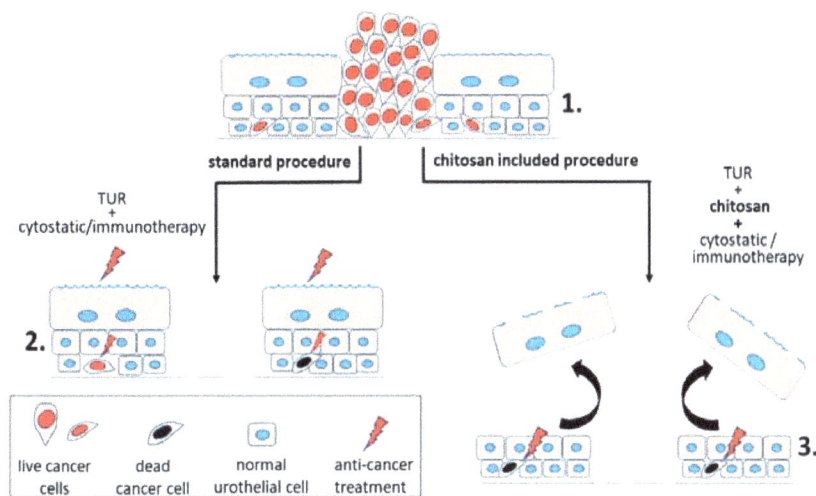

Figure 4. Two procedures to treat non-invasive urinary bladder cancer. (**1**) Urinary bladder cancer cells persist in the urothelium in tumours and as individual migrating cancer cells. (**2**) Standard procedure in treatment of non-invasive urothelial tumours is transurethral resection (TUR) of tumours and subsequent chemo- or immunotherapy. Frequently, individual cancer cells survive such treatment because of the diffusion-limiting blood-urine barrier. (**3**) Combined treatment with TUR removing tumour mass, chitosan compromising the barrier function, and immunotherapy more efficiently destroys even remote cancer cells.

4. Chitosan in Treatment of Bacterial Infection of the Urinary Bladder

Urinary tract infections (UTI) include infections of the urethra (urethritis), bladder (cystitis), ureters (ureteritis), and kidney (pyelonephritis) [46]. Uropathogenic Escherichia coli (UPEC) is the primary agent of urinary tract infections. There is an estimated annual occurrence of over 8 million UTIs in the United States [47]. Nearly all patients with UTI are prescribed a regimen of antibiotics. The annual cost of UTI treatment in the United States is estimated at $2.14 billion [48]. Despite administration of antibiotics that clear the bacteria from urine, the probability that a patient will develop a second UTI within six months is 25%. In more than 50% of recurrent UTI episodes, the bacterial strains responsible for both the initial infection and the recurrence are genetically identical [49], which indicates the involvement of the intracellular bacterial reservoirs within the bladder epithelial cells from which these recurrences originate. UPEC can invade umbrella cells, as well as the underlying less differentiated urothelial cells. UPEC can rapidly multiply in the cytosol of umbrella cells, forming a biofilm-like assembly known as an intracellular bacterial community (IBC) [50]. The development of IBCs can enhance the ability of UPEC to prosper within the urinary tract, while being sequestered away from the immune system of the host [51]. The occasional exit of UPEC from umbrella cells enables the distribution of UPEC to the new uninfected urothelial cells of all differentiation stages. Within less differentiated urothelial cells, individual UPEC can enter a dormant state [50]. The quiescent nature and intracellular localization of these bacteria protects them from most antibiotic treatments. These quiescent intracellular UPEC reservoirs (QIRs) can persist for long periods of time in the absence of any clinical symptoms and with no signs of bacterial existence in urine [51]. Differentiation of immature urothelial cells hosting quiescent bacteria can trigger the resurgent growth of UPEC, causing the development and dispersal of IBCs and the reinitiation of clinical symptoms [46]. These issues urge the need for therapeutic strategies that effectively target both active and dormant stages of UTI. By inducing the exfoliation of the superficial layer of the urothelium, chitosan was shown to stimulate rapid regenerative processes and the reactivation and efflux of quiescent intracellular UPEC reservoirs. When combined with antibiotics, chitosan treatment (chitosan hydrochloride, 86% degree of deacetylation, 30–400 kDa) significantly reduced bacterial loads within the urine and also eradicated bacteria from the bladder wall and was proposed to be of therapeutic value to individuals with chronic, recurrent UTIs [52]. However, though a single treatment of chitosan followed by ciprofloxacin administration had a marked effect on reducing UPEC titters within the bladder, this treatment failed to prevent relapsing bacterial outbursts. Our recent study elucidates that after four repeated applications of chitosan chitosan (chitosan hydrochloride, 86% degree of deacetylation, 30–400 kDa) in combination with the antibiotic ciprofloxacin, a complete eradication of UPEC from the urinary tract was achieved with no relapsing bouts of bacteriuria and no lasting harm to the urothelium [53] (Figure 5).

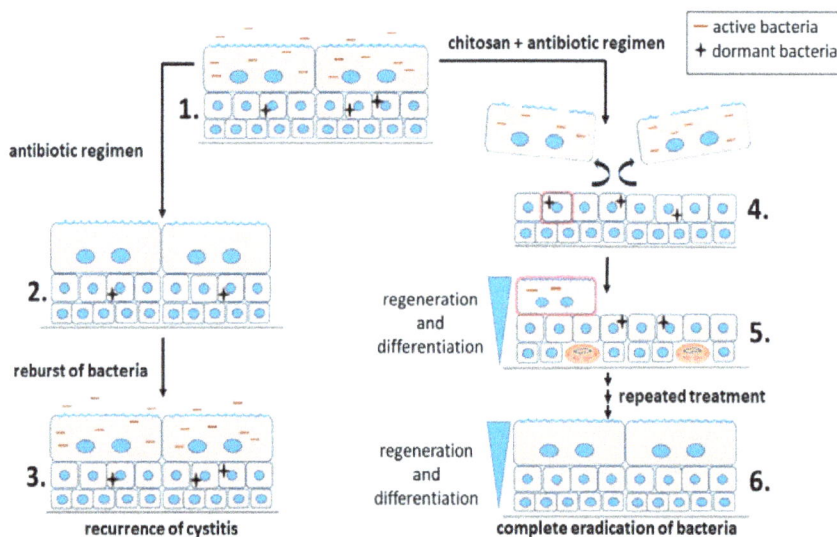

Figure 5. Two treatment regimens of infected urinary bladder epithelium. (**1**) Both active and dormant bacteria dwell in infected urothelial cells. (**2**) Antibiotic regimen efficiently clears active bacteria from both urine and host cells. However, immature urothelial cells retain dormant bacteria. (**3**) From the nests of dormant bacteria, new outbursts of bacteria appear periodically. (**4**) Treatment with a combination of chitosan and antibiotics induces a detachment of umbrella cells and activates differentiation of immature cells (red glowing cell). (**5**) Dormant bacteria reactivate in such differentiating cells (red glowing cell) and become again sensitive to antibiotics. Proliferating basal cells enable fast replacement of lost umbrella cells. (**6**) By repeated treatment with chitosan and antibiotics, bacteria become completely eliminated from urine and urothelial cells.

Polymers **2018**, *10*, 625

5. Conclusions

Chitosan offers versatile biomedical applications in urinary bladder epithelial cells due to the capability of this polymer to transitorily abolish the barrier function of urothelium and consequently to enable better penetration of specific drugs to the deeper cell layers. Chitosan has a great potential for clinical application in treatment of bladder tumours as a supplementary system in combination with cytostatic and immunotherapy, and also as an auxiliary antimicrobial drug in treatment of uroinfections. Common in both treatments of bladder cancer and bacterial infection of the urinary bladder is the benefit of local topical application, where higher dosages can be applied and the systemic side effects of the drugs otherwise used by oral administration is omitted.

Acknowledgments: The study was supported by the Slovenian Research Agency (Grant No. P3-0108). No other funds were received for covering the costs to publish in open access.

Conflicts of Interest: The authors declare no conflict of interest.

References

1. Khandelwal, P.; Abraham, S.N.; Apodaca, G. Cell biology and physiology of the uroepithelium. *Am. J. Physiol.-Ren. Physiol.* **2009**, *297*, F1477–F1501. [CrossRef] [PubMed]
2. Lewis, S.A.; Diamond, J.M. Na$^+$ transport by rabbit urinary bladder, a tight epithelium. *J. Membr. Biol.* **1976**, *28*, 1–40. [CrossRef] [PubMed]
3. Hu, P.; Meyers, S.; Liang, F.-X.; Deng, F.-M.; Kachar, B.; Zeidel, M.L.; Sun, T.-T. Role of membrane proteins in permeability barrier function: Uroplakin ablation elevates urothelial permeability. *Am. J. Physiol.-Ren. Physiol.* **2002**, *283*, F1200–F1207. [CrossRef] [PubMed]
4. Kachar, B.; Liang, F.; Lins, U.; Ding, M.; Wu, X.-R.; Stoffler, D.; Aebi, U.; Sun, T.-T. Three-dimensional analysis of the 16 nm urothelial plaque particle: Luminal surface exposure, preferential head-to-head interaction, and hinge formation 1 1Edited by W. Baumeisser. *J. Mol. Biol.* **1999**, *285*, 595–608. [CrossRef] [PubMed]
5. Hurst, R.E.; Rhodes, S.W.; Adamson, P.B.; Parsons, C.L.; Roy, J.B. Functional and structural characteristics of the glycosaminoglycans of the bladder luminal surface. *J. Urol.* **1987**, *138*, 433–437. [CrossRef]
6. Acharya, P.; Beckel, J.; Ruiz, W.G.; Wang, E.; Rojas, R.; Birder, L.; Apodaca, G. Distribution of the tight junction proteins ZO-1, occludin, and claudin-4, -8, and -12 in bladder epithelium. *Am. J. Physiol.-Ren. Physiol.* **2004**, *287*, F305–F318. [CrossRef] [PubMed]
7. Varley, C.L.; Garthwaite, M.A.E.; Cross, W.; Hinley, J.; Trejdosiewicz, L.K.; Southgate, J. PPARγ-regulated tight junction development during human urothelial cytodifferentiation. *J. Cell. Physiol.* **2006**, *208*, 407–417. [CrossRef] [PubMed]
8. Baskin, L.; Meaney, D.; Landsman, A.; Zderic, S.A.; Macarak, E. Bovine bladder compliance increases with normal fetal development. *J. Urol.* **1994**, *152*, 692–695, discussion 696–697. [CrossRef]
9. Truschel, S.T.; Wang, E.; Ruiz, W.G.; Leung, S.-M.; Rojas, R.; Lavelle, J.; Zeidel, M.; Stoffer, D.; Apodaca, G. Stretch-regulated exocytosis/endocytosis in bladder umbrella cells. *Mol. Biol. Cell* **2002**, *13*, 830–846. [CrossRef] [PubMed]
10. Baskin, L.S.; Hayward, S.W.; Young, P.F.; Cunha, G.R. Ontogeny of the rat bladder: Smooth muscle and epithelial differentiation. *Acta Anat. (Basel)* **1996**, *155*, 163–171. [CrossRef] [PubMed]
11. Veranic, P.; Jezernik, K. Trajectorial organisation of cytokeratins within the subapical region of umbrella cells. *Cell Motil. Cytoskelet.* **2002**, *53*, 317–325. [CrossRef] [PubMed]
12. Veranic, P.; Romih, R.; Jezernik, K. What determines differentiation of urothelial umbrella cells? *Eur. J. Cell Biol.* **2004**, *83*, 27–34. [CrossRef] [PubMed]
13. Veranic, P.; Jezernik, K. The response of junctional complexes to induced desquamation in mouse bladder urothelium. *Biol. Cell* **2000**, *92*, 105–113. [CrossRef]
14. Jost, S.P.; Potten, C.S. Urothelial proliferation in growing mice. *Cell Tissue Kinet.* **1986**, *19*, 155–160. [CrossRef] [PubMed]
15. Veranic, P.; Erman, A.; Kerec-Kos, M.; Bogataj, M.; Mrhar, A.; Jezernik, K. Rapid differentiation of superficial urothelial cells after chitosan-induced desquamation. *Histochem. Cell Biol.* **2009**, *131*, 129–139. [CrossRef] [PubMed]

16. Erman, A.; Kerec Kos, M.; Žakelj, S.; Resnik, N.; Romih, R.; Veranič, P. Correlative study of functional and structural regeneration of urothelium after chitosan-induced injury. *Histochem. Cell Biol.* **2013**, *140*, 521–531. [CrossRef] [PubMed]

17. Romih, R.; Koprivec, D.; Martincic, D.S.; Jezernik, K. Restoration of the rat urothelium after cyclophosphamide treatment. *Cell Biol. Int.* **2001**, *25*, 531–537. [CrossRef] [PubMed]

18. Watanabe, S.; Sasaki, J. 12-*O*-tetradecanoylphorbol-13-acetate induces selective desquamation of superficial cells in rat urinary bladder epithelium. *Cell Tissue Res.* **1992**, *268*, 239–245. [CrossRef] [PubMed]

19. Romih, R.; Jezernik, K.; Masera, A. Uroplakins and cytokeratins in the regenerating rat urothelium after sodium saccharin treatment. *Histochem. Cell Biol.* **1998**, *109*, 263–269. [CrossRef] [PubMed]

20. Jacob, J.; Hindmarsh, J.R.; Ludgate, C.M.; Chisholm, G.D. Observations on the ultrastructure of human urothelium: The response of normal bladder of elderly subjects to hyperthermia. *Urol. Res.* **1982**, *10*, 227–237. [CrossRef] [PubMed]

21. Kos, M.K.; Bogataj, M.; Veranic, P.; Mrhar, A. Permeability of pig urinary bladder wall: Time and concentration dependent effect of chitosan. *Biol. Pharm. Bull.* **2006**, *29*, 1685–1691. [CrossRef] [PubMed]

22. Singla, A.K.; Chawla, M. Chitosan: Some pharmaceutical and biological aspects—An update. *J. Pharm. Pharmacol.* **2001**, *53*, 1047–1067. [CrossRef] [PubMed]

23. Agerkvist, I.; Eriksson, L.; Enfors, S.O. Selective flocculation with chitosan in Escherichia coli disintegrates: Effects of pH and nuclease treatment. *Enzym. Microb. Technol.* **1990**, *12*, 584–590. [CrossRef]

24. Višnjar, T.; Jerman, U.D.; Veranič, P.; Kreft, M.E. Chitosan hydrochloride has no detrimental effect on bladder urothelial cancer cells. *Toxicol. In Vitro* **2017**, *44*, 403–413. [CrossRef] [PubMed]

25. Hsu, L.-W.; Ho, Y.-C.; Chuang, E.-Y.; Chen, C.-T.; Juang, J.-H.; Su, F.-Y.; Hwang, S.-M.; Sung, H.-W. Effects of pH on molecular mechanisms of chitosan-integrin interactions and resulting tight-junction disruptions. *Biomaterials* **2013**, *34*, 784–793. [CrossRef] [PubMed]

26. Pavinatto, F.J.; Pavinatto, A.; Caseli, L.; dos Santos, D.S.; Nobre, T.M.; Zaniquelli, M.E.D.; Oliveira, O.N. Interaction of Chitosan with Cell Membrane Models at the Air−Water Interface. *Biomacromolecules* **2007**, *8*, 1633–1640. [CrossRef] [PubMed]

27. Guo, J.; Ping, Q.; Jiang, G.; Huang, L.; Tong, Y. Chitosan-coated liposomes: Characterization and interaction with leuprolide. *Int. J. Pharm.* **2003**, *260*, 167–173. [CrossRef]

28. Perugini, P.; Genta, I.; Pavanetto, F.; Conti, B.; Scalia, S.; Baruffini, A. Study on glycolic acid delivery by liposomes and microspheres. *Int. J. Pharm.* **2000**, *196*, 51–61. [CrossRef]

29. Pavinatto, F.J.; Pacholatti, C.P.; Montanha, É.A.; Caseli, L.; Silva, H.S.; Miranda, P.B.; Viitala, T.; Oliveira, O.N. Cholesterol Mediates Chitosan Activity on Phospholipid Monolayers and Langmuir-Blodgett Films. *Langmuir* **2009**, *25*, 10051–10061. [CrossRef] [PubMed]

30. Artursson, P.; Lindmark, T.; Davis, S.S.; Illum, L. Effect of chitosan on the permeability of monolayers of intestinal epithelial cells (Caco-2). *Pharm. Res.* **1994**, *11*, 1358–1361. [CrossRef] [PubMed]

31. Hejazi, R.; Amiji, M. Chitosan-based gastrointestinal delivery systems. *J. Control. Release Off. J. Control. Release Soc.* **2003**, *89*, 151–165. [CrossRef]

32. Min, G.; Wang, H.; Sun, T.-T.; Kong, X.-P. Structural basis for tetraspanin functions as revealed by the cryo-EM structure of uroplakin complexes at 6-A resolution. *J. Cell Biol.* **2006**, *173*, 975–983. [CrossRef] [PubMed]

33. Hu, C.-C.A.; Liang, F.-X.; Zhou, G.; Tu, L.; Tang, C.-H.A.; Zhou, J.; Kreibich, G.; Sun, T.-T. Assembly of urothelial plaques: Tetraspanin function in membrane protein trafficking. *Mol. Biol. Cell* **2005**, *16*, 3937–3950. [CrossRef] [PubMed]

34. Parkin, D.M. The global burden of urinary bladder cancer. *Scand. J. Urol. Nephrol. Suppl.* **2008**, 12–20. [CrossRef] [PubMed]

35. Ploeg, M.; Aben, K.K.H.; Kiemeney, L.A. The present and future burden of urinary bladder cancer in the world. *World J. Urol.* **2009**, *27*, 289–293. [CrossRef] [PubMed]

36. Habuchi, T.; Marberger, M.; Droller, M.J.; Hemstreet, G.P.; Grossman, H.B.; Schalken, J.A.; Schmitz-Dräger, B.J.; Murphy, W.M.; Bono, A.V.; Goebell, P.; et al. Prognostic markers for bladder cancer: International Consensus Panel on bladder tumor markers. *Urology* **2005**, *66*, 64–74. [CrossRef] [PubMed]

37. Chamie, K.; Litwin, M.S.; Bassett, J.C.; Daskivich, T.J.; Lai, J.; Hanley, J.M.; Konety, B.R.; Saigal, C.S. Urologic Diseases in America Project Recurrence of high-risk bladder cancer: A population-based analysis. *Cancer* **2013**, *119*, 3219–3227. [CrossRef] [PubMed]

38. Nirmal, J.; Chuang, Y.-C.; Tyagi, P.; Chancellor, M.B. Intravesical therapy for lower urinary tract symptoms. *Urol. Sci.* **2012**, *23*, 70–77. [CrossRef]

39. Neutsch, L.; Wambacher, M.; Wirth, E.-M.; Spijker, S.; Kählig, H.; Wirth, M.; Gabor, F. UPEC biomimickry at the urothelial barrier: Lectin-functionalized PLGA microparticles for improved intravesical chemotherapy. *Int. J. Pharm.* **2013**, *450*, 163–176. [CrossRef] [PubMed]

40. Key, J.; Dhawan, D.; Cooper, C.L.; Knapp, D.W.; Kim, K.; Kwon, I.C.; Choi, K.; Park, K.; Decuzzi, P.; Leary, J.F. Multicomponent, peptide-targeted glycol chitosan nanoparticles containing ferrimagnetic iron oxide nanocubes for bladder cancer multimodal imaging. *Int. J. Nanomed.* **2016**, *11*, 4141–4155. [CrossRef] [PubMed]

41. Zhou, J.; Patel, T.R.; Fu, M.; Bertram, J.P.; Saltzman, W.M. Octa-functional PLGA nanoparticles for targeted and efficient siRNA delivery to tumors. *Biomaterials* **2012**, *33*, 583–591. [CrossRef] [PubMed]

42. Martin, D.T.; Steinbach, J.M.; Liu, J.; Shimizu, S.; Kaimakliotis, H.Z.; Wheeler, M.A.; Hittelman, A.B.; Mark Saltzman, W.; Weiss, R.M. Surface-Modified Nanoparticles Enhance Transurothelial Penetration and Delivery of Survivin siRNA in Treating Bladder Cancer. *Mol. Cancer Ther.* **2014**, *13*, 71–81. [CrossRef] [PubMed]

43. Jerman, U.D.; Kreft, M.E.; Veranič, P. Epithelial-Mesenchymal Interactions in Urinary Bladder and Small Intestine and How to Apply Them in Tissue Engineering. *Tissue Eng. Part B* **2015**, *21*, 521–530. [CrossRef] [PubMed]

44. Smith, S.G.; Koppolu, B.P.; Ravindranathan, S.; Kurtz, S.L.; Yang, L.; Katz, M.D.; Zaharoff, D.A. Intravesical chitosan/interleukin-12 immunotherapy induces tumor-specific systemic immunity against murine bladder cancer. *Cancer Immunol. Immunother. CII* **2015**, *64*, 689–696. [CrossRef] [PubMed]

45. Smith, S.G.; Baltz, J.L.; Koppolu, B.P.; Ravindranathan, S.; Nguyen, K.; Zaharoff, D.A. Immunological mechanisms of intravesical chitosan/interleukin-12 immunotherapy against murine bladder cancer. *OncoImmunology* **2017**, *6*, e1259050. [CrossRef] [PubMed]

46. Barber, A.E.; Norton, J.P.; Spivak, A.M.; Mulvey, M.A. Urinary tract infections: Current and emerging management strategies. *Clin. Infect. Dis. Off. Publ. Infect. Dis. Soc. Am.* **2013**, *57*, 719–724. [CrossRef] [PubMed]

47. Dielubanza, E.J.; Schaeffer, A.J. Urinary tract infections in women. *Med. Clin. N. Am.* **2011**, *95*, 27–41. [CrossRef] [PubMed]

48. Brown, P.; Ki, M.; Foxman, B. Acute pyelonephritis among adults: Cost of illness and considerations for the economic evaluation of therapy. *Pharmacoeconomics* **2005**, *23*, 1123–1142. [CrossRef] [PubMed]

49. Bower, J.M.; Eto, D.S.; Mulvey, M.A. Covert Operations of Uropathogenic Escherichia coli within the Urinary Tract. *Traffic* **2005**, *6*, 18–31. [CrossRef] [PubMed]

50. Eto, D.S.; Sundsbak, J.L.; Mulvey, M.A. Actin-gated intracellular growth and resurgence of uropathogenic Escherichia coli. *Cell. Microbiol.* **2006**, *8*, 704–717. [CrossRef] [PubMed]

51. Blango, M.G.; Mulvey, M.A. Persistence of uropathogenic Escherichia coli in the face of multiple antibiotics. *Antimicrob. Agents Chemother.* **2010**, *54*, 1855–1863. [CrossRef] [PubMed]

52. Blango, M.G.; Ott, E.M.; Erman, A.; Veranic, P.; Mulvey, M.A. Forced Resurgence and Targeting of Intracellular Uropathogenic Escherichia coli Reservoirs. *PLoS ONE* **2014**, *9*, e93327. [CrossRef] [PubMed]

53. Erman, A.; Križan Hergouth, V.; Blango, M.G.; Kerec Kos, M.; Mulvey, M.A.; Veranič, P. Repeated treatments with chitosan in combination with antibiotics completely eradicate uropathogenic *Escherichia coli* from infected mouse urinary bladders. *J. Infect. Dis.* **2017**, *216*, 375–381. [CrossRef] [PubMed]

polymers

MDPI

Review

Chitosan in Non-Viral Gene Delivery: Role of Structure, Characterization Methods, and Insights in Cancer and Rare Diseases Therapies

Beatriz Santos-Carballal [1] , Elena Fernández Fernández [2] and Francisco M. Goycoolea [3,*]

[1] ChiPro GmbH, Anne-Conway-Street 1, 28359 Bremen, Germany; bcarballal@chipro.de
[2] Lung Biology Group, Department Clinical Microbiology, RCSI, Education and Research Centre, Beaumont Hospital, Dublin 9, Ireland; elenaffernandez@rcsi.ie
[3] School of Food Science and Nutrition, University of Leeds, Leeds LS2 9JT, UK
* Correspondence: F.M.Goycoolea@leeds.ac.uk; Tel.: +44-113-343-1412

Received: 9 March 2018; Accepted: 11 April 2018; Published: 15 April 2018

Abstract: Non-viral gene delivery vectors have lagged far behind viral ones in the current pipeline of clinical trials of gene therapy nanomedicines. Even when non-viral nanovectors pose less safety risks than do viruses, their efficacy is much lower. Since the early studies to deliver pDNA, chitosan has been regarded as a highly attractive biopolymer to deliver nucleic acids intracellularly and induce a transgenic response resulting in either upregulation of protein expression (for pDNA, mRNA) or its downregulation (for siRNA or microRNA). This is explained as the consequence of a multi-step process involving condensation of nucleic acids, protection against degradation, stabilization in physiological conditions, cellular internalization, release from the endolysosome ("proton sponge" effect), unpacking and enabling the trafficking of pDNA to the nucleus or the siRNA to the RNA interference silencing complex (RISC). Given the multiple steps and complexity involved in the gene transfection process, there is a dearth of understanding of the role of chitosan's structural features (M_w and degree of acetylation, DA%) on each step that dictates the net transfection efficiency and its kinetics. The use of fully characterized chitosan samples along with the utilization of complementary biophysical and biological techniques is key to bridging this gap of knowledge and identifying the optimal chitosans for delivering a specific gene. Other aspects such as cell type and administration route are also at play. At the same time, the role of chitosan structural features on the morphology, size and surface composition of synthetic virus-like particles has barely been addressed. The ongoing revolution brought about by the recent discovery of CRISPR-Cas9 technology will undoubtedly be a game changer in this field in the short term. In the field of rare diseases, gene therapy is perhaps where the greatest potential lies and we anticipate that chitosans will be key players in the translation of research to the clinic.

Keywords: gene delivery; non-viral vectors; chitosan structure; pDNA; siRNA

1. Introduction

Modern understanding of health is based on the concept of regulation of metabolism by a complex network of molecular-based communication mechanisms known as cell signaling that governs basic cellular activities and coordinates cell responses so that they can act in concert. Cells in the body perform their life cycle functions partially by genetic programming, but also by responding to molecular signals generated within the cell. These networks respond to, are controlled by, and can be disrupted by processes that take place on the electrical, molecular, macromolecular and supramolecular scales [1]. These are also the domains of nanoscience and nanotechnology. Nanomedicine has emerged as a recent multidisciplinary field in which the manipulation of matter at such scales is used for diagnostics and therapy to tackle the current most important challenges in health under innovative, affordable and more effective approaches.

Even when extensive research is conducted focusing on advanced nanobiomaterials to be used in biomedicine or biotechnology, that their design and biophysical properties can be finely tuned, and their effectiveness in vitro or in vivo as drug and gene carriers has been demonstrated, sound fundamental understanding of the mechanistic aspects at the molecular and cell level is still lacking. Moreover, in the case of biopolymer-based biomaterials, there is a general lack of studies that enable establishing robust structure-function relationships and, hence, a more rational design of innovative biomaterials. This is particularly relevant for the development of new and more effective nanomedicines to treat cancer, autoimmune diseases, viral and antibiotic-resistant bacterial infections, and genetic rare diseases. At the same time, gene therapy has started to deliver results in clinical trials, and it is considered to be at a booming stage. Children suffering devastating diseases have seen their lives transformed. In the last two decades, gene therapy has caught significant attention as a potential method for treating genetic disorders such as cystic fibrosis [2], Parkinson's disease [3,4] as well as an alternative method for treating cancer [5]. The discovery of CRISPR (clustered regularly interspaced short palindromic repeats) sequences in DNA as part of the immune system of bacteria and archae, and the subsequent rise of CRISPR-Cas9 technology that has enabled the genome editing of plant and animal cells and to target disease-related genes in patients, has brought gene therapy to a brand new light. Indeed, an ongoing revolution in the therapeutic paradigms for many diseases, particularly genetic diseases, is currently underway, and we have only started to see the "tip of the iceberg" [6].

Another driver that accelerated the recent progress in gene therapy over the past years is the basic research into novel vectors for gene delivery. The use of genetically engineered recombinant viruses, namely adenoviruses and adeno-associated viruses, that carry the therapeutic gene payload in their viral capsid, thus protecting it from enzymatic degradation, has been the most pursued route to bring gene therapy from the laboratory bench to the bedside. This includes recent advances in vivo delivery of the Cas9 therapeutics [7]. However, this approach poses serious shortcomings and challenges due to the widespread human immunity against viruses, off-target genomic damage and their small packing size. Hence, the search for non-viral vectors has also geared substantial research in this field [8]. In the case of CRISPR-Cas9 technology, a recent paper has documented the use of gold nanoparticles conjugated with DNA and synthetic polymer as an advanced co-delivery strategy for Cas9 ribonucleoprotein and the donor DNA payload [9].

For more than two decades, chitosan has been a highly researched non-viral gene delivery biopolymer. However, it is still not fully understood what type of chitosan works best for the different type of therapeutic genes (e.g., pDNA, siRNA, miRNA, etc.) or the role of their conformations (e.g., single versus double stranded), and the shape and size of the complexes (also regarded here indistinctly as "polyplexes" or "nanocomplexes"). The breath of this review is on the utilization of chitosan as a non-viral gene delivery vector. To put this in the current context of knowledge, we give an overview of the current progress on viral and non-viral vectors for gene delivery. We review the current understanding on the role of chitosan's molecular structure and the efficacy to deliver different type of polynucleotides to mammalian cells. The focus of this paper circumscribes only to unmodified chitosans. A separate section addresses the major methods available to characterize the biophysical properties of non-viral gene delivery systems. This is followed by another two sections in which we focus on the advances on plasmid DNA and silencing RNA delivery and the current progress on gene therapy in the treatment of rare diseases of which we review the major types and their specific challenges. Even when most pre-clinical research has been geared towards viral vectors, chitosan-based non-viral systems have started to offer promising results in vitro (e.g., cystic fibrosis gene therapy).

Finally, we draw a few conclusions on the current gaps of the current knowledge, unmet challenges and future perspectives of chitosan as key players in the potential future translation of the wealth of in vitro and pre-clinical proof of concept research studies. Even though excellent comprehensive reviews that address non-viral gene delivery, including chitosan and its chemical derivatives, and combinatorial therapies, have recently been published [10–17], we have attempted in our review to bring together the current status of scattered knowledge related to chitosan non-viral gene delivery and its potential to replace viral vectors, particularly for rare diseases therapy.

2. Gene Therapy: Viral and Non-Viral Vectors in Gene Delivery

The main goal of gene therapy is to introduce new genetic material into targeted cells in the body. This approach provides several advantages compared to conventional protein therapy. Through the introduction of exogenous nucleic acids into a specific cell, it is possible to control and modulate the genomic expression. The direct transfer of a specific gene into a patient will lead to in vivo production of proteins in the target cells as "mini-bioreactors" [18]. Ideally, this process will occur at more physiological conditions than those achieved by conventional administration of therapeutic proteins [19,20]. Therefore, the simple approach of using therapeutic genes as a "pro-drug" to treat a patient may result in an alternative way to overcome the drawbacks associated with the use of recombinant proteins. The effectiveness of gene therapy relies on several feats, namely protection of the nucleotides from premature degradation in the extracellular environment, targeting of specific cells, and delivery of sufficient amounts of genetic material to produce a therapeutic effect [21]. Therefore, the major challenge related to gene therapy is the development of a non-toxic, non-immunogenic and effective intracellular delivery system. To find suitable vehicles for the efficient delivery of therapeutic oligo- and polynucleotides at the expense of minimal toxicity and immunogenicity, both viral and non-viral vectors have been developed.

After the administration of genetic material, a sequence of biochemical and physical barriers must be overcome. These are shown schematically in Figure 1, and are described as follows:

- The existence of serum nucleases in the extracellular environment results in rapid degradation of genetic material on intravenous, mucosal and intramuscular administration.
- The entry of nucleic acids (e.g., DNA, mRNA, and siRNA/miRNA) into the cell is mainly restricted to the endocytotic pathway, but the association of nucleotides with the cell surface is very low as a consequence of the high negative charge of the polynucleotides and the proteoglycans present in the cell membrane [22].
- Once internalized by the specific cells, the genetic material has to escape from the endosomal vesicle, where the low pH and enzymes present can lead to its degradation [23,24].
- DNA has to diffuse in the cytosol to get into the nucleus; this diffusion process is size-dependent and DNA larger than 3000 base pairs present highly reduced mobility [23,24].
- siRNA and miRNA must be loaded into the RNA-induced silencing complex (RISC), whereas mRNA must bind to the translational machinery [25].

To improve the efficacy of the deliver machinery of genes across these barriers, the nucleic acids can be inserted into vehicles for gene delivery to assist the transfer of exogenous genes to the specific cells. These vehicles can be divided in two groups: viral vectors and non-viral vectors.

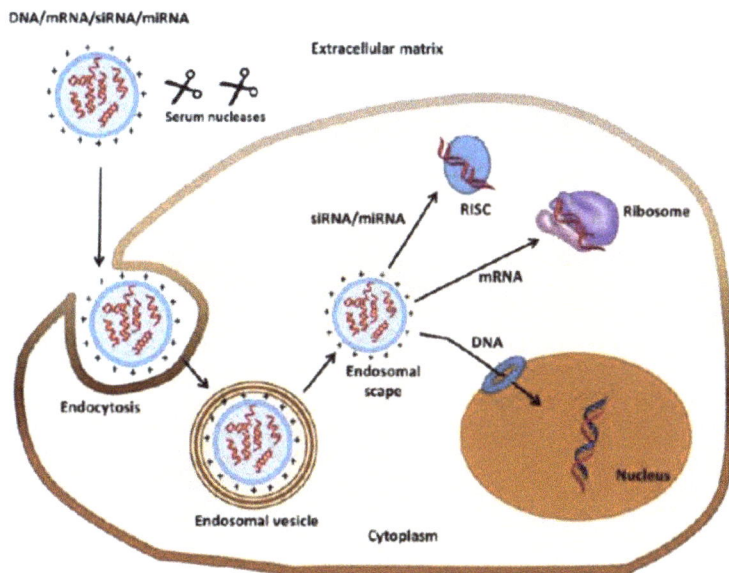

Figure 1. Barriers to successful in vivo delivery of nucleic acids.

2.1. Viral Vectors in Gene Delivery

The approach used in viral gene delivery is based on the ability of vectors to infect cells. This involves the use of genetically engineered recombinant viruses, adenoviruses and adeno-associated viruses that carry the therapeutic gene in their viral capsid, thus protecting it from enzymatic degradation [26]. Viral vectors have been developed based on a wide range of viruses and typically include strong promoters that achieve a high level of heterologous gene expression. Over the past two decades, more than 1800 clinical trials have been completed, are ongoing or have been approved worldwide in more than 30 countries [27]. However, these trials have only yielded five products globally. These products include Gendicine, Oncorine, Rexin G, Neovasculgen, and Glybera. In Europe, Glybera was approved for the treatment of patients with a familial lipoprotein lipase deficiency (LPLD) and, after some years, was discontinued; other formulations for cancer, myocardial ischemia, Duchenne muscular dystrophy, and painful diabetic neuropathy are currently in Phase III of clinical trials [28,29]. The large majority (~90%) of clinical trials utilized viral vectors, such as adenoviruses, adeno-associated viruses (AAVs), lentiviruses, or retroviruses. The low success rate of approved products underscores the many challenges that the utilization of viral vectors entails, as discussed below.

Viruses are biological particles that have evolved to transfer genetic information from one cell to another. The simplest viral particles consist of a protein coat that surrounds a strand of nucleic acid. They vary in the composition of their capside envelope, as well as in size and morphology. Hence, they occur as nanospheres of size ranging from ~20 nm (e.g., adeno-associated virus) to ~100 nm (e.g., adenovirus); short (~300 nm) nanorods (e.g., tobacco mosaic virus) and long (~700 nm) rods (e.g., Stygiolobus virus); worm-like particles (~1 μm; e.g., Ebola virus); and other shapes such as ellipsoid-, and cubic-shaped structures (e.g., *Acidianus convivator*). Moreover, viral genomes are diverse in size, structure, and nucleotide composition. They can be linear or circular chains of dsDNA, ssDNA, dsRNA or ssRNA. Given this great diversity, viruses offer a vast source of bioinspiration for bottom-up nanobiotechnology biomimetic approaches to conceiving synthetic novel vectors for gene delivery. This is particularly relevant, as the use of viral vectors in therapy raises significant safety issues, such as potential immunogenicity and reversion to pathogenicity of the vector [26]. Moreover, undesired mutations through the integration of their DNA into the genome of the introduced cells may lead to insertional mutagenesis and oncogenesis [19,30]. Other shortcomings include: limited DNA packaging capacity, complex production processes, broad tropism, and cytotoxicity. This has encouraged the development of non-viral vectors with better biosafety profiles and potential to address many of the limiting issues of viral vectors, thanks to the advances in materials science, nucleic acids chemistry and nanobiotechnology [31]. The most widely studied non-viral vectors include: cationic lipids, cell penetrating peptides, and cationic macromolecules. The major aspects of these systems are discussed in the next subsection.

2.2. Non-Viral Vectors in Gene Delivery and the "Proton Sponge" Hypothesis

The development of non-viral vectors aims to reach at least the same level of gene expression and specificity obtained when using viral vectors. The advantages of the use of non-viral vectors are related to their low cost and ease of production, their reduced immunogenicity and immunotoxicity and, therefore, greater bio-safety, in comparison to viral-mediated gene therapy. Non-viral vectors can be produced in a large scale, they are more flexible for optimization and control of the formulation and they are able to deliver large DNA sequences [23,32]. Non-viral vectors comprise naked nucleic acids or more complex systems based on the use of cationic molecules such as lipids, cell penetrating peptides or polycationic macromolecules. The nature of interaction between these non-viral vectors and the opposite negatively charged nucleic acids is mainly electrostatic, involving indeed, to transfer the genes effectively, it is desired a delivery system that carry a net positive charge to facilitate the interaction with the negatively charged cell membrane. Afterwards, it is hypothesized that the internalization into the cytoplasm will occur via endocytosis [33]. In this section, we briefly present the main type of used non-viral vectors to give the broad context of chitosan as one of the major non-viral gene delivery systems.

2.2.1. Cationic Lipids

Lipid-based vectors are among the most widely used non-viral gene carriers. Cationic lipids present amino groups to interact with nucleotides and hydrophobic groups constituted by fatty acids. The hydrophobic moieties contribute to the formation of bilayer vesicles in an aqueous medium. The first report on the use of lipid carriers is from 1987 with the introduction of *N*-[1-(2,3,-dioleyloxy)propyl]-*N*,*N*,*N*-trimethylamonium [34]. Since then, several cationic lipids have been synthesized and studied for gene delivery [35–40]. Lipocomplexes (also termed "lipoplexes") are promising candidates for in vitro and in vivo gene delivery [37,41]. However, these systems present some limitations including poor stability and rapid clearance, as well as the generation of an inflammatory response and relatively high cytotoxicity [26].

2.2.2. Cell Penetrating Peptides (CPPs)

CPPs are peptides containing domains of less than 20 amino acids and are characterized to interact specifically with receptors in the cell membrane and can transport molecules across it. They have gained popularity as non-viral transmembrane delivery vectors. CPPs are employed to enhance extracellular and intracellular internalization of relevant biomolecules including nucleotides. Gene delivery mediated by CPPs is classified in covalently bound and electrostatically bound. Either way, the CPPs are used to promote the delivery of their associated drugs and drug carriers into cells facilitated by an active transport mechanism. The most commonly used CPP is TAT peptide (TATp), derived from the transcriptional activator protein encoded by human immunodeficiency virus type 1 (HIV-1). TATp has been used for intracellular delivery in a range of cell types both in vitro and in vivo [42–45]. CPPs have recently been used to deliver Cas9 protein-RNA complex to enable rapid and timed editing with potential in human gene therapy [46].

2.2.3. Cationic Macromolecules

Cationic macromolecules have emerged as an alternative type of carriers for gene therapy. Macromolecules with functional groups able to be protonated at physiological pH, thus bearing positive charges, can be complexed with the negatively charged phosphates groups from nucleotides in nucleic acids [47–49]. The resulting systems are regarded as "polyplexes" or simply "complexes" and are self-assembled systems that exert their properties depending on the (+/−) charge ratio used.

The compaction of DNA by multivalent cations was studied by Matulis et al. using isothermal titration calorimetry (ITC) [50]. The model proposed by the authors consists of a two-stage process: (1) the cation binds to DNA through non-specific electrostatic forces, resulting in neutralization of the charges of the nucleotides, a decrease in charge repulsion and therefore an increase in the flexibility of the chains; and (2) after reaching a critical ligand concentration, DNA-DNA interactions occur and they condense into self-assembled systems, an entropically driven process [50]. In the specific case of polymeric cations (e.g., chitosan and PEI), it has been observed that the compaction of DNA occurs in a less tight manner, producing larger complexes than those formed with multivalent cations [51]. The condensation of smaller nucleic acids like siRNA by cationic polymers has been studied; it is believed that this process comprises the presence of many siRNA molecules in the formation of nanoparticles through interparticle assembly [52].

Among the main advantages of the use of cationic macromolecules are their ease for functionalization and possibility of binding specific targeting moiety, their higher stability compared to lipoplexes and lower cytotoxicity. Cationic macromolecules can be synthetic such as poly(L-lysine), polyethyleneimine and dendrimers or natural such as poly(D,L-lactic acid) and chitosan [23,26,32]. Despite the extensive reported progress on non-viral gene vectors, these systems still present deficient expression of their transgenes when transfecting mammalian cells as compared to viral systems [53]. This has demanded many research efforts on developing suitable non-viral vectors able to protect the nucleic acids against degradation, achieve specific cell targeting, promote cellular uptake, and induce minimal cytotoxicity and immunogenic rates.

Among the cationic polymers used for gene transfection, it has been observed that polymers containing amine groups with pKas around physiological pH lead to the best transgene expression. It is hypothesized that these systems exhibit "proton sponge" potential [54]. The "proton sponge" hypothesis has been described as the followed route to induce endosomal disruption and prevent nucleic acids from lysosomal degradation. Figure 2 describes the process after endocytosis of the complexes. Throughout the evolution of the endosomes, protons are translocated by ATPase proton pumps from the cytoplasm into the endosomes. This will cause a reduction in the pH of the endosomal compartments and the protonation of the cationic polymers with "proton sponge" potential. Therefore, more protons will be pumped in and chloride ions will passively accumulate into the endosomes. The increase in the ionic concentration inside the endosomes will cause water inflow, swelling and rupture of the endosomes and release of their content into the cytoplasm [33,54].

Polymers like polyethyleneimine and dendrimers containing protonable secondary or tertiary amine groups possess good rates of transfection efficiency both in vitro and in vivo [55–59]. On the other hand, chitosans are reported to have low proton sponge capacity [60,61]. However, Richard et al. have found that CS (M_w ~ 8 kDa and DA = 8%) presents similar capability as polyethyleneimine to induce proton sponge effect, and, therefore, to mediate endosomal escape. One of the main considerations has been that previous studies were carried out using mass concentration of chitosan instead of concentration on N-glucosamine units, leading to underestimation of the potential of chitosan to produce effective endosomal release [62]. Even when the proton sponge hypothesis offers a general explanation to the endolysosomal escape of polycationic non-viral vectors during intracellular trafficking, it is not the only mechanism at play that limits the rate of transit of the nucleic acid payload or the transfection efficiency [63].

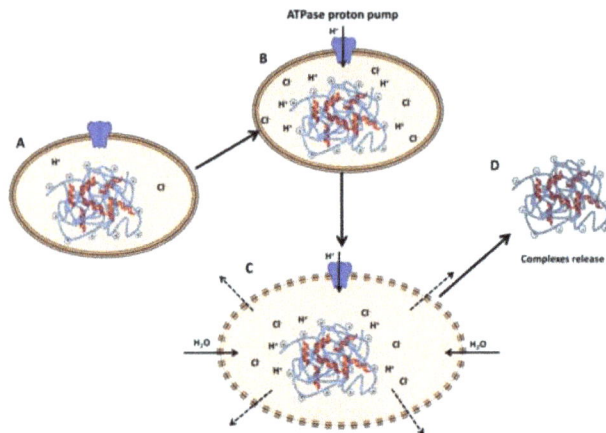

Figure 2. Schematic representation of the "proton sponge" hypothesis in which the endosomes containing the complexes with protonable polymers (**A**) evolve to late endosomes where protons are introduced by ATPase proton pumps, producing protonation of the polymer and a reduction in the pH (**B**); subsequently, chloride ions will be introduced in a non-active way causing a water inflow due to the osmotic pressure (**C**). Swelling of the endosomes leads to their rupture and finally release of their content into the cytoplasm (**D**).

3. Chitosan as a Non-Viral Gene Delivery Vector

As described above, polymers with cationic characteristics such as chitosan present enormous potential as gene delivery carriers. The primary amines in the chitosan backbone are protonated at slightly acidic pH, resulting in positive charges available to interact with nucleic acids via electrostatic forces. Mumper et al., have reported the use of chitosan as a non-viral gene delivery system for plasmid transfection for first time in 1995 [64]. In 1998, the use of chitosan in in vivo applications and its potential for the delivery of nucleic acids in mucosal epithelia (e.g., nose and lung) was documented [47]. There are several subsequent studies on the potential use of chitosan and its derivatives for the delivery of DNA [65,66]. From 2006 onwards, chitosan has been also used for condensing short interfering RNA (siRNA) [67–70]. More recently, chitosan–miRNA complexes have been investigated to target cystic fibrosis cells [71].

Thus far, only few studies have systematically investigated how the structure of chitosan, specifically the degree of acetylation (DA), and degree of polymerization (DP), affects the biophysical characteristics and biological functionality of chitosan-based systems. Indeed, attempts have been made to establish a relationship between the DP and DA, the salt form and pH, on the efficiency of transfection with plasmid DNA in vitro [72–76] and to determine the intracellular trafficking routes underlying their mode of action [77]. An ideal balance between the strength of the interaction between chitosan and plasmid DNA and the dissolution of the complex within the cell (thus conferring optimal transfection efficiency) can be achieved using chitosan molecules with specific M_w and DA [78]. The biophysical properties of chitosan–siRNA complexes and their capacity for transfection have been investigated in detail [79]. Further studies suggest that optimal properties include the use of chitosans of low molecular weight and high DA, and complexes of small particle size (~100 nm) and a moderate positive surface zeta potential along with a high (N/P) charge ratio [80]. Efforts to elucidate the role of the molecular structure of chitosan (M_w and DA) in terms of the major barriers, including internalization, endolysosomal escape, unpacking, and nuclear entry, that limit the transfection efficiency of chitosan–pDNA polyplexes, have been carried out in HEK293 cells [77]. This cell line is very popular for being easy to culture and to accept foreign DNA. The kinetics of the polyplex decondensation in relation to lysosomal sequestration and escape on chitosan–DNA systems have been suggested to be critically dependent on chitosan's structure. A conclusion that emerged from these studies was that chitosans that promote relatively stable polyplexes, but not too stable (e.g., low DA and high M_w), are optimal (DA~20% M_w 40 kDa or DA~8% M_w 10 kDa). Our own results on chitosan–microRNA polyplexes are consistent with this view, as we have evidenced that chitosans of intermediate affinity to bind miRNA (DA~20% M_w 26 kDa) have the greatest transfection efficiency in MCF-7 cells [81]. Despite the experimental evidence available, there is not yet a firm theory explaining how these factors contribute to the observed transfection efficiency in DNA and RNA systems.

4. Supramolecular Chitosan-Based Nanostructures for Gene Delivery

Chitosan based systems can be prepared following three main techniques: simple complexation, ionic gelation using crosslinkers and adsorption of DNA/siRNA onto the surface of preformed chitosan nanoparticles (Figure 3) [10,69].

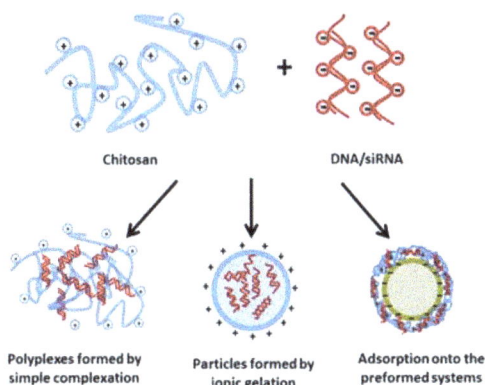

Chitosan   DNA/siRNA

Polyplexes formed by simple complexation   Particles formed by ionic gelation   Adsorption onto the preformed systems

Figure 3. Preparation of chitosan-based DNA/siRNA nanoparticles following different strategies.

The formation of self-assembled complexes with polynucleotides by direct mixing of the components in water is the simplest method. As already mentioned, the formation of such complexes is driven by electrostatic forces in aqueous solution. Despite the simplicity of the method, there are some issues that must be carefully adjusted concerning the mixing conditions, the ratio of charges used, as well as the characteristics of the chitosans. It is reported in a patent by Buschmann et al. that efficient transfecting complexes are formulated by adding chitosan over the nucleotides, pipetting up and down, tapping the tube gently and further incubating for 30 min [82].

The preparation of the delivery system using ionic gelation is based on the ability of chitosan to undergo a sol-gel transition due to the ionic interaction with a polyanion [83,84]. The addition of a third component (e.g., pentasodium tripolyphosphate, TPP) is reported to reduce the size of the particles and to increase the stability of complexes during their incubation in biological fluids [67,84,85]. Chitosan-crosslinked nanoparticles are suitable for the simultaneous encapsulation and sustained release of DNA molecules. Recently, Rafiee et al. showed the preparation of hydrogel nanoparticles with encapsulated plasmid, the system was prepared by simple complexation of chitosan and DNA and in a second step the addition of alginate to protect DNA while forming the hydrogel [86].

Adsorption of DNA on the surface of nanoparticles has been reported using preformed chitosan–TPP–hyaluronic acid nanoparticles. The particles were prepared by ionic gelation and in a further step the plasmid was added, showing efficient transfecting in ocular gene therapy [87,88].

4.1. Effect of the Degree of Acetylation

The degree of acetylation of chitosan is directly related with the density of positive charges along the chitosan chain. Chitosans of low DAs generate high density of positive charges, meaning a greater number of sites for nucleotides binding and improved capacity to interact with the cellular membrane surface and hence, to favor the uptake [89]. The average particle size diameter of complexes formed with either pDNA or siRNA is also known to be affected by the DA; as a general trend is found an increase in size with an increment of the acetylated units on the chitosan chain [10,73,90].

It has been confirmed that the chitosan's DA affects pDNA binding, release and gene transfection efficiency in vitro and in vivo [91]. Koping-Hoggard et al. reported that the DA must be lower than 35% to obtain stable complexes with DNA that transfect HEK293 cells [72]. An increase in the DA decreases the stability of the particles in presence of serum proteins and components from the medium, thus decreasing the transfection efficiency in HEK293, HeLa and SW756 cells [91]. Liu et al. studied the influence of the structural properties on chitosan–siRNA nanoparticle and its influence on gene silencing in H1299 human lung carcinoma cells. The highest gene silencing efficiency was achieved under specific characteristics: DA = 16% and high molecular weight using chitosan–siRNA nanoparticles at N/P 150 [48]. Our own work on delivering hsa-miRNA-145 to MCF-7 breast cancer cells revealed that the greatest transfection efficiency, measured in terms of the downregulation of the target gene, was observed for polyplexes of chitosan with DA 29%, when compared with a series of DA between 1.6% and 49% (M_w ~ 18–26 kDa), as shown in Figure 4 [81]. When we measured the association and dissociation affinity constants for the binding of the series of chitosans with the hsa-miRNA-145, we confirmed that the chitosan with DA 29% had an intermediate binding affinity [81,92] (see Section 5, Surface Plasmon Resonance). Recently, an exhaustive study on the uptake of chitosan–siRNA polyplexes and their transfection efficiency in vitro and in vivo was reported by Alameh et al. Their results show a predominant effect of chitosan's DA on controlling the charge density of the complexes, and the most successful in vitro knockdown rates were obtained with chitosan (DA 28% M_w 10 kDa). In agreement with previous reports, they also experienced that the degree of polymerization and the N/P ratio had a minor effect on the knockdown efficiency [93].

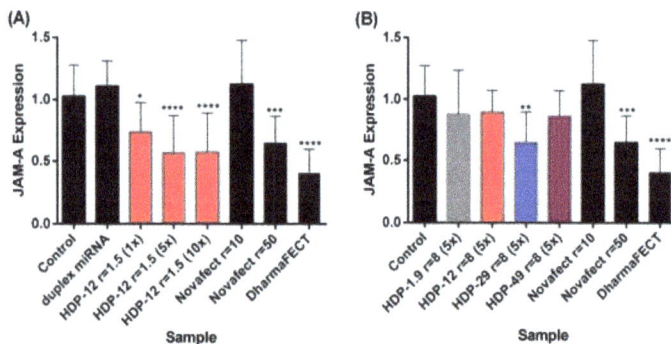

Figure 4. Transfection efficiency expressed as downregulation of JAM-A mRNA in MCF-7 cells: (**A**) complexes containing CS HDP-12 at (N/P) charge ratio = 1.5; and (**B**) complexes containing CS HDP-1.9, HDP-12, HDP-29 and HDP-49 at (N/P) charge ratio = 8. Duplex miRNA (dose 1× = 0.05 nmol/well), DharmaFECT (5 µL/well) and Novafect O 25 were used as controls. Data represent mean values (± SD) of three independent biological experiments and three technical replicates. Statistical comparisons were between each treatment and the control of untreated cells using non-parametric Kruskal–Wallis test (* $p < 0.1$; **; $p < 0.01$***; $p < 0.001$; **** $p < 0.0001$). Source Santos-Carballal et al. Scientific Reports 5, Article number: 13567 (2015) doi:10.1038/srep13567 [81], licensed under a Creative Commons Attribution 4.0 International License.

The role of the pattern of acetylation (PA) that defines the distribution of *N*-acetyl-D-glucosamine residues along blocks of charged poly-D-glucosamine in the chitosan chains, on the specific interactions between chitosan and DNA or RNA is widely unknown. Ongoing efforts in our laboratories are being made in this direction and the first enzymatically patterned chitosans have started to become available for the studies in this direction.

4.2. Effect of the Molecular Weight

The length of the chitosan chain has a prominent influence on the particle size, stability, dissociation of the complexes after internalization in the cytoplasm and therefore an impact on the final transfection efficiency [94]. In general, it has been found that the size of the complexes decreases as the molecular weight is reduced [73]. However, the complexes formed between chitosan with low M_w (10–17 kDa, DA = 12%) and pDNA tend to increase in size [73]. Our own studies with polyplexes of chitosan of very low M_w chitosan (~1.2–2.0 kDa, DA ~ 1.6–67%) and microRNA revealed that, even when the overall average size and polydispersity and zeta potential of the polyplexes did not differ much from those of the systems obtained with chitosans of low M_w (~18–26 kDa, DA ~ 1.9% to 49%), they aggregated in RPMI minimal cell culture medium [81]. Sato et al. showed that chitosans with high M_w (>100 kDa; DA = 8%) are poorly effective in transfecting DNA, whereas chitosans with low M_w (~15 and ~52 kDa, DA = 20% and 6%, respectively) largely promote pDNA luciferase expression in several cell lines (namely A549, B16 and HeLa cells) [94], in line with our results for microRNA transfection of MCF-7 breast cancer cells [81]. Recently, Bordi et al. have compared the transfection efficiency of chitosan oligomers and chitosan of M_w 50 kDa (DA = 34%) complexed with pDNA in various cell lines. In all tested cells, the chitosan of M_w 50 kDa performed better than the oligomers, thus providing evidence of higher protection of pDNA and stability of these complexes conferred by chitosan [95]. In the Laboratory of Maria J. Alonso, where nanoparticles of chitosan by ionotropic gelation with TPP were discovered, they formulated pDNA and short dsDNA oligonucleotides into chitosan/TPP prepared with chitosan of varying M_w. Low M_w chitosan (10 kDa) provided more compact nanocarriers (~100 nm) compared to high M_w chitosan (125 kDa) because of the lower viscosity of the former polymer dispersion, in consonance with previous studies [96]. Importantly, the efficiency of transfection also seemed highly dependent on the chitosan M_w, with low M_w chitosan–TPP NPs exhibiting superior gene transfer in vitro. In addition, low M_w chitosan–TPP particle displayed a marked transgene expression following intratracheal administration in mice. This was comparable with the corresponding M_w chitosan–DNA polyplexes prepared in the absence of TPP though [97].

Katas and Alpar (2006) have shown that siRNA molecules are efficiently condensed and protected by high M_w chitosan (M_w = 110 and 270 kDa and DA = 14%) [67]. Furthermore, the molecular weight will influence the capacity of chitosan to entangle siRNA and the final size of the chitosan–siRNA complexes, which seems to be yet another aspect at play in cellular uptake [48,67]. It has been suggested that only chitosan molecules that were 5–10 times the length of siRNA (M_w of 13.36 kDa) could form suitable nanocomplexes. Thus, the chitosan optimal M_w recommended to obtain nanocomplexes with siRNA is 64.8–170 kDa [66]. Evidence is consistent with the notion that the M_w of chitosan can be used to tweak the average particle size of the obtained complexes [10]. It has also been found that the M_w of chitosan has an influence on the morphology. Spherical or elongated and irregular nanoparticles were formed with chitosans of M_w 44 (DA 14%) or 143 kDa (DA 22%), respectively, as evidenced by TEM and SAXS [98]. The influence of the morphology of polyplexes on the cellular uptake, intracellular trafficking and transfection efficiency, has only been recently examined on synthetic brush polymers of methacrylamide-oligolysine–pDNA (pCMV-Luc2) polyplexes [99]. In the case of chitosan-based systems, it is known that the structure (M_w and DA) determines the morphology, but how this dictates biological activity, is a so far widely neglected aspect.

Recently, it has been seen that increased chain length increased biological performance, and that a certain M_w threshold of 10 kDa (~60−70 monomers) is required to achieve eGFP knockdown efficiency.

Particles formed with chitosan of M_w below this value (e.g., 5 kDa) are found to have a reduced stability in serum and therefore low transfection rates [93]. Most likely, the interactions governing these types of systems are highly cooperative in nature. Hence, for each type of nucleic acid, there must exist a minimum cooperative length between chitosan and the nucleic acid that favors the interaction and influences the overall molecular architecture of the assembled structures. From the practical standpoint, it is very important to carefully select the optimal M_w and DA of chitosan to obtain desirable physicochemical properties and achieve proper transfection rates. The size of the particles will be determined by the molecular mass of chitosan and this will influence the stability of chitosan–nucleic acid complexes and their biological activity in vitro, namely: cellular uptake, dissociation and processing of DNA within the cells.

4.3. Effect of the N/P Charge Ratio

The N/P charge ratio is defined as the ratio between the protonated amines from the chitosan and the negatively charged phosphate groups from nucleic acid, also used indistinctly in this review as the (+/−) charge ratio. The N/P charge ratio can assume values from 0 to 1, for negatively charged complexes, or values higher than 1 that correspond to positively charged complexes. Complexes with stoichiometric equivalent amount of charges are not desired since these systems are unstable and agglomerate and precipitate due to the absence of repulsive forces [49]. The positive surface net charge of the complexes influences directly the interaction with the negatively charged glycosaminoglycan molecules from the glycocalyx of the cell membrane [69].

N/P charge ratios below 1 imply defect of chitosan and surplus of the nucleic acid in the complex and therefore, an increase in size due to deficient condensation of the nucleotides leading to unstable systems, has been reported for complexes comprising pDNA [72,94]. By contrast, the formation of positively charged complexes, may in principle provide many advantages: Under a certain range of N/P charge ratio is observed a contraction in the size of the complexes and efficient condensation of the polynucleotides, the positive net surface charge will favor the electrostatic interaction with the cell membrane and the presence of a high number of amino groups will potentiate the mechanism of "proton sponge" for endolysosomal release and efficient unpacking of the genetic material [72].

The design of successful gene delivery systems entails to find the fine balance of the parameters described above. Indeed, complexes with optimal association and dissociation capacity that attain the highest rates of transfection, can be achieved under specific values of chitosan's M_w and DA, and at the optimal N/P charge ratio. The DA affects the charge density of chitosan and thus the binding sites to nucleotides, while the length of the chitosan influences the condensation efficiency of the cargo and together with the N/P ratio are important in balancing condensation, protection and intracellular release, to ensure efficient nucleic acid delivery and high rate of transfection [48,69,90,100]. For instance, it has been found that a high molecular weight chitosan yielded a higher transfection efficiency of DNA at a low N/P ratio, whereas a low M_w chitosan was required at higher N/P ratio to completely form the complexes [75,90]. On the other hand, chitosan of high DAs requires elevated values of N/P charge ratio to fully condense nucleotides (DNA and siRNA) [72,90,91].

Table 1 summarizes the reviewed studies on the role of different chitosans to complex with DNA or RNA and the major findings of each.

In vivo applications using chitosan based non-viral gene delivery are limited due to lower transfection efficiencies in comparison with viral systems. The conflicting evidence from the various studies regarding the role of the M_w, DA and N/P of chitosan on the transfection efficiency by different type of nucleic acids reveals the overall dearth of fundamental understanding on the mechanisms of interaction at molecular level between chitosan and nucleic acids. Computational simulations using molecular dynamics may turn out to be extremely insightful in shedding light in this regard. This knowledge is key to a more rationally-based design of effective gene delivery systems.

Table 1. Summary of chitosans used as non-viral delivery nanovehicles to associate different types of genes.

Chitosans Origin	DA (%)	M_w (kDa)	Nucleic Acid Name	In Vitro Studies	Major Findings	Reference
Chitosan Seacure Natural Biopolymer Inc., USA	10 20 18 10	102 230 540 7–92 serie	Plasmid containing a CMV promoter	Cos-1 cells	The highest level of expression in vitro was obtained using complexes prepared at a N/P = 2 and using a chitosan of M_w 102 kDa. This finding was 250-fold lower than that observed with the control of lipofectamine. Further improvement in transfection efficiency was achieved by the presence of a pH-sensitive endosomolytic peptide in complex formulation.	[47]
Chitopharm® Cognis Deutschland GmbH & Co., Germany	5 23 22 16 46	8.9 11.9 64.8 114.2 170 173	siRNA-eGFP duplex	H1299	Physicochemical properties and in vitro gene silencing of chitosan/siRNA nanoparticles are strongly dependent on chitosan M_w and DA. Chitosan/siRNA formulations (N/P = 50) prepared with low M_w (~10 kDa) showed almost no knockdown of eGFP, whereas those prepared from high M_w (64.8–170 kDa) and DA = 20% showed greater gene silencing ranging between 45% and 65%. The highest gene silencing efficiency (80%) was achieved using chitosan/siRNA nanoparticles at N:P 150 using higher M_w (114 and 170 kDa) and DA = 16%.	[48]
Molecular tailored chitosans	<0.2	8 9,8 21	DNA pWizLuc (6.7 kb)	—	Tailored chitosans include linear (LCO), trisaccharide substituted-(TCO), and self-branched trisaccharide (SB-TCO) substituted chitosan oligomers. This study revealed that, besides differences in the stability of complexes, SB-TCO and DNA formed structures with a larger height and a larger fraction of globular structures compared to the other chitosans. In addition, complexes formed by SB-TCO contained a larger fraction of unbound chitosan, which may lead to increased transfection.	[51]
Piramal Healthcare, India	8	8	eGFPLuc (6.4 kb)	—	Chitosan buffering capacity and its comparison to PEI on a molar basis revealed that chitosan possesses a higher buffering capacity than PEI in the endosomal pH range. Chitosan-DNA complex alone have an ~2-fold reduced buffering capacity as compared to free chitosan. These findings suggest that the proton sponge effect could be at least partially responsible for mediating chitosan endosomal escape.	[62]
Protasan Ultrapure, Pronova Biomedical	14 14	270 210	siRNA targeting pGL3 luciferase	CHO K1 HEK 293	Chitosan-TPP nanoparticles with entrapped siRNA are shown to be better vectors as siRNA delivery vehicles compared to chitosan-siRNA complexes possibly due to their high binding capacity and loading efficiency.	[67]
Sigma-Aldrich	1	5	DNA	–	Chitosan (5 kDa)/DNA complexes remain in B conformation, and the binding affinity of chitosan to DNA is dependent on pH of media where a great binding affinity is generated at pH 5.4, whereas at pH 12.0 a low affinity with DNA is observed. The charge ratios of chitosan to DNA strongly influence the morphology of complexes formed. At low charge ratio, not all DNA can be entrapped in the complex; at higher ratios, the complexes without free DNA evolve into spherical shape with mean size of nanoscale.	[90]
Kitomer (Marinard, Canada)	19	500	Linear calf thymus DNA (13 kbp)	—	A panel of biophysical techniques (conductivity, zeta potential, dynamic light scattering, atomic force microscopy, circular dichroism and UV/VIS spectroscopy) were used to determine the stoichiometry, net charge, dimensions, conformation and thermal stability of complexes of varying N/P ratio both in water and in 10 mM NaCl. Complexation of partially denaturated DNA in water, and double-helical DNA, showed similar electrostatic behavior and stoichiometry. The behavior for complexing was nearly independent of M_w.	[101]
Primex	4	50				
Protasan® UP G 113, Sigma-Aldrich	10–25	160	miR126	CFBE41o-	High-content analysis data indicate that miRNA-PEI nanomedicines facilitated greater uptake than miRNA-TPP-chitosan nanoparticles and the commercial transfection agent, Riboluce®. Polymeric nanoparticles can deliver premiRs effectively to CFBEs to modulate gene expression but must be tailored specifically for miRNA delivery.	[71]

Table 1. *Cont.*

Chitosans Origin	DA (%)	M_w (kDa)	Nucleic Acid Name	In Vitro Studies	Major Findings	Reference
Chitosan Aldrich Chemical Co.	12 12 12 12 12 39 54	213 98 48 17 10 213 213	pEGFP-C2 plasmid	A549	DNA condensation of N90% was achieved at the N/P charge ratio of 6, independent of the chitosan M_w and DA. NP produced with chitosan of M_w 213 kDa and DA of 12% showed the highest zeta potential (+23 mV), cellular uptake (4.1 Ag/mg protein) and transfection efficiency (12.1%), while chitosan vector with M_w of 213 kDa and DA of 54% showed the lowest cellular uptake (0.4 Ag/mg protein) and transfection efficiency (0.05%)	[73]
KITTOLIFE, Korea.	27.5	22	pSV-β-galactosi-dase	293T	The transfection efficiency of low M_w chitosan complexes (LMWC) was significantly higher than naked DNA and higher than poly-Lysine (PLL). MTT assay showed that LMWC was less cytotoxic than PLL.	[74]
Seafresh Chitosan Lab, Thailand	13 13 13 13	20 45 200 460	pcDNA3-CMV-Luc	CHO-K1	The transfection efficiency of chitosan (CS)/DNA complexes was dependent on the salt form and M_w of chitosan, and the N/P ratio of CS/DNA complexes. Of different CS, the maximum transfection efficiency was found in different N/P ratios. CS/DNA (hydrochloric acid), CS/DNA (lactic acid), CS/DNA (acetic acid) and CS/DNA (aspartic acid) and CS/DNA (d glutamic acid) complexes showed maximum transfection efficiencies at N/P ratios of 12, 12, 8, 6 and 6, respectively. Cytotoxicity results showed that all CS/DNA complexes had low cytotoxicity.	[76]
Chitosan Seacure Pronova Biopolymers, Norway	15 32 15 25	6.6 90 160 160	pcDNA3-luc	EPC cells	The in vitro transfection efficiency was affected by the polyplex (N/P) charge ratio, the DNA concentration in the complexes, the molecular weight and degree of acetylation of the chitosans. Two favorable formulations were identified: chitosan (DA-15%; 6.6 kDa) (theoretical charge ratio 10) and chitosan (DA-15%; 160 kDa) (theoretical charge ratio 3). The size of the complexes was affected by the degree of acetylation, concentration of DNA, pH, inclusion of a coacervation agent and the charge ratio.	[75]
Biosyntech, Laval, Canada	20 8 8 28	40 10 150 40	eGFPLuc	HEK293	The kinetics of decondensation in relation to lysosomal escape was a most critical structure-dependent process affecting the transfection efficiency of chitosan polyplexes. The most efficient chitosans showed an intermediate stability and a kinetics of dissociation, which occurred in synchrony with lysosomal escape. In contrast, a rapid dissociation before lysosomal escape was found for the inefficient high DA chitosan whereas the highly stable and inefficient complex formed by a high M_w and low DA chitosan did not dissociate even after 24 h.	[77]
Molecular tailored chitosans	all 0.2	146 32.9 24.8 16.4 11.6 8.0 4.7	gWiz Luc and gWiz GFP	HEK293	Maximum level of transgene expression was found with chitosan with M_w 8 and 11.6 kDa. An increase in chain length and / or the amino-phosphate (A/P) ratio reduced and delayed transgene expression. The gene transfer pattern correlated with the ability of heparin to release DNA from the polyplexes. As a tool to facilitate the unpacking of the polyplexes, we substituted the chitosans with uncharged oligosaccharides that reduced the interaction with DNA. The substitution of chitosans shorter than 4.7 kDa completely abolished transfection.	[78]

Table 1. Cont.

Chitosans Origin	DA (%)	M_w (kDa)	Nucleic Acid Name	In Vitro Studies	Major Findings	Reference
Sascha Mahtani Chitosan PVT Ltd., India	1.5 12 29 49 1.6 11 25 67	26 25 20 18 1.3 1.2 1.1 1.9	miRNA-145	MCF-7	Chitosan–miRNA nanocomplexes with degree of acetylation 12% and 29% were biologically active, showing successful downregulation of target mRNA expression in MCF-7 cells. We found no evidence that these complexes were cytotoxic towards MCF-7 cells. DA has an influence on the transfection efficiency for complexes with equivalent (+/−) charge ratio (8.0): more efficient downregulation of the target gene in the presence of intermediate DA (~30%)	[81]
Sigma-Aldrich	15–25	192	siRNAs targeting the VEGF gene	DLD-1	Particles with different cross-linkers were prepared. Chitosan–TPP nanoparticles showed better siRNA protection during storage at 4 °C. TEM micrographs revealed the assorted morphology of chitosan–TPP–siRNA nanoparticles in contrast to irregular morphology displayed by chitosan–DS–siRNA and chitosan–PGA–siRNA nanoparticles. All siRNA loaded chitosan–TPP–DS–PGA nanoparticles showed initial burst release followed by sustained release of siRNA. All the formulations showed low and concentration–dependent cytotoxicity with human colorectal cancer cells (DLD-1), in vitro. The cellular uptake studies with chitosan–TPP–siRNA nanoparticles showed successful delivery of siRNA within cytoplasm of DLD-1 cells.	[54]
Sigma-Aldrich	–	Low	pEGFPN1	HEK293	Suitable candidate for gene delivery would be alginate–chitosan nanoparticles. The effect of alginate on reducing the strength of electrostatic interactions between chitosan and pDNA, resulting in better transfection and increasing the plasmid release.	[86]
Protasan UP CL 113 FMC Biopolymers (Norway)	10–15	~110	pEGFP-C1 and pβ-gal	HCE; IOBA-NHC	Evidence of the potential of hyaluronic acid (HA)–chitosan nanoparticles, which exhibit very low cytotoxicity to the ocular surface. Potential use of HA–chitosan nanoparticles for the targeting and further transfer of genes to the ocular surface.	[87]
Ditto	10–15	~110	pSEAP	IOBA-NHC; HCE; RAW264.7	HA–chitosan oligomer (CSO)-based nanoparticles (HA–CSO NPs) were internalized by two different ocular surface cell lines by an active transport mechanism. Potential use of HA–CSO NPs to deliver genetic material to the ocular surface.	[88]
Ditto	10–15	~110	Luciferase duplex siRNA (21 bp)	A549-Luc	Chitosan–TPP nanoparticles without and with HA (CS–TPP–siRNA and CS–TPP–HA–siRNA, respectively); N/P charge ratio 5–200, diameter ~320–420 nm. Inclusion of HA reduced the cytotoxicity. Greater inhibition of luciferase expression was for CS–HA NPs (N/P = 120)–luciferase knockdown of ~85% (vs. <70% for CS–TPP–siRNA).	[102]
Ditto	10–15	~110 ~10 (depolyme-rized)	gWiz™ pSEAP	Calu-3 (in air-liquid interface)	Chitosan–TPP NPs comprising anionic β-cyclodextrins; 5% DNA loading; diameter ~264–356 nm. Slightly lower cytotoxicity for NPs comprising CDs; interaction of NPs with Calu-3 cells studied by CLSM. Pharmacokinetics of SEAP expression: NPs of chitosan–TPP, and chitosan–TPP–carboxymethyl-β-CD had greater transfection efficiency than those comprising sulfobutylether-β-CD. In NPs comprising chitosan of ~10 kDa, this effect was not observed.	[103]
Fluka	–	111	pCMV Lac–Z (7kbp)	HeLa	Chitosan–DNA polyplexes of N/P charge ratio 1–20. The chitosan of higher valence (M_w) required larger amounts to compact DNA. A 4–5-fold lower enthalpic contribution observed for the highest valence chitosan. Heterogeneous population of particles in the diameter range ~250–500 nm. Very low transfection efficiency observed for all systems, ascribed to the core-shell morphology of the polyplexes.	[104]
	– –	266 467	pCMV-Luc (6kbp)			

Table 1. *Cont.*

Chitosans Origin	DA (%)	M_w (kDa)	Nucleic Acid Name	In Vitro Studies	Major Findings	Reference
Shrimp shell	24 14	— —	—	L929 BHK21 (C13)	Biocompatibility was investigated for: (1) cell adherence and growth on the chitosan samples as substrate; (2) the effect of extract media on 2-day and 7-day growth; and (3) the presence of an inhibition zone. The results were similar for both cell lines.	[89]
Cuttlefish	19 10	— —				
UltrasanTM, Biosyntech Inc., Canada	2 8 20 28	120 200 320 220	eGFPLuc	HEK293	Results revealed an important coupling between DA and M_w of chitosan in determining transgene expression. Maximum expression was obtained with a certain combination of DA and M_w that depended on N/P charge ratio and the pH.	[90]
Vanson, USA	10 30 38	390 209 138	pcDNA encoding for Luc.	HEK293 HeLa SW756	Degree of chitosan deacetylation is an important factor in chitosan–DNA nanoparticle formulation as it affects DNA binding, release and gene transfection efficiency in vitro and in vivo.	[91]
Marinard, Canada	28 20 8 2	10 120 10 120 5 10 40 80 120 10 120	siRNA	EGFP + H1299	Highly deacetylated chitosans are superior siRNA delivery systems compared to partially acetylated chitosans. Highly deacetylated chitosans (low DA and high M_w) provide the optimal balance between biological performance and toxicity. A minimum polymer length of ~60–70 monomers (~10 kDa) was required for stability and knockdown. In vitro knockdown was equivalent to lipid control with no metabolic or genotoxicity. An inhibitory effect of serum on biological performance was dependent on DA, M_w, and N/P charge ratio. In vivo biodistribution in mice show accumulation of nanoparticles in kidney with 40–50% functional knockdown.	[93]
Yaizu Industry, Japan	20 6 8	15 52 >100	pGL3	A549 HeLa B16	Transfection efficiencies of the pGL3/chitosan complexes were dependent on pH of culture medium, stoichiometry of pGL3: chitosan, serum, and molecular mass of chitosan.	[94]
Sigma–Aldrich	34 43	50 150	pMAX-eGFP	HEK293 H441 16HBE	The morphology and the net charge of chitosan–pDNA aggregates is mainly controlled by the overall stoichiometric ratio between the positively charged (protonated) groups on chitosan chains and the negative charges on the DNA. Complexes with the higher molecular weight chitosan are more stable, and clearly demonstrate a significantly higher transfection efficiency.	[95]
KiOmedicine-CsU from *Agaricus bisporus* (3–5% glucan content)	14 16 19 22	44 63 93 144	siRNAluc G13	H1299 pGL3 (expressing Luc reporter gene)	Comparisons of biophysical and transfection efficiency properties of fungal chitosans with similar DA, and M_w from 44kDa to 143 kDa (N/P =4 and 8). Polyplexes despite very similar size (129–165 nm), zeta potential (+20–30 mV) and complex stability (K_d = 1–1.9 nM), displayed differences in particle morphology, cellular uptake and transfection efficiency. Spherical or elongated and irregular nanoparticles were formed with chitosans of M_w 44 (DA 14%) or 143 kDa (DA 22%), respectively. First study to use ITC for profiling of siRNA interactions with chitosan.	[98]
Protasan UPG Pronova Biopolymer, Norway	0.1	1.2–10	gWizTM-Luc pCMV-Luc	HEK293 HeLa	A major improvement of chitosan-mediated non-viral gene delivery to the lung was obtained by using polyplexes of well-defined chitosan oligomers. Polyplexes of oligomer fractions also had superior physicochemical properties to commonly used high-molecular-weight ultrapure chitosan (UPC).	[100]

Table 1. *Cont.*

| Chitosans | | | Nucleic Acid Name | In Vitro Studies | Major Findings | Reference |
Origin	DA (%)	M_w (kDa)				
Norwegian Biopolymer Laboratory	0	5–6	pNGVL-eGFPLuc	CFBE41o−/HEK293	The transfection efficacy of polyplexes in the CFBE41o− cell line was poor compared with that in HEK293 cells. The narrow-size-distributed chitosan at low pH shows a better transfection efficiency compared with PEI.	[105]
HMC+, Germany	30	20	pEGFP-C1 or pEGFP-C1/siRNA	CFBE41o−	Proof-of-principle that co-transfection with chitosan, as a natural non-toxic vector, might be an effective delivery system in a human CF cell line, reaching comparable levels to those achieved using lipid-based systems.	[106]
HMC+, Germany	30	20	wtCFTR-mRNA	CFBE41o−	Transfection of an immortalized CF cell line with wtCFTR-mRNA using chitosan as a carrier results in increased CFTR function	[107]
HMC+, Germany	30	30	CFTR-LNA	−	CFTR-specific locked nucleic acids (LNA) biopolymer-based nanoparticles represent a promising system for further development of new lung-targeted CF therapeutic approaches. First time the use of chitosan from a non-animal source as a potential therapeutic vector has been reported.	[108]
ChiPro, Germany	20	200				

5. Characterization Techniques of the Biophysical Properties and Biological Performance of Chitosan–Polynucleotide Complexes

Chitosan–polynucleotide supramolecular complexes are characterized in their biophysical properties, namely size distribution, zeta potential, morphology, binding affinity, buffering capacity, colloidal stability, cytotoxicity, and transfection efficiency. A wide range of characterization techniques are used for this purpose. In this section, we address the main techniques with examples of their use in the study of chitosan–polynucleotide systems.

5.1. Agarose Gel Electrophoresis

Electrophoresis is a routinely used technique to separate macromolecules based on their size and charge—especially proteins and nucleic acids. Charged molecules can migrate in presence of an electric field towards the polarity of the system used. Nucleic acids have a consistent negative charge from the phosphate groups present and they migrate towards the anode (positive pole). The migration speed is determined by the molecular weight of the nucleotides. Visualization is carried out using ethidium bromide or SYBR Green under 300 nm UV light. Agarose gel electrophoresis can be used to study binding affinity of chitosan to nucleic acids during complexation [84], release capacity, protection against endonucleases [73] and stability [84,109].

5.2. Dynamic Light Scattering (DLS)

The technique is based on probing the time fluctuations of macromolecules and colloidal particles subjected to Brownian motion when dispersed in a solvent. Then, a monochromatic light beam (HeNe) hits the solution, causing a Doppler Shift of the light. DLS measurements allow evaluating the ability of a macromolecule to diffuse in solution. This is determined by the mutual translational diffusion coefficient (or diffusivity). The diffusion coefficient, according to Fick's first law of diffusion, relates the concentration gradient of a solute in a solvent along an axis. Using an autocorrelation function, together with the diffusion coefficient, it is possible to calculate the mean hydrodynamic radius of the particle assuming spherical geometry and the polydispersity of the particle size distribution. The Stokes-Einstein equation (Equation (1)) relates the diffusion of spherical particles through a liquid and allows the determination of the hydrodynamic radius (R_H) of the scattering particles in a medium of known viscosity (η) at specific temperature (T) assuming Brownian motion (k_B) [110]:

$$D_S = \frac{k_B T}{6\pi\eta R_H} \tag{1}$$

In Equation (1), D_S defines the self-diffusion coefficient measured by DLS, and this leads to correlate that small particles will move faster than larger ones, and therefore will have high diffusion coefficients [110–112]. This technique is useful for particles with size lower than a micron. Generally, the average size diameter of chitosan-based complexes is determined by DLS. Complexes formed using a series of different chitosans range from few nanometers to less than a micron [81,93,101,113].

5.3. Zeta Potential

The zeta potential can be determined from the electrophoretic mobility and determination of the velocity of the particles using Laser Doppler Velocimetry (LDV). The electrophoretic mobility is the velocity of a charged particle relative to the liquid it is suspended in under the influence of an applied electric field. When the voltage is applied to the electrodes, the charged particles will migrate towards the oppositely charged electrode with a determined mobility. Using the Henry equation (Equation (2)) the electrophoretic mobility is converted to Zeta potential to enable comparison of materials under different experimental conditions.

$$U_E = \frac{2\varepsilon\xi f_{(K_a)}}{3\eta} \tag{2}$$

where U_E is the electrophoretic mobility ($m^2 \cdot V^{-1} \cdot s^{-1}$), ε is the dielectric constant of the medium (F/m), ξ is the zeta potential (mV), $f_{(K_a)}$ is Henrys function and η (poise) is the viscosity of the medium.

The zeta potential gives an indication of the potential stability of the particles. If the particles possess a large value of zeta potential, then they will tend to repel each other and there is no tendency to flocculate. However, if the particles have low zeta potential values the repulsive forces are very small in magnitude and the system turns unstable and inexorably precipitates. This technique has also been utilized to characterize the stoichiometry of complex formation between chitosan and DNA [95,101,114].

5.4. Transmission Electron Microscopy (TEM)

TEM operates on the same basic principles as the light microscope but with the contrast is based on the use of electrons. The electrons are emitted from a tungsten filament and travel through a 2 m column under vacuum conditions to avoid their scattering. During their trajectory, the electrons are accelerated at a high voltage (100–1000 kV) to a velocity approaching the speed of light (0.6–0.9 c); the associated electron wavelength is five orders of magnitude smaller than the light wavelength. The resolution obtained is then many orders of magnitude better than with normal microscopes, since the resolving power is directly proportional to the wavelength of irradiation; the faster the electrons are accelerated, the shorter their wavelength and higher the resolution. These characteristics enable materials imaging with the finest details of internal structure and determination at the atomic level [115,116]. TEM have been used to analyze the morphology, size and density of chitosan–DNA complexes. In the literature is described that the complexes can present different structures such as spherical shape [117], toroids and globular particles [47,118]. In Figure 5, a micrograph for chitosan–has-miRNA-145 polyplexes obtained with two distinct chitosans and at varying N/P ratios is shown. The micrographs reveal that the simple complexation of the two biopolymeric structures from aqueous solution, leads to the formation of spherical micro-heterogeneous structures, reminiscent of the morphology and size range of adenoviruses.

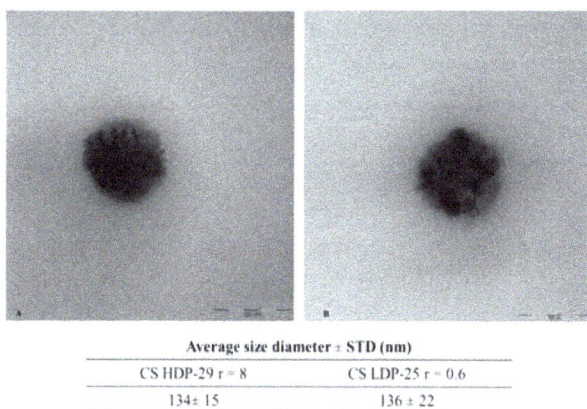

Average size diameter ± STD (nm)	
CS HDP-29 r = 8	CS LDP-25 r = 0.6
134± 15	136 ± 22

Figure 5. Representative TEM images of complexes containing: (**A**) CS HDP-29 N/P = 8; and (**B**) CS LDP-25 N/P = 0.6 stained with uranyl acetate. The embedded table shows the measured diameter of the complexes using ImageJ v1.49n (*n* = 8; mean average ± SD). Source Santos-Carballal et al. Scientific Reports 5, Article number: 13567 (2015) doi:10.1038/srep13567 [81], licensed under a Creative Commons Attribution 4.0 International License.

5.5. Atomic Force Microscopy (AFM)

AFM operates under the principle of a spring-like device called cantilever, with a spring constant weaker than the equivalent spring between atoms (<~10 N/m). This is used to sense Ångström-size displacements while a tip attached to the cantilever is lowered to the sample and moves along its surface, thus creating a deflection. Given the small force applied, it would not be enough to push the atoms out of their atomic sites. The applied force can be magnetic, electrostatic, or the result of interatomic interactions between the tip and the sample. Regardless of this, all AFM instruments have five essential components [119]: (i) a sharp tip, usually 10–20 nm made commonly of silicon nitride (but can also be of ~5 nm made of carbon nanotubes), mounted on a soft cantilever spring; (ii) a way of sensing the cantilever's deflection; (iii) a feedback system to monitor and control the deflection (and hence, the interaction force); (iv) a mechanical scanning system (usually piezoelectric) that moves the sample with respect to the tip in a raster pattern; and (v) a display system that converts the measured data into an image. Two different AFM techniques are known to generate the image. In the "contact mode", the distance between the sample and the tip is adjusted to keep the cantilever at constant deflection. The voltage supplied to the piezoelectric tube to maintain the constant deflection is used to generate the image. This technique is not adequate for imaging soft materials. A more widely used technique, known as "tapping mode" or "non-contact mode" is used to image fragile and soft materials including biological samples. In this case, the cantilever oscillates in a sinusoidal manner at its resonance frequency. The amplitude of the oscillation is dampened when the tip approaches the sample, and the data are relayed to the piezoelectric tube, to ensure the amplitude of the vibration is kept constant. The advantage of the tapping mode is that lateral forces are eliminated and it allows to image weakly absorbed samples.

AFM is a high-resolution microscopy technique that enables to image atomic-scale topologies. It has been instrumental to elucidate the morphology of diverse structures of chitosan–DNA and chitosan–siRNA complexes formed under different conditions. Tapping-mode AFM has been utilized to demonstrate that the complexation of both linear and circular pDNA (pBR322) with chitosan (M_w 162 kDa, DA 10%) yields a blend of toroidal and rodlike structures [120]. Differences in the toroid-to-rod ratios were associated to the conformation of the DNA for complexes at N/P = 1.0 (Figure 6). A larger fraction of globular structures appeared for the linear DNA than for the pDNA. Quantitative analysis of the formed structures revealed that the fraction of toroids/rods decreases with decreasing charge density of chitosan (i.e., increasing DA). High and low M_w chitosans yielded the same type of structures.

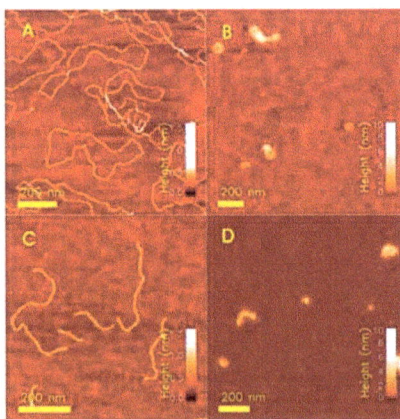

Figure 6. Tapping mode AFM height topographs of: uncomplexed pBR322 (**A**); and linear DNA (**C**); alongside with complexes of these formed when mixed with the chitosan C (0.01,162) (**B, D**) cDNA 4 µg/mL and N/P = 1. Reprinted with permission from Danielsen et al. (2004) Biomacromolecules 5, 928–936 [120]. Copyright 2004 American Chemical Society.

It has been shown by tapping-mode that chitosan (M_w 3 and 150 kDa DA 43%)–linear DNA complexes, results in the formation of "tadpole" and "question mark" structures [114]. Another study gave evidence of the formation of irregular shaped nanoparticles for chitosan (M_w 50 kDa DA 4%) linear DNA, thus confirming the capacity of chitosan to compact DNA [101]. Other studies have documented the formation of ionically gelled nanoparticles of chitosan (low M_w and DA ~15–25%) with sodium deoxycholate to associate pDNA (pCMV-GLuc, 5.7 kbp, load 5%). Spherical morphology for these systems was imaged by tapping mode AFM [121]. In addition, the same technique has been utilized to visualize the topology of complexes formed by a series of chitosans (M_w 8.9–173 kDa, DA 5–46%) with siRNA (21 bp) N/P = 50 deposited in a mica surface [4]. The formation of different type of structures spanning rod- or circle-shaped nanoparticles and open structures was evidenced for the different chitosans.

5.6. Surface Plasmon Resonance (SPR)

SPR is a technique used to characterize the affinity events between an analyte and a ligand, allowing real-time monitoring of binding kinetics. Briefly, a beam of polarized light propagates in a medium of high refractive index (e.g., a prism) with total reflection until it hits a gold-coated surface with low refractive index (n2). The electromagnetic field component penetrates the surface for few micrometers and the intensity of the resulted polarized light is attenuated. Electrons oscillate within the surface of the conductor (gold surface), and the quantization of this oscillation is called plasmon. Once the surface is irradiated with the polarized light, the surface plasmons can couple with the protons of the polarized light, and this phenomenon is called surface plasmon resonance (Figure 7). This occurs when the condition that the wavevector of the photon (k_x) is equal to the wavevector of the surface plasmon (k_{sp}) is satisfied. The k_{sp} is determined by the refractive index of the conductor and the k_x depends on the wavelength of the polarized light and its angle of incidence. SPR probes the variation in the refractive index of this transducing layer induced by the adsorption or chemical reaction of an analyte. The change in the refractive index is followed by measuring the intensity of the reflected light at different angles of incidence (Figure 7) [122–124].

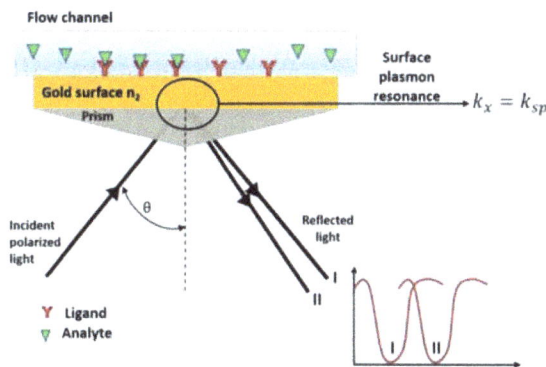

Figure 7. General principle of SPR, where n2 is the refractive index of medium with lower refractive index, E is the evanescent field amplitude, k_{sp} is the wavevector of surface plasmons, and k_x is the wavevector of photon.

The final response units (RU), related to the change in the refractive index, are directly associated to the concentration of analyte absorbed on the surface. These values can be applied to the Langmuir Adsorption Isotherm model, which describes 1:1 interaction where one ligand molecule interacts with one analyte molecule, and the chemical reaction for the monolayer adsorption can be represented as follows [125]:

$$A + B \leftrightarrow AB \tag{3}$$

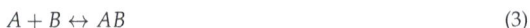

where AB represents a solute molecule bound to a surface site on B during the reaction with the analyte A. The dissociation constant K_D for this reaction is given by:

$$K_D = \frac{k_d}{k_a} = \frac{[A][B]}{[AB]} \tag{4}$$

During the reaction, the fraction of B that has reacted at certain time can be described as follows:

$$Fraction\ B = \frac{[A]}{K_D + [A]} \tag{5}$$

When the reaction has taken place to the half of its extension and the fraction of B left is only $\frac{1}{2}$, the condition of $K_D = [A]$ is satisfied and the dissociation constant can be related to the amount of analyte injected [125]. The sensitivity of SPR method to probe interactions between a given ligand-analyte system can reach the order of fM [126].

SPR has been used to study the interaction between polymer-DNA complexes based on PEI or dendrimers and hyaluronic acid as a model glycosaminoglycan [127]. González-Fernández et al. used SPR to evaluate the binding affinity of RNA anti-tobramycin aptamer with different modifications to understand its influence on affinity towards its target molecule, tobramycin [128]. More recently, SPR has been used to study the dissociation of complexes between microRNA and a family of chitosan, e.g., two molecular weights and various degrees of acetylation to comprehend the influence of these factors in the binding [81]. Figure 8 illustrates a representative SPR curve obtained between chitosan of DA 12% M_w 25.5 kDa. The KD values obtained for the different chitosans in this series ranged from 5.93 to 31.76 µM, revealed that the chitosan found with greater transfection efficiency (M_w ~29.2 DA 29%), had a KD ~11.84 µM, thus confirming that the affinity of binding must be neither too high nor too low.

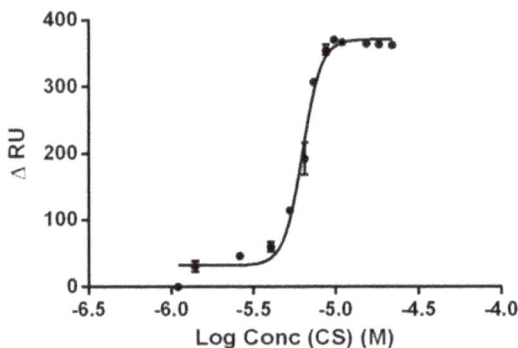

Figure 8. Saturation curve for hsa-miR-145-5p with HDP-12. Acetate buffer (35 mM, pH 5.1/10 mM NaCl) (n = 2). Source Santos-Carballal et al. Scientific Reports 5, Article number: 13567 (2015) doi:10.1038/srep13567 [81], licensed under a Creative Commons Attribution 4.0 International License.

5.7. Isothermal Titration Calorimetry (ITC)

ITC is a technique that enables to characterize the complete thermodynamic profile of molecular interactions and molecular complex formation. This profile includes conformational changes, hydrogen bonding, hydrophobic interactions and electrostatic interactions, and how these interactions are interpreted at the molecular scale. An ITC instrument is composed of two identical cells made of a good conducting material and surrounded by an adiabatic jacket (Figure 9).

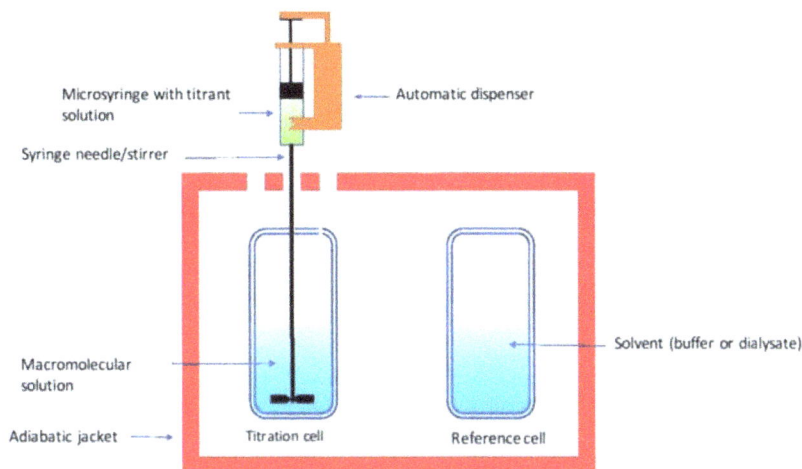

Figure 9. Schematic view of an isothermal titration calorimeter.

Very sensitive thermocouples are used to measure the differences in temperature between the reference and the titration cell. These are filled, respectively, with the solvent (e.g., 0.05% acetate buffer pH 5.0) and the measured solution (e.g., the chitosan solution). Solution with well-known concentration of the titrant (e.g., DNA solution) are placed on the automatic micro-syringe and small aliquots are injected into the titration cell. The titrant syringe functions as stirrer (e.g., 200 to 300 rpm).

For a bimolecular interaction between chitosan and a polynucleotide, the equilibrium binding constant, *Ka*, between a free chitosan molecule and a free binding site of DNA or RNA, is represented as the ratio of the molar concentration of the complex [*AB*] to the product of the molar concentrations of free binding sites in DNA and free chitosan, [*A*] and [*B*], respectively.

For the reaction $[A] + [B] \rightarrow [AB]$

$$K_a = \frac{[bound\ chitosan]}{[free\ binding\ sites\ DNA][free\ chitosan]} = \frac{[AB]}{[A][B]} \tag{6}$$

This model assumes independent binding sites. The complete characterization of a given macromolecular complex by ITC requires separating the enthalpic and entropic contributions of the binding process. Hence, ITC measures *Ka*, enthalpy change ΔH, binding stoichiometry and the entropy of change ΔS. ITC has been used to characterize the interactions of chitosan (M_w 7.4 to 153 kDa, DA 2–28%) and 6.4 kb plasmid EGFPLuc [129], as well as between chitosans of fungal origin (M_w 44 to 143 kDa and DA 14% to 22%) and siRNA luc GI3 [98]. A conclusion from these ITC studies regarding the influence of chitosan M_w and DA on binding affinity to DNA was that chitosans of low M_w bind more strongly to DNA at high charge density (i.e., low DA), whereas chitosans of high M_w bind more strongly at lower charge densities (i.e., high DA) In previous studies using ITC, it has also been possible to address the influence of the inclusion of hyaluronic acid on chitosan–siRNA complexes [102]. Both chitosan–siRNA systems (in the absence and presence of hyaluronic acid) show that the formation of the complexes is an endothermic process. This is ascribed to the delocalization of water molecules around the charge compensated regions and thus, the shuffling of condensed counterions.

5.8. Dye Displacement Titrations

Dye displacement titration is based on the quenching of fluorescence using fluorescent dyes as reporter molecules. Dyes need to exhibit relative enhancement of fluorescence upon binding to nucleotides as compared with that obtained when the dye is in solution [92,130]. Cyanines such as ethidium bromide, SYBR Green® and SYBR Gold® present high values of fluorescent quantum yield when they are intercalated into the bases of nucleic acids forming complexes [131,132]. The displacement of the dye from their complexes with nucleotides upon analyte titration will result in a decrease of fluorescence. The extent of fluorescence decrease is directly related to the binding between the ligand and the analyte. This titration assay is generally useful for establishing DNA/RNA binding affinity to different molecules, sequence selectivity, and binding stoichiometry [92,130]. The assay is not exclusive to titrations with small molecules; it has been used with a variety of ligands, including proteins and polymers [66,131–133]. Recently, a protocol based on the fluorescence determination of free pDNA with SYBR Gold® to quantify the percentage of pDNA complexation with trimethyl chitosan has been documented [134].

5.9. Circular Dichroism (CD)

Circular dichroism spectroscopy is an optical technique that evaluates the interaction between chiral molecules with circularly polarized light. Optically active molecules (e.g., asymmetric molecules) containing chromophore groups generate a deviation in circularly polarized light, resulting in a variation in the absorption of the left and right-hand components of the light. The conformational changes in the structure of macromolecules can be evaluated using CD. This spectroscopy is very sensitive to the secondary structure of polypeptides, proteins and nucleic acids [135–137]. CD of nucleic acids is generated by the asymmetry of the backbone that is composed by chiral sugars. The additional $\pi \rightarrow \pi^*$ interactions that lead to the helical arrangement of its bases generate electronic transitions within 200–320 nm, being the stacking geometry of the bases the main component of the CD spectra of nucleic acids [136,137]. Therefore, DNA and RNA may experience conformational changes as a result of the binding process to different compounds that can be tracked using CD spectroscopy [138].

5.10. Confocal Laser Scanning Microscopy (CLSM)

CLSM is a classical technique for a wide range of investigations in the biological and medical sciences. CLSM uses a focused laser beam to scan the three-dimensional volume of cells or tissues labeled with specific fluorophores. The images are obtained at a higher resolution with depth selectivity compared to conventional optical microscopy or fluorescence microscopy [139]. It has been extensively used for studying the intracellular trafficking of a broad range of nanoparticles during their interaction with cells [38,58,140] and in addition for the exhaustive comprehension of chitosan as gene delivery system [72,81]. The advantages of using CLSM include that, as the sample is exposed to the laser beam, it can be imaged many times; the ability to control depth of field; and elimination or reduction of background information. However, the main disadvantage lays on the limited number of excitation wavelengths available with common lasers [139].

5.11. Digital Holographic Microscopy (DHM)

DHM is a label-free imaging technique that allows visualization of transparent cells. The use of conventional techniques such as light bright field microscopy to observe cells can only generate small changes in the amplitude of light. In turn, DHM allows to multi-focus quantitative phase imaging of living cells. Furthermore, the cell volume and mass can be determined after determination of the refractive index of the cells. This provides real-time information about a possible cell swelling or shrinking in living cells. Based on these features, DHM has been successfully applied in several biomedical applications [141].

A basic setup for DHM is composed by an illumination source, most likely a coherent laser (monochromatic) to produce interference. This is followed by an interferometer and a digitizing camera. The monochromatic light used is divided in two beams to go through the reference and the sample. After illumination, the waves front from the sample and from the reference are superposed by a beam splitter to generate the interference pattern, e.g., the hologram. A single hologram is used to reconstruct the optical field and with that to generate the image using an appropriate algorithm to reconstruct the digital image [141,142].

DHM is an interferometric and non-invasive method able to distinguish cellular morphological changes. It is related to biophysical parameters and it can reveal absolute values of cell volume, dry mass, tissue density, transmembrane water transport and cell death [143,144]. DHM presents many advantages such as continuous cell monitoring over time and enabling to get insights about cellular processes such as migration, proliferation, death and differentiation. Ongoing studies in our laboratories are addressing the use of DHM in gene delivery by observing the motility of the cells in a migration study after treatment with chitosan particles containing miRNA.

5.12. Real-Time Quantitative Reverse Transcription Polymerase Chain Reaction (RT-qPCR)

Quantitative RT-PCR is a sensitive and powerful tool for analyzing RNA and is the method of choice for detection and quantification. The initial step in RT-PCR is the synthesis of the complementary DNA (cDNA) of the isolated RNA. The reaction is carried out using a reverse transcriptase (retroviral enzyme) and an oligonucleotide primer. The oligonucleotide primer anneals to the RNA and, therefore, will initiate the synthesis of cDNA toward the 5′ end of the mRNA through the RNA-dependent DNA polymerase activity of the reverse transcriptase. Primers can target a specific gene or be non-specific (random hexamer primers) that present all possible nucleotides combinations and bind to all RNAs sequences. The synthesis of cDNA is the source of variability in the results, since the reverse transcriptase enzyme is sensitive to salts or alcohols remaining from the RNA isolation [145]. Afterwards, the cDNA is amplified by PCR. PCR is generally constituted by steps of denaturation, annealing and elongation that are programmed in cycles. The number of cycles depends on the amount of target present and the efficiency of the reaction. The quantification and detection of the target gene can be carried out using two main techniques: "end-point" that measures the final concentration at the end of the reaction and the "real-time" that monitors the formation of the product during each cycle of the polymerase chain reaction. The quantification technique for "end-point" determinations can be done using fluorescent intercalating dyes. The "real-time" quantification uses primers and probes separately. This can be carried out using Taqman-DNA® polymerase, by using its 5′ exonuclease activity to cleave and displace the probe which leads to sequence-specific fluorogenic hydrolysis, fluorescence is no longer quenched and can be quantified [145,146]. The number of cycles required to reach a threshold of fluorescence is expressed as the Ct value. The "real-time" quantification requires the normalization to a reference gene or housekeeping gene for internally controlling the variation of the target gene. The analysis of the data is done by the relative expression of the housekeeping gene to the 2-ΔΔCt [147].

6. Plasmid DNA Delivery

At the molecular scale, pDNA can be considered a prodrug, given that, once it reaches the nucleus, it drives the synthesis of a therapeutic protein. Moreover, plasmids can also be considered as vaccines.

The Food and Drug Administration (FDA) defines DNA vaccines as "purified plasmid preparations containing one or more DNA sequences capable of inducing and/or promoting an immune response against a pathogen". A DNA vaccine consists of DNA plasmid (pDNA) containing a transgene that encodes the sequence of a target protein from a pathogen under the control of a eukaryotic promoter. Although the idea of using DNA plasmids for vaccination dates from more than 20 years ago, only four DNA veterinary vaccines are approved, and no DNA vaccines for humans are available yet [148]. One reason for this low success rate might be the weak immune response that DNA vaccines alone showed in different clinical trials. Nevertheless, genetic vaccines have many advantages over conventional ones. DNA vaccines can also induce both long-lasting cellular and immune responses but do not revert into virulence, hence they rise fewer safety concerns. Another advantage is that DNA vaccines do not integrate into the genome. pDNA is a double stranded DNA molecule, containing up to 200 kbp, and they can exist in three distinct topological configurations:

(i) compact supercoiled from; (ii) relaxed open circular form; and (iii) linearized form. The FDA requires that more than 80% of the pDNA be in the supercoiled form for application in pDNA vaccines for infectious diseases applications [149]. It has been shown that the supercoiled form of DNA has greater transfection efficiency than its open circular and linear counterparts [150,151].

pDNA cell transfection entails cellular uptake, endolysosomal escape, pDNA unpackaging, intracellular trafficking and nuclear import. The potential of chitosan and its derivatives for dDNA delivery has been widely documented. Among the pioneering studies on the use of chitosan polyplexes as an effective non-viral pDNA immunization vector gave proof of concept of the oral delivery of the plasmid pCMVArah2 encoding for the dominant peanut allergen to hypersensitive AKR/J mice. Four weeks after administration, the animals that were treated with chitosan/pDNA nanoparticles exhibited greater levels of IgA, consistent with the induction of a mucosal immunization response. In addition, the immunized mice, had less severe and delayed anaphylactic responses after intraperitoneal challenge with peanut protein Arah2 [152]. This study was the first one to show the potential of chitosan as a non-delivery vector for pDNA mucosal immunization. A pitfall in this work though, was that the chitosan used was only partially characterized as having a very high M_w (390 kDa), while the DA was not given neither in the paper nor in the preceding one.

Chitosan–pDNA polyplexes, chitosan-based nanoparticles crosslinked by ionic gelation, or either of these mixed with other biopolymers (e.g., hyaluronic acid) or beta cyclodextrins, have been reported in both in vitro and in vivo studies focused on pDNA delivery. Chitosan is a particularly attractive biopolymer for mucosal administration of pDNA vaccines. Recent studies have shown that chitosan (low M_w (undefined), DA 2%) ionically crosslinked with sodium tripolyphosphate, can be used to deliver the supercoiled topoisoform of pcDNA-FLAG-p53 plasmid [151]. To the best of our knowledge, this is the only study that has addressed the influence of chitosan on the conformation of DNA. This is an aspect that deserves much more attention in establishing the role of chitosan's structure on the conformational preference of both pDNA and siRNA.

7. RNA Interference Machinery

In cells exists a process known as RNA interference (RNAi) that regulates the degradation of RNA in a highly sequence-specific manner [153]. In general, RNAi process is activated by introduction of double stranded RNAs (dsRNAs) into cells, promoting the degradation of mRNAs, controlling mRNA translation, and therefore interrupting protein synthesis [153]. RNAi mechanism can be triggered by either small interfering RNAs (siRNAs), which can decrease gene expression through mRNA transcript cleavage, or endogenous microRNAs (miRNAs), which primarily inhibit protein translation [154]. siRNA machinery is found in plants and bacteria, and exogenous introduction in mammalian cells can target a specific mRNA with 100% complementarity. Endogenous mRNA regulation in mammalian cells is mediated by miRNAs [154,155].

In Figure 10, is shown a schematic description of miRNA biogenesis and the RNAi mechanism mediated in mammalian cells. Briefly, the gene-silencing mechanism for miRNAs is induced by endogenous dsRNA transcribed by RNA polymerase II (Pol II) in the nucleus, producing so-called pri-miRNAs. The pri-miRNAs are cleaved to yield ~70-nucleotide pre-miRNAs by the RNase III enzyme, Drosha, and its cofactor, the double-stranded RNA-binding protein, Pasha. The pre-miRNAs are imperfect stem-loop structures, which are exported into the cytoplasm via Exportin 5 (Exp 5) and its catalytic partner Ran-GTP. These pre-miRNAs are degraded by the RNase III endoribonuclease Dicer into miRNAs constituted by 21–23 nucleotides. Afterwards, they incorporate into a miRNA ribonucleoprotein complexes (miRNPs). In the case of siRNA, their exogenous introduction into mammalian cells also leads to assemble into a ribonucleoprotein complex known as the RNA-induced silencing complex (RISC). These complexes are believed to be similar, if not identical. Argonaute unwinds miRNA/siRNA within their respective functional complexes and leads to the retention of the guided strand in the complex and at the same time the complementary strand is removed and degraded. Further, exogenous siRNA mediate sequence-specific silencing by inducing mRNA cleavage and subsequent mRNA degradation of target transcripts. In the case of miRNA, it can bind either with imperfect complementarity to its target mRNA matching with the targets 3′ untranslated region (UTR) resulting in inhibition of protein translation and/or with perfect complementarity which results in degradation of the target mRNA [25,154,156–158].

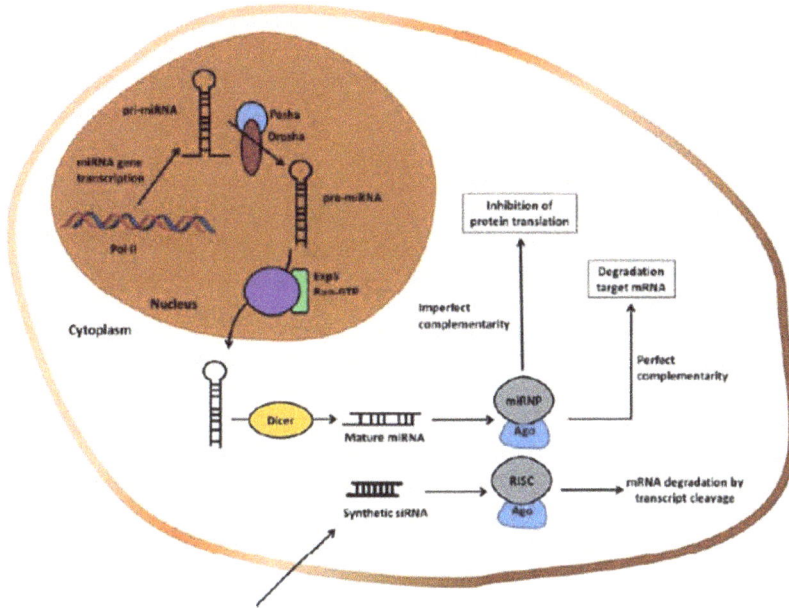

Figure 10. Simplified scheme of RNA interference mechanism in mammalian cells. Only processes mentioned in the text are illustrated.

7.1. MicroRNAs Implication in Cancer: Oncogenes and Tumor Suppressors

MicroRNAs (miRNAs) are small non-coding RNAs of 20 to 24 nucleotides found in all eukaryotic cells, playing important roles in almost all biological pathways in mammals and other multicellular organisms [158]. The first reports identifying small non-coding RNA date from 1993 and 2000, where lin-4 and let-7 genes were identified in *Caenorhabditis elegans* acting as post-transcriptional repressors of their target genes when bound to their specific sites in the 3′ untranslated region of the target mRNA [159,160]. In the last fifteen years, many reports have emerged focusing on miRNA and it is estimated that they regulate ~60% of human genes, e.g., 250–500 different mRNAs, making them very powerful gene regulators [161]. MicroRNAs influence numerous cancer-relevant processes such as proliferation, cell cycle control, apoptosis, differentiation, migration and metabolism [162–164]. This leads to huge opportunities, for researchers looking for therapeutic targets or mimics. MicroRNAs can promote or repress cell proliferation, differentiation and apoptosis during normal cell development [165,166]. Dysregulation of miRNA expression can affect a multiple number of cell signaling pathways and in that way influence cancer onset and progression. Some miRNAs may function as oncogenes or tumor suppressors. Understanding the role of miRNAs in human malignant tumors is key in the development of new therapeutic approaches.

Lu et al. demonstrated that miRNAs have different profiles in cancers compared with normal tissues, and those vary among different cancers [167]. The function of miRNAs in cancer pathogenesis can be studied by upregulation or downregulation of its expression. Upregulation of miRNA expression in tumors is regarded as oncogenes. Increased expression of miRNAs, would negatively inhibit miRNA-target tumor suppressor genes that control cell differentiation or apoptosis, contributing to tumor formation by stimulating proliferation, angiogenesis, and/or invasion. The causes of this dysregulation are not completely elucidated, but might be due to amplification of the miRNA gene, which would increase the efficiency in miRNA processing or increased stability of the miRNA [165,168–170]. The expression of some miRNAs can be as well downregulated, these miRNAs are also referred as tumor suppressor miRNAs. Tumors often present reduced levels of mature miRNAs due to genetic loss, epigenetic silencing, defects in their biogenesis pathway or widespread transcriptional repression [167]. In normal cellular process these tumor suppressor miRNAs prevent tumor development by negatively inhibiting oncogenes. Therefore, a downregulation of those consequently increases proliferation, invasiveness or angiogenesis, decreases levels of apoptosis, or undifferentiated or de-differentiated tissue, leading to tumor formation [165,168–170].

8. Rare Diseases

A rare or "Orphan" disease, as defined by the World Health Organization (WHO), is any disease or condition that affects 0.65–1% of the total global population. These diseases affect a significant proportion of the population; indeed, there have been more than 6000 types described worldwide [171]. The complex etiology of such diseases and the widely heterogeneous symptoms result in significant challenges faced by the scientific community as well as a lack of viable therapeutic options for patients suffering from these diseases. The treatment of individuals affected by rare diseases is hampered by poorly understood mechanisms slowing the development of suitable therapeutics. Advances in rare disease diagnostics, based on data from clinical trials, are the focus of gene therapy studies, enzyme replacement therapy and new drug discoveries, whose identification can be accelerated by drug repositioning. These advances have allowed the better characterization of rare diseases, especially those that are monogenic. Nonetheless, several factors have hindered therapeutic development for rare diseases such as the small market share that they represent the high cost of production new therapeutics and the potential low return on investment [172].

Addressing several rare diseases that share a common molecular etiology within a given project is especially attractive as the majority of rare diseases have an underlying genetic cause [173]. Undertaking drug repurposing screens for a variety of rare diseases that share a common molecular etiology will expedite drug discovery for these conditions. For the rare disease population, gene therapy is a hopeful inspiration.

8.1. Cystic Fibrosis

Cystic fibrosis (CF) is a genetic disease caused by mutations in the cystic fibrosis transmembrane conductance regulator (CFTR) gene [174,175]. This encodes a cyclic AMP (cAMP) dependent channel expressed in the epithelia of many exocrine tissues including the airways, lung, pancreas, liver, intestine, vas deferens and sweat gland/duct [176]. CF is the most common lethal autosomal recessive disease in people from Northern European descent affecting approximately 70,000 individuals worldwide (www.cff.org). Although medical advances in recent decades have prolonged the average life expectancy for CF patients to ~40 years, there is still no cure for this devastating disease [177].

The CFTR gene encodes a chloride and bicarbonate channel. In addition to this, CFTR is proposed to regulate the function of other membrane proteins including the epithelial sodium channel (ENaC) [178]. CFTR and ENaC play the most important role in maintaining homeostasis by controlling the movement of water through the epithelium, thus regulating the hydration of the epithelial surface in many organs, but predominantly in the airways. Therefore, epithelial CFTR dysfunction leads to airway surface liquid volume depletion due to an imbalance between CFTR–mediated Cl^- secretion and ENaC-mediated Na^+ absorption [179].

8.1.1. Pathophysiology of CF

Mutations in CFTR cause abnormal ion transport in the epithelium of several tissues, which results in the production of abnormally thick and sticky mucus that blocks the organ and is responsible for CF pathology. CF encompasses a wide range of symptoms; one of the earliest markers of the disease is the meconium ileus affecting a number of CF newborns [180]. Another symptom used for early diagnosis of CF is the salty sweat caused by defective salt reabsorption in the sweat ducts [181]. Furthermore, approximately 85% of CF patients are pancreatic insufficient whereby thick, dehydrated secretions cause occlusion of the pancreatic ducts and prevent the release of the digestive enzymes into the intestines [182]. This can be managed by controlling food intake and incorporating pancreatic enzyme supplements, minerals and fat-soluble vitamins into the diet [183,184]. Older patients can suffer from the destruction of the islets of Langerhans and reduced insulin production, leading to diabetes mellitus [185]. Liver disease and infertility are other common manifestations of CF. In fact, around 98% of male CF sufferers are infertile, mainly due to congenital bilateral absence or blockage of the vas deferens [186]. However, the main cause of morbidity and mortality in CF is lung disease, with lung malfunction and pulmonary failure [187,188]. The loss of chloride secretion from CFTR deficiency results in changes in osmotic pressures and electro-neutrality which likely lead to excessive sodium and water absorption [189]. The production of sticky mucus in the lumen of the lungs impedes mucociliary clearance [190]. Clinically, this manifests as chronic inflammation and recurrent bacterial infections, commonly by pathogens such as *Pseudomonas aeruginosa* and *Staphylococcus aureus* [188]. Mucus and inflammatory cells cause bronchiectasis and the continuous cycle of infection and inflammation lead to the progressive destruction of lung tissue [191].

8.2. CFTR Channel

Located on the long arm of chromosome 7 (7q31.2), the CFTR gene spans around 250 kb of DNA [192]. Consisting of 27 exons of various sizes and producing an mRNA of around 6.2 kb, the gene encodes a single polypeptide chain of 1480 amino acids with a predicted molecular weight of around 168 kDa [174,193]. The observed sequence of the CFTR protein has led to a proposed structure that shows similarity to proteins in the family of ATP-binding cassette (ABC) transporter [194]. The CFTR protein spans the apical membrane of epithelial cells and consists of distinct structural domains; two membranes spanning domains, two nucleotide binding domains and a regulatory domain. These nucleotide binding domains contain conserved motifs for ATP binding and hydrolysis. Both, the regulatory domain region, as well as the long N- and C-terminal extension, is unique in CFTR. The N- and C-terminal are about 80 and 30 residues in length, respectively. Phosphorylation of the R domain is carried out by PKA, PKC and CK2, a process which is dependent on the presence of intracellular ATP [195,196]. This event is important for regulating the opening of the CFTR channel. The dephosphorylation of the R region is thought to be brought about by phosphatases and has an inhibitory role on CFTR activity [197]. Small conformational changes may only be needed to bring about channel opening, and modest structural changes have been observed upon phosphorylation and the binding of ATP. Many other proteins including PDZ-interacting proteins and STAS domain interactors are believed to be important for the regulation of CFTR activity [198,199].

8.2.1. CFTR Mutation

Almost 2000 distinct CFTR mutations have been described thus far by researchers working in the field of CF genetics [200]. However, by far the most common mutation described is a three-base deletion that removes a phenylalanine at position 508 (F508del) of the CFTR protein, which is the cause of disease symptoms in 85% of patients worldwide [201]. The scientific community has organized all known CFTR mutations into six classes according to their deleterious effect on the protein [176]. Classes I–III are associated with no effective CFTR function, such as total or partial lack of production, defective processing or missense mutations. Classes IV–VI show residual function, issues with pre-mRNA splicing or stability at the plasma membrane. There is now a growing consensus that CFTR mutations may also provide a scientific basis for mutation-specific corrective therapies and are accordingly grouped into seven classes depending on their functional defect [201]. Those mutations in Class VII have been defined as any which cannot be corrected through pharmacological means. A large proportion of ongoing studies now aim to apply the same therapeutic correction to the basic defect in each class.

8.3. Cystic Fibrosis: Implication in Gene Therapy

8.3.1. Treatments for CF: Drug Therapies

CF is a disease that may manifest with various physiological abnormalities, thus creating a diverse CF population. One main challenge with treatment of CF is therefore identifying which of these abnormalities merits more attention to achieve the best outcome [202]. The major factor linked to mortality, such as pulmonary disease, nutritional deficiencies and CF related disease could be a crucial point in the optimal management of the patients. Nutritional deficiency is linked to pulmonary function decline, and early intervention to improve nutrition is associated with better clinical outcomes, including increased survival. Furthermore, nutritional supplements and pancreatic enzyme replacement therapy have been essential in patients of all ages [203]. With improved nutritional status, CF patients have improved outcomes with respect to pulmonary symptoms. Since pulmonary disease pathogenesis in CF is a complex process associated with multiple abnormalities in the respiratory tract, laboratories around the world have characterized several strategies to restore CFTR function in the airways.

Personalized protein therapy targeting specific mutation classes allows the individualization of treatment, thus anticipating which drug will be effective in one patient versus another [204]. Administration of aminoglycoside like gentamycin has shown benefits in patients with CF [205], but pre-clinical studies have demonstrated that high-dose aminoglycosides could induce a lack of potency and side-effects [206]. Therefore, a new compound named Ataluren or PTC124, with a similar mechanism of action as aminoglycosides but lacking their antibiotic properties, has been developed [207].

Chemical compounds called "correctors", agents that could prevent the degradation of the CFTR protein in the cell, have been developed [208]. The company Vertex developed VX-809 or Lumacaftor, a corrector that may achieve this goal. Studies with bronchial epithelial cells containing the F508del-CFTR mutation have shown increased chloride transport with VX-809 [209]. Other strategies also looked at agents that increased the chloride transport activity of CFTR, molecules called "potentiators" [210]. High-throughput screening methodologies identified a potentiator named VX-770 or Ivacaftor. VX-700 is an approved CFTR potentiator that increases the open probability of the CFTR channels [211]. In addition, the combination of Lumacaftor and Ivacaftor also called Orkambi has been associated with a higher increase in chloride transport than either agent used alone [209]. Recently, Phase III trials were conducted to evaluate the efficacy and safety of Orkambi in patients with CF homozygous for F508del-CFTR mutation. The results confirm an improvement in forced expiratory volume and a reduction in pulmonary exacerbations, thus providing a benefit for CF patients [212]. Another study by Leier and colleagues has shown that the activation of CFTR can be achieved by increasing the level of cAMP in the cell using phosphodiesterase (PDE) inhibitors such as sildenafil (Viagra) [213]. Leier and co-authors used sildenafil in F508del-CFTR/wt-CFTR expressing *Xenopus laevis* oocytes and in CF and non-CF human bronchial epithelial cell lines (CFBE41o-/16HBE14o-) to investigate the mode of action of this component. Unfortunately, the necessary high doses of the drug for CFTR recovery limit its use in the clinic [213].

Alternatively, ENaC activity could be inhibited to decrease the elevated Na$^+$ absorption seen in CF airways. After aerosol delivery of the ENaC inhibitor Amiloride, to the airways of CF patients in clinical trials, correction of the Na$^+$ transport defect was reported, although no long-term clinical benefit was observed [214]. Early Amiloride derivatives (e.g., Phenamil and Benzamil) had limited therapeutic use due to rapid absorption and clearance from the lungs [215]. Another strategy utilizes antisense oligonucleotides (ASOs) to inhibit ENaC activity. ASOs are short synthetic DNA molecules (15–18 bases) that are complementary to specific mRNA sequences. Some studies have focused on the ASOs strategy to down-regulate the expression of the alpha-ENaC subunit in human primary nasal epithelial cells. Sobczak et al. have shown an inhibition of Na$^+$ absorption through ENaC in CF tissue by about 75% and in non-CF tissue by about 66%. Furthermore, this ASOs strategy sustained ENaC inhibition for more than 72 h [216]. Hypertonic saline treatment (HTS) has been shown to possess mucolytic properties and aids mucociliary clearance by restoring the liquid layer lining the airways [217]. However, different studies have suggested that this treatment may not be suitable as a long-term strategy to slow disease progression [218]. Nonetheless, HTS could provide a safe, low-cost addition to the daily therapy of CF patients.

Most of the potential drug therapies mentioned above target a specific class of CFTR mutation and are effective only at high doses. Subsequently, these therapies are only relevant for a small number of CF patients and would not be suitable treatments for the clinic.

8.3.2. CFTR Gene Therapy

Gene therapy has emerged as an alternative to conventional treatment for diseases. The discovery of the CFTR gene in 1989 created excitement for the development of CF gene therapy [219]. The delivery of genetic material, DNA or RNA, can be utilized as a promising concept for heritable diseases, such as CF with the prospect of correcting many aspects of the complex pathology [220]. The vast majority of efforts over the last 20 years have focused on developing a curative treatment targeting the basic defect rather than treating CF disease manifestations [221]. In addition, gene therapy is mutation class independent also termed "mutation agnostic" and a single treatment strategy would be suitable for all patients.

One of the most powerful genetic therapeutic technologies is gene repair, where the specific mutated bases are targeted and corrected in the genome of an individual. The emergence of the engineered nucleases such as CRISPR-Cas-9 system, zinc finger nucleases (ZFN) and transcription activator-like effector nucleases (TALEN), are encouraging, but the low efficiencies and the time-consuming selection of the repaired cells is not feasible in vivo [222,223]. Subsequently, studies have mainly focused on gene complementation approaches, where an additional copy of the wtCFTR cDNA is delivered to cells homozygous for CFTR mutations [224].

Many viral and non-viral vectors have been tested for their usefulness in CF gene therapy [225]. Viral studies using recombinant adenovirus (rAd) were a promising gene delivery vector due to the high gene transfer efficiencies observed in animal models. However, the paucity of adenoviral receptors on the apical lung surface and the severity of the host-immune response to repeat viral delivery have demonstrated a strong argument against the effectivity of adenoviral CF gene therapy [226,227]. Subsequently, another promising viral vector evaluated for the treatment of chronic lung disease is based on recombinant adeno-associated virus (rAAV). Furthermore, numerous AAV serotypes have been discovered, with the aim of increasing transduction efficiency. However, results of rAAV treatment evaluated clinically were disappointing due to the antiviral immune response activated by repeated administration [228]. In addition to rAAv and rAAV, various cytoplasmic RNA viruses have been validated for airway gene transfer. The murine parainfluenza virus type 1 (SeV), the human respiratory syncital virus and the human parainfluenza virus have been shown to efficiently transfect airway epithelial cells via the apical membrane [229,230]. Only SeV has been used in animal models in vivo and has been able to correct the Cl$^-$ transport defect in the nasal epithelium of CF knockout mice [231]. However, repeated administration was not feasible and therefore the vector has not made it to clinical trials [232,233]. Most recently recombinant lentiviruses (rLV) have gained interest as they appear to evade host immunological defenses and are able to inducegene expression in non-dividing cells [234,235]. They can be modified by the addition of novel surface proteins (pseudotyping) to specifically increase the efficiency of airway gene transfer [236]. Furthermore, lentiviral vectors can be repeatedly administered to murine airways [237], which is a major requirement for the treatment of chronic diseases such as CF.

The investigation of non-viral vectors to treat chronic CF lung disease was fueled by the necessity to develop an effective long-term, repeatedly administered treatment [238]. Typically, non-viral vectors comprise circular plasmid DNA molecules manufactured from bacteria, which are then complexed with a range of cationic lipids and polymers, known as "lipoplexes" or "polyplexes", respectively [239]. The mechanism of non-viral gene transfer is poorly understood, but it is thought that lipoplexes and polyplexes bind to the cell membrane, are endocytosed and subsequently escape from endosomes by inducing rupture of the endosomal membrane [240]. Non-viral vectors are considered less efficient than viruses due to the lack of the specific components required for cell entry that are present in viruses [241]. Nevertheless, non-viral vectors do not contain viral proteins rendering them less inflammatory and immunogenic [242]. Further advantages associated with these vectors are their easy manipulation, the possibility to manufacture them in large quantities, their possible storage for extend periods and their unlimited packaging capacity [243].

Messenger ribonucleic acid (mRNA) is another potential alternative to conventional DNA therapies. The development of mRNA-based therapeutic approaches presents several important differences in comparison with other nucleic acid-based therapies. The delivery of mRNA allows direct translation in the cytoplasm, as mRNA is not required to enter the nucleus to be functional. Furthermore, mRNA does not integrate into the genome and therefore does not pose the risk of insertional mutagenesis. The production of mRNA is also relatively simple and inexpensive [244]. Approaches with mRNA-based gene therapy and delivery by non-viral vectors bypass some of the disadvantages associated with DNA delivery [245]. The delivery of mRNA-based therapeutics offers a greater safety profile than viral and pDNA-based vectors since they do not contain bacterial sequences. RNA is recognized by Toll-like receptor (TLRs), a family of receptors that trigger the innate and adaptive immune system to deal with infections by the recognition of pathogen-associated molecular patterns (PAMPs) [246]. Both dsRNA and ssRNA are often components of viral genomes or intermediates of replication, and are therefore recognized by TLR3 and TLR7/TLR8, respectively [246–248]. Future studies may show whether nucleoside modified IVT mRNA will avoid the activation of human TLRs in the clinical setting.

Recently, a proof-of-concept mRNA-based functional restoration of impaired CFTR function have been demonstrated, not only in an immortalized human bronchial CF cell line, but also in primary human nasal epithelial (HNE) cells [113,249]. Transfection of the CFBE41o- cells with CFTR-mRNA restored cAMP induced CFTR currents similar to the values seen in control cells (16HBE14o-) and an almost two-fold increase in the cAMP-stimulated CFTR current after transfection using primary cultured HNE cells. The authors demonstrated that optimized CFTR-mRNA can be reduced to a minimal dose of 0.6 μg/cm^2 in primary HNE cells, and that this dose can be persistent for a period longer than 24 h after transfection procedures [113]. These experiments successfully established a new strategy for the delivery of CFTR-mRNA directly to airway epithelial cells.

8.3.3. Different Delivery Methods for CF Lung Disease

In CF pulmonary disease the opportunity to selectively target a drug to the lungs remains a fascinating option. In fact, local drug delivery may allow maximum pharmacological targeting, and thus therapeutic efficacy. Consequently, researchers continue to apply efforts to develop new inhalation devices and advanced drug delivery [250]. CFTR-mRNA aerosol administration to airways of CF patients could be delivered as shown by Rudolph and coworkers [251]. Another recent study by Hasespunch and co-authors successfully demonstrated gene delivery with magnetized aerosol comprising iron oxide nanoparticles in the lungs of mice [252]. Therefore, administration of drugs via the inhalation route is of great interest in CF treatment. The main advantages of aerosol technologies are the limited systemic toxicity, direct drug action on target site and the suitability for home therapy [253]. Potential disadvantages include the uncertainty about drug dose at the target site as well as limited information on drug interactions in the lung [254]. Another important issue in the gene delivery method is the biocompatibility and biosafety of the nanocarrier employed in the transfection procedure [255]. Lipid based formulations like cationic lipids or cationic polymers have been proved a successful method to transfect cells and to reach adequate transfection efficiency in vitro [256].

Orientating studies demonstrated promising results using lipid-based delivery by LipofectamineTM 2000 transfection reagent in primary cultured cells. However, lipofectamine presents high cytotoxicity, compromising cell viability. Thus, despite its robust high transfection efficiency [257] it is not an appropriate carrier to assess potential clinical therapies in the treatment of CF. Therefore, alternative and stable formulations using biopolymers, such as chitosan, have been assessed for their efficacy in targeting intratracheal routes. In this work, the development and optimization of an alternative non-viral mRNA system based on a biocompatible polymer, in particular chitosan, was of great interest [258]. In general, the use of natural polymers has gained increasing interest as a safe and cost-effective delivery strategy for gene material [259]. Recent studies, including our own ones, have reported improved plasmid-based gene transfer and expression in the airways using chitosan as a nanocarrier [71,105,106]. Nydert and coworkers examined the possibility of using CFBE41o as a transfection cell model for chitosan; this was the first time chitosan polyplexes were used to transfect a CF cell line [105]. Nevertheless, the search for the optimal vector for gene therapy in the hope of finding a mutation-independent cure for CF continues and represents a major challenge in CF. Recently, we described for the first time the use of non-animal source chitosan as a gene delivery vector in the context of CF lung disease which provides an alternative to classical animal-source chitosan [108].

Rather than treating CF disease manifestations, the finding of the CFTR gene identification opened the door for targeting the basic defect to correct the mutation at a cellular level. In general, gene-based therapeutics introduce nucleic acids into cells to alter the gene expression of a pathological process [260]. Thus, the delivery of a therapeutic nucleic acid (DNA or RNA) is a promising concept for an inherited single-gene defect such as CF, with the prospect of correcting many aspects of the complex pathology [261]. In addition, one therapy might be suitable to treat subjects with a wide variety of mutations, which means that a single treatment strategy would be relevant to all patients.

Since CF lung symptoms are responsible for the majority of deaths in CF, most investigators have focused their efforts on gene therapy in airways and on targeting a reduction in lung disease [262]. Several challenges were identified in pulmonary gene delivery, including the development of efficient vectors for gene transfer concerning safety requisites, the circumvention of inflammatory responses, and the maintenance of long-term gene therapeutic expression in airways [263]. Indeed, the ciliated epithelial cells that are located in the airways are the best target for CF gene therapy [264]. The expression of the CFTR protein in the airway is mostly localized in the ciliated epithelial cells and ducts of the submucosal glands [265,266]. Furthermore, the main functions of ciliated cells include the facilitation of mucus transport and the maintenance of airway surface hydration [267,268]. Therefore, ciliated cells and their properties make them a relevant therapeutic target for CFTR gene delivery.

Although it is known that the mRNA of CFTR is in low abundance in airway epithelia [269], a minor level of gene transfer of CFTR to the airway epithelia is sufficient to correct the Cl^- transport in vitro and in vivo [270]. Furthermore, only 10% of normal cells were sufficient to normalize the main dysregulated parameters such as Cl^- or Na^+ conductance and Interleukin 8 secretion [271].

Up to now, more than 25 clinical trials which aimed to investigate the safety and sustainability of gene transfer have been completed. Most of them were of short duration and carried out with a reduced number of patients [272]. Initial approaches, which involved the direct administration of recombinant CFTR, based on conventional DNA delivery to the airway, have not been successful for several reasons, such as immunogenicity or the limited DNA packaging capacity of the vector [273,274].

Thus, no significant benefits for patients treated with viral and non-viral vectors were found [264]. Subsequently, in vitro transcribed mRNA has been pointed out as a potential new drug class to deliver genetic information [244]. Furthermore, the development of mRNA-based therapy presents several important differences in comparison with other nucleic acid-based therapies, e.g., the direct translation in the cytoplasm, the non-integration into the genome and therefore, the avoidance of the potential oncogene expression that is caused by insertional mutagenesis [260]. In addition, exogenous mRNA can be efficiently delivered into cells independently of the differentiation stage or confluence [275].

The transfer of genetic material into the appropriate target cells is the first critical step for successful gene therapy. Among the different available methods for accomplishing gene delivery, the chemical lipofection procedure has been reported to be widely efficient [34,276,277]. In the literature, it has been said that lipofection methods present high transfection efficiencies and in vitro studies showed increased levels of transgene expression in different mammalian cell types [278]. Nonetheless, a potential reduction in cell viability and immunogenicity induction restrict the use of lipofection in vivo [279].

8.4. Other Rare Diseases

- Lysosomal storage diseases (LSD)

LSDs are a group of more than 40 single-gene recessive genetic diseases that result in metabolic imbalances of the lysosome. LSDs, over time, result in a spectrum of symptoms and disabilities due to tissue-specific substrate accumulations [280]. LSDs are promising targets for gene therapy because the delivery of a single gene, into a small percentage of the appropriate target cells may be sufficient to impact the clinical course of the disease. In addition to these characteristics, another important one is the possibility for the "cross-correction phenomenon" of affected tissues in patients via the extra-cellular provision of recombinant forms of the deficient lysosomal enzyme [281]. Adeno-associated virus therapies, adenovirus therapies, and hematopoietic stem cell (HSC) transplant have overcome limitations associated with earlier clinical and preclinical trials, suggesting that gene therapy may be a reality for LSDs soon. At the same time, the first EU-approved gene therapy drug, Glybera®, has been discontinued, and other ex vivo-based therapies although approved for clinical use, have failed to be widely adapted and are no longer economically viable. In the case of aromatic L-amino acid decarboxylase (AADC) deficiency, it is also reported that some gene therapies using specific AAV serotypes, such as AAV9, could cross the blood–brain barrier to deliver genes to the central nervous system [282].

- Metachromatic leukodystrophy (MLD)

The concept of leukodystrophy (leuko-white, dystrophy-defective nutrition) refers to a very heterogeneous group of inherited diseases in which molecular abnormalities result in a defect in myelin sheath formation or maintenance, or in some cases both [283]. Metachromatic leukodystrophy (MLD) is an autosomal recessive LSD caused by the deficiency of lysosomal arylsulfatase A (ARSA) enzyme [284]. The optimal therapy for MLD would provide persistent and high level expression of ARSA in the central nervous system (CNS). Gene therapy using adeno-associated virus (AAV) is an ideal choice for clinical development as it provides the best balance of potential for efficacy with a reduced safety risk profile. Therefore, various therapeutic approaches to MLD have been tested in experimental animal models. In addition, there are several promising approaches with potential clinical translation, including: (1) enzyme-replacement therapy; (2) bone marrow transplants; (3) gene therapy by ex vivo transplantation of genetically modified HSC; and (4) AAV-mediated gene therapy directly to the CNS [285]. Initial studies with AAV-mediated gene therapy for MLD were conducted with AAV5, and then with the discovery of the non-human primate derived serotypes, second generation studies were carried out with AAVrh.10 [286–288]. A Phase I/II trial in early forms of MLD using intracerebral delivery of AAVrh10 has been instated in France (clinicaltrials.gov NCT01801709) [283] and in the United States [285]. Therefore, CNS targeted AAV gene therapy vectors have become a promising modality for treating leukodystrophies.

- Cellular Immunodeficiencies

Nowadays, more than a hundred of primary immunodeficiency syndromes have been described. These disorders involve one or more components of the immune system, including T, B, and natural killer lymphocytes; phagocytic cells; and complement proteins [289].

The Wiskott-Aldrich syndrome (WAS) is an X-linked recessive syndrome characterized by eczema, thrombocytopenic purpura with normal-appearing megakaryocytes but small defective platelets, and undue susceptibility to infection [290]. For many years, the only potential curative therapy has been allogeneic HSC transplantation (HSCT), but HLA-mismatched HSCT may still be accompanied by unacceptable risk in some cases [291]. Gene therapy for WAS was first attempted using a retroviral vector. This approach resulted in a sustained increase in the proportion of WAS-corrected cells in all patients. However, the majority of patients developed acute leukemia secondary to viral enhancer-mediated insertional mutagenesis [292]. Recently, gene therapy approaches using a self-inactivating lentiviral vector has shown promising results in children and young adults, demonstrating the feasibility of the use of new gene therapies in WAS patients [293].

Severe combined immunodeficiency (SCID) consists of a group of fatal genetic disorders characterized by profound deficiencies of T- and B-cell (and sometimes NK-cell) function. Infants with SCID are lymphopenic [294]. Twenty-five years ago, bone marrow cells genetically corrected with a retroviral vector were administered to a child suffering from adenosine deaminase-deficient severe combined immunodeficiency (ADA-SCID). This was the first approach to using genetically modified stem cells to treat a human disease [295]. In 2016, the pharmaceutical company GlaxoSmithKline announced that the European Medicines Agency had granted marketing authorization to Strimvelis, the commercial name of gene therapy for ADA-SCID [296].

- Neuronal Neuropathies

Giant axonal neuropathy (GAN) is a rare childhood onset autosomal recessive neurodegenerative disorder affecting the central and peripheral nervous system [297]. Mutations in the GAN gene cause loss of function of gigaxonin, a cytoskeletal regulatory protein, clinically leading to progressive sensorimotor neuropathy, reduced coordination, slurred speech, seizures, and progressive respiratory failure leading to death [298]. Recently, AAV9-mediated GAN gene transfer has been shown as a potential therapeutic approach for improved patient outcomes in GAN, a study which is currently being conducted as a Phase I trial at the National Institutes of Health Clinical Center (clinicaltrials.gov identifier: NCT02362438) [299,300].

- Respiratory Ciliopathies

Ciliopathies are a growing class of disorders caused by abnormal ciliary axonemal structure and function [301]. Current treatment of ciliopathies is limited to symptomatic therapy. However, the growing understanding of ciliopathy genetics coupled with recent advances in gene delivery and endogenous gene and transcript repair, demonstrated thus far in tissues of the eye, nose, and airway, offers hope for curative measures in the near future [301].

Primary ciliary dyskinesia (PCD) is a rare childhood disease, the prototype for motile ciliary dysfunction. PCD results from abnormal ciliary function, leading to neonatal respiratory distress, chronic sinopulmonary disease causing sinusitis, bronchiectasis, recurrent ear infections, and infertility. The goal for the management of PCD is to prevent exacerbations and complications as much as possible, and to slow the progression of disease [302].

A PCD-like phenotype caused by reduced generation of multiple motile cilia (RGMC) has been described, which is characterized by sinopulmonary symptoms and fertility defects similar to those observed in PCD patients [303,304]. The residual motile cilia in RGMC caused by mutations in CCNO (encoding cyclin O) show a normal ciliary beat, while the few remaining motile cilia in patients with RGMC caused by MCIDAS mutations are immotile. However, no situs defects have been observed in any patients with RGMC, thus suggesting that the function of nodal cilia is intact [305].

Current gene therapy still has many disadvantages, such as the lack of safe and effective methods to inject a permanently active gene, which further prevents its further repetitions in terms of multiple applications in genetic diseases. Besides, in some clinical trials, patients showed undesired immune responses to the treatment and decreased therapeutic effect over time. Finally, new gene-delivery vectors remain to be discovered and developed with better efficiency and safety than current viral vectors, opening the door for biopolymer-based non-viral approaches as safer and alternative to the aforementioned ones.

9. Conclusions

Our current knowledge on the use of chitosans as non-viral gene delivery vectors has still important gaps. The role of chitosan's M_w and DA on the various physicochemical and biological phenomena associated to gene transfection has been addressed in many studies, and one undisputable conclusion is that there is not "one-size-fits-all" chitosan. More systematic approaches are necessary, particularly aimed to expand our understanding on the influence of chitosan's structure on pDNA and siRNA molecular conformation, size, shape, and surface characteristics of complexes obtained either by direct electrostatic polyelectrolyte complexation, by co-crosslinking by ionic gelation (e.g., with TPP) or covalently (e.g., with genipin), co-complexed with other polyanions (e.g., hyaluronan, alginate) or with proteins or phospholipids. This knowledge is essential to aid in the rational design of virus-like particles with tailor made characteristics.

Thus far, the results accrued from techniques such as SPR, fluorescence spectroscopy and ITC to probe the interaction between chitosans and/or miRNA have been instrumental to allow measuring quantitatively the binding affinity constants of chitosans of varying M_w and DA. The evidence seems consistent with the notion that there is a narrow window of affinity for a given chitosan–nucleic acid pair, where the transfection efficiency is maximized. Although the reason behind this is not fully understood, it seems that this is due to a compromise between a binding affinity which needs being high enough to condense and protect the integrity of the gene, but, at the same time, weak enough to allow the intracellular dissociation and unpacking needed to release the gene in its functional form. Whether the nature of the molecular interactions at play is merely non-specific electrostatic or of other type is unknown. It is also not known whether specific pattern sequences of acetylated and deacetylated sugar units in chitosan have specific binding affinity for given gene sequences in DNA or RNA in each of their preferred conformations. Theoretical and computational chemistry approaches are urgently needed in these regards. Gleaning this intelligence will surely open huge possibilities for specific partially acetylated chitosan oligomers, for example, to develop riboswitches, novel ribozymes or as specific therapeutic gene targets. The ongoing revolution brought about by the recent discovery of CRISPR-Cas9 technology for surgical scar-less genome editing will undoubtedly be a game changer in this field on the short term. In the field of rare diseases, gene therapy is perhaps where the greatest potential lies at first, before the costs entailed in this technology start to decrease and it becomes amenable for the treatment of more widespread diseases. We anticipate that specific chitosans will be key players in the translation of research to the clinic. The low cytotoxicity, low biopersistance and mucoadhesive properties of chitosans are hardly shared by any other biopolymer. Chitosan chemical derivatives have lain beyond the scope of this review; however, they surely also hold enormous promise yet to be discovered and realized in future.

Acknowledgments: We dedicate this review to the memory of our dear friend and colleague Nour Eddine El Gueddari of the Department of Biology of University of Münster (Germany), who sadly and untimely passed away last 8 February.

Author Contributions: All authors contributed to writing and revising the manuscript.

Conflicts of Interest: The authors declare no conflict of interest.

Abbreviations

AADC	Aromatic L-amino acid decarboxylase
AAV	Adeno-associated virus
ADA-SCID	Adenosine deaminase-deficient severe combined immunodeficiency
AFM	Atomic force microscopy
ARSA	Deficiency of lysosomal arylsulfatase A
ASOs	Antisense oligonucleotides
cAMP	Cyclic adenosine monophosphate
CD	Circular dichroism
cDNA	Complementary deoxyribonucleic acid
CF	Cystic fibrosis
CFTR	Cystic fibrosis transmembrane conductance regulator
CLSM	Confocal laser scanning microscopy
CMV	Cytomegalovirus
CNS	Central nervous system
CPP	Cell penetrating peptide
CRISPR	Clustered regularly interspaced short palindromic repeats
CS	Chitosan
DA	Degree of acetylation
DHM	Digital holographic microscopy
DLS	Dynamic light scattering
DP	Degree of polymerization
DS	Dextran sulphate
dsRNA	Double-stranded ribonucleic acid
eGFP	Green fluorescent protein
ENaC	Epithelial sodium channel
GAN	Giant axonal neuropathy
HA	Hyaluronic acid
HDP	High degree of polymerization
HSC	Hematopoietic stem cells
ITC	Isothermal titration calorimetry
LDV	Laser Doppler velocimetry
LNA	Locked nucleic acids
LPLD	Familial lipoprotein lipase deficiency
LSD	Lysosomal storage diseases
miRNA	Micro ribonucleic acid
MLD	Methachromatic leukodistrophy
mRNA	Messenger ribonucleic acid
M_w	Molar mass
N/P	Amino to phosphate charge ratio
NPs	Nanoparticles
oc pDNA	Open chain plasmid deoxyribonucleic acid
PA	Pattern of acetylation
PAMPs	Pathogen-associated molecular patterns
PCD	Primary ciliary dyskinesia
PDE	Phosphodiesterase
pDNA	Plasmid deoxyribonucleic acid
PEI	Polyethylene imine
PGA	Poly-D-glutamic acid
PLL	Poly-Lysine

rAAV	Recombinant adeno-associated virus
rAD	Recombinant adenovirus
RGMC	Reduced generation of multiple motile cilia
RISC	RNA interference silencing complex
rLV	Recombinant lentivirus
RT-PCR	Real-time quantitative reverse transcription polymerase chain reaction
SAXS	Small angle X-ray scattering
SCID	Severe combined immunodeficiency
SeV	Murine parainfluenza virus type 1
siRNA	Small interfering ribonucleic acid
sc pDNA	Supercoiled plasmid deoxyribonucleic acid
SPR	Surface plasmon resonance
ssRNA	Single-stranded ribonucleic acid
TALEN	Transcription activator-like effector nucleases
TEM	Transmission electron microscopy
TLR	Toll-like receptor
TPP	Pentasodium tripolyphosphate
WAS	Wiskott-Aldrich syndrome
ZFN	Zinc finger nucleases

References

1. Tibbals, H.F. *Medical Nanotechnology and Nanomedicine*; CRC Press: Boca Raton, FL, USA, 2010; ISBN 1439808767.
2. Bigger, B.; Coutelle, C. Perspectives on Gene Therapy for Cystic Fibrosis Airway Disease. *BioDrugs* **2001**, *15*, 615–634. [CrossRef] [PubMed]
3. Coune, P.G.; Schneider, B.L.; Aebischer, P. Parkinson's Disease: Gene Therapies. *Cold Spring Harb. Perspect. Med.* **2012**, *2*, a009431. [CrossRef] [PubMed]
4. Azzouz, M.; Martin-Rendon, E.; Barber, R.D.; Mitrophanous, K.A.; Carter, E.E.; Rohll, J.B.; Kingsman, S.M.; Kingsman, A.J.; Mazarakis, N.D. Multicistronic Lentiviral Vector-Mediated Striatal Gene Transfer of Aromatic L-Amino Acid Decarboxylase, Tyrosine Hydroxylase, and GTP Cyclohydrolase I Induces Sustained Transgene Expression, Dopamine Production, and Functional Improvement in a Rat Model. *J. Neurosci.* **2002**, *22*, 10302–10312. [CrossRef] [PubMed]
5. Hanahan, D.; Weinberg, R.A. Hallmarks of Cancer: The Next Generation. *Cell* **2011**, *144*, 646–674. [CrossRef] [PubMed]
6. Doudna, J.A.; Charpentier, E. The new frontier of genome engineering with CRISPR-Cas9. *Science* **2014**, *346*, 1258096. [CrossRef] [PubMed]
7. Lin, S.; Staahl, B.T.; Alla, R.K.; Doudna, J.A. Enhanced homology-directed human genome engineering by controlled timing of CRISPR/Cas9 delivery. *eLife* **2014**, *3*, e04766. [CrossRef] [PubMed]
8. Yin, H.; Song, C.-Q.; Dorkin, J.R.; Zhu, L.J.; Li, Y.; Wu, Q.; Park, A.; Yang, J.; Suresh, S.; Bizhanova, A.; et al. Therapeutic genome editing by combined viral and non-viral delivery of CRISPR system components in vivo. *Nat. Biotechnol.* **2016**, *34*, 328–333. [CrossRef] [PubMed]
9. Lee, K.; Conboy, M.; Park, H.M.; Jiang, F.; Kim, H.J.; Dewitt, M.A.; Mackley, V.A.; Chang, K.; Rao, A.; Skinner, C.; et al. Nanoparticle delivery of Cas9 ribonucleoprotein and donor DNA in vivo induces homology-directed DNA repair. *Nat. Biomed. Eng.* **2017**, *1*, 889–901. [CrossRef]
10. Ragelle, H.; Vandermeulen, G.; Préat, V. Chitosan-based siRNA delivery systems. *J. Control. Release* **2013**, *172*, 207–218. [CrossRef] [PubMed]
11. Raemdonck, K.; Martens, T.F.; Braeckmans, K.; Demeester, J.; De Smedt, S.C. Polysaccharide-based nucleic acid nanoformulations. *Adv. Drug Deliv. Rev.* **2013**, *65*, 1123–1147. [CrossRef] [PubMed]
12. Buschmann, M.D.; Merzouki, A.; Lavertu, M.; Thibault, M.; Jean, M.; Darras, V. Chitosans for delivery of nucleic acids. *Adv. Drug Deliv. Rev.* **2013**, *65*, 1234–1270. [CrossRef] [PubMed]
13. Babu, A.; Munshi, A.; Ramesh, R. Combinatorial therapeutic approaches with RNAi and anticancer drugs using nanodrug delivery systems. *Drug Dev. Ind. Pharm.* **2017**, *43*, 1391–1401. [CrossRef] [PubMed]

14. Mokhtarzadeh, A.; Alibakhshi, A.; Hashemi, M.; Hejazi, M.; Hosseini, V.; de la Guardia, M.; Ramezani, M. Biodegradable nano-polymers as delivery vehicles for therapeutic small non-coding ribonucleic acids. *J. Control. Release* **2017**, *245*, 116–126. [CrossRef] [PubMed]

15. Gomes, C.P.; Ferreira Lopes, C.D.; Duarte Moreno, P.M.; Varela-Moreira, A.; Alonso, M.J.; Pêgo, A.P. Translating chitosan to clinical delivery of nucleic acid-based drugs. *MRS Bull.* **2014**, *39*, 60–70. [CrossRef]

16. Shi, B.; Zheng, M.; Tao, W.; Chung, R.; Jin, D.; Ghaffari, D.; Farokhzad, O.C. Challenges in DNA Delivery and Recent Advances in Multifunctional Polymeric DNA Delivery Systems. *Biomacromolecules* **2017**, *18*, 2231–2246. [CrossRef] [PubMed]

17. Kritchenkov, A.S.; Andranovitš, S.; Skorik, Y.A. Chitosan and its derivatives: Vectors in gene therapy. *Russ. Chem. Rev.* **2017**, *86*, 231–239. [CrossRef]

18. Rolland, A. Gene medicines: The end of the beginning? *Adv. Drug Deliv. Rev.* **2005**, *57*, 669–673. [CrossRef] [PubMed]

19. Anderson, W.F. Human gene therapy. *Nature* **1998**, *392*, 25–30. [CrossRef] [PubMed]

20. Friedman, T.; Roblin, R. Gene Therapy for Human Genetic Disease. *Science* **1972**, *175*, 949. [CrossRef]

21. Mahato, R.I.; Smith, L.C.; Rolland, A. Pharmaceutical perspectives of nonviral gene therapy. *Adv. Genet.* **1999**, *41*, 95–156. [PubMed]

22. Pouton, C.; Seymour, L. Key issues in non-viral gene delivery. *Adv. Drug Deliv. Rev.* **1998**, *34*, 3–19. [CrossRef]

23. Yin, H.; Kanasty, R.L.; Eltoukhy, A.A.; Vegas, A.J.; Dorkin, J.R.; Anderson, D.G. Non-viral vectors for gene-based therapy. *Nat. Rev. Genet.* **2014**, *15*, 541–555. [CrossRef] [PubMed]

24. Mintzer, M.A.; Simanek, E.E. Nonviral vectors for gene delivery. *Chem. Rev.* **2009**, *109*, 259–302. [CrossRef] [PubMed]

25. Ha, M.; Kim, V.N. Regulation of microRNA biogenesis. *Nat. Rev. Mol. Cell Biol.* **2014**, *15*, 509–524. [CrossRef] [PubMed]

26. Lundstrom, K.; Boulikas, T. Viral and Non-viral Vectors in Gene Therapy: Technology Development and Clinical Trials. *Technol. Cancer Res. Treat.* **2003**, *2*, 471–485. [CrossRef] [PubMed]

27. Konishi, M.; Kawamoto, K.; Izumikawa, M.; Kuriyama, H.; Yamashita, T. Gene transfer into guinea pig cochlea using adeno-associated virus vectors. *J. Gene Med.* **2008**, *10*, 610–618. [CrossRef] [PubMed]

28. Andtbacka, R.H.I.; Kaufman, H.L.; Collichio, F.; Amatruda, T.; Senzer, N.; Chesney, J.; Delman, K.A.; Spitler, L.E.; Puzanov, I.; Agarwala, S.S.; et al. Talimogene Laherparepvec Improves Durable Response Rate in Patients With Advanced Melanoma. *J. Clin. Oncol.* **2015**, *58*, 3377. [CrossRef] [PubMed]

29. Hill, A.B.; Chen, M.; Chen, C.-K.; Pfeifer, B.A.; Jones, C.H. Overcoming Gene-Delivery Hurdles: Physiological Considerations for Nonviral Vectors. *Trends Biotechnol.* **2016**, *34*, 91–105. [CrossRef] [PubMed]

30. Misra, S. Human gene therapy: A brief overview of the genetic revolution. *J. Assoc. Physicians India* **2013**, *61*, 127–133. [PubMed]

31. Schlenk, F.; Grund, S.; Fischer, D. Recent developments and perspectives on gene therapy using synthetic vectors. *Ther. Deliv.* **2013**, *4*, 95–113. [CrossRef] [PubMed]

32. Liu, F.; Huang, L. Development of non-viral vectors for systemic gene delivery. *J. Control. Release* **2002**, *78*, 259–266. [CrossRef]

33. Pack, D.W.; Hoffman, A.S.; Pun, S.; Stayton, P.S. Design and development of polymers for gene delivery. *Nat. Rev. Drug Discov.* **2005**, *4*, 581–593. [CrossRef] [PubMed]

34. Felgner, P.L.; Gadek, T.R.; Holm, M.; Roman, R.; Chan, H.W.; Wenz, M.; Northrop, J.P.; Ringold, G.M.; Danielsen, M. Lipofection: A highly efficient, lipid-mediated DNA-transfection procedure. *Proc. Natl. Acad. Sci. USA* **1987**, *84*, 7413–7417. [CrossRef] [PubMed]

35. De Jesus, M.B.; Zuhorn, I.S. Solid lipid nanoparticles as nucleic acid delivery system: Properties and molecular mechanisms. *J. Control. Release* **2015**, *201*, 1–13. [CrossRef] [PubMed]

36. Ozpolat, B.; Sood, A.K.; Lopez-Berestein, G. Liposomal siRNA nanocarriers for cancer therapy. *Adv. Drug Deliv. Rev.* **2014**, *66*, 110–116. [CrossRef] [PubMed]

37. McLendon, J.M.; Joshi, S.R.; Sparks, J.; Matar, M.; Fewell, J.G.; Abe, K.; Oka, M.; McMurtry, I.F.; Gerthoffer, W.T. Lipid nanoparticle delivery of a microRNA-145 inhibitor improves experimental pulmonary hypertension. *J. Control. Release* **2015**, *210*, 67–75. [CrossRef] [PubMed]

38. Serikawa, T.; Suzuki, N.; Kikuchi, H.; Tanaka, K.; Kitagawa, T. A new cationic liposome for efficient gene delivery with serum into cultured human cells: A quantitative analysis using two independent fluorescent probes. *Biochim. Biophys. Acta* **2000**, *1467*, 419–430. [CrossRef]

39. Stephan, D.J.; Yang, Z.Y.; San, H.; Simari, R.D.; Wheeler, C.J.; Felgner, P.L.; Gordon, D.; Nabel, G.J.; Nabel, E.G. A new cationic liposome DNA complex enhances the efficiency of arterial gene transfer in vivo. *Hum. Gene Ther.* **1996**, *7*, 1803–1812. [CrossRef] [PubMed]

40. Bochicchio, S.; Dalmoro, A.; Barba, A.; Grassi, G.; Lamberti, G. Liposomes as siRNA Delivery Vectors. *Curr. Drug Metab.* **2015**, *15*, 882–892. [CrossRef]

41. Berchel, M.; Le Gall, T.; Haelters, J.-P.; Lehn, P.; Montier, T.; Jaffrès, P.-A. Cationic lipophosphoramidates containing a hydroxylated polar headgroup for improving gene delivery. *Mol. Pharm.* **2015**, *12*, 1902–1910. [CrossRef] [PubMed]

42. Alhakamy, N.A.; Nigatu, A.S.; Berkland, C.J.; Ramsey, J.D. Noncovalently associated cell-penetrating peptides for gene delivery applications. *Ther. Deliv.* **2013**, *4*, 741–757. [CrossRef] [PubMed]

43. Torchilin, V.P. Cell penetrating peptide-modified pharmaceutical nanocarriers for intracellular drug and gene delivery. *Biopolymers* **2008**, *90*, 604–610. [CrossRef] [PubMed]

44. Margus, H.; Padari, K.; Pooga, M. Cell-penetrating peptides as versatile vehicles for oligonucleotide delivery. *Mol. Ther.* **2012**, *20*, 525–533. [CrossRef] [PubMed]

45. Munyendo, W.L.; Lv, H.; Benza-Ingoula, H.; Baraza, L.D.; Zhou, J. Cell penetrating peptides in the delivery of biopharmaceuticals. *Biomolecules* **2012**, *2*, 187–202. [CrossRef] [PubMed]

46. Kim, S.; Kim, D.; Cho, S.W.; Kim, J.; Kim, J.-S. Highly efficient RNA-guided genome editing in human cells via delivery of purified Cas9 ribonucleoproteins. *Genome Res.* **2014**, *24*, 1012–1019. [CrossRef] [PubMed]

47. MacLaughlin, F.C.; Mumper, R.J.; Wang, J.; Tagliaferri, J.M.; Gill, I.; Hinchcliffe, M.; Rolland, A.P. Chitosan and depolymerized chitosan oligomers as condensing carriers for in vivo plasmid delivery. *J. Control. Release* **1998**, *56*, 259–272. [CrossRef]

48. Liu, X.; Howard, K.A.; Dong, M.; Andersen, M.Ø.; Rahbek, U.L.; Johnsen, M.G.; Hansen, O.C.; Besenbacher, F.; Kjems, J. The influence of polymeric properties on chitosan/siRNA nanoparticle formulation and gene silencing. *Biomaterials* **2007**, *28*, 1280–1288. [CrossRef] [PubMed]

49. De Smedt, S.C.; Demeester, J.; Hennink, W.E. Cationic polymer based gene delivery systems. *Pharm. Res.* **2000**, *17*, 113–126. [CrossRef] [PubMed]

50. Matulis, D.; Rouzina, I.; Bloomfield, V.A. Thermodynamics of DNA binding and condensation: Isothermal titration calorimetry and electrostatic mechanism. *J. Mol. Biol.* **2000**, *296*, 1053–1063. [CrossRef] [PubMed]

51. Reitan, N.K.; Maurstad, G.; de Lange Davies, C.; Strand, S.P. Characterizing DNA condensation by structurally different chitosans of variable gene transfer efficacy. *Biomacromolecules* **2009**, *10*, 1508–1515. [CrossRef] [PubMed]

52. Scholz, C.; Wagner, E. Therapeutic plasmid DNA versus siRNA delivery: Common and different tasks for synthetic carriers. *J. Control. Release* **2012**, *161*, 554–565. [CrossRef] [PubMed]

53. Takakura, Y.; Nishikawa, M.; Yamashita, F.; Hashida, M. Influence of physicochemical properties on pharmacokinetics of non-viral vectors for gene delivery. *J. Drug Target.* **2002**, *10*, 99–104. [CrossRef] [PubMed]

54. Liang, W.; Lam, J.K.W. Endosomal Escape Pathways for Non-Viral Nucleic Acid Delivery Systems. In *Molecular Regulation of Endocytosis*; Ceresa, B., Ed.; InTech: Rijeka, Croatia, 2012; Chapter 17; ISBN 978-953-51-0662-3.

55. Urban-Klein, B.; Werth, S.; Abuharbeid, S.; Czubayko, F.; Aigner, A. RNAi-mediated gene-targeting through systemic application of polyethylenimine (PEI)-complexed siRNA in vivo. *Gene Ther.* **2005**, *12*, 461–466. [CrossRef] [PubMed]

56. Pfeifer, C.; Hasenpusch, G.; Uezguen, S.; Aneja, M.K.; Reinhardt, D.; Kirch, J.; Schneider, M.; Claus, S.; Friess, W.; Rudolph, C. Dry powder aerosols of polyethylenimine (PEI)-based gene vectors mediate efficient gene delivery to the lung. *J. Control. Release* **2011**, *154*, 69–76. [CrossRef] [PubMed]

57. Merkel, O.M.; Beyerle, A.; Librizzi, D.; Pfestroff, A.; Behr, T.M.; Sproat, B.; Barth, P.J.; Kissel, T. Nonviral siRNA delivery to the lung: Investigation of PEG-PEI polyplexes and their in vivo performance. *Mol. Pharm.* **2009**, *6*, 1246–1260. [CrossRef] [PubMed]

58. Spassova, C. Characterization of Peptide Dendrimers for DNA Delivery in Living Cells. Doctoral Thesis, Universität Heidelberg, Heidelberg, Germany, 2014. [CrossRef]

59. Lakshminarayanan, A.; Ravi, V.K.; Tatineni, R.; Rajesh, Y.B.R.D.; Maingi, V.; Vasu, K.S.; Madhusudhan, N.; Maiti, P.K.; Sood, A.K.; Das, S.; et al. Efficient dendrimer-DNA complexation and gene delivery vector properties of nitrogen-core poly(propyl ether imine) dendrimer in mammalian cells. *Bioconjug. Chem.* **2013**, *24*, 1612–1623. [CrossRef] [PubMed]

60. Wagner, E.; Kloeckner, J. Gene Delivery Using Polymer Therapeutics. In *Polymer Therapeutics I*; Satchi-Fainaro, R., Duncan, R., Eds.; Advances in Polymer Science; Springer: Berlin/Heidelberg, Germany, 2006; Volume 192, ISBN 978-3-540-29210-4.

61. Jiang, H.-L.; Kim, T.-H.; Kim, Y.-K.; Park, I.-Y.; Cho, M.-H.; Cho, C.-S. Efficient gene delivery using chitosan-polyethylenimine hybrid systems. *Biomed. Mater.* **2008**, *3*, 25013. [CrossRef] [PubMed]

62. Richard, I.; Thibault, M.; De Crescenzo, G.; Buschmann, M.D.; Lavertu, M. Ionization behavior of chitosan and chitosan-DNA polyplexes indicate that chitosan has a similar capability to induce a proton-sponge effect as PEI. *Biomacromolecules* **2013**, *14*, 1732–1740. [CrossRef] [PubMed]

63. Benjaminsen, R.V.; Mattebjerg, M.A.; Henriksen, J.R.; Moghimi, S.M.; Andresen, T.L. The Possible "Proton Sponge" Effect of Polyethylenimine (PEI) Does Not Include Change in Lysosomal pH. *Mol. Ther.* **2013**, *21*, 149–157. [CrossRef] [PubMed]

64. Mumper, R.; Wang, J.; Claspell, J.; Rolland, A. Novel polymeric condensing carriers for gene delivery. *Proc. Control. Release Soc.* **1995**, *22*, 178–179.

65. Mansouri, S.; Lavigne, P.; Corsi, K.; Benderdour, M.; Beaumont, E.; Fernandes, J.C. Chitosan-DNA nanoparticles as non-viral vectors in gene therapy: Strategies to improve transfection efficacy. *Eur. J. Pharm. Biopharm.* **2004**, *57*, 1–8. [CrossRef]

66. Liu, W.; Sun, S.; Cao, Z.; Zhang, X.; Yao, K.; Lu, W.W.; Luk, K.D.K. An investigation on the physicochemical properties of chitosan/DNA polyelectrolyte complexes. *Biomaterials* **2005**, *26*, 2705–2711. [CrossRef] [PubMed]

67. Katas, H.; Alpar, H.O. Development and characterisation of chitosan nanoparticles for siRNA delivery. *J. Control. Release* **2006**, *115*, 216–225. [CrossRef] [PubMed]

68. Harish Prashanth, K.V.; Tharanathan, R.N. Chitin/chitosan: Modifications and their unlimited application potential—An overview. *Trends Food Sci. Technol.* **2007**, *18*, 117–131. [CrossRef]

69. Mao, S.; Sun, W.; Kissel, T. Chitosan-based formulations for delivery of DNA and siRNA. *Adv. Drug Deliv. Rev.* **2010**, *62*, 12–27. [CrossRef] [PubMed]

70. Vauthier, C.; Zandanel, C.; Ramon, A.L. Chitosan-based nanoparticles for in vivo delivery of interfering agents including siRNA. *Curr. Opin. Colloid Interface Sci.* **2013**, *18*, 406–418. [CrossRef]

71. McKiernan, P.J.; Cunningham, O.; Greene, C.M.; Cryan, S.-A. Targeting miRNA-based medicines to cystic fibrosis airway epithelial cells using nanotechnology. *Int. J. Nanomed.* **2013**, *8*, 3907–3915. [CrossRef]

72. Köping-Höggård, M.; Tubulekas, I.; Guan, H.; Edwards, K.; Nilsson, M.; Vårum, K.M.; Artursson, P. Chitosan as a nonviral gene delivery system. Structure-property relationships and characteristics compared with polyethylenimine in vitro and after lung administration in vivo. *Gene Ther.* **2001**, *8*, 1108–1121. [CrossRef] [PubMed]

73. Huang, M.; Fong, C.-W.; Khor, E.; Lim, L.-Y. Transfection efficiency of chitosan vectors: Effect of polymer molecular weight and degree of deacetylation. *J. Control. Release* **2005**, *106*, 391–406. [CrossRef] [PubMed]

74. Lee, M.; Nah, J.W.; Kwon, Y.; Koh, J.J.; Ko, K.S.; Kim, S.W. Water-soluble and low molecular weight chitosan-based plasmid DNA delivery. *Pharm. Res.* **2001**, *18*, 427–431. [CrossRef] [PubMed]

75. Romøren, K.; Pedersen, S.; Smistad, G.; Evensen, Ø.; Thu, B.J. The influence of formulation variables on in vitro transfection efficiency and physicochemical properties of chitosan-based polyplexes. *Int. J. Pharm.* **2003**, *261*, 115–127. [CrossRef]

76. Weecharangsan, W.; Opanasopit, P.; Ngawhirunpat, T.; Apirakaramwong, A.; Rojanarata, T.; Ruktanonchai, U.; Lee, R.J. Evaluation of chitosan salts as non-viral gene vectors in CHO-K1 cells. *Int. J. Pharm.* **2008**, *348*, 161–168. [CrossRef] [PubMed]

77. Thibault, M.; Nimesh, S.; Lavertu, M.; Buschmann, M.D. Intracellular Trafficking and Decondensation Kinetics of Chitosan-pDNA Polyplexes. *Mol. Ther.* **2010**, *18*, 1787–1795. [CrossRef] [PubMed]

78. Strand, S.P.; Lelu, S.; Reitan, N.K.; de Lange Davies, C.; Artursson, P.; Vårum, K.M. Molecular design of chitosan gene delivery systems with an optimized balance between polyplex stability and polyplex unpacking. *Biomaterials* **2010**, *31*, 975–987. [CrossRef] [PubMed]

79. Gary, D.J.; Puri, N.; Won, Y.-Y. Polymer-based siRNA delivery: Perspectives on the fundamental and phenomenological distinctions from polymer-based DNA delivery. *J. Control. Release* **2007**, *121*, 64–73. [CrossRef] [PubMed]

80. Rudzinski, W.E.; Aminabhavi, T.M. Chitosan as a carrier for targeted delivery of small interfering RNA. *Int. J. Pharm.* **2010**, *399*, 1–11. [CrossRef] [PubMed]

81. Santos-Carballal, B.; Aaldering, L.J.; Ritzefeld, M.; Pereira, S.; Sewald, N.; Moerschbacher, B.M.; Götte, M.; Goycoolea, F.M. Physicochemical and biological characterization of chitosan-microRNA nanocomplexes for gene delivery to MCF-7 breast cancer cells. *Sci. Rep.* **2015**, *5*, 13567. [CrossRef] [PubMed]

82. Buschmann, M.D.; Lavertu, M.; Methot, S. Composition and Method for Efficient Delivery of Nucleic Acids to Cells Using Chitosan. US Application WO2007059605 A1, 31 May 2007.

83. Calvo, P.; Remuñán-López, C.; Vila-Jato, J.L.; Alonso, M.J. Novel hydrophilic chitosan-polyethylene oxide nanoparticles as protein carriers. *J. Appl. Polym. Sci.* **1997**, *63*, 125–132. [CrossRef]

84. Abdul Ghafoor Raja, M.; Katas, H.; Jing Wen, T. Stability, Intracellular Delivery, and Release of siRNA from Chitosan Nanoparticles Using Different Cross-Linkers. *PLoS ONE* **2015**, *10*, e0128963. [CrossRef] [PubMed]

85. Vimal, S.; Abdul Majeed, S.; Taju, G.; Nambi, K.S.N.; Sundar Raj, N.; Madan, N.; Farook, M.A.; Rajkumar, T.; Gopinath, D.; Sahul Hameed, A.S. Chitosan tripolyphosphate (CS/TPP) nanoparticles: Preparation, characterization and application for gene delivery in shrimp. *Acta Trop.* **2013**, *128*, 486–493. [CrossRef] [PubMed]

86. Rafiee, A.; Riazi-Rad, F.; Alimohammadian, M.H.; Gazori, T.; Fatemi, S.M.R.; Havaskary, M. Hydrogel nanoparticle encapsulated plasmid as a suitable gene delivery system. *TSitol. Genet.* **2015**, *49*, 16–20. [CrossRef]

87. De la Fuente, M.; Seijo, B.; Alonso, M.J. Novel hyaluronic acid-chitosan nanoparticles for ocular gene therapy. *Investig. Ophthalmol. Vis. Sci.* **2008**, *49*, 2016–2024. [CrossRef] [PubMed]

88. Contreras-Ruiz, L.; de la Fuente, M.; Párraga, J.E.; López-García, A.; Fernández, I.; Seijo, B.; Sánchez, A.; Calonge, M.; Diebold, Y. Intracellular trafficking of hyaluronic acid-chitosan oligomer-based nanoparticles in cultured human ocular surface cells. *Mol. Vis.* **2011**, *17*, 279–290. [PubMed]

89. Prasitsilp, M.; Jenwithisuk, R.; Kongsuwan, K.; Damrongchai, N.; Watts, P. Cellular responses to chitosan in vitro: The importance of deacetylation. *J. Mater. Sci. Mater. Med.* **2000**, *11*, 773–778. [CrossRef] [PubMed]

90. Lavertu, M.; Méthot, S.; Tran-Khanh, N.; Buschmann, M.D. High efficiency gene transfer using chitosan/DNA nanoparticles with specific combinations of molecular weight and degree of deacetylation. *Biomaterials* **2006**, *27*, 4815–4824. [CrossRef] [PubMed]

91. Kiang, T.; Wen, J.; Lim, H.W.; Leong, K.W. The effect of the degree of chitosan deacetylation on the efficiency of gene transfection. *Biomaterials* **2004**, *25*, 5293–5301. [CrossRef] [PubMed]

92. Tse, W.C.; Boger, D.L. A fluorescent intercalator displacement assay for establishing DNA binding selectivity and affinity. *Acc. Chem. Res.* **2004**, *37*, 61–69. [CrossRef] [PubMed]

93. Alameh, M.; Lavertu, M.; Tran-Khanh, N.; Chang, C.-Y.; Lesage, F.; Bail, M.; Darras, V.; Chevrier, A.; Buschmann, M.D. siRNA Delivery with Chitosan: Influence of Chitosan Molecular Weight, Degree of Deacetylation, and Amine to Phosphate Ratio on in Vitro Silencing Efficiency, Hemocompatibility, Biodistribution, and in Vivo Efficacy. *Biomacromolecules* **2018**, *19*, 112–131. [CrossRef] [PubMed]

94. Sato, T.; Ishii, T.; Okahata, Y. In vitro gene delivery mediated by chitosan. Effect of pH, serum, and molecular mass of chitosan on the transfection efficiency. *Biomaterials* **2001**, *22*, 2075–2080. [CrossRef]

95. Bordi, F.; Chronopoulou, L.; Palocci, C.; Bomboi, F.; Di Martino, A.; Cifani, N.; Pompili, B.; Ascenzioni, F.; Sennato, S. Chitosan–DNA complexes: Effect of molecular parameters on the efficiency of delivery. *Colloids Surf. A Physicochem. Eng. Asp.* **2014**, *460*, 184–190. [CrossRef]

96. Malhotra, M.; Kulamarva, A.; Sebak, S.; Paul, A.; Bhathena, J.; Mirzaei, M.; Prakash, S. Ultrafine chitosan nanoparticles as an efficient nucleic acid delivery system targeting neuronal cells. *Drug Dev. Ind. Pharm.* **2009**, *35*, 719–726. [CrossRef] [PubMed]

97. Csaba, N.; Köping-Höggård, M.; Alonso, M.J. Ionically crosslinked chitosan/tripolyphosphate nanoparticles for oligonucleotide and plasmid DNA delivery. *Int. J. Pharm.* **2009**, *382*, 205–214. [CrossRef] [PubMed]

98. Holzerny, P.; Ajdini, B.; Heusermann, W.; Bruno, K.; Schuleit, M.; Meinel, L.; Keller, M. Biophysical properties of chitosan/siRNA polyplexes: Profiling the polymer/siRNA interactions and bioactivity. *J. Control. Release* **2012**, *157*, 297–304. [CrossRef] [PubMed]

99. Shi, J.; Choi, J.L.; Chou, B.; Johnson, R.N.; Schellinger, J.G.; Pun, S.H. Effect of Polyplex Morphology on Cellular Uptake, Intracellular Trafficking, and Transgene Expression. *ACS Nano* **2013**, *7*, 10612–10620. [CrossRef] [PubMed]

100. Köping-Höggård, M.; Vårum, K.M.; Issa, M.; Danielsen, S.; Christensen, B.E.; Stokke, B.T.; Artursson, P. Improved chitosan-mediated gene delivery based on easily dissociated chitosan polyplexes of highly defined chitosan oligomers. *Gene Ther.* **2004**, *11*, 1441–1452. [CrossRef] [PubMed]

101. Bravo-Anaya, L.M.; Soltero, J.F.A.; Rinaudo, M. DNA/chitosan electrostatic complex. *Int. J. Biol. Macromol.* **2016**, *88*, 345–353. [CrossRef] [PubMed]

102. Al-Qadi, S.; Alatorre-Meda, M.; Zaghloul, E.M.; Taboada, P.; Remunán-López, C. Chitosan–hyaluronic acid nanoparticles for gene silencing: The role of hyaluronic acid on the nanoparticles' formation and activity. *Colloids Surf. B Biointerfaces* **2013**, *103*, 615–623. [CrossRef] [PubMed]

103. Teijeiro-Osorio, D.; Remuñán-López, C.; Alonso, M.J. Chitosan/cyclodextrin nanoparticles can efficiently transfect the airway epithelium in vitro. *Eur. J. Pharm. Biopharm.* **2009**, *71*, 257–263. [CrossRef] [PubMed]

104. Alatorre-Meda, M.; Taboada, P.; Hartl, F.; Wagner, T.; Freis, M.; Rodríguez, J.R. The influence of chitosan valence on the complexation and transfection of DNA: The weaker the DNA–chitosan binding the higher the transfection efficiency. *Colloids Surf. B Biointerfaces* **2011**, *82*, 54–62. [CrossRef] [PubMed]

105. Nydert, P.; Dragomir, A.; Hjelte, L. Chitosan as a carrier for non-viral gene transfer in a cystic-fibrosis cell line. *Biotechnol. Appl. Biochem.* **2008**, *51*, 153–157. [CrossRef] [PubMed]

106. Fernández Fernández, E.; Santos-Carballal, B.; Weber, W.-M.; Goycoolea, F.M. Chitosan as a non-viral co-transfection system in a cystic fibrosis cell line. *Int. J. Pharm.* **2016**, *502*, 1–9. [CrossRef] [PubMed]

107. Weber, W.-M.; Kolonko, A.K.; Fernández Fernández, E.; Santos-Carballal, B.; Goycoolea, F.M. Functional Restoring of Defect CFTR by Transfection of CFTR-mRNA Using Chitosan. *JSM Genet. Genom.* **2016**, *3*, 1016.

108. Fernández Fernández, E.; Santos-Carballal, B.; de Santi, C.; Ramsey, J.; MacLoughlin, R.; Cryan, S.-A.; Greene, C.M. Biopolymer-Based Nanoparticles for Cystic Fibrosis Lung Gene Therapy Studies. *Materials* **2018**, *11*, 122. [CrossRef] [PubMed]

109. Liu, C.; Zhu, Q.; Wu, W.; Xu, X.; Wang, X.; Gao, S.; Liu, K. Degradable copolymer based on amphiphilic N-octyl-N-quatenary chitosan and low-molecular weight polyethylenimine for gene delivery. *Int. J. Nanomed.* **2012**, *7*, 5339–5350. [CrossRef]

110. Schärtl, W. *Light Scattering from Polymer Solutions and Nanoparticle Dispersions*; Springer: Berlin/Heidelberg, Germany, 2007; ISBN 978-3-540-71950-2.

111. Lorber, B.; Fischer, F.; Bailly, M.; Roy, H.; Kern, D. Protein analysis by dynamic light scattering: Methods and techniques for students. *Biochem. Mol. Biol. Educ.* **2012**, *40*, 372–382. [CrossRef] [PubMed]

112. Fankhauser, F. Application of static and dynamic light scattering—A review. *Klin. Monbl. Augenheilkd.* **2010**, *227*, 194–198. [CrossRef] [PubMed]

113. Fernández Fernández, E.; Bangel-Ruland, N.; Tomczak, K.; Weber, W.-M. Optimization of CFTR-mRNA transfection in human nasal epithelial cells. *Transl. Med. Commun.* **2016**, *1*, 5. [CrossRef]

114. Amaduzzi, F.; Bomboi, F.; Bonincontro, A.; Bordi, F.; Casciardi, S.; Chronopoulou, L.; Diociaiuti, M.; Mura, F.; Palocci, C.; Sennato, S. Chitosan–DNA complexes: Charge inversion and DNA condensation. *Colloids Surf. B Biointerfaces* **2014**, *114*, 1–10. [CrossRef] [PubMed]

115. Sigle, W. Analytical transmission electron microscopy. *Annu. Rev. Mater. Res.* **2005**, *35*, 239–314. [CrossRef]

116. Williams, D.B.; Carter, C.B. Transmission electron microscopy: A textbook for materials science. In *Transmission Electron Microscopy: A Textbook for Materials Science*; Springer: Boston, MA, USA, 2009; pp. 1–760. ISBN 9780387765006.

117. Corsi, K.; Chellat, F.; Yahia, L.; Fernandes, J.C. Mesenchymal stem cells, MG63 and HEK293 transfection using chitosan-DNA nanoparticles. *Biomaterials* **2003**, *24*, 1255–1264. [CrossRef]

118. Erbacher, P.; Zou, S.; Bettinger, T.; Steffan, A.-M.; Remy, J.-S. Chitosan-Based Vector/DNA Complexes for Gene Delivery: Biophysical Characteristics and Transfection Ability. *Pharm. Res.* **1998**, *15*, 1332–1339. [CrossRef] [PubMed]

119. Rugar, D.; Hansma, P. Atomic Force Microscopy. *Phys. Today* **1990**, *43*, 23–30. [CrossRef]

120. Danielsen, S.; Varum, K.; Stokke, B. Structural Analysis of Chitosan Mediated DNA Condensation by AFM: Influence of Chitosan Molecular Parameters. *Biomacromolecules* **2004**, *5*, 928–936. [CrossRef] [PubMed]

121. Cadete, A.; Figueiredo, L.; Lopes, R.; Calado, C.C.R.; Almeida, A.J.; Gonçalves, L.M.D. Development and characterization of a new plasmid delivery system based on chitosan–sodium deoxycholate nanoparticles. *Eur. J. Pharm. Sci.* **2012**, *45*, 451–458. [CrossRef] [PubMed]

122. Homola, J.; Yee, S.S.; Gauglitz, G. Surface plasmon resonance sensors: Review. *Sens. Actuators B Chem.* **1999**, *54*, 3–15. [CrossRef]

123. Daghestani, H.N.; Day, B.W. Theory and applications of surface plasmon resonance, resonant mirror, resonant waveguide grating, and dual polarization interferometry biosensors. *Sensors* **2010**, *10*, 9630–9646. [CrossRef] [PubMed]

124. Ritzefeld, M.; Sewald, N. Real-Time Analysis of Specific Protein-DNA Interactions with Surface Plasmon Resonance. *J. Amino Acids* **2012**, *2012*, 816032. [CrossRef] [PubMed]

125. De-Mol, N.J.; Fischer, M.J.E. Kinetic and Thermodynamic Analysis of Ligand-receptor Interactions: SPR Applications in Drug Development. In *Handbook of Surface Plasmon Resonance*; Schasfoort, R.B.M., Tudos, A., Eds.; Royal Society of Chemistry: London, UK, 2008; pp. 123–172.

126. Homola, J. Surface plasmon resonance sensors for detection of chemical and biological species. *Chem. Rev.* **2008**, *108*, 462–493. [CrossRef] [PubMed]

127. Dubruel, P.; Urtti, A.; Schacht, E.H. Surface plasmon resonance as an elegant technique to study polyplex-GAG interactions. *J. Control. Release* **2006**, *116*, E77–E79. [CrossRef] [PubMed]

128. González-Fernández, E.; de-los-Santos-Álvarez, N.; Miranda-Ordieres, A.J.; Lobo-Castañón, M.J. SPR evaluation of binding kinetics and affinity study of modified RNA aptamers towards small molecules. *Talanta* **2012**, *99*, 767–773. [CrossRef] [PubMed]

129. Ma, P.L.; Lavertu, M.; Winnik, F.M.; Buschmann, M.D. New Insights into Chitosan-DNA Interactions Using Isothermal Titration Microcalorimetry. *Biomacromolecules* **2009**, *10*, 1490–1499. [CrossRef] [PubMed]

130. Eggleston, A.K.; Rahim, N.A.; Kowalczykowski, S.C. A Helicase Assay Based on the Displacement of Fluorescent, Nucleic Acid-Binding Ligands. *Nucleic Acids Res.* **1996**, *24*, 1179–1186. [CrossRef] [PubMed]

131. Tuma, R.S.; Beaudet, M.P.; Jin, X.; Jones, L.J.; Cheung, C.Y.; Yue, S.; Singer, V.L. Characterization of SYBR Gold nucleic acid gel stain: A dye optimized for use with 300-nm ultraviolet transilluminators. *Anal. Biochem.* **1999**, *268*, 278–288. [CrossRef] [PubMed]

132. Wiethoff, C.M.; Gill, M.L.; Koe, G.S.; Koe, J.G.; Middaugh, C.R. A fluorescence study of the structure and accessibility of plasmid DNA condensed with cationic gene delivery vehicles. *J. Pharm. Sci.* **2003**, *92*, 1272–1285. [CrossRef] [PubMed]

133. Santos-Carballal, B.; Swamy, M.J.; Moerschbacher, B.M.; Goycoolea, F.M.; Swamy, M.J.; Goycoolea, F.M. SYBR Gold fluorescence quenching is a sensitive probe of chitosan-microRNA interactions. *J. Fluoresc.* **2015**, *26*. [CrossRef] [PubMed]

134. Santos, J.C.C.; Moreno, P.M.D.; Mansur, A.A.P.; Leiro, V.; Mansur, H.S.; Pêgo, A.P. Functionalized chitosan derivatives as nonviral vectors: Physicochemical properties of acylated *N,N,N*-trimethyl chitosan/oligonucleotide nanopolyplexes. *Soft Matter* **2015**, *11*, 8113–8125. [CrossRef] [PubMed]

135. Kypr, J.; Kejnovská, I.; Renciuk, D.; Vorlíčková, M. Circular dichroism and conformational polymorphism of DNA. *Nucleic Acids Res.* **2009**, *37*, 1713–1725. [CrossRef] [PubMed]

136. Ranjbar, B.; Gill, P. Circular Dichroism Techniques: Biomolecular and Nanostructural Analyses—A Review. *Chem. Biol. Drug Des.* **2009**, *74*, 101–120. [CrossRef] [PubMed]

137. Vorlíčková, M.; Kejnovská, I.; Bednářová, K.; Renčiuk, D.; Kypr, J. Circular dichroism spectroscopy of DNA: From duplexes to quadruplexes. *Chirality* **2012**, *24*, 691–698. [CrossRef] [PubMed]

138. Chang, Y.-M.; Chen, C.K.-M.; Hou, M.-H. Conformational Changes in DNA upon Ligand Binding Monitored by Circular Dichroism. *Int. J. Mol. Sci.* **2012**, *13*, 3394–3413. [CrossRef] [PubMed]

139. Zhang, L.W.; Monteiro-Riviere, N.A. Use of confocal microscopy for nanoparticle drug delivery through skin. *J. Biomed. Opt.* **2013**, *18*, 61214. [CrossRef] [PubMed]

140. Li, Q.; Liu, W.; Dai, J.; Zhang, C. Synthesis of polysaccharide-block-polypeptide copolymer for potential co-delivery of drug and plasmid DNA. *Macromol. Biosci.* **2015**, *15*, 756–764. [CrossRef] [PubMed]

141. Marquet, P.; Rappaz, B.; Magistretti, P.J.; Cuche, E.; Emery, Y.; Colomb, T.; Depeursinge, C. Digital holographic microscopy: A noninvasive contrast imaging technique allowing quantitative visualization of living cells with subwavelength axial accuracy. *Opt. Lett.* **2005**, *30*, 468. [CrossRef] [PubMed]

142. Kim, M.K. Principles and techniques of digital holographic microscopy. *J. Photonics Energy* **2010**, 018005. [CrossRef]

143. Rappaz, B.; Breton, B.; Shaffer, E.; Turcatti, G. Digital holographic microscopy: A quantitative label-free microscopy technique for phenotypic screening. *Comb. Chem. High Throughput Screen.* **2014**, *17*, 80–88. [CrossRef] [PubMed]

144. Bettenworth, D.; Lenz, P.; Krausewitz, P.; Brückner, M.; Ketelhut, S.; Domagk, D.; Kemper, B. Quantitative stain-free and continuous multimodal monitoring of wound healing in vitro with digital holographic microscopy. *PLoS ONE* **2014**, *9*, e107317. [CrossRef] [PubMed]

145. Freeman, W.M.; Walker, S.J.; Vrana, K.E. Quantitative RT-PCR: Pitfalls and potential. *Biotechniques* **1999**, *26*, 112–125. [PubMed]

146. Bustin, S. Quantification of mRNA using real-time reverse transcription PCR (RT-PCR): Trends and problems. *J. Mol. Endocrinol.* **2002**, *29*, 23–39. [CrossRef] [PubMed]

147. Huggett, J.; Dheda, K.; Bustin, S.; Zumla, A. Real-time RT-PCR normalisation; strategies and considerations. *Genes Immun.* **2005**, *6*, 279–284. [CrossRef] [PubMed]

148. Ingolotti, M.; Kawalekar, O.; Shedlock, D.J.; Muthumani, K.; Weiner, D.B. DNA vaccines for targeting bacterial infections. *Expert Rev. Vaccines* **2010**, *9*, 747–763. [CrossRef] [PubMed]

149. USDHHS/FDA/CBER. *Guidance for Industry: Considerations for Plasmid DNA Vaccines for Infectious Disease Indications*; Food and Drug Administration: Rockville, MD, USA, 2007. Available online: http://www.fda.gov/cber/guidelines.htm. (accessed on 15 April 2018).

150. Remaut, K.; Sanders, N.N.; Fayazpour, F.; Demeester, J.; De Smedt, S.C. Influence of plasmid DNA topology on the transfection properties of DOTAP/DOPE lipoplexes. *J. Control. Release* **2006**, *115*, 335–343. [CrossRef] [PubMed]

151. Gaspar, V.M.; Correia, I.J.; Sousa, Â.; Silva, F.; Paquete, C.M.; Queiroz, J.A.; Sousa, F. Nanoparticle mediated delivery of pure P53 supercoiled plasmid DNA for gene therapy. *J. Control. Release* **2011**, *156*, 212–222. [CrossRef] [PubMed]

152. Roy, K.; Mao, H.-Q.; Huang, S.-K.; Leong, K.W. Oral gene delivery with chitosan–DNA nanoparticles generates immunologic protection in a murine model of peanut allergy. *Nat. Med.* **1999**, *5*, 387–391. [CrossRef] [PubMed]

153. Agrawal, N.; Dasaradhi, P.V.N.; Mohmmed, A.; Malhotra, P.; Bhatnagar, R.K.; Mukherjee, S.K. RNA interference: Biology, mechanism, and applications. *Microbiol. Mol. Biol. Rev.* **2003**, *67*, 657–685. [CrossRef] [PubMed]

154. Martin, S.E.; Caplen, N.J. Applications of RNA interference in mammalian systems. *Annu. Rev. Genom. Hum. Genet.* **2007**, *8*, 81–108. [CrossRef] [PubMed]

155. Carthew, R.W.; Sontheimer, E.J. Origins and Mechanisms of miRNAs and siRNAs. *Cell* **2009**, *136*, 642–655. [CrossRef] [PubMed]

156. Esquela-Kerscher, A.; Slack, F.J. Oncomirs—MicroRNAs with a role in cancer. *Nat. Rev. Cancer* **2006**, *6*, 259–269. [CrossRef] [PubMed]

157. Acunzo, M.; Romano, G.; Wernicke, D.; Croce, C.M. MicroRNA and cancer—A brief overview. *Adv. Biol. Regul.* **2014**, *57*, 1–9. [CrossRef] [PubMed]

158. Jansson, M.D.; Lund, A.H. MicroRNA and cancer. *Mol. Oncol.* **2012**, *6*, 590–610. [CrossRef] [PubMed]

159. Lee, R.C.; Feinbaum, R.L.; Ambros, V. The *C. elegans* heterochronic gene lin-4 encodes small RNAs with antisense complementarity to lin-14. *Cell* **1993**, *75*, 843–854. [CrossRef]

160. Reinhart, B.J.; Slack, F.J.; Basson, M.; Pasquinelli, A.E.; Bettinger, J.C.; Rougvie, A.E.; Horvitz, H.R.; Ruvkun, G. The 21-nucleotide let-7 RNA regulates developmental timing in *Caenorhabditis elegans*. *Nature* **2000**, *403*, 901–906. [CrossRef] [PubMed]

161. Bartel, D.P. MicroRNAs: Target Recognition and Regulatory Functions. *Cell* **2009**, *136*, 215–233. [CrossRef] [PubMed]

162. Calin, G.A.; Dumitru, C.D.; Shimizu, M.; Bichi, R.; Zupo, S.; Noch, E.; Aldler, H.; Rattan, S.; Keating, M.; Rai, K.; et al. Frequent deletions and down-regulation of micro-RNA genes miR15 and miR16 at 13q14 in chronic lymphocytic leukemia. *Proc. Natl. Acad. Sci. USA* **2002**, *99*, 15524–15529. [CrossRef] [PubMed]

163. Götte, M.; Mohr, C.; Koo, C.-Y.; Stock, C.; Vaske, A.-K.; Viola, M.; Ibrahim, S.A.; Peddibhotla, S.; Teng, Y.H.-F.; Low, J.-Y.; et al. MiR-145-dependent targeting of Junctional Adhesion Molecule A and modulation of fascin expression are associated with reduced breast cancer cell motility and invasiveness. *Oncogene* **2010**, *29*, 6569–6580. [CrossRef] [PubMed]

164. Adammek, M.; Greve, B.; Kässens, N.; Schneider, C.; Brüggemann, K.; Schüring, A.N.; Starzinski-Powitz, A.; Kiesel, L.; Götte, M. MicroRNA miR-145 inhibits proliferation, invasiveness, and stem cell phenotype of an in vitro endometriosis model by targeting multiple cytoskeletal elements and pluripotency factors. *Fertil. Steril.* **2013**, *99*, 1346–1355.e5. [CrossRef] [PubMed]

165. Zhang, B.; Pan, X.; Cobb, G.P.; Anderson, T.A. microRNAs as oncogenes and tumor suppressors. *Dev. Biol.* **2007**, *302*, 1–12. [CrossRef] [PubMed]

166. Miska, E.A. How microRNAs control cell division, differentiation and death. *Curr. Opin. Genet. Dev.* **2005**, *15*, 563–568. [CrossRef] [PubMed]

167. Lu, J.; Getz, G.; Miska, E.A.; Alvarez-Saavedra, E.; Lamb, J.; Peck, D.; Sweet-Cordero, A.; Ebert, B.L.; Mak, R.H.; Ferrando, A.A.; et al. MicroRNA expression profiles classify human cancers. *Nature* **2005**, *435*, 834–838. [CrossRef] [PubMed]

168. Kent, O.A.; Mendell, J.T. A small piece in the cancer puzzle: MicroRNAs as tumor suppressors and oncogenes. *Oncogene* **2006**, *25*, 6188–6196. [CrossRef] [PubMed]

169. Shenouda, S.K.; Alahari, S.K. MicroRNA function in cancer: Oncogene or a tumor suppressor? *Cancer Metastasis Rev.* **2009**, *28*, 369–378. [CrossRef] [PubMed]

170. Wang, D.; Qiu, C.; Zhang, H.; Wang, J.; Cui, Q.; Yin, Y. Human microRNA oncogenes and tumor suppressors show significantly different biological patterns: From functions to targets. *PLoS ONE* **2010**, *5*. [CrossRef] [PubMed]

171. Zhao, M.; Wei, D.-Q. Rare Diseases: Drug Discovery and Informatics Resource. *Interdiscip. Sci. Comput. Life Sci.* **2017**, 1–10. [CrossRef] [PubMed]

172. Melnikova, I. Rare diseases and orphan drugs. *Nat. Rev. Drug Discov.* **2012**, *11*, 267–268. [CrossRef] [PubMed]

173. Sun, W.; Zheng, W.; Simeonov, A. Drug discovery and development for rare genetic disorders. *Am. J. Med. Genet. Part A* **2017**, *173*, 2307–2322. [CrossRef] [PubMed]

174. Riordan, J.R.; Rommens, J.M.; Kerem, B.; Alon, N.; Rozmahel, R.; Grzelczak, Z.; Zielenski, J.; Lok, S.; Plavsik, N.; Chou, J.L.; et al. Identification of the cystic fibrosis gene: Cloning and characterization of complementary DNA. *Science* **1989**, *245*, 1066–1073. [CrossRef] [PubMed]

175. Kerem, B.; Rommens, J.M.; Buchanan, J.A.; Markiewicz, D.; Cox, T.K.; Chakravarti, A.; Buchwald, M.; Tsui, L.C. Identification of the cystic fibrosis gene: Genetic analysis. *Science* **1989**, *245*, 1073–1080. [CrossRef] [PubMed]

176. Welsh, M.J.; Smith, A.E. Molecular mechanisms of CFTR chloride channel dysfunction in cystic fibrosis. *Cell* **1993**, *73*, 1251–1254. [CrossRef]

177. Elborn, J.S. Cystic fibrosis. *Lancet* **2016**, *388*, 2519–2531. [CrossRef]

178. Briel, M.; Greger, R.; Kunzelmann, K. Cl⁻ transport by cystic fibrosis transmembrane conductance regulator (CFTR) contributes to the inhibition of epithelial Na⁺ channels (ENaCs) in *Xenopus* oocytes co-expressing CFTR and ENaC. *J. Physiol.* **1998**, *508*, 825–836. [CrossRef] [PubMed]

179. Hartl, D.; Gaggar, A.; Bruscia, E.; Hector, A.; Marcos, V.; Jung, A.; Greene, C.; McElvaney, G.; Mall, M.; Döring, G. Innate immunity in cystic fibrosis lung disease. *J. Cyst. Fibros.* **2012**, *11*, 363–382. [CrossRef] [PubMed]

180. Giglio, L.; Candusso, M.; D'Orazio, C.; Mastella, G.; Faraguna, D. Failure to thrive: The earliest feature of cystic fibrosis in infants diagnosed by neonatal screening. *Acta Paediatr.* **1997**, *86*, 1162–1165. [CrossRef] [PubMed]

181. Di Sant Agnese, P.A. Cystic Fibrosis of the pancreas. *Am. J. Med.* **1956**, *21*, 406–422. [CrossRef]

182. Strandvik, B. Fatty acid metabolism in cystic fibrosis. *Prostaglandins. Leukot. Essent. Fat. Acids* **2010**, *83*, 121–129. [CrossRef] [PubMed]

183. Maqbool, A.; Stallings, V.A. Update on fat-soluble vitamins in cystic fibrosis. *Curr. Opin. Pulm. Med.* **2008**, *14*, 574–581. [CrossRef] [PubMed]

184. Baker, S.S.; Borowitz, D.; Duffy, L.; Fitzpatrick, L.; Gyamfi, J.; Baker, R.D. Pancreatic enzyme therapy and clinical outcomes in patients with cystic fibrosis. *J. Pediatr.* **2005**, *146*, 189–193. [CrossRef] [PubMed]

185. Barrio, R. Management of endocrine disease: Cystic fibrosis-related diabetes: Novel pathogenic insights opening new therapeutic avenues. *Eur. J. Endocrinol.* **2015**, *172*, R131–R141. [CrossRef] [PubMed]

186. Jarzabek, K.; Zbucka, M.; Pepinski, W.; Szamatowicz, J.; Domitrz, J.; Janica, J.; Wolczynski, S.; Szamatowicz, M.; Pepiński, W.; Szamatowicz, J.; et al. Cystic fibrosis as a cause of infertility. *Reprod. Biol.* **2004**, *4*, 119–129. [PubMed]

187. Ratjen, F.; Döring, G. Cystic fibrosis. *Lancet* **2003**, *361*, 681–689. [CrossRef]

188. Davis, P.B. Cystic Fibrosis since 1938. *Am. J. Respir. Crit. Care Med.* **2006**, *173*, 475–482. [CrossRef] [PubMed]

189. Cantin, A.M. Cystic Fibrosis Transmembrane Conductance Regulator. Implications in Cystic Fibrosis and Chronic Obstructive Pulmonary Disease. *Ann. Am. Thorac. Soc.* **2016**, *13*, S150–S155. [CrossRef] [PubMed]

190. Boucher, R.C. Evidence for airway surface dehydration as the initiating event in CF airway disease. *J. Intern. Med.* **2007**, *261*, 5–16. [CrossRef] [PubMed]

191. Hoenderdos, K.; Condliffe, A. The neutrophil in chronic obstructive pulmonary disease. *Am. J. Respir. Cell Mol. Biol.* **2013**, *48*, 531–539. [CrossRef] [PubMed]

192. Zielenski, J.; Bozon, D.; Kerem, B.; Markiewicz, D.; Durie, P.; Rommens, J.M.; Tsui, L.C. Identification of mutations in exons 1 through 8 of the cystic fibrosis transmembrane conductance regulator (CFTR) gene. *Genomics* **1991**, *10*, 229–235. [CrossRef]

193. Rommens, J.M.; Iannuzzi, M.C.; Kerem, B.; Drumm, M.L.; Melmer, G.; Dean, M.; Rozmahel, R.; Cole, J.L.; Kennedy, D.; Hidaka, N. Identification of the cystic fibrosis gene: Chromosome walking and jumping. *Science* **1989**, *245*, 1059–1065. [CrossRef] [PubMed]

194. Dean, M.; Hamon, Y.; Chimini, G. The human ATP-binding cassette (ABC) transporter superfamily. *J. Lipid Res.* **2001**, *42*, 1007–1017. [CrossRef] [PubMed]

195. Sheppard, D.N.; Welsh, M.J. Structure and function of the CFTR chloride channel. *Physiol. Rev.* **1999**, *79*, S23–S45. [CrossRef] [PubMed]

196. Riordan, J.R. Assembly of functional CFTR chloride channels. *Annu. Physiol.* **2005**, *67*, 701–718. [CrossRef] [PubMed]

197. Csanady, L.; Seto-Young, D.; Chan, K.W.; Cenciarelli, C.; Angel, B.B.; Qin, J.; McLachlin, D.T.; Krutchinsky, A.N.; Chait, B.T.; Nairn, A.C.; et al. Preferential phosphorylation of R-domain Serine 768 dampens activation of CFTR channels by PKA. *J. Gen. Physiol.* **2005**, *125*, 171–186. [CrossRef] [PubMed]

198. Venerando, A.; Franchin, C.; Cant, N.; Cozza, G.; Pagano, M.A.; Tosoni, K.; Al-Zahrani, A.; Arrigoni, G.; Ford, R.C.; Mehta, A.; et al. Detection of phospho-sites generated by protein kinase CK2 in CFTR: Mechanistic aspects of Thr1471 phosphorylation. *PLoS. ONE* **2013**, *8*, e74232. [CrossRef] [PubMed]

199. Bozoky, Z.; Krzeminski, M.; Muhandiram, R.; Birtley, J.R.; Al-Zahrani, A.; Thomas, P.J.; Frizzell, R.A.; Ford, R.C.; Forman-Kay, J.D. Regulatory R region of the CFTR chloride channel is a dynamic integrator of phospho-dependent intra- and intermolecular interactions. *Proc. Natl. Acad. Sci. USA* **2013**, *110*, E4427–E4436. [CrossRef] [PubMed]

200. Cystic Fibrosis Mutation Database: Statistics. 2015. Available online: Genet.sickkids.on.ca (accessed on 15 January 2018).

201. Boeck, D.; Amaral, M. Classification of CFTR mutation classes. *Lancet Respir. Med.* **2016**. [CrossRef]

202. Ashlock, M.A.; Olson, E.R. Therapeutics development for cystic fibrosis: A successful model for a multisystem genetic disease. *Annu. Med.* **2011**, *62*, 107–125. [CrossRef] [PubMed]

203. Matel, J.L.; Milla, C.E. Nutrition in cystic fibrosis. *Semin. Crit. Care Med.* **2009**, *30*, 579–586. [CrossRef] [PubMed]

204. Corvol, H.; Thompson, K.E.; Tabary, O.; le Rouzic, P.; Guillot, L. Translating the genetics of cystic fibrosis to personalized medicine. *Transl. Res.* **2015**. [CrossRef] [PubMed]

205. Wilschanski, M.; Famini, C.; Blau, H.; Rivlin, J.; Augarten, A.; Avital, A.; Kerem, B.; Kerem, E. A pilot study of the effect of gentamicin on nasal potential difference measurements in cystic fibrosis patients carrying stop mutations. *Am. J. Respir. Crit. Care Med.* **2010**, *161*, 860–865. [CrossRef] [PubMed]

206. Wilschanski, M.; Yahav, Y.; Yaacov, Y.; Blau, H.; Bentur, L.; Rivlin, J.; Aviram, M.; Bdolah-Abram, T.; Bebok, Z.; Shushi, L.; et al. Gentamicin-induced correction of CFTR function in patients with cystic fibrosis and CFTR stop mutations. *N. Engl. J. Med.* **2003**, *349*, 1433–1441. [CrossRef] [PubMed]

207. Sermet-Gaudelus, I.; De Boeck, K.; Casimir, G.J.; Vermeulen, F.; Leal, T.; Mogenet, A.; Roussel, D.; Fritsch, J.; Hanssens, L.; Hirawat, S.; et al. Ataluren (PTC124) induces cystic fibrosis transmembrane conductance regulator protein expression and activity in children with nonsense mutation cystic fibrosis. *Am. J. Respir. Crit. Care Med.* **2010**, *182*, 1262–1272. [CrossRef] [PubMed]

208. Kreindler, J.L. Cystic fibrosis: Exploiting its genetic basis in the hunt for new therapies. *Pharmacol. Ther.* **2010**, *125*, 219–229. [CrossRef] [PubMed]

209. Van, G.F.; Hadida, S.; Grootenhuis, P.D.; Burton, B.; Stack, J.H.; Straley, K.S.; Decker, C.J.; Miller, M.; McCartney, J.; Olson, E.R.; et al. Correction of the F508del-CFTR protein processing defect in vitro by the investigational drug VX-809. *Proc. Natl. Acad. Sci. USA* **2011**, *108*, 18843–18848.

210. Hwang, T.-C.; Sheppard, D.N. Molecular pharmacology of the CFTR Cl^- channel. *Trends Pharmacol. Sci.* **1999**, *20*, 448–453. [CrossRef]

211. Yu, H.; Burton, B.; Huang, C.J.; Worley, J.; Cao, D.; Johnson, J.; Urrutia, A.; Joubran, J.; Seepersaud, S.; Sussky, K.; et al. Ivacaftor potentiation of multiple CFTR channels with gating mutations. *J. Cyst. Fibros.* **2012**, *11*, 237–245. [CrossRef] [PubMed]

212. Wainwright, C.E.; Elborn, J.S.; Ramsey, B.W.; Marigowda, G.; Huang, X.; Cipolli, M.; Colombo, C.; Davies, J.C.; De Boeck, K.; Flume, P.A.; et al. Lumacaftor–Ivacaftor in Patients with Cystic Fibrosis Homozygous for Phe508del CFTR. *N. Engl. J. Med.* **2015**, *373*, 220–231. [CrossRef] [PubMed]

213. Leier, G.; Bangel-Ruland, N.; Sobczak, K.; Knieper, Y.; Weber, W.M. Sildenafil acts as potentiator and corrector of CFTR but might be not suitable for the treatment of CF lung disease. *Cell Physiol. Biochem.* **2012**, *29*, 775–790. [CrossRef] [PubMed]

214. Noone, P.G.; Regnis, J.A.; Liu, X.; Brouwer, K.L.; Robinson, M.; Edwards, L.; Knowles, M.R. Airway deposition and clearance and systemic pharmacokinetics of amiloride following aerosolization with an ultrasonic nebulizer to normal airways. *Chest* **1997**, *112*, 1283–1290. [CrossRef] [PubMed]

215. Hirsh, A.J.; Sabater, J.R.; Zamurs, A.; Smith, R.T.; Paradiso, A.M.; Hopkins, S.; Abraham, W.M.; Boucher, R.C. Evaluation of second generation amiloride analogs as therapy for cystic fibrosis lung disease. *J. Pharmacol. Exp. Ther.* **2004**, *311*, 929–938. [CrossRef] [PubMed]

216. Sobczak, K.; Segal, A.; Bangel-Ruland, N.; Semmler, J.; Van Driessche, W.; Lindemann, H.; Heermann, R.; Weber, W.-M. Specific inhibition of epithelial Na$^+$ channels by antisense oligonucleotides for the treatment of Na$^+$ hyperabsorption in cystic fibrosis. *J. Gene Med.* **2009**, *11*, 813–823. [CrossRef] [PubMed]

217. Reeves, E.P.; Molloy, K.; Pohl, K.; McElvaney, N.G. Hypertonic saline in treatment of pulmonary disease in cystic fibrosis. *Sci. World J.* **2012**, *2012*, 465230. [CrossRef] [PubMed]

218. Elkins, M.R.; Robinson, M.; Rose, B.R.; Harbour, C.; Moriarty, C.P.; Marks, G.B.; Belousova, E.G.; Xuan, W.; Bye, P.T. A controlled trial of long-term inhaled hypertonic saline in patients with cystic fibrosis. *N. Engl. J. Med.* **2006**, *354*, 229–240. [CrossRef] [PubMed]

219. Griesenbach, U.; Alton, E.W.F.W. Recent advances in understanding and managing cystic fibrosis transmembrane conductance regulator dysfunction. *F1000Prime Rep.* **2015**, *7*, 64. [CrossRef] [PubMed]

220. Jones, C.H.; Chen, C.K.; Ravikrishnan, A.; Rane, S.; Pfeifer, B.A. Overcoming nonviral gene delivery barriers: Perspective and future. *Mol. Pharm.* **2013**, *10*, 4082–4098. [CrossRef] [PubMed]

221. Clancy, J.P.; Jain, M. Personalized medicine in cystic fibrosis: Dawning of a new era. *Am. J. Respir. Crit. Care Med.* **2012**, *186*, 593–597. [CrossRef] [PubMed]

222. Gaj, T.; Gersbach, C.A.; Barbas, C.F. ZFN, TALEN, and CRISPR/Cas-based methods for genome engineering. *Trends Biotechnol.* **2013**, *31*, 397–405. [CrossRef] [PubMed]

223. Ikpa, P.T.; Bijvelds, M.J.C.; De Jonge, H.R. Cystic fibrosis: Toward personalized therapies. *Int. J. Biochem. Cell Biol.* **2014**, *52*, 192–200. [CrossRef] [PubMed]

224. Oakland, M.; Sinn, P.L.; McCray, P.B., Jr. Advances in Cell and Gene-based Therapies for Cystic Fibrosis Lung Disease. *Mol. Ther.* **2012**, *20*, 1108–1115. [CrossRef] [PubMed]

225. Conese, M.; Ascenzioni, F.; Boyd, A.C.; Coutelle, C.; De Fino, I.; De Smedt, S.; Rejman, J.; Rosenecker, J.; Schindelhauer, D.; Scholte, B.J. Gene and cell therapy for cystic fibrosis: From bench to bedside. *J. Cyst. Fibros.* **2011**, *10* (Suppl. 2), S114–S128. [CrossRef]

226. Griesenbach, U.; Alton, E.W. Expert opinion in biological therapy: Update on developments in lung gene transfer. *Expert Opin. Biol. Ther.* **2013**, *13*, 345–360. [CrossRef] [PubMed]

227. Prickett, M.; Jain, M. Gene therapy in cystic fibrosis. *Transl. Res.* **2013**, *161*, 255–264. [CrossRef] [PubMed]

228. Griesenbach, U.; Alton, E.W. Progress in gene and cell therapy for cystic fibrosis lung disease. *Curr. Pharm. Des.* **2012**, *18*, 642–662. [CrossRef] [PubMed]

229. Ferrari, S.; Griesenbach, U.; Shiraki-Iida, T.; Shu, T.; Hironaka, T.; Hou, X.; Williams, J.; Zhu, J.; Jeffery, P.K.; Geddes, D.M.; et al. A defective nontransmissible recombinant Sendai virus mediates efficient gene transfer to airway epithelium in vivo. *Gene Ther.* **2004**, *11*, 1659–1664. [CrossRef] [PubMed]

230. Zhang, L.; Peeples, M.E.; Boucher, R.C.; Collins, P.L.; Pickles, R.J. Respiratory Syncytial Virus Infection of Human Airway Epithelial Cells Is Polarized, Specific to Ciliated Cells, and without Obvious Cytopathology. *J. Virol.* **2002**, *76*, 5654–5666. [CrossRef] [PubMed]

231. Griesenbach, U.; Inoue, M.; Hasegawa, M.; Alton, E.W. Sendai virus for gene therapy and vaccination. *Curr. Opin. Mol. Ther.* **2005**, *7*, 346–352. [PubMed]

232. Griesenbach, U.; McLachlan, G.; Owaki, T.; Somerton, L.; Shu, T.; Baker, A.; Tennant, P.; Gordon, C.; Vrettou, C.; Baker, E.; et al. Validation of recombinant Sendai virus in a non-natural host model. *Gene Ther.* **2011**, *18*, 182–188. [CrossRef] [PubMed]

233. Alton, E.W.F.W.; Armstrong, D.K.; Ashby, D.; Bayfield, K.J.; Bilton, D.; Bloomfield, E.V.; Boyd, A.C.; Brand, J.; Buchan, R.; Calcedo, R.; et al. Repeated nebulisation of non-viral CFTR gene therapy in patients with cystic fibrosis: A randomised, double-blind, placebo-controlled, phase 2b trial. *Lancet Respir. Med.* **2015**, *3*, 684–691. [CrossRef]

234. Nishida, K.; Smith, Z.; Rana, D.; Palmer, J.; Gallicano, G.I.I. Cystic fibrosis: A look into the future of prenatal screening and therapy. *Birth Defects Res. Part C Embryo Today Rev.* **2015**, *105*, 73–80. [CrossRef] [PubMed]

235. Yu, Z.-Y.; McKay, K.; van Asperen, P.; Zheng, M.; Fleming, J.; Ginn, S.L.; Kizana, E.; Latham, M.; Feneley, M.P.; Kirkland, P.D.; et al. Lentivirus vector-mediated gene transfer to the developing bronchiolar airway epithelium in the fetal lamb. *J. Gene Med.* **2007**, *9*, 429–439. [CrossRef] [PubMed]

236. Copreni, E.; Penzo, M.; Carrabino, S.; Conese, M. Lentivirus-mediated gene transfer to the respiratory epithelium: A promising approach to gene therapy of cystic fibrosis. *Gene Ther.* **2004**, *11* (Suppl. 1), S67–S75. [CrossRef] [PubMed]

237. Sinn, P.L.; Arias, A.C.; Brogden, K.A.; McCray, P.B., Jr. Lentivirus vector can be readministered to nasal epithelia without blocking immune responses. *J. Virol.* **2008**, *82*, 10684–10692. [CrossRef] [PubMed]

238. Armstrong, D.K.; Cunningham, S.; Davies, J.C.; Alton, E.W.F.W. Gene therapy in cystic fibrosis. *Arch. Dis. Child.* **2014**, *99*, 465–468. [CrossRef] [PubMed]

239. Nayerossadat, N.; Maedeh, T.; Ali, P.A. Viral and nonviral delivery systems for gene delivery. *Adv. Biomed. Res.* **2012**, *1*, 27. [CrossRef] [PubMed]

240. Pichon, C.; Billiet, L.; Midoux, P. Chemical vectors for gene delivery: Uptake and intracellular trafficking. *Curr. Opin. Biotechnol.* **2010**, *21*, 640–645. [CrossRef] [PubMed]

241. Griesenbach, U.; Alton, E.W. Moving forward: Cystic fibrosis gene therapy. *Hum. Mol. Genet.* **2013**, *22*, R52–R58. [CrossRef] [PubMed]

242. Nishikawa, M.; Hashida, M. Nonviral Approaches Satisfying Various Requirements for Effective in Vivo Gene Therapy. *Biol. Pharm. Bull.* **2002**, *25*, 275–283. [CrossRef] [PubMed]

243. Ratko, T.A.; Cummings, J.P.; Blebea, J.; Matuszewski, K.A. Clinical gene therapy for nonmalignant disease. *Am. J. Med.* **2003**, *115*, 560–569. [CrossRef]

244. Sahin, U.; Karikó, K.; Türeci, Ö. mRNA-based therapeutics—Developing a new class of drugs. *Nat. Rev. Drug Discov.* **2014**, *13*, 759–780. [CrossRef] [PubMed]

245. Lutz, J.F.; Zarafshani, Z. Efficient construction of therapeutics, bioconjugates, biomaterials and bioactive surfaces using azide-alkyne "click" chemistry. *Adv. Drug Deliv. Rev.* **2008**, *60*, 958–970. [CrossRef] [PubMed]

246. Akira, S. Toll receptor families: Structure and function. *Semin. Immunol.* **2004**, *16*, 1–2. [CrossRef] [PubMed]

247. Diebold, S.S.; Kaisho, T.; Hemmi, H.; Akira, S.; Sousa, R.E. Innate antiviral responses by means of TLR7-mediated recognition of single-stranded RNA. *Science* **2004**, *303*, 1529–1531. [CrossRef] [PubMed]

248. Alexopoulou, L.; Holt, A.C.; Medzhitov, R.; Flavell, R.A. Recognition of double-stranded RNA and activation of NF-kappaB by Toll-like receptor 3. *Nature* **2001**, *413*, 732–738. [CrossRef] [PubMed]

249. Bangel-Ruland, N.; Tomczak, K.; Fernández Fernández, E.; Leier, G.; Leciejewski, B.; Rudolph, C.; Rosenecker, J.; Weber, W.-M. Cystic fibrosis transmembrane conductance regulator-mRNA delivery: A novel alternative for cystic fibrosis gene therapy. *J. Gene Med.* **2013**, *15*, 414–426. [CrossRef] [PubMed]

250. Daniels, T.; Mills, N.; Whitaker, P. Nebuliser systems for drug delivery in cystic fibrosis. *Cochrane Database Syst. Rev.* **2013**, *4*, CD007639. [CrossRef]

251. Rudolph, C.; Ortiz, A.; Schillinger, U.; Jauernig, J.; Plank, C.; Rosenecker, J. Methodological optimization of polyethylenimine (PEI)-based gene delivery to the lungs of mice via aerosol application. *J. Gene Med.* **2005**, *7*, 59–66. [CrossRef] [PubMed]

252. Hasenpusch, G.; Geiger, J.; Wagner, K.; Mykhaylyk, O.; Wiekhorst, F.; Trahms, L.; Heidsieck, A.; Gleich, B.; Bergemann, C.; Aneja, M.K.; et al. Magnetized aerosols comprising superparamagnetic iron oxide nanoparticles improve targeted drug and gene delivery to the lung. *Pharm. Res.* **2012**, *29*, 1308–1318. [CrossRef] [PubMed]

253. Rubin, B.K.; Williams, R.W. Emerging aerosol drug delivery strategies: From bench to clinic. *Adv. Drug Deliv. Rev.* **2014**, *75*, 141–148. [CrossRef] [PubMed]

254. Heijerman, H.; Westerman, E.; Conway, S.; Touw, D.; Doring, G. Inhaled medication and inhalation devices for lung disease in patients with cystic fibrosis: A European consensus. *J. Cyst. Fibros.* **2009**, *8*, 295–315. [CrossRef] [PubMed]

255. Xia, E.; Munegowda, M.A.; Cao, H.; Hu, J. Lung gene therapy-How to capture illumination from the light already present in the tunnel. *Genes Dis.* **2014**, *1*, 40–52. [CrossRef] [PubMed]

256. Kariko, K.; Muramatsu, H.; Keller, J.M.; Weissman, D. Increased erythropoiesis in mice injected with submicrogram quantities of pseudouridine-containing mRNA encoding erythropoietin. *Mol. Ther.* **2012**, *20*, 948–953. [CrossRef] [PubMed]

257. Zhong, Y.Q.; Wei, J.; Fu, Y.R.; Shao, J.; Liang, Y.W.; Lin, Y.H.; Liu, J.; Zhu, Z.H. Toxicity of cationic liposome Lipofectamine 2000 in human pancreatic cancer Capan-2 cells. *Nan Fang Yi Ke Da Xue Xue Bao* **2008**, *28*, 1981–1984. [PubMed]

258. Jin, L.; Zeng, X.; Liu, M.; Deng, Y.; He, N. Current progress in gene delivery technology based on chemical methods and nano-carriers. *Theranostics* **2014**, *4*, 240–255. [CrossRef] [PubMed]

259. Ojea-Jiménez, I.; Tort, O.; Lorenzo, J.; Puntes, V.F. Engineered nonviral nanocarriers for intracellular gene delivery applications. *Biomed. Mater.* **2012**, *7*, 54106. [CrossRef] [PubMed]

260. Kay, M.A. State-of-the-art gene-based therapies: The road ahead. *Nat. Rev. Genet.* **2011**, *12*, 316–328. [CrossRef] [PubMed]

261. Gill, D.R.; Hyde, S.C. Delivery of genes into the CF airway. *Thorax* **2014**, *69*, 962–964. [CrossRef] [PubMed]

262. Davis, P.B.; Drumm, M.L.; Konstan, M.W. Cystic fibrosis. *Am. J. Respir. Crit. Care Med.* **1996**, *154*, 1229–1256. [CrossRef] [PubMed]

263. Bell, S.C. A new phase of CFTR treatment for cystic fibrosis? *Lancet. Respir. Med.* **2015**. [CrossRef]

264. Griesenbach, U.; Alton, E.W. Cystic fibrosis gene therapy: Successes, failures and hopes for the future. *Expert Rev. Respir. Med.* **2009**, *3*, 363–371. [CrossRef] [PubMed]

265. Engelhardt, J.F.; Zepeda, M.; Cohn, J.A.; Yankaskas, J.R.; Wilson, J.M. Expression of the cystic fibrosis gene in adult human lung. *J. Clin. Investig.* **1994**, *93*, 737–749. [CrossRef] [PubMed]

266. Kreda, S.M.; Mall, M.; Mengos, A.; Rochelle, L.; Yankaskas, J.; Riordan, J.R.; Boucher, R.C. Characterization of wild-type and deltaF508 cystic fibrosis transmembrane regulator in human respiratory epithelia. *Mol. Biol. Cell* **2005**, *16*, 2154–2467. [CrossRef] [PubMed]

267. Zhang, L.; Button, B.; Gabriel, S.E.; Burkett, S.; Yan, Y.; Skiadopoulos, M.H.; Dang, Y.L.; Vogel, L.N.; McKay, T.; Mengos, A.; et al. CFTR delivery to 25% of surface epithelial cells restores normal rates of mucus transport to human cystic fibrosis airway epithelium. *PLoS Biol.* **2009**, *7*, e1000155. [CrossRef] [PubMed]

268. Cotton, C.U.; Stutts, M.J.; Knowles, M.R.; Gatzy, J.T.; Boucher, R.C. Abnormal apical cell membrane in cystic fibrosis respiratory epithelium. An in vitro electrophysiologic analysis. *J. Clin. Investig.* **1987**, *79*, 80–85. [CrossRef] [PubMed]

269. Ramachandran, S.; Krishnamurthy, S.; Jacobi, A.M.; Wohlford-Lenane, C.; Behlke, M.A.; Davidson, B.L.; McCray, P.B. Efficient delivery of RNA interference oligonucleotides to polarized airway epithelia in vitro. *AJP Lung Cell. Mol. Physiol.* **2013**, *305*, L23–L32. [CrossRef] [PubMed]

270. Ramalho, A.S.; Beck, S.; Meyer, M.; Penque, D.; Cutting, G.R.; Amaral, M.D. Five percent of normal cystic fibrosis transmembrane conductance regulator mRNA ameliorates the severity of pulmonary disease in cystic fibrosis. *Am. Respir. Cell Mol. Biol.* **2002**, *27*, 619–627. [CrossRef] [PubMed]

271. Dannhoffer, L.; Blouquit-Laye, S.; Regnier, A.; Chinet, T. Functional properties of mixed cystic fibrosis and normal bronchial epithelial cell cultures. *Am. Respir. Cell Mol. Biol.* **2009**, *40*, 717–723. [CrossRef] [PubMed]

272. Griesenbach, U.; Ferrari, S.; Geddes, D.M.; Alton, E.W.F.W. Gene therapy progress and prospects: Cystic fibrosis. *Gene Ther.* **2002**, *9*, 1344–1350. [CrossRef] [PubMed]

273. Thomas, C.E.; Ehrhardt, A.; Kay, M.A. Progress and problems with the use of viral vectors for gene therapy. *Nat. Rev. Genet.* **2003**, *4*, 346–358. [CrossRef] [PubMed]

274. Bessis, N.; GarciaCozar, F.J.; Boissier, M.-C. Immune responses to gene therapy vectors: Influence on vector function and effector mechanisms. *Gene Ther.* **2004**, *11* (Suppl. 1), S10–S17. [CrossRef] [PubMed]

275. Hansson, M.L.; Albert, S.; González Somermeyer, L.; Peco, R.; Mejía-Ramírez, E.; Montserrat, N.; Izpisua Belmonte, J.C. Efficient delivery and functional expression of transfected modified mRNA in human embryonic stem cell-derived retinal pigmented epithelial cells. *J. Biol. Chem.* **2015**, *290*, 5661–5672. [CrossRef] [PubMed]

276. Parker, A.L.; Newman, C.; Briggs, S.; Seymour, L.; Sheridan, P.J. Nonviral gene delivery: Techniques and implications for molecular medicine. *Expert Rev. Mol. Med.* **2004**, *5*, 1–15. [CrossRef]

277. Mars, T.; Strazisar, M.; Mis, K.; Kotnik, N.; Pegan, K.; Lojk, J.; Grubic, Z.; Pavlin, M. Electrotransfection and lipofection show comparable efficiency for in vitro gene delivery of primary human myoblasts. *J. Membr. Biol.* **2015**, *248*, 273–283. [CrossRef] [PubMed]

278. Dalby, B.; Cates, S.; Harris, A.; Ohki, E.C.; Tilkins, M.L.; Price, P.J.; Ciccarone, V.C. Advanced transfection with Lipofectamine 2000 reagent: Primary neurons, siRNA, and high-throughput applications. *Methods* **2004**, *33*, 95–103. [CrossRef] [PubMed]

279. Nguyen, L.T.; Atobe, K.; Barichello, J.M.; Ishida, T.; Kiwada, H. Complex formation with plasmid DNA increases the cytotoxicity of cationic liposomes. *Biol. Pharm. Bull.* **2007**, *30*, 751–757. [CrossRef] [PubMed]

280. Futerman, A.H.; van Meer, G. The cell biology of lysosomal storage disorders. *Nat. Rev. Mol. Cell Biol.* **2004**, *5*, 554–565. [CrossRef] [PubMed]

281. Rastall, D.P.W.; Amalfitano, A. Current and Future Treatments for Lysosomal Storage Disorders. *Curr. Treat. Options Neurol.* **2017**, *19*, 45. [CrossRef] [PubMed]

282. Rastall, D.P.; Amalfitano, A. Recent advances in gene therapy for lysosomal storage disorders. *Appl. Clin. Genet.* **2015**, *8*, 157–169. [CrossRef] [PubMed]

283. Aubourg, P. Gene therapy for leukodystrophy: Progress, challenges and opportunities. *Expert Opin. Orphan Drugs* **2016**, *4*, 359–367. [CrossRef]

284. Gieselmann, V. Metachromatic leukodystrophy: Genetics, pathogenesis and therapeutic options. *Acta Paediatr.* **2008**, *97*, 15–21. [CrossRef] [PubMed]

285. Rosenberg, J.B.; Kaminsky, S.M.; Aubourg, P.; Crystal, R.G.; Sondhi, D. Gene therapy for metachromatic leukodystrophy. *J. Neurosci. Res.* **2016**, *94*, 1169–1179. [CrossRef] [PubMed]

286. Piguet, F.; Sondhi, D.; Piraud, M.; Fouquet, F.; Hackett, N.R.; Ahouansou, O.; Vanier, M.-T.; Bieche, I.; Aubourg, P.; Crystal, R.G.; et al. Correction of Brain Oligodendrocytes by AAVrh.10 Intracerebral Gene Therapy in Metachromatic Leukodystrophy Mice. *Hum. Gene Ther.* **2012**, *23*, 903–914. [CrossRef] [PubMed]

287. Rosenberg, J.B.; Sondhi, D.; Rubin, D.G.; Monette, S.; Chen, A.; Cram, S.; De, B.P.; Kaminsky, S.M.; Sevin, C.; Aubourg, P.; et al. Comparative Efficacy and Safety of Multiple Routes of Direct CNS Administration of Adeno-Associated Virus Gene Transfer Vector Serotype rh.10 Expressing the Human Arylsulfatase A cDNA to Nonhuman Primates. *Hum. Gene Ther. Clin. Dev.* **2014**, *25*, 164–177. [CrossRef] [PubMed]

288. Sondhi, D.; Johnson, L.; Purpura, K.; Monette, S.; Souweidane, M.M.; Kaplitt, M.G.; Kosofsky, B.; Yohay, K.; Ballon, D.; Dyke, J.; et al. Long-Term Expression and Safety of Administration of AAVrh.10hCLN2 to the Brain of Rats and Nonhuman Primates for the Treatment of Late Infantile Neuronal Ceroid Lipofuscinosis. *Hum. Gene Ther. Methods* **2012**, *23*, 324–335. [CrossRef] [PubMed]

289. Buckley, R.H. Primary cellular immunodeficiencies. *J. Allergy Clin. Immunol.* **2002**, *109*, 747–757. [CrossRef] [PubMed]

290. Baptista, M.A.P.; Keszei, M.; Oliveira, M.; Sunahara, K.K.S.; Andersson, J.; Dahlberg, C.I.M.; Worth, A.J.; L021 L021 L021 L021 L021 L021 L021 L021 L021 L021 L021 L021 L021 L021 L021 L021 L021 L021
L021 L021 L021 L021 L021 L021 L021 L021 L021 L021 L021 L021 L021 L021 L021 L021 L021 L021 L021 L021 L021 L021 L021 L021 L021 L021 L021 L021 L021 L021 L021 L021 L021 L021 L021 L021 L021 L021 L021 L021 L021 L021 L021 L021 L021 L021 L021 L021 L021 L021 L021 L021 L021 L021 L021 L021 L021 L021 L021 L021 L021 L021 L021 L021 L021 L021 L021 L021 L021 L021 L021 L021 L021 L021 L021 L021 L021 L021 L021 L021 L021 L021
Lidén, A.; Kuo, I.-C.; Wallin, R.P.A.; et al. Deletion of Wiskott–Aldrich syndrome protein triggers Rac2 activity and increased cross-presentation by dendritic cells. *Nat. Commun.* **2016**, *7*, 12175. [CrossRef] [PubMed]

291. Morris, E.C.; Fox, T.; Chakraverty, R.; Tendeiro, R.; Snell, K.; Rivat, C.; Grace, S.; Gilmour, K.; Workman, S.; Buckland, K.; et al. Gene therapy for Wiskott-Aldrich syndrome in a severely affected adult. *Blood* **2017**, *130*, 1327–1335. [CrossRef] [PubMed]

292. Braun, C.J.; Boztug, K.; Paruzynski, A.; Witzel, M.; Schwarzer, A.; Rothe, M.; Modlich, U.; Beier, R.; Gohring, G.; Steinemann, D.; et al. Gene Therapy for Wiskott-Aldrich Syndrome—Long-Term Efficacy and Genotoxicity. *Sci. Transl. Med.* **2014**, *6*, 227ra33. [CrossRef] [PubMed]

293. Hacein-Bey Abina, S.; Gaspar, H.B.; Blondeau, J.; Caccavelli, L.; Charrier, S.; Buckland, K.; Picard, C.; Six, E.; Himoudi, N.; Gilmour, K.; et al. Outcomes Following Gene Therapy in Patients With Severe Wiskott-Aldrich Syndrome. *JAMA* **2015**, *313*, 1550. [CrossRef] [PubMed]

294. Fischer, A. Severe combined immunodeficiencies (SCID). *Clin. Exp. Immunol.* **2000**, *122*, 143–149. [CrossRef] [PubMed]

295. Abbott, A. Italians first to use stem cells. *Nature* **1992**, *356*, 465. [CrossRef] [PubMed]

296. Mavilio, F. Developing gene and cell therapies for rare diseases: An opportunity for synergy between academia and industry. *Gene Ther.* **2017**, *24*, 590–592. [CrossRef] [PubMed]

297. Asbury, A.K.; Gale, M.K.; Cox, S.C.; Baringer, J.R.; Berg, B.O. Giant axonal neuropathy—A unique case with segmental neurofilamentous masses. *Acta Neuropathol.* **1972**, *20*, 237–247. [CrossRef] [PubMed]

298. Kuhlenbäumer, G.; Timmerman, V.; Bomont, P. Giant Axonal Neuropathy. In *GeneReviews*; University of Washington: Seattle, WA, USA, 1993.

299. Bharucha-Goebel, D.; Jain, M.; Waite, M.; Lehky, T.; Foley, R.; Marra, J.D.; Zein, W.; Bonnemann, C.G. 715. Giant Axonal Neuropathy—The Role of Natural History Studies in Gene Transfer Therapy Trial Design. *Mol. Ther.* **2016**, *24*, S282. [CrossRef]

300. Bailey, R.M.; Armao, D.; Nagabhushan Kalburgi, S.; Gray, S.J. Development of Intrathecal scAAV9 Gene Therapy for Giant Axonal Neuropathy. *Mol. Ther.* **2015**, *23*, S6–S7. [CrossRef]

301. McIntyre, J.C.; Williams, C.L.; Martens, J.R. Smelling the roses and seeing the light: Gene therapy for ciliopathies. *Trends Biotechnol.* **2013**, *31*, 355–363. [CrossRef] [PubMed]

302. Lobo, L.J.; Zariwala, M.A.; Noone, P.G. Ciliary dyskinesias: Primary ciliary dyskinesia in adults. In *Bronchiectasis*; European Respiratory Society: Lausanne, Switzerland, 2011; pp. 130–149. [CrossRef]

303. Wallmeier, J.; Al-Mutairi, D.A.; Chen, C.-T.; Loges, N.T.; Pennekamp, P.; Menchen, T.; Ma, L.; Shamseldin, H.E.; Olbrich, H.; Dougherty, G.W.; et al. Mutations in CCNO result in congenital mucociliary clearance disorder with reduced generation of multiple motile cilia. *Nat. Genet.* **2014**, *46*, 646–651. [CrossRef] [PubMed]

304. Boon, M.; Wallmeier, J.; Ma, L.; Loges, N.T.; Jaspers, M.; Olbrich, H.; Dougherty, G.W.; Raidt, J.; Werner, C.; Amirav, I.; et al. MCIDAS mutations result in a mucociliary clearance disorder with reduced generation of multiple motile cilia. *Nat. Commun.* **2014**, *5*, 4418. [CrossRef] [PubMed]

305. Praveen, K.; Davis, E.E.; Katsanis, N. Unique among ciliopathies: Primary ciliary dyskinesia, a motile cilia disorder. *F1000Prime Rep.* **2015**, *7*, 36. [CrossRef] [PubMed]

polymers

MDPI

Article

Reversible pH-Sensitive Chitosan-Based Hydrogels. Influence of Dispersion Composition on Rheological Properties and Sustained Drug Delivery

Nieves Iglesias [1], Elsa Galbis [1] ⓘ, Concepción Valencia [2,3], M.-Violante de-Paz [1,*] ⓘ and Juan A. Galbis [1]

[1] Departamento de Química Orgánica y Farmacéutica, Facultad de Farmacia, Universidad de Sevilla, 41012 Sevilla, Spain; nievesiglesias@us.es (N.I.); elsa@us.es (E.G.); jgalbis@us.es (J.A.G.)
[2] Departamento de Ingeniería Química, Campus de "El Carmen", Universidad de Huelva, 21071 Huelva, Spain; barragan@diq.uhu.es
[3] Pro2TecS—Chemical Process and Product Technology Research Center, Universidad de Huelva, 21071 Huelva, Spain
* Correspondence: vdepaz@us.es; Tel.: +34-954-556-740

Received: 27 February 2018; Accepted: 29 March 2018; Published: 1 April 2018

Abstract: The present work deals with the synthesis of micro-structured biomaterials based on chitosan (CTS) for their applications as biocompatible carriers of drugs and bioactive compounds. Twelve dispersions were prepared by means of functional cross-linking with tricarballylic acid (TCA); they were characterized by Fourier transform infrared spectroscopy (FT-IR), modulated temperature differential scanning calorimetry (MTDSC) and scanning electron microscopy (SEM), and their rheological properties were studied. To the best of the authors' knowledge, no study has been carried out on the influence of CTS concentration, degree of cross-linking and drug loading on chitosan hydrogels for drug delivery systems (DDS) and is investigated herein for the first time. The influence of dispersion composition (polymer concentration and degree of cross-linking) revealed to exert a marked impact on its rheological properties, going from liquid-like to viscoelastic gels. The release profiles of a model drug, diclofenac sodium (DCNa), as well as their relationships with polymer concentration, drug loading and degree of cross-linking were evaluated. Similar to the findings on rheological properties, a wide range of release profiles was encountered. These formulations were found to display a well-controlled drug release strongly dependent on the formulation composition. Cumulative drug release under physiological conditions for 96 h ranged from 8% to 67%. For comparative purpose, Voltaren emulgel® from Novartis Pharmaceuticals was also investigated and the latter was the formulation with the highest cumulative drug release (85%). Some formulations showed similar spreadability values to the commercial hydrogel. The comparative study of three batches confirmed the reproducibility of the method, leading to systems particularly suitable for their use as drug carriers.

Keywords: ionic cross-linking; eco-friendly formulations; thermal transition sol-gel; drug delivery systems; MTDSC; DSC

1. Introduction

Hydrogels are currently under investigation as matrices for the controlled release of bioactive molecules, in particular drugs and proteins, and for the encapsulation of living cells. In general terms, hydrogels are constituted by cross-linked polymer networks that have a high amount of hydrophilic domains with affinity for water. Among the favorable features of these systems, the similarity between their physical properties and those of living tissues, such as low interfacial tension with water or biological fluids, can be highlighted [1,2]. Likewise, the elastic nature of hydrated hydrogels minimizes irritation of surrounding tissues after implantation. Moreover, the low interfacial tension between the hydrogel surface and body fluid lessens protein adsorption and cell adhesion, which reduces the chances of a negative immune reaction [3]. For biomedical applications, gels are often required to degrade under physiological conditions and lead to the disintegration of the three-dimensional structure, preferably in harmless products, to ensure good biocompatibility of the hydrogel [2].

Being a non-toxic, biocompatible and biodegradable polymer, chitosan (CTS) and its hydrogels have been widely used as biomaterials for drug delivery, gene delivery, and tissue engineering and, in recent years, numerous scientific papers have been published on a wide variety of biomedical applications [4,5]. The polysaccharide CTS, a weak cationic polysaccharide composed of randomly distributed β-(1–4)-linked D-glucosamine and N-acetyl-D-glucosamine repeating units, is a copolymer prepared from renewable resources. It can be obtained from the partial deacetylation of the second most important natural polymer in the world: chitin or poly(N-acetyl-β-D-glucosamine). Its abundance in marine crustacean such as shrimp and crabs makes it a commercial product with global impact in polymer science.

CTS hydrogels can be prepared via physical association (ionic crosslinking) [5–9], coordination with metal ions [10], or irreversible/chemical cross-linking between chitosan and the crosslinker [11–13]. The chemical and/or physical linkages will prevent the networks from dissolving. Interestingly, and regarding the release of bioactive molecules, the reversible nature of ionically cross-linked networks is useful for their potential applications as drug delivery systems. Therefore, once the release of the drug in the medium have been accomplished, the formulations can subsequently disintegrate into biocompatible components that will then be metabolized and eliminated from the body [2].

Few anions such as citrate salts, dextran sulfate, glycerol-mono or diphosphates, polyphosphates, glucose-phosphates and other polyol-phosphates have been tested in the formation of ionically crosslinked chitosan hydrogels [2,6–8,14]. However, a lack of study on carboxylate derivatives as cross-linkers is found apart from few polyanionic materials such as poly(methacrylic acid) and alginates, in which an additional secondary force, the chain entanglement is involved [7,10,15–17].

We describe herein a highly efficient strategy for the preparation of a new batch of pH-sensitive reversible chitosan hydrogels, which are endowed with controlled release properties under physiological conditions. The prepared chitosan hydrogels are stabilized by ionic crosslinking with propane-1,2,3-tricarboxylic acid (tricarballylic acid, TCA). To the best of the authors' knowledge, no study has been carried out on the influence of the CTS concentration, degree of cross-linking and drug loading on the rheological properties and drug release profiles of CTS-based hydrogels and this is explored herein for the first time. The influence of the concentration of CTS (from 2% to 4%) and the degree of cross-linking (from 0% to 15%) on the rheological properties of the systems is investigated. The anionic model drug, diclofenac sodium, is loaded in CTS formulations in which the concentration of CTS (from 2% to 4%), the degree of cross-linking (from 0% to 15%) and the drug loading (from 0.5% to 2%) are set to investigate the effect of those parameters on the drug release. Thus, a wide range of systems is studied and marked differences encountered, not only in their rheological properties but also in their release profiles.

2. Materials and Methods

All the chemicals used were purchased from Sigma-Aldrich (Madrid, Spain) and used as received. A commercial chitosan (CTS) from Sigma-Aldrich with a deacetylation degree of 75% was chosen. The molecular weight of the CTS used in the present work was determined by viscometric analysis. Its viscosity was measured in a buffered solution of 0.5 M acetic acid—0.5 M sodium acetate solution at $25.0 \pm 0.1\ ^\circ$C using an Anton Paar AMVn automated microviscometer (Ashland, VA, US). The viscometric constants a and K in the Mark-Houwink equation were previously determined for this solvent—CTS system and found to be $a = 0.59$ and $K = 0.119$ cm$^3 \cdot$g^{-1} [18]. The weight of the CTS used in the present work was calculated means of the Mark-Houwink equation ([η] = 3.385 dL/g) and its value was 299 kDa. The acetate buffer at pH 5.5 (25 °C) for release assays was prepared in-house, with pH variations vs. temperature of ±0.1 from 5 to 50 °C. Voltaren emulgel® (Novartis Farmaceutica, Barcelona, Spain) was purchased from a licensed drugstore (Seville, Spain).

To investigate the chemical interaction between chitosan and tricarballylic acid (TCA) in aqueous media, IR spectra of chitosan, tricarballylic acid and CTS-TCA conjugate were recorded on a Jasco FT/IR 4200 spectrometer (Great Dunmow, Essex, UK) equipped with attenuated total single reflection (ATR) accessory in the range between 4000 and 600 cm^{-1}.

Measurement of UV and visible light absorbance was performed with an Agilent 8453 UV-Visible spectrophotometer (Palo Alto, CA, USA), equipped with diode array detection (DAD); the data were the result of at least three measurements.

The phase transitions exhibited by the new prepared cross-linked hydrogels were examined by modulated temperature differential scanning calorimetry (MTDSC), using a TA DSC Q-200 Instrument (calibrated with indium, Cerdanyola del Valles, Spain) and a refrigerated cooling system (RCS) to ensure proper temperature cycling. Accurately weighed 3–6 mg samples were then hermetically sealed into aluminum DSC pans, and the equipment was operated in modulation mode. Calorimetric scans were carried out at a scanning rate of 2 °C/min under nitrogen atmosphere over an appropriate temperature range (from 0 to 140 °C). Modulation amplitude was ±0.159 °C every 30 s.

The morphologies of selected samples were examined by Scanning Electron Microscopy (SEM). Before SEM observations, the hydrogel scaffold was directly frozen at −20 °C for 3 h, then at −80 °C for 24 h. The samples were then lyophilized by freeze drying for 24 h. Finally, the dry hydrogel was fixed on aluminum stubs, coated with a thickness of about 25 nm of gold, and analyzed at the Electron Microscopy Division of the Scientific Integrated Services (SC-ICYT) of the University of Cádiz (Spain) using a field emission scanning electron microscope FEI Nova NanoSEM 450 (Hillsboro, Oregon, US) operated at 5 kV.

To determine of spreadability of a specific formulation, the selected hydrogel (300 mg) was placed in the center of a glass plate (20 cm × 20 cm) and the sample was covered by another plate of similar dimensions. A weight of 30 g was carefully placed in the center of the upper cover avoiding the sliding between the plates. The spread area (diameter, in cm) of the gel was measured after 1 min and 30 min.

2.1. Preparation of Hydrogels from Crosslinked Chitosan-Tricarballylic Acid (CTS$_x$-TCA$_y$)

Twelve systems named **CTS$_x$-TCA$_y$** were prepared with CTS of molecular weight ranging from 190 to 375 kDa, (based on viscosity values). The targeted final CTS concentrations were 2%, 3% or 4% w/w and the degree of crosslinking 0%, 5%, 10% or 15%. The later parameter was fixed by the amount of tricarballylic acid (TCA) added to the formulation. In Table 1 and along the text "x" denotes CTS concentration (% w/w) and "y" denotes the degree of cross-linking in the hydrogel.

Table 1. Concentration and degree of cross-linking of the 12 CTS-based hydrogels prepared. Rheological parameters.

Hydrogel	CTS		TCA			Rheological Properties [1]			
CTS$_x$-TCA$_y$	mg CTS	Conc. (% *w/w*)	mg TCA	Degree Xr (%)	pH	η_0 (Pa.s)	K (s)	m	Tan δ (1 rad/s)
CTS$_2$-TCA$_0$	200	2%	-	0	5.2	5	0.13	0.6	2.57
CTS$_3$-TCA$_0$	300	3%	-	0	5.3	496	20.18	0.6	1.19
CTS$_4$-TCA$_0$	400	4%	-	0	5.3	5616	17.44	0.85	0.62
CTS$_2$-TCA$_5$	200	2%	2.6	5	5.3	9	0.095	0.55	3.08
CTS$_3$-TCA$_5$	300	3%	3.8	5	5.2	1123	8.57	0.7	0.81
CTS$_4$-TCA$_5$	400	4%	5.1	5	5.2	2061	7.77	0.82	0.59
CTS$_2$-TCA$_{10}$	200	2%	5.1	10	5.3	5	0.47	0.37	5.14
CTS$_3$-TCA$_{10}$	300	3%	7.7	10	5.3	155	4.57	0.6	1.47
CTS$_4$-TCA$_{10}$	400	4%	10.3	10	5.2	3745	13.24	0.83	0.68
CTS$_2$-TCA$_{15}$	200	2%	7.7	15	5.3	3	0.08	0.59	2.88
CTS$_3$-TCA$_{15}$	300	3%	11.5	15	5.2	100	10.97	0.49	1.77
CTS$_4$-TCA$_{15}$	400	4%	15.4	15	5.2	2151	8.41	0.82	0.86

[1] Williamson's model parameters and loss tangent at 1 rad/s for CTS$_x$-TCA$_y$ hydrogels studied.

A typical procedure for the preparation of aqueous cross-linked chitosan-carballylic acid conjugates at 3% *w/w* polymer concentration and 5% of degree of crosslinking (**CTS$_3$-TCA$_5$**) is summarized next: chitosan with a deacetylation degree of 75% (chitosan, 0.3 g, 1.31 mmol of free amine groups) was charged in a round-bottom flask provided with a stirrer bar; then, an aqueous solution of tricarballylic acid (TCA, 0.38 mL, 10mg/mL, 0.02 mmol), a solution of acetic acid (HAc, 0.1 mL, 52% *w/v*) and double-distilled water (up to a final weight of 10 g, and final polymer concentration of 3% *w/w*) were added in sequence. The mixture was stirred to homogenization at 40 °C during 1.5 h. The solution was cooled and the stirring proceeded overnight at 25 °C. Three different batches of these conjugates were synthesized for comparative purposes.

CTS$_x$-TCA$_y$ hydrogels were rheologically characterized in a controlled-strain (ARES, Rheometric Scientific, Surrey, UK) rheometer, using a serrated plate-plate (25 mm diameter, 1 mm gap) geometry. Small amplitude oscillatory shear (SAOS) tests were carried out inside the linear viscoelastic region in a frequency range of 0.03–100 rad/s at 25 °C. Stress sweep tests were previously performed to determine the linear viscoelastic regime. Viscous flow tests were also made by applying a stepped shear rate ramp in a shear rate range of 0.06–100 s^{-1} at 25 °C. Each fresh sample was tested at least in duplicate.

2.2. Preparation of Diclofenac Sodium Loaded Formulations from Crosslinked Chitosan-Conjugates

Eight systems named **CTS$_x$-TCA$_y$-DCNa$_z$** were prepared with targeted final CTS concentrations 2%, 3% or 4% *w/w* and degree of crosslinking 0%, 5%, 10% or 15%. Diclofenac sodium was loaded into the hydrogel formulation so that the final *w/w* concentrations ranged from 0.5% to 2%. In Table 2 and along the text "x" denotes CTS concentration (% *w/w*), "y" denotes the degree of cross-linking and "z" denotes DCNa concentration (% *w/w*).

Table 2. Composition and evaluation of drug formulations.

Formulation Code	CTS Conc (% w/w)	Degree Xr (%)	Drug Content (%)	Spreadability (Diameter, cm)			Drug Release (%)		
				$t = 1$ min	$t = 30$ min	Diameter (%)	$t = 5$ h	$t = 24$ h	$t = 96$ h
CTS_3-TCA_0.$DCNa_1$	3	–	1	4.4	6	36	11	17	33
CTS_3-TCA_5.$DCNa_1$	3	5	1	4.2	7.3	74	18	35	46
CTS_3-TCA_{10}.$DCNa_1$	3	10	1	4.4	5.2	18	9	18	22
CTS_3-TCA_{15}.$DCNa_1$	3	15	1	4.1	5.70	39	4	16	31
CTS_2-TCA_{10}.$DCNa_1$	2	10	1	5.4	10.6	96	15	31	51
CTS_3-TCA_{10}.$DCNa_1$	3	10	1	4.4	5.2	18	9	18	22
CTS_4-TCA_{10}.$DCNa_1$	4	10	1	3.6	4.80	33	8	16	26
CTS_3-TCA_{10}.$DCNa_{0.5}$	3	10	0.5	3.9	6.4	64	15	44	67
CTS_3-TCA_{10}.$DCNa_1$	3	10	1	4.4	5.2	18	9	18	22
CTS_3-TCA_{10}.$DCNa_2$	3	10	2	2.3	2.5	9	0	1	9
Voltaren emulgel	–	–	1	4.30	5.70	33	46	79	85

2.3. Diclofenac Sodium Release Studies

Prior to the release analyses, a calibration curve of diclofenac sodium was made with DCNa standard solutions (in buffered acetate solutions at pH 5.5) at 280 nm (Figure 1). For the calibration, a stock solution of sodium diclofenac at 100 µg/mL concentration was used. By dilution, 4 solutions were prepared with the following concentrations: 75, 50, 25, and 12.5 µg/mL.

The selected hydrogel (0.1 g) was transferred to a dialysis bag (molecular weight cut-off: 8000–14,000 Da) and then immersed in 20 mL of acetate buffer solution (pH = 5.5). Experiments were performed at 37 °C in a shaker incubator (Heidolph Unimax 1010-Heidolph Inkubator 1000, Schwabach, Germany). At pre-designed time intervals, aliquots of 1 mL were taken from the release medium and the amount of DCNa released was determined by UV-Vis spectroscopy at 280 nm. 1 mL of pre-heated buffer solution was added to the release medium to maintain a constant volume. DCNa release experiments were performed in triplicates. Analogous experiments were performed under the same conditions with Voltaren emulgel® (0.1 g).

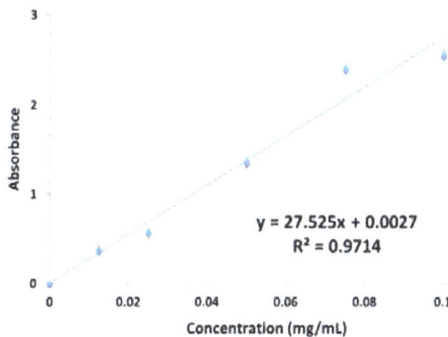

$$y = 27.525x + 0.0027$$
$$R^2 = 0.9714$$

Figure 1. Calibration curve of diclofenac sodium at 280 nm (UV-Vis spectroscopy) at 25 °C.

3. Results and Discussion

3.1. Cross-Linked Chitosan-Tricarballylic Acid (CTSx TCAy) Hydrogels

To form stabilizing linkages, CTS-based hydrogels have amino and hydroxyl functional groups that allow linking between the chain and the cross-linker to prevent gel dissolution. Hydrogel binding could be accomplished by reversible cross-linking reactions, for example by means of ionic cross-linkers negatively ionized at physiological pH such as citrate salts [6], glycerol-mono or diphosphates [14], polyphosphates [8], glucose-phosphates, and other polyol-phosphates [7]. This method can sufficiently restrain hydrogel structure, but the physical associations are reversible bonds, whereas the covalent cross-linkages between polymer chains are not. This distinction is relevant for the drug release kinetics and further biodegradation of drug-delivery-system (DDS) hydrogels [3].

In the present work, the preparation of a batch of new DDS from CTS-based reversible cross-linked hydrogels has been carried out. Propane-1,2,3-tricarboxylic acid (tricarballylic acid, TCA) has been the cross-linker of choice due to its hydrophilic character, biocompatibility, and symmetry that can facilitate the formation of homogeneous aqueous dispersions with reproducible physical properties. In CTS/TCA systems, several types of interactions may be involved during the gelation process: (1) electrostatic attraction between CTS ammonia groups and TCA carboxylate groups: since pKa of CTS and TCA are 6.5 and 4.1, respectively, the amine and carboxylic acid groups will be mainly ionized at the pH of the formulations (pH $\approx$ 5.2–5.3), i.e., when the pH is more than one unit under or one unit over their pKa, respectively. Thus, electrostatic attractions between the ammonia groups from chitosan and the carboxylate groups from the tricarballylic acid are expected to be significant leading to the stability of the hydrogel systems at those pHs; (2) Other non-covalent physical associations or secondary forces such as hydrogen bonding as a consequence of reduced electrostatic repulsion after neutralization of CTS with TCA and physical entanglements [19].

pH Values from 5.5 to 7.0 are within the so-called physiological values in human beings; for example, slightly acidic microenvironments are found in mucous membranes and other topical areas. However, maintaining acid-base balance is critical for the survival of living species since cellular processes are highly sensitive to changes in proton concentrations. Although in humans, pH varies within a narrow range (in the blood between pH 7.35 and 7.45), local deviations from the systemic pH are often caused by pathologies, such as cancer, inflammation, infection, ischemia, renal failure or pulmonary disease [20]. Consequently, drug delivery systems active at acidic pH, such as the formulations studied herein, could find significant applications in a wide field of pathologies and locations.

Twelve systems named **CTS$_x$-TCA$_y$** were prepared with CTS of molecular weight ranging from 190 to 375 kDa, (based on viscosity values). The targeted final CTS concentrations, "x", were 2%, 3% or 4% w/w and the degree of cross-linking, "y" 0%, 5%, 10% or 15% (Table 1). The degree of cross-linking was calculated based on the number of mequiv of free amine groups present in the CTS used and the number of mequiv of carboxylic acid groups present in the amount of the cross-linker added. All the samples displayed a clear and slightly-colored appearance with the absence of solid particles in suspension and the pH of the dispersions ranged from 5.2 to 5.3. An in-depth analysis on their rheological properties was conducted to determine the influence of the hydrogel composition on their viscoelastic properties and viscosities. Likewise, some additional studies were carried out by Fourier transform infrared spectroscopy (FT-IR), scanning electron microscopy (SEM), and differential scanning calorimetry (DSC) and the results are discussed below.

3.1.1. Fourier Transform Infrared Spectroscopy (FT-IR)

To examine the chemical interactions between CTS and the cross-linker in the hydrogels, IR spectra of CTS, TCA, and freeze-dried CTS-TCA conjugates were recorded. Figure 2 shows the FTIR spectra of CTS, TCA and the freeze-dried hydrogel **CTS$_3$-TCA$_{10}$** and they were in concurrence with their chemical structure. Thus, from the CTS spectra, the O-H bonds from hydroxyl groups provide a broad band centered at 3288 cm^{-1}, which is overlapped with the stretching bands corresponding to the N–H bonds from amine and amide groups. The absorption band centered at 1064 cm^{-1} can also be observed (st C–OH). The acetamido groups in the N-acetyl-D-glucosamine repeating units are responsible of the characteristic band at 1664 cm^{-1}, attributed to the stretching vibration of the C=O amide (amide I band). From the TCA FTIR spectrum two significant bands correlated to the carboxylic acid groups of the cross-linker can be found: the first one, a broad band centered at 2937 cm^{-1} displays the characteristic feature of those associated to the stretching of O–H bonds from carboxylic acids, and the second one, an intense stretching band due to the C=O group, appeared at 1687 cm^{-1}. The most intense band in the spectrum is due to the stretching vibration of OC–O bond from the carboxylic acid groups at

1238 cm^{-1}. None of them are present in sample **CTS$_3$-TCA$_{10}$** in which the acid-base reaction between CTS and TCA had taken placed. Consequently, a new band at 1555 cm^{-1} (stretching band of C=O in carboxylate ions), has emerged as well as a shift and gain in intensity of st N–H band, in this case correlated with ammonia ions. These findings were consistent with the effective preparation of ionic cross-linked CTS-based hydrogels.

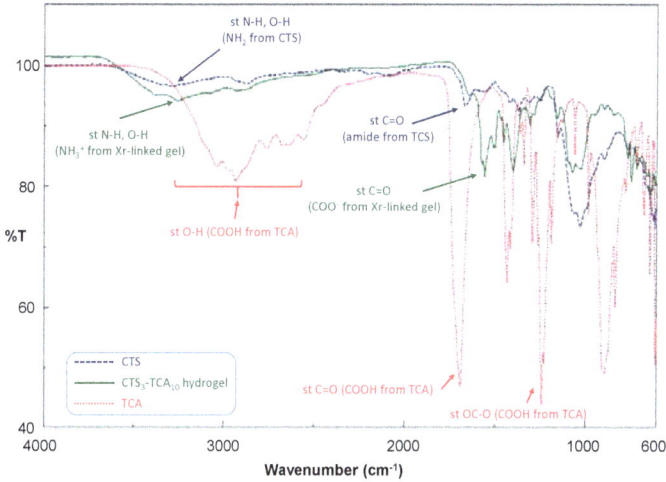

Figure 2. FT-IR spectra of CTS (blue); freeze-dried CTS$_3$-TCA$_{10}$ hydrogel (green) and TCA (red).

3.1.2. Rheological Characterization of CTS$_x$-TCA$_y$ Hydrogels

Figure 3a–d illustrate the evolution of SAOS functions with frequency, at 25 °C, inside the linear viscoelastic range for **CTS$_x$-TCA$_y$** hydrogels as function of CTS concentration and degree of cross-linking based on TCA, respectively. As can be observed, different responses were obtained depending on CTS concentration and degree of cross-linker. Regarding the influence of CTS concentration, at low CTS concentration (**CTS$_2$-TCA$_{10}$**) the values of the loss modulus (G'') are significantly higher than those found for the storage modulus (G'), and the formulations are not really resulting in gels but liquid-like viscoelastic dispersions. A tendency to reach a crossover point between these moduli was observed at high frequencies. This corresponds to the dynamic characteristics of a viscoelastic polymer fluid without entanglements. At an intermediate CTS concentration (**CTS$_3$-TCA$_{10}$**), G' coincides with G'' over a wide frequency range. This power law behavior suggests a gel-like behavior quite close to a critical gel [21]. On the other hand, at high CTS concentration (**CTS$_4$-TCA$_{10}$**), G' becomes higher than G'' and parallel in the whole frequency range. This indicates the formation of viscoelastic gels.

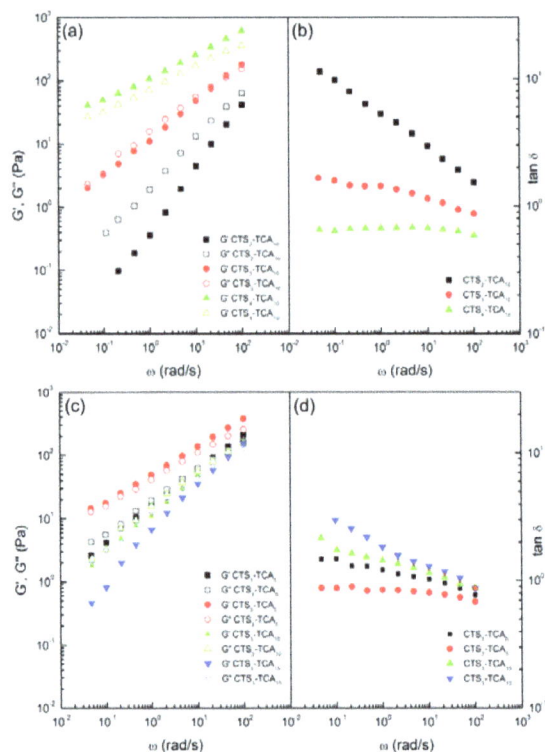

Figure 3. Frequency dependence of the storage, G', and loss, G'', moduli and the loss tangent for CTS_x-TCA_y hydrogels as function of CTS (**a,b**) and degree of cross-linking (**c,d**).

As well known [20], the gel strength of biopolymer dispersed systems, from dilute solutions to crosslinked gels, can be quantified from SAOS measurements as a function of the G' and G'' frequency dependence, i.e., the slopes of G' and G'' versus frequency plots, and the relative values of both viscoelastic functions, i.e., the relative elasticity, expressed in terms of the loss tangent (tan δ = G''/G'). Regarding the evolution of the loss tangent with frequency (Figure 3b), the **CTS$_4$-TCA$_{10}$** hydrogel shows the lowest values of the loss tangent indicating a higher relative elasticity due to the high level of cross-linking and physical entanglements, followed by **CTS$_3$-TCA$_{10}$** hydrogel. On the other hand, **CTS$_2$-TCA$_{10}$** displays the highest values of the loss tangent, higher than 1 in the whole frequency range indicating a essentially viscous behavior characteristic of polymer fluids without significant formation of physical entanglements [22]. The influence of cross-linker concentration on the mechanical spectra, obtained from SAOS measurements and loss tangent of **CTS$_x$-TCA$_y$** hydrogels are presented in Figure 3c,d, respectively. At high TCA concentration (**CTS$_3$-TCA$_{15}$**), the values of G'' are higher than those found for G' and a tendency to reach a crossover point between G' and G'' was found at high frequencies, which mainly exhibits a viscous response. This fact is reflected in the much higher loss tangent values (Figure 3d) On the other hand, at low TCA concentration (**CTS$_3$-TCA$_5$**), G' becomes slightly higher than G'' in the whole frequency range studied, showing viscoelastic response. Moreover, a typical sol-gel transition response, with a crossover at medium frequencies, was exhibited by the **CTS$_3$-TCA$_0$** hydrogel. **CTS$_3$-TCA$_0$** and **CTS$_3$-TCA$_{10}$** present similar relative elasticity and **CTS$_3$-TCA$_5$** the lowest loss tangent values at low frequency (Figure 3d).

Viscosity values depend on CTS and TCA concentration in the same way as G' and G'' do. Figure 4a,b depict the viscous flow behavior exhibited by selected hydrogels as a function of CTS and

TCA concentration, respectively. The Williamson model fits fairly well this flow behavior in the shear rate range studied ($R^2 > 0.995$) [23].

$$\eta = \frac{\eta_0}{1 + (K\dot{\gamma})^m} \tag{1}$$

where η is the non-Newtonian viscosity; η_0 is the zero-shear rate limiting viscosity; $\dot{\gamma}$ is the shear rate; m is a parameter related to the slope of the shear-thinning region; and K is a constant whose reciprocal coincides with the shear rate at which $\eta = \eta_0/2$.

The values of these fitting parameters are displayed in Table 1 for **CTS$_x$-TCA$_y$** hydrogels studied. Figure 4a depicts that viscosity clearly increases with CTS concentration, yielding higher values of η_0, K and m. Moreover, regarding to TCA concentrations, **CTS$_3$-TCA$_5$** shows the highest values of viscosity and **CTS$_3$-TCA$_{15}$** the lowest ones.

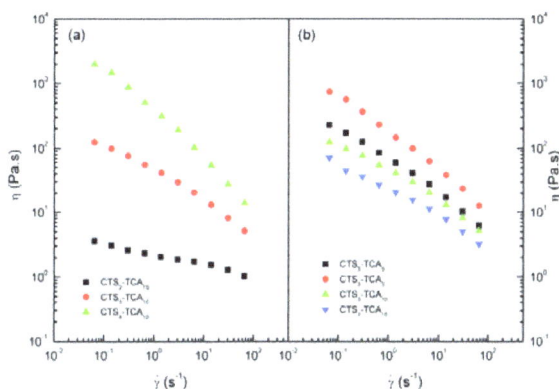

Figure 4. Viscous flow curves for **CTS$_x$-TCA$_y$** hydrogels studied, at 25 °C, as a function of (**a**) CTS concentration and (**b**) TCA concentration.

The values of the fitting parameters to the Williamson model and loss tangent are shown in Table 2 for all **CTS$_x$-TCA$_y$** studied. As can be seen, viscosity values increase by increasing the CTS concentration and generally decrease by increasing the degree of crosslinking. Moreover, m values augment with CTS concentration indicating less pronounced shear-thinning characteristics of these hydrogels at low CTS concentrations but their flow behavior is almost unaffected by the degree of cross-linking. On the other hand, loss tangent decreases by increasing the CTS concentration and generally follows a similar trend with the degree of crosslinking.

3.1.3. Modulated Temperature Differential Scanning Calorimetry (MTDSC) Studies of CTS-TCA Hydrogels

Modulated temperature differential scanning calorimetry (MTDSC) is the general term for DSC techniques where a non-linear heating or cooling rate is applied to the sample to separate the kinetic data from the thermodynamic ones. This technique finds numerous applications in the polymer field: for example, Higginbotham et al. [24] developed some thermosensitive polymer matrices based on *N*-isopropylacrylamide (NIPAAm). The thermosensitive hydrogels synthesized belong to the negative temperature-sensitive hydrogels category, with a lower critical solution temperature (LCST), and contract upon heating above the LCST. In general terms, the reversing heat-flow signal provided great sensitivity in the determination of LCST when compared with conventional DSC. In addition, we have studied polyHEMA hydrogels by thermal analysis. The transitions gel-sol were explored by MTDSC and the study was centered on the reversible heat-flow plot corresponding to transitions where the HEMA systems experienced a transformation from an arrangement with a higher heat capacity to one with a lower heat capacity [25]. As the phase transition temperature is a fully reversible transition [26,27], the reversible heat-flow curves reflect true gel-sol transitions and are, consequently, a better measurement of physical changes in these materials. Similarly we explored herein the phase transition temperature of the CTS-based hydrogels by MTDSC.

Some of the prepared CTS-based samples exhibited the typical transition gel-sol on thermo-reversible hydrogels and both, peak onset and peak maximum values were recorded in Table 3.

Table 3. Thermal properties of selected CTS-based hydrogels from modulated temperature differential scanning calorimetry (MTDSC).

Sample	MTDSC		
	Peak Temp. [a]	Onset Temp. [b]	Rev. Heat Flow [c]
CTS_2-TCA_{15}	112.71	112.09	28.68
CTS_3-TCA_{10}	113.10	112.85	96.19
CTS_3-TCA_5	–	–	–
CTS_4-TCA_0	109.01	107.71	25.39

[a] Peak Temp.: Phase transition temperature from the temperature at the peak of the transition endotherm (°C); [b] Onset Temp.: Phase transition temperature from the onset of the transition endotherm (°C); [c] Rev. Heat Flow: Reversible heat flow (J/g).

Figure 5 displays the phase transition temperatures (°C) found at the peak of the transition endotherm and from the onset of the transition endotherm for several hydrogels. For samples **CTS_3-TCA_{10}** and **CTS_2-TCA_{15}** the transitions gel-sol took place in a very narrow interval confirming the thermo-reversible nature of **CTS_3-TCA_{10}** and **CTS_2-TCA_{15}** hydrogels. However, when the degree of crosslinking was reduced (for example, sample **CTS_3-TCA_5** compared to sample **CTS_3-TCA_{10}**), this transition was not observed.

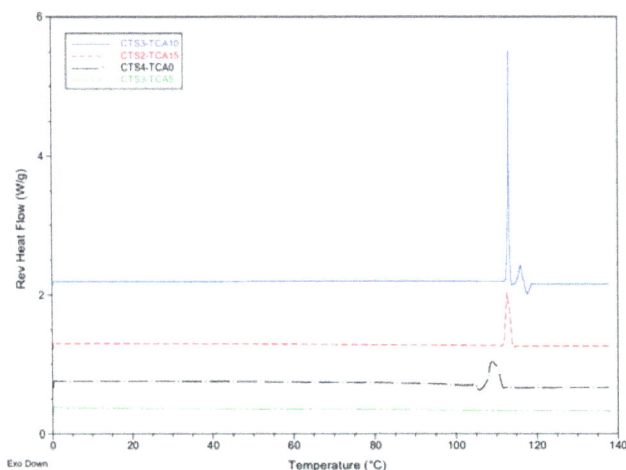

Figure 5. Reversible heat-flow plot and phase transition temperature from the temperature at the peak of the transition endotherm (°C) of **CTS_3-TCA_{10}** (solid line, in blue), **CTS_2-TCA_{15}** (short dash line, in red), and **CTS_4-TCA_0** (dash-dot line, in black) hydrogels established using modulated temperature DSC (MTDSC) overlaid with the reversible heat-flow versus temperature of **CTS_3-TCA_5** system (dash-dot line, in green).

Additionally, an increase in CTS concentration helps the formation of gel-like systems since it decreases the amount of TCA needed as the CTS concentration increases. Nonetheless, if cross-linking is absent in the formulation (sample **CTS₄-TCA₀**), the transition endotherm shifts to lower values and does not show the typical sharp peak of truly cross-linked samples.

3.2. Diclofenac Loaded (CTS$_x$-TCA$_y$-DCNa$_Z$) Formulations

Among the multitude of materials used to load and release drugs or biologically active species, positively charged polymers play an important role as nanocarriers with enhanced cellular uptake, improved encapsulation efficiency and high stability. Cationic polymershave been proven to be promising for the encapsulation and delivery of various anionic therapeutic compounds. Their charged nature allows complex formation with anionic molecules, such as diclofenac, DNA, and more [28]. Likewise, in drug delivery, the cationic polysaccharide CTS has been of special relevance as a carrier for various bioactive molecules due to its physico-chemical and biocompatible properties. Being hydro-soluble and positively charged, CTS interacts with negatively charged species in aqueous media and exhibits the singular feature of adhering to mucosal surfaces, a fact that makes it a useful biomaterial for mucosal drug delivery. CTS is able to promote transmucosal absorption of small polar drugs, including peptides, inducing a transient opening of the tight junctions of the cell membrane [29]. In addition, networks formed by non-covalent cross-linked CTS are largely used as DDS due to the possibility of drug diffusion [30].

Eight DDS named **CTS$_x$-TCA$_y$-DCNa$_z$** were prepared with targeted final CTS concentrations 2%, 3% or 4% w/w and degree of crosslinking 0%, 5%, 10% or 15%. Diclofenac sodium (DCNa) was loaded into the hydrogel formulation so that the final w/w concentrations ranged from 0.5 to 2 %. In Table 2 and along the text "x" denotes CTS concentration (% w/w), "y" denotes the degree of cross-linking and "z" denotes DCNa concentration (% w/w). The influence of drug content, degree of cross-linking and CTS concentration in the drug release profiles and spreadability of the formulations were investigated.

3.2.1. Scanning Electron Microscopy (SEM)

SEM is a valuable tool to evaluate the scaffold characteristics (pore size and morphology) of hydrogels and biological systems. In a previous study in which freeze dried method has been used with cells, it was demonstrated that this technique preserved the scaffold structure [31]. In this method the freezing step is critical since it fixes and determines the size of ice crystal in the scaffold structure. In the present work the hydrogels were freeze-dried following the procedure described by Nedel and co-workers [32] so that the water movement could be prevented by cryofixation.

Figure 6 shows the SEM micrographs from the hydrogel **CTS₃-TCA₁₀** and DCNa-loaded hydrogel **CTS₃-TCA₁₀-DCNa₁**. **CTS₃-TCA₁₀** did not show a porous scaffold but a well-defined lamellar structure probably due to the homogeneity of system (Figure 6a–c). Nonetheless, these results are opposite to those obtained by other authors; for example, when a bulky ionic cross-linker was used such as guanosine 5′-diphosphate (GDP), the SEM images showed a porous micro-structure [7] The packing of the bulky cross-linking agent in the hydrogel structure can cause some micro-irregularities responsible for the porous microstructure encountered. In addition, the method used for the preparation of the samples also exerts a great impact on the final microstructures. For example, and in contrast with the results presented herein, when the liquid-nitrogen fracture surfaces of some hydrogels prepared by ionic cross-linking were characterized using SEM after lyophilization, highly porous surfaces were encountered [5].

The spatial architecture of the prepared loaded CTS-hydrogels (Figure 6d–f), displayed highly porous scaffolds in contrast with non-loaded systems. Leroux and coworkers also found some porous structures in CTS-based DDS by SEM. They observed that drug loaded chitosan/glycerophosphate gels presented a highly porous structure after 24 h of exposure to a continuous flow of phosphate buffered saline in drug release experiments [19]. Unfortunately, no SEM image of drug-loaded formulations prior to release trials was taken to corroborate the origin of the highly porous scaffolds in their formulations.

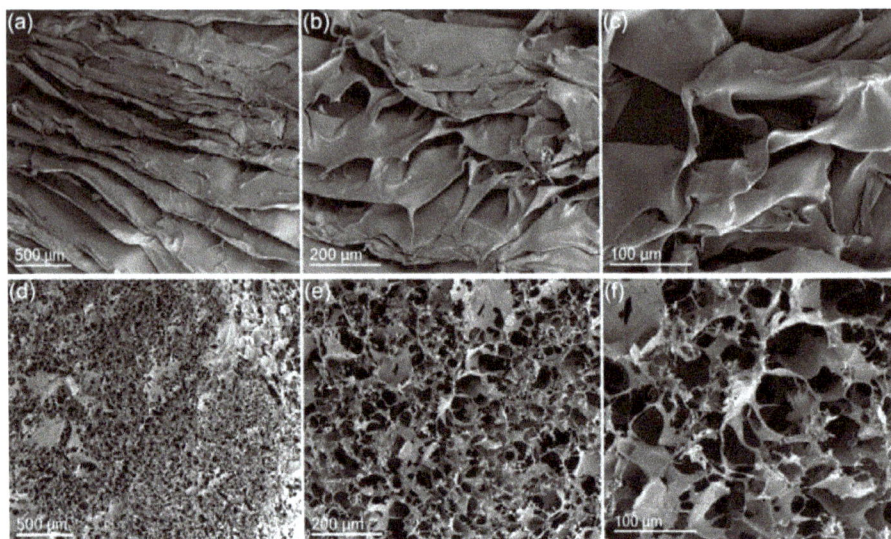

Figure 6. SEM images showing morphology of the as-synthesized unloaded chitosan-based hydrogel **CTS$_3$-TCA$_{10}$** (a–c) and loaded hydrogel **CTS$_3$-TCA$_{10}$-DCNa$_1$** (d–f).

In summary, the freeze-drying method used has proved to be a great method for fixation of CTS-based scaffolds for SEM studies since with the described lyophilization method, the pores, if present, were clearly distinguishable and the scaffold morphologies remained intact.

3.2.2. Spreadability

In the design of polymeric pharmaceutical formulations for topical applications, several required product characteristics may be defined. Of those, spreadability of the product on the surface (skin or mucosal epithelium) and adhesion of the formulation when it is dealing with a mucosal surface are included. Such properties contribute to the final clinical efficacy of the product [33]. Mucoadhesion is controlled by the affinity of the DDS for the mucin glycoproteins of the mucus. Polysaccharides are very good mucoadhesive polymers because of their non-toxic nature, and capacity to bind to mucins through either electrostatic or hydrophobic interactions. The amine and hydroxyl groups of CTS are involved in its excellent mucoadhesive properties, leading, for example, to prolonged residence time in the gastrointestinal tract [3].

Evaluation of the spreadability of the formulations and voltaren emulgel® was conducted for comparative purposes. Some formulations showed similar values of spreadability to the commercial hydrogel (see Table 2, **CTS$_3$-TCA$_{15}$-DCNa$_1$** and **CTS$_4$-TCA$_{10}$-DCNa$_1$**). It was observed that a reduction in CTS concentration led to low-structured formulations (**CTS$_2$-TCA$_{10}$-DCNa$_1$**) and hydrogels, in accordance with their poor rheological properties. Conversely, with an increasing concentration of the anionic drug, a significant reduction in spreadability was observed. This could be explained by means of the pseudo-cross-linking effect of DCNa in CTS-based hydrogels.

3.2.3. Diclofenac Sodium Release Studies

The exceptional biomedical properties of CTS have made it an ideal polymer for the preparation of muco-adhesive preparations that have been industrialized for ocular, nasal, buccal, gastrointestinal, and vaginal administrations with a great variety of drugs. Among them, DCNa, has been extensively used as a model anionic drug in numerous DDS formulations, particularly in CTS-based systems such as tablets, microspheres, microparticles and gels, due to its ionic linkage with CTS at acidic pH [29]. For example, the preparation of chitosan hydrogels bearing diclofenac, acetylsalicylic acid, and hydrocortisone acetate as anti-inflammatory drugs has been reported [34].

Additionally, macroporous CTS-nanographene oxide hybrid hydrogels have been shown to be effective adsorbent of this anti-inflammatory drug in wastewater [30]. However, as far as the authors are aware, an in-depth study of the influence of various parameters such as CTS concentration and drug content on the release profiles of an anionic drug in CTS-based formulations is still lacking in the literature.

The formulations prepared in the present work are recorded in Table 2. From the samples studied, it was observed that all the hydrogels were able to control the DCNa release for long periods of time at 37 °C in acetate buffer at pH 5.5 (Figure 7), close to the acidic pH which corresponds to the reduced pH microenvironment typical of cancerous cells (pH 6.5) [35]. The drug was released in a slow and sustained manner, ranging from 9% to 67% after 96 h. The influence of several factors, such as (a) the degree of crosslinking, (b) the CTS concentration, and (c) the drug content, on the release profiles has been studied.

It was first investigated the degree of cross-linking. The other two variables, concentration of CTS and drug content, were fixed at 3% and 1%, respectively. The results are recorded in Figure 7a. Regarding the degree of cross-linking, the percentage of DCNa delivered (after 96 h) ranged from 22% to 47% (samples **CTS$_3$-TCA$_{10}$-DCNa$_1$** and **CTS$_3$-TCA$_5$-DCNa$_1$**, respectively), displaying the formulation with a degree of crosslinking of 10% the most controlled release pattern. Surprisingly, a degree of crosslinking greater than or less than 10% unequally affected the release of DCNa. Thus, for example, when the crosslinking was 5%, a boost in the drug release was observed, being the latter even greater than those results obtained from CTS-based dispersions without TCA in the formulation. It is hypothesized that the presence of TCA alters, to a certain extent, the hydrogen bonds that keep the CTS chains tight together. Thus, more free volume can be found, which allows a quicker diffusion of the drug to the medium.

When the concentration of CTS was investigated (Figure 7b), the degree of cross-linking was set at 10% for all samples and the concentration of the drug in the formulation was fixed at 1%. It was observed that the hydrogels displayed improved control over the drug release when the CTS concentration was 3% or higher (with percentages of DCNa release below 25% after 96 h). In contrast, CTS concentrations inferior to 3% conducted to a lower drug-controlled efficiency, causing a two-fold increase in drug release.

Lastly, the amount of drug in the formulation exerted a significant impact on the kinetics of drug release (Figure 7c). For the samples studied, the CTS concentration and degree of cross-linking were set at 3% and 10%, respectively and the drug concentration ranged from 0.5% to 2%. For comparative purpose, Voltaren emulgel® from Novartis Pharmaceuticals was investigated, with a 1% concentration of diclofenac diethylammonium in its formulation. In all cases, except for the commercial product with a boosted release from the first few hours, the higher the diclofenac content in the prepared samples, the lower the DCNa released from the formulations, generating well-controlled delivery formulations. With a pKa of 4.1, the non-steroidal anti-inflammatory agent diclofenac is anticipated that would mostly be in its anionic form under the experimental conditions, interacting ionically with the ammonia residues of CTS. It should be noted that the ionized carboxylate group equally distributes its negative charge between the two oxygen atoms from the functional group and, therefore, ionic attractions with two ammonia groups could be possible, acting the carboxylate group as a pseudo-cross-linker in the dispersion. These hypotheses can also explain the reduction in spreadability of samples with the same CTS content and degree of crosslinking (see Table 2, samples **CTS$_3$-TCA$_{10}$-DCNa$_{0.5}$,** **CTS$_3$-TCA$_{10}$-DCNa$_1$** and **CTS$_3$-TCA$_{10}$-DCNa$_2$**). Consequently, at the highest DCNa concentration studied, the release of the drug is almost totally impeded, with values lower than 10% in 96 h.

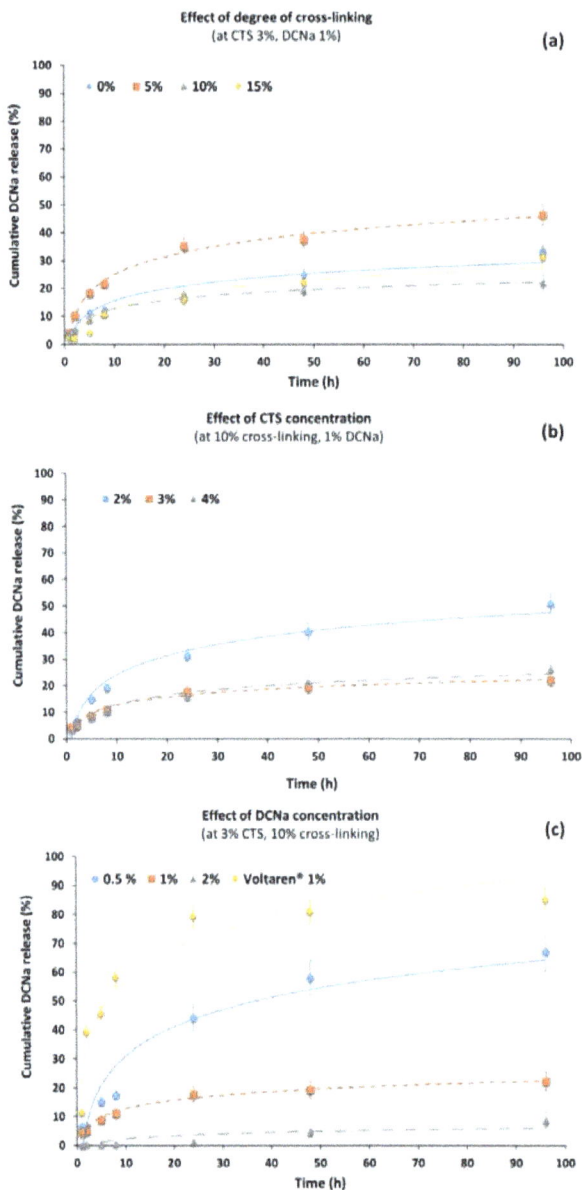

Figure 7. In vitro release profiles of diclofenac sodium (DCNa) from chitosan hydrogels in acetate buffer solution at pH 5.5 at 37 °C. Data were obtained from UV-Vis spectroscopy at 280 nm and reported as mean ± S.D from five independent experiments. (**a**) Effect of degree of cross-linking from non-cross-linked samples to 15% of cross-linked (CTS concentration 3%; DCNa concentration: 1%). (**b**) Effect of CTS concentration from 2% to 4% CTS concentration (degree of cross-linking: 10%; DCNa concentration: 1%). (**c**) Effect of DCNa concentration from 0.5% to 2% DCNa concentration (degree of cross-linking: 10%; CTS concentration: 3%).

4. Conclusions

This study reports the preparation and use as drug carriers of a new family of eco-friendly reversible chitosan hydrogels highly compatible with biological compounds. The hydrogels have been successfully prepared by ionic cross-linking with propane-1,2,3-tricarboxylic acid (tricarballylic acid, TCA).

The rheological properties of CTS_x-TCA_y hydrogels could be modulated by modifying the CTS and TCA concentrations; thus, low CTS concentrations produced liquid-like viscoelastic dispersions in contrast to the strong gel-like features found at high CTS, whereas an intermediate response was observed at 3% w/w CTS. On the contrary, at high TCA concentration the dispersions mainly exhibited a viscous response, whereas at an intermediate TCA concentration, G' coincided with G'' or typical sol-gel transition responses were found. Viscosity values increased with increasing the CTS concentration and generally decreased with increasing the degree of crosslinking.

The thermal transitions gel-sol were investigated by MTDSC and the study was focused on the reversible heat-flow plot. Some transitions gel-sol took place in a very narrow interval confirming the thermo-reversible nature of **CTS_3-TCA_{10}** and **CTS_2-TCA_{15}** hydrogels. The presence of a broader peak and the shift of the transition endotherm to lower values was observed when the cross-linker was lacking in the formulation (sample **CTS_4-TCA_0**). The spatial architecture of non-loaded and loaded CTS-hydrogels were studied and they markedly differed in their microstructure, displaying the latter highly-porous scaffolds in contrast with non-loaded systems.

Eight diclofenac sodium formulations were prepared and the relationships between polymer concentration, drug loading and degree of cross-linking with the release profiles were evaluated. They displayed a well-controlled drug release strongly dependent on the formulation composition and cumulative drug release under physiological conditions for 96 h varied from 8% to 67%. For comparative purpose, a commercial formulation, Voltaren emulgel® from Novartis Pharmaceuticals, with 1% of diclofenac concentration, was also investigated. In conclusion, the preparation method of CTS-based dispersions presented herein provides a simple and easy method to tailor-made controlled-release systems with improved rheological properties.

Acknowledgments: The authors would like to thank the Ministerio de Economía y Competitividad, Spain (Grant MAT2016-77345-C3-2-P) and the Junta de Andalucía, Spain (Grant P12-FQM-1553) of Spain for financial support. The authors also thank Francisco Miguel Morales and Bertrand Lacroix for their contribution in the study of CTS-crosslinked hydrogels micro-morphologies by means of scanning electron microscope.

Author Contributions: All the authors of the present work have been involved in the different stages of the research carried out, i.e., they have conceived and designed the experiments, performed them, analyzed the data and wrote the paper.

Conflicts of Interest: The authors declare no conflict of interest.

References

1. Hamidi, M.; Azadi, A.; Rafiei, P. Hydrogel nanoparticles in drug delivery. *Adv. Drug Deliv. Rev.* **2008**, *60*, 1638–1649. [CrossRef] [PubMed]
2. Hennink, W.E.; Van Nostrum, C.F. Novel crosslinking methods to design hydrogels. *Adv. Drug Deliv. Rev.* **2012**, *64*, 223–236. [CrossRef]
3. Bhattarai, N.; Gunn, J.; Zhang, M. Chitosan-based hydrogels for controlled, localized drug delivery. *Adv. Drug Deliv. Rev.* **2010**, *62*, 83–99. [CrossRef] [PubMed]
4. Sahoo, D.; Sahoo, S.; Mohanty, P.; Sasmal, S.; Nayak, P.L. Chitosan: A New Versatile Bio-polymer for Various Applications. *Des. Monomers Polym.* **2009**, *12*, 377–404. [CrossRef]
5. Xu, Y.; Han, J.; Lin, H. Fabrication and characterization of a self-crosslinking chitosan hydrogel under mild conditions without the use of strong bases. *Carbohydr. Polym.* **2017**, *156*, 372–379. [CrossRef] [PubMed]
6. Vivek, R.; Thangam, R.; Nipunbabu, V.; Ponraj, T.; Kannan, S. Oxaliplatin-chitosan nanoparticles induced intrinsic apoptotic signaling pathway: A "smart" drug delivery system to breast cancer cell therapy. *Int. J. Biol. Macromol.* **2014**, *65*, 289–297. [CrossRef] [PubMed]

7. Liu, L.; Gao, Q.; Lu, X.; Zhou, H. In situ forming hydrogels based on chitosan for drug delivery and tissue regeneration. *Asian J. Pharm. Sci.* **2016**, *11*, 673–683. [CrossRef]

8. Rafique, A.; Zia, K.M.; Zuber, M.; Tabasum, S. Chitosan functionalized poly(vinyl alcohol) for prospects biomedical and industrial applications: A review. *Int. J. Biol. Macromol.* **2016**, *87*, 141–154. [CrossRef] [PubMed]

9. Chenite, A.; Chaput, C.; Wang, D.; Combes, C.; Buschmann, M.; Hoemann, C.; Leroux, J.; Atkinson, B.; Binette, F.; Selmani, A. Novel injectable neutral solutions of chitosan form biodegradable gels in situ. *Biomaterials* **2000**, *21*, 2155–2161. [CrossRef]

10. Xu, Y.; Zhan, C.; Fan, L.; Wang, L.; Zheng, H. Preparation of dual crosslinked alginate-chitosan blend gel beads and in vitro controlled release in oral site-specific drug delivery system. *Int. J. Pharm.* **2007**, *336*, 329–337. [CrossRef] [PubMed]

11. Ferris, C.; Casas, M.; Lucero, M.J.; De Paz, M.V.; Jimenez-Castellanos, M.R. Synthesis and characterization of a novel chitosan-*N*-acetyl-homocysteine thiolactone polymer using MES buffer. *Carbohydr. Polym.* **2014**, *111*, 125–132. [CrossRef] [PubMed]

12. Lucero, M.J.; Ferris, C.; Sanchez-Gutierrez, C.A.; Jimenez-Castellanos, M.R.; De-Paz, M.-V. Novel aqueous chitosan-based dispersions as efficient drug delivery systems for topical use. Rheological, textural and release studies. *Carbohydr. Polym.* **2016**, *151*, 692–699. [CrossRef] [PubMed]

13. Dash, M.; Chiellini, F.; Ottenbrite, R.M.; Chiellini, E. Chitosan—A versatile semi-synthetic polymer in biomedical applications. *Prog. Polym. Sci.* **2011**, *36*, 981–1014. [CrossRef]

14. Wolinsky, J.B.; Colson, Y.L.; Grinstaff, M.W. Local drug delivery strategies for cancer treatment: Gels, nanoparticles, polymeric films, rods, and wafers. *J. Control. Release* **2012**, *159*, 14–26. [CrossRef] [PubMed]

15. Anitha, A.; Sowmya, S.; Kumar, P.T.S.; Deepthi, S.; Chennazhi, K.P.; Ehrlich, H.; Tsurkan, M.; Jayakumar, R. Chitin and chitosan in selected biomedical applications. *Prog. Polym. Sci.* **2014**, *39*, 1644–1667. [CrossRef]

16. Giri, T.K.; Thakur, A.; Alexander, A.; Ajazuddin; Badwaik, H.; Tripathi, D.K. Modified chitosan hydrogels as drug delivery and tissue engineering systems: Present status and applications. *Acta Pharm. Sin. B* **2012**, *2*, 439–449. [CrossRef]

17. Wu, W.; Shen, J.; Banerjee, P.; Zhou, S. Chitosan-based responsive hybrid nanogels for integration of optical pH-sensing, tumor cell imaging and controlled drug delivery. *Biomaterials* **2010**, *31*, 8371–8381. [CrossRef] [PubMed]

18. Yomota, C.; Miyazaki, T.; Okada, S. Determination of the viscometric constants for chitosan and the application of universal calibration procedure in its gel permeation chromatography. *Colloid Polym. Sci.* **1993**, *271*, 76–82. [CrossRef]

19. Ruel-Gariépy, E.; Chenite, A.; Chaput, C.; Guirguis, S.; Leroux, J.C. Characterization of thermosensitive chitosan gels for the sustained delivery of drugs. *Int. J. Pharm.* **2000**, *203*, 89–98. [CrossRef]

20. Düwel, S.; Hundshammer, C.; Gersch, M.; Feuerecker, B.; Steiger, K.; Buck, A.; Walch, A.; Haase, A.; Glaser, S.J.; Schwaiger, M.; et al. Imaging of pH in vivo using hyperpolarized ^{13}C-labelled zymonic acid. *Nat. Commun.* **2017**, *8*, 15126. [CrossRef] [PubMed]

21. Lu, L.; Liu, X.; Tong, Z. Critical exponents for sol-gel transition in aqueous alginate solutions induced by cupric cations. *Carbohydr. Polym.* **2006**, *65*, 544–551. [CrossRef]

22. Gupta, R.K. *Polymer and Composite Rheology*, 2nd ed.; Marcel Dekker, Inc.: New York, NY, USA, 2000; ISBN 0-8247-9922-4.

23. Macosko, C. *Rheology: Principles, Measurements and Applications*; VCH Publishers Inc.: New York, NY, USA, 1994; ISBN 978-0-471-18575-8.

24. Geever, L.M.; Mínguez, C.M.; Devine, D.M.; Nugent, M.J.D.; Kennedy, J.E.; Lyons, J.G.; Hanley, A.; Devery, S.; Tomkins, P.T.; Higginbotham, C.L. The synthesis, swelling behaviour and rheological properties of chemically crosslinked thermosensitive copolymers based on *N*-isopropylacrylamide. *J. Mater. Sci.* **2007**, *42*, 4136–4148. [CrossRef]

25. Galbis, E.; De Paz, M.V.; McGuinness, K.L.; Angulo, M.; Valencia, C.; Galbis, J.A. Tandem ATRP/Diels-Alder synthesis of polyHEMA-based hydrogels. *Polym. Chem.* **2014**, *5*, 5391–5402. [CrossRef]

26. Boutris, C.; Chatzi, E.G.; Kiparissides, C. Characterization of the LCST behaviour of aqueous poly(*N*-isopropylacrylamide) solutions by thermal and cloud point techniques. *Polymer* **1997**, *38*, 2567–2570. [CrossRef]

Polymers **2018**, *10*, 392

27. Otake, K.; Inomata, H.; Konno, M.; Saito, S. Thermal Analysis of the Volume Phase Transition with *N*-Isopropylacrylamide Gels. *Macromolecules* **1990**, *23*, 283–289. [CrossRef]

28. Dinu, I.A.; Duskey, J.T.; Car, A.; Palivan, C.G.; Meier, W. Engineered non-toxic cationic nanocarriers with photo-triggered slow-release properties. *Polym. Chem.* **2016**, *7*, 3451–3464. [CrossRef]

29. Enescu, D.; Olteanu, C.E. Functionalized Chitosan and Its Use in Pharmaceutical, Biomedical, and Biotechnological Research. *Chem. Eng. Commun.* **2008**, *195*, 1269–1291. [CrossRef]

30. Aiedeh, K.M.; Taha, M.O.; Al-Hiari, Y.; Bustanji, Y.; Alkhatib, H.S. Effect of Ionic Crosslinking on the Drug Release Properties of Chitosan Diacetate Matrices. *J. Pharm. Sci.* **2007**, *96*, 38–43. [CrossRef] [PubMed]

31. Favi, P.M.; Benson, R.S.; Neilsen, N.R.; Hammonds, R.L.; Bates, C.C.; Stephens, C.P.; Dhar, M.S. Cell proliferation, viability, and in vitro differentiation of equine mesenchymal stem cells seeded on bacterial cellulose hydrogel scaffolds. *Mater. Sci. Eng. C* **2013**, *33*, 1935–1944. [CrossRef] [PubMed]

32. Palma, S.B.; Nedel, F.; Perelló, F.C.; Marques, S.R.; Fernandes, S.A.; Fernando, D.; Flávio, V.C.N.L. Comparing different methods to fix and to dehydrate cells on alginate hydrogel scaffolds using scanning electron microscopy. *Microsc. Res. Tech.* **2015**, *78*, 553–561. [CrossRef]

33. Ferrari, F.; Bertoni, M.; Caramella, C.; La Manna, A. Description and validation of an apparatus for gel strength measurements. *Int. J. Pharm.* **1994**, *109*, 115–124. [CrossRef]

34. Dreve, S.; Kacso, I.; Popa, A.; Raita, O.; Dragan, F.; Bende, A.; Borodi, G.; Bratu, I. Structural investigation of chitosan-based microspheres with some anti-inflammatory drugs. *J. Mol. Struct.* **2011**, *997*, 78–86. [CrossRef]

35. Sun, T.; Zhang, Y.S.; Pang, B.; Hyun, D.C.; Yang, M.; Xia, Y. Engineered nanoparticles for drug delivery in cancer therapy. *Angew. Chem. Int. Ed.* **2014**, *53*, 12320–12364. [CrossRef] [PubMed]

polymers

MDPI

Article

Chitosan Composites Synthesized Using Acetic Acid and Tetraethylorthosilicate Respond Differently to Methylene Blue Adsorption

Thomas Y. A. Essel [1], Albert Koomson [1], Marie-Pearl O. Seniagya [1], Grace P. Cobbold [1], Samuel K. Kwofie [1,3], Bernard O. Asimeng [1], Patrick K. Arthur [2,3], Gordon Awandare [2,3] and Elvis K. Tiburu [1,3,*]

[1] Department of Biomedical Engineering, University of Ghana, P.O. Box LG 25, Legon, Ghana; tyaessel@st.ug.edu.gh (T.Y.A.E.); albert.koomson@stu.ucc.edu.gh (A.K.); moseniagya@st.ug.edu.gh (M.-P.O.S.); gracecobbold2016@gmail.com (G.P.C.); skwofie2000@gmail.com (S.K.K.); boasimeng@ug.edu.gh (B.O.A.)

[2] Department of Biochemistry, Cell and Molecular Biology, University of Ghana, P.O. Box LG 25, Legon, Ghana; parthur14@gmail.com (P.K.A.); gawandare@ug.edu.gh (G.A.)

[3] West Africa Center for Cell Biology of Infectious Pathogens (WACCBIP), University of Ghana, P.O. Box LG 25, Legon, Ghana

* Correspondence: etiburu@ug.edu.gh; Tel.: +233-559585194

Received: 20 February 2018; Accepted: 11 April 2018; Published: 24 April 2018

Abstract: The sol-gel and cross-linking processes have been used by researchers to synthesize silica-based nanostructures and optimize their size and morphology by changing either the material or the synthesis conditions. However, the influence of the silica nanostructures on the overall physicochemical and mechanistic properties of organic biopolymers such as chitosan has received limited attention. The present study used a one-step synthetic method to obtain chitosan composites to monitor the uptake and release of a basic cationic dye (methylene blue) at two different pH values. Firstly, the composites were synthesized and characterized by Fourier Transform Infrared Spectroscopy (FTIR) and X-ray Diffraction (XRD) to ascertain their chemical identity. Adsorption studies were conducted using methylene blue and these studies revealed that Acetic Acid-Chitosan (AA-CHI), Tetraethylorthosilicate-Chitosan (TEOS-CHI), Acetic Acid-Tetraethylorthosilicate-Chitosan (AA-TEOS-CHI), and Acetic Acid-Chitosan-Tetraethylorthosilicate (AA-CHI-TEOS) had comparatively lower percentage adsorbances in acidic media after 40 h, with AA-CHI adsorbing most of the methylene blue dye. In contrast, these materials recorded higher percentage adsorbances of methylene blue in the basic media. The release profiles of these composites were fitted with an exponential model. The R-squared values obtained indicated that the AA-CHI at pH ~2.6 and AA-TEOS-CHI at pH ~7.2 of methylene blue had steady and consistent release profiles. The release mechanisms were analyzed using Korsmeyer-Peppas and Hixson-Crowell models. It was deduced that the release profiles of the majority of the synthesized chitosan beads were influenced by the conformational or surface area changes of the methylene blue. This was justified by the higher correlation coefficient or Pearson's R values ($R \geq 0.5$) computed from the Hixson-Crowell model. The results from this study showed that two of the novel materials comprising acetic acid-chitosan and a combination of equimolar ratios of acetic acid-TEOS-chitosan could be useful pH-sensitive probes for various biomedical applications, whereas the other materials involving the two-step synthesis could be found useful in environmental remediation of toxic materials.

Keywords: TEOS; methylene blue; chitosan; modelling; cross-linking; interpenetrating; XRD; FTIR

Polymers **2018**, *10*, 466

1. Introduction

Biopolymers are macromolecules produced by living organisms; these include chitin, gelatin, cellulose, and a plethora of other diverse chemical entities. The properties of macromolecules are usually modified to make them suitable for most in vivo and in vitro studies [1,2]. Chitosan, the deacetylated derivative of chitin, is a linear cationic biopolymer of glucosamine residues, specifically N-acetyl-D-glucosamine and N-D-glucosamine, linked through β-(1–4)-glyosidic bonds. Due to its abundance, biocompatibility, and low cost, chitosan has been extensively used either in its natural form or as a derivative with other organic or inorganic molecules in various applications including bioseparation, drug delivery systems, environmental remediation, and imaging [3,4].

However, the high crystallinity, minimum surface area, high molecular weight, excessive swelling, and the degree of acetylation of chitosan impact negatively on its solubility, material forming capacity, biodegradability, and diverse bioactive attributes [5]. In addition, organic-based materials such as chitosan possess reactive functional groups including amines and hydroxyls, which are sites for attachment of other macromolecules and metal ions. For decades, efforts have been targeted at the modification of these biopolymers with suitable molecules for a wide range of applications [6]. For example, chemical cross-linkers such as acetic acid, glutaraldehyde, epichlorohydrin, and tetraethylorthosilicate (TEOS) have been used as alternatives to modify the structures of certain biopolymers including chitosan and, as a result, new functional materials with improved physicochemical properties have been produced [7]. Although these cross-linkers improve the mechanical strength, thermal stability, swelling ability, and pH sensitivity of the chitosan nanocomposite materials, glutaraldehyde, for example, tends to polymerize upon addition to the reaction mixture leading to loss of adsorption sites, thus diminishing the capacity of the resulting end product to be used in drug uptake and delivery [8]. Besides this, most cross-linkers are toxic and their usage in most biomedical applications is limited [9].

The sol-gel process has been used by researchers to synthesize silica nanostructures and optimize their size and morphology by changing either the materials or synthesis conditions [10–12]. It has been established that a mixture of an inorganic strong acid such as TEOS with a weak organic acid (acetic acid) can produce nanostructures with unique properties [13]. For example, chitosan can be cross-linked with TEOS to form an interpenetrative network (IPN), thus enhancing drug permeation as well as controlled release of food nutrients. Similarly, chitosan, when mixed with acetic acid, enhances gelation as well as drug entrapment. A hybrid polymer derived from siloxane and chitosan has also been obtained by the sol-gel technique using tetraethylorthosilicate (TEOS) as a precursor to immobilize enzymes for studying a wide range of applications including bio-sensing and bio-catalysis [14–16].

However, the influence of the silica nanostructures on the overall physicochemical properties of biopolymers such as chitosan has received limited attention. In most reaction conditions, a two-step reaction scheme is adopted: chitosan is first activated by acetic acid, followed by cross-linking with a suitable molecule [6,7,10,17]. The underlying question that the current work seeks to investigate is whether a one-step reaction pathway, using a mixture of equimolar concentrations of precursor molecules such as TEOS and acetic acid, can influence the overall structure and properties of the parent chitosan molecule. The present study therefore reports a one-step synthetic method of using chitosan composites to monitor the uptake and release of methylene blue at two different pH values, with a combination of equimolar mixture of acetic acid and TEOS. The materials' adsorption and release capacities of methylene blue is compared to the results of similarly synthesized chitosan composites using a two-step method at the acidic and basic conditions. Methylene blue was used as a model cationic dye to monitor the physicochemical properties of the materials because it has been established that anionic and cationic dyes, have differential properties depending on the pH conditions of the media. This work seeks to propose that the dye adsorptive properties towards the chitosan composites are dependent on pH conditions [18–20].

2. Materials and Methods

2.1. Materials

All analytical-grade chemicals were purchased from Sigma-Aldrich, St. Louis, MO, USA with the exception of TEOS, which was obtained from Philip Harris Ltd., Birmingham, UK. Chitosan (medium molecular weight M_w = 141 kDa, and degree of acetylation DA = 15–25%) was used as the parent biomolecule for the formation of the composites. Tetraethylorthosilicate (TEOS, 98% grade), a strong inorganic acid, and glacial acetic acid, a weak organic acid, were used as a cross-linker and an activator, respectively, to modify the chitosan.

2.2. Methods

2.2.1. Synthesis of Chitosan Beads Using Acetic Acid

Chitosan beads were synthesized with acetic acid according to methods described in previous publications [21,22] with slight modifications.

2.2.2. Synthesis of Chitosan Beads Using TEOS

Chitosan beads were synthesized with TEOS according to the above modified method [6]. Briefly, 2 g of chitosan was added to 2% *v/v* TEOS (2 mL TEOS, 6 mL 0.5 M HCl, and 92 mL deionized water) and the resulting mixture stirred until a uniform solution was obtained. The resulting gel was aspirated with a syringe and added in drops to 250 mL of 2 M NaOH solution. The beads formed in the NaOH solution were stirred at room temperature for 24 h using a magnetic stirrer. The beads were then filtered with deionized water until a neutral pH was achieved and then air dried.

2.2.3. Synthesis of Chitosan Beads Using Double Cross-Linking (Acetic Acid Followed by TEOS)

Chitosan beads were synthesized with acetic acid followed by TEOS according to the established method [23].

2.2.4. Synthesis of Chitosan Beads Using One-Step Method (Equimolar Concentrations of Acetic Acid and TEOS)

The chitosan beads were produced using the sol-gel method with slight modifications [17]. Briefly, 50 mL of 2% *v/v* acetic acid was added to 50 mL of 2% *v/v* TEOS (1 mL TEOS, 3 mL 0.5 M HCl, and 46 mL deionized water). A quantity of 2 g of chitosan was then added and the mixture was stirred until a uniform solution was obtained. The resulting gel was aspirated and added in drops to 250 mL of 2 M NaOH solution. The beads formed in the NaOH solution were stirred at room temperature for 24 h using a magnetic stirrer. The beads were then filtered with deionized water until a neutral pH was reached, and then air dried.

2.2.5. Sample Characterization

A PANanalytical Empyrean X-ray Diffractometer, Almelo, The Netherlands with Cu Kα (λ = 1.5406 Å) was used to acquire the XRD patterns for chitosan and its derivatives. The scan rate was $2°$ per minute in a range of 5–90°. The FTIR spectra were acquired over the region of 500–4000 cm^{-1} using a PerkinElmer spectrometer, Waltham, Massachusetts, USA. The FTIR scans were processed using the PerkinElmer Spectrum Version 10.03.09.I.

2.3. Analysis

2.3.1. Point of Zero Charge

The point of zero charge of each composite was estimated by using an established drift method [24]. Briefly, a fixed amount of each of the synthesized beads was added to 10 mL of deionized water with

Polymers **2018**, *10*, 466

varying pH values of 2, 4, 6, 8, 10, and 12. The samples were left to sit for 48 h after which pH readings were taken and a graph of the two pH values was recorded. The points of intersection of the two pH readings for each sample were recorded.

2.3.2. Swelling Ratio

Fixed amounts of the various synthesized beads were immersed in separate 10 mL buffers (pH ~2.6 and 7.2) and left overnight. The swelling ratio was calculated using the following equation by Park et al. [6].

$$Swelling\ ratio = \frac{Weight\ of\ wet\ or\ swelled\ beads - Weight\ of\ dried\ beads}{Weight\ of\ dried\ beads} \tag{1}$$

The following equation was used to estimate the %ε (percentage porosity) as described by Chatterjee et al. [5]:

$$\%\varepsilon = \frac{(W_W - W_D)/\rho_W}{\frac{W_D}{\rho_{Mat}} + (W_W - W_D)/\rho_W} \times 100\% \tag{2}$$

where W_W is the weight of the wet beads in grams before drying; W_D is the weight of the dry beads in grams; ρ_W is the density of water, 1.0 g/mL; and ρ_{Mat} (g/mL) is the material density of the dry bead.

2.3.3. Methylene Blue Adsorption and Release Properties of the Composites

Diluted methylene blue solutions adjusted to pH ~2.6 and 7.2 were prepared from 1 g/L stock methylene blue solution. Adsorbance values were recorded using a JENWAY, 6705 UV-Vis Spectrophotometer, Stone, Staffordshire, UK, at a wavelength of 661 nm. A quantity of 1.5 mL of the diluted dye was added separately to 50 mg of chitosan beads. The resulting solution was gently agitated for even dispersion of beads in the dye. The solution was left to stand for 16, 20, 24, or 40 h, after which 1 mL of the supernatant was used for UV-Vis adsorbance readings. Percentage changes in adsorbances at each time interval were calculated using the formula

$$\%\ change\ in\ adsorbance\ (methylene\ blue) = \frac{A_t - A_o}{A_o} \times 100 \tag{3}$$

where A_o and A_t are adsorbances at times 0 and t, respectively.

The release studies were estimated over 240 min using the adsorbent capacity formula as used by Chatterjee et al. [5]:

$$q = \frac{(C_o - C_{eq})}{W} \times V \tag{4}$$

where q is the adsorbent capacity, mg/g; C_o is the initial concentration of methylene blue, g/L; C_{eq} is the final or equilibrium concentration of methylene blue, mg/L; V is the volume of the experimental solution, L; and W is the dry weight of the hydrogel beads, g. An exponential fit was used to analyze the best release profile. Models used to describe the release mechanism were Korsmeyer-Peppas and Hixson-Crowell models [25].

3. Results

The functionalization of chitosan was undertaken using two cross-linkers at room temperature, and the reaction scheme is outlined in Table 1. The key difference between AA-CHI-TEOS and AA-TEOS-CHI is that the later was generated using a one-step scheme.

Table 1. Different reaction schemes for the synthesis of chitosan composites.

	Chitosan (CHI)	Acetic acid (AA)	TEOS
CHI	√		
AA-CHI	√	√	
TEOS-CHI	√		√
[a] AA-CHI-TEOS	√	√	√
[b] AA-TEOS-CHI	√	√	√

[a] AA-CHI-TEOS are beads synthesized by first activating chitosan with acetic acid and cross-linked with tetraethylorthosilicate (TEOS). [b] AA-TEOS-CHI are beads which were synthesized with a mixture of equimolar concentrations of both acetic acid (weak organic acid) and TEOS (strong inorganic acid).

In Figure 1, the X-ray diffraction patterns of the chitosan and its synthesized composites from acetic acid and TEOS using the different synthetic routes are displayed. The signature peaks of chitosan (CHI) revealed two unique broad peaks at 2θ degrees at positions 9.1° and 20.1°, indicating the presence of chitosan [26–28]. The additional peaks at 2θ degrees of 44.5, 64.9, and 78.0 are probably residual chitin peaks because of the incomplete deacetylation of the parent chitin. Except for AA-CHI-TEOS, all the other composites have similar XRD patterns. Also, additional peaks appeared at about 2θ degrees of 32, as well as the reduction of the background (broad peaks) in the AA-CHI-TEOS pattern, indicating the introduction of a crystalline phase into the sample.

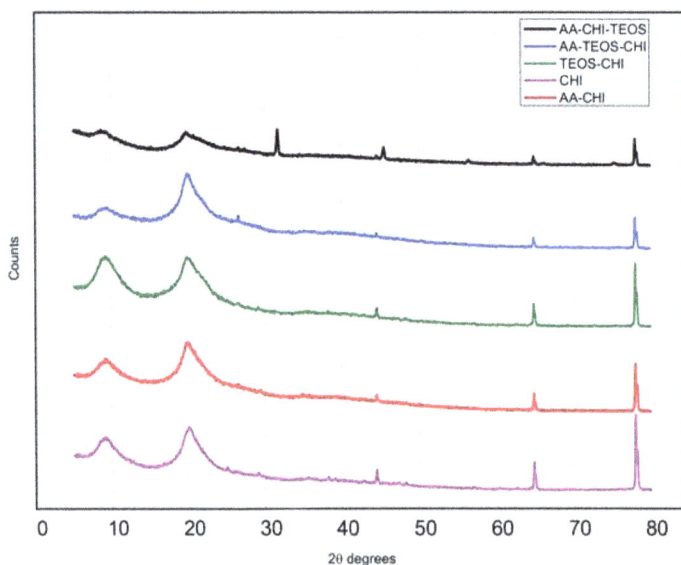

Figure 1. XRD patterns of chitosan and its derivatized composites obtained from acetic acid and TEOS.

The FTIR spectra were obtained within the range of 500–4000 cm^{-1}, but the major fingerprints of chitosan as shown in Figure 2 are between 1000 and 3345 cm^{-1}. Chitosan and its derivatives all showed broad percentage transmittances greater than 3000 cm^{-1} due to the stretching vibrations of the O–H and N–H bonds. The relative wavenumbers remained unchanged, but the percentage transmittances of the broad peaks differed based on the treatment of the parent chitosan with the various additives. The doublet peaks at 2921/2874 cm^{-1} are characteristic fingerprints of C–H due to symmetric and antisymmetric stretching vibrations. It is also interesting to note that the peaks below 2000 cm^{-1} have different percentage transmittances with AA-CHI having the highest, indicating that acetic acid treatment of chitosan was very efficient in cross-linking the organic additives with chitosan. Except for TEOS-CHI, all the composites showed higher percentage transmittances at wavenumbers below 2000 cm^{-1} compared with the pure chitosan. When chitosan was treated with only TEOS as in the TEOS-CHI sample, the percentage transmittances of the peaks seemed to diminish. The wavenumbers in the IR spectra of pure chitosan and the derivatized product clearly indicated that the parent chitosan was not chemically different from the product but rather that its percent transmittance was influenced by the organic acids.

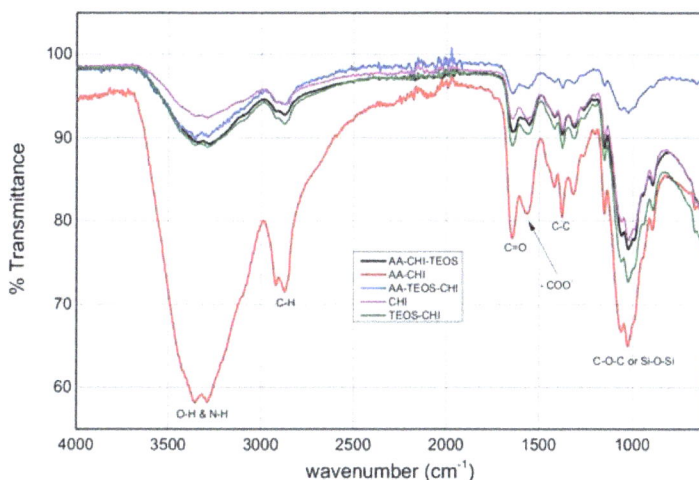

Figure 2. FTIR spectra of pure chitosan and its derivatized composites obtained from acetic acid and TEOS.

The points of zero charge (pH$_{pzc}$) of each of the materials as well as their water swelling capacities were estimated at different pH values; the results are displayed in Table 2. The percentage porosities (%ε) of all the synthesized materials are recorded in Table 2. Almost all the samples at pH = 2.6 revealed similar porosities of 86%, 87%, and 88% for AA-CHI, AA-CHI-TEOS, and AA-TEOS-CHI, respectively, except for TEOS-CHI which recorded a %ε of 92. However, different %ε values were obtained at pH = 7.2 for AA-CHI (83%), AA-TEOS-CHI (70%), AA-TEOS-CHI (77%), and TEOS-CHI (89%).

Table 2. Swelling studies conducted at different pH values. The point of zero charge (pH$_{pzc}$) of each sample is also indicated.

Sample	Swelling ratio		% Porosity (%ε)		pH$_{pzc}$
	pH = 2.6	pH = 7.2	pH = 2.6	pH = 7.2	
AA-CHI	3.0	2.4	86	83	7.6
TEOS-CHI	4.0	2.6	92	89	3.5
[a] AA-CHI-TEOS	4.0	2.0	87	77	7.1
[b] AA-TEOS-CHI	3.0	1.0	88	70	7.5

[a] AA-CHI-TEOS are beads synthesized by first activating chitosan with acetic acid and cross-linked with tetraethylorthosilicate (TEOS). [b] AA-TEOS-CHI are beads which were synthesized with a mixture of equimolar concentrations of both acetic acid (weak organic acid) and TEOS (strong inorganic acid).

The effect of pH on the adsorption behavior of methylene blue was investigated in both acidic and basic media in different time frames; the results are as shown in Figures 3 and 4. It was observed that after 16 h of adsorption, the trend seemed to favour AA-CHI, with AA-CHI-TEOS exhibiting the least adsorption at pH = 2.6, as shown in Figure 3. After 20 h, TEOS-CHI performed best among all the materials studied. AA-CHI-TEOS recorded a net negative percentage, indicating water uptake from the methylene blue solution. Similar trends were observed after 24 h with slight variation in the adsorption behavior between AA-CHI and AA-TEOS-CHI. The pattern seemed to favour TEOS-CHI, with AA-CHI-TEOS exhibiting the least adsorption behavior at pH = 2.6, as shown in Figure 3. However, after 40 h, all the materials, including AA-CHI-TEOS, showed significant amounts of dye adsorption. It was concluded that after 40 h, AA-CHI had a better adsorption capacity among all materials studied at pH = 2.6. Unlike at the acidic pH conditions, however, there was an overall increase in dye adsorption at pH = 7.2 with abrupt changes to the adsorption behavior of AA-CHI-TEOS after 40 h, as shown in Figure 4. The pH effect on the methylene blue released was also monitored at different time intervals, as shown in Figures 5 and 6.

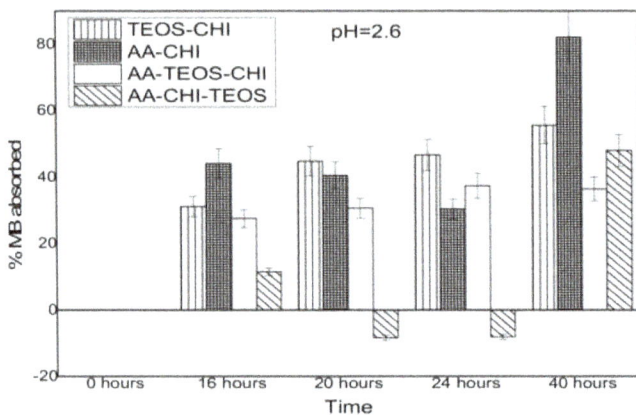

Figure 3. Adsorbance profile of methylene blue at pH = 2.6 as a function of time (hours).

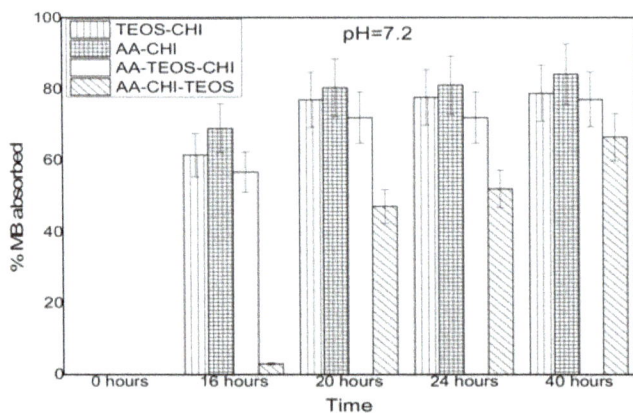

Figure 4. Adsorbance profile of methylene blue at pH = 7.2 as a function of time (hours).

The release profile plots of methylene blue over 240 min are shown in Figures 5 and 6 at pH = 2.6 and 7.2, respectively. At pH = 2.6, the release profile of AA-CHI follows an exponential order whereas the other materials fluctuate over the same time, as shown in Figure 5. However, at pH = 7.2, AA-TEOS-CHI exhibited a unique exponential order as opposed to the other materials as shown in Figure 6. It is interesting to note that AA-CHI, which exhibited progressive exponential order at a lower pH value in Figure 5, now behaves erratically at high pH conditions, and vice versa for AA-TEOS-CHI. This could be due to changes in the pore structures either through shrinkage in the pore size or the modification of certain functional groups within the pore matrix under different pH conditions. An exponential curve fitting model was used to extract R-squared values for comparison, as shown in Table 3. As observed, AA-CHI performs best, revealing R^2 values at pH = 2.6 and 7.2 of 0.989 and 0.757, respectively. Similarly, AA-TEOS-CHI also showed reasonable R^2 values of 0.658 and 0.977 at acidic and basic pH, respectively. Thus, the other functionalized materials could not follow the exponential order profile and were subjected to other mathematical models to explain whether the dye conformational properties could influence the materials' behavior. Nonetheless, as shown in Figure 7, AA-CHI released the dye efficiently at lower pH, whereas AA-TEOS-CHI performed better at higher pH with significant degree of accuracy based on the exponential fit model.

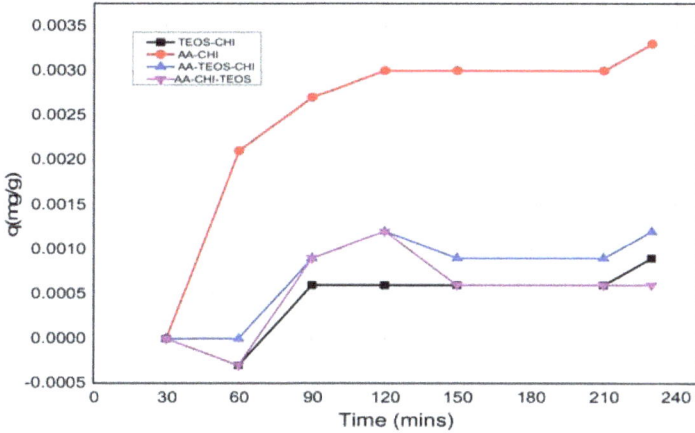

Figure 5. The release profiles of methylene blue at pH = 2.6 as a function of time.

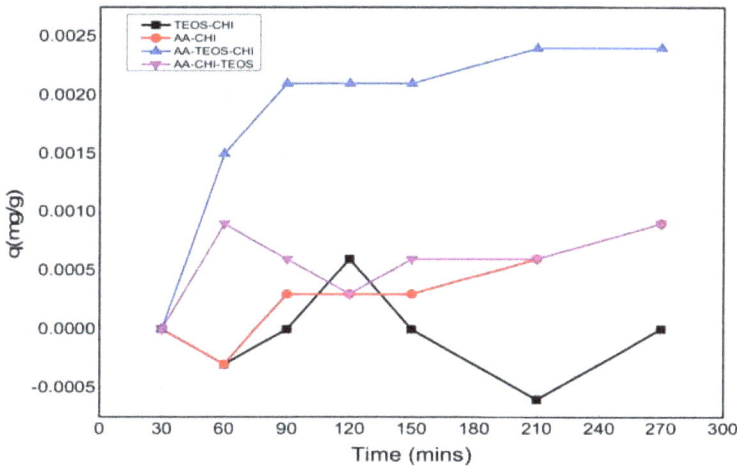

Figure 6. The release profiles of methylene blue at pH = 7.2 as a function of time.

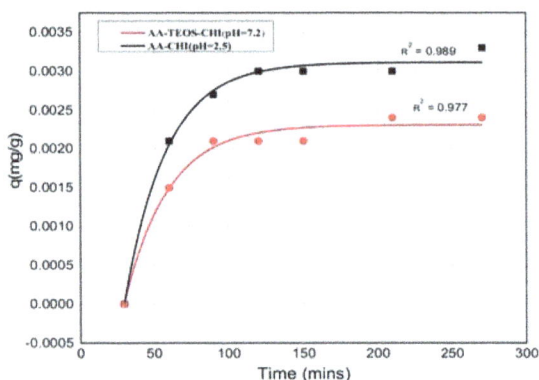

Figure 7. The exponential fit release profile of methylene blue from chitosan composites at two different pH values. The experimental and fitted values are represented by the scatter plots and the solid lines, respectively.

Table 3. A table of R-squared values obtained from an exponential fit of the release profile.

Samples	R^2 Values	
	pH = 2.6	pH = 7.2
AA-CHI	0.989	0.757
TEOS-CHI	0.539	−0.5
[a] AA-CHI-TEOS	0.156	0.380
[b] AA-TEOS-CHI	0.658	0.977

[a] AA-CHI-TEOS are beads synthesized by first activating chitosan with acetic acid and cross-linking with tetraethylorthosilicate (TEOS). [b] AA-TEOS-CHI are beads which were synthesized with a mixture of equimolar concentrations of both acetic acid (weak organic acid) and TEOS (strong inorganic acid).

4. Discussion

The X-ray diffraction patterns revealed the same positions but different intensities, which probably suggests the introduction of crystalline phase into the chitosan when the biopolymer was treated with the acids. This was especially true for AA-CHI-TEOS, where the background broad peaks had reduced drastically, leading to the appearance of additional peaks. The differences in the FTIR transmittance clearly indicated the modification of the parent chitosan through amide formation and other physicochemical changes caused by the two acids. To facilitate the interpretation of the materials under consideration, parameters such as pH_{pzc} and percentage porosity were carefully determined in order to provide the surface charge and porosity information that are crucial determinants of material behavior. All the materials synthesized except for TEOS-CHI were neutral in acidic media based on the point of zero charge (pH_{pzc}) values. In similar acidic conditions, methylene blue also existed as a cationic molecule and, as a result, only simple diffusion of methylene blue would be permissible for the dye uptake into these materials. However, approximately 30–45% of methylene blue had permeated AA-TEOS-CHI and AA-CHI-TEOS after 40 h, whereas within the same time frame, AA-CHI with comparable pH_{pzc} and %ε (% porosity) recorded almost 80% methylene blue adsorption. These observations underscored the importance of surface charge neutralization and porosity, which are the major determinants of dye adsorption into these materials. As a result, we attributed this behavior to the influence of the silica framework in the adsorption behavior of AA-TEOS-CHI and AA-CHI-TEOS. As previously reported, the swollen state of chitosan within the silica nanostructures at high pH, the insolubility of the chitosan composite, and the barrier created by the silica framework could impact the dye's diffusion into the interpenetration polymer pore network of these two materials, causing a significant reduction in methylene blue's uptake [6]. Such physicochemical factors were not available in AA-CHI to impede the simple diffusion of the dye into the activated chitosan. Based on the observations, it was speculated that inherent properties requiring energy for active transport of the methylene blue into AA-CHI-TEOS could be the driving force for the low uptake of the dye. This was especially true for AA-CHI-TEOS where water uptake was the preferred choice for the material as indicated in the negative adsorption profile.

At basic pH, there was rapid response (70–80%) in methylene blue uptake by all the materials studied after 40 h. It was observed that the percentage porosity of the chitosan composites drastically reduced in basic pH due to shrinkage of the chitosan within the silica framework, yet dye uptake was significantly higher compared with the values at high pH (acidic medium). This could be due to the biopolymer shrinkage within the silica framework that might be responsible for the elevated amount of dye uptake at low pH (basic medium). Again, AA-CHI seems to have the highest adsorption capacity among all the materials because of increased surface area at low pH.

To probe the release kinetics of entrapped methylene blue at the two pH conditions, the release isotherm model was used to monitor the release of the dye as a function of time in phosphate-buffered saline. The release kinetics of the entrapped methylene blue were evaluated using the Korsmeyer-Peppas ($\frac{M_t}{M_\infty} = Kt^n$) model, where $\frac{M_t}{M_\infty}$ is the fraction of the dye released at time t, K is the release rate constant, and n is the release exponent. The n values corresponding to the released mechanism of the dye from the material at different pH values were determined. The boundary conditions set by the above model to determine the type of release mechanisms of the dye were based on the n values. From our determination, the computed n values from this work concluded that the transport mechanism did not follow the Fickian diffusion mechanism, or other non-Fickian transport such as the Case II transport or Super Case II transport.

Our attention was then directed towards using the Hixson-Crowell Model to evaluate whether conformation changes of the dye could have an influence on the release process. The formula used to evaluate this assertion is given by $C_0^{\frac{1}{3}} - C_t^{\frac{1}{3}} = K_{HC}t$, where C_0 = initial amount of drug adsorbed, C_t = amount of drug remaining at time t, and K_{HC} = rate constant. A graph of ($C_0^{\frac{1}{3}} - C_t^{\frac{1}{3}}$) was plotted against t for the release profile of each composite synthesized and Pearson's R was recorded based on the linear fit constructed for each release profile. The boundary conditions for this model were that if the absolute value of the correlation coefficient or Pearson's R of the above equation was higher, then the release mechanism was as a result of a change in surface area/conformation of the dye. The Pearson's R for the release profile of each chitosan bead is reported in Table 4. From the table, it was concluded that change in surface area/conformation of the dye has a significant effect on the respective release profiles of AA-CHI, AA-TEOS-CHI, and AA-CHI-TEOS at pH = 2.6, while in the slightly basic pH, the release of TEOS-CHI, AA-CHI, and AA-TEOS-CHI was also strongly influenced by the change in surface area or conformation of the dye.

Table 4. Evaluation of the release mechanism of methylene blue using Hixson-Crowell model.

	Sample	TEOS-CHI	AA-CHI	AA-TEOS-CHI	AA-CHI-TEOS
Pearson's R	pH = 2.6	−0.18	0.92	0.76	0.52
	pH = 7.2	0.77	0.745	0.748	0.38

However, it is important to note that the steady rise in the release profile just before the plateau region suggests an asymptotic behavior which is due to the exponential release pattern. In the acidic medium, only AA-CHI revealed a release profile that follows an exponential order but none of the materials' behaviors were able to fit that model. As expected, the adsorption studies in acidic media wherein AA-CHI entrapped most of the dye after 40 h followed an exponential release curve with an R^2 value close to 1. The other composites at the same pH exhibited a release profile deviant from the exponential model and this is evident from the low R^2 values. In terms of the release profile in

Polymers **2018**, *10*, 466

basic medium, AA-TEOS-CHI followed an exponential profile reminiscent of what is observed at pH = 2.6 for AA-CHI. In terms of the practical applications of the materials synthesized, AA-CHI and AA-TEOS-CHI could be useful for wider biomedical applications including drug entrapment and release due to their steady mechanistic behavior, whereas the other materials could find use in environmental remediation of toxic materials.

5. Conclusions

The current study highlights a unique opportunity to utilize a combination of organic and inorganic precursors in either a one-step or two-step reaction mechanism to produce new functional materials that could either be used for targeted drug delivery or for environmental remediation. It has been revealed that two of the materials comprising acetic acid-chitosan and a combination of equimolar ratios of acetic acid-TEOS-chitosan could be useful pH-sensitive probes for various biomedical applications, whereas the other materials from the two-step synthesis could be useful in environmental remediation of toxic materials.

Acknowledgments: We would like to acknowledge support from the West African Centre for Cell Biology of Infectious Pathogens (WACCBIP). The authors also thank the Office of Research, Innovation and Development (ORID) of University of Ghana for funding. We would also like to thank Beatrice Agyapomah at the XRD Facility in University of Ghana as well as John Kutor for the expert advice. We would also like to thank Robert E. Armah-Sekum, Issah Ibrahim and Nadia K. Amoateng for proof reading the manuscript. We also thank the Office of Research, Innovation and Development for supporting this work, Grant #: UGRF/9/LMG-011(EKT). The work also received support from grants from the Wellcome Trust (WACCBIP DELTAS grant; 107755/Z/15/Z awarded to GAA), and the World Bank Africa Centres of Excellence project (ACE02-WACCBIP) awarded to (GAA).

Author Contributions: Thomas Y. A. Essel, Albert Koomson, Marie-Pearl O. Seniagya and Grace P. Cobbold synthesized the composites and conducted swelling test, dye uptake, release studies and other physicochemical analysis. Samuel K. Kwofie, part of the interpretation and manuscript editing. Bernard O. Asimeng, part of the interpretation and manuscript editing. Patrick K. Arthur did data interpretation and edit the paper. Gordon Awandare contributed analysis tools, laboratory resources and structure of the manuscript. Elvis K. Tiburu conceived the idea, designed the experiments and wrote the manuscript.

Conflicts of Interest: The authors declare no conflict of interest.

References

1. Algul, D.; Gokce, A.; Onal, A.; Servet, E.; Dogan Ekici, A.I.; Yener, F.G. In vitro release and In vivo biocompatibility studies of biomimetic multilayered alginate-chitosan/beta-TCP scaffold for osteochondral tissue. *J. Biomater. Sci. Polym. Ed.* **2016**, *27*, 431–440. [CrossRef] [PubMed]

2. Koc, A.; Finkenzeller, G.; Elcin, A.E.; Stark, G.B.; Elcin, Y.M. Evaluation of adenoviral vascular endothelial growth factor-activated chitosan/hydroxyapatite scaffold for engineering vascularized bone tissue using human osteoblasts: In vitro and in vivo studies. *J. Biomater. Appl.* **2014**, *29*, 748–760. [CrossRef] [PubMed]

3. Bernkop-Schnurch, A.; Dunnhaupt, S. Chitosan-based drug delivery systems. *Eur. J. Pharm. Biopharm.* **2012**, *81*, 463–469. [CrossRef] [PubMed]

4. Cabral, J.D.; McConnell, M.A.; Fitzpatrick, C.; Mros, S.; Williams, G.; Wormald, P.J.; Moratti, S.C.; Hanton, L.R. Characterization of the in vivo host response to a bi-labeled chitosan-dextran based hydrogel for postsurgical adhesion prevention. *J. Biomed. Mater. Res. A* **2015**, *103*, 2611–2620. [CrossRef] [PubMed]

5. Chatterjee, S.; Chatterjee, T.; Lim, S.R.; Woo, S.H. Adsorption of a cationic dye, methylene blue, on to chitosan hydrogel beads generated by anionic surfactant gelation. *Environ. Technol.* **2011**, *32*, 1503–1514. [CrossRef] [PubMed]

6. Park, S.B.; You, J.O.; Park, H.Y.; Haam, S.J.; Kim, W.S. A novel pH-sensitive membrane from chitosan-TEOS IPN; preparation and its drug permeation characteristics. *Biomaterials* **2001**, *22*, 323–330. [CrossRef]

7. Jothimani, B.; Sureshkumar, S.; Venkatachalapathy, B. Hydrophobic structural modification of chitosan and its impact on nanoparticle synthesis—A physicochemical study. *Carbohydr. Polym.* **2017**, *173*, 714–720. [CrossRef] [PubMed]

8. Cheng, M.; Han, J.; Li, Q.; He, B.; Zha, B.; Wu, J.; Zhou, R.; Ye, T.; Wang, W.; Xu, H.; et al. Synthesis of galactosylated chitosan/5-fluorouracil nanoparticles and its characteristics, in vitro and in vivo release studies. *J. Biomed. Mater. Res. B Appl. Biomater.* **2012**, *100*, 2035–2043. [CrossRef] [PubMed]

9. Delval, F.; Crini, G.; Vebrel, J. Removal of organic pollutants from aqueous solutions by adsorbents prepared from an agroalimentary by-product. *Bioresour. Technol.* **2006**, *97*, 2173–2181. [CrossRef] [PubMed]

10. Xin, C.; Zhao, N.; Zhan, H.; Xiao, F.; Wei, W.; Sun, Y. Phase transition of silica in the TMB-P123-H₂O-TEOS quadru-component system: A feasible route to different mesostructured materials. *J. Colloid Interface Sci.* **2014**, *433*, 176–182. [CrossRef] [PubMed]

11. Budnyak, T.M.; Pylypchuk, I.V.; Tertykh, V.A.; Yanovska, E.S.; Kolodynska, D. Synthesis and adsorption properties of chitosan-silica nanocomposite prepared by sol-gel method. *Nanoscale Res. Lett.* **2015**, *10*, 87. [CrossRef] [PubMed]

12. Toskas, G.; Cherif, C.; Hund, R.D.; Laourine, E.; Mahltig, B.; Fahmi, A.; Heinemann, C.; Hanke, T. Chitosan(PEO)/silica hybrid nanofibers as a potential biomaterial for bone regeneration. *Carbohydr. Polym.* **2013**, *94*, 713–722. [CrossRef] [PubMed]

13. Pipattanawarothai, A.; Suksai, C.; Srisook, K.; Trakulsujaritchok, T. Non-cytotoxic hybrid bioscaffolds of chitosan-silica: Sol-gel synthesis, characterization and proposed application. *Carbohydr. Polym.* **2017**, *178*, 190–199. [CrossRef] [PubMed]

14. Wang, D.; Liu, W.; Feng, Q.; Dong, C.; Liu, Q.; Duan, L.; Huang, J.; Zhu, W.; Li, Z.; Xiong, J.; et al. Effect of inorganic/organic ratio and chemical coupling on the performance of porous silica/chitosan hybrid scaffolds. *Mater. Sci. Eng. C Mater. Biol. Appl.* **2017**, *70*, 969–975. [CrossRef] [PubMed]

15. Matsuhisa, H.; Tsuchiya, M.; Hasebe, Y. Protein and polysaccharide-composite sol-gel silicate film for an interference-free amperometric glucose biosensor. *Colloids Surf. B Biointerfaces* **2013**, *111*, 523–529. [CrossRef] [PubMed]

16. Li, F.; Li, J.; Zhang, S. Molecularly imprinted polymer grafted on polysaccharide microsphere surface by the sol-gel process for protein recognition. *Talanta* **2008**, *74*, 1247–1255. [CrossRef] [PubMed]

17. Elsagh, A. Synthesis of Silica Nanostructures and Optimization of their Size and Morphology by Use of Changing in Synthesis Conditions. *J. Chem.* **2011**, *9*, 659–668. [CrossRef]

18. Gurses, A.; Dogar, C.; Yalcin, M.; Acikyildiz, M.; Bayrak, R.; Karaca, S. The adsorption kinetics of the cationic dye, methylene blue, onto clay. *J. Hazard. Mater.* **2006**, *131*, 217–228. [CrossRef] [PubMed]

19. Dos Santos, C.C.; Mouta, R.; Junior, M.C.C.; Santana, S.A.A.; Silva, H.; Bezerra, C.W.B. Chitosan-edible oil based materials as upgraded adsorbents for textile dyes. *Carbohydr. Polym.* **2018**, *180*, 182–191. [CrossRef] [PubMed]

20. Ju, H.; Zhou, J.; Cal, C.; Chen, H. The Electrochemical Behaviour of Methylene Blue at a Microcylinder Carbon Fiber Electrode. *Electroanalysis* **1994**, *7*, 1165–1169. [CrossRef]

21. Kamari, A.; Ngah, W.S.W.; Liew, L.K. Chitosan and Chemically Modified Beads for acid dyes sorption. *J. Environ. Sci.* **2009**, *21*, 296–302.

22. Tiburu, E.K.; Fleischer, H.N.A.; Aidoo, E.O.; Salifu, A.A.; Asimeng, B.O.; Zhou, H. Crystallization of Linde Type A nanomaterials at two Temperatures Exhibit Differential Inhibition of HeLa Cervical Cancer Cells In vitro. *J. Biomim. Biomater. Biomed. Eng.* **2016**, *28*, 66–77. [CrossRef]

23. Copello, J.G.; Villanueva, M.E.; Gonzalez, J.A.; Eguez, L.S.; Diaz, L.E. TEOS as an Improved Alternative for Chitosan Beads Cross-Linking: A Comparative Adsorption Study. *Appl. Polym.* **2014**, *131*, 41005–41012. [CrossRef]

24. Lopez-Ramon, M.V.; Stoeckli, F.; Moreno-Castilla, C.; Carrasco-Marin, F. On the characterization of acidic and basic surface sites on carbons by various technoques. *Carbon* **1999**, *37*, 1215. [CrossRef]

25. Gouda, R.; Baishya, H.; Qing, Z. Application of Mathematical Models in Drug Release Kinetics of Carbidopa and Levodopa ER Tablets. *J. Dev. Drugs* **2017**, *6*, 171.

26. Tiburu, E.K.; Salifu, A.A.; Aidoo, E.O.; Fleischer, H.N.A.; Manu, G.; Yaya, A.; Zhou, H.; Efavi, J.K. Formation of Chitosan Nanoparticles Using Deacetylated Chitin Isolated from Freshwater Algae and Locally Synthesized Zeolite A and their Influence on Cancer Cell Growth. *J. Nanopart. Res.* **2017**, *48*, 156. [CrossRef]

27. Argüelles-Monal, W.M.; Lizardi-Mendoza, J.; Fernández-Quiroz, D.; Recillas-Mota, M.T. Montiel-Herrera, M. Chitosan Derivatives: Introducing New Functionalities with a Controlled Molecular Architecture for Innovative Materials. *Polymers* **2018**, *10*, 342. [CrossRef]
28. Gohi, B.F.C.A.; Zeng, H.-Y.; Pan, A.D.; Han, J.; Yuan, J. pH Dependence of Chitosan Enzymolysis. *Polymers* **2017**, *9*, 174. [CrossRef]

polymers

MDPI

Review

Application of Chitin/Chitosan and Their Derivatives in the Papermaking Industry

Zhaoping Song *, Guodong Li, Feixiang Guan and Wenxia Liu

Key Laboratory of Pulp and Paper Science and Technology (Ministry of Education), Qilu University of Technology (Shandong Academy of Sciences), Jinan 250353, China; lgd@qlu.edu.cn (G.L.); 1043117338@stu.qlu.edu.cn (F.G.); liuwenxia@qlu.edu.cn (W.L.)
* Correspondence: zsong@qlu.edu.cn; Tel.: +86-158-6676-9018

Received: 5 February 2018; Accepted: 26 March 2018; Published: 27 March 2018

Abstract: Chitin/chitosan and their derivatives have become of great interest as functional materials in many fields within the papermaking industry. They have been employed in papermaking wet-end, paper surface coating, papermaking wastewater treatment, and other sections of the papermaking industry due to their structure and chemical properties. The purpose of this paper is to briefly discuss the application of chitin/chitosan and their derivatives in the papermaking industry. The development of their application in the papermaking area will be reviewed and summarized.

Keywords: chitosan; papermaking; wet-end; coating; wastewater

1. Introduction

Chitin is the second most abundant natural polymer in the world. The main sources are from two marine crustaceans, shrimp and crabs [1]. Chitin and chitosan are β(1–4) glycans whose chains are formed by 2-acetamide-2-deoxy-D-glucopyranose and 2-amino-2-deoxy-D-glucopyranose units, respectively. Chitosan is generally prepared by the deacetylation of chitin. Having a unique set of biological properties including biocompatibility, biodegradability, and low to absent toxicity [2], chitin and chitosan, as well as their derivatives, have been found to be attractive materials for some high value-added products, including: cosmetics, food additives, drugs carriers, pharmaceutics, and semi-permeable membranes [1–4].

There are some very detailed review papers on the introduction, processing, and application of chitin and chitosan, addressing the varied application of chitosan in many fields [1,5]. This paper aims to give a short review on the application of chitin/chitosan as well as their derivatives in the papermaking industry. In recent years, studies focused on investigating the applications of chitosan as a papermaking additive, for both internal and surface applications [6], improving the wet and dry strength of paper [7,8], demonstrating the compatibility of chitosan with paper stock components, and its ability to work as a retention and drainage additive [9,10], or as dye fixative in producing coloured paper [11,12]. Meanwhile, the inherent antibacterial properties and the film forming ability of chitosan are also studied for potential applications in papermaking, laying a foundation for fabricating functional papers such as antibacterial paper and greaseproof paper [13,14]. In addition, chitin/chitosan and their derivatives are also used widely as a chelating and coagulating agent for wastewater treatments [15], due to the sorption of dyes, humic acids, metallic ions, bacterial cells, and xenobiotics on chitin/chitosan in wastewater from papermaking and other industries, according to their unique characteristics and properties [16–20]. In this short review, the recent applications of chitin/chitosan and their derivatives in the papermaking industry will be summarized, and the development of their application in the papermaking field will be discussed.

2. Wet-End Application

Wet-end is where the slurry of fibers forms a wet paper web on the paper machine. At the wet-end, there is a continuous water phase and a dispersed phase of cellulose fibers. Wet-end chemistry is very complex since there are many components, including cellulose fibers, paper additives, fines, water, etc. Examples of paper additives used at the wet-end are retention aids, strength resins, internal sizing agents, fillers, and so on. The additives are used to achieve specific paper properties and enhance the paper machine efficiency. Due to the complex of the components in papermaking wet-end, the interactions between the cellulose fibers and the additives are very complex. Chitosan can be used in the wet-end of papermaking, for retention and drainage agents, strength agents, or sizing promoters. As for the application of chitosan in wet-end of papermaking, chitosan and their derivatives are first dissolved in water or acid solution, then the solution is added to the pulp suspension before the formation of paper sheets.

2.1. As Retention and Drainage Agents

Retention and drainage agents are very important wet-end additives in the papermaking process. They are used to promote the aggregation of fillers, fibers, and other fines in wet-end by electrostatic interactions. Retention and drainage agents are added in the papermaking wet-end to generate flocs by flocculation or coagulation which improves the retention of fillers and fines. If the flocs are formed by coagulation, the addition of retention and drainage aid also impart better drainability to the pulp suspension which allows higher paper machine speed. For retention and drainage agents, they can be natural or synthetic polymers, single- or multi-component systems [21–23]. Currently, natural polymers have become more popular than synthetic ones due to increasing environmental awareness [24]. Chitosan and its derivatives appear to be good candidates to be used as retention and drainage agents in papermaking industry and numerous application strategies of chitosan have already been investigated [10,25–33]. Nicu et al. compared the ability of three chitosans with different molecular weights to flocculate grounded calcium carbonate (GCC) and pulp/GCC suspensions in papermaking [34]. Chitosans with a higher molecular weight (MW) showed greater flocculation efficiency since it had a stronger affinity towards cellulose fibers [34]. Since the MW and degree of substitution (DS) of quaternary chitosan have a great influence on the properties of chitosan as a retention and drainage aid [34,35], N-(2-Hydroxyl-3-trimethylammonio)-propyl chitosan chloride with varying DS and MW were prepared and the effects of the DS and MW of the quaternary chitosan on its adsorption and flocculation properties in alkaline papermaking were studied [28]. Compared with a commercial cationic starch, the results showed that quaternary chitosan had a lower flocculation concentration and a higher flocculation performance, when used to induce the flocculation of $CaCO_3$ fillers in alkaline papermaking. It was also found that the absorption of chitosan on $CaCO_3$ fillers as well as the flocculation of $CaCO_3$ dispersion were significantly improved by increasing the DS of quaternary chitosan from 43% to 93% due to the enhanced electrostatic interactions between quaternary chitosan and negatively charged $CaCO_3$ particles [28]. Meanwhile, quaternary chitosan with lower MW demonstrated higher efficiency in inducing the flocculation of $CaCO_3$ particles when chitosan with high DS was used. Similarly, Chi et al. [10] reported the retention and drainage-aid behavior of quaternary chitosan(N-(2-Hydroxy-3-trimethylammonio)-propyl chitosan chlorider (C-CS)) as retention and drainage-aids for peroxide bleached reed kraft pulp in papermaking system. They found that the drainage rate of pulp suspension was increased significantly upon addition of C-CS, although, there was a small decrease in the mechanical propertiesshen of finished paper due to the improved retention of $CaCO_3$ fillers and fibrous fines. The reason for the higher retention rate of quaternary chitosan can be attributed to the electrostatic interaction between the quaternary chitosan and the cellulosic substrates or mineral fillers in the wet-end of a papermaking system [28]. When the chitosan was modified by introducing a quaternary ammonium group, the modified chitosan became soluble in both neutral and alkaline solutions and obtained good mineral binding properties, which is required for anchoring the mineral to the fibers. Furthermore, dual- or multi-component retention systems involving chitosan, such as a chitosan/bentonite particulate retention system, chitosan/silica (SiO$_2$) retention system [36,37], and chitosan/cellulose nanofiber (CNF) retention system, have also been used in the papermaking process [38,39]. Quaternary chitosan (QCS)/nano-SiO$_2$ retention aid system has been used in flocculation of reed pulp suspension [36]. The results showed that the flocculation was increased with the increasing of SiO$_2$ when the fiber substrates surfaces was net positively charged by an adsorbed QCS layer. The effect on the fiber flocculation of electrolyte concentration in the QCS-nano-SiO$_2$ system was governed not only by the adsorption of QCS onto the substrate surface but also by the interaction between polyelectrolyte and nano-SiO$_2$ particles [36]. Chitosan with a nano-silica retention/drainage system used in recycled cellulosic fibers was studied [37] and the results showed that retention and drainage in recycled waste office pulp was significantly improved in comparison to the control sample. The effect of chitosan/bentonite particulate retention system on the retention and drainage performance of the tobacco pulp, which was used to make tobacco sheets using papermaking technology, was studied [38]. The results showed that the chitosan/bentonite particle retention and drainage system can improve retention and drainage performance of the tobacco slurry significantly [38]. The employment of CNF combined with chitosan as a dual retention system in the papermaking process showed that the introduction of CNF in the presence of chitosan reduced drainage times [39].

2.2. As Paper Strength Agents

Strength properties, both dry and wet, are very important to the paper sheets and paper-based packaging materials. However, the hydrophilic nature of cellulose fibers limits the application of paper products, especially for those products used in humid surroundings that require high strength properties. There are two kinds of paper strength agents, including synthetic polymers and natural renewable polymers, used most frequently in the papermaking industry. Among the bio-based renewable polymers, chitosan with important functional groups such as hydroxyl, amino, and even acetamido groups has been found to be effective as a dry strength agent in the papermaking industry [28,37,40–48]. The structure of chitosan is similar as cellulose, making it possible to have strong bonding with fibers thus giving a dry and wet strength in papermaking. It was also observed that the amino groups on chitosan could react with cellulose's aldehydes and subsequently produce covalent bonds [49].

Chitosan can be used alone as a strength agent in papermaking. For example, the effect of chitosan on properties of handsheets made from bleached eucalypt pulp has been quantified [50], and the research concluded that chitosan has the potential to be used as a dry strength additive in neutral, acidic, or alkaline conditions depending on process requirements. Meanwhile, the effects of shrimp chitosan on the physical properties of handsheets were investigated by Khantayanuwong et al. [51]. The results showed that most of the mechanical properties of shrimp-chitosan-treated handsheets such as the bursting index, folding endurance, tensile index, modulus of elasticity, and tensile energy absorption were greatly increased with the addition of chitosan at 0.25–0.5 o.d. wt. % of pulp, except that there was no change in the tearing strength.

In order to achieve satisfying strength properties of paper sheets, chitosan is often used in combination with other products. For example, synergy of carboxymethyl cellulose (CMC) and modified chitosan [52], and synthesized chitosan-complexed starch nanoparticles [53] have been adopted to enhance the strength properties of a cellulosic fiber network. The combination of chitosan with bentonite microparticles to act as a wet-end additive system for paper reinforcement has been studied [54]. It was suggested that the bentonite may make a bridging between different chitosan molecules, making them act like a higher-mass cationic polymer and chitosan showed potential as a dry strength additive in mixed hardwood chemical-mechanical pulp in acidic pH. Additionally, chitosan was used in combination with cationic starch as dry-strength agents to improve the strength properties of bagasse paper [55]. The results showed that the dry-strength agent acted as a protecting film or glazed on the surfaces of bagasse paper handsheets, which had a positive impact on the pulp properties. This work showed the feasibility of using chitosan and cationic starch as dry-strength additives for application of non-wood materials in the paper industry. In another study, chitosan, cationic starch and poly vinyl alcohol (PVA) were used in various sequences to find out the optimal combination for improving both wet and dry tensile strength of old corrugated containerboard (OCC) furnish and the best results in wet and dry tensile strengths were simultaneously achieved using sequential PVA-chitosan-cationic starch [56]. Moreover, the synthesis and application of chitosan-complexed starch nanoparticles for improving the physical properties of recycled paper furnish (OCC), was also studied [53] and provided a uniquely renewable and useful approach to enhance the mechanical properties of pulp while maintaining environmental compatibility, industrial compatibility, and paper qualities. Other complexes such as xylan/chitosan complex [44], CMC/chitosan complex [57], maleic anhydride-acylated chitosan [58], and soy flour combined chitosan dual system complexes [59] have been prepared and used to enhance paper strength. The results indicated that those complexes are potential strength agents used in the paper industry. Soy protein flour—DTPA (Diethylenetriaminepentaacetic acid)–chitosan agent [60] and nanocellulose-DTPA-chitosan agent [61] were prepared by Salam et al. and the performance of those complexes used as dry-strength agents in papermaking was investigated. The results revealed that those agents provided increased tensile and burst strength for the modified OCC pulp sheet and significantly increased gloss and water repellency with diminished surface roughness.

Chitosan and its derivatives have been identified as the potential dry and wet strengthening additive for papermaking. The potential advantages can be illustrated in the film forming property of chitosan improves the surface properties of paper [62], the formation of hydrogen bonds [63] and the imine [64]. The paper strength is a function of fiber-fiber bond strength, fiber strength, and sheet formation. However, there are always air voids between the fibers in fiber networks after sheet formation. In recent years, research found that fiber-fiber hydrogen bonding can be greatly enhanced by the use of strength additives. When chitosan is added to the wet-end of papermaking, a film covering the fiber crossing areas could lead to stronger bonds by welding the surfaces together. Meanwhile, the hydroxyl groups of chitosan could form hydrogen bonds with weakly polar areas of fiber surfaces, therefore contributing to paper strength development if the fibers come sufficiently close in order to meet the required geometry conditions. Therefore, the film-forming potential of chitosan not only facilitates the formation of van der Waals forces between the fibers but also provides suitable conditions for hydrogen bonds to occur [63]. Moreover, the formation of imine has been proposed as an important contribution to the ability of chitosan to increase wet strength [64] and studies into the nature of chitosan promoted discoloration of paper have proposed and provided evidence of imine formation [65].

2.3. As ASA Sizing Promoter

Chitosan has also been reported to act as a paper sizing promoter to improve the alkenyl succinic anhydride (ASA) emulsion stability. Liu's group reported the capability of chitosan promoting the sizing performance of ASA emulsion stabilized by montmorillonite [66] or laponite [67] due to its large amount of amino groups. In this role, chitosan, with low molecular weight, could significantly improve the sizing performance of ASA emulsion without inducing the flocculation of the ASA droplets at low charge amounts.

2.4. As the Other Wet-End Additives

In addition to the strength properties, chitosan and some derivatives can bring about other properties to paper sheets such as electrical properties [68], antibacterial properties [69,70], and barrier properties [71]. Work done by Nada et al. [68] focused on studying the dependence of paper sheet strength properties on the composition of additives—chitosan and its derivatives, which enhanced the strength properties and the dielectric properties of unaged and aged paper sheets. Research on the synergistic effects of chitosan–guanidine complexes used as functional additives for paper on enhancing wet-strength and antimicrobial activity of paper was carried out, and the chitosan-guanidine complexes synergistically improved wet-strength and antimicrobial activities [69]. Similarly, the chitosan-cellulose blends provided mechanical, antibacterial and water barrier properties [70]. Another study on the pulp-fiber-chitosan sheets investigated the effects of incorporating chitosan or chitosan-acetic acid salt as oxygen-barrier or air-barrier components on the packaging-related mechanical and barrier properties. The results showed that the addition of the chitosan solution to the pulp slurry led to a substantial loss of fiber and chitosan through the wire screen and consequently a low grammage and high sheet porosity and air permeance [71].

3. Surface Coating Application

As a linear carbohydrate biopolymer, chitosan has a very similar chemical structure to cellulose. It is easily absorbed onto the cellulosic surface of fibers due to its chemical affinity. Chitosan has been applied widely to improve some properties of cellulose-based materials, especially those of cellulose fibers and paper sheets [72]. Chitosan coatings on the surface of cellulose fiber network or paper sheets have been considered for antimicrobial [73,74] or antibacterial purpose [75,76], as well as enhancing water vapor barrier properties [77], oxygen barrier properties, grease barrier properties [14,77–79], anti-electrostatic effects, dyeability promoter of paperboard [80], and increasing the mechanical strength [64,81,82] or the surface property of paper products [83]. As for the surface coating application, chitosan can be used in the form of aqueous solution [76] or emulsion [84], via rod coater [14], bar coater [76], wire bar coater [84], multicoated [85] or size press [79], to be transferred onto the paper surface to endue paper with specific characteristics.

3.1. Antibacterial and Antimicrobial Properties

Antibacterial paper is highly significant to living environments and health condition, while also being widely used as food wrappers, hospital paper, indoor environmental protection paper, and sanitary paper, etc. [86]. Antibacterial paper can be produced by coating chitosan and its derivatives or chitosan complex systems on the paper surface, since chitosan possesses the antibacterial properties. Chitin and chitosan have been investigated as an antimicrobial material against a wide range of target organisms like algae, bacteria, yeasts, and fungi in experiments involving in vivo and in vitro interactions with chitosan in different forms (solutions, film, and composites), and three possible and accepted antimicrobial mechanisms for chitosan discussed by Goy et al. [87].

Research on antimicrobial properties of chitosan-coated paper by Vartiainen et al. showed that chitosan dissolved in 1.6, 3.2, and 6.4% lactic acid showed antimicrobial activity against *Bacillus subtilis* [73]. Janjic et al. developed biologically active cellulose-chitosan fibers by oxidizing lyocell fibers with potassium periodate followed by a chitosan coating [88]. Chitosan-coated lyocell fibers were prepared by subsequent treatment of oxidized lyocell fibers with a solution of chitosan in aqueous acetic acid. The free amino group of chitosan reacts with an aldehyde to give the corresponding Schiff base with high degrees of substitution. The antibacterial activity of the cellulose-chitosan fibers against different pathogens including *Staphylococcus aureus* and *Escherichia coli* was confirmed in their experiments. Novel antibacterial paper was fabricated by a surface coating based on modified chitosan and organic montmorillonite/Ag nanocomposites complex and the results proved that this study provided basic data for an efficient and safe chitosan-contained antibacterial agent that can be applied in the paper industry [89]. The combination of propolis and chitosan was also used to impart antimicrobial as well as antioxidant capacity to paper and cellulosic packaging materials, improving some fundamental features of paper and food packaging materials [90,91].

3.2. Strength and Barrier Properties

Chitosan is selected as the coating material to enhance paper strength and barrier properties due to its good film forming property, and the reactive amino and hydroxyl groups of chitosan have the potential to form hydrogen bonds with fiber surfaces, therefore contributing to paper strength development. Wang et al. [92] discussed the film formation of chitosan coated on the surface of Kraft paper and they found that there was no chitosan penetration through the Kraft paper, indicating the good film-forming property of chitosan. The resulting paper obtained strongest water vapor barrier properties when there was a higher concentration of chitosan solution at the optimum pH, stirring speed, and those with a thicker coating on the Kraft paper. Gandini [93] reported that the deposition of chitosan films of different thicknesses on uncoated paper sheets not only improved the optical properties of the ensuing surfaces and their printability, but also brought about useful modifications of certain mechanical and permeability properties. In addition, production of coated papers with a water-soluble chitosan derivative was discussed by Fernandes et al. They claimed that paper coated with such chitosan derivative presented superior optical properties, printability, and had better results on aging measurements than the pristine chitosan-coated papers [94]. However, the hydrophobic property of water-soluble chitosan coated paper was limited.

It was said that chitosan can be used as a pre-coating on paper to provide better bonding and a more uniform surface for other processing steps like, e.g., the application of an additional biopolymer layer by extrusion coating [72,95]. This is similar to the layer-by-layer (LBL) assembly technique, which was used to build coating multilayers on the paper surface and employed in improving the properties of paper in the papermaking field. Despond et al. [78] carried out the experiment focusing on the barrier properties of paper coated with chitosan and carnauba wax. Chitosan was first coated to obtain a dense polymer layer at the paper surface, which gave interesting gas barrier properties in the anhydrous state of coated paper. This was followed by the coating of carnauba wax forming a bilayer which led to a hydrophobic surface of the treated paper. Similarly, carboxymethyl cellulose-chitosan complex LBL treatment on cellulose fiber networks was carried to enhance the wet and dry tensile strength of cellulose fibre networks [57]. Another study showed that the incorporation of sodium alginate in a chitosan formulation significantly improved the fat resistance of the coated paper in comparison with a pure chitosan coating; however, the introduction of cellulose ethers in the chitosan formulations did not improve the fat resistance of coated papers [79]. Zhang et al. [85] used chitosan in combination with beeswax to create a high water vapour barrier property and grease resistance of coated paper. The results showed that as the concentration of chitosan solution increased from 1.0 to 3.0 wt. %, its water vapour transport rate (WVTR) decreased from 171.6 to 52.8 $g/m^2/d$ but using reduced beeswax coating weight (from 10.1 to 4.9 g/m^2). It also displayed an enhanced performance of grease resistance. Study on chitosan-caseinate bilayer coatings for paper packaging materials was also reported [96] and the results showed that chitosan significantly increased the elongation at break of coated paper while caseinate led to a decrease in water vapor permeability. Moreover, LBL self-assembly deposition of the chitosan lactate-carboxymethyl cellulose complex previously modified with metal oxide for reinforcement of aged papers was studied [97] and shown to have an excellent improvement in the mechanical properties of the treated paper. Nanofibrillated cellulose-chitosan nanocomposite films were prepared and used for paper coating [98]. The coating of nanocomposite films improved the tensile strength properties and grease-proof properties of the coated paper while decreasing the porosity and water absorption of paper. However, the water vapor permeability was not affected.

3.3. Other Coating Applications

In addition to the applications of chitosan coating on paper to create oxygen gas barriers, water vapor barriers, and grease or fat barriers of coated paper, another interesting topic should be the preparation of some intelligent chitosan coated paper based materials [6,99,100]. For example, an interesting study showed that an intelligent and biodegradable temperature indicator packaging material could be developed through incorporating a heat-sensitive pigment (anthocyanin-ATH) into chitosan-acetic acid dispersion that was applied as surface coating on card paper. This smart packaging materials can indicate temperature variations in a specific range by irreversible visual colour changes [100]. Chitosan as a paperboard coating additive for use in HVAC (heating, ventilation, and air conditioning) applications has been reported [6]. Commercial chitosan and fungal chitosan solution were coated onto unbleached Kraft paper to be an alternative to phenolic resin coatings and the results indicated the potential of chitosan coated paper for manufacturing evaporative cooling pads used in livestock enclosures.

4. Wastewater Treatment

A variety of pollutants are generated from pulp and papermaking mills depending on which process is used. The high amount of water and various chemicals used in the complicated processes in papermaking industry generates large amounts of contaminated wastewater. The pulp and paper industry is considered as a big polluter in the world. Therefore, learning how to deal with the papermaking wastewater is a big issue. Advanced wastewater treatment technologies are mandatory to reduce fresh water consumption with minimum detrimental effects on papermaking operations and paper quality [101]. Pokhrel et al. listed considerable methods for dealing with the wastewater in the papermaking industry, including physicochemical, biological, fungal, and integrated treatment processes. Chitosan was mentioned in the coagulation and flocculation method, which is normally employed in the tertiary treatment in the case of pulp and paper mill wastewater treatment [17].

Coagulation and flocculation is an important secondary treatment procedure in the removal of turbidity, colloids, and natural organic matter during water treatment processes [102]. Chitosan has the characteristics of both a coagulant and a flocculant with high cationic charge density, long polymer chains, and acts as a bridge for aggregates and precipitation. It may be considered as one of the most promising bioflocculants for environmental and purification purposes [16,103]. Many studies showed that chitosan-based flocculants have many advantages, including their widespread availability, environmental friendliness, biodegradability, and prominent structural features when they are used in the wastewater treatment [104]. Wang et al. [105] reported a novel cationic chitosan-based flocculant with a high water-solubility for pulp mill wastewater treatment. The flocculate was synthesized through grafting (2-methacryloyloxyethyl) trimethyl ammonium chloride (DMC) onto chitosan initiated by potassium persulphate. The results showed that this chitosan-based flocculant had an excellent flocculation capacity and its flocculation efficiency was greater than that of polyacrylamide. In another study, chitosan dissolved in acetic acid and was used as a flocculating agent in the flocculation of cardboard industry wastewater treated by a biological process in an aerated lagoon [16]. Compared with commercial grade polyaluminium chloride (PAC), an extensively used flocculant in wastewater treatment, chitosan induced a more efficient flocculating process. Chitosan lowered the chemical oxygen demand (COD) by over 80% and turbidity by more than 85% which were much higher than PAC did. Moreover, using chitosan as flocculant generated bigger flocs makes settling faster than in the case of using PAC. Meanwhile, chitosan-induced flocculation removed more residual colour and led to a significant decrease in the amount of heavy metals present in the effluent.

In order to investigate the effects of molecular weights on the chitosan performance, Miranda et al. evaluated two native chitosans with different molecular weights on a laboratory scale for their effectiveness in the removal of contaminants from papermaking process waters by dissolved air flotation [106]. The use of chitosan quaternary derivatives and the use of the native chitosans in combination with anionic bentonite microparticles have also been tested. The results demonstrated a high efficiency of the native chitosan products at intermediate dosages and their efficiency was enhanced by the combined addition of bentonite. Quaternary derivatives obtained lower efficiency than the base chitosan used. The main reason for this was the lower charge density of the quaternary derivatives compared to the native chitosans at the operational conditions.

Many studies proved that chitosan can also be used in combination with other polymers for wastewater treatment. Tong et al. [107] and Ganjidoust et al. [108] carried out a comparative study of horseradish peroxidase (chitosan) and other synthetic polymers, including hexamethylene diamine epichlorohydrin polycondensate (HE), polyethylenimine(PEI), and polyacrylamide(PAM), to remove lignin and other kinds of chlorinated organic compounds from pulp and paper industrial wastewater. The results showed that modified chitosan was far more effective in removing these pollutants than other coagulants. Zeng et al. [109] prepared a composite flocculant that consisted of chitosan, polymerized ferrous sulfate (PFS), and PAM to treat papermaking wastewater. This composite flocculant has economic and environmental benefits due to its lower price and higher efficiency. Liu et al. reported a macroporous resin with a methyl acrylate matrix and coated with chitosan of various molecular weights through glutaraldehyde crosslinking for the treatment of whitewater from papermaking after pectinase and lipase were immobilized on the resin coated with chitosan [110]. The macroporous resin, immobilized with dual-enzymes, was proved useful for the treatment of whitewater in the papermaking industry by reducing the cationic demand and pitch deposits in whitewater by 58% and 74%, respectively. Another study reported the preparation of cross-linked chitosan beads with immobilized pectinase, which were used to investigate the effects of enzymes in lowering pectins or polygalacturonic acids (PGA) concentration in papermaking industries. The results showed that the PGA-absorption capability of chitosan beads was greatly affected by its cross-linking degree [19]. This revealed the potential for cross-linked chitosan beads which lowers canionic demand of PGA by solute adsorption and pectinase immobilization for potential use in water treatment of the papermaking industry. Petzold et al. studied the removal of dissolved and colloidal substances (DCS) in paper cycling water with modified starch and chitosan compared with a control [111]. Results revealed that turbidity and total organic carbon (TOC) were lowered especially due to charge interaction, whereas the increase in surface tension is mainly caused by the hydrophobic character of the modified natural polymers.

5. Other Applications

Other applications of chitosan in the papermaking or papermaking-related industries have also been studied, such as modification of cellulose fibers [65,112], blending with cellulose to prepare chitosan/cellulose blend beads [113], crosslinking with cellulose nanofibers to form nanopaper with water-resistant and transparent properties [114], etc. As such, photochromic paper from wood pulp modified via LBL assembly chitosan-spiropyran on pulp fibers has been studied [112]. The LBL–treated fibers were compatible with pulp fibers, which gave a highly effective method to impart the photochromic characteristic to paper. Chitosan with two different molecular weights were employed as flocculant to recover the dissolved lignocellulosic materials of industrially produced pre-hydrolysis liquor (PHL). The addition of chitosan causes the precipitation of dissolved lignocellulosic materials. Chitosan with higher molecular weight induces more precipitation of dissolved lignocellulosic materials at a lower concentration [115].

6. Conclusions

Chitin and chitosan and their derivatives have a wide range of applications in the papermaking industry. They can be employed to solve numerous problems in wet-end chemistry and wastewater treatment such as improving the efficiency of the paper machine, enhancing paper strength, or to prepare antibacterial, high barrier, intelligent paper-based packaging materials. Together, with the biodegradable nature of chitosan, it appears that chitosan can be an interesting and promising candidate for environmentally-friendly high value-added paper production. Up to now, the economy of the chitosan application in large scale in the papermaking industry has not been considered.

Acknowledgments: The authors of this work would like to thank the financial support from National Natural Science Foundation of China (No. 21506105 and 31270625). The funding from NSFC No. 21506105 and 31270625 support this work and will cover the costs to publish this paper in open access. The authors would also like to thank Alex Wang of Texas A&M Health Science Center, for English correction of this manuscript.

Polymers **2018**, *10*, 389

Author Contributions: Zhaoping Song and Wenxia Liu conceived and designed the review; Feixiang Guan contributed the references collection; Zhaoping Song wrote the paper, Guodong Li and Wenxia Liu went over and revised the manuscript.

Conflicts of Interest: The authors declare no conflict of interest. The founding sponsors had no role in the design of the study; in the collection, analyses, or interpretation of data; in the writing of the manuscript, and in the decision to publish the results.

References

1. Rinaudo, M. Chitin and chitosan: Properties and applications. *Prog. Polym. Sci.* **2006**, *31*, 603–632. [CrossRef]
2. Baldrick, P. The safety of chitosan as a pharmaceutical excipient. *Regul. Toxicol. Pharmacol.* **2010**, *56*, 290–299. [CrossRef] [PubMed]
3. Casariego, A.; Souza, B.W.S.; Cerqueira, M.A.; Teixeira, J.A.; Cruz, L.; Díaz, R.; Vicente, A.A. Chitosan/clay films' properties as affected by biopolymer and clay micro/nanoparticles' concentrations. *Food Hydrocoll.* **2009**, *23*, 1895–1902. [CrossRef]
4. Azuma, K.; Izumi, R.; Osaki, T.; Ifuku, S.; Morimoto, M.; Saimoto, H.; Minami, S.; Okamoto, Y. Chitin, chitosan, and its derivatives for wound healing: Old and new materials. *J. Funct. Biomater.* **2015**, *6*, 104–142. [CrossRef] [PubMed]
5. Hamed, I.; Özogul, F.; Regenstein, J.M. Industrial applications of crustacean by-products (chitin, chitosan, and chitooligosaccharides): A review. *Trends Food Sci. Technol.* **2016**, *48*, 40–50. [CrossRef]
6. Atkinson, J.; Mondala, A.; Senger, Y.D.S.; Al-Mubarak, R.; Young, B.; Pekarovic, J.; Joyce, M. Chitosan as a paperboard coating additive for use in HVAC (heating, ventilation and air conditioning) applications. *Cellul. Chem. Technol.* **2017**, *51*, 477–481.
7. Agusnar, H.; Nainggolan, I. Mechanical properties of paper from oil palm pulp treated with chitosan from horseshoe crab. *Adv. Environ. Biol.* **2013**, *7*, 28–29.
8. Habibie, S.; Hamzah, M.; Anggaravidya, M.; Kalembang, E. The effect of chitosan on physical and mechanical properties of paper. *J. Chem. Eng. Mater. Sci.* **2016**, *7*, 1–10.
9. Bobu, E.; Ciolacu, F.; Anghel, N. Prevention of colloidal material accumulation in short circulation of paper machine. *Wochenbl. Pap.* **2002**, *130*, 576–582.
10. Chi, H.; Li, H.; Liu, W.; Zhan, H. The retention- and drainage-aid behavior of quaternary chitosan in papermaking system. *Colloids Surf. Physicochem. Eng. Asp.* **2007**, *297*, 147–153. [CrossRef]
11. Anthonsen, T.; Sandford, P.A.; Skjåk-Bræk, G. *Chitin and Chitosan: Sources, Chemistry, Biochemistry, Physical Properties, and Applications*; Elsevier Applied Science: London, UK, 1989.
12. Ali, N.F.; El-Mohamedy, R.S.R. Microbial decolourization of textile waste water. *J. Saudi Chem. Soc.* **2012**, *16*, 117–123. [CrossRef]
13. Bobu, E.; Ciolacu, F.; Parpalea, R. Effective use of FWAs in papermaking by controlling their interactions with other chemicals. *Wochenbl. Pap.* **2002**, *130*, 1510–1523.
14. Kjellgren, H.; Gällstedt, M.; Engström, G.; Järnström, L. Barrier and surface properties of chitosan-coated greaseproof paper. *Carbohydr. Polym.* **2006**, *65*, 453–460. [CrossRef]
15. Varma, A.J.; Deshpande, S.V.; Kennedy, J.F. Metal complexation by chitosan and its derivatives: A review. *Carbohydr. Polym.* **2004**, *55*, 77–93. [CrossRef]
16. Renault, F.; Sancey, B.; Charles, J.; Morin-Crini, N.; Badot, P.-M.; Winterton, P.; Crini, G. Chitosan flocculation of cardboard-mill secondary biological wastewater. *Chem. Eng. J.* **2009**, *155*, 775–783. [CrossRef]
17. Pokhrel, D.; Viraraghavan, T. Treatment of pulp and paper mill wastewater—A review. *Sci. Total Environ.* **2004**, *333*, 37–58. [CrossRef] [PubMed]
18. You, L.; Lu, F.; Li, D.; Qiao, Z.; Yin, Y. Preparation and flocculation properties of cationic starch/chitosan crosslinking-copolymer. *J. Hazard. Mater.* **2009**, *172*, 38–45. [CrossRef] [PubMed]
19. Liu, K.; Li, X.-F.; Li, X.-M.; He, B.-H.; Zhao, G.-L. Lowering the cationic demand caused by PGA in papermaking by solute adsorption and immobilized pectinase on chitosan beads. *Carbohydr. Polym.* **2010**, *82*, 648–652. [CrossRef]
20. Wan, M.; Wang, C.; Chen, C. The adsorption study of copper removal by chitosan-coated sludge derived from water treatment plant. *Int. J. Environ. Sci. Technol. Dev.* **2013**, *4*, 545–551. [CrossRef]

21. Cadotte, M.; Tellier, M.E.; Blanco, A.; Fuente, E.; van de Ven, T. G.; Paris, J. Flocculation, retention and drainage in papermaking: A comparative study of polymeric additives. *Can. J. Chem. Eng.* **2007**, *85*, 240–248. [CrossRef]

22. Hubbe, M.A.; Nanko, H.; Mcneal, M.R. Retention aid polymer interactions with cellulosic surfaces and suspensions: A review. *BioResources* **2009**, *4*, 850–906.

23. Honig, D.S.; Farinato, R.S.; Jackson, L.A. Design and development of the micropolymer system: An "organic microparticle" retention/drainage system. *Nordic Pulp Pap. Res. J.* **2000**, *15*, 536–544. [CrossRef]

24. Fatehi, P.; Shen, J. A review on the use of lignocellulose-derived chemicals in wet-end application of papermaking. *Curr. Org. Chem.* **2013**, *17*, 1647–1654.

25. Diab, M.; Curtil, D.; El-shinnawy, N.; Hassan, M.L.; Zeid, I.F.; Mauret, E. Biobased polymers and cationic microfibrillated cellulose as retention and drainage aids in papermaking: Comparison between softwood and bagasse pulps. *Ind. Crops Prod.* **2015**, *72*, 34–45. [CrossRef]

26. Dryabina, S.; Fotina, K.; Navrotskii, A.; Novakov, I. The flocculation of kaolin aqueous dispersion by two cationic polyelectrolytes. *Colloids Surf. A Physicochem. Eng. Asp.* **2017**, *515*, 12–21. [CrossRef]

27. Prado, H.J.; Matulewicz, M.C. Cationization of polysaccharides: A path to greener derivatives with many industrial applications. *Eur. Polym. J.* **2014**, *52*, 53–75. [CrossRef]

28. Li, H.; Du, Y.; Wu, X.; Zhan, H. Effect of molecular weight and degree of substitution of quaternary chitosan on its adsorption and flocculation properties for potential retention-aids in alkaline papermaking. *Colloids Surf. A Physicochem. Eng. Asp.* **2004**, *242*, 1–8. [CrossRef]

29. Gao, Z.; Zhai, X.; Liu, F.; Zhang, M.; Zang, D.; Wang, C. Fabrication of TiO_2/EP super-hydrophobic thin film on filter paper surface. *Carbohydr. Polym.* **2015**, *128*, 24–31. [CrossRef] [PubMed]

30. Shen, J.; Song, Z.; Qian, X.; Song, C. Chitosan-coated papermaking grade PCC filler prepared by alkali precipitation: Properties and application. In Proceedings of the 2nd International Papermaking and Environment Conference, Tianjin, China, 1–3 May 2008; pp. 659–664.

31. Kuutti, L.; Haavisto, S.; Hyvarinen, S.; Mikkonen, H. Properties and flocculation efficiency of cationized biopolymers and their applicability in papermaking and in conditioning of pulp and paper sludge. *Bioresources* **2011**, *6*, 2836–2850.

32. Fukuda, S.; Chaussy, D.; Belgacem, M.N.; Reverdy-Bruas, N.; Thielemans, W. Characterization of oil-proof papers containing new-type of fluorochemicals part 1: Surface properties and printability. *Appl. Surf. Sci.* **2013**, *277*, 57–66. [CrossRef]

33. Ian, R.; Michelle, R. Pectinase in papermaking: Solving retention problems in mechanical pulps bleached with hydrogen peroxide. *Enzym. Microb. Technol.* **2000**, *26*, 115–123.

34. Nicu, R.; Bobu, E.; Miranda, R.; Blanco, A. Flocculation efficiency of chitosan for papermaking applications. *BioResources* **2013**, *8*, 768–784. [CrossRef]

35. Fredheim, G.E.; Christensen, B.E. Polyelectrolyte complexes: Interactions between lignosulfonate and chitosan. *Biomacromolecules* **2003**, *4*, 232–239. [CrossRef] [PubMed]

36. Zhang, X.; Gu, W.J.; Li, H.; Chi, H.; Chen, L. Flocculation of reed pulp suspensions by quaternary chitosan-nanoparticle SiO_2 retention aid systems. *J. Appl. Polym. Sci.* **2010**, *117*, 742–749. [CrossRef]

37. Sabazoodkhiz, R.; Rahmaninia, M.; Ramezani, O. Interaction of chitosan biopolymer with silica nanoparticles as a novel retention/drainage and reinforcement aid in recycled cellulosic fibers. *Cellulose* **2017**, *24*, 3433–3444. [CrossRef]

38. Hu, H.; Shi, S. Application of chitosan/bentonite particle retention and drainage system in tobacco sheet manufacture by papermaking process. *China Pulp Pap.* **2011**, *30*, 30–33.

39. Balea, A.; Merayo, N.; De La Fuente, E.; Negro, C.; Blanco, Á. Assessing the influence of refining, bleaching and tempo-mediated oxidation on the production of more sustainable cellulose nanofibers and their application as paper additives. *Ind. Crop. Prod.* **2017**, *97*, 374–387. [CrossRef]

40. Ashori, A.; Harun, J.; Wan, M.Z.; Yusoff, N.M. Enhancing dry-strength properties of kenaf (*Hibiscus cannabinus*) paper through chitosan. *J. Macromol. Sci. Part D Rev. Polym. Process.* **2006**, *45*, 125–129. [CrossRef]

41. Dutta, P.K.; Dutta, J.; Tripathi, V.S. Chitin and chitosan: Chemistry, properties and applications. *J. Sci. Ind. Res.* **2004**, *63*, 20–31.

42. Kumar, M.N.R. A review of chitin and chitosan applications. *React. Funct. Polym.* **2000**, *46*, 1–27. [CrossRef]

43. Li, H.; Du, Y.; Xu, Y.; Zhan, H.; Kennedy, J.F. Interactions of cationized chitosan with components in a chemical pulp suspension. *Carbohydr. Polym.* **2004**, *58*, 205–214. [CrossRef]

44. Mocchiutti, P.; Schnell, C.N.; Rossi, G.D.; Peresin, M.S.; Zanuttini, M.A.; Galvan, M.V. Cationic and anionic polyelectrolyte complexes of xylan and chitosan. Interaction with lignocellulosic surfaces. *Carbohydr. Polym.* **2016**, *150*, 89–98. [CrossRef] [PubMed]

45. Myllytie, P.; Salmi, J.; Laine, J. The influence of pH on the adsorption and interaction of chitosan with cellulose. *Bioresources* **2009**, *4*, 9–17.

46. Nicu, R. Synthesis and characterization of n-alkyl chitosan for papermaking applications. *Cellul. Chem. Technol.* **2011**, *45*, 619–625.

47. Nikolaeva, M. Measurement and Improvement of Wet Paper Web Strength. Master's Thesis, Lappeenranta University of Technology, Lappeenranta, Finland, 2010.

48. Rohi, M.; Ramezani, O.; Rahmaninia, M.; Zabihzadeh, S.M.; Hubbe, M.A. Influence of pulp suspension pH on the performance of chitosan as a strength agent for hardwood CMP paper. *Cellul. Chem. Technol.* **2016**, *50*, 873–878.

49. Laleg, M.; Pikulik, I.I. Strengthening of mechanical pulp webs by chitosan. *Nord. Pulp Pap. Res. J.* **1992**, *7*, 174–180. [CrossRef]

50. Lertsutthiwong, P.; Chandrkrachang, S.; Nazhad, M.M.; Stevens, W.F. Chitosan as a dry strength agent for paper. *Appita J.* **2002**, *55*, 208–212.

51. Khantayanuwong, S.; Khemarom, C.; Salaemae, S. Effects of shrimp chitosan on the physical properties of handsheets. *Agric. Natl. Resour.* **2017**, *51*, 53–56. [CrossRef]

52. Fatehi, P.; Kititerakun, R.; Ni, Y.; Xiao, H. Synergy of CMC and modified chitosan on strength properties of cellulosic fiber network. *Carbohydr. Polym.* **2010**, *80*, 208–214. [CrossRef]

53. Salam, A.; Lucia, L.A.; Jameel, H. Synthesis, characterization, and evaluation of chitosan-complexed starch nanoparticles on the physical properties of recycled paper furnish. *ACS Appl. Mater. Interfaces* **2013**, *5*, 11029. [CrossRef] [PubMed]

54. Rahmaninia, M.; Rohi, M.; Hubbe, M.A.; Zabihzadeh, S.M.; Ramezani, O. The performance of chitosan with bentonite microparticles as wet-end additive system for paper reinforcement. *Carbohydr. Polym.* **2018**, *179*, 328–332. [CrossRef] [PubMed]

55. Ashori, A.; Cordeiro, N.; Faria, M.; Hamzeh, Y. Effect of chitosan and cationic starch on the surface chemistry properties of bagasse paper. *Int. J. Biol. Macromol.* **2013**, *58*, 343–348. [CrossRef] [PubMed]

56. Hamzeh, Y.; Sabbaghi, S.; Ashori, A.; Abdulkhani, A.; Soltani, F. Improving wet and dry strength properties of recycled old corrugated carton (OCC) pulp using various polymers. *Carbohydr. Polym.* **2013**, *94*, 577–583. [CrossRef] [PubMed]

57. Wu, T.; Farnood, R. Cellulose fibre networks reinforced with carboxymethyl cellulose/chitosan complex layer-by-layer. *Carbohydr. Polym.* **2014**, *114*, 500–505. [CrossRef] [PubMed]

58. Chen, Z.; Zhang, H.; Song, Z.; Qian, X. Preparation and application of maleic anhydride-acylated chitosan for wet strength improvement of paper. *Bioresources* **2013**, *8*, 3901–3911. [CrossRef]

59. Arboleda, J.C.; Niemi, N.; Kumpunen, J.; Lucia, L.A.; Rojas, O.J. Soy protein-based polyelectrolyte complexes as biobased wood fiber dry strength agents. *ACS Sustain. Chem. Eng.* **2014**, *2*, 2267–2274. [CrossRef]

60. Salam, A.; Lucia, L.A.; Jameel, H. A new class of biobased paper dry strength agents: Synthesis and characterization of soy-based polymers. *ACS Sustain. Chem. Eng.* **2015**, *3*, 524–532. [CrossRef]

61. Salam, A.; Lucia, L.A.; Jameel, H. A novel cellulose nanocrystals-based approach to improve the mechanical properties of recycled paper. *ACS Sustain. Chem. Eng.* **2013**, *1*, 1584–1592. [CrossRef]

62. Ashori, A.; Raverty, W.D.; Harun, J. Effect of chitosan addition on the surface properties of kenaf (*Hibiscus cannabinus*) paper. *Fibers Polym.* **2005**, *6*, 174–179. [CrossRef]

63. Adel, A.M.; Dupont, A.L.; Abou-Yousef, H.; El-Gendy, A.; Paris, S.; El-Shinnawy, N. A study of wet and dry strength properties of unaged and hygrothermally aged paper sheets reinforced with biopolymer composites. *J. Appl. Polym. Sci.* **2014**, *131*, 14–20. [CrossRef]

64. Laleg, M.; Pikulik, I.I. Wet-web strength increase by chitosan. *Nordic Pulp Pap. Res. J.* **1991**, *6*, 99–103. [CrossRef]

65. Urreaga, J.M.; De la Orden, M.U. Chemical interactions and yellowing in chitosan-treated cellulose. *Eur. Polym. J.* **2006**, *42*, 2606–2616. [CrossRef]

66. Lu, P.; Liu, W.; Wang, H.; Wang, Z. Using chitosan as sizing promoter of ASA emulsion stabilized by montmorillonite. *Bioresources* **2013**, *8*, 4923–4936. [CrossRef]

67. Zhang, W.; Liu, W.X.; Han, J.M. Application of chitosan to improve sizing performance of as a emulsion stabilized by laponite. *Chung Kuo Tsao Chih/China Pulp Pap.* **2013**, *32*, 18–23.

68. Nada, A.M.A.; El-Sakhawy, M.; Kamel, S.; Eid, M.A.M.; Adel, A.M. Mechanical and electrical properties of paper sheets treated with chitosan and its derivatives. *Carbohydr. Polym.* **2006**, *63*, 113–121. [CrossRef]

69. Sun, S.; An, Q.; Li, X.; Qian, L.; He, B.; Xiao, H. Synergistic effects of chitosan-guanidine complexes on enhancing antimicrobial activity and wet-strength of paper. *Bioresour. Technol.* **2010**, *101*, 5693–5700. [CrossRef] [PubMed]

70. Wu, Y.-B.; Yu, S.-H.; Mi, F.-L.; Wu, C.-W.; Shyu, S.-S.; Peng, C.-K.; Chao, A.-C. Preparation and characterization on mechanical and antibacterial properties of chitsoan/cellulose blends. *Carbohydr. Polym.* **2004**, *57*, 435–440. [CrossRef]

71. Gällstedt, M.; Hedenqvist, M.S. Packaging-related mechanical and barrier properties of pulp–fiber–chitosan sheets. *Carbohydr. Polym.* **2006**, *63*, 46–53. [CrossRef]

72. Kuusipalo, J. Chitosan as a coating additive in paper and paperboard. *Tappi J.* **2005**, *4*, 17–21.

73. Vartiainen, J.; Motion, R.; Kulonen, H.; Rättö, M.; Skyttä, E.; Ahvenainen, R. Chitosan-coated paper: Effects of nisin and different acids on the antimicrobial activity. *J. Appl. Polym. Sci.* **2004**, *94*, 986–993. [CrossRef]

74. Alonso, D.; Gimeno, M.; Olayo, R.; Vázquez-Torres, H.; Sepúlveda-Sánchez, J.D.; Shirai, K. Cross-linking chitosan into UV-irradiated cellulose fibers for the preparation of antimicrobial-finished textiles. *Carbohydr. Polym.* **2009**, *77*, 536–543. [CrossRef]

75. Bordenave, N.; Grelier, S.; Coma, V. Hydrophobization and antimicrobial activity of chitosan and paper-based packaging material. *Biomacromolecules* **2010**, *11*, 88–96. [CrossRef] [PubMed]

76. Zakaria, S.; Chia, C.H.; Wan, H.W.A.; Kaco, H.; Chook, S.W.; Chi, H.C. Mechanical and antibacterial properties of paper coated with chitosan. *Sains Malays.* **2015**, *44*, 905–911. [CrossRef]

77. Kittur, F.S.; Kumar, K.R.; Tharanathan, R.N. Functional packaging properties of chitosan films. *Zeitschrift für Lebensmitteluntersuchung und-Forschung A* **1998**, *206*, 44–47. [CrossRef]

78. Despond, S.; Espuche, E.; Cartier, N.; Domard, A. Barrier properties of paper–chitosan and paper–chitosan–carnauba wax films. *J. Appl. Polym. Sci.* **2005**, *98*, 704–710. [CrossRef]

79. Hampichavant, F.; Sebe, G.; Pardon, P.; Coma, V. Fat resistance properties of chitosan-based paper packaging for food applications. *Carbohydr. Polym.* **2005**, *61*, 259–265. [CrossRef]

80. Makino, Y.; Hirata, T. Modified atmosphere packaging of fresh produce with a biodegradable laminate of chitosan-cellulose and polycaprolactone. *Postharvest Biol. Technol.* **1997**, *10*, 247–254. [CrossRef]

81. Liu, X.D.; Nishi, N.; Tokura, S.; Sakairi, N. Chitosan coated cotton fiber: Preparation and physical properties. *Carbohydr. Polym.* **2001**, *44*, 233–238. [CrossRef]

82. Gandini, A.; Lacerda, T.M. From monomers to polymers from renewable resources: Recent advances. *Prog. Polym. Sci.* **2015**, *48*, 1–39. [CrossRef]

83. Lertsutthiwong, P.; Nazhad, M.M.; Chandrkrachang, S.; Stevens, W.F. Chitosan as a surface sizing agent for offset printing paper. *World Pulp Pap.* **2005**, *57*, 274–280.

84. Reis, A.B.; Yoshida, C.M.P.; Reis, A.P.C.; Franco, T.T. Application of chitosan emulsion as a coating on kraft paper. *Polym. Int.* **2011**, *60*, 963–969. [CrossRef]

85. Zhang, W.; Xiao, H.; Qian, L. Enhanced water vapour barrier and grease resistance of paper bilayer-coated with chitosan and beeswax. *Carbohydr. Polym.* **2014**, *101*, 401–406. [CrossRef] [PubMed]

86. Nassar, M.A.; Youssef, A.M. Mechanical and antibacterial properties of recycled carton paper coated by PS/Ag nanocomposites for packaging. *Carbohydr. Polym.* **2012**, *89*, 269. [CrossRef] [PubMed]

87. Goy, R.C.; Britto, D.D.; Assis, O.B.G. A review of the antimicrobial activity of chitosan. *Polímeros* **2009**, *19*, 241–247. [CrossRef]

88. Janjic, S.; Kostic, M.; Vucinic, V.; Dimitrijevic, S.; Popovic, K.; Ristic, M.; Skundric, P. Biologically active fibers based on chitosan-coated lyocell fibers. *Carbohydr. Polym.* **2009**, *78*, 240–246. [CrossRef]

89. Ling, Y.; Luo, Y.; Luo, J.; Wang, X.; Sun, R. Novel antibacterial paper based on quaternized carboxymethyl chitosan/organic montmorillonite/Ag NP nanocomposites. *Ind. Crop. Prod.* **2013**, *51*, 470–479. [CrossRef]

90. Coma, R. Polysaccharide-based biomaterials with antimicrobial and antioxidant properties. *Polimeros* **2013**, *23*, 287–297. [CrossRef]

91. Coma, V.; Freire, C.S.R.; Silvestre, A.J.D. *Recent Advances on the Development of Antibacterial Polysaccharide-Based Materials*; Springer: Berlin, Germany, 2015.

92. Wang, S.; Jing, Y. Effects of a chitosan coating layer on the surface properties and barrier properties of kraft paper. *Bioresources* **2016**, *11*, 1868–1881. [CrossRef]

93. Gandini, A. Polymers from renewable resources: A challenge for the future of macromolecular materials. *Macromolecules* **2008**, *41*, 37–59. [CrossRef]

94. Fernandes, S.C.M.; Freire, C.S.R.; Silvestre, A.J.D.; Desbrières, J.; Gandini, A.; Neto, C.P. Production of coated papers with improved properties by using a water-soluble chitosan derivative. *Ind. Eng. Chem. Res.* **2010**, *49*, 6432–6438. [CrossRef]

95. Rastogi, V.K.; Samyn, P. Bio-based coatings for paper applications. *Coatings* **2015**, *2015*, 887–930. [CrossRef]

96. Khwaldia, K.; Basta, A.H.; Aloui, H.; El-Saied, H. Chitosan-caseinate bilayer coatings for paper packaging materials. *Carbohydr. Polym.* **2014**, *99*, 508–516. [CrossRef] [PubMed]

97. Jiang, F.; Yang, Y.; Weng, J.; Zhang, X. Layer-by-layer self-assembly for reinforcement of aged papers. *Ind. Eng. Chem. Res.* **2016**, *55*, 10544–10554. [CrossRef]

98. Hassan, E.A.; Hassan, M.L.; Abou-Zeid, R.E.; El-Wakil, N.A. Novel nanofibrillated cellulose/chitosan nanoparticles nanocomposites films and their use for paper coating. *Ind. Crop. Prod.* **2016**, *93*, 219–226. [CrossRef]

99. Wang, S.; Ge, L.; Song, X.; Yu, J.; Ge, S.; Huang, J.; Zeng, F. Paper-based chemiluminescence elisa: Lab-on-paper based on chitosan modified paper device and wax-screen-printing. *Biosens. Bioelectron.* **2012**, *31*, 212–218. [CrossRef] [PubMed]

100. Maciel, V.B.V.; Yoshida, C.M.P.; Franco, T.T. Development of a prototype of a colourimetric temperature indicator for monitoring food quality. *J. Food Eng.* **2012**, *111*, 21–27. [CrossRef]

101. Lakhdhar, I.; Mangin, P.; Chabot, B. Copper (II) ions adsorption from aqueous solutions using electrospun chitosan/peo nanofibres: Effects of process variables and process optimization. *J. Water Process Eng.* **2015**, *7*, 295–305. [CrossRef]

102. Rong, H.; Gao, B.; Zhao, Y.; Sun, S.; Yang, Z.; Wang, Y.; Yue, Q.; Li, Q. Advanced lignin-acrylamide water treatment agent by pulp and paper industrial sludge: Synthesis, properties and application. *J. Environ. Sci.* **2013**, *25*, 2367–2377. [CrossRef]

103. Kurita, K. Chitin and chitosan: Functional biopolymers from marine crustaceans. *Mar. Biotechnol.* **2006**, *8*, 203. [CrossRef] [PubMed]

104. Yang, R.; Li, H.; Huang, M.; Yang, H.; Li, A. A review on chitosan-based flocculants and their applications in water treatment. *Water Res.* **2016**, *95*, 59–89. [CrossRef] [PubMed]

105. Wang, J.P.; Chen, Y.Z.; Yuan, S.J.; Sheng, G.P.; Yu, H.Q. Synthesis and characterization of a novel cationic chitosan-based flocculant with a high water-solubility for pulp mill wastewater treatment. *Water Res.* **2009**, *43*, 5267. [CrossRef] [PubMed]

106. Miranda, R.; Nicu, R.; Latour, I.; Lupei, M.; Bobu, E.; Blanco, A. Efficiency of chitosans for the treatment of papermaking process water by dissolved air flotation. *Chem. Eng. J.* **2013**, *231*, 304–313. [CrossRef]

107. Zhang, T.; Wada, S.; Yamagishi, T.; Hiroyasu, I.; Tatsumi, K.; Zhao, Q.X. Treatment of bleaching wastewater from pulp paper plants in china using enzymes and coagulants. *J. Environ. Sci.* **1999**, *11*, 480–484.

108. Ganjidoust, H.; Tatsumi, K.; Yamagishi, T.; Gholian, R.N. Effect of synthetic and natural coagulant on lignin removal from pulp and paper wastewater. *Water Sci. Technol.* **1997**, *35*, 291–296.

109. Zeng, D.F.; Hu, D.; Cheng, J. Experimental study on chitosan composite flocculant for treating papermaking wastewater. *J. Water Chem. Technol.* **2012**, *34*, 35–41. [CrossRef]

110. Liu, K.; Zhao, G.; He, B.; Chen, L.; Huang, L. Immobilization of pectinase and lipase on macroporous resin coated with chitosan for treatment of whitewater from papermaking. *Bioresour. Technol.* **2012**, *123*, 616–619. [CrossRef] [PubMed]

111. Petzold, G.; Petzold-Welcke, K.; Qi, H.; Stengel, K.; Schwarz, S.; Heinze, T. The removal of stickies with modified starch and chitosan—Highly cationic and hydrophobic types compared with unmodified ones. *Carbohydr. Polym.* **2012**, *90*, 1712–1718. [CrossRef] [PubMed]

112. Tian, X.; Wang, B.; Li, J.; Zeng, J.; Chen, K. Photochromic paper from wood pulp modification via layer-by-layer assembly of pulp fiber/chitosan/spiropyran. *Carbohydr. Polym.* **2017**, *157*, 704–710. [CrossRef] [PubMed]

113. Twu, Y.K.; Huang, H.I.; Chang, S.Y.; Wang, S.L. Preparation and sorption activity of chitosan/cellulose blend beads. *Carbohydr. Polym.* **2003**, *54*, 425–430. [CrossRef]

114. Toivonen, M.S.; Kurki-Suonio, S.; Schacher, F.H.; Hietala, S.; Rojas, O.J.; Ikkala, O. Water-resistant, transparent hybrid nanopaper by physical cross-linking with chitosan. *Biomacromolecules* **2015**, *16*, 1062–1071. [CrossRef] [PubMed]

115. Saeed, A.; Fatehi, P.; Ni, Y. Chitosan as a flocculant for pre-hydrolysis liquor of kraft-based dissolving pulp production process. *Carbohydr. Polym.* **2011**, *86*, 1630–1636. [CrossRef]

polymers

MDPI

Article

Biomaterials Based on Electrospun Chitosan. Relation between Processing Conditions and Mechanical Properties

Christian Enrique Garcia Garcia [1], Félix Armando Soltero Martínez [1] [iD], Frédéric Bossard [2] [iD] and Marguerite Rinaudo [3,*]

[1] Departamento de Ingeniería Química, Universidad de Guadalajara, Blvd. M. García Barragán #1451, C.P. Guadalajara 44430, Jalisco, Mexico; christian0309@hotmail.com (C.E.G.G.); jfasm@hotmail.com (F.A.S.M.)
[2] Institute of Engineering Grenoble Alpes, CNRS, Grenoble INP, LRP, 38000 Grenoble, France; frederic.bossard@univ-grenoble-alpes.fr
[3] Biomaterials Applications, 6 Rue Lesdiguières, 38000 Grenoble, France
[*] Correspondence: marguerite.rinaudo38@gmail.com or marguerite.rinaudo@sfr.fr; Tel.: +33-611-434-806

Received: 25 January 2018; Accepted: 26 February 2018; Published: 1 March 2018

Abstract: In this paper, it is shown that pure chitosan nanofibers and films were prepared with success in 0.5 M acetic acid as solvent using poly (ethylene oxide) (PEO) at different yields, allowing electrospinning of the blends. After processing, a neutralization step of chitosan followed by water washing is performed, preserving the initial morphology of chitosan materials. The influence of the yield in PEO in the blend on the degree of swelling and hydrophilicity of films and nanofibers is demonstrated. Then, the mechanical behavior of blended nanofibers and films used as reference are determined for small stress applied in the linear domain by DMA and by uniaxial traction up to rupture. The dried and wet states are covered for the first time. It is shown that the mechanical properties are increased when electrospinning is performed in the presence of PEO up to a 70/30 chitosan/PEO weight ratio even after PEO extraction. This result can be explained by a better dispersion of the chitosan in the presence of PEO.

Keywords: PEO/chitosan blend; swelling; mechanical properties; wet and dried states

1. Introduction

Processing of chitosan and chitosan blended with different polymers using electrospinning is often proposed in the literature to produce new biomaterials, especially developed for biomedical applications [1–10]. The advantage of chitosan is that it is obtained from a natural polymer, chitin, after controlled deacetylation. Chitosan is soluble in aqueous medium in acidic conditions due to $-NH_2$ protonation as soon as its degree of acetylation is lower than 0.5 [11]. Then, processing of chitosan is relatively easy, and it can be used under fiber, nanofiber, film, capsule, bead, sponge, gel, powder, tablet based on its insolubility in neutral medium.

Additionally, chitosan is an interesting biodegradable and biocompatible polymer with antibacterial and antifungal properties often described in the literature [12]. In addition, chitosan is stabilized by H–bond network in the solid state providing good mechanical properties under film or fiber materials. The main applications proposed for chitosan in the biomedical domain are: drug delivery, gene delivery vehicle, encapsulation of sensitive drugs, medical textiles, guided bone regeneration, scaffold for nerve tissue regeneration, and wound healing.

In a previous work, the conditions for electrospinning of pure chitosan nanofibers were optimized considering conditions published in the literature [13]. Electrospinning of chitosan has drawn a lot of attention in several scientific studies and a wide range of methods have been used to produce chitosan-based nanofibrous materials in the presence of poly (ethylene oxide) (PEO) as mentioned previously [13–19]. The optimum condition for a good spinnability and solubility of chitosan was obtained by using 0.5 M acetic acid as solvent and blending with PEO either in solution or in powder. Compatibility of chitosan and PEO was also shown and it was demonstrated that interaction between PEO and chitosan favored the processing. Nanofibers were obtained in the presence of 20 up to 40% (*w/w*) PEO yield solutions. In addition, it was confirmed that a chitosan sample with moderate molar mass (*MW* ~100,000) is more convenient to control the viscosity of the spinnable solution. In these conditions, nanofibers made of blend PEO/chitosan have average diameters under dried state between 100 and 150 nm [13].

The advantage of electrospinning is that materials are easily produced with high porosity and large surface area which favors the cell development [20–22]. Due to its good adhesive characteristic, chitosan is usually blended with other polymers [21]. One reference introduces dibasic sodium phosphate as ionic cross-linker [23]. In these conditions, based on its mechanical characteristics, the chitosan mats are adapted for soft tissue regeneration [23]. The chitosan characteristics are recognized to be used as potential wound dressing for skin lesions due to mechanical properties in the range of that of normal skin [24]. It is very recently that the mechanical properties of these electrospun chitosan-based nanofibers were discussed in literature. It was separately demonstrated that lower degree of acetylation increases the strength of the fibers and decreases the elongation at break for a chitosan/polyvynylalcohol (PVA) (ratio 1/1) electrospun membrane. These authors found a tensile strength at break between 3.5 and 5.2 MPa with 8% and 4% elongation at break when the degrees of acetylation are 16 and 8% respectively [25]. The content of chitosan in the blend chitosan/PVA has also shown to increase the Young modulus and stress at break and to decrease the elongation at break [20]. The values given at 30% of chitosan in the material gives respectively E = 201.7 Pa, σ_{break} = 4.15 MPa and ε_{break} = 3.96% [26]. For a chitosan/gelatin system (weight ratio 1/1), these parameters are respectively 48 MPa, 0.478 MPa and 1.3% [27]. Up to now, to our knowledge, no paper concerning the mechanical properties of pure chitosan electrospun nanofibers in dry and wet states are available in the literature. The only published results on wet state concern poly (L-lactide) nanofibers immersed in chitosan and treated with polydopamine at pH = 8.5. Chitosan decreases the stress at break and the elongation. At dried state, in presence of chitosan, the stress at break is around 2 MPa and the elongation is between 40% and 80%. The characteristics of the same samples in the wet state decreases for σ_{break} in the range of 0.6–1.0 MPa and increases slowly for ε_{break} up to 50% to 120% [21]. Nevertheless, the conditions for determination of these important parameters are usually not described.

In the present work, the electrospinning of pure chitosan-based nanofibers was extended to prepare chitosan nanofiber mats and determine their mechanical properties. As reference, films were prepared from the same solutions as used for electrospinning by casting allowing us to compare their physical properties with that of the nanofibers. It will be interesting to test the main differences between unorganized casted and fibrous materials.

For fibers and films production, the two different techniques previously proposed were used: (i) solutions of PEO and solutions of chitosan were mixed and casted in a mold or electrospun; (ii) powdered PEO was added into chitosan solutions [13]. Acetic acid at 0.5 M concentration was adopted as solvent. The change of PEO and chitosan weight ratio and PEO molar mass in the chitosan blends were investigated and related with the electrospun nanofibers and films physical properties under dried and wet conditions.

2. Materials and Methods

2.1. Materials

Chitosan (CS) sample from Northern cold-water shrimp, *Pandalus borealis* is a gift from Primex Cy (Batch TM4778, code 42010, Siglufjordur, Iceland). Its molecular weight (*MW*) is around 200 kg/mol and its degree of acetylation determined using ^{1}H NMR is degree of acetylation (DA) = 0.05. The viscosity of 1% solution is 86 mPa in 1% acetic acid at 25 °C. Poly (ethylene oxide) (PEO) with different molecular weight MW (5×10^3 and 1×10^3 kg/mol), acetic acid ($\geq$99.7%), ethanol and K_2CO_3 were purchased from Sigma-Aldrich (Saint Quentin Fallavier, France). Deionized water was used as solvent to make up the solutions. All reagents and polymers were used as received without further purification.

2.2. Sample Preparation

Chitosan (CS) solutions were prepared separately at 5% (w/v) in 0.5 M acetic acid. These solutions were prepared at room temperature with slow stirring for 4 days to obtain homogeneous solutions. In the same manner, poly (ethylene oxide) with different MW (5×10^3 and 1×10^3 kg/mol) were solubilized at 5% and 3% w/v in 0.5 M acetic acid on rotating stirrer. Chitosan (CS) solutions were mixed with the solution of PEO at chitosan/PEO weight ratio (in %) of 95/5, 90/10, 80/20, 70/30 and 60/40. Similarly, the same concentration of CS was blended with powdered PEO. The weight ratios were expressed as weight of chitosan or PEO divided by the total polymer weight for each system tested. This addition of powder (PEO) in the solution of chitosan was selected to avoid the dilution of the final chitosan concentration in the polymer blend solution when mixing the two polymers solutions. Viscosities and ionic conductivities of the blended CS/PEO solutions were measured at room temperature.

2.3. Chitosan Stabilization

Weighted initial nanofiber mats or films cut in pieces were immersed in alkaline ethanol/water (70/30) mixture dissolving K_2CO_3 at pH = 12 to neutralize the chitosan. Further, nanofibers membranes or films were washed for 3 days four times in a day with deionized water until neutral pH to obtain removal of the salt formed from chitosan solutions (potassium acetate), K_2CO_3 excess and PEO. At last, the membranes were dried at room temperature for further determination of the swelling capacity or rehydration after a first drying of the materials.

2.4. Casting of Chitosan/PEO Films

A certain amount of each of the CS/PEO mixtures showing good spinnability was placed in a Teflon mold of known volume to obtain a uniform polymer film. The probes were stored at room temperature for 3 days until complete evaporation of the solvent.

Different samples of a regular shape were taken from the films obtained for future measurements of their mechanical properties and degree of swelling.

2.5. Electrospinning

The prepared solutions were placed in a 5 mL plastic syringe fitted with a 21-gauge stainless steel needle. The syringe pump delivers solutions at specified flow rate vertically (model: KDS Legato 200, KD Scientific, Holliston, MA, USA), and electrospinning is realized with an applied voltage around 20 kV between the electrodes using a homemade dual high voltage power supplier ($\pm$30 kV, iseq GmbH, Radeberg, Germany). Then, the nanofibers were recovered on a microstructured collector with a regular pattern supporting an aluminum film used as collector and kept from 10 to 17 cm from the tip of the needle. The flow rates vary from 0.7 to 1.5 mL/h. The experiments were carried out at room temperature in closed Plexiglas® box with relative humidity ranging between 40% and 60%. The produced nanofibers matrices were left in ambient conditions to evaporate excess of acetic acid and water prior to further analyses.

2.6. Characterization of Nanofibers

2.6.1. Morphology of the Nanofibers Membranes

The Scanning electron microscopy (SEM) analyses of the samples were performed at CMTC-INP, Grenoble, France. The morphology of electrospun nanofiber membranes including the washed samples were observed with a scanning electron microscope (ultra 55 SEM FEG, Zeiss, Jena, Germany) operated at 3 kV. The nanofibers samples were coated with 10 nm carbon layer prior to SEM imaging. The average fiber diameter (AFD) was calculated by randomly selected diameter of 500 nanofibers from each sample.

2.6.2. Determination of Swelling Capacity

The swelling of the nanofibrous membranes or films were examined in terms of water loss between swollen state in water at neutral pH and final dried weight at room temperature. The wet swollen samples were weighed after blotting with tissue paper to remove excess surface water (W_w). Accordingly, the dried samples were also weighted repeatedly until the mass became constant (W_d). The measurements were carried out three times each. These values correspond to the first swelling. The average data were taken for the determination of swelling ratio S using the following equation:

$$S(\text{g/g}) = \left(\frac{W_w - W_d}{W_d} \right) \qquad (1)$$

where W_w (g) is the weight of the swollen nanofibrous mat or film and W_d (g) is the weight of the samples after drying at room temperature.

After drying, rehydration of dried samples was tested under wet form using the same conditions after two days in water at room temperature.

2.6.3. NMR Characterization of Nanofibers

The composition of the nanofibers in chitosan, PEO and acetic acid (or acetate) remaining in the fibers were determined by ^{1}H NMR at 80 °C on a Bruker Avance III 400 spectrometer (Billerica, MA, USA). Selected samples were used to analyze the NMR spectrum and exemplify the change in the nanofibers compositions [13]. Nanofiber samples of 7 to 10 mg were dissolved in 1 mL D_2O in presence of stoichiometric amount of DCl. Analysis of spectra allows the determination of the amount of acetic acid remaining in the dried samples and yield of PEO remaining in the samples after washing in different conditions.

2.6.4. Rheological Behavior

Rheological characteristics were studied using a ARG2 rheometer from TA Instruments (New Castle, DE, USA) with a cone-plate geometry; the cone has a diameter of 25 mm, a 4° angle and a 107 μm gap. The temperature is controlled at 20 °C by a Peltier plate. Steady-state flow experiments were performed in the range of 0.01 to 10 s^{-1}.

2.6.5. Ionic Conductivity of Blend Solutions

The conductivity of solution was determined at 20 °C using a Conductimeter sensION+ EC7 from Hach Lange (Loveland, CO, USA) equipped with a titanium electrode Crison 5073. The solvent 0.5 M acetic acid has a conductivity of 1.211 mS/cm.

2.6.6. Mechanical Characterization

The measurements were carried out using ARES-G2 rheometer (TA Instruments, New Castle, DE, USA) equipped with a rectangular geometry, used for axial tension consisting of two axial clamps that hold the material when the force is applied. Samples were taken from the nanofibrous electrospun matrices or films maintaining a length/width ratio around 2.69 (suggested value in the rheometer procedure). The results are expressed as the Stress σ = Force applied/section area in Pa. In addition, to compare the samples which have not the same morphology, results of tensile tests are expressed by the reduced force in $N/(kg/m^2)$ including the mass of the sample divided by its area.

Break tests were performed using the same geometry, starting from a zero-applied force until the material presents a breaking point, with a deformation rate of 0.01 mm/s. The experiments were carried out at constant temperature around 20 °C and a special device was adopted to maintain the relative humidity in the sample environment.

The rheometer also allowed obtaining of the thickness of the samples by measuring the gap between the two plates when they approach film or fiber mat as close as possible until the detector perceives zero axial force during compression. This measurement was repeated with a micrometer (Mitutoyo Digimatic micrometer; −25 mm with precision of 0.001 mm) giving very close values. Both techniques used to determine the thickness are in good agreement with a precision of 1 µm.

The dynamic mechanical analysis Dynamic mechanical analysis (DMA) tests were performed using an initial force applied of 0.02 N with deformations between 0.01% and 0.1% strain imposed by the length/width ratio of the sample. For analysis, the storage moduli E observed were normalized taking into consideration the sample weight per unit of surface with the expression:

$$E_s(\text{Specific modulus}) = \frac{\text{Experimental Storage Modulus (Pa)}}{\frac{\text{Sample weight (kg)}}{\text{Sample surface (m}^2)}} \tag{2}$$

3. Results and Discussion

To establish the influence of the morphology of chitosan in the materials produced, in a first step, film of chitosan/PEO blends are prepared by casting using the different compositions of the blends. Then, the same solutions were used for electrospinning and production of chitosan nanofibers.

3.1. Characterization of Chitosan/PEO Solutions

3.1.1. Viscosity

The viscosity of the different solutions used to produce films and fibers was determined in steady state conditions at low shear rate. The values at zero shear rate are given in Figure 1a for mixing of the two initial solutions (*w*/*v* 5% chitosan/5% PEO *MW* = 5 × 10⁶) and in Figure 1b for addition of the same PEO in solid state into the 5% chitosan solution.

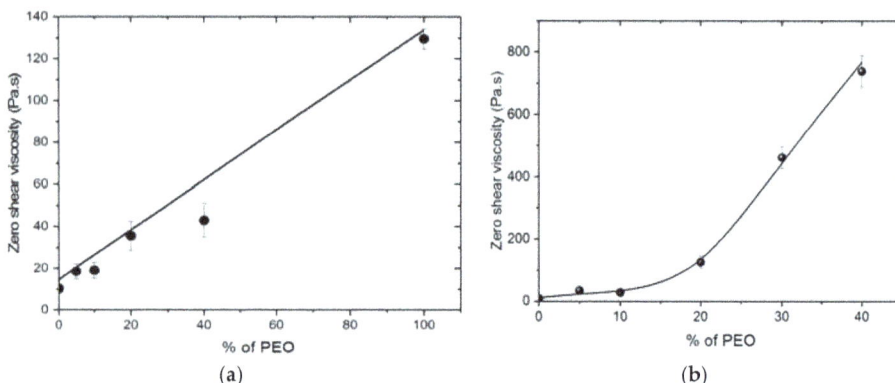

Figure 1. Zero shear viscosity at 20 °C for solutions prepared for films and fibers production as a function of w/w % of PEO in the blend. $T = 20$ °C. (**a**) By mixing 5% solutions of chitosan and 5% solution PEO in 0.5 M acetic acid; (**b**) by addition of powder PEO in 5% chitosan solution.

In Figure 1a, the total polymer concentration remains constant and equal to 5% w/v. These results indicate good compatibility between the two polymers in the chosen solvent used as mentioned in the literature [13,15,28,29].

As shown in Figure 1b, the viscosity of chitosan increases progressively with increasing PEO ratio. In these conditions, the chitosan concentration remains constant (5% w/v). Up to the ratio 70/30, the solutions are Newtonian at low shear rate corresponding to electrospinning conditions. These data confirm the compatibility of the chitosan and PEO. The same conclusions were previously obtained with PEO $MW = 1 \times 10^6$ [13].

3.1.2. Conductivity

The conductivity was also determined for the different solutions: all the values vary from 4.5 up to 8.2 mS/cm. At a given chitosan/PEO ratio, the conductivity decreases when the neutral PEO weight fraction increases whatever the blend conditions (mixture of solutions or PEO powder added in chitosan solution). However, the conductivity of mixed solutions remains slightly lower than when powder is added in chitosan solution due to a larger content of chitosan in this case (5% weight concentration). A small increase of conductivity with decrease of the viscosity is also observed when conductivity of chitosan/PEO blend with 5% $MW = 1 \times 10^6$ is compared with that of blend with 5% $MW = 5 \times 10^6$ at the same chitosan/PEO ratio.

3.2. Casting and Film Characterization

3.2.1. Degree of Swelling of Films

After casting and dying in atmospheric conditions, the chitosan in the blended films is neutralized in non-solvent basic conditions (ethanol/water 70/30 *V/V* containing CO_3K_2) and washed with fresh water until neutrality. The chitosan becomes insoluble at pH > 6.5 and PEO is solubilized in water. To control this step, the dried weight of the sample was determined in the same time as the degree of swelling (using the wet weight after washing step). All the data are given in Table 1. Taking these results into consideration, after drying, the samples are stabilized in water to determine the degree of rehydration of the chitosan materials, the hydrophilic character being important for biological application.

Table 1. Degree of swelling of blended film and composition after chitosan neutralization.

Systems	Chitosan Weight Ratio %	Film Thickness (mm)	Density (g/cm^3)	Initial Swelling Degree after PEO Extraction (g/g)	Remaining Polymer Fraction after First Neutralization	Swelling Degree after Rehydration (g/g)	Remaining Polymer Fraction after Rehydration	Theoretical Chitosan Fraction in the Material
	100	0.108	0.781	1.893	0.99	1.291	0.98	1.0
Powdered PEO added	90	0.092	0.767	1.852	0.96	1.372	0.95	0.9
	80	0.078	0.923	1.786	0.93	1.589	0.92	0.8
	70	0.085	0.978	2.439	0.86	1.966	0.85	0.7
	60	0.124	1.041	2.954	0.72	2.544	0.70	0.6
Blend of CS and PEO solutions	90	0.079	0.967	1.410	0.95	1.222	0.94	0.9
	80	0.083	0.817	1.767	0.90	1.466	0.89	0.8
	70	0.078	0.757	2.809	0.69	1.564	0.68	0.7
	60	0.068	1.004	1.939	0.64	1.860	0.64	0.6

From Table 1, it is clearly shown that the dried weight after rehydration remains the same as after the first step of drying indicating extraction of PEO. Nevertheless, the values remain slightly higher than the predicted value based on the initial weight fraction of chitosan especially when the PEO is added in the powder form. This is probably due to localization of PEO in the chitosan compact matrix. A small PEO fraction within the film may be not accessible for the washing process. The final composition of the sample was examined by ^{1}H NMR as previously described [13]. On films prepared at 80/20 chitosan/PEO ratio, it is confirmed that a small amount of PEO was not extracted in the process adopted especially when PEO is added under powder form.

The degree of swelling of the films is directly related with the composition of the blend as shown in Figure 2. The larger is the initial PEO yield, after extraction of the PEO, the larger is the swelling degree in water i.e., also of the specific area. In addition, it is shown that the hydrophilic character reflected by the degree of swelling remains high even after complete drying of the PEO extracted films. Then, the role of the yield in PEO on the properties of the material prepared remains important.

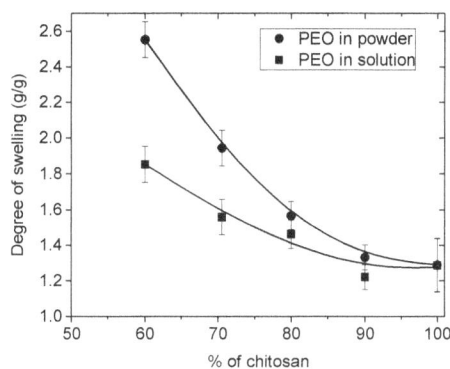

Figure 2. Influence of the initial PEO content in rehydration of chitosan films as a function of w/w % of chitosan in the blend (● powder form; ■ solution 5%, $MW = 1 \times 10^6$).

It may be added also that when PEO is added under powder form, the chitosan concentration is larger in the electrospun blend than in case of mixed solutions. After extraction of PEO and drying the pure chitosan film swelling is larger in relation to a larger porosity.

3.2.2. Mechanical Characteristics of Films

Firstly, DMA was performed under a low force (0.02 N) in the linear domain giving the elastic modulus E expressed in Pa. To compare results from one sample to another, moduli were normalized by the mass per unit area of each samples and expressed as E_s in Pa/(kg/m^2). Mechanical tests on membranes based on chitosan/powder PEO were performed both on the same sample just after the electrospinning process and after PEO extraction (Figure 3, Table 2).

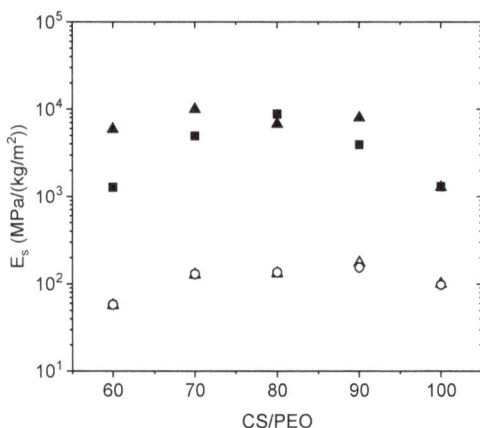

Figure 3. Specific modulus E_s expressed in MPa/(kg/m^2) obtained by DMA on initial blended film for powdered PEO added in 5% chitosan solution (dried state ▲, ■) and after PEO ($MW = 1 \times 10^6$) extraction, after drying and rehydration (wet state Δ, ○). Standard deviation is lower than 10%.

Table 2. Experimental values of E_s obtained in DMA on initial blended films and after extraction of PEO with different MW and concentrations being added in powder or solution.

Systems	$\frac{CS}{PEO}$	$E_s \left(\frac{MPA}{kg/m^2}\right)$	$E_s \left(\frac{MPA}{kg/m^2}\right)$	$\frac{E_{initial}}{E_{swollen}}$	$\frac{CS}{PEO}$	$E_s \left(\frac{MPA}{kg/m^2}\right)$	$E_s \left(\frac{MPA}{kg/m^2}\right)$	$\frac{E_{initial}}{E_{swollen}}$
		Initial State	Wet State after Rehydration			Initial State	Wet State after Rehydration	
		PEO $MW = 1 \times 10^6$ g/mol				PEO $MW = 5 \times 10^6$ g/mol		
PEO powder	60	5970	55	108				
	70	9970	126	79				
	80	6760	134	50.6	80	3830	166	22.99
	90	7820	176	44.5	90	2350	183	12.84
	100	1300	98	13.3	95	1590	201	7.92
5% PEO solution	60	1310	-	-				
	70	5080	-	-	80	3430	306	11.19
	80	8930	142	62.8	90	3370	294	11.46
	90	3990	166	24.0	95	2230	244	9.15
3% PEO solution					70	1070	170	6.26
					80	1320	260	4.99
					90	1010	240	4.19

The modulus of the film in the initial dried state is reinforced by the presence of PEO, nevertheless for biological application the properties under the chitosan insoluble form in aqueous medium must be determined. For that purpose, the films were neutralized and the PEO was extracted to perform the test under the swollen state at equilibrium in aqueous medium. Even after PEO extraction, the elastic modulus is reinforced when the chitosan was blended with PEO at least up to 70/30 chitosan/PEO ratio. Considering the data given in Table 2, it is shown that there is a slight influence of the conditions of PEO mixing (powder or solution) but it appears that the stronger performance in the wet state are obtained with solution of the higher molar mass PEO.

An interesting characteristic is the ratio between initial modulus in the dried state and the modulus in the wet state after PEO extraction. This ratio is very large when PEO is added in the powder form and it decreases when the content in PEO decreases. It indicates a larger porosity of the films formed in presence of PEO. The ratio remains lower when solutions are mixed corresponding to a larger homogeneity of the materials as well as a lower degree of swelling (see Figure 2).

Tests of uniaxial traction stress were performed on the wet films after PEO extraction. The blending with powder and solution PEO are compared in Figure 4a. To be able to compare all the experimental results and especially because the different materials prepared are porous, the performances are expressed by the force in Newton (N) reduced by the surfacic mass (kg/m^2) are represented in Figure 4b.

From Figure 4b, it is clear that the mechanical properties of the films in the wet state are directly related to the chitosan content in the initial casted film. The density of chitosan is larger when the yield of chitosan in the blend is larger.

Down to a ratio 60/40 chitosan/PEO, the strain at break remains nearly constant around 40%. It must be pointed out that the wet films have a high stress at beak (σ ~7 MPa) when prepared at high chitosan content (Table 3). This value is clearly higher than values given in the literature for different blends under dried state.

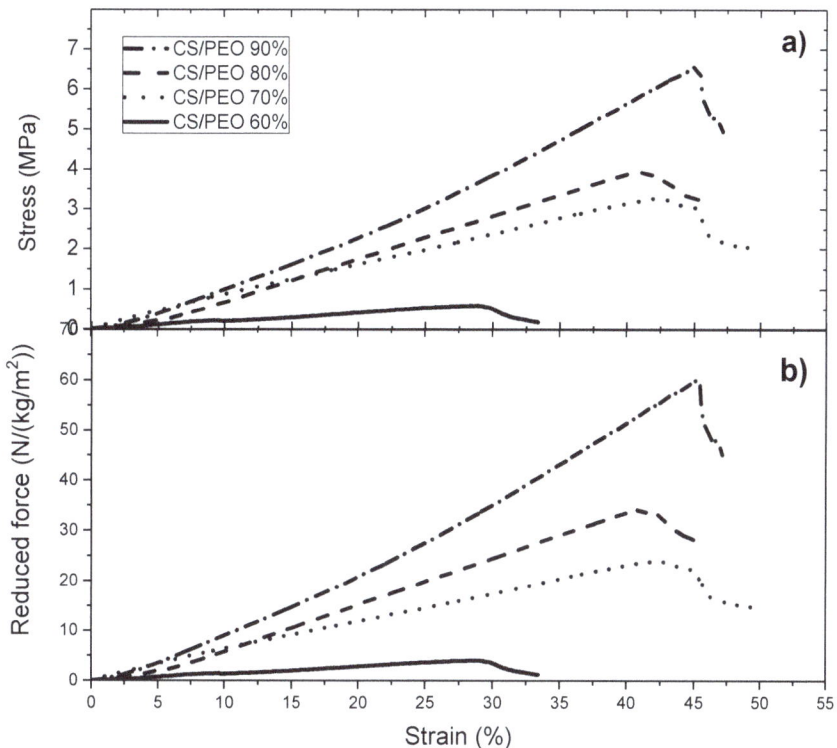

Figure 4. (**a**) Stress under uniaxial traction σ (expressed in MPa) and (**b**) Reduced force (expressed in N/(kg/m^2)) on films under rehydrated form (wet state) after chitosan neutralization, PEO (under powder form) extraction and drying. PEO *MW* = 1 × 10^6.

Table 3. Experimental data obtained in uniaxial tensile test on blended films after PEO extraction (PEO 1×10^6) and drying under wet state.

CS/PEO Powder	σ_{Break}		ε_{Break}	CS/5%PEO Solution	σ_{Break}		ε_{Break}
%	MPa	$\frac{N}{kg/m^2}$	%	%	MPa	$\frac{N}{kg/m^2}$	%
100	1.41	14	33.2	100	1.41	14	33.2
90	6.57	60	45.1	90	5.12	43	44.5
80	3.95	34	41.2	80	3.91	38	19.4
70	3.28	24	42.6				
60	0.66	4	29.2				

From this table, it is shown that the stress at break is increased when the films are prepared in the presence PEO (up to 70% CS/PEO) as well as the degree of extension and that there is no important difference in dependence of the conditions of PEO addition (powder or solution). It is suggested that mixing chitosan and PEO allows better dispersion of chitosan molecules and more homogeneous films with stronger H bond stabilization.

3.3. Electrospinning of Blends

3.3.1. Conditions of Electrospinning

Different experimental conditions were explored to optimize the production of nanofibers at high yield in chitosan using two PEO with different molar masses. The average parameters determined to get fibers in absence of spray or beads are given in Table 4.

Table 4. Experimental conditions for electrospinning on the different systems studied

CS/PEO	Pump Rate (mL/h)	Tip to Collector Distance (cm)	Applied Voltage (kV)	Electrospun Products
PEO MW = 1×10^6				
Powdered PEO Added into the Chitosan Solution				
70/30	0.2–0.4 *	11 *–12	24–27 *	Fibers
80/20	0.22 *–0.25	13	25–27 *	Fibers
Mixture of the Polymer Solutions				
60/40	0.05–0.2 *	14	22–25 *	Fibers
70/30	0.09–0.1 *	14	27	Fibers
PEO MW = 5×10^6				
Powdered PEO Added into the Chitosan Solution				
80/20	1.0–1.15	13–15	21–25	Fibers
90/10	1–1.1	12–15	20–24	Fibers
95/5	0.7–1.5	14 *–17	21–24 *	Fibers, few beads
Mixture of the Polymer Solutions (5%)				
70/30	1.4–1.5	15	24–27	Fibers
80/20	0.6 *–1.0	15	21–24 *	Fibers
90/10	0.65–0.7	15	20	Fibers
95/5	1.2	15	22	Fibers, few beads
Mixture of the Polymer Solutions (3%)				
60/40	0.7 *–1.4	15	20–24	Fibers, few beads
70/30	0.8 *–1.3	14 *–17	21–22 *	Fibers
80/20	0.7–0.8 *	13 *–14	20–22 *	Fibers
90/10	0.7 *–0.9	14 *–16	21 *–24	Few beads, fibers

* Indicates the optimum conditions.

In our electrospinning process, the collector was designed permitting to recover the fibers matrix from a metal plate avoiding sticking on the support (Figure 5). Aluminum foils cut in cross or rectangle have been chosen as support to remove the mat after processing as shown in Figure 5. In these conditions, the probes for mechanical tests are easy to take out.

Figure 5. Experimental devices for nanofibers production.

Considering the experimental conditions given in Table 4, it must be noticed that they are different, especially considering the flow rates, compared with our previous work even if the solvent is the same as well as one of the PEO ($MW = 1 \times 10^6$) [13]. In this previous work, the flow rates varied between 20 and 100 μL/h and the viscosities of systems were much lower than in the present study. It is due to the higher molar mass of the chitosan used in the present work as well as PEO molar mass of 5×10^6. It is important to point out the importance of the molar mass of the chitosan used as well probably of its degree of acetylation which controls the solubility. PEO with $MW = 5 \times 10^6$ was preferred to get stronger fibers in the following part of the work.

3.3.2. Degree of Swelling of the Nanofibers

From the yield in polymer remaining after extraction of PEO, Table 5 indicates that PEO is nearly totally extracted. ^{1}H analysis confirmed this information by small signal located at 3.8 ppm corresponding to the superposition of one H of chitosan and $-CH_2-$ of PEO left.

Table 5. Degree of rehydration of nanofibers after neutralization, extraction of PEO ($MW = 5 \times 10^6$) and drying.

Systems	Fiber Mat Thickness (mm)	Chitosan Weight Concentration Ratio	Density (g/cm^3)	Rehydration Degree (g/g)	Remaining Polymer Fraction after Rehydration	Theoretical Chitosan Fraction in the Material
Powdered PEO added	0.16	95	0.117	3.13	0.93	0.95
	0.156	90	0.132	4.29	0.91	0.90
	0.127	80	0.141	4.50	0.84	0.80
Blend of CS and 5% PEO solution	0.128	95	0.173	3.58	0.90	0.95
	0.122	90	0.163	3.71	0.88	0.90
	0.102	80	0.200	3.95	0.84	0.80
	0.115	70	0.164	4.19	0.73	0.70
Blend of CS and 3% PEO solution	0.115	90	0.169	3.80	0.95	0.90
	0.120	80	0.192	3.93	0.83	0.80
	0.112	70	0.176	4.09	0.73	0.70
	0.114	60	0.167	5.33	0.65	0.60

Due to porosity of the mat of fibers, the water content corresponds to the swelling of the fibers but also to water included in the porous materials between hydrophilic fibers. Then, the values are larger than values obtained on films prepared from the same solutions. In addition, it is found that the dried weight fraction obtained compared to the initial chitosan dried weight fraction indicates nearly complete PEO extraction as confirmed by NMR experiments. Then, these data demonstrate that accessibility for extraction is much better under fiber morphology compared to the films (compare with Table 1).

The influence of the PEO content in the blend controls the degree of rehydration which increases when PEO content increases as represented in Figure 6. The values are nearly the same for the two PEO concentrations used during electrospinning.

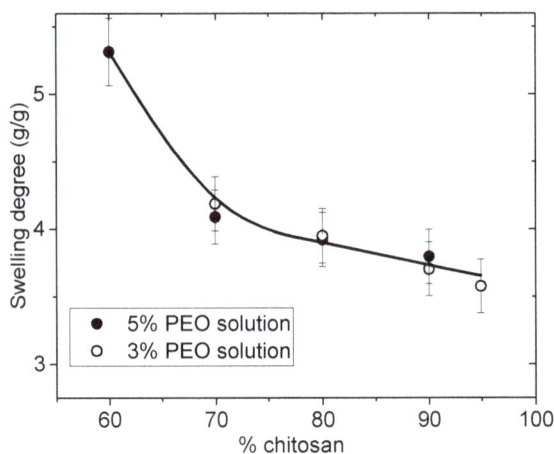

Figure 6. Degree of swelling of nanofiber mats after extraction of PEO ($MW = 5 \times 10^6$) and rehydration as a function of the initial chitosan content w/w % in the initial blend.

The swelling of nanofibers in the presence of PEO powder in the blend was found larger than when using solutions as observed previously with films. The nanofibers are better dispersed when blended with PEO solutions giving a more homogeneous network. The hydration is larger than that of films due to the lower chitosan density of the nanofiber mats compared with the casted films. This result indicates a larger porosity and larger accessible surface of the fiber mat compared with film. This point should be important for biological applications and especially cell developments.

3.3.3. Mechanical Characterization of Nanofibers

The characterization of the mechanical properties of the matrices obtained by electrospinning and their relationship with those of the casted non-structured materials is the fundamental part of this research work. Therefore, different stages in the measurement of properties were carried out. For this purpose, a sequence of dynamic mechanical analysis (DMA) was chosen where small deformations are applied to the material without affecting its structure, nor causing significant elongation, being able to recover the sample for subsequent treatments (Table 6). This sequence ends with material break tests using uniaxial deformation under the wet state.

Table 6. Experimental values of E_s obtained in DMA on initial blended fibers and after PEO extraction ($MW = 5 \times 10^6$) used in powder form and in 5% and 3% solutions under swollen state after drying and rehydration.

Systems		$E_s\left(\frac{\text{MPA}}{\text{kg/m}^2}\right)$ Initial State	$E_s\left(\frac{\text{MPA}}{\text{kg/m}^2}\right)$ Wet State after Rehydration	$\frac{E_{initial}}{E_{swollen}}$
CS/PEO powder	80	1650	466	3.53
	90	1460	434	3.36
CS (5%)/PEO solution	70	939	421	2.23
	80	1490	652	2.28
	90	1370	602	2.28
CS (3%)/PEO solution	60	2430	432	5.63
	70	2760	510	5.41
	80	2580	546	4.73

From this table, it comes that the ratio between initial modulus and swollen wet modulus are much lower than for films (compare with Table 2). This indicates a larger stability of the fibrillary network compared with films. In addition, modulus in the hydrated form remains higher when PEO was initially introduced in 5% solution even if in the initial dried state, it is the opposite due to the degree of distribution of the fibers and larger interfaces but lower cross section. It must be pointed that the E_s values in the swollen state are much higher than on film for the same composition even if the chitosan density is lower in the mat.

The data for uniaxial traction up to break are analyzed in Table 7.

Table 7. The uniaxial stress/strain properties on nanofibers mat under hydrated form after neutralization and PEO extraction, drying and rehydration. ($MW = 5 \times 10^6$).

[CS]	σ_{Break}		ε_{Break}	[CS]	σ_{Break}		ε_{Break}	[CS]	σ_{Break}		ε_{Break}
%	MPa	$\frac{\text{N}}{\text{kg/m}^2}$	%	%	MPa	$\frac{\text{N}}{\text{kg/m}^2}$	%	%	MPa	$\frac{\text{N}}{\text{kg/m}^2}$	%
Powder PEO				5% Solution				3% Solution			
								60	0.4	11	30.4
				70	0.42	18	44.9	70	0.64	23	34.5
80	0.83	34	53.8	80	1.43	40	46.0	80	0.72	26	32.8
90	0.61	23	41.2	90	1.19	47	31.9				

From these data, it is shown that in the initial dried state, the performance of the material is better when the blend is prepared by mixing the solutions even if the yield in chitosan during preparation is lower in this case (due to mixing of the two solutions) compared to addition of solid PEO. The higher is the chitosan content the higher is σ_{Break} and lower is the elongation (ε_{Break} %). From these data, it is concluded that normalized σ_{break} values are slightly larger than for the corresponding chitosan/PEO composition obtained with the films even if the density of the mat is lower (Table 3).

An example of mechanical properties of chitosan/PEO nanofibers blends is given in Figure 7a,b, in which the mechanical behavior is compared when PEO is added under powder form and in solution respectively. The elongation at break increases with increasing the ratio of PEO. These results point out that PEO interacts with chitosan as shown previously [13]. Then, PEO plays the role of a plasticizer and the material becomes less brittle.

Then, the performances were compared under wet state after PEO extraction and rehydration. Few results are given in Figure 8 in which the influence of the PEO concentration is tested.

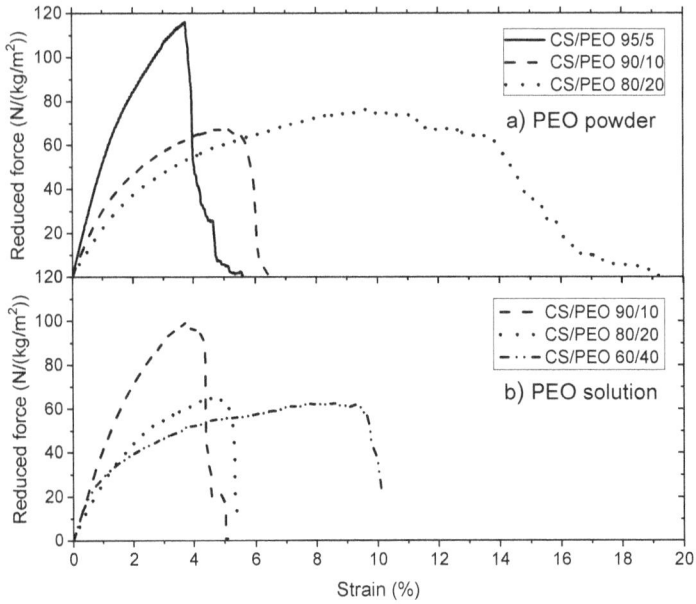

Figure 7. Uniaxial extension for chitosan/PEO nanofibers blends in the initial state. (**a**) PEO $MW = 5 \times 10^6$ added under powder form; (**b**) PEO mixed in solution from $MW = 5 \times 10^6$ at 3%.

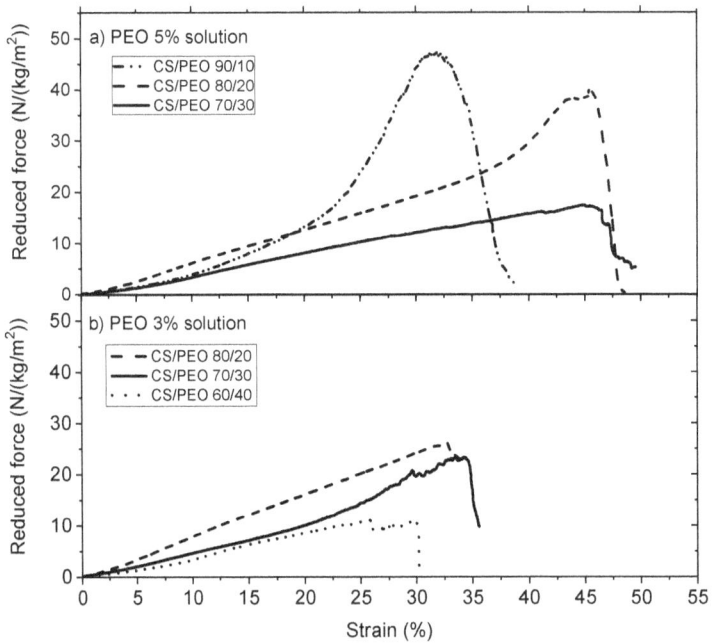

Figure 8. Uniaxial extension for chitosan nanofibers in the wet state after PEO extraction (**a**) PEO mixed in solution from $MW = 5 \times 10^6$ at 5% (**b**) PEO mixed in solution from PEO 5×10^6 solution at 3%.

From Figure 8, it is clearly shown that the mechanical behavior of the chitosan nanofibers is stronger when the yield of chitosan increases with a relatively high elongation around 30% to 35% under wet state. In the presence of 5% PEO solution having a higher viscosity but also a higher chitosan concentration for the same chitosan/PEO ratio, the stress at break is higher probably due to a lower dispersion of chitosan fibers with higher degree of packing of the chitosan molecules causing H-bonding in the intermediate drying of the materials.

It must be pointed out that the influence of PEO in the blend on the reinforcement of the mechanical properties is lower than for the films and the values of σ_{Break} are of the same order of magnitude with a lower density of chitosan (compare Tables 1 and 5).

3.3.4. Morphology of the Nanofibers

Different states of the fibers mats were examined by scanning electron microscopy to point out the role of the blend composition and the role of PEO extraction on the morphology of the fibers.

Some examples are given in the following pictures in which nanofibers produced at 70% and 95% chitosan are compared (Figure 9). From their analysis, it is shown that the average diameters of 70% chitosan fibers produced with PEO powder are larger than when 5% solution is used with PEO $MW = 1 \times 10^6$ (Figure 9a,b).

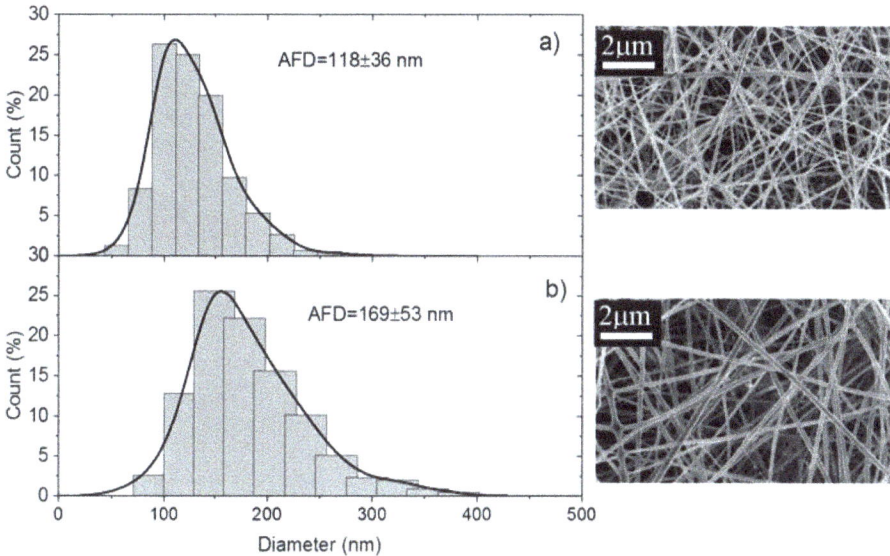

Figure 9. Diameter distribution of chitosan/PEO fibers expressed as a function of the count of fibers in % for (**a**) PEO $MW = 10^6$ at the concentration of 5% mixed with chitosan solution and (**b**) PEO $MW = 10^6$ added in powder form.

In addition, the diameters of the fibers are lower for 70% chitosan when PEO is added as 5% PEO solution ($MW = 1 \times 10^6$) (Figure 9a) compared with PEO ($MW = 5 \times 10^6$) (see Figure 10a). This influence of MW was previously shown for cellulose acetate and poly(vinyl chloride) nanofibers [30]. SEM indicates that fibers are thinner with 3% solution than with 5% with also a broader distribution causing lower mechanical properties (Figure 10a,b). These results are directly related with the viscosity of the electrospun systems and consequently with the flow rate necessary to optimize the process.

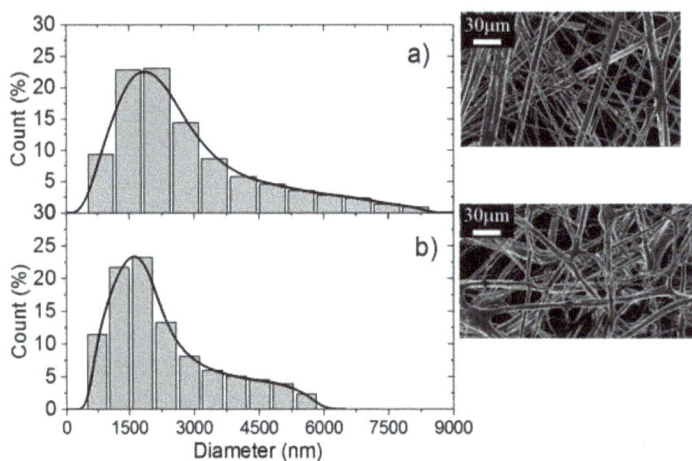

Figure 10. Diameter distribution of chitosan/PEO fibers expressed as a function of the count of fibers in % for PEO *MW* = 5 × 10^6 at the concentration of (**a**) 5% and (**b**) 3%.

Concerning the influence of the chitosan/PEO ratio, using PEO *MW* = 5 × 10^6 at 5% mixed with 5% chitosan solution, the fiber diameters for 70/30 are 10 times higher than for the 95/5 ratio (Figures 10a and 11a). Such larger diameter distribution is essentially due to much higher viscosity of the 70/30 system (see Figure 1a,b), all the parameters for electrospinning being equivalent.

The last parameter tested is the washing step as shown in Figure 11.

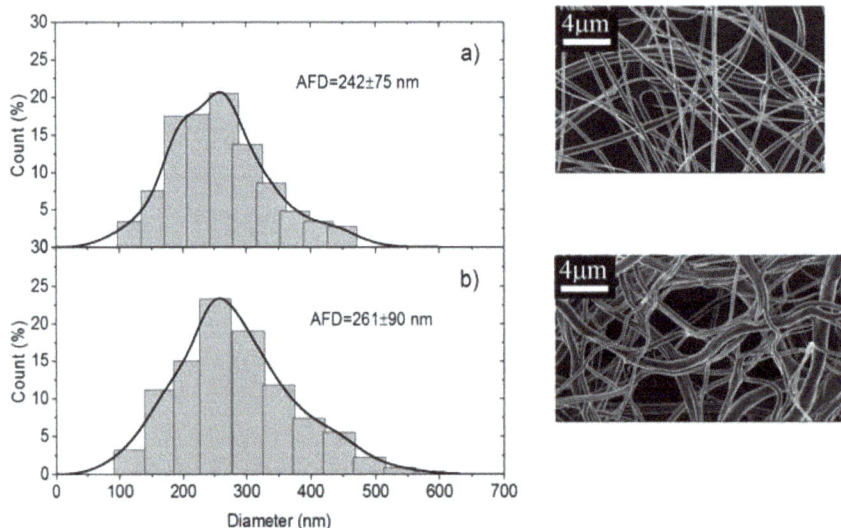

Figure 11. Diameter distribution of chitosan/PEO fibers expressed as a function of the count of fibers in % for PEO *MW* = 5 × 10^6 at the concentration of 5% (**a**) before washing and (**b**) after washing.

Comparison of Figure 11a,b indicates that washing does not modify the fiber morphology. Nevertheless, the distribution of diameters is a little broader than that obtained for 70/30 in presence of PEO $MW = 1 \times 10^6$ but the use of PEO $MW = 5 \times 10^6$ was necessary to get convenient conditions for electrospinning even if the viscosity was larger for the PEO solution.

4. Conclusions

In this paper, the influence of the structure of materials based on chitosan was examined. Firstly, films were casted from PEO/chitosan blends in different conditions. It was clearly shown that the degrees of swelling and mechanical properties of these chitosan films are controlled by the content and conditions of PEO addition even after extraction of PEO. The degree of swelling increases when PEO content increased up to 60% in the blend. In the same time, the specific modulus E_s from DMA increases in the initial dried state as well as in the wet state after PEO extraction corresponding to a better dispersion of chitosan and stronger inter-chain interactions (involving H-bond stabilization). Such chitosan films prepared with 90% chitosan in the initial blend with PEO powder ($MW = 1 \times 10^6$) and after PEO extraction have stress at beak around 6.6 MPa or 60 N/(kg/m$^2)$ with an elongation at beak larger than 40% in the wet state. These results confirm the good film forming character of chitosan recognized for a long time. The films are less brittle when casted from the blend due to their more porous and hydrophilic structure.

Secondly, using the same solutions, electrospinning was performed in experimental conditions optimized to get good nanofibers which were tested in the initial state but also after PEO extraction in the wet state. This study allows discussion of the advantage of preparing chitosan nanofibers among unstructured films. The PEO extraction is more efficient under fiber morphology as shown by NMR. The density decreases and the degree of hydration of nanofiber mats increases much when processed in presence of PEO compared with film. The ratio between E_s values in the initial composite state and in the wet state on pure chitosan remains lower than on films corresponding to a larger stability of the network formed. In the same time, the specific modulus E_s is much larger on the mats in the wet state than under film structure even if the density is lower. This is attributed to the larger porosity and higher dispersion and higher degree of connection between the chitosan chains. This aspect is certainly interesting for cell development. At 90% chitosan blended with 5% PEO solution ($MW = 5 \times 10^6$), after extraction of PEO in the wet state, the stress at break is around 1.2 MPa or 47 N/(Kg/m$^2)$ and 32% elongation at break which are of the same order of magnitude than for films even with a lower density of fibrous material. One advantage of the nanofibers is the large porosity of membranes as well as its large specific surface allowing a possible high surface adsorption.

To conclude, for the first time, chitosan nanofibers mats are prepared and characterized completely for their structural and mechanical characteristics in the dried and wet state allowing the use of these materials for biological applications in the near future. In addition, it must be remembered that these fibers are stable in aqueous medium at pH higher than chitosan pK_0 for years.

Acknowledgments: This work has been financially supported by Mexican CONACYT grant 453169. The Primex Cy (Iceland) is recognized for gifting us the pure chitosan sample used in this work. The Centre de Recherche sur les Macromolécules Végétales (CERMAV, CNRS) is also acknowledge for providing us technical assistance for NMR experiments.

Author Contributions: Marguerite Rinaudo and Frédéric Bossard conceived and designed the experiments; Christian Enrique Garcia Garcia performed the experiments; Félix Armando Soltero Martínez; Marguerite Rinaudo, Christian Enrique Garcia Garcia and Frédéric Bossard analyzed the data and wrote the paper.

Conflicts of Interest: The authors declare no conflict of interest.

References

1. Haider, A.; Haider, S.; Kang, I.K. A comprehensive review summarizing the effect of electrospinning parameters and potential applications of nanofibers in biomedical and biotechnology. *Arab. J. Chem.* **2015**. [CrossRef]
2. Chen, J.P.; Chang, G.Y.; Chen, J.K. Electrospun collagen/chitosan nanofibrous membrane as wound dressing. *Colloids Surf. A* **2008**, *313*, 183–188. [CrossRef]

3. Chen, Z.G.; Wang, P.W.; Wei, B.; Mo, X.M.; Cui, F.Z. Electrospun collagen–chitosan nanofiber: A biomimetic extracellular matrix for endothelial cell and smooth muscle cell. *Acta Biomater.* **2010**, *6*, 372–382. [CrossRef] [PubMed]

4. Zhang, Y.; Reddy, V.J.; Wong, S.Y.; Li, X.; Su, B.; Ramakrishna, S.; Lim, C.T. Enhanced biomineralization in osteoblasts on a novel electrospun biocomposite nanofibrous substrate of hydroxyapatite/collagen/chitosan. *Tissue Eng. Part A* **2010**, *16*, 1949–1960. [CrossRef] [PubMed]

5. Chen, J.P.; Chen, S.H.; Lai, G.J. Preparation and characterization of biomimetic silk fibroin/chitosan composite nanofibers by electrospinning for osteoblasts culture. *Nanoscale Res. Lett.* **2012**, *7*, 1–11. [CrossRef] [PubMed]

6. Zhang, Y.; Venugopal, J.R.; El-Turki, A.; Ramakrishna, S.; Su, B.; Lim, C.T. Electrospun biomimetic nanocomposite nanofibers of hydroxyapatite/chitosan for bone tissue engineering. *Biomaterials* **2008**, *29*, 4314–4322. [CrossRef] [PubMed]

7. Jalaja, K.; Naskar, D.; Kundu, S.C.; James, N.R. Potential of electrospun core–shell structured gelatin-chitosan nanofibers for biomedical applications. *Carbohydr. Polym.* **2016**, *136*, 1098–1107. [CrossRef] [PubMed]

8. Jayakumar, R.; Prabaharan, M.; Nair, S.V.; Tamura, H. Novel chitin and chitosan nanofibers in biomedical applications. *Biotechnol. Adv.* **2010**, *28*, 142–150. [CrossRef] [PubMed]

9. Torres-Giner, S.; Ocio, M.J.; Lagaron, J.M. Development of active antimicrobial fiber-based chitosan polysaccharide nanostructures using electrospinning. *Eng. Life Sci.* **2008**, *8*, 303–314. [CrossRef]

10. Subramanian, A.; Vu, D.; Larsen, G.F.; Lin, H.Y. Preparation and evaluation of the electrospun chitosan/PEO fibers for potential applications in cartilage tissue engineering. *J. Biomater. Sci. Polym. Ed.* **2005**, *16*, 861–873. [CrossRef] [PubMed]

11. Rinaudo, M. Chitin and chitosan: Properties and applications. *Prog. Polym. Sci.* **2006**, *31*, 603–632. [CrossRef]

12. Younes, I.; Rinaudo, M. Chitin and chitosan preparation from marine sources. Structure, properties and applications. *Mar. Drugs* **2015**, *13*, 1133–1174. [CrossRef] [PubMed]

13. Mengistu Lemma, S.; Bossard, F.; Rinaudo, M. Preparation of pure and stable chitosan nanofibers by electrospinning in the presence of poly(ethylene oxide). *Int. J. Mol. Sci.* **2016**, *17*, 1790. [CrossRef] [PubMed]

14. Desai, K.; Kit, K.; Li, J.; Zivanovic, S. Morphological and surface properties of electrospun chitosan nanofibers. *Biomacromolecules* **2008**, *9*, 1000–1006. [CrossRef] [PubMed]

15. Pakravan, M.; Heuzey, M.C.; Ajji, A. A fundamental study of chitosan/PEO electrospinning. *Polymer* **2011**, *52*, 4813–4824. [CrossRef]

16. Ojha, S.S.; Stevens, D.R.; Hoffman, T.J.; Stano, K.; Klossner, R.; Scott, M.C.; Gorga, R.E. Fabrication and characterization of electrospun chitosan nanofibers formed via templating with polyethylene oxide. *Biomacromolecules* **2008**, *9*, 2523–2529. [CrossRef] [PubMed]

17. Duan, B.; Dong, C.; Yuan, X.; Yao, K. Electrospinning of chitosan solutions in acetic acid with poly(ethylene oxide). *J. Biomater. Sci. Polym. Ed.* **2004**, *15*, 797–811. [CrossRef] [PubMed]

18. Spasova, M.; Manolova, N.; Paneva, D.; Rashkov, I. Preparation of chitosan-containing nanofibres by electrospinning of chitosan/poly(ethylene oxide) blend solutions. *e-Polymers* **2004**, *4*, 624–635. [CrossRef]

19. Cheng, F.; Gao, J.; Wang, L.; Hu, X. Composite chitosan/poly(ethylene oxide) electrospun nanofibrous mats as novel wound dressing matrixes for the controlled release of drugs. *J. Appl. Polym. Sci.* **2015**, *132*. [CrossRef]

20. Balagangadharan, K.; Dhivya, S.; Selvamurugan, N. Chitosan based nanofibers in bone tissue engineering. *Int. J. Biol. Macromol.* **2017**, *104*, 1372–1382. [CrossRef] [PubMed]

21. Liu, H.; Li, W.; Wen, W.; Luo, B.; Liu, M.; Ding, S.; Zhoua, C. Mechanical properties and osteogenic activity of poly(L-lactide) fibrous membrane synergistically enhanced by chitosan nanofibers and polydopamine layer. *Mater. Sci. Eng. C* **2017**, *81*, 280–290. [CrossRef] [PubMed]

22. Bhattarai, N.; Edmondson, D.; Veiseh, O.; Matsen, F.A.; Zhang, M. Electrospun chitosan-based nanofibers and their cellular compatibility. *Biomaterials* **2005**, *26*, 6176–6184. [CrossRef] [PubMed]

23. Tonda-Turoa, C.; Ruini, F.; Ramella, M.; Boccafoschi, F. Non-covalently crosslinked chitosan nanofibrous mats prepared byelectrospinning as substrates for soft tissue regeneration. *Carbohydr. Polym.* **2017**, *162*, 82–92. [CrossRef] [PubMed]

24. Trinca, R.B.; Westin, C.B.; da Silva, J.A.F.; Moraes, A.M. Electrospun multilayer chitosan scaffolds as potential wound dressings for skin lesions. *Eur. Polym. J.* **2017**, *88*, 161–170. [CrossRef]

25.	Habiba, U.; Siddique, T.A.; Talebian, S.; Lee, J.J.L.; Salleh, A.; Ang, B.C.; Afifi, A.M. Effect of deacetylation on property of electrospun chitosan/PVA nanofibrous membrane and removal of methyl orange, Fe(III) and Cr(VI) ions. *Carbohydr. Polym.* **2017**, *177*, 32–39. [CrossRef] [PubMed]

26.	Gonçalvesa, R.P.; Ferreira, W.H.; Gouvêa, R.F.; Andrade, C.T. Effect of chitosan on the properties of electrospun fibers from mixed poly (vinyl alcohol)/chitosan solutions. *Mater. Res.* **2017**, *20*, 984–993. [CrossRef]

27.	Cai, N.; Hou, D.; Luo, X.; Han, C.; Fu, J.; Zen, H.; Yu, F. Enhancing mechanical properties of polyelectrolyte complex nanofibers with graphene oxide nanofillers pretreated by polycation. *Compos. Sci. Technol.* **2016**, *135*, 128–136. [CrossRef]

28.	Nikolova, A.; Manolova, N.; Rashkov, I. Rheological characteristics of aqueous of solutions mixtures of chitosan and polyoxyethylene. *Polym. Bull.* **1998**, *41*, 115–121. [CrossRef]

29.	Rošic, R.; Pelipenko, J.; Kocbek, P.; Baumgartner, S.; Bešter-Rogač, M.; Kristl, J. The role of rheology of polymer solutions in predicting nanofiber formation by electrospinning. *Eur. Polym. J.* **2012**, *48*, 1374–1384. [CrossRef]

30.	Tarus, B.; Fadel, N.; Al-Oufi, A.; El-Missiry, M. Effect of polymer concentration on the morphology and mechanical characteristics of electrospun cellulose acetate and poly(vunyl chloride) nanofiber mats. *Alex. Eng. J.* **2016**, *55*, 2975–2984. [CrossRef]

polymers

MDPI

Communication

Nanosphere Lithography of Chitin and Chitosan with Colloidal and Self-Masking Patterning

Rakkiyappan Chandran, Kyle Nowlin and Dennis R. LaJeunesse *

Department of Nanoscience, Joint School of Nanoscience and Nanoengineering,
University of North Carolina Greensboro, Greensboro, NC 27401, USA;
rakkiyappanchandran@gmail.com (R.C.); ksnowlin@gmail.com (K.N.)
* Correspondence: drlajeun@uncg.edu; Tel.: +1-336-285-2866

Received: 15 January 2018; Accepted: 15 February 2018; Published: 23 February 2018

Abstract: Complex surface topographies control, define, and determine the properties of insect cuticles. In some cases, these nanostructured materials are a direct extension of chitin-based cuticles. The cellular mechanisms that generate these elaborate chitin-based structures are unknown, and involve complicated cellular and biochemical "bottom-up" processes. We demonstrated that a synthetic "top-down" fabrication technique—nanosphere lithography—generates surfaces of chitin or chitosan that mimic the arrangement of nanostructures found on the surface of certain insect wings and eyes. Chitin and chitosan are flexible and biocompatible abundant natural polymers, and are a sustainable resource. The fabrication of nanostructured chitin and chitosan materials enables the development of new biopolymer materials. Finally, we demonstrated that another property of chitin and chitosan—the ability to self-assemble nanosilver particles—enables a novel and powerful new tool for the nanosphere lithographic method: the ability to generate a self-masking thin film. The scalability of the nanosphere lithographic technique is a major limitation; however, the silver nanoparticle self-masking enables a one-step thin-film cast or masking process, which can be used to generate nanostructured surfaces over a wide range of surfaces and areas.

Keywords: chitin; chitosan; nanostructured biomaterial; polymer; self-masking nanosphere lithography; cicada

1. Introduction

The surfaces of many insects are decorated with nanostructure topographies that control and determine specific physical and chemical properties. The wings and eyes of many insects—including cicadas and moths—have arrays of multimodal nanoscale cones and cylinders that are anti-reflective, anti-wetting, self-cleaning, and anti-microbial [1–5]. Insect cuticles are complex natural composite materials that are composed of a polysaccharide chitin fiber network and a matrix of proteins and lipids, and can potentially form through complex interactions during cuticle deposition [6,7]. While many of these nanostructures in insect cuticles are surface features composed of a lipid–protein matrix, in the case of the dog day cicada (*Tibicens* species), the nanostructured surface is, in part, controlled by deeper extensions of the underlying chitin exoskeleton. This demonstrates that, in some cases, the nanoscale organization of chitin assists in the formation of nanoscale topology [4]. Many insect nanostructured surfaces demonstrate a hexagonally close-packed array nanostructure [4,6,8]. While the mechanisms that control this type of nanoscale chitin organization are unresolved, in this paper, we take a direct approach to organize chitin and the related polysaccharide chitosan at the nanoscale, using a technique called nanosphere lithography (NSL), which utilizes an hexagonal close packed (HCP) patterning method [9–13]. NSL is a two-step fabrication technique that enables the generation of micro- and nanoscale topographies that are composed of hexagonally close-packed arrays of cylindrical, conical, or hemispherical structures [9–13]. The initial step of the NSL process involves the masking of a substrate with an HCP monolayer nanosphere film, which is followed by the processing of the masked substrate using either standard deposition (e.g., plasma vapor deposition), or etching techniques (e.g., reactive ion etching). NSL has been used to generate micro- and nanostructured surfaces in hard semiconductor and synthetic polymeric materials [12,13]. In this work, we demonstrate the generation of a nanostructured surface in a biopolymer—in this case, either the chitin-mats derived from an insect wing, or a thin film composed of chitosan. The nanostructured surface (NSS) generated by NSL mimics the surfaces of cicada wings, however, these synthetic biopolymer surfaces are dynamic and change their morphology in an aqueous environment. One of the major issues with NSL as a large-scale processing technique is its scalability. This limitation is largely due to the difficulty of generating large areas of the nanosphere mask [9,10]. In this paper, we utilize a property of chitin and chitosan—the ability of these polymers to generate silver nanoparticles (AgNP) [14]—as a means to bypass the application of the nanosphere of the substrate in the traditional NSL process. Using a drop-cast chitosan/AgNP as a self-masking thin-film etching substrate, we generated a nanostructured AgNP/chitosan surface. This self-masking technique has great application potential for the large-scale fabrication of nanostructured polymeric surfaces, especially for large and non-uniform areas.

2. Methods

Chitin and chitosan sources: For these experiments, chitin was prepared from insect wing cuticles from the periodic 17-year cicada Brood II, *Magicicada septendecim*, collected locally in Greensboro, North Carolina, during the June 2013 emergence. The chitin from these wings was prepared as previously described [4]. Chitosan (molecular weight: 150,000, 1.5% w/v), acetic acid, NaOH, and NaCl were purchased from Sigma-Aldrich (St. Louis, MO, USA).

Synthesis of silver nanoparticle in chitosan solution: AgNO$_3$ (>99%) was purchased from Sigma Aldrich chemicals, and was prepared as a 10^{-2} M solution. Chitosan (0.5 g, dissolved in 10 mL of 1% v/v acetic acid solution) and 0.2 M NaCl in a 5 mL solution were added dropwise. Mixtures of chitosan and AgNO$_3$ solution were prepared in a 1:5 ratio (by volume). For the formation of monodispersed nanoparticles, the mixed sample solutions underwent ultra-sonication for 3–4 h. The samples were then drop-casted and dried in oven for about 10 min at 40 °C to make films, and were then assessed for various analytical characteristics.

Colloidal/nanosphere lithography: We fabricated 2D HCP monolayer nanosphere (NS) crystals composed of four polystyrene NSs using an indirect method, via assembly at an air-liquid interface, as previously described [13]. We etched the NS-masked and self-masking chitin and chitosan substrates using the South Bay Technology Model PC-2000 Plasma Cleaners (South Bay Technology, San Clemente, CA, USA), with oxygen as the process gas. We performed a 90 s reactive ion etching (RIE) of the surface using the following process parameters: 100 W forward power, −700 V DC bias, and 200 mT chamber pressures.

SEM imaging: Scanning electron micrographs were obtained using a Zeiss Auriga FIB/SEM (Carl Zeiss Microscopy, LLC, Thornwood, NY, USA) with accelerating voltage 2–4 kV (30 μm or 7.5 μm aperture), and an Inlens SESI detector. Samples were sputter-coated with an approximately 5 nm gold layer using a Leica EM ACE200 (Leica Microsystems, Buffalo Grove, IL, USA), with real-time thickness monitoring using a quartz crystal microbalance.

3. Results

To determine whether a purified chitin scaffold could serve as a substrate for colloidal/nanosphere lithography, we prepared a purified chitin scaffold from a wing of the periodic 17-year cicada Brood II *Magicicada septendecim* (Figure 1A). The wing of the Brood II cicada is decorated by a low-aspect ratio nanostructured surface with a hexagonal close-packed arrangement (Figure 1C) [5]. Previous work has demonstrated that these nanoscale hemispherical structures are predominately protein or wax in composition, and do not retain any nanostructures after the in situ chitin purification procedure [4]. The in situ-purified Brood II cicada wing retained the general appearance of a cicada wing, but lacked any pigmentation or nanostructure (Figure 1C,D). This purified surface was then masked with a 300 nm polystyrene nanosphere monolayer as previously described [13], and used as the etching target for NSL (Figure 2).

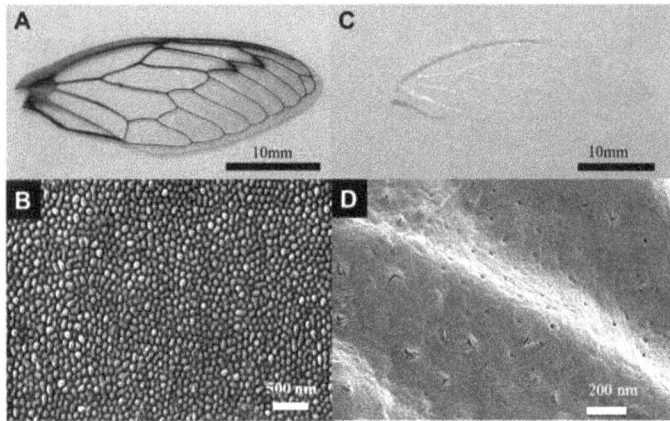

Figure 1. The source of chitin for the colloidal/nanosphere lithography substrate. (**A**) A forewing from the periodic 17-year cicada Brood II *Magicicada septendecim*; (**B**) SEM micrograph of the inter-vein (clear) regions of a wing showing an array of hemispherical nano-features, which is a rough hexagonal close-packed arrangement; (**C**) A forewing from the periodic 17-year cicada Brood II *Magicicada septendecim* after in situ chitin purification. Note the loss of all pigment and color from the wing; (**D**) The surface of the inter-vein region showing the loss of the array of nanofeatures, and the presence of a nanoscale fibrous network.

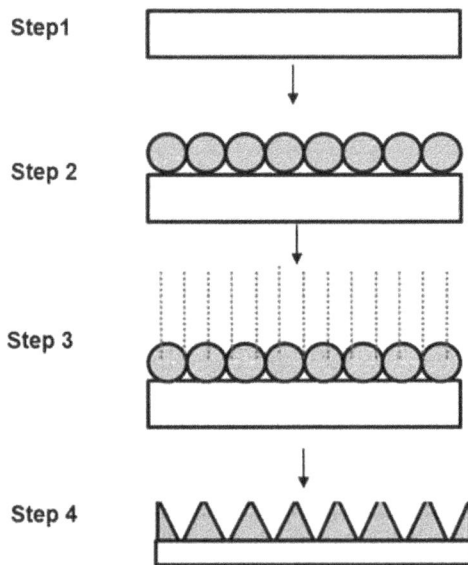

Figure 2. Schematic of the colloidal/nanosphere lithographic process. Step (1) preparation of biopolymer substrate; Step (2) nanosphere masking of substrate; Step (3) Reactive ion etching of masked substrate; and Step (4) completed NSS.

Unlike previous masking substrates, both polymeric and semiconductor [9,13], chitin nanofibers interacted with the polystyrene nanospheres within the mask, showing both projections from the surface and the nanofiber coating of the spheres themselves (Figure 3A, arrows). Chitin substrates prepared from the wings of cicadas have a random organization of nanofibers [4]. Although we do not know what controls this type of interaction (chitin has a low solubility in aqueous solutions [15] and the masking step is performed within an aqueous bath [13]), we suspect that the polystyrene beads are redepositing chitin from the substrate onto the surface. A 90 s reactive ion etch of the nanosphere-masked chitin substrate resulted in a nanostructure surface dominated by high aspect ratio nanocones arranged in a hexagonal close-packed arrangement (Figure 3B).

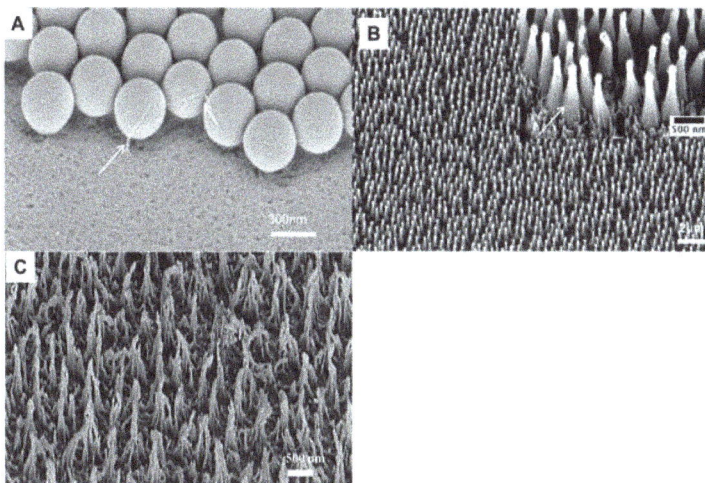

Figure 3. Colloidal/nanosphere lithography on chitin substrate derived from the wing of the periodic 17-year cicada Brood II *Magicicada septendecim*. (**A**) A nanosphere mask on the chitin substrate. Notice the filaments emanating from the chitin surface to the nanospheres, and the presence of a fibrous network on the nanospheres (as noted by the arrows); (**B**) An array of chitin nanocones after the etching process. The inset shows the presence of longitudinal ridges along the nanocones; (**C**) The array of chitin nanocones after incubation in an aqueous solution; note the loss of organization in the array and the structure of the nanocones.

The surface of the chitin derived from the wing of the periodic 17-year cicada Brood II *Magicicada septendecim* now resemble the wings of cicadas that have surfaces with higher-aspect ratio nanofeatures [4,5,16]. Closer inspection of these chitin nanocones shows these structures—akin to the previously-described synthetic polymer structures—have secondary finer nanoscale structures (longitudinal ridges) along the length of the nanocones (Figure 3B, inset arrows). We suspect that like the synthetic polymers, these longitudinal nano-ridges are a product of the interaction of the reactive ions and nanoscale structure within the chitin fiber [14]. Chitin nanofibers in natural systems range in size from 2 nm to 10 nm, depending on the context [4,15,17]. In synthetic systems, chitin nanofibers have been demonstrated to self-assemble into nanoscale fibers ranging from 5 nm to 20 nm, depending on solvent conditions [18,19]; the filamentous ridges fall within these ranges. Unlike the nanostructured surfaces generated from synthetic polymers, chitin nanostructured surfaces are extremely labile in aqueous solutions, and unravel to form random tangles of chitin nanofibers (Figure 3C). The dynamic nature of chitin nanocones may serve in applications like drug delivery, in which a dynamic surface with nanoscale features could allow for response to a changing environment.

We also performed colloidal/nanosphere lithography on drop-cast thin films of chitosan. Drop-cast chitosan thin films (~1 μm thick) demonstrated a finer organization than chitin (Figure 4A) thin film, and displayed a stronger interaction with the polystyrene nanospheres, literally stretching or distorting the nanospheres across its surface (Figure 4B). Post-etching, the chitosan substrates displayed a similar arrangement of nanocones as the chitin surfaces (Figure 4C); the nanostructured chitosan thin films were flexible and could be contorted post-etch (Figure 4D).

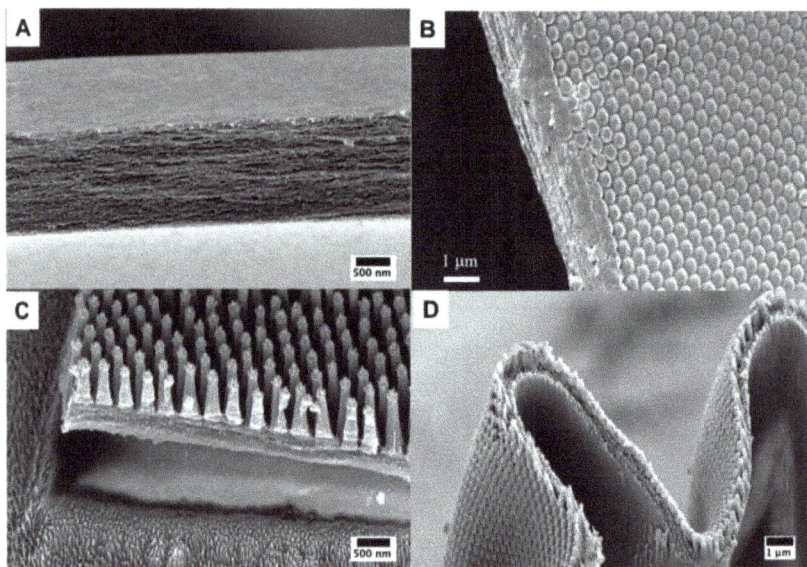

Figure 4. Colloidal/nanosphere lithography on drop-cast chitosan thin films. (**A**) A drop-cast chitosan thin film; note the layers of chitosan nanoscale fibers; (**B**) 300 nm nanosphere mask on chitosan thin film—note the distortion of the polystyrene nanospheres as they are stretched across the surface of the chitosan substrate; (**C**) Post-reactive ion etch of a nanosphere-masked chitosan substrate—note the high-aspect ratio nanocone arrays; (**D**) Flexible nanostructured chitosan thin film.

Due to the presence of stabilizing and reducing components within their composition, chitin and chitosan have been used for the generation of silver nanoparticles (AgNP) [14]. One of the scaling limitations of colloidal/nanosphere lithography is the masking step. Typically, this step involves either the self-assembly of a nanosphere monolayer at a gas–liquid interface, or the mechanical generation of a nanosphere monolayer via centrifugation [9]; in either case, the area masked is limited. We hypothesized that a substrate with spatially distributed components and inherent differential etching rates would enable a simplified colloidal/nanosphere lithographic procedure. We tested this hypothesis by generating a composite chitosan–silver nanoparticle thin film (Figure 5A). To do this, we used a chitosan solution doped with silver nitrate; this solution generated cuboidal silver nanoparticles and acquired a characteristic purple hue (Figure 5B). We characterized the presence and size of the silver nanoparticles in the chitosan solution using EDX, DLS, and UV-Vis spectroscopy. SEM imaging of the dried thin film produced by drop-casting revealed the presence of fairly evenly-distributed cuboidal AgNP, between 40–60 nm (Figure 5A, arrows). A 90 s reactive ion etch of the chitosan–AgNP composite thin film resulted in a nanostructured surface (Figure 5B). This nanostructured surface was similar in appearance to those generated using a polystyrene nanosphere mask, although the self-masked AgNP-nanostructured surfaces demonstrated far less organization compared to the neatly-arranged hexagonally close-packed nanocones generated from the nanosphere-masked surface, as expected. The silver nanoparticles accumulated on the peaks of the rough chitin nanocones (Figure 5B). As with nanostructured chitosan surfaces, these self-masked chitosan surfaces were flexible (Figure 5C,D), and thus this procedure or modifications of this procedure can be used to coat a variety of materials that are in mechanically dynamic environments (e.g., cell scaffolds, tissue implants).

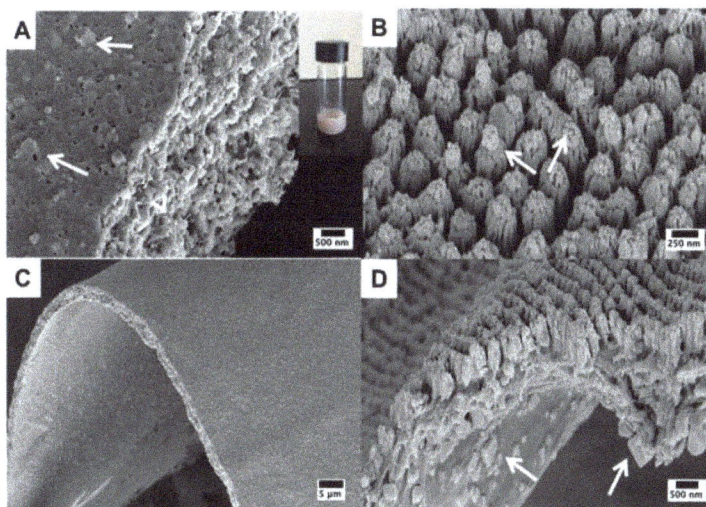

Figure 5. Nanostructured surfaces generated on a self-masking AgNP–chitosan thin film by colloidal-nanosphere lithography. (**A**) Chitosan–AgNP composite thin film generated by a chitosan–AgNP solution (inset); note the presence of cuboidal AgNP embedded within the film (denoted by arrows) (**B**) Post-reactive ion etch of the self-masked chitosan–AgNP composite thin film; note that the array is less-organized without a nanosphere mask compared to with one, and note the presence of colloidal AgNP at the nanocone apex. (**C,D**) Demonstration of the flexibility of the nanostructured chitosan–AgNP composite thin films.

4. Discussion

Nanostructured surfaces are a powerful means of controlling and changing the surface properties of a material [20,21]. Naturally-structured chitin in insects and in other invertebrates has been demonstrated to have compelling physical properties [4,7]. Among the most interesting are the silica–chitin composite spicules of the glass sponge, *Sericolophus hawaiicus*, which have been shown to be the first biomaterial to display supercontinuum generation [22]. While much of the previous efforts in colloidal/nanosphere lithography has been focused on the modification or application of surface topography to semiconductor materials (and most recently synthetic polymers) [9,13], no work has been done using biopolymers. Chitin and chitosan have long been used as a template for nanoparticle synthesis and the deposition of nanoscale materials [23–27]. Many of the chitin–metal oxide nanocomposite materials are generated by a hydrothermal process, and display enhanced nanoscale topography that synergizes with compositional contributions to the properties and the functionality of these new and interesting materials [23,25]. However, little work has been done to directly structure chitin using standard etching techniques. This work demonstrates that biopolymers like chitin and chitosan are not only compatible with the nanosphere lithographic process, but also provide unique and powerful opportunities for this fabrication technique—especially for the generation of antimicrobial surfaces and biomimetic optical materials. The chemical and structural properties of chitin and chitosan (e.g., their nanoscale fiber size and organization) enables the generation of dynamic

Polymers **2018**, *10*, 218

and responsive nanostructure surfaces which can be used to release drugs or growth factors in response to changes in the local environment. The self-assembly of metallic nanoparticles by chitin and chitosan has enabled the generation of a self-masking substrate for a colloidal/nanosphere lithographic process. While further work is needed to refine the self-assembly of the metallic nanomaterials and their distribution within the chitosan thin film, using chitosan or other polymers enables the large-scale application of nanostructure surfaces to a broad range of targets via a colloidal/nanosphere lithographic-like approach.

Acknowledgments: This work was supported by the North Carolina Biotechnology Center (NCBC) Biotechnology Research Grant to Dennis LaJeunesse (213-BRG-1209), NIH grant to Dennis LaJeunesse (1R15EB024921-01), and through generous support of Dean James Ryan, the Joint School of Nanoscience and Nanoengineering, and the State of North Carolina. This work was performed at the JSNN, a member of Southeastern Nanotechnology Infrastructure Corridor (SENIC) and National Nanotechnology Coordinated Infrastructure (NNCI), which is supported by the National Science Foundation (ECCS-1542174).

Author Contributions: Rakkiyappan Chandran, Kyle Nowlin and Dennis R. LaJeunesse conceived and designed the experiments; Rakkiyappan Chandran and Kyle Nowlin performed the experiments; Rakkiyappan Chandran analyzed the data; Kyle Nowlin contributed reagents/materials/analysis tools; Dennis R. LaJeunesse wrote the paper.

Conflicts of Interest: The authors declare no conflict of interest.

References

1. Sergeev, A.; Timchenko, A.A.; Kryuchkov, M.; Blagodatski, A.; Enin, G.A.; Katanaev, V.L. Origin of order in bionanostructures. *RSC Adv.* **2015**, *5*, 63521–63527. [CrossRef]
2. Kryuchkov, M.; Lehmann, J.; Schaab, J.; Fiebig, M.; Katanaev, V.L. Antireflective nanocoatings for UV-sensation: The case of predatory owlfly insects. *J. Nanobiotechnol.* **2017**, *15*, 52. [CrossRef] [PubMed]
3. Hasan, J.; Webb, H.K.; Truong, V.K.; Pogodin, S.; Baulin, V.A.; Watson, G.S.; Watson, J.A.; Crawford, R.J.; Ivanova, E.P. Selective bactericidal activity of nanopatterned superhydrophobic cicada Psaltoda claripennis wing surfaces. *Appl. Microbiol. Biotechnol.* **2013**, *97*, 9257–9262. [CrossRef] [PubMed]
4. Chandran, R.; Williams, L.; Hung, A.; Nowlin, K.; LaJeunesse, D. SEM characterization of anatomical variation in chitin organization in insect and arthropod cuticles. *Micron* **2016**, *82*, 74–85. [CrossRef] [PubMed]
5. Nowlin, K.; Boseman, A.; Covell, A.; LaJeunesse, D. Adhesion-dependent rupturing of Saccharomyces cerevisiae on biological antimicrobial nanostructured surfaces. *J. R. Soc. Interface* **2015**, *12*, 20140999. [CrossRef] [PubMed]
6. Blagodatski, A.; Sergeev, A.; Kryuchkov, M.; Lopatina, Y.; Katanaev, V.L. Diverse set of Turing nanopatterns coat corneae across insect lineages. *Proc. Natl. Acad. Sci. USA* **2015**, *112*, 10750–10755. [CrossRef] [PubMed]
7. Vincent, J.F.; Wegst, U.G. Design and mechanical properties of insect cuticle. *Arthropod Struct. Dev.* **2004**, *33*, 187–199. [CrossRef] [PubMed]
8. Lee, K.C.; Erb, U. Remarkable crystal and defect structures in butterfly eye nano-nipple arrays. *Arthropod Struct. Dev.* **2015**, *44*, 587–594. [CrossRef] [PubMed]
9. Colson, P.; Henrist, C.; Cloots, R. Nanosphere lithography: A powerful method for the controlled manufacturing of nanomaterials. *J. Nanomater.* **2013**, *2013*. [CrossRef]
10. Zhang, Y.; Wang, X.; Wang, Y.; Liu, H.; Yang, J. Ordered nanostructures array fabricated by nanosphere lithography. *J. Alloys Compd.* **2008**, *452*, 473–477. [CrossRef]
11. Wang, D.; Zhao, A.; Li, L.; He, Q.; Guo, H.; Sun, H.; Gao, Q. Bioinspired ribbed hair arrays with robust superhydrophobicity fabricated by micro/nanosphere lithography and plasma etching. *RSC Adv.* **2015**, *5*, 96404–96411. [CrossRef]
12. Xu, X.; Yang, Q.; Wattanatorn, N.; Zhao, C.; Chiang, N.; Jonas, S.J.; Weiss, P.S. Multiple-Patterning Nanosphere Lithography for Fabricating Periodic Three-Dimensional Hierarchical Nanostructures. *ACS Nano* **2017**, *11*, 10384–10391. [CrossRef] [PubMed]
13. Nowlin, K.; LaJeunesse, D.R. Fabrication of hierarchical biomimetic polymeric nanostructured surfaces. *Mol. Syst. Des. Eng.* **2017**, *2*, 201–213. [CrossRef]

14. Marpu, S.; Kolailat, S.S.; Korir, D.; Kamras, B.L.; Chaturvedi, R.; Joseph, A.; Smith, C.M.; Palma, M.C.; Shah, J.; Omary, M.A. Photochemical formation of chitosan-stabilized near-infrared-absorbing silver Nanoworms: A "Green" synthetic strategy and activity on Gram-negative pathogenic bacteria. *J. Colloid Interface Sci.* **2017**, *507*, 437–452. [CrossRef] [PubMed]

15. Mikhailov, G.M.; Lebedeva, M.F. Procedures for preparing chitin-based fibers. *Russ. J. Appl. Chem.* **2007**, *80*, 685–694. [CrossRef]

16. Sun, M.; Watson, G.S.; Zheng, Y.; Watson, J.A.; Liang, A. Wetting properties on nanostructured surfaces of cicada wings. *J. Exp. Biol.* **2009**, *212*, 3148–3155. [CrossRef] [PubMed]

17. Ehrlich, H.; Simon, P.; Carrillo-Cabrera, W.; Bazhenov, V.V.; Botting, J.P.; Ilan, M.; Ereskovsky, A.V.; Muricy, G.; Worch, H.; Mensch, A.; et al. Insights into chemistry of biological materials: Newly discovered silica-aragonite-chitin biocomposites in demosponges. *Chem. Mater.* **2010**, *22*, 1462–1471. [CrossRef]

18. Salaberria, A.M.; Labidi, J.; Fernandes, S.C.M. Different routes to turn chitin into stunning nano-objects. *Eur. Polym. J.* **2015**, *68*, 503–515. [CrossRef]

19. Nogi, M.; Kurosaki, F.; Yano, H.; Takano, M. Preparation of nanofibrillar carbon from chitin nanofibers. *Carbohydr. Polym.* **2010**, *81*, 919–924. [CrossRef]

20. Huang, Y.F.; Jen, Y.J.; Chen, L.C.; Chen, K.H.; Chattopadhyay, S. Design for approaching cicada-wing reflectance in low- and high-index biomimetic nanostructures. *ACS Nano* **2015**, *9*, 301–311. [CrossRef] [PubMed]

21. Wang, Z.; Zhao, J.; Bagal, A.; Dandley, E.C.; Oldham, C.J.; Fang, T.; Parsons, G.N.; Chang, C.H. Wicking Enhancement in Three-Dimensional Hierarchical Nanostructures. *Langmuir* **2016**, *32*, 8029–8033. [CrossRef] [PubMed]

22. Ehrlich, H.; Maldonado, M.; Parker, A.R.; Kulchin, Y.N.; Schilling, J.; Köhler, B.; Skrzypczak, U.; Simon, P.; Reiswig, H.M.; Tsurkan, M.V.; et al. Supercontinuum Generation in Naturally Occurring Glass Sponges Spicules. *Adv. Opt. Mater.* **2016**, *4*, 1608–1613. [CrossRef]

23. Wysokowski, M.; Motylenko, M.; Stöcker, H.; Bazhenov, V.V.; Langer, E.; Dobrowolska, A.; Czaczyk, K.; Galli, R.; Stelling, A.L.; Behm, T.; et al. An extreme biomimetic approach: Hydrothermal synthesis of β-chitin/ZnO nanostructured composites. *J. Mater. Chem. B* **2013**, *1*, 6469–6476. [CrossRef]

24. Wysokowski, M.; Szalaty, T.J.; Jesionowski, T.; Motylenko, M.; Rafaja, D.; Koltsov, I.; Stöcker, H.; Bazhenov, V.V.; Ehrlich, H.; Stelling, A.L.; et al. Extreme biomimetic approach for synthesis of nanocrystalline chitin-(Ti,Zr)O$_2$ multiphase composites. *Mater. Chem. Phys.* **2017**, *188*, 115–124. [CrossRef]

25. Wysokowski, M.; Motylenko, M.; Beyer, J.; Makarova, A.; Stöcker, H.; Walter, J.; Galli, R.; Kaiser, S.; Vyalikh, D.; Bazhenov, V.V.; et al. Extreme biomimetic approach for developing novel chitin-GeO$_2$ nanocomposites with photoluminescent properties. *Nano Res.* **2015**, *8*, 2288–2301. [CrossRef]

26. Wysokowski, M.; Motylenko, M.; Walter, J.; Lota, G.; Wojciechowski, J.; Stöcker, H.; Galli, R.; Stelling, A.L.; Himcinschi, C.; Niederschlag, E.; et al. Synthesis of nanostructured chitin-hematite composites under extreme biomimetic conditions. *RSC Adv.* **2014**, *4*, 61743–61752. [CrossRef]

27. Brião, G.V.; Jahn, S.L.; Foletto, E.L.; Dotto, G.L. Adsorption of crystal violet dye onto a mesoporous ZSM-5 zeolite synthetized using chitin as template. *J. Colloid Interface Sci.* **2017**, *508*, 313–322. [CrossRef] [PubMed]

polymers

MDPI

Article

Aerogels from Chitosan Solutions in Ionic Liquids

Gonzalo Santos-López [1], Waldo Argüelles-Monal [2] , Elizabeth Carvajal-Millan [1],
Yolanda L. López-Franco [1] , Maricarmen T. Recillas-Mota [2] and Jaime Lizardi-Mendoza [1,*]

[1] Grupo de Investigación en Biopolímeros—CTAOA. Centro de Investigación en Alimentación y Desarrollo, A.C., Hermosillo, Sonora 83304, Mexico; gonzalosantos@estudiantes.ciad.mx (G.S.-L.); ecarvajal@ciad.mx (E.C.-M.); lopezf@ciad.mx (Y.L.L.-F.)

[2] Polímeros Naturales. Centro de Investigación en Alimentación y Desarrollo, A.C., Unidad Guaymas, Guaymas, Sonora 85480, Mexico; waldo@ciad.mx (W.A.-M.); mrecillas@ciad.mx (M.T.R.-M.)

* Correspondence: jalim@ciad.mx; Tel.: +52-662-289-2400

Received: 16 November 2017; Accepted: 14 December 2017; Published: 16 December 2017

Abstract: Chitosan aerogels conjugates the characteristics of nanostructured porous materials, i.e., extended specific surface area and nano scale porosity, with the remarkable functional properties of chitosan. Aerogels were obtained from solutions of chitosan in ionic liquids (ILs), 1-butyl-3-methylimidazolium acetate (BMIMAc), and 1-ethyl-3-methyl-imidazolium acetate (EMIMAc), in order to observe the effect of the solvent in the structural characteristics of this type of materials. The process of elaboration of aerogels comprised the formation of physical gels through anti-solvent vapor diffusion, liquid phase exchange, and supercritical CO_2 drying. The aerogels maintained the chemical identity of chitosan according to Fourier transform infrared spectrophotometer (FT-IR) spectroscopy, indicating the presence of their characteristic functional groups. The internal structure of the obtained aerogels appears as porous aggregated networks in microscopy images. The obtained materials have specific surface areas over $350 \ m^2/g$ and can be considered mesoporous. According to swelling experiments, the chitosan aerogels could absorb between three and six times their weight of water. However, the swelling and diffusion coefficient decreased at higher temperatures. The structural characteristics of chitosan aerogels that are obtained from ionic liquids are distinctive and could be related to solvation dynamic at the initial state.

Keywords: aerogels; chitosan; ionic liquids; ionogels

1. Introduction

Chitosan (Cs) is a natural linear polysaccharide generated from the deacetylation of chitin and is composed of β-(1-4)-D-glucosamine units and β-(1-4)-N-acetyl-glucosamine distributed along the polymeric chain. The physicochemical characteristics and functional properties of Cs, such as its polycationic character, biocompatibility, low toxicity, and structural capacity, make it a polysaccharide of interest in different fields. The mechanical and structural properties of chitosan allow for different types of materials to be obtained from chitosan solutions, e.g., nanostructured porous materials. Aerogels are a specific type of nanostructured porous material that are characterized by mesopores (pore diameter between 2 and 50 nm) and large surface areas. Obtaining aerogels of chitosan could improve the availability of the functional groups of chitosan [1]. Together, the aerogel characteristics with the functional properties of chitosan, potentially provides materials with multiple applications, e.g., adsorption, transport, and controlled release of bioactive molecules, toxics, and pollutants removers, among others.

Ionic liquids (IL) are an alternative medium for chitosan dissolution. When compared to the traditional aqueous acid solvents used to dissolve Cs, IL have different physicochemical characteristics because they consist of only ions and water is not needed to dissolve chitosan. Most of the studies of chitosan with IL have been focused on the formation of solutions. Imidazolium based IL, as 1-butyl-3-methylimidazolium acetate (BMIMAc), and 1-ethyl-3-methyl-imidazolium acetate (EMIMAc), have been highlighted because they are able to dissolve chitosan at high concentrations (up to 10% *w/w*). Other reports have focused on forming materials from chitosan solutions in IL e.g., fibers, blends, films, membranes, hydrogels, and ionogels [2–12]. Aerogels from chitin with BMIMAc solutions has been reported [13], but the information about chitosan ionogels and aerogels from IL solutions is limited.

Supercritical CO_2 drying, unlike other types of drying, keeps most of the internal structure formed at gelation because the effects of surface tension on the three-dimensional macromolecular network are minimized. Therefore, chitosan aerogels that are generated from solutions of Cs in IL (BMIMAc and EMIMAc) were produced in order to study the physicochemical characteristics of these novel materials. Concurrently, observations on the structural features of the aerogels could be related to the network formation and molecular conformation of chitosan in ionic liquids at the physical gelling process.

2. Materials and Methods

2.1. Materials

Shrimp shells (*Pandalus borealis*) Cs was acquired from Primex (batch No. TM 1961, Siglufjörður, Iceland). The degree of acetylation (DA) of Cs was 16.3%, as determined by solid-state cross-polarization magic angle spinning ^{13}C nuclear magnetic resonance spectroscopy (CP/MAS ^{13}C-NMR) performed in a Bruker Avance TM 400WB (Bruker Biospin, Wissembourg, France, 9.4 Tesla, 1 ms contact time, 2000 scans) [14,15]. The weight average molecular weight (M_w) of Cs was 2.01×10^5 g/mol, determined by static light scattering, as described previously [16]. Chitosan was purified, as described by Rinaudo et al. (1999), before use it [17]. Commercial grade IL, BMIMAc and EMIMAc (BASF, Steinheim, Germany), were used as received just prior each use they were stored in desiccator and kept at 105° C for 24 h in vacuum to minimize moisture. High grade supercritical drying liquid CO_2 (99.99% purity) was supplied by Infra (Hermosillo, México). All of the compounds and solvents that were used were reagent grade acquired from recognized commercial chemical distributors. The water used for the experiments was deionized type I (resistivity of 18.2 MΩ·cm at 25 °C) unless stated otherwise.

2.2. Solutions of Chitosan in Ionic Liquids

The moisture-free IL was added into a round-bottom flask with a sufficient amount of chitosan to obtain a 2% (*w/w*) concentration. This solution was heated in an oil bath at 105 °C under magnetic stirring at least 6 hours under a nitrogen atmosphere. Upon complete Cs dissolution it was stored in a desiccator at room temperature until use. Two types of solutions were obtained: Cs in EMIMAc (CsEMIM) and Cs in BMIMAc (CsBMIM).

2.3. Formation of Ionogels

The physical gels were formed using approximately 0.1 g of the Cs-IL solution that was casted in cylindrical containers (0.4 cm diameter), with the bottom being covered with plastic paraffin film. Gelation was induced by vapor diffusion of an antisolvent, such as ethanol or water, in a closed chamber [18]. The gels were recovered after 48 h and were transferred to a 70% ethanol aqueous mixture. These gels that contain IL as liquid phase confined within a polymer matrix are defined as ionogels [19].

2.4. Formation of Aerogels

The ionogels were thoroughly rinsed with aqueous ethanol (70%) until the IL was completely removed. The presence of IL in the rinsing media was monitored with UV/VIS spectroscopy, scanning between 190 to 300 nm using the ethanol-water mixture as reference. Afterwards, the liquid phase was replaced with acetone, which has higher miscibility with supercritical CO_2. For this, the gels were repeatedly rinsed with a gradient of acetone-water mixtures that ended with two rinses with pure acetone. The acetogels (gels containing acetone as fluid phase) were dried with supercritical CO_2 (>32 °C and 73 atm) in a pressurized reactor. At the end of the process, the dried aerogels were stored in a desiccator at room temperature.

2.5. Characterization of Aerogels

2.5.1. Chemical Identity

The chemical composition of the samples was determined analyzing the characteristic bands in the infrared spectrum obtained by a Fourier transform infrared spectrophotometer (FT-IR, Thermo Scientific, Nicolet iS-50, Madison, WI, USA), using the attenuated total reflection (ATR) mode. All of the measurements were performed at room temperature collecting 32 scans at 4 cm^{-1} resolution.

2.5.2. Structural Analysis

The surface area, pore volume, and nitrogen adsorption and desorption isotherms of the aerogels were determined with the surface area analyzer Nova 2200e (Quantachrome Instruments, Boynton Beach, FL, USA), and analysis of the data with the software NovaWin version 11.02 (Quantachrome Instruments, Boynton Beach, FL, USA). The morphological characteristics of the aerogels were thoroughly observed by field emission scanning electron microscopy (Hitachi SU8000, Tokyo, Japan). The images were obtained using an accelerating voltage of 1.0 KeV.

2.5.3. Degree of Swelling

The swelling capacity of the aerogels at equilibrium was determined from their immersion in water at different temperatures (20, 25, 30, and 40 °C). The weight gain of the samples was periodically monitored by removing the excess of water with filter paper. Equation (1) was used to estimate the degree of swelling (*W*), as follows:

$$W = \frac{P - P_0}{P_0} = \frac{P}{P_0} - 1 \tag{1}$$

where P_0 is the weight of the dry aerogel and P is the weight of the aerogel in the wet state [20].

3. Results and Discussion

The vapor diffusion of a non-solvent agent was useful to produce ionogels from Cs solutions in IL. By this way, it was possible to obtain three different types of physical chitosan ionogels (Table 1). Using ethanol as non-solvent agent gels were obtained from both Cs-IL solutions. These gels were clear, rigid, and brittle; the only noticeable difference among them was the color tone, darker yellow for the gels from CsBMIM. It has been indicated that the main effect of IL on polysaccharides is to disrupt the hydrogen bonds and promoting their dissolution. Low molecular weight alcohols are miscible with imidazolium based ionic liquids [21], but chitosan does not dissolve in alcohols. Therefore, when ethanol diffuses into a Cs-IL solution, the solvation effect of the ionic liquid over chitosan decreases, favoring the interactions between chitosan chains and subsequently leading to the generation of a gel. Conversely, water vapor was only useful to produce gels from the CsEMIM solution. It has been noticed that BMIMAc has a lower affinity for water than EMIMAc [22], this could be related to lower diffusion rates that do not decrease its chitosan solvation capacity in preventing the gel formation. It should be taken in account that chitosan do not dissolve in water, but it is hygroscopic. As result, the obtained gels from water diffusion were weak and difficult to manipulate without compromising their integrity. Hence, subsequent procedures and analysis were performed using only ionogels that are produced by ethanol treatment.

Table 1. Outcome of the treatment of Cs-IL solutions with non-solvent agents vapor diffusion.

Cs-IL Solution	Non-Solvent Agent	
	Water	Ethanol
CsEMIM	Ionogel (soft)	Ionogel (CsE)
CsBMIM	dilution	Ionogel (CsB)

When the ionogels were rinsed with an ethanol-water mixture they became more translucent, reducing their yellow color (Figure 1). The continuous rinsing of the gels gradually eliminated the ionic liquid from inside the gels. The ethanol concentration that was used allowed for keeping the ionogels volume without causing drastic swelling or shrinkage. In the subsequent fluid phase replacement with acetone, the CsE ionogels decrease 25% their volume, and in the case of CsB, the reduction was 42%. Apparently, the chitosan chains underwent rearrangement within the formed network to a more stable configuration as a result of the interaction with acetone [23].

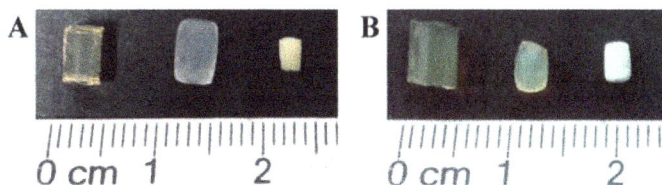

Figure 1. Materials formed from the Cs-IL solutions: (**A**) CsE and (**B**) CsB. From left to right: ionogel, acetogel and aerogel.

The obtained aerogels were rigid and brittle cylinders with opaque white color (Figure 1). The volume reduction when compared to the starting ionogels was of 73% for CsE and 82% for CsB. This behavior is similar to that reported for aerogels obtained from chitosan and κ-carrageenan [24]. Such volume reduction has been associated with the rearrangements of chitosan chains due to their lower affinity for acetone. This molecular movement does not cease until all of the acetone has been removed, even when the supercritical CO_2 drying reduce the effects of surface tension in the material [25].

3.1. Characterization of Chitosan Aerogels

The infrared spectra (2000–500 cm^{-1}) of chitosan and the obtained aerogels are shown in Figure 2. The main characteristic bands of chitosan are also observed in the aerogels spectra. At 1652 cm^{-1} is observed the stretching vibration band of the C=O bond that is associated to the amide I; the amide II $-NH_2$ deformation is related to the band at 1580 cm^{-1}; the band at 1424 cm^{-1} is associated to the CH_2 bending; at 1380 cm^{-1}, the symmetrical vibration deformation of the CH_3 group is observed; the band at 1318 cm^{-1} is associated to the amide III; the antisymmetric tension mode of the COC bridge is observed at 1150 cm^{-1}; finally, the fingerprint zone, between 1075 and 1026 cm^{-1}, is characteristic of the polysaccharides [26,27]. There is no evident modification on the spectra that indicate chemical changes in the chitosan as result of the gel formation or the drying process. Furthermore, there are not absorbance bands that could be related to the presence of residual IL in the aerogels.

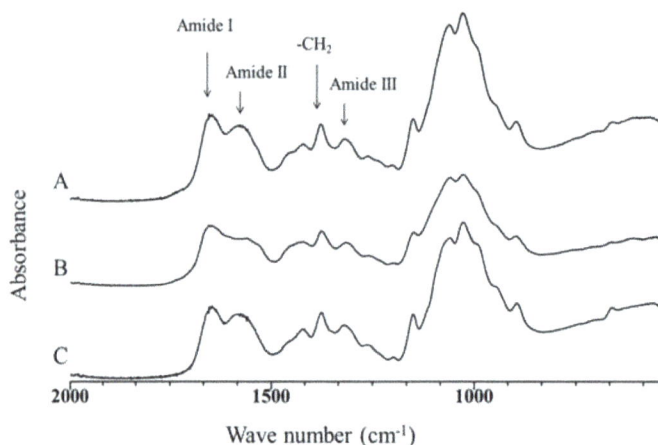

Figure 2. Fourier transform infrared spectrophotometer (FT-IR) spectra of (**A**) CsE aerogel; (**B**) CsB aerogel and (**C**) chitosan.

The N_2 adsorption and desorption isotherms of both types of aerogels are classified as type IV according to the IUPAC conventions (Figure 3). The observed hysteresis of N_2 desorption at high relative pressures is indicative of a mesoporous dry material [28]. The specific surface area (S_{BET}) and pore volume that are calculated from the adsorption and desorption isotherms are included in Table 2. The specific surface areas of the aerogels are in the higher rank when compared with other pure polysaccharide aerogels [24,29,30]. The pore size obtained was within the range of mesopores, which is characteristic of aerogels.

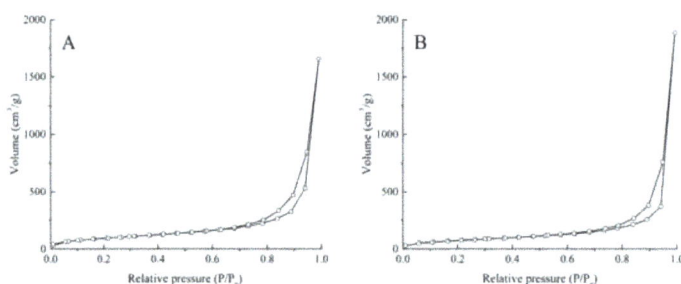

Figure 3. N_2 adsorption and desorption BET isotherms of (**A**) CsE and (**B**) CsB aerogels.

Table 2. Specific surface area (S_{BET}), pore volume (V_p) and pore diameter (D_p) of the CsE and CsB aerogels.

Parameters	CsE	CsB
S_{BET} (m^2/g)	358 ± 79	478 ± 264
V_p (cm^3/g)	0.0733 ± 0.016	0.236 ± 0.083
D_p (nm)	31.0 ± 1.4	46.0 ± 3.6

The scanning electron microscopy (SEM) images of the aerogels are shown in Figures 4 and 5. Both of the aerogels appear as uniform materials with some imperfections, which could be caused by fracture events. At the highest magnification available ($\times 10,000$), the internal structure of the aerogels looks as aggregated clumps forming a compact network with heterogeneous pores. The appearance of the pores is consistent with the mesoporous characteristics of the aerogels. The main difference between both types of aerogels is that the internal structure of CsB appears to be denser. The internal structure of the aerogels produced from ionogels is different to previously reported chitosan aerogels [24,29]. These differences could be related to the solvent-polymer interaction. Electrostatic repulsions dominate in the aqueous acid chitosan solutions. In such conditions, the polysaccharide molecules adopt an extended hydrodynamic volume conformation [17]. Conversely, these repulsive forces are absent in Cs-IL solutions; thus, the chitosan molecules have relatively smaller dimensions, generating more compact structures in their aerogels.

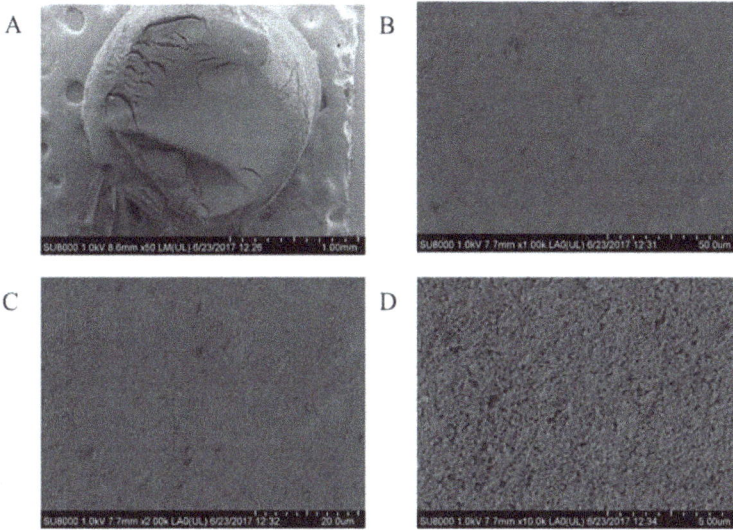

Figure 4. SEM images of the CsE aerogels (**A**) 50×; (**B**) 1000×; (**C**) 2000×; and, (**D**) 10,000× magnification.

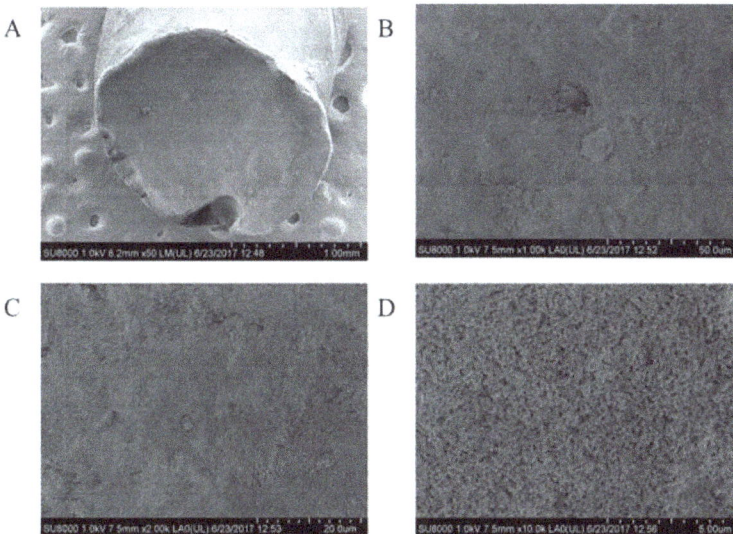

Figure 5. SEM images of the CsB aerogels. (**A**) 50×; (**B**) 1000×; (**C**) 2000×; and, (**D**) 10,000× magnification.

3.2. Diffusion Properties of Aerogels

The physical characteristics of aerogels, such as their large surface area and mesoporosity, endow these materials with a large capacity to adsorb certain compounds. The diffusion properties are key to evaluating the performance of these materials for important applications in pharmacy and biotechnology, among others. For this purpose, dried aerogels that are obtained from chitosan solutions in EMIMAc and BMIMAc, as described previously, were swollen in water at different temperatures between 20 and 40 °C and their kinetics was followed. In Figures 6 and 7, the experimental data is presented.

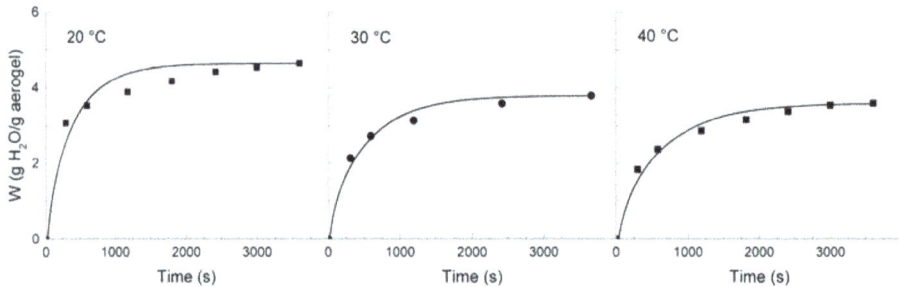

Figure 6. Swelling curves of the CsE aerogels in water at different temperatures (20, 30, and 40 °C). Experimental data (points) and adjustment (lines) with Equation (2) are included.

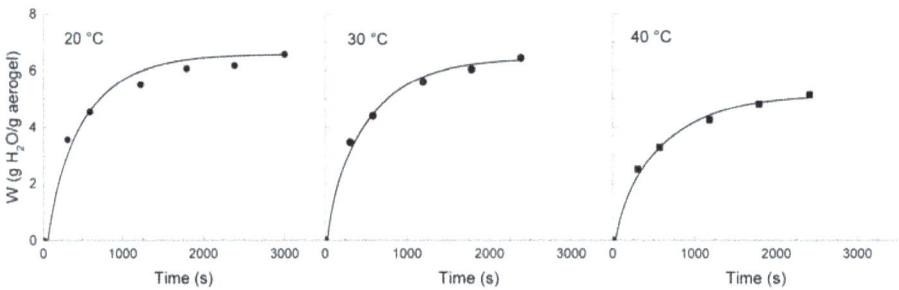

Figure 7. Swelling curves of the CsB aerogels in water at different temperatures (20, 30, and 40 °C). Experimental data (points) and adjustment (lines) with Equation (2) are included.

The CsE aerogels absorbed between three and four times their weight and exhibited a decreasing swelling capacity with an increasing temperature. In contrast, the CsB aerogels showed a greater capacity of absorption (between five and six times their weight), and the effect of temperature on the swelling capacity seems to be less marked. When compared with the aerogels of chitosan that was obtained from aqueous acid solutions, the aerogels from chitosan in IL showed higher W_∞. Previous studies with chitosan or chitosan-polyelectrolyte complex matrices that were prepared from aqueous media have shown a similar tendency to decrease the swelling capacity with temperature [20,31].

The Fick's law equation resolved for diffusion through a circular cylinder of radius r, keeping the diffusant concentration constant, becomes [32]:

$$\frac{W}{W_\infty} = 1 - \sum_{n=1}^{\infty} \frac{4}{r^2\alpha_n^2} \exp\left(-D\alpha_n^2 t\right), \tag{2}$$

in which W is the swelling degree at time t, and W_∞ is the corresponding quantity at equilibrium, α_n^2 are the n first positive roots of the Bessel function of the first-kind, and D is the diffusion coefficient.

Equation (3) was solved for the first 15 terms of the summation and the diffusion coefficients that were adjusted through a non-linear least square fitting process. In all situations, satisfactory adjustments were obtained from Equation (2), as could be appreciated from the curves that are traced in Figures 6 and 7. The estimated values of D are summarized in Table 3. The diffusion coefficients are similar for both types of aerogels and compare to those that are reported for swelling of other polymeric materials [20,33]. It should be remarked, however, the unusual trend that exhibits the diffusion coefficient, decreasing with the increase in temperature.

Table 3. Values of the degree of swelling, diffusion coefficient, second-order kinetic constant and the release exponent of CsE and CsB aerogels of swelled in water at different temperatures.

T (°C)	CsE				CsB			
	W_∞ (g H_2O/g gel)	$D \times 10^{10}$ (m²/s) [a]	$k \times 10^4$ (s⁻¹) [b]	n [c]	W_∞ (g H_2O/g gel)	$D \times 10^{10}$ (m²/s) [a]	$k \times 10^4$ (s⁻¹) [b]	n [c]
20	4.64	3.99	3.011	0.16	6.56	3.36	1.111	0.26
25	3.96	3.55	3.426	0.20	5.22	3.32	1.367	0.31
30	3.79	3.01	3.464	0.23	6.43	3.09	1.703	0.35
40	3.59	2.54	4.481	0.29	5.15	2.68	2.738	0.37

[a] Evaluated according to Equation (2), [b] Evaluated according to Equations (8) and (10), [c] Dimensionless release exponent from the power law relation.

It is well known that diffusion coefficients have a dependence on temperature that is similar to the Arrhenius equation:

$$D = D_0 exp\left(-\frac{E_D}{RT}\right) \tag{3}$$

where E_D is the apparent activation energy for the diffusion process. For both aerogels, the expected linear dependence was obtained, but as a result of the inverse tendency that is shown by the diffusion coefficients with the temperature (Table 3), negative activation energy values were obtained. From a physical point of view, this fact indicates that there is another process competing with the diffusion, giving rise to negative values of this parameter. On the other hand, an analysis of the swelling values as a function of time, for each material at each temperature, according to the power law relation $W = kt^n$, allowed for finding the values of the release exponent, n, which are shown in the Table 3. Here, again, values are obtained that have no physical sense (for cylinders, n should be between 0.46 and 1) [34]. As seen above, both of the aerogels have a porous structure, with pore sizes ranging between 15 and 23 nm. According to these morphological characteristics, it should be expected that a Fickean-type diffusion kinetics would be fulfilled, since the relaxation of the polymer chains is not the limiting step for swelling. Consequently, the experimental data had to show a linear dependence between the swelling and $t^{1/2}$. However, this is not the case either.

All of this analysis shows that along with the diffusion, another process besides swelling is taking place. In this sense, the Eyring equation [35,36] can provide information about the diffusion mechanism that takes place and sheds light on the causes of the observed negative values for the activation energies of the global diffusion process:

$$k = \frac{k_B T}{h} exp\left(-\frac{\Delta G^\ddagger}{RT}\right) \tag{4}$$

here k_B, h, and R are the Boltzmann, Planck, and the gas constants, respectively. The activation Gibbs free energy, $\Delta G^\ddagger$, is related to the activation enthalpy and entropy ($\Delta H^\ddagger$, $\Delta S^\ddagger$), according to the following expression:

$$\Delta G^\ddagger = \Delta H^\ddagger - T\Delta S^\ddagger \tag{5}$$

Then, substituting (5) in (4):

$$k = \frac{k_B T}{h} \, exp\left(\frac{\Delta S^{\ddagger}}{R}\right) exp\left(-\frac{\Delta H^{\ddagger}}{RT}\right) \tag{6}$$

which can be linearized as follows:

$$ln\frac{k}{T} = ln\frac{k_B}{h} + \frac{\Delta S^{\ddagger}}{R} - \frac{\Delta H^{\ddagger}}{R} \cdot \frac{1}{T} \tag{7}$$

This kinetic treatment describes the dependence of the rate of a chemical reaction with temperature when the concepts of statistical mechanics are applied. Even though this analysis is based on the theory of absolute reaction rates developed to treat ordinary chemical reactions, it was demonstrated that this model could be successfully used to describe the kinetic treatment of viscosity and diffusion [37,38].

In order to elucidate the mechanism that takes place during the diffusion of water through these aerogels using the statistical approach of Eyring, it is necessary to estimate the rate constants of the swelling process. The experimental data showed an excellent fit to the following Equation proposed by Schott [39]:

$$\frac{t}{W} = A + Bt \tag{8}$$

which describes a second-order kinetics with respect to the unrealized swelling:

$$\frac{dW}{dt} = k(W_{\infty} - W)^2 \tag{9}$$

and the kinetic constant becomes equal to:

$$k = \frac{1}{AW_{\infty}^2} \tag{10}$$

The calculated values of the second-order rate constant, k, are included in Table 3, and were used to perform a thermo-kinetic analysis according to the Eyring Equation (7), as shown in Figure 8. The excellent adjustment that is obtained is evident. In Table 4, the values of $\Delta H^{\ddagger}$, $\Delta S^{\ddagger}$, and $\Delta G^{\ddagger}$ are summarized.

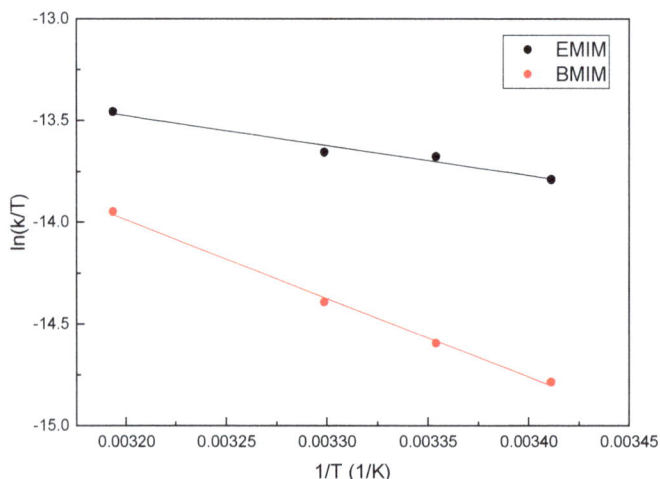

Figure 8. Plots of $\ln(k/T)$ versus $1/T$ using the second-order rate constants obtained from Equation (10) for the two aerogels.

Table 4. Activation energies calculated from D by the Arrhenius Equation (3). Thermo-kinetic parameters: $\Delta H^{\ddagger}$ and $\Delta S^{\ddagger}$ estimated from the Eyring Equation (7) and $\Delta G^{\ddagger}$ calculated at 298.15 K using the Equation (5).

Aerogel	Ea (kJ/mol)	$\Delta H^{\ddagger}$ (kJ/mol)	$\Delta S^{\ddagger}$ (J/mol·K)	$-T\Delta S^{\ddagger}$ [a] (kJ/mol)	$\Delta G^{\ddagger}$ [a] (kJ/mol)
CsE	−17.5	12.1	−271	80.7	92.9
CsB	−9.2	32.1	−211	63.0	95.1

[a] Evaluated at T = 298.15 K.

When analyzing the thermo-kinetic parameters, it is observed that the activation enthalpies are positive due to the endothermic change from starting intermediates to the transition state during the overall diffusion process. According to these activation parameters, at 298.15 K, the change of the activation Gibbs free energy is largely controlled by the activation entropy term (e.g., $-T\Delta S^{\ddagger}$ = 81 kJ/mol for the CsE aerogel), while the contribution of the activation enthalpy to the transition state is not so significant ($\Delta H^{\ddagger}$ = 12 kJ/mol). In this sense, it is important to note that these data indicate a significant increase in the order of the atoms that are involved during the pass to the transition state, with a small absorption of heat.

The values of these activation parameters are very similar to those that are observed during the formation of hydrogen bonds [40–42]. Their magnitudes coincide with the chemical nature of hydrogen bonding. Their formation involves relatively low energy and implicates the rearrangement of water molecules around the polymer, creating a hydration shell that is spatially oriented towards the hydroxyl and amino groups of chitosan.

At this point, it is important to remember that during the dissolution of chitosan in ionic liquids, the intra- and intermolecular hydrogen bonds of chitosan chains are broken. Although at the moment of forming the physical cross-linking the formation of some amount of hydrogen bonds is propitiated, these are relatively few in comparison with the large number of groups that are able to form hydrogen bonds in the polymer. In this way, during the swelling of the aerogels in water, along with the diffusion process, the generation of a huge amount of hydrogen bonds takes place with the consequent formation of a hydration shell. This analysis unambiguously supports the observed complex nature of the swelling processes of these aerogels, whose hydrated structure is apparently organized through hydrogen bonds. For this reason, a large contribution of $\Delta S^{\ddagger}$ to the transition state is appreciated, when compared to that of $\Delta H^{\ddagger}$.

4. Conclusions

Physical ionogels were obtained from chitosan solutions in EMIMAc and BMIMAc by non-solvent agent vapor diffusion (ethanol or water). The gels that were formed with ethanol treatment were rigid and brittle. Conversely, water vapor treatment only produce gels from the EMIMAc solution, these gels were too soft and brittle to be processed. Aerogels were produced by supercritical CO_2 drying using ionogels that were produced by ethanol treatment. Such aerogels were low density mesoporous materials with surface area between 350 and 480 m^2/g. There is neither spectroscopic evidence of changes in the chemical identity of the chitosan after aerogels production nor evidence of residual IL in the obtained materials. The internal structures of both types of aerogels were similar, appearing as mesoporous materials that are formed by agglomerated clumps.

The swelling of these aerogels followed a second-order kinetics. The application of the Eyring equation to the dependence of the rate constants on temperature allowed for clarifying the characteristics of the diffusion of water. The change of the activation Gibbs energy is mainly controlled by the activation entropy, rather than by the activation enthalpy at the tested temperatures. The kinetics of swelling seems to be controlled by the formation of hydrogen bonding, instead of the diffusion of water itself.

Polymers **2017**, *9*, 722

Aerogels with such characteristics have the potential for applications in the pharmaceutical industry as materials for the encapsulation, retention, and transport of model molecules that have affinities for chitosan. In the environmental field, these aerogels can be used as materials for the removal of pollutants from the effluents of industries in an extractive process by picking up minerals or compounds that are considered to be contaminants with affinity for chitosan. Another potential use is in the food industry or biotechnology, where components such as enzymes, proteins, or compounds that are part of a process could be immobilized inside the aerogels.

Acknowledgments: The authors want acknowledge the financial support of CONACYT through the Project CB-2011-01-169626 and the fellowship DC 2010-778 for GSL. Similarly, it is recognized the valuable input of the technical staff of the Biopolymer Research Group of CIAD.

Author Contributions: Jaime Lizardi-Mendoza and Waldo Argüelles-Monal conceived and designed the experiments; Gonzalo Santos-López performed the experiments; Gonzalo Santos-López, Waldo Argüelles-Monal and Jaime Lizardi-Mendoza analyzed the data; Maricarmen T. Recillas-Mota, Yolanda L. López-Franco and Elizabeth Carvajal-Millan contributed with reagents, materials and analytical advice; Gonzalo Santos-López, Waldo Argüelles-Monal and Jaime Lizardi-Mendoza wrote the paper and all the authors contributed to discuss and review the paper.

Conflicts of Interest: The authors declare no conflict of interest. The founding sponsors had no role in the design of the study; in the collection, analyses, or interpretation of data; in the writing of the manuscript, and in the decision to publish the results.

References

1. Valentin, R.; Bonelli, B.; Garrone, E.; Di Renzo, F.; Quignard, F. Accessibility of the Functional Groups of Chitosan Aerogel Probed by FT-IR-Monitored Deuteration. *Biomacromolecules* **2007**, *8*, 3646–3650. [CrossRef] [PubMed]

2. Pohako-Esko, K.; Bahlmann, M.; Schulz, P.S.; Wasserscheid, P. Chitosan Containing Supported Ionic Liquid Phase Materials for CO_2 Absorption. *Ind. Eng. Chem. Res.* **2016**, *55*, 7052–7059. [CrossRef]

3. Guyomard-Lack, A.; Buchtová, N.; Humbert, B.; Bideau, J.L. Ion segregation in an ionic liquid confined within chitosan based chemical ionogels. *Phys. Chem. Chem. Phys.* **2015**, *17*, 23947–23951. [CrossRef] [PubMed]

4. Trivedi, T.J.; Rao, K.S.; Kumar, A. Facile preparation of agarose-chitosan hybrid materials and nanocomposite ionogels using an ionic liquid via dissolution, regeneration and sol-gel transition. *Green Chem.* **2014**, *16*, 320–330. [CrossRef]

5. Ma, B.; Li, X.; Qin, A.; He, C. A comparative study on the chitosan membranes prepared from glycine hydrochloride and acetic acid. *Carbohydr. Polym.* **2013**, *91*, 477–482. [CrossRef] [PubMed]

6. Kuzmina, O.; Heinze, T.; Wawro, D. Blending of Cellulose and Chitosan in Alkyl Imidazolium Ionic Liquids. *ISRN Polym. Sci.* **2012**, *2012*, 251950. [CrossRef]

7. Xiong, Y.; Wang, H.; Wu, C.; Wang, R. Preparation and characterization of conductive chitosan–ionic liquid composite membranes. *Polym. Adv. Technol.* **2012**, *23*, 1429–1434. [CrossRef]

8. Silva, S.S.; Santos, T.C.; Cerqueira, M.T.; Marques, A.P.; Reys, L.L.; Silva, T.H.; Caridade, S.G.; Mano, J.F.; Reis, R.L. The use of ionic liquids in the processing of chitosan/silk hydrogels for biomedical applications. *Green Chem.* **2012**, *14*, 1463–1470. [CrossRef]

9. Liu, Z.; Wang, H.; Liu, C.; Jiang, Y.; Yu, G.; Mu, X.; Wang, X. Magnetic cellulose–chitosan hydrogels prepared from ionic liquids as reusable adsorbent for removal of heavy metal ions. *Chem. Commun.* **2012**, *48*, 7350–7352. [CrossRef] [PubMed]

10. Naseeruteen, F.; Hamid, N.S.A.; Suah, F.B.M.; Ngah, W.S.W.; Mehamod, F.S. Adsorption of malachite green from aqueous solution by using novel chitosan ionic liquid beads. *Int. J. Biol. Macromol.* **2017**. [CrossRef] [PubMed]

11. Muzzarelli, R.A.A. Biomedical Exploitation of Chitin and Chitosan via Mechano-Chemical Disassembly, Electrospinning, Dissolution in Imidazolium Ionic Liquids, and Supercritical Drying. *Mar. Drugs* **2011**, *9*, 1510–1533. [CrossRef] [PubMed]

12. Silva, S.S.; Mano, J.F.; Reis, R.L. Ionic liquids in the processing and chemical modification of chitin and chitosan for biomedical applications. *Green Chem.* **2017**, *19*, 1208–1220. [CrossRef]

13. Silva, S.S.; Duarte, A.R.C.; Carvalho, A.P.; Mano, J.F.; Reis, R.L. Green processing of porous chitin structures for biomedical applications combining ionic liquids and supercritical fluid technology. *Acta Biomater.* **2011**, *7*, 1166–1172. [CrossRef] [PubMed]

14. Kasaai, M.R. Various Methods for Determination of the Degree of N-Acetylation of Chitin and Chitosan: A Review. *J. Agric. Food Chem.* **2009**, *57*, 1667–1676. [CrossRef] [PubMed]

15. Ottøy, M.H.; Vårum, K.M.; Smidsrød, O. Compositional heterogeneity of heterogeneously deacetylated chitosans. *Carbohydr. Polym.* **1996**, *29*, 17–24. [CrossRef]

16. Scherließ, R.; Buske, S.; Young, K.; Weber, B.; Rades, T.; Hook, S. In vivo evaluation of chitosan as an adjuvant in subcutaneous vaccine formulations. *Vaccine* **2013**, *31*, 4812–4819. [CrossRef] [PubMed]

17. Rinaudo, M.; Pavlov, G.; Desbrières, J. Solubilization of Chitosan in Strong Acid Medium. *Int. J. Polym. Anal. Charact.* **1999**, *5*, 267–276. [CrossRef]

18. Montembault, A.; Viton, C.; Domard, A. Rheometric study of the gelation of chitosan in aqueous solution without cross-linking agent. *Biomacromolecules* **2005**, *6*, 653–662. [CrossRef] [PubMed]

19. Le Bideau, J.; Viau, L.; Vioux, A. Ionogels, ionic liquid based hybrid materials. *Chem. Soc. Rev.* **2011**, *40*, 907–925. [CrossRef] [PubMed]

20. Cárdenas, A.; Argüelles-Monal, W.; Goycoolea, F.M.; Higuera-Ciapara, I.; Peniche, C. Diffusion through Membranes of the Polyelectrolyte Complex of Chitosan and Alginate. *Macromol. Biosci.* **2003**, *3*, 535–539. [CrossRef]

21. Seddon, K.R. Ionic Liquids for Clean Technology. *J. Chem. Technol. Biotechnol.* **1997**, *68*, 351–356. [CrossRef]

22. Chen, Y.; Sun, X.; Yan, C.; Cao, Y.; Mu, T. The Dynamic Process of Atmospheric Water Sorption in [EMIM][Ac] and Mixtures of [EMIM][Ac] with Biopolymers and CO_2 Capture in These Systems. *J. Phys. Chem. B* **2014**, *118*, 11523–11536. [CrossRef] [PubMed]

23. Vachoud, L.; Domard, A. Physicochemical properties of physical chitin hydrogels: Modeling and relation with the mechanical properties. *Biomacromolecules* **2001**, *2*, 1294–1300. [CrossRef] [PubMed]

24. Quignard, F.; Valentin, R.; Renzo, F.D. Aerogel materials from marine polysaccharides. *New J. Chem.* **2008**, *32*, 1300–1310. [CrossRef]

25. Subrahmanyam, R.; Gurikov, P.; Dieringer, P.; Sun, M.; Smirnova, I. On the Road to Biopolymer Aerogels—Dealing with the Solvent. *Gels* **2015**, *1*, 291–313. [CrossRef]

26. Brugnerotto, J.; Lizardi, J.; Goycoolea, F.M.; Argüelles-Monal, W.; Desbrieres, J.; Rinaudo, M. An infrared investigation in relation with chitin and chitosan characterization. *Polymer* **2001**, *42*, 3569–3580. [CrossRef]

27. Socrates, G. *Infrared and Raman Characteristic Group Frequencies: Tables and Charts*; John Wiley & Sons: Chichester, NY, USA, 2004; ISBN 978-0-470-09307-8.

28. Sing, K.S.W.; Everett, D.H.; Haul, R.A.W.; Moscou, L.; Pierotti, R.A.; Rouquerol, J.; Siemieniewska, T. Reporting Physisorption Data for Gas/Solid Systems. In *Handbook of Heterogeneous Catalysis*; Wiley-VCH Verlag GmbH & Co. KGaA: Weinheim, German, 2008; ISBN 978-3-527-61004-4.

29. Chang, X.; Chen, D.; Jiao, X. Chitosan-Based Aerogels with High Adsorption Performance. *J. Phys. Chem. B* **2008**, *112*, 7721–7725. [CrossRef] [PubMed]

30. Tsioptsias, C.; Stefopoulos, A.; Kokkinomalis, I.; Papadopoulou, L.; Panayiotou, C. Development of micro- and nano-porous composite materials by processing cellulose with ionic liquids and supercritical CO_2. *Green Chem.* **2008**, *10*, 965–971. [CrossRef]

31. Goycoolea, F.M.; Fernández-Valle, M.E.; Aranaz, I.; Heras, Á. pH- and Temperature-Sensitive Chitosan Hydrogels: Swelling and MRI Studies. *Macromol. Chem. Phys.* **2011**, *212*, 887–895. [CrossRef]

32. Crank, J. *The Mathematics of Diffusion*, 2nd ed.; Clarendon Press: Oxford, UK, 1979; ISBN 0-19-853411-6.

33. Vázquez, B.; San Roman, J.; Peniche, C.; Cohen, M.E. Polymeric Hydrophilic Hydrogels with Flexible Hydrophobic Chains. Control of the Hydration and Interactions with Water Molecules. *Macromolecules* **1997**, *30*, 8440–8446. [CrossRef]

34. Ritger, P.L.; Peppas, N.A. A simple equation for description of solute release I. Fickian and non-fickian release from non-swellable devices in the form of slabs, spheres, cylinders or discs. *J. Control. Release* **1987**, *5*, 23–36. [CrossRef]

35. Eyring, H. The Activated Complex in Chemical Reactions. *J. Chem. Phys.* **1935**, *3*, 107–115. [CrossRef]

36. Wynne-Jones, W.F.K.; Eyring, H. The Absolute Rate of Reactions in Condensed Phases. *J. Chem. Phys.* **1935**, *3*, 492–502. [CrossRef]

37. Eyring, H. Viscosity, Plasticity, and Diffusion as Examples of Absolute Reaction Rates. *J. Chem. Phys.* **1936**, *4*, 283–291. [CrossRef]

38. Kincaid, J.F.; Eyring, H.; Stearn, A.E. The Theory of Absolute Reaction Rates and its Application to Viscosity and Diffusion in the Liquid State. *Chem. Rev.* **1941**, *28*, 301–365. [CrossRef]

39. Schott, H. Swelling kinetics of polymers. *J. Macromol. Sci. Part B* **1992**, *31*, 1–9. [CrossRef]

40. Pauling, L. *The Nature of the Chemical Bond and the Structure of Molecules and Crystals: An Introduction to Modern Structural Chemistry*; Cornell University Press: Ithaca, NY, USA, 1960; ISBN 978-0-8014-0333-0.

41. Prabhumirashi, L.S.; Jose, C.I. Infra-red studies and thermodynamics of hydrogen bonding in ethylene glycol monoalkyl ethers. Evidence for a ten membered ring dimer. *J. Chem. Soc. Faraday Trans. 2 Mol. Chem. Phys.* **1975**, *71*, 1545–1554. [CrossRef]

42. Ishizuka, T.; Ohzu, S.; Kotani, H.; Shiota, Y.; Yoshizawa, K.; Kojima, T. Hydrogen atom abstraction reactions independent of C–H bond dissociation energies of organic substrates in water: Significance of oxidant–substrate adduct formation. *Chem. Sci.* **2014**, *5*, 1429–1436. [CrossRef]

polymers

MDPI

Article

Rheo-Kinetic Study of Sol-Gel Phase Transition of Chitosan Colloidal Systems

Piotr Owczarz [1], Patryk Ziółkowski [1,*], Zofia Modrzejewska [2], Sławomir Kuberski [3] and Marek Dziubiński [1]

[1] Department of Chemical Engineering, Lodz University of Technology, Lodz 90-924, Poland; piotr.owczarz@p.lodz.pl (P.O.); marek.dziubinski@p.lodz.pl (M.D.)
[2] Department of Environmental Engineering, Lodz University of Technology, Lodz 90-924, Poland; zofia.modrzejewska@p.lodz.pl
[3] Department of Molecular Engineering, Lodz University of Technology, Lodz 90-924, Poland; slawomir.kuberski@p.lodz.pl
* Correspondence: patryk.ziolkowski@dokt.p.lodz.pl; Tel.: +48-42-631-3975

Received: 4 October 2017; Accepted: 2 January 2018; Published: 5 January 2018

Abstract: Chitosan colloidal systems, created by dispersing in aqueous solutions of hydrochloric acid, with and without the addition of disodium β-glycerophosphate (β-NaGP), were prepared for the investigation of forming mechanisms of chitosan hydrogels. Three types of chitosan were used in varying molecular weights. The impacts of the charge and shape of the macromolecules on the phase transition process were assessed. The chitosan system without the addition of β-NaGP was characterized by stiff and entangled molecules, in contrast to the chitosan system with the addition of β-NaGP, wherein the molecules adopt a more flexible and disentangled form. Differences in molecules shapes were confirmed using the Zeta potential and thixotropy experiments. The chitosan system without β-NaGP revealed a rapid nature of phase transition—consistent with diffusion-limited aggregation (DLA). The chitosan system with β-NaGP revealed a two-step nature of phase transition, wherein the first step was consistent with reaction-limited aggregation (RLA), while the second step complied with diffusion-limited aggregation (DLA).

Keywords: chitosan; hydrogel; phase transition; gelation mechanism

1. Introduction

Chitosan is a derivative of chitin that constitutes a building component of crustaceans. Obtained in the deacetylation process, chitosan is a linear polysaccharide consisting of acetylglucosamine and glucosamine molecules, in which the ratio defines the degree of acetylation. The interest in chitosan for biomedical applications results from the biocompatible, biodegradable and non-toxic properties of this compound [1–8]. The main aspects of chitosan use in biomedicine are smart drug delivery systems and scaffolds in tissue engineering [9–13]. Both of these uses employ the phenomenon of sol-gel phase transition of chitosan hydrogels.

Due to the mutual association of chitosan molecules, this compound is not water-soluble [14]. Obtaining a soluble system is possible using water solutions of organic and non-organic acids. Solubility in such a system is caused by the protonation phenomenon of free amino groups $H^+ + -NH_2 \leftrightarrow -NH^{3+}$ [15,16]. Chitosan molecule properties change from hydrophobic to hydrophilic. The association forces are overcome, which enables the hydration of polymer chains. A colloidal system (sol) is formed in which the polymer chains become dispersed in a continuous medium—a water solution of the acid. As a result of the changes in physical or chemical parameters (e.g., pH, temperature, polymer concentration) of the chitosan solution, the system undergoes a phase transition to the form of a hydrogel [17–20].

Chitosan macromolecules may form various shape configurations in solutions specified with parameter *a* in Mark Houwink's equation. The shapes of macromolecules assume the form of stiff, impenetrable spheres ($a < 0.5$), flexible balls enabling the flow of the solvent ($0.5 < a < 1.2$) and disentangled, stiff rods ($a > 1.2$). Using this type of solvent and the related differences in the ionic power and pH of the solutions lead to various shape configurations of the polysaccharide macromolecules in the solution [21–24]. This results from the equilibrium between the intermolecular interactions, mainly electrostatic repulsion, hydrogen bonds and mutual association of the polymer molecules. The molecular weight of chitosan also influences the configurations of the chain structures, as the charge distribution of the functional groups changes with the molecular weight [23].

The necessity to use aqueous acid solutions as a chitosan solvent and the connected acid reaction of the chitosan systems excluded certain biomedical uses. What turned out to be an answer was systems with an addition of substances neutralizing the reaction of the solutions and allowing to maintain a pH close to the physiological value of a human body (pH~7). In the literature, the most frequently used is disodium β-glycerophosphate (β-NaGP). These systems are characterized by a neutral reaction and a thermo-induced gelation mechanism [18,25,26].

When temperature increases, a reduction of positively charged amino groups occurs, which is caused by a decrease of chitosan pK_a. The released H^+ protons are captured by the dissociated β-GP^{2-} groups. The charge neutralization of the chitosan molecules results in a vanishing of electrostatic repulsion and polymer molecules change their character from hydrophilic to hydrophobic. Coagulation of polymer molecules begins, induced by hydrophobic effects and hydrogen interactions and an initiation of the gelation process occurs. The addition of β-NaGP functions as a buffer and acceptor for the protons released in the gelation process in the system. It was also stated that β-NaGP does not participate in building a crosslinked structure. This was proven by experiments in which β-NaGP freely diffused from a prepared gel structure [27].

In the literature [28], chitosan gels are considered as physical hydrogels, in which a lattice is created by hydrogen bonds among amino groups. However, there is no explicit answer as to which processes take place, when the samples are heated and a structure is formed in them in the case of acid reaction solutions (chitosan salt solutions created with ions of an acid solvent) and what processes take place in the case when a buffering substance is added to the solution.

The aim of the paper was to specify the mechanism of forming chitosan hydrogels from aqueous solutions of hydrochloric acid with and without the addition of disodium β-glycerophosphate (β-NaGP). Characteristic parameters of the process of phase transition, induced by a temperature increase, were determined.

For this purpose, the changes in storage and loss moduli as a function of temperature were determined using oscillation rheology methods. The influence of shearing on the deformation of polymer chains was tested. The pH values and the Zeta potential of the tested chitosan solutions were specified.

2. Materials and Methods

2.1. Materials

For the present study, three types of chitosan were used varying in molecular weight: Chitosan from the Fluka company of unknown origin (sample 1), product No. 50949, lot No. 1078112; chitosan from the Sigma-Aldrich® company (Sigma-Aldrich Sp. z o.o., Poznan, Poland) obtained from crab shells (sample 2), product No. 50494, lot No. 0001424218; chitosan from the Sigma-Aldrich® company obtained from shrimp shells (sample 3), product No. 50494, lot No. BCBB5837.

2.2. Preparation of Chitosan Solutions

Chitosan solutions (2.5% *w/w*) were prepared by dispersing 0.4 g chitosan in 16 g 0.1 M HCl solution. The obtained solutions were characterized by an equimolar number of H^+ ions and $-NH_2$

amino groups in chitosan molecules. After dispersing, the container with the sample was covered (in order to prevent vaporization) and left for 24 h at room temperature. After that time, the sample was left at the temperature of 5 °C for 2 h. The above procedure was repeated twice for each type of chitosan in order to investigate two sets of solutions: Chitosan-solvent and chitosan-solvent-buffer. The second set was prepared by addition of disodium β-glycerophosphate (β-NaGP). First, the buffer in the amount of 2 g was dissolved in 2 g of distilled water and then the resulting solution was gradually added to a cooled solution of chitosan.

2.3. Materials Characteristics

The degree of acetylation (DA) was determined by titrimetric method. The pH measurements were made in a heated bath with the use of a pH-meter ELMETRON CP-401 (ELMETRON Sp. j., Zabrze, Poland), equipped with an electrode for viscous liquids ERH-12-6. The molecular weight (M_w and M_n) of the tested samples was determined by gel permeation chromatography (GPC/SEC) with the use of the high-performance liquid chromatography (HPLC) on a KnauerSmartline company (Berlin, Germany) apparatus, equipped with an analytical isocratic Pump1000 and DRI detector (S-2300/2400, Knauer). The Zeta potential was measured on the basis of electrophoretic mobility of the molecules in an electric field by means of Malvern ZetaSizer Nano ZS (Malvern Instruments Ltd., Malvern, UK). The measurements were conducted at a steady temperature of 30 °C using a folded capillary cell DTS1060 type. The changes in the Zeta potential during the heating of the sample were also established. In order to do so, measurements were made on one sample placed in a cell at the temperature of 20 °C and the value of the Zeta potential was established every 2 °C while heating the sample to 40 °C. The basic properties of chitosan solutions samples are presented in Table 1.

Table 1. Properties of chitosan samples.

Chitosan Samples	Weight Average Molar Mass M_w (g/mol)	Number Average Molar Mass M_n (g/mol)	Polydispersity Index PDI (–)	Degree of Acetylation DA (%)	Zeta Potential ζ (mV) [1]	
					No Addition of β-NaGP	Addition of β-NaGP
Sample 1	463,000	79,000	5.9	16.8	48.5 (±1.9)	4.74 (±5.3)
Sample 2	680,000	110,000	6.1	18.2	42.3 (±4.4)	1.76 (±4.2)
Sample 3	862,000	145,000	5.9	16.6	37.4 (±2.4)	3.23 (±4.5)

[1] Zeta potential values expressed as an average of 30 repeated measurements on the same sample at constant temperature 30 °C. Some of a Zeta potential values for samples with β-NaGP were below zero, hence standard deviation values are higher than average values.

2.4. Rheological Measurements

The rheological properties of the chitosan solutions were measured using rotational rheometer Anton Paar Physica series MCR 301 (Anton Paar, Warszawa, Poland). The viscous properties of the tested samples were established in a cone-plate system, of 50 mm diameter, 1° cone angle and 47 μm cone truncation. The flow curves in the range of shear rates $0.01 \text{ s}^{-1} < \gamma < 500 \text{ s}^{-1}$ were established. A sample, placed at the temperature of 5 °C, was heated at the rate of 1 deg/min to 20 °C. After that, subsequent flow curves were made at steady temperatures of 20, 25 30, 35 and 40 °C using the same heating rate between measurement intervals.

The viscoelastic properties of chitosan systems were established in the same measurement system. The measurement was conducted in the linear range of viscoelasticity at low amplitude of deformations $\gamma = 1\%$ and constant angular frequency $\omega = 5$ rad/s. The gelation process was conducted in non-isothermal conditions, at a constant heating rate of 1 deg/min, from the temperature of 5 °C (storage temperature) to 60 °C. Data of temperature ramp tests are presented as an average from three measurements for each sample. The free surface of the sample was covered with thin layer of low-viscosity silicone oil to prevent sample evaporation during temperature sweep tests [18,20].

The impact of the shear rate on the deformation of a sample was conducted at the temperature of 5 °C and the method of a three-interval thixotropy test was applied (three-interval thixotropy test (3ITT) cillations—rotations—oscillations). In the middle interval, the sample was deformed at a constant shear rate (10^1, 10^2 and 10^3 s^{-1}). In the initial and final interval, the sample was tested by means of oscillatory shear at a constant deformation amplitude $\gamma = 1\%$ and angular frequency $\omega = 5$ rad/s in order to establish the properties before and after the rotational deformation process.

3. Results

3.1. The pH and the Zeta Potential of Chitosan Solutions

The pH values of chitosan solutions in the temperature function are presented in Figure 1. A significant influence of the addition of β-NaGP on the pH value was observed. Dissociation of this compound in the solution creates the conditions of equilibrium between β-GP^{2-} ions and positively charged –NH^{3+} amino groups, increasing the pH value of the solutions. For all samples containing the buffer, the pH values of the solutions oscillate around a neutral reaction (pH~7) and are independent of the temperature, excluding the pH values for T < 10 °C (Figure 1).

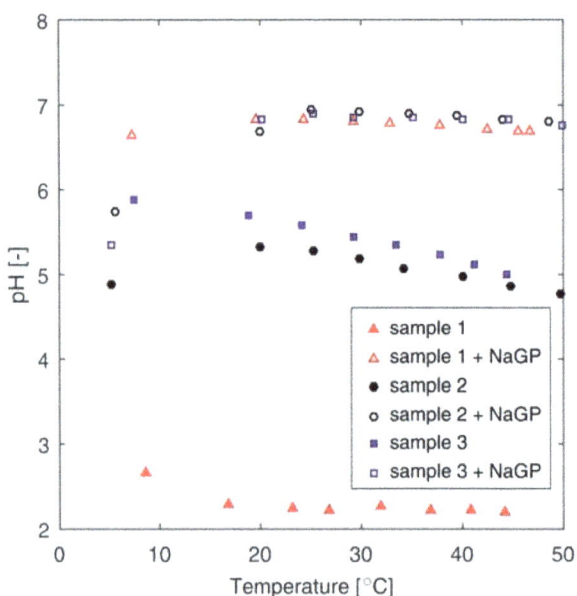

Figure 1. Influence of the temperature on the pH value for different chitosan solutions.

Chitosan solutions without the addition of the buffer show an acidic pH. The lowest pH value occurs for sample 1 and is approximate to the pH of a pure solvent (0.1 M HCl). Correlations between the molecular weight of chitosan and the pH of the solutions for systems without the addition of β-NaGP can be observed. For lower molecular weight of chitosan, the pH values of the solution are lower. An acid reaction of the solution suggests that a major part of the –NH^2 amino group was not protonated, despite an equimolar number of these groups and H$^+$ ions. This can be justified with the configuration taken by chitosan macromolecules in a solution. In a case when a macromolecule has the form of a stiff, entangled ball, the access to the amino groups is difficult. Protonation occurs only outside the ball. In a case when the configuration of a macromolecule has a more disentangled (flexible) form, protonation can take place inside the macromolecule.

The presence of the above-mentioned configurations in the tested solutions is confirmed by the values of the Zeta potential presented in Table 1. For the samples without the addition of β-NaGP, the Zeta potential assumes values above 30 mV (the verge of colloidal stability). Thus, in these solutions the chitosan macromolecules have the form of a ball suspended in a dispersion medium. The hydrophobic core and the positively charged external layer of the macromolecule enable the solvation by water molecules. On the other hand, values of the Zeta potential for solutions with an addition of β-NaGP are ca. 10 times lower and are far below the verge of stability. This probably results from the more flexible and disentangled form of the macromolecule, allowing solvent penetration into the core and as a consequence this results in distribution of the charge not only on the external layer but also inside the macromolecule.

Analysis of the changes in the Zeta potential values obtained during the heating of the samples containing β-NaGP (Figure 2) shows that during the test, the samples showed unstable properties in a transitional form between a sol and a gel. At the temperature below 30 °C, the measured values of the Zeta potential gradually increase. With the time increasing, a decrease of ionic strength of the amino groups takes place and thus in the polymer chains properties change from hydrophilic to hydrophobic, combined with a disentangling of the chitosan macromolecules. After exceeding the temperature of 30 °C, the sol-gel equilibrium moved towards the creation of a polymer network—a gel structure. Due to this, any further measurements of the Zeta potential are encumbered with errors. This is a consequence of the measuring technique of the equipment—the measurement of the electrophoretic mobility of the molecules. Due to a limited mobility of the molecules trapped in the lattice structure, erroneous readings of the measured values occur. It is visible in Figure 2 as a cloud of scattered spots—high standard deviation. This also explains the high values of measurement errors of the Zeta potential values for all samples containing β-NaGP, which is presented in Table 1.

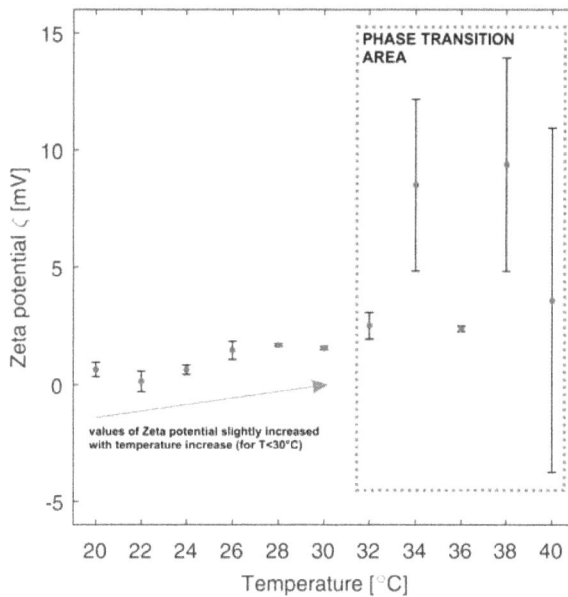

Figure 2. Zeta potential values determined during heating of sample 2 with addition of β-NaGP. Error bars represent the standard deviation.

3.2. Influence of Temperature on the Viscous Properties of the Chitosan Solutions

The flow curves of the chitosan solutions allow us to state that in the investigated temperature range, all the tested samples show properties of non-Newtonian shear thinning fluid. In Figure 3, changes in the values of flow behaviour index n_o of Ostwald-de Waele power law model were presented for samples containing β-NaGP and without the addition of a buffering substance. For all the tested samples, an increase of temperature decreases the viscosity of the solvent. Furthermore, an increase of non-Newtonian shear thinning properties can be observed—the value of the n_o coefficient decreases. However, the character of those changes depends on the molecular weight of the polymer and the presence of a buffer.

In the case of samples without β-NaGP, the values of the flow behaviour index n_o clearly depends on the molecular weight of the polymer; the lower the M_w value, the higher the n_o value. It could also be observed that the polymer chains assume the configurations of entangled balls. It is confirmed by the Newtonian character of some tested solutions, which is supported by the n_o values that are slightly lower than one (Figure 3, sample 1). An increase in the molecular weight and consequently larger balls of more flexible structure [23,29], cause a significant decrease of the value of the n_o parameter. During the heating of these samples it was observed that the values of the n_o coefficient reveal rapid changes after exceeding the temperature of 30 °C. This is clearly visible for the slopes of the curves assigned to samples 2 and 3 (Figure 3).

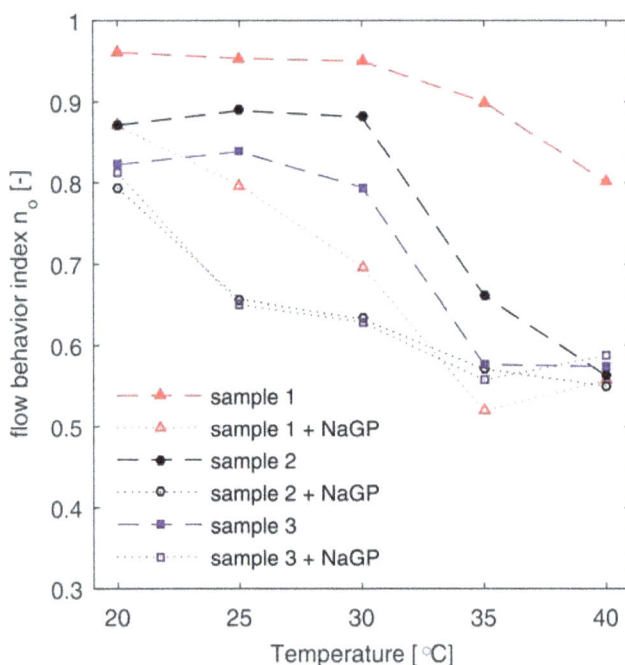

Figure 3. Influence of the temperature on the flow behaviour index n_o values for different chitosan solutions (the lines are only for visual aid).

The curves characterizing changes in the flow behaviour index of the samples containing β-NaGP have a different slope. These samples show significantly lower values of the n_o parameter, already in the temperature of 20 °C. It must be assumed that the addition of the buffer (pH increase, Figure 1) intensifies the disentangling of the polysaccharide chains [22,23] which during shearing arrange more easily in parallel to the shear flow direction—thus, causing an increase in the non-Newtonian properties of the tested liquid.

3.3. Influence of Temperature on the Viscoelastic Properties of the Chitosan Solutions

The evolution of storage modulus G' and loss modulus G'' as functions of temperature is shown in Figure 4, for systems with and without the addition of β-NaGP. At the initial stage, both systems reveal sol-like behaviour, showing a typical behaviour for a viscoelastic liquid. The loss modulus dominates over the storage modulus. At the same time the values of both moduli and thus the complex viscosity decreases with the increase of temperature and prove the general dependency of liquid viscosity decrease with the temperature increase. In accordance with the Stokes-Einstein equation [30], with a temperature increase the diffusion coefficient increases and an intensification of Brownian motion takes place. Chitosan solutions at this stage are characterised by charged polymer chains. This charge stabilizes the systems and prevents crosslinking [20,31].

With a temperature increase the phase transition occurs—the values of the storage and loss moduli increase. This is a result of the neutralization of the charge of chitosan and formation of a crosslinked structure (gel). Comparing the two types of systems (with and without β-NaGP), a difference in the dynamics of the moduli changes can be noticed. A more rapid increase of moduli for solutions without the addition of β-NaGP compared with the solutions containing the buffer can be observed. Additionally, in the chitosan-solvent-β-NaGP systems, two regions of phase transition can be distinguished in contrast to chitosan-solvent systems where there is one rapid increase of storage modulus.

In the final stage, dominance of storage modulus over loss modulus indicate the gel-like behaviour. However, the plateau values of moduli in chitosan-solvent systems are one decade lower compared to second systems. Furthermore, this plateau is not stable after exceeding a temperature of 50 °C in contrast to chitosan-solvent-β-NaGP systems. The other difference between two systems is revealed as a slightly higher phase transition temperature for chitosan-solvent systems.

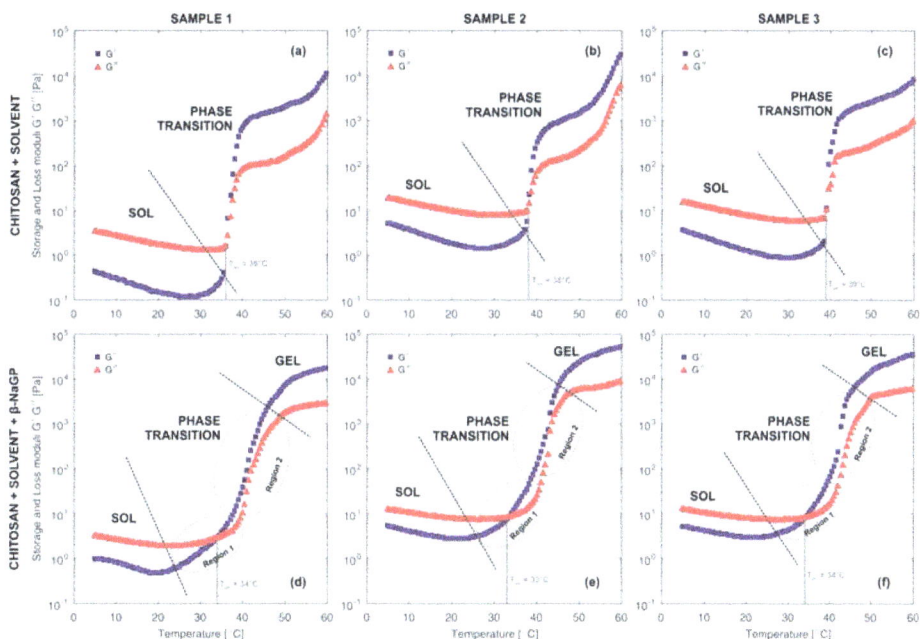

Figure 4. The storage modulus G′ and loss modulus G″ as a function of temperature. (**a**–**c**) Samples without addition of β-NaGP. (**d**–**f**) Samples with addition of β-NaGP.

In the initial phase of the temperature ramp experiments, the stability of the solutions is conditioned by the hydrophilic character of the chitosan molecules (Chit) related with the presence of positively charged –NH^{3+} groups. The protonation reactions of the amino groups in both solution types are presented below:

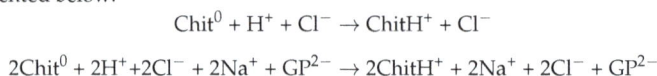

$$Chit^0 + H^+ + Cl^- \rightarrow ChitH^+ + Cl^-$$

$$2Chit^0 + 2H^+ + 2Cl^- + 2Na^+ + GP^{2-} \rightarrow 2ChitH^+ + 2Na^+ + 2Cl^- + GP^{2-}$$

Analysing the systems of chitosan-β-NaGP [25,26], a decrease in chitosan pK_a was proved with an increase in temperature and at the same time an impact of temperature on pK_a of β-glycerophosphate was not observed. Accordingly, an increase in temperature (supplying energy) causes a release of protons and a reduction in the charge on the chitosan molecule. The released H$^+$ protons are connected to a negatively charged GP^{2-} ion [26]. Repulsive electrostatic interactions among the chitosan molecules vanish. Chitosan interchain attractive interactions resulting from hydrophobic effects and hydrogen interactions begin to dominate. The reactions taking place during phase transition are presented below:

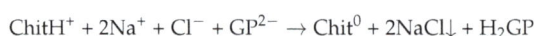

$$ChitH^+ + 2Na^+ + Cl^- + GP^{2-} \rightarrow Chit^0 + 2NaCl\downarrow + H_2GP$$

The basic factor leading to the gelation of chitosan systems is a reduction of the charge on the polymer molecule, induced by supplying energy. The presence of β-NaGP in the role of an acceptor of the protons released from the chitosan causes lowering of the phase transition temperature. In acid systems, with no buffer present, phase transition is heavily dependent on the diffusion of the polymer molecules. For this reason, the phase transition process requires a greater amount of supplied energy (a higher temperature).

3.4. Non-Isothermal Gelation Kinetics

An analysis of the changes in the values of the storage modulus G′ and the use of kinetics model of polymer crystallization which is based on the Arrhenius equation, allows to determine the activation energy for the gelation process [32–34]. An equation describing the kinetics of a gelation process taking place with a temperature increase takes the following form [32]:

$$\ln\left(\frac{1}{G'^{n_a}}\frac{dG'}{dt}\right) = \ln k_0 + \left(\frac{E_a}{RT}\right) \tag{1}$$

where exponent n_a in the Equation (1), described with the dependency $n_a = r + s$, specifies the order of the polymerization reaction, where r represents the dimension of the growing crystals and s the nucleation type. Parameter r assumes values 1, 2 or 3 respectively for one-, two- or three-dimension structures (rod, disk, sphere), whereas s assumes value 0 for predetermined nucleation (nuclei already present), or 1 for sporadic nucleation (nuclei arise and their number increases linearly with time). In the calculations parameter $n_a = 2$ was assumed on the basis of research presented by authors [18,35], regarding similar polymer systems. The value of dG′/dt in the Equation (1) means the so-called structure development rate. By drawing the dependency $\ln(1/G'^{n_a} dG'/dt)$ vs. 1/T from the curve slope, the value of activation energy E_a was determined.

A graphic interpretation of Equation (1) and determined values of activation energies are presented in Figure 5. Comparing the systems with and without an addition of β-NaGP, differences in the amounts of energy inducing the phase transition can be observed. Chitosan-solvent systems required greater energy compared to chitosan-solvent-β-NaGP systems. Furthermore, samples with addition of β-NaGP reveal a two-step nature of phase transition.

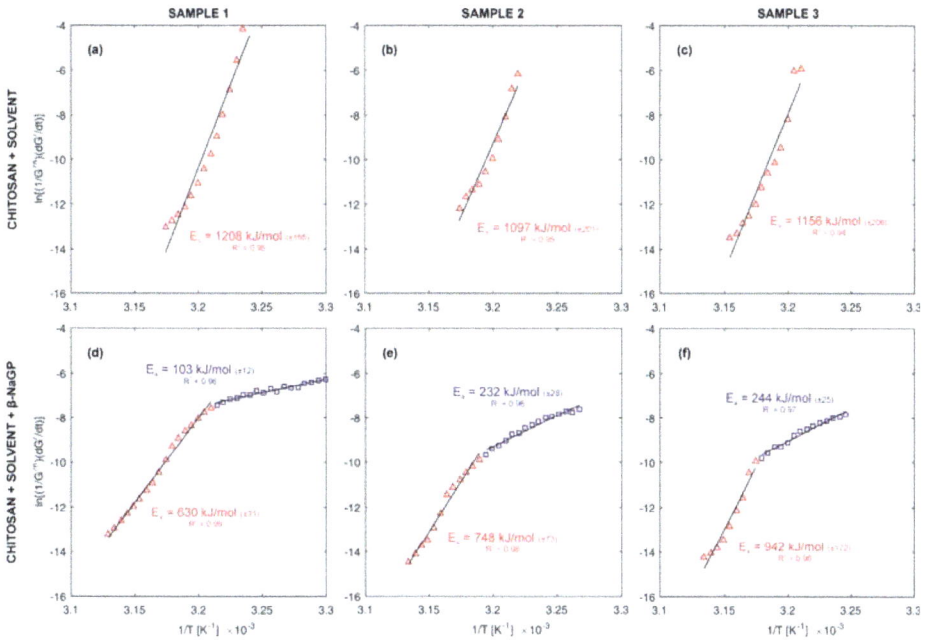

Figure 5. The time-temperature relationship from Equation (1) on Arrhenius plot enabling to determination of activation energy. The straight lines represent the linear approximation of the Equation (1). (**a–c**) Samples without addition of β-NaGP. The red triangles represent experimental data from phase transition stage (see Figure 4). (**d–f**) Samples with addition of β-NaGP. The blue squares and red triangles represent experimental data from region 1 and region 2, respectively (see Figure 4).

3.5. Influence of Shear Rate on Polymer Deformations

The impact of shearing deformation on the behaviour of samples in time was specified based on the three-interval thixotropy test (3ITT). The samples were deformed at three different shearing deformations: 10^1, 10^2 and 10^3 s^{-1}. On the basis of the obtained measurement results, deformation parameters Def were determined from Equation (2) [36]:

$$\text{Def} = \frac{G_i - G_0}{G_i} \qquad (2)$$

where G_i—value of the storage modulus in the initial stage of the experiment (interval 1), G_0—value of the storage modulus immediately after the deformation stage (t = 0 s interval 3), see Figure 6a.

Changes in the storage modulus G' in the recovery stage were fitted to a 2-order kinetic structural model [37,38], assuming G_e as an equilibrium value of the storage modulus when t → ∞, k as a constant of the recovery speed and parameter $n_a = 2$ (2-order of the kinetic model):

$$\left[\frac{G' - G_e}{G_0 - G_e} \right]^{1-n_a} = (n_a - 1)kt + 1 \qquad (3)$$

Table 2. Parameters of second order structural model for all investigated samples.

Chitosan Samples	Shear Rate (s^{-1})	G_0 (Pa)		G_e (Pa)		k ($\times 10^2$)		G_e/G_0		R^2	
		$-\beta$*	$+\beta$*	$-\beta$	$+\beta$	$-\beta$	$+\beta$	$-\beta$	$+\beta$	$-\beta$	$+\beta$
	10^1	- **	0.89	-	2.36	-	0.12	-	2.64	-	0.96
Sample 1	10^2	0.31	0.75	0.36	1.96	1.23	0.38	1.13	2.64	0.79	0.98
	10^3	0.28	0.46	0.35	2.13	5.99	0.34	1.31	4.68	0.91	0.99
	10^1	-	5.09	-	7.57	-	0.67	-	1.49	-	0.90
Sample 2	10^2	4.06	4.25	4.61	8.18	4.08	1.09	1.13	1.92	0.96	0.96
	10^3	3.94	3.41	4.73	8.16	5.49	0.96	1.20	2.40	0.94	0.97
	10^1	-	4.93	-	8.87	-	0.68	-	1.80	-	0.93
Sample 3	10^2	3.31	3.80	3.76	7.71	1.76	1.35	1.14	2.03	0.96	0.96
	10^3	3.35	3.27	3.93	7.80	3.21	1.26	1.17	2.38	0.95	0.97

* Samples without addition of β-NaGP ($-\beta$), samples with addition of β-NaGP ($+\beta$). ** Not possible to fitted data because of the weak phenomenon of deformation.

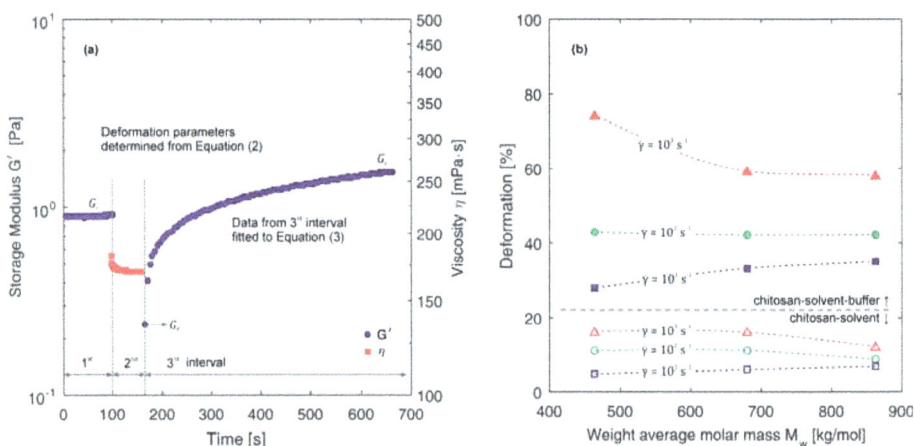

Figure 6. (**a**) Storage modulus and shear viscosity during the three interval thixotropy test for a sample 1 with addition of β-NaGP. (**b**) Deformation parameters as a function of molecular weight for all investigate samples (the lines are only for visual aid).

Figure 6a presents the result of the thixotropy test with a marked characteristic point. The obtained values of deformation parameters for all investigated samples are shown in Figure 6b. Parameters of second order structural model determined by Equation (3) are presented in Table 2. Comparison of the two types of systems indicate that chitosan-solvent-β-NaGP systems reveal stronger thixotropic properties in contrast to chitosan-solvent systems. These differences can result from a shape of molecules. As mentioned above, molecules of chitosan in systems without β-NaGP take the form of stiff, entangled balls. Thus, a shear deformation effect on polymer chains is insignificant. Whereas disentanglement and the more flexible form of chitosan molecules in systems with β-NaGP lead to disentangling of polymer chains due to shear deformation which is revealed by the thixotropic properties.

Furthermore, a correlation between the molecular weight and the degree of deformation can be observed. With a decrease of the value of molecular weight, the degree of deformation of the polymer increases. However, this correlation weakens or vanishes with the decrease of applied shear rate. The effect of molecular weight on the range of deformation parameters values can be also observed. For sample 1 (lowest molecular weight), the difference between maximum and minimum value of deformation parameter is significantly higher compared to sample 3 (highest molecular weight).

4. Discussion

Analysis of the phase transition process of chitosan hydrogels enables to observe two different mechanisms of this phenomenon. The differences in the composition of the solutions cause the formation of systems varying in pH reaction which influences the shape that the polymer macromolecules take in the solution [22–24]. A schematic comparison of the gelation mechanisms occurring in the tested chitosan systems is presented in Figure 7. The figure shows the differences and indicates the characteristic features of the process of structure forming for chitosan-solvent and chitosan-solvent-buffer systems.

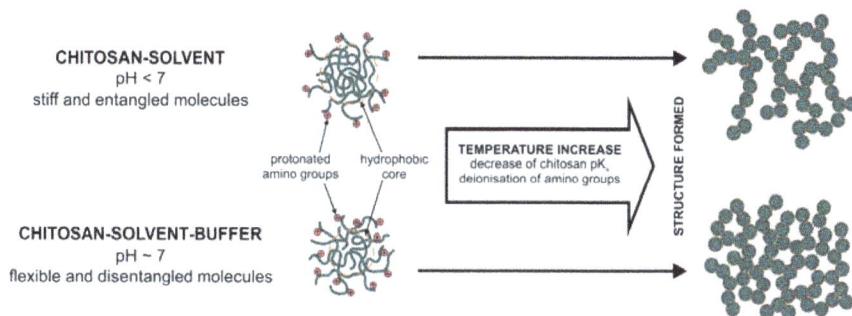

Figure 7. Comparison of the two types of chitosan systems. Shape configuration of chitosan macromolecule and proposed forming of structure.

Systems of acid reaction (chitosan-solvent) are characterized by macromolecules with a structure of stiff, entangled balls. Protonation of amino groups takes place largely on the external layer of the macromolecules. The determined values of the flow behaviour index n_0 and the Zeta potential suggest the presence of these macromolecules in the form of dispersed balls surrounded by solvent molecules and the shorter the polymer chain (lower molecular weight), the stiffer the structure of the ball [23,29]. The phase transition temperatures determined in oscillatory tests are included in the range of 36–39 °C. The dynamics of these transitions is a rapid increase in the value of modulus G' after reaching a certain critical temperature (Figure 4a–c).

Systems of neutral pH reaction (chitosan-solvent-buffer) are characterized by a flexible structure of balls with a possibility of the solvent penetrating inside. The charge distribution, resulting from the protonation of amino groups, takes place both on the external layer and inside the ball. This is indicated by the determined values of the Zeta potential, proving the lack of a double layer. The phase transition temperatures determined in oscillatory and Zeta potential tests are included in the range 33–34 °C. After reaching the critical temperature the value of modulus G' increases much more slowly than in the case of solutions without β-NaGP (Figure 4d–f). The activation energy of the gelation process indicates higher values for systems of acid reaction in comparison with systems of neutral reaction (Figure 5).

The theory of aggregation processes distinguishes two mechanisms of the phenomenon: diffusion limited aggregation (DLA) and reaction limited aggregation (RLA) [39]. At the initial stage of gelation phenomena, both investigated systems are present in the sol form. Charged chitosan molecules cause a repulsive force between particles [20]. Therefore, the aggregation is limited by repulsive barrier and systems reveal RLA mechanism. As the temperature increases, the motion of particles intensified due to thermal activation. This enables overcoming repulsive barrier. The simultaneous heating of systems reduces the charge of chitosan molecules what follows from neutralization of $-NH^{3+}$ groups [26]. When a certain critical point is reached, the phase transition occurs. Repulsive barrier is overcome by the attractive forces between chitosan molecules.

Comparison of the phase transition stages of both systems indicates differences between gelation mechanism. The chitosan-solvent systems show a rapid increase of storage modulus in a short period of time (Figure 4a–c). It can be assumed that this rapid increase of dynamic moduli results from a vanishing of the repulsive barrier after reaching a critical point. Weakness of repulsive forces lead to an aggregation limited by diffusion (DLA). The chitosan-solvent-buffer systems show a two-step phase transition. The dynamic moduli increase is smoother and occurs in a longer period of time compared with chitosan-solvent system. It can be assumed that in the first period (Region 1 from Figure 4d–f) of phase transition there is a still repulsive barrier between chitosan particles. Thus, the aggregation mechanism is defined by reaction limited aggregation (RLA). However, as time proceeded the repulsive barrier becomes negligible and the aggregation is limited by diffusion (DLA), which can be observed as a rapid increase of dynamic moduli (Region 2 from Figure 4d–f).

5. Conclusions

The conducted research revealed that there is a significant impact of the molecular weight on the crosslinked structure of chitosan in an acid solution. The solubility of chitosan depends on amount of H^+ ions bonding to amino groups [15,16]. In the case of polymers with short chains—low molecular weight—the protonation process inside molecule is inhibited due to the chitosan chains entangling into a stiff ball. Penetration of ions inside the ball is inhibited (sample 1 without β-NaGP). The longer the polymer chain (the higher the molecular weight), the looser is its entanglement. This is proved by both of the results of the conducted tests of the Zeta potential—indicating a decrease of the Zeta potential with an increase of molecular weight (Table 1) and the results of the rheological research—indicating a decrease of the flow behaviour index (transition from Newtonian to non-Newtonian properties) with an increase of molecular weight (Figure 2).

In the case of systems without β-NaGP, a solvation shell is formed around the polymer macromolecules which influences the increase of stability of the system (high values of the Zeta potential). Such a system is hydrophilic and initiation of the aggregation process is significantly inhibited by the solvation shell around macromolecules and existence of repulsive barrier. After exceeding the energy barrier, the phase transition occurs rapidly, which can be explained by the mechanism of diffusion limited aggregation (DLA).

In the case of systems with β-NaGP, two-step phase transition is observed in which the first step is reaction limited aggregation and the second step is diffusion limited aggregation. Occurrence of a first step of the phase transition is a result of buffer addition. The presence of the –GP^{2-} ions, which interact with chitosan amino groups, probably contributes to the formation of the barrier. This barrier needs to be first overcome to enable aggregation between chitosan molecules. After overcoming this barrier, the mechanism of gelation changes from RLA to DLA.

Acknowledgments: The authors would like to acknowledge the financial support from National Science Centre (Poland)—research grant (NCN UMO-2014/15/B/ST8/02512).

Author Contributions: Piotr Owczarz, Patryk Ziółkowski, Marek Dziubiński conceived and designed the experiments; Piotr Owczarz, Patryk Ziółkowski performed the experiments, analysed the data and wrote the paper; Zofia Modrzejewska supervised and directed the project; Sławomir Kuberski consulted on technical and theoretical issues; and all authors reviewed, edited and approved the manuscript.

Conflicts of Interest: The authors declare no conflict of interest.

References

1. Ravi Kumar, M.N.V. A review of chitin and chitosan applications. *React. Funct. Polym.* **2000**, *46*, 1–27. [CrossRef]
2. El-hefian, E.A. Rheological study of chitosan and its blends: An overview. *Maejo Int. J. Sci. Technol.* **2010**, *4*, 210–220.
3. Shigemasa, Y.; Minami, S. Applications of chitin and chitosan for biomaterials. *Biotechnol. Genet. Eng. Rev.* **1996**, *13*, 383–420. [CrossRef] [PubMed]

4.   Yi, H.; Wu, L.-Q.; Bentley, W.E.; Ghodssi, R.; Rubloff, G.W.; Culver, J.N.; Payne, G.F. Biofabrication with chitosan. *Biomacromolecules* **2005**, *6*, 2881–2894. [CrossRef] [PubMed]

5.   Ravichandran, R.; Sundarrajan, S.; Venugopal, J.R.; Mukherjee, S.; Ramakrishna, S. Advances in polymeric systems for tissue engineering and biomedical applications. *Macromol. Biosci.* **2012**, *12*, 286–311. [CrossRef] [PubMed]

6.   Li, Q.; Dunn, E.T.; Grandmaison, E.W.; Goosen, M.F.A. Applications and properties of chitosan. *J. Bioact. Compat. Polym.* **1992**, *7*, 370–397. [CrossRef]

7.   Kurita, K. Chemistry and application of chitin and chitosan. *Polym. Degrad. Stab.* **1998**, *59*, 117–120. [CrossRef]

8.   Noble, L.; Gray, A.I.; Sadiq, L.; Uchegbu, I.F. A non-covalently cross-linked chitosan based hydrogel. *Int. J. Pharm.* **1999**, *192*, 173–182. [CrossRef]

9.   Bhattarai, N.; Gunn, J.; Zhang, M. Chitosan-based hydrogels for controlled, localized drug delivery. *Adv. Drug Deliv. Rev.* **2010**, *62*, 83–99. [CrossRef] [PubMed]

10. Bernkop-Schnürch, A.; Dünnhaupt, S. Chitosan-based drug delivery systems. *Eur. J. Pharm. Biopharm. Off. J. Arbeitsgemeinsch. Pharm. Verfahrenstech. EV* **2012**, *81*, 463–469. [CrossRef] [PubMed]

11. Croisier, F.; Jérôme, C. Chitosan-based biomaterials for tissue engineering. *Eur. Polym. J.* **2013**, *49*, 780–792. [CrossRef]

12. Teimouri, A.; Azadi, M. Preparation and characterization of novel chitosan/nanodiopside/nanohydr oxyapatite composite scaffolds for tissue engineering applications. *Int. J. Polym. Mater. Polym. Biomater.* **2016**, *65*, 917–927. [CrossRef]

13. Yang, B.; Li, X.; Shi, S.; Kong, X.; Guo, G.; Huang, M.; Luo, F.; Wei, Y.; Zhao, X.; Qian, Z. Preparation and characterization of a novel chitosan scaffold. *Carbohydr. Polym.* **2010**, *80*, 860–865. [CrossRef]

14. Sogias, I.A.; Khutoryanskiy, V.V.; Williams, A.C. Exploring the Factors affecting the solubility of chitosan in water. *Macromol. Chem. Phys.* **2010**, *211*, 426–433. [CrossRef]

15. Park, J.-W.; Choi, K.-H.; Park, K.K. Acid-base equilibria and related properites of chitosan. *Bull. Korean Chem. Soc.* **1983**, *4*, 68–72.

16. Vachoud, L.; Zydowicz, N.; Domard, A. Formation and characterisation of a physical chitin gel. *Carbohydr. Res.* **1997**, *302*, 169–177. [CrossRef]

17. Cho, J.; Heuzey, M.-C.; Bégin, A.; Carreau, P.J. Viscoelastic properties of chitosan solutions: Effect of concentration and ionic strength. *J. Food Eng.* **2006**, *74*, 500–515. [CrossRef]

18. Cho, J.; Heuzey, M.-C.; Bégin, A.; Carreau, P.J. Physical gelation of chitosan in the presence of beta-glycerophosphate: The effect of temperature. *Biomacromolecules* **2005**, *6*, 3267–3275. [CrossRef] [PubMed]

19. Chenite, A.; Chaput, C.; Wang, D.; Combes, C.; Buschmann, M.D.; Hoemann, C.D.; Leroux, J.C.; Atkinson, B.L.; Binette, F.; Selmani, A. Novel injectable neutral solutions of chitosan form biodegradable gels in situ. *Biomaterials* **2000**, *21*, 2155–2161. [CrossRef]

20. Chenite, A.; Buschmann, M.; Wang, D.; Chaput, C.; Kandani, N. Rheological characterisation of thermogelling chitosan/glycerol-phosphate solutions. *Carbohydr. Polym.* **2001**, *46*, 39–47. [CrossRef]

21. Struszczyk, H. Chitin and Chitosan. Part I. Properties and production. *Polimery* **2002**, *47*, 316–325.

22. Kasaai, M.R. Calculation of Mark–Houwink–Sakurada (MHS) equation viscometric constants for chitosan in any solvent–temperature system using experimental reported viscometric constants data. *Carbohydr. Polym.* **2007**, *68*, 477–488. [CrossRef]

23. Chen, R.H.; Tsaih, M.L. Effect of temperature on the intrinsic viscosity and conformation of chitosans in dilute HCl solution. *Int. J. Biol. Macromol.* **1998**, *23*, 135–141. [CrossRef]

24. Christensen, B.E.; Vold, I.M.N.; Vårum, K.M. Chain stiffness and extension of chitosans and periodate oxidised chitosans studied by size-exclusion chromatography combined with light scattering and viscosity detectors. *Carbohydr. Polym.* **2008**, *74*, 559–565. [CrossRef]

25. Filion, D.; Lavertu, M.; Buschmann, M.D. Ionization and solubility of chitosan solutions related to thermosensitive chitosan/glycerol-phosphate systems. *Biomacromolecules* **2007**, *8*, 3224–3234. [CrossRef] [PubMed]

26. Lavertu, M.; Filion, D.; Buschmann, M.D. Heat-induced transfer of protons from chitosan to glycerol phosphate produces chitosan precipitation and gelation. *Biomacromolecules* **2008**, *9*, 640–650. [CrossRef] [PubMed]

27. Filion, D.; Buschmann, M.D. Chitosan–glycerol-phosphate (GP) gels release freely diffusible GP and possess titratable fixed charge. *Carbohydr. Polym.* **2013**, *98*, 813–819. [CrossRef] [PubMed]

28. Van Tomme, S.R.; Storm, G.; Hennink, W.E. In situ gelling hydrogels for pharmaceutical and biomedical applications. *Int. J. Pharm.* **2008**, *355*, 1–18. [CrossRef] [PubMed]

29. Ferguson, J.; Kemblowski, Z. *Applied Fluid Rheology*; Elsevier Applied Science: London, UK; New York, NY, USA, 1991; ISBN 978-1-85166-588-4.

30. Einstein, A. Über die von der molekularkinetischen Theorie der Wärme geforderte Bewegung von in ruhenden Flüssigkeiten suspendierten Teilchen. *Ann. Phys.* **1905**, *322*, 549–560. [CrossRef]

31. Montembault, A.; Viton, C.; Domard, A. Physico-chemical studies of the gelation of chitosan in a hydroalcoholic medium. *Biomaterials* **2005**, *26*, 933–943. [CrossRef] [PubMed]

32. Da Silva, J.A.; Gonçalves, M.P.; Rao, M.A. Kinetics and thermal behaviour of the structure formation process in HMP/sucrose gelation. *Int. J. Biol. Macromol.* **1995**, *17*, 25–32. [CrossRef]

33. Fu, J.-T.; Rao, M.A. Rheology and structure development during gelation of low-methoxyl pectin gels: The effect of sucrose. *Food Hydrocoll.* **2001**, *15*, 93–100. [CrossRef]

34. Ross-Murphy, S.B. The estimation of junction zone size from geltime measurements. *Carbohydr. Polym.* **1991**, *14*, 281–294. [CrossRef]

35. Cho, J.; Heuzey, M.-C.; Bégin, A.; Carreau, P.J. Chitosan and glycerophosphate concentration dependence of solution behaviour and gel point using small amplitude oscillatory rheometry. *Food Hydrocoll.* **2006**, *20*, 936–945. [CrossRef]

36. Toker, O.S.; Karasu, S.; Yilmaz, M.T.; Karaman, S. Three interval thixotropy test (3ITT) in food applications: A novel technique to determine structural regeneration of mayonnaise under different shear conditions. *Food Res. Int.* **2015**, *70*, 125–133. [CrossRef]

37. Butler, F.; McNulty, P. Time dependent rheological characterisation of buttermilk at 5 °C. *J. Food Eng.* **1995**, *25*, 569–580. [CrossRef]

38. Abu-Jdayil, B. Modelling the time-dependent rheological behavior of semisolid foodstuffs. *J. Food Eng.* **2003**, *57*, 97–102. [CrossRef]

39. Lin, M.Y.; Lindsay, H.M.; Weitz, D.A.; Ball, R.C.; Klein, R.; Meakin, P. Universality in colloid aggregation. *Nature* **1989**, *339*, 360–362. [CrossRef]

polymers

MDPI

Article

Sorption of Hg(II) and Pb(II) Ions on Chitosan-Iron(III) from Aqueous Solutions: Single and Binary Systems

Byron Lapo [1,2,*] , Hary Demey [3,4,*] , Jessenia Zapata [1], Cristhian Romero [1] and Ana María Sastre [4]

[1] School of Chemical Engineering, Universidad Técnica de Machala, UACQS, BIOeng, 070151 Machala, Ecuador; jmzapata_est@utmachala.edu.ec (J.Z.); caromeros_est@utmachala.edu.ec (C.R.)

[2] Department of Chemical Engineering, Universitat Politècnica de Catalunya, EPSEVG, Av. Víctor Balaguer, s/n, 08800 Vilanova i la Geltrú, Spain

[3] Commissariat à l'Energie Atomique et aux Energies Alternatives, CEA/DRT/LITEN/DTBH/LTB, 17 rue des Martyrs, 38054 Grenoble, France

[4] Department of Chemical Engineering, Universitat Politècnica de Catalunya, ETSEIB, Diagonal 647, 08028 Barcelona, Spain; ana.maria.sastre@upc.edu

* Correspondence: blapo@utmachala.edu.ec (B.L.); hary.demey@upc.edu (H.D.); Tel.: +59-3987949667 (B.L.); +34-938-937-778 (H.D.)

Received: 28 February 2018; Accepted: 23 March 2018; Published: 24 March 2018

Abstract: The present work describes the study of mercury Hg(II) and lead Pb(II) removal in single and binary component systems into easily prepared chitosan-iron(III) bio-composite beads. Scanning electron microscopy and energy-dispersive X-ray (SEM-EDX) analysis, Fourier transform infrared spectroscopy (FTIR), thermogravimetric analysis (TGA) and point of zero charge (pH_{pzc}) analysis were carried out. The experimental set covered pH study, single and competitive equilibrium, kinetics, chloride and sulfate effects as well as sorption–desorption cycles. In single systems, the Langmuir nonlinear model fitted the experimental data better than the Freundlich and Sips equations. The sorbent material has more affinity to Hg(II) rather than Pb(II) ions, the maximum sorption capacities were 1.8 mmol·g^{-1} and 0.56 mmol·g^{-1} for Hg(II) and Pb(II), respectively. The binary systems data were adjusted with competitive Langmuir isotherm model. The presence of sulfate ions in the multicomponent system [Hg(II)-Pb(II)] had a lesser impact on the sorption efficiency than did chloride ions, however, the presence of chloride ions improves the selectivity towards Hg(II) ions. The bio-based material showed good recovery performance of metal ions along three sorption–desorption cycles.

Keywords: binary; chitosan; desorption; iron; lead; mercury; salt effects; single; sorption competition

1. Introduction

Heavy metals are potentially toxic to humans; mercury and lead are two of the most harmful metals present in wastewater [1], some studies label these metals as relevant and very toxic elements [2,3]. Even nowadays, the presence of these metals, particularly in water resources is reported [4,5], hence it represents a potential risk to our ecosystems. Typical ways of being released into the environment are mining, smelters, coal burning, hydropower plants, agriculture system, etc. [6], particularly, these contaminants can reach the natural waters degrading its quality. Nevertheless, the scientific community is constantly developing removal methods, such as adsorption, membrane filtration, electrodialysis, ionic liquids [7,8], towards a suitable solution to this problem.

Sorption and particularly biosorption are some of the most promising methods to remove toxic metals. Particular focus is given to innovative adsorbents based on biomaterials. One of the major potential substances is the chitosan, a well-known biopolymer derived from alkaline *N*-deacetylation of chitin [9], which can surpass its natural limitations in aqueous media throughout targeted modifications, driving to enhance its adsorption properties as well as its selectivity for targeted ions and mechanical properties. Several chitosan-based materials have been reported to provide excellent sorption of Hg(II) and Pb(II) [10–13]. Although, many reported chitosan-composite materials have shown excellent performance in terms of sorption capacity, several limitations make the industrial scaling-up difficult. Some restraints are: Very low particle size, which can produce blockage in hydraulic systems, low mechanical stability that does not enable the use in columns, extended use of chemicals for chitosan matrix modification, not to mention in many cases the related toxicity of the chemical compounds used is overlooked.

Iron(III) represents an alternative to making this possible. It presents many advantages such as being abundant, cheap and easy to manage. It could also provide better mechanical properties to chitosan. However, the major studies are based on magnetic iron, which involves complex procedures and high energy consumption to achieve the desired magnetic properties. Nevertheless, few researchers have proposed chitosan-iron composite without further process in adsorbing metals and metalloids from water. Example of these are: As(III) and As(V) with Fe(III) immobilized on chitosan beads [14], and boron from seawater with chitosan iron(III) hydroxide beads [15]. However, two of the most toxic and highly relevant heavy metals, Hg(II) and Pb(II) have not been studied with this promising bio-based material.

On the other hand, the major of studies are targeted toward single element studies, neglecting the synergic effect of other ions in real water, which is definitively important for assessing industrial wastewater treatment. Focusing on developing a potential material with industrial insight, which means the convergence between cost, good sorption–desorption performance and being environmentally friendly, the present study targets a single, binary component, desorption, salt interference on the sorption of Hg(II) and Pb(II) using beads based on chitosan-iron(III) composite.

2. Materials and Methods

2.1. Chemicals

Chloride(II) nitrate (HgCl$_2$, 99%, (Probus, Barcelona, Spain), lead(II) nitrate (Pb(NO$_3$)$_2$; 98%, Panreac, Barcelona, Spain), chitosan (Aber Technologies, Lannilis, France, M$_W$ = 125,000 g·mol^{-1} was determined by gel permeation chromatography technique, and the degree of acetylation DA = 0.13 was obtained by Fourier transform infrared spectroscopy [16,17]), acetic acid (CH$_3$COOH, 99.7%, Panreac, Barcelona, Spain), hydrochloric acid (HCl, 37.4%, J.T. Baker, Phillipsburg, NJ, USA), sodium hydroxide (NaOH, 97%, Probus, Barcelona, Spain), Sodium Borohydride (NaBH$_4$, ≥96%, Sigma-Aldrich, St. Louis, MO, USA), nitric acid (HNO$_3$, 64.9%, Phillipsburg, NJ, USA), sodium sulfate (Na$_2$SO$_4$, 99.5%, Prolabo, Fontenay-sous-Bois CEDEX, France), sodium chloride (NaCl, 99.5%, Prolabo, Fontenay-sous-Bois CEDEX, France), Iron(III) chloride (FeCl$_3$·6H$_2$O, 99–102%, Fluka, Buchs, Switzerland, Thiourea (SC(NH$_2$)$_2$, 99%, Panreac, Barcelona, Spain), Ethylenediaminetetraacetic acid disodium salt dihydrate (EDTA, 99% Panreac, Barcelona, Spain) and deionized water type II laboratory water were used.

2.2. Preparation of Composite Beads

The composite beads were prepared with a slight improvement on our previous sorbent material reported by Demey et al. [15]. Neat chitosan was dissolved in acetic acid (1% *w/w*). Parallelly, a previously prepared solution of FeCl$_3$·6H$_2$O (30% *w/w*) was mixed with chitosan solution. Then, the mixture was homogenized for 2 h at 600 rpm. This solution was added drop-by-drop into a solution of NaOH 1 M through a thin nozzle (Ø 2.0 mm), assisted by a peristaltic pump. For the beads manufacturing, the mixture was first washed with high quantities of type II laboratory water to remove the excess of iron, and dried in a laboratory freeze drier (LyoQuest-55, Telstar equipment, São Paulo, Brazil) at 218 K and 0.05 mbar.

2.3. Characterization

Infrared spectrum was performed from 450 to 4000 cm^{-1} in a FTIR Thermo Scientific Nicolet 6700 (Madison, WI, USA); the samples were crushed and blended with potassium bromide (KBr) (2 mg of material in 100 mg of KBr) to make the pellets, prior to FTIR analysis. Thermogravimetric analysis was carried out with a TGA/SDTA 851e/LF/1100 thermobalance (Mettler Toledo, Mississauga, ON, Canada). Samples with mass of 6 mg were degraded between 30 and 800 °C at a heating rate of 10 °C·min^{-1} in N$_2$ atmosphere. The pH$_{pzc}$ was evaluated according to the methodology of Yazdani et al. [18] with a Bante 901 Benchtop pH meter instruments (Bante, Shanghai, China). The morphological observations, and energy dispersive X-ray (EDX) probe analysis of composite beads was done before and after metal sorption, using a Phenom XL SEM-EDX (PhenomWorld, Rotterdam, The Netherlands).

2.4. pH Study

The optimum pH to perform the sorption experiments was evaluated prior to the equilibrium, kinetics, salt effects and desorption studies. In 25 mL of separated solutions of 0.2 mmol·L^{-1} Hg(II), 0.2 mmol·L^{-1} Pb(II), and 0.15 mmol·L^{-1} of mixed [Hg(II)-Pb(II)] solutions with adjusted pH of 2.0, 3.0, 3.5, 4.0, 4.5, 5.0, 5.5 and 6.0 were added to around 20 mg of adsorbent material. After 48 h of agitation (180 rpm), the initial and the final pH values were recorded. Solutions of diluted HNO$_3$ and NaOH were used to conveniently adjust the pH. The analysis of Hg(II) was carried out using an atomic absorption spectrophotometer (AAS) Shimazdu AA6300 (Shimadzu Corporation, Kyoto, Japan), equipped with a hydride vapor generator, and the Pb(II) analysis was carried out in an Agilent Technologies 4100 microwave plasma atomic emission spectrometer (MP–AES) (Agilent Technologies, Melbourne, Australia). The sorption capacity (q_e) versus pH was also reported.

2.5. Equilibrium Study

Sorption experiments and data analysis were carried out on single and binary component systems separately. For the single system, in 25 mL of individual solutions of Hg(II) and Pb(II) of 0.2 mmol·L^{-1} were added 25 mg of ChiFer(III) and agitated for 48 h at 180 rpm in a laboratory orbital shaker. The pH of the solution was adjusted to 4.5, before mixing it with the sorbent material.

The adsorption capacity was calculated by the Equation (1). Furthermore, to investigate the better fitting of the equilibrium parameters, Langmuir [19], Freundlich [20] and Sips [21] models were evaluated according to non-linear Equations (2)–(4) respectively.

Sorption capacity equation:

$$q_e = \frac{V(C_o - C_e)}{w} \tag{1}$$

Langmuir equation:

$$q_e = \frac{q_{max}bC_e}{1 + bC_e} \tag{2}$$

Freundlich equation:

$$q_e = K_F C_e^{1/n} \tag{3}$$

Sips equation:

$$q_e = \frac{q_{ms}K_sC_e^{1/ms}}{1 + K_sC_e^{1/ms}} \tag{4}$$

where q_e is the amount of metal adsorbed in (mmol·g^{-1}), C_o and C_e are the initial and equilibrium concentrations respectively in (mmol·L^{-1}), q_{max} is the Langmuir maximum capacity in monolayer expressed in (mmol·g^{-1}), b is the Langmuir constant in (L·mmol^{-1}), K_F is the Freundlich constant, n is sorption intensity, q_{ms} the Sips maximum adsorption capacity (mmol·g^{-1}), K_s is the Sips equilibrium constant in (L·mmol^{-1}) and ms is the Sips model exponent.

For binary component systems, solutions of 25 mL of mixed Hg(II)-Pb(II) at equimolar concentrations of 0.15 mmol·L^{-1} with 25 mg of sorbent material, under the same operation conditions of reaction time and agitation speed (48 h and 180 rpm). In binary component systems, the sorption mechanism can be explained using multi-component models [22], based on Langmuir competitive isotherm (Equation (5)), and the corresponding Equations (6) and (7) were used:

$$q_{e,i} = \frac{K_i q_m C_{e,i}}{1 + \sum_{j=1}^{N} K_j C_{e,j}} \qquad (5)$$

For Hg(II) equation:

$$q_{e,Hg(II)} = \frac{K_{Hg(II)} q_m C_{e,Hg(II)}}{1 + K_{Hg(II)} C_{e,Hg(II)} + K_{Pb(II)} C_{e,Pb(II)}} \qquad (6)$$

For Pb(II) equation:

$$q_{e,Pb(II)} = \frac{K_{Pb(II)} q_m C_{e,Pb(II)}}{1 + K_{Pb(II)} C_{e,Pb(II)} + K_{Hg(II)} C_{e,Hg(II)}} \qquad (7)$$

where, K_1 and K_2 are the constants of the model (mmol·g^{-1})

2.6. Kinetics

The kinetics parameters were evaluated both in single and mixed solutions; 50 mg of ChiFer(III) were added to each 500 mL in 0.25 mmol·L^{-1} of Hg(II) and Pb(II) solutions (for single experiments), the agitation velocity was kept constant at 180 rpm. Pseudo-first order (PFORE) and pseudo second order (PSORE) models were assessed to fit the experimental data and to obtain the kinetics parameters according to Equations (5) and (6):

Pseudo-first order rate Equation (PFORE):

$$\frac{dq_t}{dt} = K_1(q_1 - q_t) \qquad (5)$$

Pseudo-second order rate Equation (PSORE):

$$\frac{dq_t}{(q_{eq} - q_t)^2} = K_2 dt \qquad (6)$$

where q_{eq} is the equilibrium sorption capacity (mmol·g^{-1}), q_t is the sorption capacity (mmol·g^{-1}) at any time t (h) and K_2 is the pseudo-second order rate constant (g·mg^{-1}·min^{-1}). The parameters q_{eq} and K_2 parameters are pseudo-constants.

2.7. Salt Effects

The effect on the sorption of Hg(II) and Pb(II) by the presence of sulfate and chloride salts was evaluated. The sulfate concentrations were chosen based on real concentrations of sulfate in wastewater found around gold mining zones (maximum found 223.68 mg·L^{-1} [23]). Twenty milligrams of sorbent material was added to 100 mL of 0.1 mmol·L^{-1} of Hg(II) and Pb(II) binary solutions previously charged with 0.001, 0.05, 0.1 and 0.2 mmol·L^{-1} of sodium sulfate and the initial pH was set at 4.5.

Moreover, a huge range of sodium chloride concentrations were evaluated in the sorption of a binary solution of Hg(II) and Pb(II); these evaluations resembled sodium chloride concentrations normally found in rivers, underground water and seawater. Four different concentrations of NaCl (1.0, 10.0, 100.0 and 500.0 mmol·L^{-1}) in 0.1 mmol·L^{-1} of Hg(II) and Pb(II); 20 mg of sorbent material in 100 mL of each solution at pH 4.5 were constantly agitated at 180 rpm for 48 h and the remaining concentrations of Hg(II) and Pb(II) were measured. In order to avoid metal precipitation in the mix solution, a theoretical diagram of chemical species was evaluated; all chemical species diagrams were created using Medusa free software (KTH Royal Institute of Technology, Stockholm, Sweden, version 2013) which are provided in the Supplementary Materials (Figures S1–S3).

2.8. Desorption Cycles

To know the possibility of reusing the composite material, sorption–desorption cycles were done in two stages; the first stage, to choose a proper eluent, and the second to assess the reusability across sorption/desorption cycles. In previous sorption/desorption cycles, seven eluents were assessed in one desorption cycle, HNO_3 (pH = 3.5), HCl (pH = 3.5), NaOH (pH = 11), NaOH (pH = 13), Thiourea 0.1 M (pH = 3.5), Thiourea 0.05 M (pH = 3.5) and EDTA 0.05 M (pH = 10). The binary Hg(II)-Pb(II) solutions were prepared at 0.15 mmol·L^{-1}, 25 mg of sorbent material were added to 25 mL of solution at pH = 4.5 for sorption. Then, three sorption–desorption cycles were performed with the selected eluents; the desorption efficiency was calculated according to Equation (7). All tests were made in duplicated.

$$\% \text{ desorption} = \frac{m_A - m_D}{m_A} \times 100 \tag{7}$$

where m_A and m_D are the sorbed and eluted mass of the metals (mg) at each sorption/desorption cycle.

3. Results and Discussion

3.1. Characterization

3.1.1. SEM-EDX Analysis

Figure 1 shows the morphology and elemental analysis of the composite material before and after contact with Hg(II) and Pb(II). In Figure 1a the sphericity and roughness of the material is clearly observed.

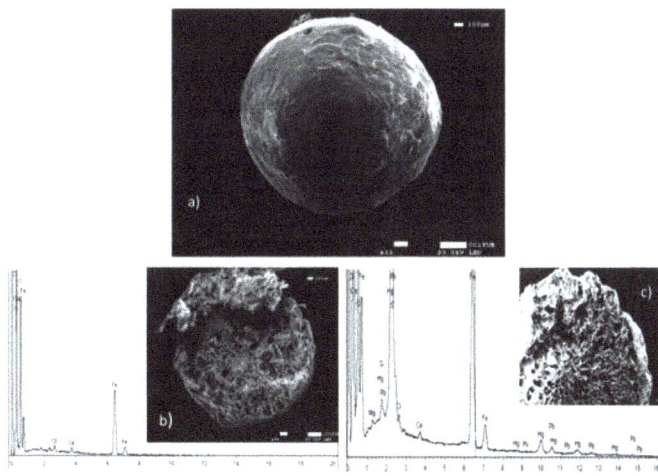

Figure 1. SEM-EDX images. (**a**) ChiFer(III) bead, (**b**) segmented/cross sectional view and EDX of ChiFer(III) bead, (**c**) segmented view and EDX of ChiFer(III) bead after sorption of Hg(II) and Pb(II).

In Figure 1b a longitudinal slit shows images of the inner porosity, with holes of around 50–100 μm, which can be compared with glutaraldehyde-cross-linked-chitosan-spheres reported by [24] in terms of porosity and size. In addition, in the same Figure 1b EDX analysis shows the presence of iron(III) in its structure. Furthermore, the addition of iron(III) does not affect the inner morphology of the beads. The beads are observed after sorption of Hg(II) and Pb(II) in Figure 1c. The porosity inside the composite did not change. The EDX analysis confirmed the presence of Hg(II) and Pb(II) as a result of the sorption process.

3.1.2. FTIR and TGA Analysis

Figure 2 shows the FTIR spectra (Figure 2a) and TGA analysis (Figure 2b) of neat chitosan (neat CS) and chitosan-iron(III) composite ChiFer(III). In FTIR both samples show a broad band at 3400–3600 cm^{-1} (hydroxyl groups) for neat CS and stretching vibrations and Fe–OH for ChiFer(III), C–H stretching vibration at 2925 cm^{-1} and O–H group of polysaccharides in 1639 cm^{-1} in neat CS [25,26], and slight change at 1644 cm^{-1} for ChiFer(III). As Sipos et al. [27] reports, well know bands of FeOOH are 1620 cm^{-1}, 1500 cm^{-1} and 1340 cm^{-1}, the first two peaks are overlapped in the ChiFer(III) specter, while the latter peak at 1384 cm^{-1} is well noted. The more notable difference in both spectra is the peak at 796 cm^{-1} in ChiFer(III), which is attributed to iron species reported by Ruan et al. [28]. The major iron species are overlapped in the main characteristic bands of neat chitosan, which could indicate the good cohesion of the composite. Also, FTIR analysis were performed before and after sorption (provided in Supplementary Materials Figure S4). In ChiFer(III) material the mean peak of amino and hydroxyl groups (3400 cm^{-1}) are overlapped by the presence of iron. Besides, after Hg(II) and Pb(II) sorption, the FTIR specter present the same behavior, it is to say that the metals were mainly bound by remaining amine groups, particularly at stretching vibrations of 3400 cm^{-1}, 1631 cm^{-1} (C = N bond), 1376 cm^{-1} and 1073 cm^{-1} peaks correspond to stretching vibration of C–OH. Moreover, in this study was not evidenced the typical peaks of 580 cm^{-1} and 759 cm^{-1} of Fe–O complex (indicative of metal-metal complexes).

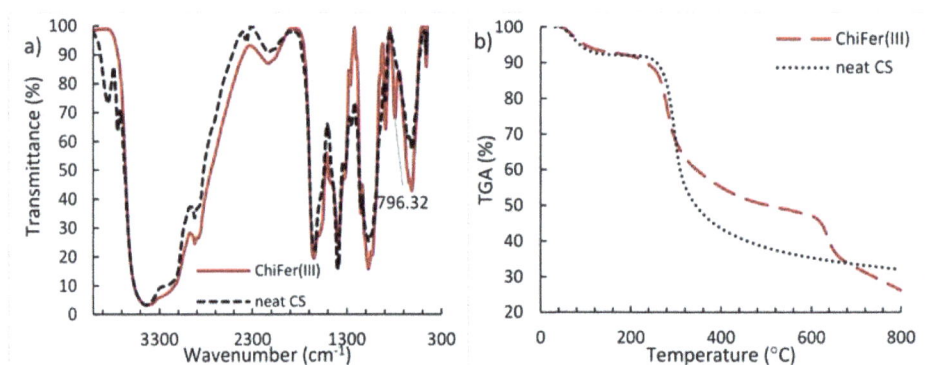

Figure 2. (a) FTIR and (b) TGA curves of neat CS and ChiFer(III).

Additionally, TGA analysis of neat CS and ChiFer(III) were conducted. The results of both samples are showed in Figure 2b; a first weight loss at 90 °C is produced due to the volatilization of low-molecular weight compounds, such as water [29]. The drop of the curves between 250–300 °C is related to the oxidation and degradation of chitosan, this is in concordance with Yu et al. [30]. Above 300 °C is observed a significant difference of loss weight for CS and ChiFer(III), it can be attributed to the thermal degradation of the polysaccharides chains of chitosan. According to Hong et al. [29], the kinetic decomposition follows the Ozawa–Flynn–Wall method [31].

It is noteworthy that the amount of iron into the composite corresponds to 33% (dried weight, d.w.), so the 67% is neat chitosan; evidently, for 6 mg samples ChiFer(III) material has less chitosan content than pure CS samples. Between 300–600 °C the weight loss of chitosan is higher than ChiFer(III), as expected (i.e., the weight loss for chitosan is 64% and for ChiFer(III) is 52%); it means that the incorporation of iron into the chitosan matrix provides high thermal stability to the resulting material (at temperatures between 300–600 °C). However, at 600 °C occurs a steep decomposition of ChiFer(III) which is attributed by Ziegler et al. [32] to the conversion of the inorganic core of iron(III). The iron species can be converted to magnetite, maghemite or wüstite Fe1-xO, consequently a weight loss is produced, which is accompanied with the evaporation of volatiles sub-products. This complex phenomenon depends on the structure and binding species in the composite. However, in sorption applications, such temperatures are not reached, moreover, TGA analysis in this study enhances the knowledge of the thermal stability of the materials.

3.2. pH Study

The pH plays a transcendental role when it is used in chitosan-based composites in metal sorption procedures [33]. Sorption experiments at different pHs were evaluated from pH = 2.0 to pH = 7.0 in single and binary solutions. It is noteworthy that at pH < 3.0 the ChiFer(III) composite is less stable (due to the chitosan hydrolysis and the consequent dissolution of the organic and inorganic content in the beads), while pH > 6 insoluble hydroxide precipitates would be formed (the images of the material stability at different pHs are provided in Supplementary Materials (Figure S5)). All these considerations were taken into account when performing the experiments and to avoid the precipitation phenomenon.

Figure 3 shows the adsorption capacity at various pHs. The best pH observed was between 4.5 and 5.0. This accords with other researchers [34–36] who tested chitosan-based composites. It is remarkable that Hg(II) sorption is greater than Pb(II) sorption in both single and binary component systems. In binary system, it is clearly shown the variations in pH, which do not surpass pH 7 (due to the buffer effect of chitosan).

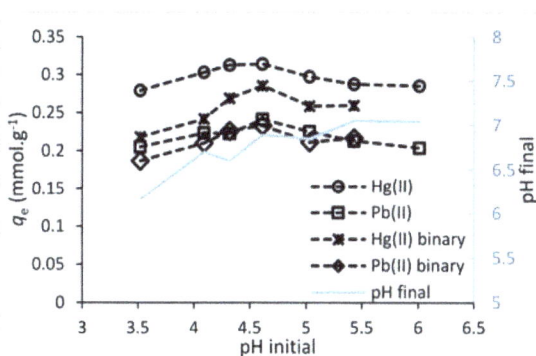

Figure 3. Adsorption capacity at different pHs. (*T*: 20 °C; sorbent dosage: 1 g·L^{-1}; agitation speed: 180 rpm; contact time: 48 h; $C_{0(single)}$: 0.2 mmol·L^{-1}; $C_{0(binary)}$: 0.1 mmol·L^{-1}).

However, the pH where the majority of surface sites are neutral, and the net charge on the surface is zero, is known as the point of zero charge (pH$_{pzc}$) [37], this value was evaluated and is presented in Figure 4. Zero net surface charge density does not imply the absence of any charges, but rather the presence of equal amounts of positive and negative charge [38]. In general, ligand exchange is favored at pH levels less than the pH$_{pzc}$ [39], as below pH$_{pzc}$ more sites are able to be protonated or, failing that, be able to be occupied by cations, depending on the predomination of electrostatic forces or chelating bonding [35]. The pH$_{pzc}$ of ChiFer(III) was recorded at 7.40 which is depicted in Figure 4, this is in agreement with findings in the literature [40]. The pH$_{pzc}$ of neat chitosan was reported as 7.1 and pH$_{pzc}$ of Fe(OH)$_3$ was 6.9 [38]. Consequently, the pH$_{pzc}$ value of ChiFer(III) corresponds somewhere between neat chitosan and its ferric form.

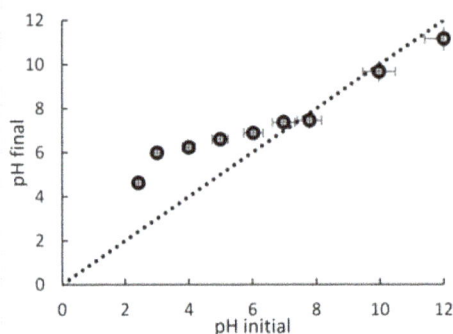

Figure 4. pH_{pzc} of ChiFer(III). (*T*: 20 °C; sorbent dosage: 1 g·L^{-1}; agitation speed: 180 rpm; contact time: 24 h; 0.01 M NaCl solution).

The optimal pH for Hg(II) and Pb(II) sorption, match with the pH values under pH_{pzc}. These accords with several authors who reported behaviour below pH_{pzc} for Hg(II) [41] and Pb(II). Moreover, the main variations in pH were observed between pH > 3.0 < pH_{pzc}, due to the protonation of the amino groups into the ChiFer(III) material. This means that in this pH range a competition between metals and protons for the active sites in the sorbent could be produced; which is confirmed by the "buffering effect" at pH 4–6 (Figure 4). Consequently, the sorbate/sorbent interactions to these systems are mainly based on chelation bonding of metals with nitrogen atoms and hydroxides iron species, and to a lesser extent on electrostatic interactions. The optimum operational pH was found as pH 4.5. Henceforth, the experiments were performed at this initial pH.

3.3. Equilibrium

The correlation of data by theoretical Equations is fundamental for the engineering design and scaling-up of sorption systems. Several models are proposed in the literature for fitting the experimental data, it includes Langmuir, Freundlich and Sips Equations. This fitting does not mean that the principles of the models are verified, but it could improve the interpretation of the sorption mechanisms [42]. The impact of metal concentration on sorption uptake is demonstrated by a progressive increase until a saturation plateau is reached. Figures 5 and 6 show the equilibrium data for Hg(II) and Pb(II) for single and binary component systems respectively, from aqueous solutions at initial pH of 4.5 and 20 °C. The obtained parameters of the models for single component are summarized in Table 1.

In single component, Sips model shows the best fitting in terms of r^2, however it presents a large standard error in the K_s parameter. Likewise, the Langmuir model gets good $r^2 > 0.98$, but much less standard (Std.) error values than the Sips model. Therefore, Langmuir model is taken as reference for the interface analysis. Nonetheless, the characteristic asymptotic shape of the isotherm is consistent with the Langmuir Equation. The maximum sorption capacity (q_{max}) of Hg(II) and Pb(II) were 1.80 and 0.56 mmol·g^{-1}, respectively. Thus, the sorption capacity for mercury ions is three times higher than that for lead ions.

In terms of sorption capacity performance, ChiFer(III) is competitive compared with other chitosan composites; e.g., Dhanapal et al. [43] tested acryloylated chitosan, 2-acrylamido-2-methyl-1-propansulfonic acid, 2-(diethylamino) ethylmethacrylate and *N,N'*-methylene bisacrylamide as a crosslinker (ACAD), obtaining a sorption removal of 2.26 mmolHg(II)·g^{-1}. Similarly, Zhang et al. [44] manufactured the cobalt ferrite/chitosan grafted with graphene composite (MCGS) material and the sorption capacity obtained was 0.67 mmolHg(II)·g^{-1}.

In the case of Pb(II), chitosan/magnetite [45], and thiolated chitosan [12] materials were tested, and the sorption uptake was in the order of 0.30–0.53 mmolPb(II)·g^{-1}. Table 2 shows additional studies regarding to chitosan-based composites for single sorption of Hg(II) and Pb(II). It is noteworthy that the ChiFer(III) material is configured in the form of beads, and this could contribute to the scale-up for future industrial manufacturing.

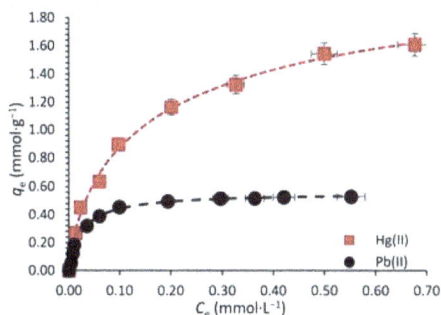

Figure 5. Single component sorption isotherms. (*T*: 20 °C; sorbent dosage: 1 g·L^{-1}; agitation speed: 180 rpm; contact time: 48 h).

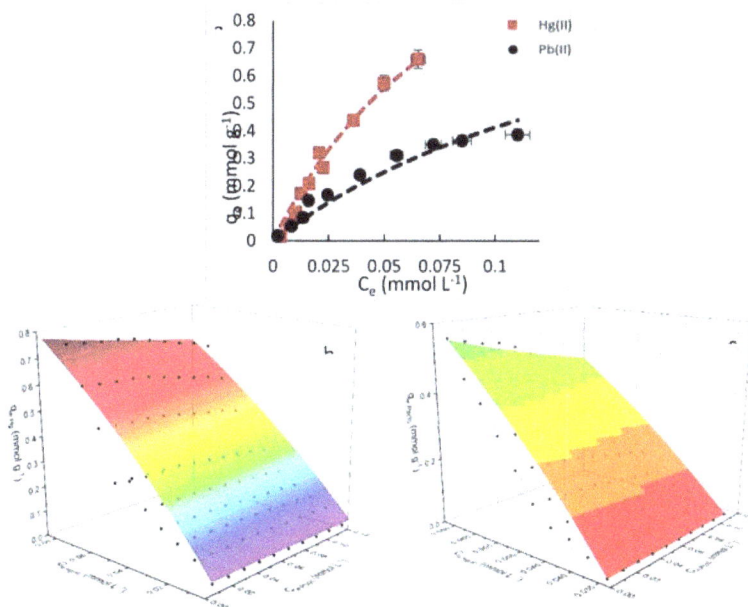

Figure 6. (**a**) Binary component sorption isotherms, (**b**) 3D surface for Hg(II) sorption capacity response vs. Hg(II) and Pb(II) interactions, and (**c**) 3D surface for Pb(II) sorption capacity response vs. Hg(II) and Pb(II) interactions. (*T*: 20 °C; sorbent dosage: 1 g·L^{-1}; agitation speed: 180 rpm; contact time: 48 h, pH$_o$ = 4.5, C$_o$ = 0.15 mmol·L^{-1}).

Table 1. Single and binary component isotherm constants of ChiFer(III) material.

			Single Component System				
				Hg(II)		**Pb(II)**	
	Parameter	Unit		Value	Error	Value	Error
Langmuir	q_{exp}	$(mmol \cdot g^{-1})$		1.61		0.52	
	q_{max}	$(mmol \cdot g^{-1})$		1.80	0.06	0.56	0.007
	b	$(L \cdot mmol^{-1})$		10.17	1.31	38.8	2.49
	r^2			0.988		0.996	
	$q_{max} \times b$	$(L \cdot g^{-1})$		18.3		21.44	
Freundlich	K_F	$(mmol^{1-1/n} \cdot g^{-1} \cdot L^{1/n})$		1.97	0.07	0.69	0.05
	n			2.63	0.19	3.66	0.54
	r^2			0.982		0.91	
Sips	q_{max}	$(mmol \cdot g^{-1})$		2.3	0.26	0.55	0.01
	K_s	$(L \cdot mmol^{-1})$		3.14	1.18	52.41	15.04
	n_s			1.4	0.15	0.93	0.05
	r^2			0.995		0.998	
			Binary Component System				
				Value	Error		
Langmuir competitive model	K_m	$(mmol \cdot g^{-1})$		2.87	0.57		
	K_1	$(mmol \cdot g^{-1})$		5.68	1.60		
	K_2	$(mmol \cdot g^{-1})$		2.24	0.63		
	r^2			0.96			

Regarding to the sorption affinity, it is clearly seen that Hg(II) ions are more sorbed onto ChiFer(III) material than Pb(II) ions: Hg(II) > Pb(II). This could be explained by the ionic radii differences between Hg(II) and Pb(II), which is 1.02 and 1.19 respectively. Therefore, Hg(II) ions can enter into the material pores easier than Pb(II) ions. However, more than one mechanism is presented to explicate in deep how the metals are bonded onto the material. This affinity accords with that reported by Zhu et al. [46], who carried out sorption experiments with Hg(II) and Pb(II) onto chitosan with thiourea groups where the sorbent had more affinity towards Hg(II) than to Pb(II) ions. Several experiments with the raw chitosan particles were carried out in single system (with 0.2 mmol·L^{-1} solutions of Hg(II) and Pb(II)) for comparing the sorption efficiency of ChiFer(III) beads. Figure S6 shows that the sorption capacity for Hg(II) is almost similar for both materials (i.e., 1.02 mmol·g^{-1} for chitosan and 0.90 mmol·g^{-1} for ChiFer(III)); the performance of chitosan particles is slightly higher than ChiFer(III); this can be attributed to the small particle size of chitosan (<0.5 mm), which in comparison with ChiFer(III) beads (2 mm), has a greater impact on the resistance to the film diffusion; making the active sites of chitosan easier accessible for mercury ions. In addition, ChiFer(III) material is more efficient for lead removal than chitosan at the same operation conditions (i.e., 0.2 mmol·g^{-1} for chitosan and 0.45 mmol·g^{-1} for ChiFer(III)); it means that the introduction of hydroxyl groups of iron(III) hydroxide improves the sorption uptake of metal ions and could improve the stability of the resulting beads [17].

Table 2. Adsorption capacity of Hg(II) and Pb(II) with different chitosan-based materials.

Modification	Metal	pH	T (°C)	q_{max} $(mmol \cdot M^+ \cdot g^{-1})$	Isotherm fitting	Ref.
Microspheres chitosan grafted with chlorosulfonic acid (CSSULF) or ethylenimine (CSPEI)	Hg(II)	6	-	0.32	Langmuir/Freundlich	[35]
Cross-linked aminated chitosan beads	Hg(II)	7		2.23	Langmuir	[33]
Chitosan/Graphene oxide imprinted Pb^{2+}	Pb(II)	5	30	0.38	Langmuir	[47]
Polyaniline grafted cross-linked chitosan beads	Pb(II)		45	0.55	Langmuir	[48]
Present study: ChiFer(III)	Hg(II) Pb(II)	4.5	Room	1.80 0.56	Langmuir	

Many published results in the literature are based in single systems, however the evaluations of the binary components had more interest for future applications in real effluents [49]. In this present study, the equilibrium analysis was done under the concept that one binding site was only available for one sorbate, supported by the competitive Langmuir isotherm model, according to Equation (5) and combination of Equations (6) and (7), which were simultaneously solved using Origin 9.0 software (OriginLab Inc., Northampton, MA, USA, 2012). Many authors have taken into account the simple models and have omitted the competitive effect on secondary species. Thus, recent publications reported the use of the competitive equilibrium Equations [50–52].

The constant values of the bi-component model were $K_{Hg(II)} = 5.68$, $K_{Pb(II)} = 2.24$ and $q_m = 2.87$ as shown in Table 1. Furthermore, the determination coefficient was acceptably fitted at $r^2 = 0.95$. Figure 6 illustrates the isotherms related to the competition of Hg(II) and Pb(II) at $pH_o = 4.5$ and the same initial concentrations of 0.15 mmol·L^{-1}. It is noted that the material has the same trend as in single systems. In other words, the affinity stays stronger in Hg(II) ions rather than in Pb(II) ions. To better illustrate of the two metal ions on the sorption capacity of each metal ion, 3D surfaces (as seen in Figure 6b), which shows a more marked decrease as the concentration of Pb(II) ions increase.

According to Mohan et al. [49], the effect of ionic interactions on sorption may be represented by the ratio of the sorption capacity for one metal ion in the presence of another metal ion (Q^{mix}), to the sorption capacity for the same metal when it is present (Q^0). Although this relation was used with $Q^0 = q_{max}$ (for this nomenclature) and calculated by models. In this work we used the same relation for an equilibrium concentration at a single point in the isotherm q_e instead of q_{max}, as it is not possible to reach the isotherm plateau due to the imminent precipitation at concentrations over 0.15 mmol·L^{-1} at pH = 5.5. It is also possible to note the ratio differences at different equilibrium concentrations. Thus, the effect of ionic interactions by each metal is calculated by the sorption capacity ratio (scr), scr = $q_{emix,i}/q_{e,i}$; so, for Hg(II) the scr$_{Hg(II)}$ = $q_{emix,Hg}/q_{e,Hg}$ and scr$_{Pb(II)0}$ = $q_{emix,Pb}/q_{e,Pb}$. Figure 7 show the scr behaviour in both metals at different C_e concentrations. It is noted that difference between the ratio is not marked in Hg(II) ions as scr changes, which means that although Pb(II) ions are in the solution, these ions do not suppress the sorption of Hg(II) ions. On the other hand, Pb(II) ions capacity decreases as long as the C_e increases, which means that there is a suppression of Pb(II) ions is achieved towards the isotherm plateau is achieved. It also indicates preference for Hg(II) ions in this Hg(II)-Pb(II) system.

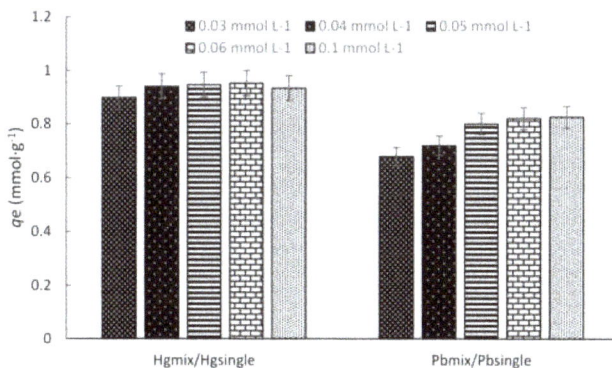

Figure 7. Sorption capacity ratio (scr) changes. (*T*: 20 °C; sorbent dosage: 1 g·L^{-1}; agitation speed: 180 rpm; contact time: 48 h, pH_o = 4.5, C_o = 0.03–0.1 mmol·L^{-1}).

3.4. Kinetics Studies

The sorption kinetics studies are important for determining the required contact time of the sorbate/sorbent systems. In industrial applications the time for achieving the maximum saturation plateau is a very relevant parameter for design of reactors [53]. Figure 8a,b show the kinetic profiles for lead and mercury from aqueous solutions. It is noteworthy that the kinetic uptakes are controlled by different mechanisms including: (i) bulk diffusion; (ii) external diffusion (so-called film diffusion); (iii) intraparticle diffusion; and (iv) reaction rate. Demey et al. [17] reported that maintaining a continuous agitation speed of 150–200 rpm (and a sorbent dosage of 1 g·L^{-1}) is enough to avoid the settling of the sorbent and neglecting the contribution to the bulk diffusion.

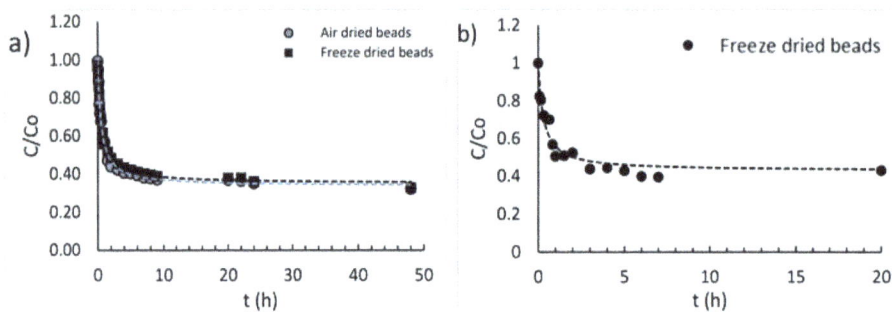

Figure 8. Kinetics profiles: (**a**) Hg(II), (**b**) Pb(II). (*T*: 20 °C; sorbent dosage: 1 g·L^{-1}; agitation speed: 180 rpm; contact time: 48 h; C_0: 0.2 mmol·L^{-1}; dashed line: fitting with pseudo-second order rate equation).

The kinetics studies of mercury (Figure 8) were performed with two types of drying beads configurations (i.e., freeze-dried beads, which is the standard material synthesized in this work, and the air dried beads). A comparison of the drying method was performed to evaluate the accessibility of the metal species into the porous network of the sorbent. Both curves shown in Figure 8 are overlapped, confirming not significant influence of the drying techniques on the metal uptake. These surprising results are in contradiction with those obtained recently by Demey et al. [17], in which ChiFer(III) material for neodymium recovery from aqueous solutions, and the accessibility of the metal into the active sites was affected according to the drying method and, as a consequence, the equilibrium time was impacted.

In this work, important differences between the drying techniques were not found, probably due to several causes: (i) the structure of the polymeric matrix remained relatively open after air drying and it did not affect the accessibility into the pores (this was not completely evidenced in the SEM images of the composite); (ii) the covalent radius of neodymium is bigger than that of the mercury ions, and these latter ions had not difficulties in transporting into the entrance of the pores (i.e., the covalent radius of mercury is 132 pm and the covalent radius of neodymium is 201 pm). Rorrer et al. [54] concluded that a pore blockage may occur at low metal concentrations. The migration of sorbed species is low, which in conjunction with the larger size of the neodymium ions, results in an accumulation at the entrance of the pore. Therefore, the low-cost air drying technique does not have strong influence in the sorption of mercury. This opens the door to manufacturing the materials on a higher scale so as to conduct future evaluations with real industrial effluents.

Regardless of the metal, (mercury or lead), the curves in Figure 8a,b are characterized by three progressive pseudo-steps: (i) an initial step that takes around 50–60 min; (ii) a second step that takes around 4–5 h, and (iii) a third step that takes around 60 min. The differences of each step are related to the gradient mass that is progressively reduced as a function of the contact time. Table 3 reports the comparison of the experimental sorption capacities at equilibrium with calculated values for both pseudo-first order (PFORE) and the pseudo-second order rate equations (PSORE). The correlations confirm much better fitting with PSORE.

Table 3. Kinetic parameters of sorbent materials.

	Experimental			PFORE			PSORE		
Metal	Sorbent	q_{exp} (mmol·g^{-1})	K_1 (h^{-1})	q_1 (mmol·g^{-1})	r^2	K_2 (g·mmol^{-1}·h^{-1})	q_2 (mmol·g^{-1})	r^2
Hg(II)	Air dried beads (AD)	1.71	1.44	1.57	0.983	1.18	1.69	0.991
	Freeze dried beads (FD)	1.70	1.36	1.53	0.977	1.12	1.65	0.995
Pb(II)	Freeze dried beads (FD)	0.64	2.09	0.63	0.930	4.94	0.679	0.945

3.5. Salt Effects

Figure 9a,b show the sorption behaviour when sodium sulfate and sodium chloride are added in binary Hg(II)-Pb(II) systems. In the case of sulfate ions, the material sorption capacity slightly decreases as the sulfate concentration increases. This effect is more pronounced in Hg(II) than in Pb(II) ions. However, it is more noticeable that even at high sulfate concentration (0.2 mmol·L^{-1}), the sorption capacity decreases at around 35% and 15% for Hg(II) and Pb(II) respectively. This means that the material sorption capacity was lightly affected by the sulfate ions, which in turn is beneficial for the treatment of wastewater.

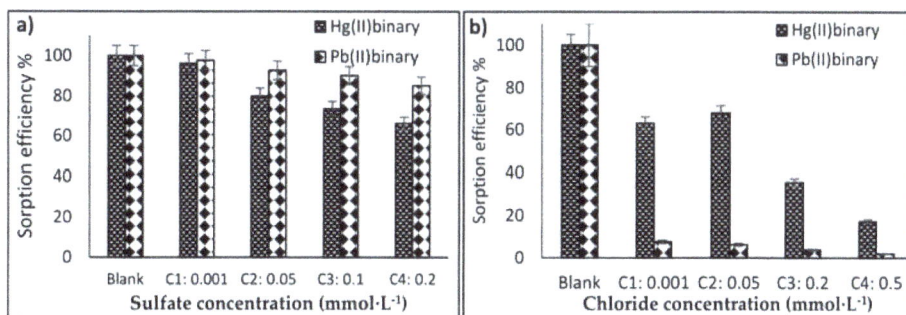

Figure 9. Salt effects: (**a**) Sulfate effect, (**b**) Chloride effect. (*T*: 20 °C; sorbent dosage: 0.2 g·L^{-1}; agitation speed: 180 rpm; contact time: 48 h; C_0: 0.1 mmol·L^{-1}).

The chloride effect in sorption efficiency of the material was then assessed, covering weak and strong chloride concentrations (0.001 M similar to (s.t.) potable water, 0.05 M s.t. river water, 0.2 M s.t. underground water and 0.5 M s.t. seawater). In all cases, sodium chloride suppresses the sorption capacity, even at low chloride concentrations. The sorption of Pb(II) was affected more strongly than the Hg(II) ones, around 97–98% and 40–83% for Pb(II) and Hg(II) respectively. This can be explained as follows: the formation of complexes with chloride ions is easier than sulfate ions [55]; i.e., chloride may form complexes with Pb(II) and Hg(II). According to Kinniburgh et al. [56] who studied the adsorption of Hg(II) on Fe gel, found that in the presence of chloride the adsorption of Hg(II) is considerably reduced. However, the biggest suppressing affect was on the Pb(II) sorption, which could indicate that the chloride-complex is more easily formed with Pb(II) rather than Hg(II). On the other hand, this factor could impact on the selectivity of the material, which in the presence of chloride, the Hg(II) is successfully adsorbed but the Pb(II) sorption has been substantially reduced (Figure 9b).

Comparing the effects of sulfates and chlorides on the sorption of Hg(II) and Pb(II), it is noted that chlorides suppress the adsorption much more than sulfates. This is in agreement with Mitani et al. [57] who found that sulfates were adsorbed in greater quantities than chlorides when this act as counter ions with metals over chitosan-based gel. That is to say that sulfates have less effect on sorption than chlorides with chitosan. In this work we found the affinity over our iron/chitosan composite as: Hg(II) > Pb(II).The possibility of regeneration, as well as the recover.

3.6. Desorption

The possibility of regeneration, as well as the recovery of metals from the loaded sorbent, is an important parameter for evaluating the feasibility of the sorption processes. In the first stage, seven eluents in one sorption/desorption cycle were tested, the results of which are illustrated in Figure 10a. The eluents were: HNO_3 (pH = 3.5), HCl (pH = 3.5), NaOH (pH = 11), NaOH (pH = 13), Thiourea 0.1 M (pH = 3.5), Thiourea 0.05 M (pH = 3.5) and EDTA 0.05 M (pH = 10). Eluents HNO_3, HCl, NaOH and thiourea 0.05 M shows recoveries below 60% and 50% for Hg(II) ions and Pb(II) ions respectively. HCl at pH = 3.5 represent a good eluent, this is in concordance with the Section 3.5 of this article, which describes the salt effects, the impact of protons was verified, as a result of a higher concentration of H^+, a strong competition for the active sites is produced, and this effect is taken in advantage for desorption procedures. Moreover, thiourea 0.1 M present 80% of recovery for Hg(II), but 32% of elution recovery for Pb(II) ions in the first sorption/desorption cycle, but after this, thiourea directly impacts on the stability of the sorbent and the beads become mechanically fragile; thus, their original brown color turns to green. It means that thiourea acts as a reducing agent and consequently iron(III) is reduced to iron(II); it is in agreement with the findings of Zhu, 1992 [58] who reported that the redox interactions of ferric ions and thiourea follows a first order reaction.

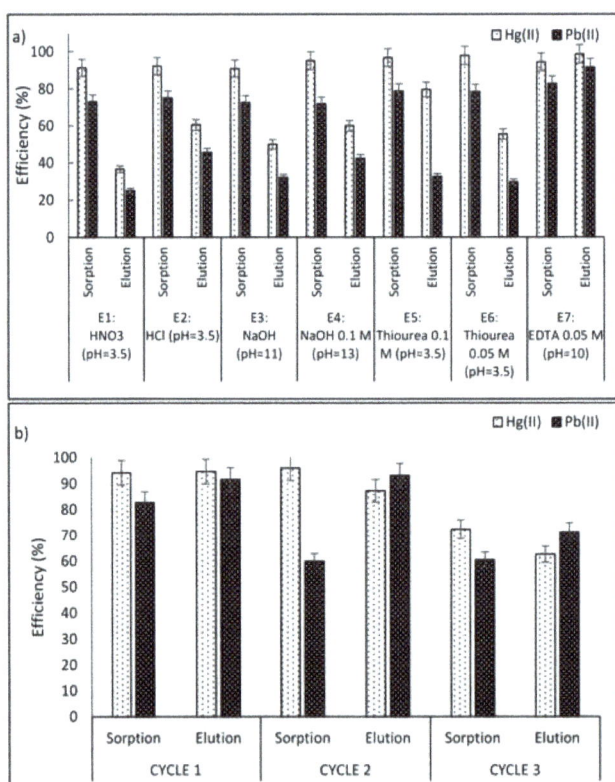

Figure 10. Desorption: (**a**) Eluent selection, (**b**) Cycles with the selected eluent: EDTA solution pH 10. (*T*: 20 °C; sorbent dosage: 1 g·L^{-1}; agitation speed: 180 rpm; contact time: 24 h; C_0: 0.15 mmol·L^{-1}).

Polymers **2018**, *10*, 367

Nevertheless, EDTA 0.05 M set at pH 10 achieved recoveries of 98% for Hg(II) and 91% for Pb(II), so, in order to assess the performance of reusability three sorption/desorption cycles were carried out with EDTA 0.05 M, which is illustrated in Figure 10b.

Along three sorption/desorption cycles, alkaline EDTA shows excellent recovery for both metals, with average elution efficiencies of 81.43% for (Hg(II)) and 81.43% for (Pb(II)), it was up to 87% for the first two cycles, followed a lightly drop in the third cycle with recoveries of 62.54% and 71.23% for Hg(II) and Pb(II) respectively. Mass of the sorbed and eluted metals of the three sorption/desorption cycles are provided in Figure S7 of Supplementary Materials section. The stability of the beads after the third cycle was good, mainly due to the alkaline media, in which the hydrogel is more stable. Desorption studies reported by [59–61], who used EDTA for desorption of heavy metals showed comparable results with the obtained in this study, EDTA at alkaline medium is deprotonated, which in conjunction with its chelating properties, represents an excellent desorbent for heavy metals strongly bonded onto the sorbent matrix.

4. Conclusions

ChiFer(III) bio-based sorbent showed good sorption capacity towards the removal of Hg(II) and Pb(II) ions in single and binary systems. There was remarkable affinity for Hg(II) even in strong chloride conditions. Besides, the stability and the performance of the material was maintained during all three sorption–desorption cycles. Langmuir and competitive Langmuir models fitted the equilibrium in single and binary component systems, respectively. The maximum sorption capacities were 1.8 mmol·g^{-1} and 0.56 mmol·g^{-1} for Hg(II) and Pb(II), respectively. Pseudo second order rate equation adjusted accurately to kinetics data. The main advantage of using this material is the very simple manufacturing procedure and cheap cost, added to the high sorption capacity compared with existing similar bio-materials.

Supplementary Materials: The following are available online at http://www.mdpi.com/2073-4360/10/4/367/s1, Figure S1: Species diagram system of Hg(II), Figure S2: Species diagram system $SO_4{}^{2-}$, Hg(II), Pb(II), Figure S3: Species diagram system Cl^-, Hg(II), Pb(II), Figure S4. FTIR spectres before and after metals sorption, Figure S5: Stability of sorbent material at different pHs, Figure S6: Metal species removal by neat chitosan and ChiFer(III) as sorbents and Figure S7: Sorbent and eluted mass in sorption/desorption cycles.

Acknowledgments: This work was supported by the Spanish Ministry of Economy and Competitiveness, MINECO (Project CTM2017-83581-R). The authors would like to thank Trifon Trifonov, Montserrate Dominguez, Agustín Fortuny, Xavier Ramis and Gema Gonzalez for their technical assistance. Byron Lapo would like to thank to Universidad Técnica de Machala for financial support of his research stay at Universitat Politècnica de Catalunya, under approved research project HCU resolution 539/2017.

Author Contributions: Byron Lapo, Hary Demey and Ana María Sastre conceived and designed the experiments; Byron Lapo, Jessenia Zapata and Cristhian Romero performed the tests and collected the experimental data; the manuscript was written by Byron Lapo and Hary Demey. The revision of the results interpretations was performed by Hary Demey and Ana María Sastre.

Conflicts of Interest: The authors declare no conflict of interest.

References

1. Morais, S.; Costa, F.G.; Lourdes Pereira, M. Heavy Metals and Human Health. In *Environmental Health—Emerging Issues and Practice*; InTech: London, UK, 2012; pp. 227–244.

2. Agency for Toxic Substances and Disease Registry. *Toxicological Profile for Mercury*; ATSDR: Atlanta, GA, USA, 1999.

3. Agency for Toxic Substances and Disease Registry. *Toxicological Profile for Lead*; ATSDR: Atlanta, GA, USA, 2007.

4. Marín, A.; Gonzalez, V.; Lapo, B.; Molina, E.; Lemus, M. Mercury levels in sediments from the coast of El Oro-Ecuador. *Gayana* **2016**, *80*, 147–153. [CrossRef]

5. Safruk, A.M.; McGregor, E.; Whitfield Aslund, M.L.; Cheung, P.H.; Pinsent, C.; Jackson, B.J.; Hair, A.T.; Lee, M.; Sigal, E.A. The influence of lead content in drinking water, household dust, soil, and paint on blood lead levels of children in Flin Flon, Manitoba and Creighton, Saskatchewan. *Sci. Total Environ.* **2017**, *593–594*, 202–210. [CrossRef] [PubMed]

6. Pandey, R.; Dwivedi, M.K.; Singh, P.K.; Patel, B.; Pandey, S.; Patel, B.; Patel, A.; Singh, B. Effluences of Heavy Metals, Way of Exposure and Bio-toxic Impacts: An Update. *J. Chem. Chem. Sci.* **2016**, *66*, 2319–7625.

7. Borras Atienzar, C.; Lapo Calderón, B.; Fernandéz Martínez, L. *Electroquímica en el Tratamiento de Efluentes (Electrochemical in Wastewater Treatment)*, 1st ed.; Universidad Técnica de Machala: Machala, Ecuador, 2015; ISBN 978-9978-316-96-2.

8. Barakat, M.A. New trends in removing heavy metals from industrial wastewater. *Arab. J. Chem.* **2011**, *4*, 361–377. [CrossRef]

9. Zhang, L.; Zeng, Y.; Cheng, Z. Removal of heavy metal ions using chitosan and modified chitosan: A review. *J. Mol. Liq.* **2016**, *214*, 175–191. [CrossRef]

10. Li, N.; Bai, R.; Liu, C. Enhanced and Selective Adsorption of Mercury Ions on Chitosan Beads Grafted with Polyacrylamide via Surface-Initiated Atom Transfer Radical Polymerization. *Langmuir* **2005**, *21*, 11780–11787. [CrossRef] [PubMed]

11. Shekhawat, A.; Kahu, S.; Saravanan, D.; Jugade, R. Removal of Cd(II) and Hg(II) from effluents by ionic solid impregnated chitosan. *Int. J. Biol. Macromol.* **2017**, *104*, 1556–1568. [CrossRef] [PubMed]

12. Kong Yong, S.; Bolan, N.; Lombi, E.; Skinner, W. Enhanced Zn(II) and Pb(II) removal from wastewater using thiolated chitosan beads (ETB). *Malays. J. Anal. Sci.* **2015**, *19*, 586–594.

13. Liu, B.; Chen, W.; Peng, X.; Cao, Q.; Wang, Q.; Wang, D.; Meng, X.; Yu, G. Biosorption of lead from aqueous solutions by ion-imprinted tetraethylenepentamine modified chitosan beads. *Int. J. Biol. Macromol.* **2016**, *86*, 562–569. [CrossRef] [PubMed]

14. Marques Neto, J.D.O.; Bellato, C.R.; Milagres, J.L.; Pessoa, K.D.; Alvarenga, E.S. Preparation and evaluation of chitosan beads immobilized with Iron(III) for the removal of As(III) and As(V) from water. *J. Braz. Chem. Soc.* **2013**, *24*, 121–132. [CrossRef]

15. Demey, H.; Vincent, T.; Ruiz, M.; Nogueras, M.; Sastre, A.M.; Guibal, E. Boron recovery from seawater with a new low-cost adsorbent material. *Chem. Eng. J.* **2014**, *254*, 463–471. [CrossRef]

16. Ruiz, M.; Sastre, A.M.; Zikan, M.C.; Guibal, E. Palladium sorption on glutaraldehyde-crosslinked chitosan in fixed-bed systems. *J. Appl. Polym. Sci.* **2001**, *81*, 153–165. [CrossRef]

17. Demey, H.; Lapo, B.; Ruiz, M.; Fortuny, A.; Marchand, M.; Sastre, A. Neodymium Recovery by Chitosan/Iron(III) Hydroxide [ChiFer(III)] Sorbent Material: Batch and Column Systems. *Polymers (Basel)* **2018**, *10*, 204. [CrossRef]

18. Yazdani, M.R.; Virolainen, E.; Conley, K.; Vahala, R. Chitosan-Zinc(II) complexes as a bio-sorbent for the adsorptive abatement of phosphate: Mechanism of complexation and assessment of adsorption performance. *Polymers (Basel)* **2017**, *10*, 25. [CrossRef]

19. Jin, L.; Bai, R. Mechanisms of lead adsorption on chitosan/PVA hydrogel beads. *Langmuir* **2002**, *18*, 9765–9770. [CrossRef]

20. Freundlich, H.M.F. Over the Adsorption in Solution. *J. Phys. Chem.* **1906**, *5*, 385–471.

21. Sips, R. On the Structure of a Catalyst Surface. *J. Phys. Chem.* **1948**, *16*, 490–495. [CrossRef]

22. Allen, S.J.; McKay, G.; Porter, J.F. Adsorption isotherm models for basic dye adsorption by peat in single and binary component systems. *J. Colloid Interface Sci.* **2004**, *280*, 322–333. [CrossRef] [PubMed]

23. Valverde Armas, P.; Galarza Romero, B. Caracterización geoquímica e isotópica del agua superficial y subterránea en el área de influencia del río Siete y de las actividades mineras en el distrito minero de Ponce Enriquez, Escuela Superior Politécnica del Litoral. *Rev. Derecho Priv.* **2013**, *62*, 465–490.

24. Vieira, R.S.; Beppu, M.M. Dynamic and static adsorption and desorption of Hg(II) ions on chitosan membranes and spheres. *Water Res.* **2006**, *40*, 1726–1734. [CrossRef] [PubMed]

25. Shigemasa, Y.; Matsuura, H.; Sashiwa, H.; Saimoto, H. Evaluation of different absorbance ratios from infrared spectroscopy for analyzing the degree of deacetylation in chitin. *Int. J. Biol. Macromol.* **1996**, *18*, 237–242. [CrossRef]

26. Dimzon, I.K.D.; Knepper, T.P. Degree of deacetylation of chitosan by infrared spectroscopy and partial least squares. *Int. J. Biol. Macromol.* **2015**, *72*, 939–945. [CrossRef] [PubMed]

27. Sipos, P.; Berkesi, O.; Tombacz, E.; St Pierre, T.G.; Webb, J.; Kalman Burger, P. Formation of spherical iron(III) oxyhydroxide nanoparticles sterically stabilized by chitosan in aqueous solutions. *J. Inorg. Biochem.* **2003**, *95*, 55–63. [CrossRef]

28. Ruan, H.D.; Frost, R.L.; Kloprogge, J.T.; Duong, L. Infrared spectroscopy of goethite dehydroxylation: III. FT-IR microscopy of in situ study of the thermal transformation of goethite to hematite. *Spectrochim. Acta Part A Mol. Biomol. Spectrosc.* **2002**, *58*, 967–981. [CrossRef]

29. Hong, P.-Z.; Li, S.-D.; Ou, C.-Y.; Li, C.-P.; Yang, L.; Zhang, C.-H. Thermogravimetric Analysis of Chitosan. *J. Appl. Electrochem.* **2007**, 449–456. [CrossRef]

30. Yu, Z.; Zhang, X.; Huang, Y. Magnetic Chitosan–Iron(III) Hydrogel as a Fast and Reusable Adsorbent for Chromium(VI) Removal. *Ind. Eng. Chem. Res.* **2013**, *52*, 11956–11966. [CrossRef]

31. Lyon, R.E. An integral method of nonisothermal kinetic analysis. *Thermochim. Acta* **1997**, *297*, 117–124. [CrossRef]

32. Ziegler-Borowska, M.; Chełminiak, D.; Kaczmarek, H.; Kaczmarek-Kędziera, A. Effect of side substituents on thermal stability of the modified chitosan and its nanocomposites with magnetite. *J. Therm. Anal. Calorim.* **2016**, *124*, 1267–1280. [CrossRef]

33. Jeon, C.; Oll, W.H.H. Chemical modification of chitosan and equilibrium study for mercury ion removal. *Water Res.* **2003**, *37*, 4770–4780. [CrossRef]

34. Zhou, L.; Wang, Y.; Liu, Z.; Huang, Q. Characteristics of equilibrium, kinetics studies for adsorption of Hg(II), Cu(II), and Ni(II) ions by thiourea-modified magnetic chitosan microspheres. *J. Hazard. Mater.* **2009**, *161*, 995–1002. [CrossRef] [PubMed]

35. Kyzas, G.Z.; Kostoglou, M. Swelling-adsorption interactions during mercury and nickel ions removal by chitosan derivatives. *Sep. Purif. Technol.* **2015**, *149*, 92–102. [CrossRef]

36. Shawky, H.A.; El-Aassar, A.H.M.; Abo-Zeid, D.E. Chitosan/carbon nanotube composite beads: Preparation, characterization, and cost evaluation for mercury removal from wastewater of some industrial cities in Egypt. *J. Appl. Polym. Sci.* **2012**, *125*, E93–E101. [CrossRef]

37. Franks, G.V.; Meagher, L. The isoelectric points of sapphire crystals and alpha-alumina powder. *Colloids Surf. Physicochem. Eng. Asp.* **2003**, *214*, 99–110. [CrossRef]

38. Kosmulski, M. *Surface Charging and Points of Zero Charge*; CRC Press: Boca Raton, FL, USA, 2009; ISBN 9781420051896.

39. Krešić, N. *Hydrogeology and Groundwater Modeling*; CRC Press: Boca Raton, FL, USA, 2007; ISBN 9780849333484.

40. Kamari, A.; Wan Saime, W.N.; Lai Ken, L. Chitosan and chemically modified chitosan beads for acid dyes sorption. *J. Environ. Sci.* **2009**, *21*, 296–302. [CrossRef]

41. Sharma, R.; Singh, N.; Tiwari, S.; Tiwari, S.K.; Dhakate, S.R. Cerium functionalized PVA–chitosan composite nanofibers for effective remediation of ultra-low concentrations of Hg(II) in water. *RSC Adv.* **2015**, *5*, 16622–16630. [CrossRef]

42. Demey, H.; Vincent, T.; Ruiz, M.; Sastre, A.M.; Guibal, E. Development of a new chitosan/Ni(OH)$_2$-based sorbent for boron removal. *Chem. Eng. J.* **2014**, *244*, 576–586. [CrossRef]

43. Dhanapal, V.; Subramanian, K. Modified chitosan for the collection of reactive blue 4, arsenic and mercury from aqueous media. *Carbohydr. Polym.* **2015**, *117*, 123–132. [CrossRef] [PubMed]

44. Zhang, Y.; Yan, T.; Yan, L.; Guo, X.; Cui, L.; Wei, Q.; Du, B. Preparation of novel cobalt ferrite/chitosan grafted with graphene composite as effective adsorbents for mercury ions. *J. Mol. Liq.* **2014**, *198*, 381–387. [CrossRef]

45. Tran, H.V.; Tran, L.D.; Nguyen, T.N. Preparation of chitosan/magnetite composite beads and their application for removal of Pb(II) and Ni(II) from aqueous solution. *Mater. Sci. Eng. C* **2010**, *30*, 304–310. [CrossRef]

46. Zhu, Y.; Zheng, Y.; Wang, W.; Wang, A. Highly efficient adsorption of Hg(II) and Pb(II) onto chitosan-based granular adsorbent containing thiourea groups. *J. Water Process Eng.* **2015**, *7*, 218–226. [CrossRef]

47. Wang, Y.; Li, L.; Luo, C.; Wang, X.; Duan, H. Removal of Pb^{2+} from water environment using a novel magnetic chitosan/graphene oxide imprinted Pb^{2+}. *Int. J. Biol. Macromol.* **2016**, *86*, 505–511. [CrossRef] [PubMed]

48. Igberase, E.; Osifo, P. Equilibrium, kinetic, thermodynamic and desorption studies of cadmium and lead by polyaniline grafted cross-linked chitosan beads from aqueous solution. *J. Ind. Eng. Chem.* **2015**, *26*, 340–347. [CrossRef]

49. Mohan, D.; Pittman, C.U.; Steele, P.H. Single, binary and multi-component adsorption of copper and cadmium from aqueous solutions on Kraft lignin-a biosorbent. *J. Colloid Interface Sci.* **2006**, *297*, 489–504. [CrossRef] [PubMed]

50. Wang, S.; Vincent, T.; Faur, C.; Guibal, E. Modeling competitive sorption of lead and copper ions onto alginate and greenly prepared algal-based beads. *Bioresour. Technol.* **2017**, *231*, 26–35. [CrossRef] [PubMed]

51. Medellin-Castillo, N.A.; Padilla-Ortega, E.; Regules-Martínez, M.C.; Leyva-Ramos, R.; Ocampo-Pérez, R.; Carranza-Alvarez, C. Single and competitive adsorption of Cd(II) and Pb(II) ions from aqueous solutions onto industrial chili seeds (*Capsicum annuum*) waste. *Sustain. Environ. Res.* **2017**, *27*, 61–69. [CrossRef]

52. Padilla-Ortega, E.; Leyva-Ramos, R.; Flores-Cano, J.V. Binary adsorption of heavy metals from aqueous solution onto natural clays. *Chem. Eng. J.* **2013**, *225*, 536–546. [CrossRef]

53. Demey, H.; Vincent, T.; Guibal, E. A novel algal-based sorbent for heavy metal removal. *Chem. Eng. J.* **2018**, *332*, 582–595. [CrossRef]

54. Rorrer, G.L.; Hsien, T.Y.; Way, J.D. Synthesis of Porous-Magnetic Chitosan Beads for Removal of Cadmium Ions from Wastewater. *Ind. Eng. Chem. Res.* **1993**, *32*, 2170–2178. [CrossRef]

55. Kawamura, Y.; Mitsuhashi, M.; Tanibe, H.; Yoshida, H. Adsorption of Metal Ions on Polyaminated Highly Porous Chitosan Chelating Resin. *Znd. Eng. Chem. Res.* **1993**, *32*, 386–391. [CrossRef]

56. Kinniburgh, D.G.; Jackson, M.L. Adsorption of Mercury(II) by Iron Hydrous Oxide Gel. *Soil Sci. Soc. Am. J.* **1978**, *42*, 45. [CrossRef]

57. Mitani, T.; Fukumuro, N.; Yoshimoto, C.; Ishii, H.; Inoue, K.; Baba, Y.; Yoshizawa, K.; Noguchi, H.; Yoshizaki, M.; Lett, C. Effects of Counter Ions (Sulfate and Chloride) on the Adsorption of Copper and Nickel Ions by Swollen Chitosan Beads. *Chem. Biol. Technol. Agric.* **1991**, *55*, 2419.

58. Zhu, T. The redox reaction between thiourea and ferric iron and catalysis of sulphide ores. *Hydrometallurgy* **1992**, *28*, 381–397. [CrossRef]

59. Monier, M.; Abdel-Latif, D.A. Preparation of cross-linked magnetic chitosan-phenylthiourea resin for adsorption of Hg(II), Cd(II) and Zn(II) ions from aqueous solutions. *J. Hazard. Mater.* **2012**, *209–210*, 240–249. [CrossRef] [PubMed]

60. Wan Ngah, W.; Endud, C.; Mayanar, R. Removal of copper(II) ions from aqueous solution onto chitosan and cross-linked chitosan beads. *React. Funct. Polym.* **2002**, *50*, 181–190. [CrossRef]

61. Badruddoza, Z.; Zakir, Z.; Tay, J.; Hidajat, K.; Uddin, M.S. Fe$_3$O$_4$/cyclodextrin polymer nanocomposites for selective heavy metals removal from industrial wastewater. *Carbohydr. Polym.* **2013**, *91*, 322–332. [CrossRef] [PubMed]

polymers

MDPI

Article

Chitosan–Zinc(II) Complexes as a Bio-Sorbent for the Adsorptive Abatement of Phosphate: Mechanism of Complexation and Assessment of Adsorption Performance

Maryam Roza Yazdani [1,*], Elina Virolainen [1], Kevin Conley [2] and Riku Vahala [1]

[1] Water and Wastewater Engineering Research Group, School of Engineering, Aalto University, P.O. Box 15200, FI-00076 Aalto, Finland; elina.virolainen@gmail.com (E.V.); riku.vahala@aalto.fi (R.V.)

[2] COMP Centre of Excellence, Department of Applied Physics, School of Science, Aalto University, FI-00076 Aalto, Finland; kevin.conley@aalto.fi

* Correspondence: roza.yazdani@aalto.fi; Tel.: +358-4074-2014

Received: 21 October 2017; Accepted: 22 December 2017; Published: 25 December 2017

Abstract: This study examines zinc(II)–chitosan complexes as a bio-sorbent for phosphate removal from aqueous solutions. The bio-sorbent is prepared and is characterized via Fourier Transform Infrared Spectroscopy (FT-IR), Scanning Electron Microscopy (SEM), and Point of Zero Charge (pH_{PZC})–drift method. The adsorption capacity of zinc(II)–chitosan bio-sorbent is compared with those of chitosan and ZnO–chitosan and nano-ZnO–chitosan composites. The effect of operational parameters including pH, temperature, and competing ions are explored via adsorption batch mode. A rapid phosphate uptake is observed within the first three hours of contact time. Phosphate removal by zinc(II)–chitosan is favored when the surface charge of bio-sorbent is positive/or neutral e.g., within the pH range inferior or around its pH_{PZC}, 7. Phosphate abatement is enhanced with decreasing temperature. The study of background ions indicates a minor effect of chloride, whereas nitrate and sulfate show competing effect with phosphate for the adsorptive sites. The adsorption kinetics is best described with the pseudo-second-order model. Sips ($R^2 > 0.96$) and Freundlich ($R^2 \geq 0.95$) models suit the adsorption isotherm. The phosphate reaction with zinc(II)–chitosan is exothermic, favorable and spontaneous. The complexation of zinc(II) and chitosan along with the corresponding mechanisms of phosphate removal are presented. This study indicates the introduction of zinc(II) ions into chitosan improves its performance towards phosphate uptake from 1.45 to 6.55 mg/g and provides fundamental information for developing bio-based materials for water remediation.

Keywords: zinc–chitosan complexes; characterization; bio-sorbent; phosphate; adsorption; mechanism; thermodynamic

1. Introduction

Phosphate is an important element for many natural organisms, yet in high concentrations, it can cause serious eutrophication in natural waters [1–5]. In eutrophic waters, the exceeding amount of nutrients leads to the excessive growth of plants and algae. This phenomenon reduces the dissolved oxygen in the water, which disturbs the natural balance of organisms, and causes, e.g., mass fish death [1,6]. Eutrophication raises the costs of water treatment, weakens the recreational use of waters and enables the growth of algal blooms that produce harmful cyanotoxins [7]. Phosphorus dissolves into natural waters from weathered rocks, peat land and forests, and it gets into the communal wastewaters through agriculture, human settlement and industry [7,8]. Miettinen et al. [8] have studied the connection of phosphorus and bacterial growth in drinking water sources and showed that in both surface and ground waters the addition of phosphorus strongly contributes to the growth of heterotrophic bacteria.

While the maximum phosphate concentration set by US Environmental Protection Agency (EPA) is 0.05 mg/L [9], even a concentration of 0.02 mg/L can cause major eutrophication [6]. Preserving the aquatic life from phosphate contamination requires new phosphate removal techniques [2,10]. The commonly practiced methods for phosphate abatement include chemical precipitation and biological removal, but they are problematic in different ways. These treatment processes usually fail to meet the standard levels set for phosphate or even to decrease it to below 10 mg/L [11]. In addition, chemical precipitation is a relatively expensive method, requiring the storage and transportation of chemical reagents and producing considerable sludge waste. Biological phosphorus removal on the other hand is relatively sensitive to water conditions, which decreases it reliability. Compared to these techniques, adsorption technology offers a simple and low cost option [7]. Recently, engineered adsorbents have attracted a great deal of attention as alternatives for common yet expensive adsorbents, e.g., activated carbon. This new type of adsorbents usually comprises two or more constituents, one of which acts as a support matrix [12]. For instance, impregnated polymers with metal oxides [3] and biomass modified with mesoporous materials [6] have previously been studied as phosphate adsorbents.

Chitosan (CTS) is a biopolymer emerging in the adsorption process and is derived from chitin, the second most plentiful natural polymer after cellulose [13–15]. It has many advantages as a bio-sorbent, such as good adsorption capacity, biodegradability, and biocompatibility [15–19]. As it is extracted from crustacean waste, it is eco-friendly and cost-effective compared to commonly employed adsorbents. Despite its good adsorption capacity for a wide range of pollutants, CTS provides a low affinity towards oxyanions mainly because of the pK_a value of its electron-donor functional sites, viz. $-NH_2$ [12]. To overcome this obstacle, the introduction of metal ions to CTS has recently been practiced [15]. Metal ions, e.g., zinc(II) [20] and copper(II) [12], are capable of forming complexes with the functional groups on the CTS chain, and consequently the complexed metal ions on the CTS structure can link with other ligands including oxyanions. Among different metal ions, zinc (Zn) can easily chelate with CTS, which has made it the focus of many studies on its application for metal ions separation [20], waste management [21], antibacterial, and medical aspects [20]. However, to date, there is limited information on Zn(II)–CTS complexes as adsorptive media for the removal of oxyanions like phosphate. Along with the research conducted on the antimicrobial activity, thermal degradation and pyrolysis characteristics of CTS complexation with zinc(II) [20,21], a study on its application as a bio-sorbent provides deeper knowledge of the complexation and enable a better performance in the application stage towards specific goals, e.g., phosphate removal. Thus, a prospect of using Zn(II)–CTS bio-sorbent for phosphate removal was conceptualized in this study.

Here, we develop Zn(II)–CTS bio-sorbents for the removal of phosphate and study their response to adsorbent dose, phosphate concentration, solution pH, water temperature, contact time and competing ions (Cl^-, NO_3^-, SO_4^{2-}) effects. The bio-sorbent dosage is optimized and compared with those of plain CTS and the composites produced from both zinc oxide (ZnO) and nano-sized zinc oxide (nano-ZnO) embedded in CTS to indicate its better performance towards phosphate abatement. The composition, morphology and interface of the Zn(II)–CTS are characterized by Fourier transform infrared spectroscopy (FT-IR), scanning electron microscopy (SEM) and determination of the pH of Point of Zero Charge (pH_{PZC}).

2. Experimental

2.1. Materials

Chitosan was supplied by Acros Organics of Thermo Fisher Scientific Inc., Geel, Belgium. The degree of deacetylation (DDA) of chitosan was $84 \pm 1\%$ determined via acid–base titration [22,23] and its average molecular weight ($\overline{M_W}$) was 153.3 kD measured via determination of intrinsic viscosity as described previously [24,25]. Zinc oxide (>99%) and zinc chloride (>98%) were purchased from Merck chemicals (Darmstadt, Germany) and zinc oxide nano-powder (<50 nm particle size (BET), >97%) was supplied by Sigma Aldrich, Darmstadt, Germany. A stock solution with the concentration of 1000 mg/L was made via weighing an accurate amount of potassium dihydrogen phosphate (KH_2PO_4) and dissolving it in reverse osmosis water. Different dilutions in the range of 1–15 mg/L were prepared daily before each adsorption set.

2.2. Analysis Methods and Instruments

The pH of phosphate solutions was measured according to the SFS-EN ISO 10523 (dated 2012) with WTW inoLab pH 720-meter and probe Sentix 81 Plus. The pH meter was calibrated before each use. Phosphate concentration was measured according to the SFS-EN ISO 15681-1 (dated 2005) via flow analysis (FIA) and spectrometric detection using tin chloride method on a FOSAS Tecator, FIAstar 5000 Analyzer and Sampler 5027.

2.3. Adsorbent Preparation and Characterization

This study investigated three different bio-sorbents developed from CTS and zinc compounds; Zn(II)–CTS, ZnO–CTS and nano-ZnO–CTS. The Zn(II)–CTS bio-sorbent was prepared as follows: dissolved CTS (1 g) into 0.1 M acetic acid (100 mL) was agitated for 24 h at room temperature at a speed of 180 rpm to achieve a thoroughly homogenous solution. Then 0.34 g of zinc chloride dissolved into 50 mL of reverse osmosis water was slowly added into the CTS solution while continuously stirring until a homogenous solution was achieved. This solution was heated to 80 °C for 1 h with continuous stirring. After cooling to room temperature, the solution was pumped into 0.5 M NaOH solution (50 mL per 250 mL NaOH) using a syringe pump to develop the bio-sorbent in the form of beads. The product was immersed in the NaOH solution overnight. The final solid product was separated, rinsed with reverse osmosis water to reach neutral pH, and dried in an oven at 25 °C. ZnO–CTS was prepared in a similar way. The suspension of ZnO (0.2 g) in 0.1 M acetic acid (100 mL) was agitated for 24 h at room temperature and at the speed of 180 rpm before mixing with the CTS. The ZnO suspension was mixed with CTS solution and agitated for another 24 h at room temperature. The final ZnO–CTS was pumped into NaOH solution. Nano-ZnO–CTS was prepared as ZnO–CTS, but the pH of nano-ZnO/acetic acid suspension was set to 4. The mass of zinc–compounds, $ZnCl_2$ and ZnO, was determined in a way that the bio-sorbent products contained an equal amount of Zn.

The FT–IR analysis of the developed bio-sorbent was performed on a Thermo scientific Nicolet iS50 FT-IR spectrometer with a PIKE Gladi-ATR. SEM was performed with ZEISS Sigma VP (Jena, Germany) using 2 kV acceleration voltage, detecting secondary electrons. Samples were attached to an aluminum stub using carbon tape, and sputter coated with platinum using Emitech K100X for 90 s and 30 mA coating current to prevent charging effects. The pH drift method was employed to determine the pH_{PZC} of the Zn(II)–CTS surface using 20 mL of 0.1 M NaCl in a series of solutions for which pH was adjusted within the range of 3 to 12. After the initial pH of NaCl solutions was adjusted using NaOH and HCl, 0.01 g of the bio-sorbent was added to each of them and their final pH was measured after 24 h. The pH_{PZC} was noted at the pH where the final pH equals the initial pH [26].

2.4. Adsorption Experiments

All batch experiments were conducted on a shaker at the speed of 180 rpm using 50 mL phosphate solutions and known amount of bio-sorbent. The pH of phosphate solution was set to 4 unless otherwise mentioned. After the required contact time, the solutions were filtrated with Sartorius Minisart 45 μm filters and analyzed for final phosphate concentration. All experiments excluding the isotherm tests were performed at room temperature for 24 h. The optimum adsorbent dose was determined by conducting the experiment with five different doses in range of 0.1–2 g/L. The pH effect was studied with adjusting the pH at different values between 4 and 12 using HCl and NaOH. The effect of contact time was studied at different time intervals between 1 min to 48 h. Phosphate solutions with initial concentration from 1 to 15 mg/L were used to study adsorption isotherm at 20, 25, and 30 °C. NO_3^-, SO_4^{2-} or Cl^- was added to investigate the effect of competing ions on phosphate adsorption. All the adsorption experiments were carried out in two or three replicates when high standard deviation values were noticed for the removal percentage. The average of the replicates and the standard deviation values are reported accordingly. Phosphate removal and adsorption capacity were calculated with the following equations:

$$Removal\% = \frac{(C_0 - C_t)}{C_0} \times 100 \tag{1}$$

$$q_t \ (mg/g) = \frac{C_0 - C_t}{m} \times V \tag{2}$$

where C_0 (mg/L) is the initial phosphate concentration, C_t (mg/L) the phosphate concentration at time t, V (L) the solution volume, and m (g) the adsorbent mass.

3. Results and Discussion

3.1. Characterization of Zn(II)–CTS Bio-Sorbent

3.1.1. FT-IR Analysis

The FT–IR spectra of Zn(II)–CTS complexes and plain CTS were analyzed to determine the compositional differences as seen in Figure 1. The broad characteristic peak in the region 3100–3600 cm^{-1}, related to the stretching vibration of –NH$_2$ and –OH groups of CTS, are shifted to lower wavenumbers on the spectrum of Zn(II)–CTS. The O–H stretching band at 3740 cm^{-1} on the spectrum of CTS has moved to 3760 cm^{-1} with a lower intensity on the spectrum of Zn(II)–CTS. The peak at 1640 cm^{-1}, related to the –NH$_2$ bending vibration of CTS, is shifted to higher frequencies (1660 cm^{-1}) for Zn(II)–CTS indicating the –NH$_2$ and –OH groups on the CTS backbone have complexed with Zn(II). The band at 1090 cm^{-1} representing the secondary –OH of CTS is moved to 1080 cm^{-1} along with a higher intensity on the spectrum of Zn(II)–CTS. The band shift from 1090 cm^{-1} to 1080 cm^{-1} is characteristic of the coordination of –OH with Zn [20]. The peaks at 2920, 2880, 1600, 1380, 1080 and 620 cm^{-1} are assigned to methylene C–H$_2$ and methyl C–H stretching vibrations, N–H groups, C–H asymmetric bending, C–O alcohol stretching and O–H bending (out-of-plane), respectively. The vibrational bands at 1780 and 898 cm^{-1} correspond to the C=O band and glucopyranose ring of CTS, respectively. The peak at 620 cm^{-1}, assigned to the hydroxyl groups of CTS, has moved to 650 cm^{-1} on the spectrum of Zn(II)–CTS. Moreover, the appearance of peaks around 450 and 570 cm^{-1} on the spectrum of Zn(II)–CTS are characteristics of the stretching vibrations of Zn–O and Zn–N [20,27].

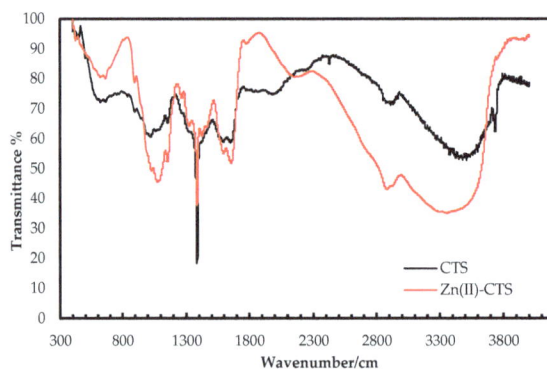

Figure 1. Fourier transform infrared (FT-IR) spectra of CTS and Zn(II)–CTS.

3.1.2. SEM Analysis

The surface morphology of Zn(II)–CTS was studied via scanning electron microscopy. Figure 2 indicates that the developed Zn(II)–CTS bio-sorbent shows an irregular and rough surface including micro-pores and small fractures, into which the oxyanions can penetrate and better access the internal functional adsorptive sites. Porosity enables the second phase adsorption according to the intra-particle diffusion theory, which is further discussed in kinetics modelling. The bio-sorbent particles have an irregular bead-like shape (granular) as seen in Figure 2.

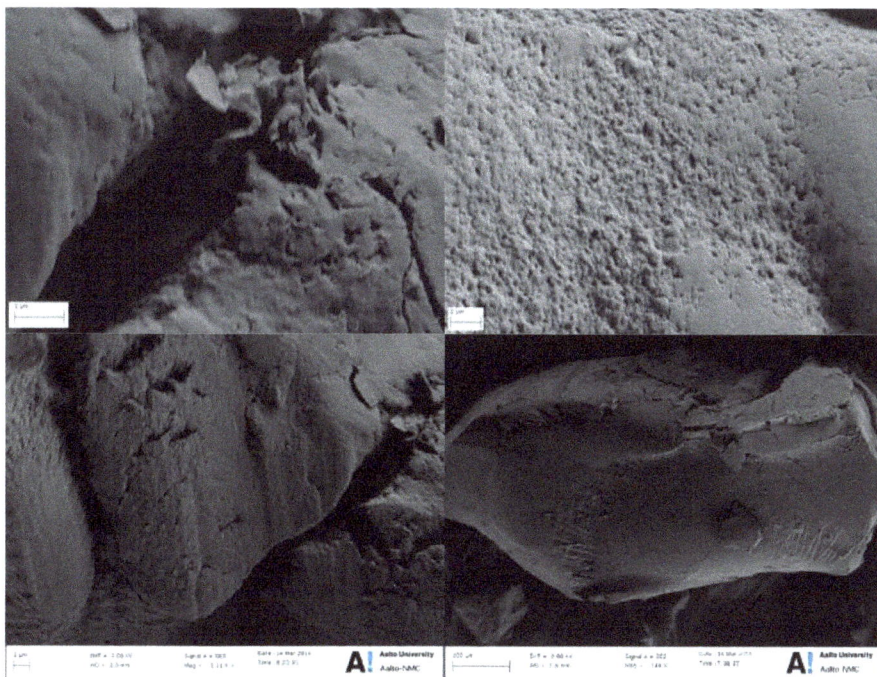

Figure 2. Scanning electron microscopy (SEM) images of Zn(II)–CTS beads at different magnifications (2 μm and 100 μm).

3.1.3. Determination of pH_{PZC}

Point of zero charge plays a key role in the surface science of environmental interfaces, where it indicates how easily adsorptive materials are able to adsorb the ions of target pollutants. The pH_{PZC} is the pH when the charge on the adsorbent surface is zero. Figure 3a depicts the results of the pH_{PZC} determination via drift method (NaCl solutions). The pH_{PZC} of Zn(II)–CTS bio-sorbent was determined to be approximately 7, which is a reasonable pH_{PZC}, given that the pK_a of CTS ranges from 6.3 to 7.2. At solution pH above the pH_{PZC}, the surface of the bio-sorbent is negatively charged, whereas at pH values below pH_{PZC}, the surface becomes positively charged and oxyanions adsorption happens due to electrostatic interaction of anions with the positively charged surface of the bio-sorbent. Hence, there is an increase in phosphate adsorption when the solution pH is lower than pH_{PZC} [3,26]. A similar experimental set was conducted with PO_4^{3-} solutions (Figure 3b). The final pH converged to around 7 when the initial pH was set from 4 to 11. For the solution with the initial pH of 3, the final pH remained below the pH_{PZC} which can be due to the abundant H^+ ions preventing the pH raise induced by the bio-sorbent. For the sample with initial pH of 12, the final pH was 10.1, higher than pH$_{PZC}$, which happens due to the excessive amount of OH$^-$ ions. Above pH$_{PZC}$, the adsorption of anions is hindered by the negatively charged surface of Zn(II)–CTS.

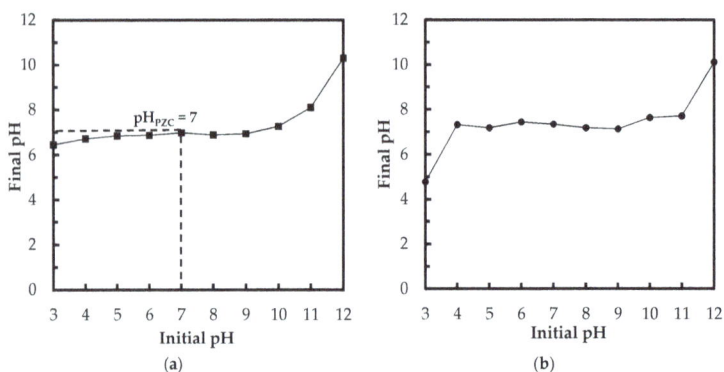

Figure 3. (**a**) The determination of point of zero charge (pH$_{PZC}$) using 0.1 M NaCl solutions (drift method); and (**b**) change of pH in the 5 mg/L PO$_4{}^{3-}$/Zn(II)–CTS solutions (Experimental conditions: 0.01 g of Zn(II)–CTS; 20 mL solution volume; contact time 24 h).

3.2. Phosphate Adsorption Studies

3.2.1. Effect of Bio-Sorbent Dose

Adsorbent dose was optimized for three zinc compound–CTS bio-sorbents, Zn(II)–CTS, ZnO–CTS and nano-ZnO–CTS, and additionally for a plain CTS sample. Different adsorbent doses varying between 0.1 to 2 g/L were studied. Similar trends in the results of phosphate removal percentage and adsorption capacity (mg/g) were observed for all three composites. As the amount of absorbent was increased, the removal percentage increases (Figure 4). This is because the growing adsorption surface area increases the active adsorptive sites on the surface. With the adsorbent dose of 0.1 g/L, all three bio-sorbents reached a removal percentage in the range of 19.7–21.7%. After this, the removal efficiency of nano-ZnO–CTS grew less than those of ZnO–CTS and Zn(II)–CTS. At the maximum adsorbent dose, 2 g/L, the removal percentage of nano-ZnO–CTS was 71.5% whereas for ZnO–CTS and Zn(II)–CTS 97.75% and 94.65%, respectively. When comparing the nano-ZnO–CTS to the other two bio-sorbets, the lower adsorption performance, e.g., removal percentages and corresponding capacities in Figure 4, along with higher standard deviations may be caused by the aggregated nano ZnO particles in the nano-ZnO–CTS beads. In comparison, the plain CTS sample (Figure 4a) proved to be relatively unsuccessful with only 10% average removal, which is mainly due to the pK_a value of its electron-donor functional groups, viz. –NH$_2$. The formation of new binding sites by Zn(II) ions within the CTS matrix is followed by the coordination of the phosphate anions to zinc(II). These chelating sites are unavailable to phosphate anions in the plain CTS, proved by the low performance of CTS in Figure 4. This observation confirms the improving effect of zinc(II) complexes in the adsorption performance of CTS towards phosphate oxyanions.

As seen in Figure 4b, the adsorption capacity q_t decreases for all four adsorbents with increasing dose. The capacity decreased for ZnO–CTS from 12.37 to 2.43 mg/g, for Zn(II)–CTS from 10.99 to 2.37 mg/g, and for nano-ZnO–CTS from 10.21 to 1.78 mg/g. This reduction can happen due to (I) a gap in the flux of phosphate concentration gradient between the concentrations in the liquid phase and on the solid surface, causing the amount of phosphate adsorbed onto the unit weight of adsorbent to decrease with the increasing dose [28], and/or (II) the increase in the adsorbent dose for a given amount of phosphate in the solution results in the unsaturation of adsorbent sites. In addition, more adsorptive sites become occupied, hence creating an increasing repulsion between the molecules in the solution and adsorbed on the surface [4].

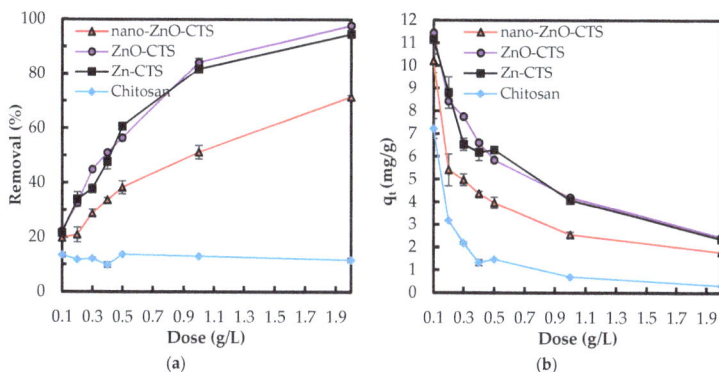

Figure 4. (**a**) Effect of adsorbent dose on (**a**) phosphate removal percentage; and (**b**) phosphate adsorption capacity (mg/g) (Experimental conditions: 5 mg/L phosphate concentration; natural pH; 50 mL solution).

The bio-sorbent/liquid ratio was set to 0.5 g/L for the rest of adsorption studies to account for the balance between the lower removal percentage at lower dosage and lower adsorption capacity (q) at higher dosage.

3.2.2. Effect of pH

The role of pH in phosphate uptake by the developed Zn(II)–CTS bio-sorbent was studied in five different pH values. The binding of phosphate oxyanions to the bio-sorbent occurred more efficiently in acidic medium, as seen in Figure 5. This enhancement of the removal percentage is consistent with previous studies of adsorptive abatement of phosphate [3].

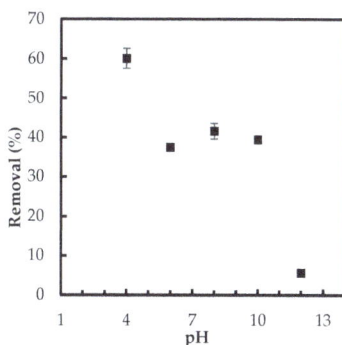

Figure 5. Effect of pH on phosphate removal by Zn(II)–CTS complex (Experimental conditions: 0.5 g/L adsorbent dose; 5 mg/L initial phosphate concentration).

The decrease in the phosphate adsorption onto Zn(II)–CTS by shifting pH from acidic to basic conditions can be attributed to the change in the distribution of phosphate species ($H_2PO_4^-$, HPO_4^{2-}, and PO_4^{3-}) and the decrease in surface protonation of the bio-sorbent. Additionally, in the basic solution, the hydroxyl ions may compete with phosphate adsorption [4]. High pH alter the aqueous form of phosphate species from $H_2PO_4^-$ to HPO_4^{2-}. The latter species is more resistant to the uptake by the surface hydroxyl groups. The limited adsorption at higher pH can also be caused by the excess amount of OH^- ions and deprotonation of Zn(II)–CTS surface, leading to a decrease in the interaction between the surface of Zn(II)–CTS and phosphate anions [28]. On the other hand, lower acidic condition protonates the surface functional groups of Zn(II)–CTS complexes, as discussed in the section related to pH_{PZC}, resulting in higher phosphate uptake due to the electrostatic attraction. As showed in Figure 3b, there is a shift in the final pH of the phosphate solutions from the initial pH. The solution with an initial pH of 12, for instance, had a final pH of 10, which is higher than pH_{PZC} of the bio-sorbent. The surface is negatively charged above pH_{PZC} and consequently repels phosphate oxyanions. For the initial pH ranging 6–10, the final pH of the solutions shifted to around pH_{PZC}, 7, resulting in a neutral surface of the bio-sorbent. Therefore, the anions of phosphate were able to easily move to the surface and finally chelate with the Zn(II) sites of the bio-sorbent. Because the surface is neutral between pH 6 and 8 (Figure 5), the slightly higher removal at pH 8 may be due to more readily accessible Zn(II) adsorptive sites. Furthermore, considering the kinetics of transport into a charged pore, salt will screen the charges in the pores allowing for faster transport into the pores. Therefore, there is a faster/better adsorption. The solution with initial pH 4 showed the highest phosphate removal implying a positive surface charge of the bio-sorbent, which favored the oxyanion removal.

3.2.3. Adsorption Kinetics

When it comes to the economical design of an adsorption process, the time for the system to reach equilibrium is an essential parameter [4]. The contact time was studied within 13 different time intervals from 1 min up to 48 h for three different phosphate concentrations (1, 5, 10 mg/L). The adsorption equilibrium and maximum capacities of phosphate adsorption were obtained within 180 min. All three sets of experiments increased the removal percentage during the first 180 min, which was followed by a plateau. The concentration gradient of phosphate in the solution and the existence of the higher number of available active sites on the surface drive the increase in adsorption. The maximum phosphate removal percentages for 10, 5 and 1 mg/L were 54.84%, 75.70% and 97.63%, respectively (Figure 6). The corresponding adsorption capacity (mg/g) for those percentages were 10.99, 7.39 and 2.00 mg/g, respectively. The decrease in removal by increasing concentration from 1 to 10 mg/L is expected as there are more molecules of adsorbate in the solution when the concentration increases so that the denominator of the percentage is greatly increased, resulting in a lower percentage. Furthermore, the adsorbent surface has a saturation stage (Langmuir monolayer coverage 5.33 mg/g discussed in the adsorption isotherm section). Once the surface is covered, the adsorption kinetics will change and the removal percent will decrease.

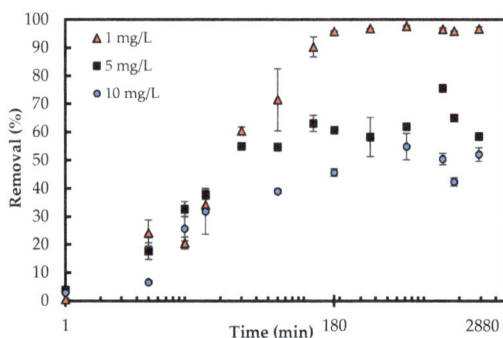

Figure 6. Effect of contact time on phosphate removal percentage (Experimental conditions: 0.5 g/L Zn(II)–CTS dose; 1, 5 and 10 mg/L initial phosphate concentration; pH 4).

The pseudo-first-order [29] and pseudo-second-order [30] kinetic models were employed to gather information on the adsorption dynamics [31]; the equations presented in Table 1. The applicability of the models via linear regression can be estimated by presenting the experimental data in plots: a linear plot of $\log(q_e - q_t)$ against t indicates better fit with pseudo-first-order model while a linear plot of t/q_t versus t presents better fit with pseudo-second-order model (Figure S1) [7]. The experimental q_t and the corresponding theoretical values, which were obtained by inserting the linearized parameters from Table 1 in the kinetic models are displayed in Figure S2. The kinetic constants and the linear R^2 values from the models are compiled in Table 1. The linear R^2 values corresponding to pseudo-first-order model all fall under 0.8 whereas the R^2 values corresponding to pseudo-second-order model are all above 0.99 (Table 1 and Figure S1). This suggests that the pseudo-second-order is more applicable than pseudo-first-order model in presenting the kinetics of phosphate adsorption onto Zn(II)–CTS.

Table 1. Calculated parameters for kinetic models via linear and nonlinear regression.

C_0 (mg/L)	$q_{e(exp)}$	Pseudo-first-order				Pseudo-second-order				Intra-particle diffusion	
		Linear regression									
		$\ln(q_e - q_t) = \ln q_e - k_1 t$				$\frac{t}{q_t} = \frac{1}{k_2 q_e^2} + \frac{t}{q_e}$				$q_t = k_p t^{\frac{1}{2}} + C$	
		$q_{e(cal)}$	k_1	R^2	Δq	$q_{e(cal)}$	k_2	R^2	Δq	k_p	C
1	1.99	0.61	0.003	0.66	3.85	2.01	0.029	1.00	0.90	0.030	0.875
5	6.34	1.85	0.004	0.55	3.27	5.85	1.476	0.998	1.03	0.067	3.541
10	10.44	4.64	0.002	0.76	3.22	10.46	0.004	1.00	0.48	0.150	4.312
		Nonlinear regression									
1	1.99	1.97	0.029	0.98	0.34	2.08	0.020	0.981	0.17	-	-
5	6.34	5.87	0.067	0.99	0.06	6.16	0.016	0.992	0.15	-	-
10	10.44	9.59	0.048	0.91	0.25	10.13	0.007	0.984	0.20	-	-

Note: k_1 pseudo-first-order rate constant (min^{-1}); t time (min); q_e and q_t phosphate adsorbed per adsorbent mass at equilibrium and at time t (mg/g); k_2 pseudo-second-order rate constant (g/mg·min); k_p intraparticle diffusion constant (mg/g·min$^{1/2}$); C boundary layer thickness constant (mg/g).

In addition, the normalized standard deviation Δq is determined for quantitative comparison of the model applicability, given in Table 1.

$$\Delta q = \sqrt{\frac{\sum\left[(q_{exp} - q_{cal})/q_{exp}\right]^2}{n-1}} \tag{3}$$

where n is the number of data points. Even though the linear regression provided high R^2 values, the corresponding Δq values (Table 1) and Figure S2 indicate a poor performance of linear regression for determination of best fitting kinetic models. Therefore, a nonlinear regression has been employed to further explore the kinetic data and confirm the best fitting kinetic model (Table 1 and Figure 7). The coefficient of determination R^2 is employed to find out the best-fitting model via nonlinear regression.

$$R^2 = \frac{\sum_{i=1}^{p}\left(q_{exp} - \overline{q_{calc}}\right)^2}{\sum_{i=1}^{p}\left(q_{exp} - \overline{q_{calc}}\right)^2 + \sum_{i=1}^{p}\left(q_{exp} + \overline{q_{calc}}\right)^2} \tag{4}$$

where $\overline{q_{calc}}$ is the average of q_{calc}. The experimental data (q_{exp}) aligned better with the theoretical data (q_{cal}) obtained by nonlinear regression of pseudo-second-order model (Figure 7) that is consistent with corresponding nonlinear R^2 and Δq values. This indicates that the adsorption may be involved in chemisorption [6] and is influenced by the characteristics of both the adsorbent and adsorbate [28]. The phosphate adsorption also followed pseudo-second-order kinetics in other studies [3–5,32].

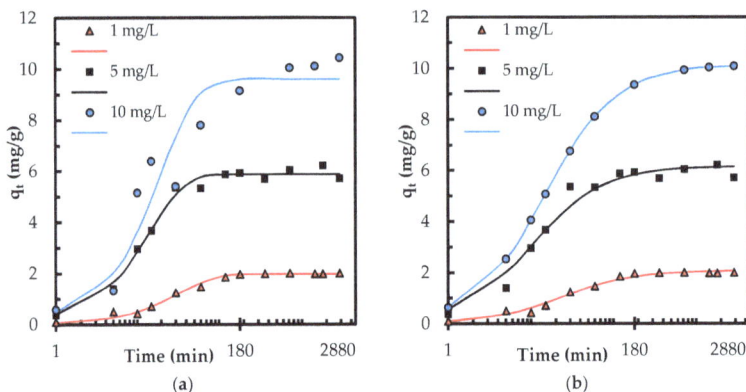

Figure 7. (**a**) Theoretical q_t (lines) by pseudo-first-order kinetic model; and (**b**) theoretical q_t (lines) by pseudo-second-order kinetic model determined by nonlinear regression compared to experimental q_t (markers).

Intra-particle diffusion mechanism was explored according to Weber and Morris model [33] using the equation given in Table 1. In this model, the C constant represents the thickness of the boundary layer [7]. The results are compiled in Table 1. The multi-linear plot of q_t versus $t^{1/2}$ (Figure 8) suggests several phases occur in the adsorption process and the pore diffusion is not the only rate-limiting stage [7,33]. During the first phase, the phosphate oxyanions are transferred onto the surface of Zn(II)–CTS complexes from the solution (boundary layer phase). During the second and slower phase, the anions are transported to the pores of Zn(II)–CTS particles where intra-particle diffusion is the rate-limiting step. As discussed in earlier sections, the surface charge of the bio-sorbent is mainly of positive or neutral charge in the studied pH range, therefore the transfer of phosphate anions from the solution to the surface of the bio-sorbent can take place faster and more easily. In some cases, there is a third and final stage which can be considered as the equilibrium phase in which intra-particle diffusion starts slowing down because of the very dilute concentration of adsorbate remained in the liquid phase [34,35]. In Figure 8, the first two phases of adsorption are visible.

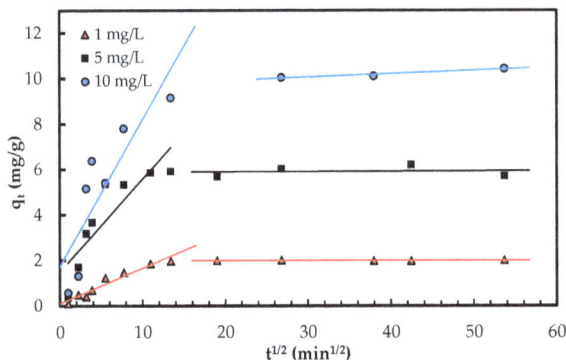

Figure 8. Phosphate adsorption steps onto Zn(II)–CTS bio-sorbent determined with the intra-particle diffusion model.

3.2.4. Adsorption Isotherm

The isotherm tests were conducted in five phosphate concentrations of 1, 2, 5, 10 and 15 mg/L at 20, 25, and 30 °C. When comparing the results of varying temperature, both the removal percentage and the adsorption capacity improved with a decrease in temperature (Figure 9). For instance, at concentration 5 mg/L, adsorption capacity increased from 3.9 to 5 mg/g and the removal percentage increased from 36% to 47.4% with decreasing temperature from 30 to 20 °C, respectively. This indicates that lower temperatures favor the adsorption process. When comparing the results at a constant temperature, while the adsorption capacity increased with increasing PO_4^{3-} concentration, the removal percentage diminished. For instance, at 20 °C, adsorption capacity increased from 1.8 to 7.2 mg/g while the removal percentage decreased from 91% to 23% with increasing phosphate concentration from 1 to 15 g/L, respectively. Lower removal percentage at higher phosphate concentrations can be caused from the saturation of adsorptive sites. The equilibrium data were examined with Langmuir, Freundlich, and Sips isotherms. While the Langmuir model hypothesizes a homogenous monolayer adsorption without molecule interaction, Freundlich isotherm assumes multilayer adsorption with molecule interaction [36]. The Sips isotherm, combining the basics of Langmuir and Freundlich isotherms, represents systems where an adsorbed molecule can be involved with more than one adsorptive site [15]. At lower adsorbate concentrations, this model turns to a Freundlich model, while, at higher concentrations, it gives the monolayer coverage characteristic of Langmuir isotherm. Giles et al. [37] have presented four main types of capacity curves, C, L, H and S. Herein, the adsorption capacity curve resembled the "L" isotherm without a strict plateau, i.e., it does not reach a point of limited adsorption capacity. For the "L" capacity curve, the Freundlich isotherm was found to be the most suitable [37].

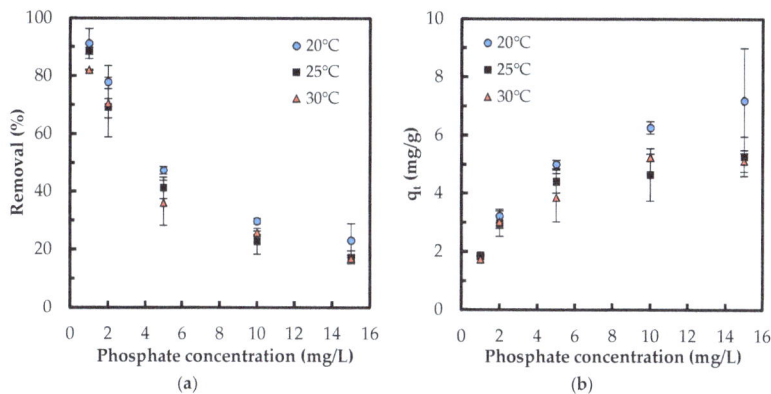

Figure 9. Effect of temperature on: (**a**) phosphate removal percentage; and (**b**) adsorption capacity (mg/g) (Experimental conditions: 0.5 g/L Zn(II)–CTS dose; pH 4; 24 h contact time).

The calculated parameters via linear regression along with the linearized isotherm equations are compiled in Table 2. The unitless Langmuir separation factor R_L shows the favorability of the adsorption; the values between 1 and 0 suggest favorable adsorption [7]. The linear R^2 values for Langmuir model were slightly higher than those of Freundlich model. However, the calculated values for q_e with linearized Freundlich isotherm showed a better fit with experimental q_e (Δq values in Table 2 and Figure S3).

$$R_L = \frac{1}{1 + K_L C_0} \tag{5}$$

Table 2. Calculated isotherm parameters via linear regression.

T (K)	Langmuir $\frac{C_e}{q_e} = \frac{1}{q_{max}}C_e + \frac{1}{K_Lq_{max}}$					Freundlich $\log q_e = \log K_F + \frac{1}{n}\log C_e$			
	q_{max}	K_L	R_L	R^2	Δq	K_F	$1/n$	R^2	Δq
293	7.37	1.32	0.05–0.43	0.99	0.31	3.72	0.27	0.99	0.04
298	5.33	1.85	0.03–0.34	0.99	0.24	3.12	0.22	0.98	0.06
303	5.40	1.56	0.04–0.38	0.99	0.18	2.92	0.25	0.95	0.1

Note: C_e equilibrium phosphate concentration; K_L Langmuir thermodynamic constant (L/mg); R_L Langmuir separation factor; q_e adsorption capacity in equilibrium; q_{max} calculated maximum adsorption capacity at each temperature; K_F Freundlich thermodynamic constant ((mg/g)·(L/mg)$^{1/n}$); $1/n$ Freundlich's intensity factor.

Even though linear regression is frequently employed to determine the best fitting isotherm, the error structure can alter upon linearizing the nonlinear equations. Based on the way the isotherm is linearized, the error distribution may change for either the worse or the better. Nonlinear regression is more suitable for the determination of the isotherm parameters, which can prevent such errors. Furthermore, the linear regression is inapplicable for isotherms with more than two adjustable parameters, e.g., Sips [15]. Therefore, the isotherm parameters were also determined by nonlinear regression. For nonlinear method, a trial and error approach was employed by minimizing the error between experimental data and calculated values. The calculated isotherm parameters via nonlinear regression are compiled in Table 3. The nonlinear R^2 values for Freundlich isotherm via nonlinear regression were higher when compared with those of the linear R^2 values. This indicates the error distribution altered to the worse while fitting the experimental data in linearized Freundlich model. The Freundlich R^2 values were higher when compared with those of Langmuir R^2 values (nonlinear regression). Figure 10 depicts the experimental q_e and the predicted q_{calc} by the isotherm models via nonlinear modeling. The Δq values based on Freundlich model were also lower compared with those of Langmuir isotherm via both linear and nonlinear regressions. Figure 10 and Table 3 clearly suggest that the adsorption isotherm of phosphate onto Zn(II)–CTS is better fit to a Freundlich model than a Langmuir model, yet this is the Sips model indicating the best fit. The fitting of the experimental data to the Sips model indicates that phosphate adsorption takes place on homogeneous–heterogeneous surface of the bio-sorbent.

Table 3. Calculated isotherm parameters via nonlinear regression.

T (K)	Langmuir $q_e = \frac{q_{max}K_LC_e}{1 + K_LC_e}$				Freundlich $q_e = K_FC_e^{\frac{1}{n}}$				Sips $q_e = \frac{q_sK_sC_e^{ns}}{1 + K_sC_e^{ns}}$				
	q_{max}	K_L	R^2	Δq	K_F	$1/n$	R^2	Δq	q_s	K_s	n_S	R^2	Δq
293	6.75	2.11	0.92	0.23	3.77	0.26	1.00	0.05	17.1	0.29	0.34	1.00	0.03
298	4.98	3.19	0.93	0.16	3.17	0.20	0.97	0.07	6.94	0.94	0.33	0.99	0.04
303	5.12	2.25	0.93	0.11	3.01	0.23	0.95	0.12	7.26	0.77	0.48	0.96	0.08

Note: q_s Sips maximum capacity (mg/g); K_S Sips equilibrium constant ((L/mg)^{ns}); n_S Sips exponent.

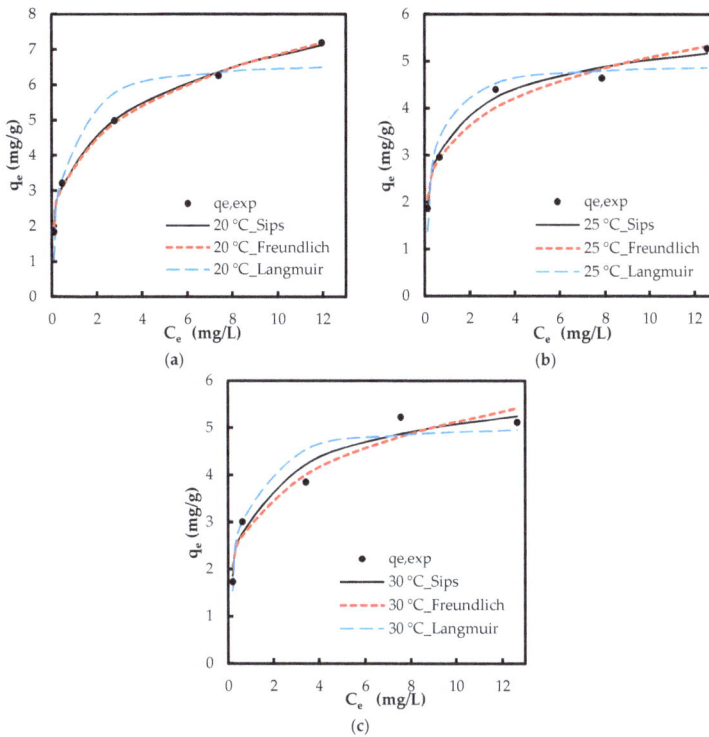

Figure 10. Experimental q_e (markers) and theoretical q_e (lines) via nonlinear regression of isotherm models at: (**a**) 20 °C; (**b**) 25 °C; and (**c**) 30 °C.

3.2.5. Adsorption Thermodynamics

The adsorption thermodynamics, the change of Gibb's free energy (ΔG), the change of entropy (ΔS), and the change in enthalpy (ΔH) [38] are defined by:

$$\Delta G = -RT \ln K_0 \tag{6}$$

$$\Delta G = \Delta H - T\Delta S \tag{7}$$

where R is the universal gas constant (8.314 J/K·mol), T (K) the temperature, and K_0 the unitless thermodynamic equilibrium constant [39]. The K_0 is extrapolated from the plot of $\ln(q_e/C_e)$ versus C_e (Figure S4) [39]. The thermodynamic parameters are tabulated in Table 4.

The negative values of ΔG indicate the spontaneous adsorption of phosphate onto Zn(II)–CTS. Table 4 shows the negative value of ΔG increases with decreasing temperature, which is consistent with the isotherm results showing a more favorable adsorption at lower temperatures. The negative enthalpy change confirms the exothermic characteristic of adsorption. The heat evolved in the adsorption reveals the physical and chemical nature of the reaction. The value of ΔH (-37.86 kJ/mol) suggests the phosphate adsorption onto Zn(II)–CTS is a physicochemical process rather than a purely physical or chemical adsorption [26,40]. The negative ΔS indicates the randomness decreases at the adsorption interface. Similar results have been reported in previous studies [41,42]. Worch [38] argues that a negative change in entropy is caused by the immobilization of adsorbate in the system. It can also indicate the fast adsorption of phosphate on the adsorptive sites [41].

Table 4. Calculated thermodynamic parameters.

T (K)	ΔG (kJ/mol)	ΔH (kJ/mol)	ΔS (J/K·mol)
293	−5.27	−37.86	−111.32
298	−4.60		
303	−4.15		

3.2.6. Effect of Coexisting Ions

Natural waters usually contain other anions competing for the adsorptive sites with phosphate. The competing effect of coexisting anions on phosphate adsorption by the developed bio-sorbent was studied with NO_3^-, SO_4^{2-} and Cl^- ions. Of these ions, chloride showed a minor effect, while nitrate and sulfate showed competing effect on the adsorption of phosphate (Figure 11). The equilibrium capacity of the bio-sorbent for phosphate in the absence of other ions was 7.63 mg/g while the phosphate adsorption in the presence of NO_3^-, SO_4^{2-} and Cl^- were 4.26, 4.79 and 7.15 mg/g, respectively. These results agree with previously reported studies on phosphate adsorption [3,4]. The sensitivity of the phosphate adsorption to the presence of background ions, e.g., SO_4^{2-} and NO_3^-, can indicate the outer-sphere complexation of phosphate with the bio-sorbent. Liu and Zhang [3] reported a notable reduction in phosphate uptake by modified chitosan beads in the presence of SO_4^{2-}, while Cl^- and NO_3^- slightly reduced phosphate adsorption. It was explained that SO_4^{2-} ions could more likely link with the functional groups on the surface of the composite, which in turn reduced the available active sites and hindered phosphate uptake. In addition, the accumulation of sulfate ions on the surface of the composite could contribute in forming a negatively charged surface, leading to an increasing repulsive force against phosphate oxyanions and hence a decreasing phosphate uptake. In the case of competing effect of nitrate, Seliem et al. [6] reported that the presence of NO_3^- significantly reduced the phosphate adsorption onto the composite of MCM-41 silica with rice husk and they attributed this effect to the initial pH of the solutions (pH 9.56). Their study showed that by maintaining a constant pH of 6, no obvious shift was noticed in the removal of phosphate, especially in the presence of nitrate molecules. Herein the nitrate molecules might hinder the adsorption of phosphate through mechanisms including the accumulation of nitrate anions on the surface of Zn(II)–CTS and formation of a negatively charged surface. Based on the results observed in this study and those previously reported, it can be inferred that: (I) The binding affinity of these ions for the adsorptive sites of the developed bio-sorbent are comparable with those of phosphate. (II) These ions can compete with phosphate for the adsorptive sites. (III) The mechanisms involved in the adsorption of phosphate and the other two ions onto the developed bio-sorbent are similar [43].

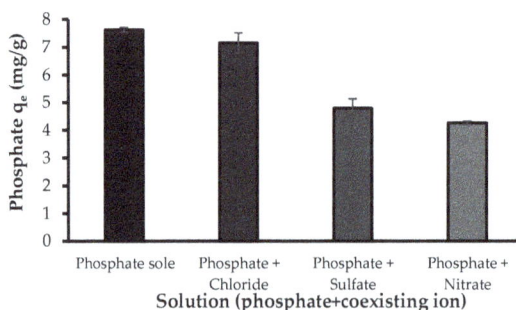

Figure 11. The effect of coexisting ions on phosphate adsorption (Experimental conditions: 5 mg/L phosphate concentration; 5 mg/L coexisting ion concentration; 0.5 g/L Zn(II)–CTS dose; pH 4; 24 contact time).

3.3. Mechanisms of Zn(II)–CTS Complexation and Phosphate Adsorption

CTS provided limited affinity for the oxyanions of phosphate as showed in Figure 4, which can be due to the pK_a values of its electron-donating groups. Metal cations, e.g., Zn(II), however, are able to bind to CTS at different electron-donating sites, e.g., amine $-NH_2$, on its chain. The complexed metal ion can then coordinate other ligands, providing adsorptive sites for phosphate on the CTS matrix that previously were unavailable, which is confirmed with the results presented in Figure 4. The mechanisms of metal ions complexation with CTS can be categorized in two categories of: (I) monodentate pattern; and (II) bidendate pattern [12,20]. In the former mechanism (I), the metal ions bind to one functional groups on CTS chain, while in the latter type (II) the metal ions bind to two or more functional groups, e.g., amino and hydroxyl groups, on one or more CTS chains so that act as a bridge between the chains [20]. Wang et al. [20] developed Zn(II)–CTS complexes for antibacterial application and noted that the complexes prepared with different zinc/CTS ratios were found to coordinate different amount of the metal ion and indicated diverse characteristics. The importance of pH and metal ion/CTS ratio in coordinating the metal ions on CTS structure was also reported for the Cu(II)–CTS complexes. Large Cu(II) loading changes the complexation mechanism from solely type (I) to a combination of types (I) and (II). It was also reported that the pH range <~5.5 mainly led to the formation of type (I) and exceeding pH values created more of type (II) [44]. The formation of metal ion–CTS complexation can be explained through Lewis acid–base theory, where the metal ion (M^{2+}) plays the role of the acid by accepting the electron pairs provided by CTS as the base. The reaction equations are written as follows:

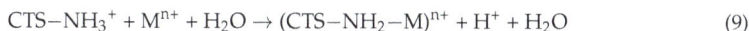

$$CTS-NH_2 + H_3O^+ \rightarrow CTS-NH_3^+ + H_2O \tag{8}$$

$$CTS-NH_3^+ + M^{n+} + H_2O \rightarrow (CTS-NH_2-M)^{n+} + H^+ + H_2O \tag{9}$$

where $(CTS-NH_2-M)^{n+}$ represents the metal ion–CTS complexes. Herein, M^{n+} is Zn^{2+} (Zn(II)). The FT-IR analysis showed that the $-NH_2$ and $-OH$ functional groups of CTS are involved in the complexation of CTS with Zn(II) in the bio-sorbent. The potential molecular structures of Zn(II)–CTS complexation are depicted in Figure 12.

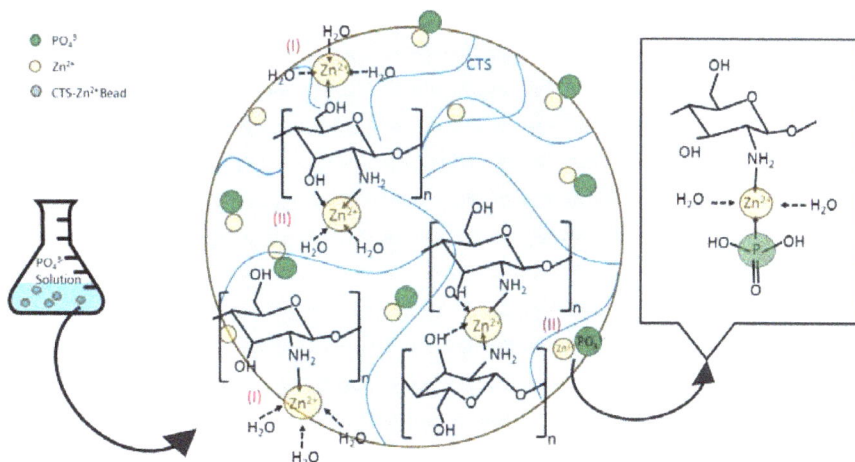

Figure 12. Possible structures of Zn(II)–CTS complexation and corresponding mechanism for phosphate uptake.

The type (I) complexation has been reported as the main mechanism favoring the uptake of phosphate Cu(II)–CTS complexes [12]. Some extent of phosphate uptake might also occur through the type (II) mechanism, yet, the magnitude of this uptake is minor in comparison with those of type (I) [12]. One explanation for this preference may lie in electrostatic interactions: electrons are more likely in a "diffused state" in complexation type (II) as they are distributed between different monomers and chains of CTS, whereas, in complexation type (I), electrons are configured in a more "concentrated state" so that electrostatic forces are strong enough to attract phosphate anions [12,45]. Besides, the involvement of two CTS chains in complexation type (II) may obstruct phosphate from adsorbing. Accordingly, the mechanism of phosphate uptake by the Zn(II)–CTS bio-sorbent proposed here is illustrated in Figure 12.

3.4. Evaluation of Zn(II)–CTS Bio-Sorbent

The adsorption capacities achieved by Zn(II)–CTS are compared with those of earlier reports on engineered adsorbents including zirconium–modified chitosan (ZCB) [3], zinc ferrite [46], magnetic iron oxide nanoparticles [47], $ZnCl_2$–activated carbon [48], titanium dioxide [49], synthetic iron oxide coated sand [50] and magnetic illite clay [51], which are compiled in Table 5. The adsorption capacity of Zn(II)–CTS was superior or comparable to the other adsorbents. For instance, magnetic iron, $ZnCl_2$–activated carbon, zinc ferrite, and synthetic iron oxide coated sand provided lower maximum adsorption capacities compared with those of observed in the present study. When comparing different adsorbents studied for phosphate removal, the differences in operational conditions need to be taken into account. Unlike the studies conducted with higher phosphate concentrations, e.g., ZCB (Table 5), Zn(II)–CTS was studied with relatively lower, yet, more environmentally relevant phosphate concentrations, which most likely explains the difference in maximum capacities.

Table 5. Phosphate adsorption on different engineered adsorbents.

Adsorbent	C_0 (mg/L)	q_{max} (mg/g)	pH	Reference
Zn(II)–CTS	5	7.37	4	Present
Zn(II)–CTS	5	6.3	natural	Present
Chitosan	5	1.45	natural	Present
Chitosan	5	4.75	4	Present
Naturally iron oxide coated sand	5–30	0.88	5	[50]
Synthetic iron oxide coated sand	5–30	1.50	5	[50]
Zinc ferrite	5	5.23	-	[46]
Magnetic illite clay	10–100	5.48	-	[51]
ZCB	5–50	60.60	4	[3]
Quaternized chitosan beads	1000	59.00	-	[5]
TiO_2	-	2.63	5	[49]
$ZnCl_2$–carbon	10	5.10	-	[48]
Magnetic iron	2–20	5.03	-	[47]

4. Conclusions

This study developed zinc(II)–chitosan bio-sorbents for phosphate removal from aqueous solution. The effect of operational parameters, e.g., pH, dose and contact time, on the adsorption phosphate by the complexes was explored in detail. The optimum dose of the bio-sorbent was found to be 0.5 mg/L. It was observed that lower pH values favor the adsorption of phosphate. A rapid adsorption was observed within the first 3 h of contact time for all three studied concentrations. The temperature study revealed that the adsorption process was more successful in lower temperatures and it was exothermic and spontaneous in nature. The study of co-existing ions revealed that Cl^- shows minor effect on phosphate removal, whereas NO_3^- and SO_4^{2-} show competing effect. Pseudo-second-order model was more applicable to kinetics study. The intra-particle diffusion study revealed that at least two steps were involved in the adsorption process, the boundary layer stage and the intra-particle diffusion

Polymers **2018**, *10*, 25

stage. The adsorption isotherm was well fitted with Freundlich and Sips isotherms. CTS is a unique adsorbent, a completely natural and biodegradable polymer. Along with studies on the antimicrobial activity, thermal degradation, and pyrolysis characteristics of CTS complexes with zinc(II) [20,21], this study provides deeper understanding of the zinc(II)–CTS as a bio-sorbent for more accurate application towards specific target, e.g., phosphate removal.

Supplementary Materials: The following are available online at www.mdpi.com/2073-4360/10/1/25/s1, Figure S1: The linear plot of t/q_t versus t for the determination of psodo-second-order kinetic model parameters via linear regression. Figure S2: (a) Theoretical q_t (lines) by pseudo-first-order kinetic model; and (b) theoretical q_t (lines) by pseudo-second-order kinetic model compared to experimental q_t (markers). Figure S3: (a) Theoretical values of q_e (lines) from Langmuir isotherm model; and (b) theoretical values of q_e (lines) from Freundlich isotherm model compared to experimental q_e (markers). Figure S4: The plot of $\ln(q_e/C_e)$ versus C_e for determination of K_0.

Acknowledgments: This work was conducted at Aalto University's analytical water laboratory. The electron microscopy was carried out at Aalto University's nanomicroscopy center. The research was funded by the MVTT organization (Maa– ja vesitekniikan tuki) (grant reference number 32900), which is gratefully acknowledged. The first author would like to thank the Doctoral School of Aalto University and Foundation for Aalto University Science and Technology for their financial support. The authors would like to acknowledge Laboratory Manager Ari Järvinen, Senior Laboratory Technician Aino Peltola and Laboratory Technician Marina Sushko, Aalto University, for their helps and fast phosphate measurements. We would also like to thank Maryam Borghei, Bio-based Colloids and Materials Group-Aalto University, for the assistance with the intrinsic viscosity measurements and Taneli Tiittanen, Inorganic Chemistry Laboratory-Aalto University, for conducting the FT-IR analysis. Sincere acknowledgement to three anonymous reviewers for their valuable comments towards improvement of this work.

Author Contributions: Maryam Roza Yazdani planned the research, designed the study, and instructed the experiments. Elina Virolainen performed the experiments. Maryam Roza Yazdani, Elina Virolainen, and Kevin Conley analyzed the data and wrote the paper. Riku Vahala reviewed and commented on the paper.

Conflicts of Interest: The authors declare no conflict of interest.

References

1. Conley, D.J.; Paerl, H.W.; Howarth, R.W.; Boesch, D.F.; Seitzinge, S.P.; Havens, K.E.; Lancelot, C.; Likens, G.E. Controlling eutrophication: Nitrogen and phosphorus. *Sci. Total Environ.* **2009**, *323*, 1014–1015. [CrossRef] [PubMed]

2. Morse, G.K.; Brett, S.W.; Guy, J.A.; Lester, J.N. Review: Phosphorus removal and recovery technologies. *Sci. Total Environ.* **1998**, *212*, 69–81. [CrossRef]

3. Liu, X.; Zhang, L. Removal of phosphate anions using the modified chitosan beads: Adsorption kinetic, isotherm and mechanism studies. *Powder Technol.* **2015**, *277*, 112–119. [CrossRef]

4. Ding, L.; Wu, C.; Deng, H.; Zhang, X. Adsorptive characteristics of phosphate from aqueous solutions by miex resin. *J. Colloid Interface Sci.* **2012**, *376*, 224–232. [CrossRef] [PubMed]

5. Sowmya, A.; Meenakshi, S. An efficient and regenerable quaternary amine modified chitosan beads for the removal of nitrate and phosphate anions. *J. Environ. Chem. Eng.* **2013**, *1*, 906–915. [CrossRef]

6. Seliem, M.K.; Komarneni, S.; Abu Khadra, M.R. Phosphate removal from solution by composite of MCM-41 silica with rice husk: Kinetic and equilibrium studies. *Microporous Mesoporous Mater.* **2016**, *224*, 51–57. [CrossRef]

7. Lalley, J.; Han, C.; Li, X.; Dionysiou, D.D.; Nadagouda, M.N. Phosphate adsorption using modified iron oxide-based sorbents in lake water: Kinetics, equilibrium, and column tests. *Chem. Eng. J.* **2016**, *284*, 1386–1396. [CrossRef]

8. Miettinen, I.T.; Vartiainen, T.; Martikainen, P.J. Phosphorus and bacterial growth in drinking water. *Appl. Environ. Microbiol.* **1997**, *63*, 3242–3245. [PubMed]

9. U.S. Environmental Protection Agency (USEPA). *Quality Criteria for Water*; EPA 440/5-86-001; Office of Water Regulation and Standard: Washington, DC, USA, 1986.

10. De-Bashan, L.E.; Bashan, Y. Recent advances in removing phosphorus from wastewater and its future use as fertilizer (1997–2003). *Water Res.* **2004**, *38*, 4222–4246. [CrossRef] [PubMed]

11. Awual, M.R.; Jyo, A.; Ihara, T.; Seko, N.; Tamada, M.; Lim, K.T. Enhanced trace phosphate removal from water by zirconium(IV) loaded fibrous adsorbent. *Water Res.* **2011**, *45*, 4592–4600. [CrossRef] [PubMed]

12. Yamani, J.S.; Lounsbury, A.W.; Zimmerman, J.B. Towards a selective adsorbent for arsenate and selenite in the presence of phosphate: Assessment of adsorption efficiency, mechanism, and binary separation factors of the chitosan-copper complex. *Water Res.* **2016**, *88*, 889–896. [CrossRef] [PubMed]
13. Wan Ngah, W.S.; Teong, L.C.; Hanafiah, M.A.K.M. Adsorption of dyes and heavy metal ions by chitosan composites: A review. *Carbohydr. Polym.* **2011**, *83*, 1446–1456. [CrossRef]
14. Madill, E.A.W.; Garcia-Valdez, O.; Champagne, P.; Cunningham, M.F. CO$_2$-responsive graft modified chitosan for heavy metal (nickel) recovery. *Polymers* **2017**, *9*, 39. [CrossRef]
15. Yazdani, M.R.; Bhatnagar, A.; Vahala, R. Synthesis, characterization and exploitation of nano-TiO$_2$/feldspar embedded chitosan beads towards UV-assisted adsorptive abatement of aqueous arsenic (As). *Chem. Eng. J.* **2017**, *316*, 370–382. [CrossRef]
16. Prabhu, S.M.; Meenakshi, S. A dendrimer-like hyper branched chitosan beads toward fluoride adsorption from water. *Int. J. Biol. Macromol.* **2015**, *78*, 280–286. [CrossRef] [PubMed]
17. Yazdani, M.; Bahrami, H.; Arami, M. Feldspar/titanium dioxide/chitosan as a biophotocatalyst hybrid for the removal of organic dyes from aquatic phases. *J. Appl. Polym. Sci.* **2014**, *131*, 40247–40256. [CrossRef]
18. Mahaninia, M.H.; Wilson, L.D. Cross-linked chitosan beads for phosphate removal from aqueous solution. *J. Appl. Polym. Sci.* **2016**, *133*. [CrossRef]
19. Cao, J.; Cao, H.; Zhu, Y.; Wang, S.; Qian, D.; Chen, G.; Sun, M.; Huang, W. Rapid and effective removal of Cu^{2+} from aqueous solution using novel chitosan and laponite-based nanocomposite as adsorbent. *Polymers* **2017**, *9*, 5. [CrossRef]
20. Wang, X.; Du, Y.; Liu, H. Preparation, characterization and antimicrobial activity of chitosan-Zn complex. *Carbohydr. Polym.* **2004**, *56*, 21–26. [CrossRef]
21. Ou, C.; Chen, S.; Liu, Y.; Shao, J.; Li, S.; Fu, T.; Fan, W.; Zheng, H.; Lu, Q.; Bi, X. Study on the thermal degradation kinetics and pyrolysis characteristics of chitosan-Zn complex. *J. Anal. Appl. Pyrolysis* **2016**, *122*, 268–276. [CrossRef]
22. Yuan, Y.; Chesnutt, B.M.; Haggard, W.O.; Bumgardner, J.D. Deacetylation of chitosan: Material characterization and in vitro evaluation via albumin adsorption and pre-osteoblastic cell cultures. *Materials* **2011**, *4*, 1399–1416. [CrossRef] [PubMed]
23. Jia, Z.; Shen, D. Effect of reaction temperature and reaction time on the preparation of low-molecular-weight chitosan using phosphoric acid. *Carbohydr. Polym.* **2002**, *49*, 393–396. [CrossRef]
24. Kasaai, M.R. Calculation of Mark–Houwink–Sakurada (MHS) equation viscometric constants for chitosan in any solvent–temperature system using experimental reported viscometric constants data. *Carbohydr. Polym.* **2007**, *68*, 477–788. [CrossRef]
25. Costa, C.N.; Teixeira, V.G.; Delpech, M.C.; Souza, J.V.S.; Costa, M.A.S. Viscometric study of chitosan solutions in acetic acid/sodium acetate and acetic acid/sodium chloride. *Carbohydr. Polym.* **2015**, *133*, 245–250. [CrossRef] [PubMed]
26. Yazdani, M.R.; Tuutijärvi, T.; Bhatnagar, A.; Vahala, R. Adsorptive removal of arsenic(V) from aqueous phase by feldspars: Kinetics, mechanism, and thermodynamic aspects of adsorption. *J. Mol. Liq.* **2016**, *214*, 149–156. [CrossRef]
27. Anandhavelu, S.; Thambidurai, S. Preparation of chitosan-zinc oxide complex during chitin deacetylation. *Carbohydr. Polym.* **2011**, *83*, 1565–1569. [CrossRef]
28. Pandiselvi, K.; Thambidurai, S. Synthesis of porous chitosan-polyaniline/ZnO hybrid composite and application for removal of reactive orange 16 dye. *Colloids Surf. B Biointerfaces* **2013**, *108*, 229–238.
29. Lagergren, S. About the theory of so-called adsorption of soluble substances. *Kungliga Sven. Vetenskapsakademiens Handl.* **1898**, *24*, 1–39.
30. Ho, Y.S.; McKay, G. Pseudo-second order model for sorption processes. *Process Biochem.* **1999**, *34*, 451–465. [CrossRef]
31. Shi, X.; Li, Q.; Wang, T.; Lackner, K.S. Kinetic analysis of an anion exchange absorbent for CO$_2$ capture from ambient air. *PLoS ONE* **2017**, *12*, e0179828. [CrossRef] [PubMed]
32. Su, Y.; Yang, W.; Sun, W.; Li, Q.; Shang, J.K. Synthesis of mesoporous cerium-zirconium binary oxide nanoadsorbents by a solvothermal process and their effective adsorption of phosphate from water. *Chem. Eng. J.* **2015**, *268*, 270–279. [CrossRef]
33. Weber, W.J.; Morris, J.C. Kinetics of adsorption on carbon from solution. *J. Sanit. Eng. Div. Proc. Am. Soc. Civ. Eng.* **1963**, *89*, 31–60.

34. Lin, K.A.; Liu, Y.; Chen, S. Adsorption of fluoride to UiO-66-NH_2 in water: Stability, kinetic, isotherm and thermodynamic studies. *J. Colloid Interface Sci.* **2016**, *461*, 79–87. [CrossRef] [PubMed]

35. Cheung, W.H.; Szeto, Y.S.; McKay, G. Intraparticle diffusion processes during acid dye adsorption onto chitosan. *Bioresour. Technol.* **2007**, *98*, 2897–2904. [CrossRef] [PubMed]

36. Liu, Q.; Hu, P.; Wang, J.; Zhang, L.; Huang, R. Phosphate adsorption from aqueous solutions by zirconium (IV) loaded cross-linked chitosan particles. *J. Taiwan Inst. Chem. Eng.* **2016**, *59*, 311–319. [CrossRef]

37. Giles, C.H.; Smith, D.; Huitson, A. A general treatment and classification of the solute adsorption isotherm. I. Theoretical. *J. Colloid Interface Sci.* **1974**, *47*, 755–765. [CrossRef]

38. Worch, E. *Adsorption Technology in Water Treatment: Fundamentals, Processes, and Modeling*; Walter de Gruyter: Munchen, Germany, 2012.

39. Demirbas, A.; Sari, A.; Isildak, O. Adsorption thermodynamics of stearic acid onto bentonite. *J. Hazard. Mater.* **2006**, *135*, 226–231. [CrossRef] [PubMed]

40. Smith, J.M. *Chemical Engineering Kinetics*, 3rd ed.; McGraw-Hill: New York, NY, USA, 1981.

41. Raji, F.; Pakizeh, M. Kinetic and thermodynamic studies of Hg(II) adsorption onto MCM-41 modified by $ZnCl_2$. *Appl. Surf. Sci.* **2014**, *301*, 568–575. [CrossRef]

42. Tan, Z.; Peng, H.; Liu, H.; Wang, L.; Chen, J.; Lu, X. Facile preparation of EDTA-functionalized chitosan magnetic adsorbent for removal of Pb(II). *J. Appl. Polym. Sci.* **2015**, *132*. [CrossRef]

43. Guan, X.; Dong, H.; Ma, J.; Jiang, L. Removal of arsenic from water: Effects of competing anions on As(III) removal in $KMnO_4$–Fe(II) process. *Water Res.* **2009**, *43*, 3891–3899. [CrossRef] [PubMed]

44. Rhazi, M.; Desbrières, J.; Tolaimate, A.; Rinaudo, M.; Vottero, P.; Alagui, A. Contribution to the study of the complexation of copper by chitosan and oligomers. *Polymer* **2001**, *43*, 1267–1276. [CrossRef]

45. Shi, X.; Xiao, H.; Chen, X.; Lackner, K.S. The effect of moisture on the hydrolysis of basic salts. *Chem. Eur. J.* **2016**, *22*, 18326–18330. [CrossRef] [PubMed]

46. Gu, W.; Xie, Q.; Qi, C.; Zhao, L.; Wu, D. Phosphate removal using zinc ferrite synthesized through a facile solvothermal technique. *Powder Technol.* **2016**, *301*, 723–729. [CrossRef]

47. Yoon, S.-Y.; Lee, C.-G.; Park, J.-A.; Kim, J.-H.; Kim, S.-B.; Lee, S.-H.; Choi, J.-W. Kinetic, equilibrium and thermodynamic studies for phosphate adsorption to magnetic iron oxide nanoparticles. *Chem. Eng. J.* **2014**, *236*, 341–347. [CrossRef]

48. Namasivayam, C.; Sangeetha, D. Equilibrium and kinetic studies of adsorption of phosphate onto $ZnCl_2$ activated coir pith carbon. *J. Colloid Interface Sci.* **2004**, *280*, 359–365. [CrossRef] [PubMed]

49. Darcy, M.; Weiss, D.; Bluck, M.; Vilar, R. Adsorption kinetics, capacity and mechanism of arsenate and phosphate on a bifunctional TiO_2-Fe_2O_3 bi-composite. *J. Colloid Interface Sci.* **2011**, *364*, 205–212. [CrossRef] [PubMed]

50. Boujelben, N.; Bouzid, J.; Elouear, Z.; Feki, M.; Jamoussi, F.; Montiel, A. Phosphorus removal from aqueous solution using iron coated natural and engineered sorbents. *J. Hazard. Mater.* **2008**, *151*, 103–110. [CrossRef] [PubMed]

51. Chen, J.; Yan, L.; Yu, H.; Li, S.; Qin, L.; Liu, G.; Li, Y.; Du, B. Efficient removal of phosphate by facile prepared magnetic diatomite and illite clay from aqueous solution. *Chem. Eng. J.* **2016**, *287*, 162–172. [CrossRef]

polymers

MDPI

Article

Neodymium Recovery by Chitosan/Iron(III) Hydroxide [ChiFer(III)] Sorbent Material: Batch and Column Systems

Hary Demey [1,2,*], Byron Lapo [1,3], Montserrat Ruiz [4], Agustin Fortuny [4], Muriel Marchand [2] and Ana M. Sastre [1]

1   Department of Chemical Engineering, Universitat Politècnica de Catalunya, ETSEIB, Diagonal 647, 08028 Barcelona, Spain; Byron.lapo@upc.edu (B.L.); Ana.maria.sastre@upc.edu (A.M.S.)
2   Commissariat à l'Energie Atomique et aux Energies Alternatives, CEA/DRT/LITEN/DTBH/LTB, 17 rue des Martrys, 38054 Grenoble, France; muriel.marchand@cea.fr
3   School of Chemical Engineering, Universidad Técnica de Machala, UACQS, 070151 Machala, Ecuador
4   Department of Chemical Engineering, Universitat Politècnica de Catalunya, EPSEVG, Av. Víctor Balaguer, s/n, 08800 Vilanova i la Geltrú, Spain; Montserrat.ruiz@upc.edu (M.R.); agustin.fortuny@upc.edu (A.F.)
*   Correspondence: hary.demey@upc.edu; Tel.: +34-938-937-778

Received: 15 January 2018; Accepted: 16 February 2018; Published: 19 February 2018

Abstract: A low cost composite material was synthesized for neodymium recovery from dilute aqueous solutions. The in-situ production of the composite containing chitosan and iron(III) hydroxide (ChiFer(III)) was improved and the results were compared with raw chitosan particles. The sorbent was characterized using Fourier transform infrared spectroscopy (FTIR) and scanning electron microscopy-energy dispersive X-ray analyses (SEM-EDX). The equilibrium studies were performed using firstly a batch system, and secondly a continuous system. The sorption isotherms were fitted with the Langmuir, Freundlich, and Sips models; experimental data was better described with the Langmuir equation and the maximum sorption capacity was 13.8 mg·g^{-1} at pH 4. The introduction of iron into the biopolymer matrix increases by four times the sorption uptake of the chitosan; the individual sorption capacity of iron (into the composite) was calculated as 30.9 mg Nd/g Fe. The experimental results of the columns were fitted adequately using the Thomas model. As an approach to Nd-Fe-B permanent magnets effluents, a synthetic dilute effluent was simulated at pH 4, in order to evaluate the selectivity of the sorbent material; the overshooting of boron in the column system confirmed the higher selectivity toward neodymium ions. The elution step was carried out using MilliQ-water with the pH set to 3.5 (dilute HCl solution).

Keywords: boron; chitosan; iron(III) hydroxide; neodymium; sorption

1. Introduction

Rare earth elements (REEs) are critical due to their importance for many technological applications. The greater reserves of REEs are located in China; being the main world producer (since the 1990s it has been producing roughly 90% of the world's supply). REEs are massively used in the manufacturing of high-technology devices; the growing demand of the last decade has forced the mining industry of the European Union to search for new mineral sources outside of China (in order to reduce the strong dependence to this country). Additionally, the popular concept of "urban mining" has been generalized in all European countries over the last five years: society is recognizing that resources contained in wastes should be recovered and reused as much as possible [1].

REEs are distributed in moderated concentration in the earth's crust; the term "rare" originated from the difficult isolation processes on their discovery. Their abundance follows the Oddo–Harking rules: elements with even-atomic number are more stable and more abundant than adjacent odd-atomic elements [2,3]. The most abundant REEs in the earth crust are yttrium (28–70 mg·kg^{-1}) and cerium (20–70 mg·kg^{-1}) and the less abundant is thulium (0.2–1.0 mg·kg^{-1}). Special attention must be paid to neodymium (Nd), which is a key element in the high-tech industries and its applications include the manufacturing of the high strength permanent magnets (Nd-Fe-B), the fabrication of electrical motors for hybrid vehicles, and wind turbine generators [4].

Although the concentration of Nd in the earth's crust is moderate (12–41.5 mg·kg^{-1}), it is also considered as a critical element due to the possible shortage in Chinese export to the EU and USA [5,6]. Therefore, the development of novel technologies for recycling and recovery of neodymium is an important task for government authorities. Currently, there is no a simple method for Nd separation from aqueous effluents. Nevertheless, several methodologies have been used for metal removal from aqueous solutions, such as chemical precipitation [7,8], electrochemical treatment [9,10], membrane technology [11,12], solvent extraction [13], ion-exchange [14,15], and biosorption [16–19].

Recently, our research group [20] suggested liquid-liquid extraction technique with Cyanex-272 and Cyanex-572 as a good alternative for Nd separation from waste magnets effluents. This process is particularly useful and efficient for the treatment of wastewaters with high metal concentrations; otherwise, it could not be cost-effective for separating traces of metals, due to the involvement of large amounts of organic solvents and a further step for eluent treatment.

Biosorption could be regarded as the most suitable and economical method for recovering metal ions from dilutes effluents [21,22]. Furthermore, the designing of sorbents is crucial for reaching a high performance; the combination of several reactive groups may improve the stability of the resulting material and improve the removal uptake. To achieve this goal, a composite consisting of chitosan and iron(III) hydroxide [ChiFer(III)] was manufactured (in-situ) and evaluated for Nd sorption from aqueous solutions. Previous studies [23] have reported the use of ChiFer(III) material for boron removal from Mediterranean seawater, the sorbent is stable for several sorption–desorption cycles and can be used under conditions with high ion-strength (i.e., seawater); this information is advantageous for the treatment of waste magnet effluents (e.g., Nd-Fe-B).

In comparison with heavy metals, the literature for REEs recovery through the sorption process is still scarce. Several authors have reported the possibility of using biopolymer-based materials; e.g., Wang et al. [24] evaluated the performance of calcium alginate for Nd sorption, and Galhoum et al. [25] used cysteine-functionalized chitosan magnetic nano-based particles for removal of light and heavy rare earth elements from aqueous solutions: cationic species La(III), Nd(III), and Yb(III) can be sorbed by a combination of chelating and anion-exchange mechanisms; and the thermodynamic constants demonstrated the spontaneous and endothermic nature of sorption. Additionally, Zhao et al. [26,27] synthesized two innovative materials based on: i) EDTA-cross-linked β-cyclodextrin (EDTA-β-CD); and ii) polyethylenimine-cross-linked cellulose nanocrystals (PEI-CNC). The sorbents were demonstrated to be selective for Eu(III) and Er(III) recovery from waters, respectively (i.e., the selectivity of EDTA-β-CD follows: Eu(III) over Ce(III) and La(III); and PEI-CNC follows: Er(III) over La(III) and Eu(III)).

Chitosan is a copolymer of glucosamine and *N*-acetyl-D-glucosamine linked by β(1→4) glycosidic bonds [28,29]; its applications are derived from its easy availability (as a renewable resource) and its easy transformation into different configurations: chitosan is considered as the second most abundant biopolymer in the environment after cellulose. This study reports the preparation of chitosan-based composite to improve the handling of Fe(OH)$_3$ as sorbent to recover neodymium in a continuous system. The regeneration of the ChiFer(III) material was evaluated with dilute HCl solution (demineralized water at pH 3.5), and the Thomas model [30] was used to fit the experimental data.

2. Materials and Methods

2.1. Materials

Neodymium solutions were prepared using $Nd(NO_3)_3 \cdot 6H_2O$ (molecular weight 438.35 g·mol^{-1}) provided by Sigma-Aldrich (St. Louis, MO, USA). Iron(III) chloride hexahydrate used for the sorbent preparation was provided by Panreac (Barcelona, Spain). Chitosan was supplied by Aber Technologies (Lannilis, France), and its molecular weight (125,000 g·mol^{-1}) was previously reported by Ruiz et al. [31] using size exclusion chromatography (SEC) coupled with light scattering and refractometry. The degree of acetylation determined by Fourier transform infrared (FTIR) spectroscopy was found to be 0.13 (i.e., the deacetylation degree of chitin is 0.87) [32].

2.2. Preparation of ChiFer(III) Microspheres

The manufacturing of the sorbent material was slightly improved from that of Demey et al. [23]; the chitosan solution with a concentration of 2.2% *w/w*, was prepared by dissolving 30 g of chitosan in 2.2% *w/w* acetic acid solution (1350 mL) and stirring for 5 h. Thirty grams of $FeCl_3 \cdot 6H_2O$ powder were mixed in 120 mL of HCl solution (0.5 M) until complete dissolution. The chitosan solution (1350 mL, 2.2% *w/w*) was then mixed with the iron(III) solution under vigorous stirring (500 rpm) for 2 h.

The chitosan-iron(III) mixture was added drop-by-drop with a peristaltic pump through a thin nozzle (Ø 2.0 mm) into an aqueous solution of 1 M sodium hydroxide under magnetic stirring to produce microspheres of the ChiFer(III) composite. The resulting beads were kept under stirring for 8 h at room temperature (25 °C) and were then filtered and washed intensively with distilled water to remove the excess of sodium hydroxide from the surface of the sorbent. To compare the effect of the drying method on the kinetic profiles, wet samples of beads were air-dried in a laboratory oven at 45 °C (AD beads), and freeze dried (using a LyoQuest-55, Telstar equipment, São Paulo, Brazil) at 218 K and 0.01 mbar (FD beads). The standard sorbent used in this work was the freeze dried beads with an average diameter of 2.0 mm. The average particle size (Sp) of chitosan was 0.5 mm < Sp < 1 mm (Figure S1 in the supplementary materials section); it was verified through a MASTERSIZER 3000™ equipment from Malvern Instruments Ltd., Worcestershire, UK.

2.3. Characterization of Sorbents

2.3.1. Scanning Electron Microscopy

The samples were analyzed using a JEOL JSM 7100F field emission scanning electron microscope (JEOL Ltd., Peabody, MA, USA), a specialized high-performance scanning electron microscope (SEM) with low-vacuum, and high-vacuum modes, capable of analyzing samples under pressures of up to 9.4×10^{-4} Pa. The objective lens of the JEOL JSM-7100F does not create a magnetic field around the samples and a high magnification is easily obtained (which improves the study of several types of micro/nano-structures). Thus, magnetic samples can be observed and analyzed without restriction. The microscope is also equipped with an Energy Dispersive X-ray (EDX) spectrometer (INCA 250, Oxford instruments, Oxford, UK) for chemical analysis.

The sorbent samples were analyzed before and after sorption of neodymium from aqueous solutions; EDX-analysis technique was used to detect the main elements present at the surface of the sorbent particles.

2.3.2. FTIR Analyses

Fourier transform infrared (FTIR) analyses were performed using a BRUKER IFS 66 FTIR spectrophotometer (Bruker Optik GmbH, Ettlingen, Germany) equipped with a reflection diamond accessory (platinum ATR), and the spectra were recorded in the range of 4000–400 cm^{-1} with a sample amount of 2 mg of ChiFer(III).

2.4. pH Effect

The study of pH-influence on neodymium sorption was performed by mixing 25 mL of dilute metal solution (9.4 mg·L^{-1}; i.e., 0.07 mmol·L^{-1}) with 25 mg of sorbent (i.e., sorbent dosage, SD 1 g·L^{-1}) in 50 mL polyethylene flasks. Proton concentration was adjusted using 0.1 M HCl or 0.1 M NaOH (as required); the evaluated range was set at pH 1–6 (to prevent metal precipitation in all experimental series) and the stirring speed was set at 150 rpm at 20 °C, using an agitator Rotabit, J.P. Selecta (Barcelona, Spain). After 72 h of agitation, the final pH was measured, and 5 mL of solution was filtered and analyzed with 4100 MP-AES instrument (Microwave plasma-atomic emission spectrometer from Agilent technologies, Melbourne, Australia) at the wavelength of 430.35 nm (for neodymium detection).

2.5. Equilibrium Sorption

Sorption isotherms were obtained by mixing a known volume of solution (25 mL) at different metal concentrations at selected pH 4 and a fixed mass of sorbent (25 mg; i.e., SD 1 g·L^{-1}). After 72 h of contact, the pHs of the solutions were measured and the initial and equilibrium metal concentrations were systematically determined using 4100 MP-AES equipment (Agilent technologies, Melbourne, Australia).

The Langmuir, the Freundlich, and the Sips equations were used to describe the experimental sorption isotherm data [33–36]:

$$q = \frac{q_{max}bC_{eq}}{1 + bC_{eq}} \tag{1}$$

$$q = K_F C_{eq}^{1/n} \tag{2}$$

$$q = \frac{q_{max}K_S C_{eq}^{1/ns}}{1 + K_S C_{eq}^{1/ns}} \tag{3}$$

where q is the amount of sorbed metal per gram of sorbent at equilibrium (mg·g^{-1}), qmax is the maximum adsorption capacity of the sorbent (mg·g^{-1}), and Ceq is the equilibrium concentration of the solution (mg·L^{-1}). In the Langmuir model (equation 1), b is related to the energy of adsorption (L·mg^{-1}), whereas KF and n are the Freundlich adsorption constants, indicative of the relative capacity and the adsorption intensity, respectively; Ks (L·mg^{-1}) and ns are the constants of the Sips model [36]. The experiments of equilibrium studies were performed in duplicate to ensure the accuracy of the results (the relative standard deviation obtained was in the order of ±5%).

2.6. Influence of Contact Time

The uptake kinetics experiments were performed by adding (under continuous stirring) a known amount of sorbent (SD 0.5 g·L^{-1}) to 500 mL of metal solution (i.e., 10–20 mg·L^{-1}) at pH 4. Aliquots of solution were withdrawn at different times and filtered over 72 h of contact with the beads. The residual concentration was determined by the 4100 MP-AES instrument. The kinetic profiles were compared by using sorbent beads under the same conditioning (freeze-dried beads; Ø 2.0 mm). The models such as PFORE and PSORE (pseudo-first order and pseudo-second order rate equations) were evaluated to fit the experimental data; the contribution of the resistance to intraparticle diffusion was evaluated by the equation of Weber and Morris, [37]:

Pseudo-first order rate equation (PFORE) [38,39]:

$$\frac{dq_t}{dt} = K_1(q_1 - q_t) \tag{4}$$

Pseudo-second order rate equation (PSORE) [40]:

$$\frac{dq_t}{\left(q_{eq} - q_t\right)^2} = K_2 dt \tag{5}$$

where q_{eq} is the equilibrium sorption capacity (mg·g^{-1}), q_t is the sorption capacity (mg·g^{-1}) at any time t (min), and k_2 is the pseudo-second order rate constant (g·mg^{-1}·min^{-1}).The parameters q_{eq} and k_2 are pseudo-constants depending on the experimental conditions.

Equation of Weber and Morris, [37]:

$$q_t = K_p t^{1/2} + C \tag{6}$$

where C is the intercept, and K_p is the intraparticle diffusion rate constant.

2.7. Dynamic Column Testing

The applicability of a dynamic system was tested for evaluating the sorption performance of ChiFer(III) material on neodymium sorption. The glass columns (i.d., Ø 1.8 cm) were filled with hydrated beads (Ø 2.0 mm; 2.4 g in d.w.) until a bed depth of 23 cm (i.e., bed-volume: 58.53 mL). In order to avoid a poor arrangement of the packed-bed, the sorbent was carefully introduced into the columns from the top in a water suspension (typically 1 g·L^{-1}). Then, the bottom of the column was kept open (over 2 h) to evacuate the water content.

The metal solution was delivered by up-flow at 20 °C using a peristaltic pump at a flow rate of 0.01 and 0.02 L·h^{-1} (i.e., superficial velocity: 3.77 × 10^{-2} m·h^{-1} and 8.01 × 10^{-2} m·h^{-1} respectively). The tests were performed at pH 4.5 and inlet concentration of 10 mg·L^{-1}. After saturation, the packed-bed was eluted upward using dilute HCl solution (demineralized water at pH 3.5) and a flow rate of 0.01 L·h^{-1}. Samples of 5 mL were periodically collected for analyzing during sorption and desorption steps using an automatic fraction collector (Gilson FC-203 B, Dunstable, UK).

It is noteworthy that, the breakthrough curve allows the behavior of recovered metal into the sorbent bed to be represented; it is typically plotted as the ratio of effluent concentration to inlet concentration (C_t/C_0) as a function of the operational time t, or also as a function of the ratio volume of the influent/volume of the packed-bed (so called bed-volume, BV). The breakthrough and the exhaustion points were set as C_t/C_0 = 0.05 and C_t/C_0 = 0.95, respectively. The experimental sorption capacity can be obtained from Equation (7):

$$q_{exp} = \int_0^{V_{total}} \frac{(C_0 - C_t)}{m} dV \tag{7}$$

The Thomas model equation [30] was used for describing the theoretical performance of the columns due to its simplicity and adequate accuracy in predicting breakthrough curves. The model can be represented by the following equation for favorable sorption process [41]:

$$\frac{C_t}{C_0} = \frac{1}{1 + \exp[K_T(q_T m - C_0 V)/Q]} \tag{8}$$

where K_T is the Thomas rate constant (L·h^{-1}·mg^{-1}), m is the mass of sorbent (g), Q is the volumetric flow rate (L·h^{-1}), V is the volume of the solution into the column (L), C_t is the metal concentration (mg·L^{-1}), and q_T is the Thomas sorption capacity (mg·g^{-1}). The constants of the non-linearized form were obtained by origin 9.0 software (OriginLab Inc., Northampton, MA, USA, 2012).

In addition, a selectivity study was developed by using a binary system (in columns) with an equimolar solution of neodymium and boron (0.1 mmol·L^{-1}), in order to approach the real effluent of permanent magnets (Nd-Fe-B). Sorption systems are usually employed as polishing steps at pH 4–6 (as a complement of traditional physical–chemical techniques); therefore metal concentrations are very dilute at these operating conditions, especially iron concentration which is almost negligible and it was not considered in this study.

The separation coefficients of neodymium and boron ($R_{Nd/B}$) were calculated as follows and plotted versus the bed-volume [42]:

$$Kd_i = \int_0^{V_{total}} \frac{(C_0 - C_t)}{C_0 m} dV \tag{9}$$

$$R_{Nd/B} = \frac{Kd_{Nd}}{Kd_B} \tag{10}$$

where $R_{Nd/B}$ is the separation coefficient of Nd and B; K_d is the distribution coefficient ($L \cdot g^{-1}$); C_o is initial metal concentration ($mg \cdot L^{-1}$); C_t is the metal concentration at the time t ($mg \cdot L^{-1}$); m is the mass of the sorbent (g); V is the volume of the solution passed into the column (L).

3. Results and Discussion

3.1. Characterization of Sorbent

3.1.1. Scanning Electron Microscopy

SEM micrographs were performed for exploring the topography of the sorbent surface, as well as to examine the external and internal structure of the beads configuration. Figure 1a shows the spherical shape of the standard freeze-dried material (average diameter, Ø 2.0 mm); some cavities are found around the entire external surface, which is an indication of the easy accessibility for sorbate molecules into the volume of the beads. The images performed on the cross-section area, corroborate that the internal topography is relatively open (Figure 1b); this is a relevant feature for enhancing the diffusion phenomenon.

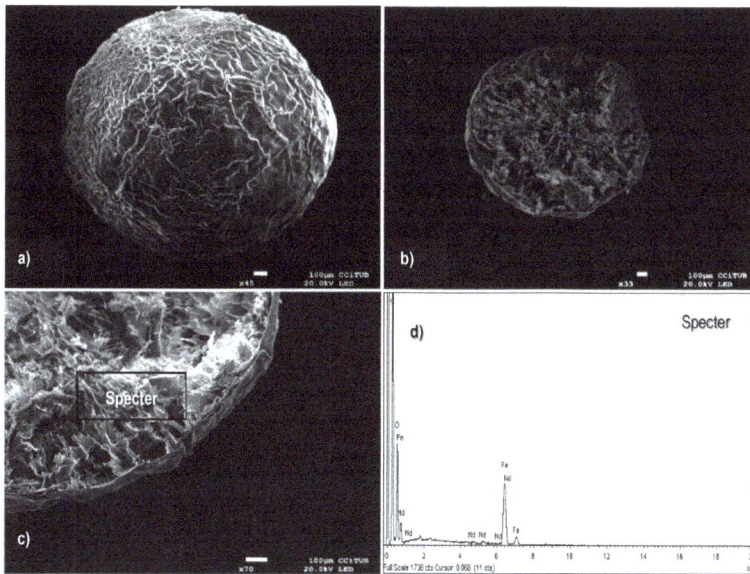

Figure 1. Scanning electron microscopy (SEM) images of the freeze dried (FD) sorbent. (**a**) External surface. (**b**) Cross-section area. (**c,d**) Energy dispersive X-ray (EDX) analysis of the cross-section area of the beads.

EDX-analyses on different zones of the cross-section area (Figure 1c,d) suggest that the main elements of the composite (i.e., oxygen, carbon, and iron) are homogenously distributed in the whole volume of the sorbent. The total amount of iron distributed throughout the dried material was quantified by 4100 MP-AES technique (a known amount of beads was kept under contact with 10 M HNO_3 solution over 10 h until complete digestion of the organic and mineral components): it was found to be 330 mg of iron per gram of ChiFer(III).

The micrographs presented in Figure 1 are in complete agreement with the previous findings reported by Demey et al. [23]; the manufacturing of iron(III) hydroxide through the in-situ coagulation of chitosan solution provides an innovative technique for immobilization of active mineral materials over a polymeric and well distributed network. A second configuration of the sorbent, the so-called xerogel material, was examined for verifying the impact of the drying method on the resulting structure; Figure 2a,b show that the uncontrolled air-drying of the hydrogels leads to shrinkage of the surfaces with depleted performance of mass transfer [43]. The air-dried particles have an average diameter of 0.90 mm; some folds over the entire surface are formed due to water evaporation at ambient pressure conditions, and the cavities or pores (around the folds) appear not to be easy accessible for sorbate ions. Figure 2c shows the cross-section area of the xerogels which differ widely from the freeze-dried beads: it seems to be a closed network; however, the EDX-analysis (Figure 2d) confirmed the content of the main elements, as expected (i.e., oxygen, carbon, and iron).

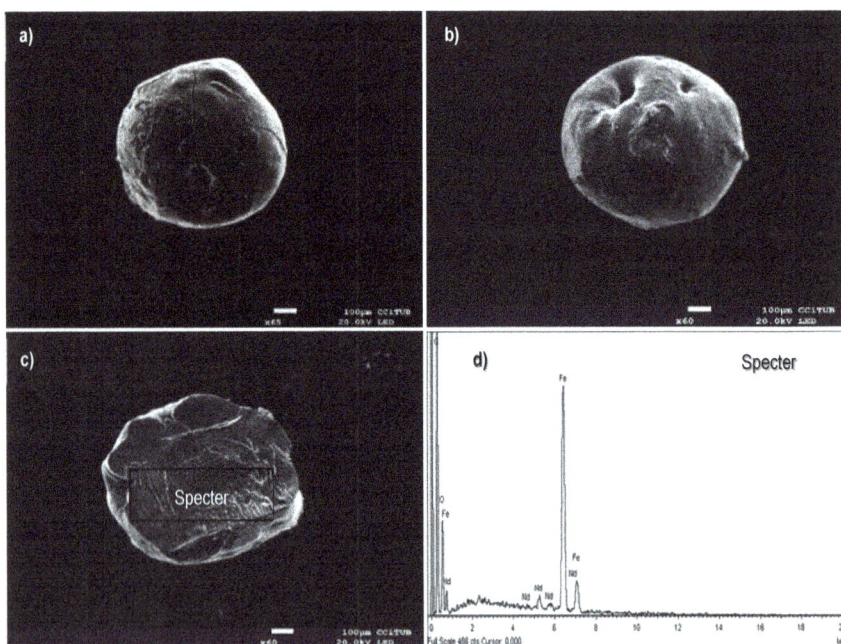

Figure 2. Scanning electron microscopy (SEM) images of the air dried (AD) sorbent. (**a**) External surface. (**b**) Cross-section area. (**c,d**) Energy dispersive X-ray (EDX) analysis of the cross-section area of the beads.

In this work, the freeze-drying method was preferred over the air-drying technique since the resulting beads could provide a network structure with more easy accessibility, and the regular spherical shape of the particles may improve packaging into the fixed bed (improving the mass transfer of the system). According to Borisova et al. [44] a better storage/conservation of the structures of the hydrogels can be performed by a novel technique based on the supercritical conditions of CO_2

(scCO$_2$), nevertheless the freeze-drying technique is less expensive than scCO$_2$ and the resulting dried materials adequately conserve their original structures [45].

Additionally, EDX-analysis were used for examining the cartography on the cross-section area of ChiFer(III) after neodymium removal from aqueous solutions. Figure 3 gives a qualitative representation of the elements contained in the volume of sorbent (such as carbon, oxygen, iron, and neodymium); neodymium appears to be homogenously distributed in the entire volume of the beads, which is useful information for interpreting the kinetic data.

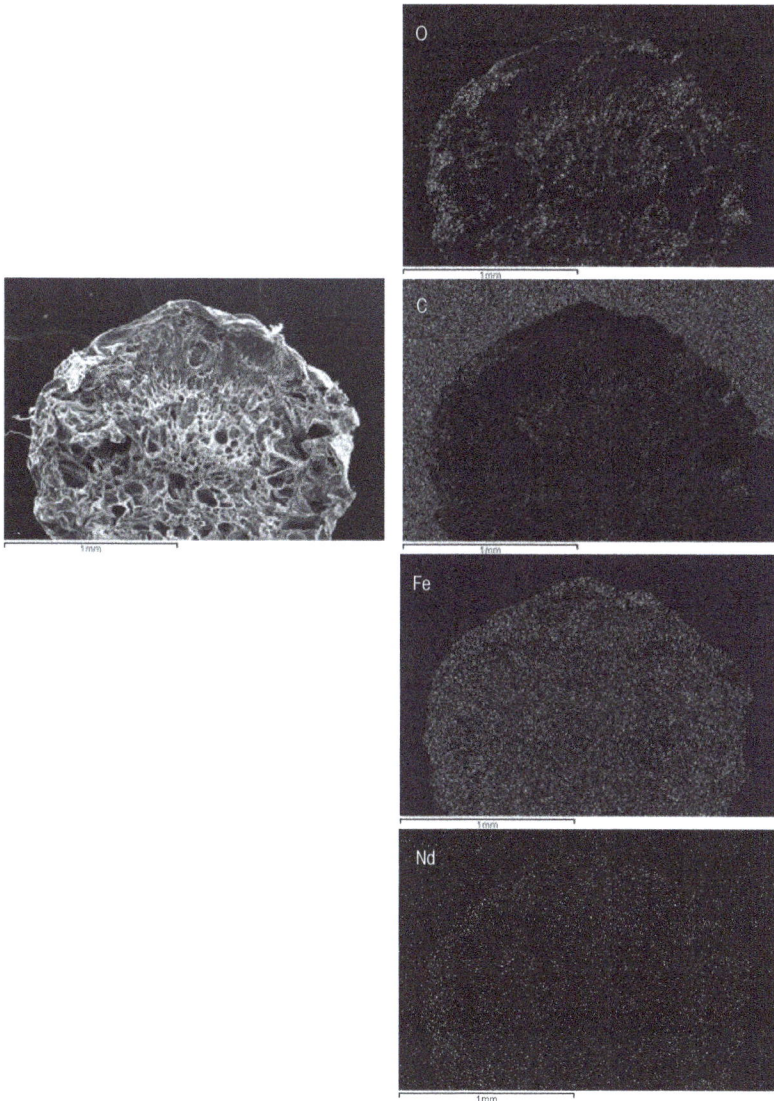

Figure 3. Distribution of the main components on the cross-section area of the ChiFer(III) material after sorption. (Right side: Original structure; left side: Distribution of O: Oxygen; C: Carbon; Fe: Iron; Nd: Neodymium).

Figure S2 (in the Supplementary Materials Section) shows the characteristics peaks (FTIR analyses) of the ChiFer(III) sorbent; the broad band centered at 3000–3600 cm^{-1} is due mainly to the hydroxyl groups (stretching vibrations of C–OH and Fe–OH) and to the extension vibration of N–H group of the primary amine (–NH$_2$) of chitosan [42]. The band at 2885 cm^{-1} is usually assigned to symmetric –CH$_2$ groups [28,46]; the bands at 1408 cm^{-1} and 1022 cm^{-1} are attributed to the C–O–C and to the C–O stretching vibrations, respectively [25,47], and the band centered at 1630 cm^{-1} is attributed to the stretching vibration of C=O (carboxylate groups of polysaccharides, [42]). The peak around the band at 898 cm^{-1} was attributed by Wang et al. [28] to the β-D-glucose unit of chitosan. The absorption band at 586 cm^{-1} is due to the Fe–O stretching vibration of Fe$_3$O$_4$, Shinde et al. [48] found that this band is specially shifted with an increase of Nd concentration (in contact with nickel ferrites material); in this study, the dilute concentrations of neodymium solutions (<0.01 mol·L^{-1}) did not lead to verification of this finding.

3.2. pH Effect

The pH is one of the most important parameters in the sorption process. The proton concentration influences metal speciation which impacts on metal/sorbent interactions, affecting the sorption performance. The plot of metal distribution of neodymium and iron species at dilute concentrations (1 mmol·L^{-1}) is represented in the Supplementary Materials Section (Figure S3); the experiments were conducted to avoid the precipitation phenomenon of neodymium.

Sorption performances of ChiFer(III) were compared with raw chitosan particles in the range of pH 1 to 6 (Figure 4a); at very acidic conditions the sorption is not favored since at pH below pH 3 some traces of iron are released into the solution, from pH < 2 the sorbents are completely dissolved (Figure S4), due to the natural hydrolysis of chitosan in acidic medium [49]. The best results and stability were obtained in the range of pH 4–6; the material is sufficiently stable and the introduction of iron into the chitosan matrix may enhance the sorption capacity of raw chitosan: at pH 5 the ChiFer(III) material increased by two times the removal achieved by chitosan particles (i.e., 98.8% and 49.3% respectively).

Figure 4b,c show the pH variation after metal sorption: when the initial pH was in the range 3–4, the equilibrium pH increased up to 4.4–5.0; for initial values of pH 5.0–6.5 the equilibrium pH tends to stabilize around pH 4.8–5.5. This type of "buffering effect" can be attributed to the acid–basic properties of chitosan; according to Sorlier et al. [50], the pKa of chitosan depends on parameters such as the deacetylation degree and the ionization extent of the polymer. The pKa for standard samples varies between 6.3 and 6.8; when the solution pH is below the pKa, chitosan tends to bind protons increasing the pH. Hence, at low pH values, the coordinating atoms in the sorbent (amino groups) are protonated which in contact with Nd species lead to repulsive electrostatic forces, limiting the sorption onto the sorbents materials.

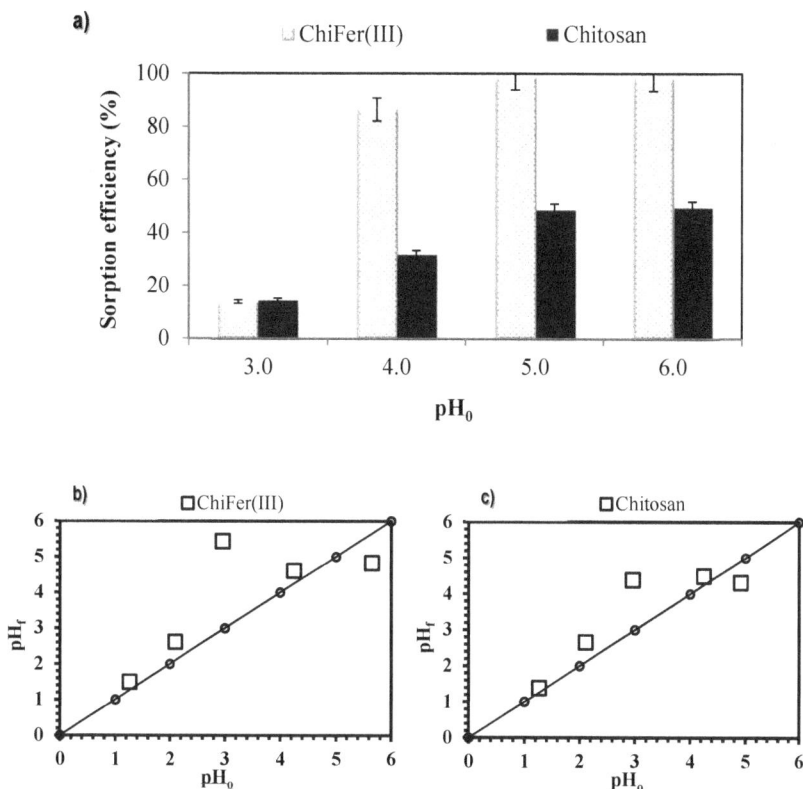

Figure 4. Influence of pH on neodymium removal. (**a**) Sorption efficiency. (**b**) Variation in pH using ChiFer(III) as sorbent. (**c**) Variation in pH using chitosan as sorbent. (T: 20 °C; sorbent dosage, SD: 1 g·L^{-1}; agitation speed: 150 rpm; contact time: 72 h; C_0: 9.4 mg·L^{-1}).

3.3. Equilibrium Studies

Isotherms are critical in optimizing sorption processes since they help to describe the distribution of the sorbate between the liquid and solid phases at equilibrium. The correlation of data by theoretical or empirical equations is essential for the practical design of the sorption systems; different models have been proposed in the literature including Langmuir, Freundlich, and Sips equations, nevertheless, the fit of the experimental data does not mean that the principles of the models are verified, although it leads to a better interpretation of the sorption mechanism.

Figure 5 shows the impact of the metal concentration on the sorption uptake; increasing the neodymium concentration increases the metal removal progressively until a saturation plateau is reached (beyond which no further sorption can take place) [51]. At dilute metal concentrations (i.e., 5 mg·L^{-1}) the sorbents allow removing of more than 60% of Nd; it meets the characteristics of a typical favorable sorption profile (asymptotic shape of the isotherm is consistent with the Langmuir equation), making these materials interesting for the treatment of dilute effluents.

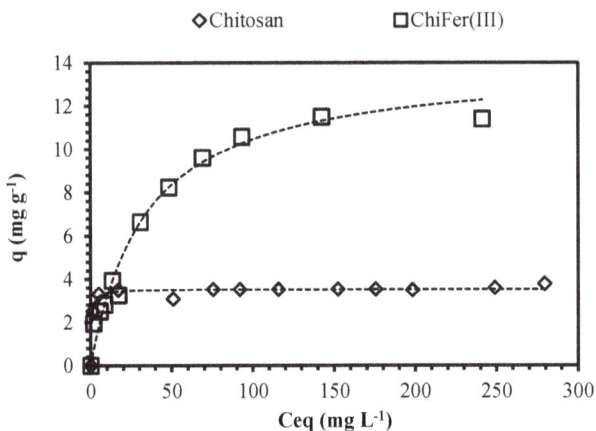

Figure 5. Isotherm plots for neodymium removal. (Solid line: Langmuir model; T: 20 °C; sorbent dosage, SD: 1 g·L^{-1}; agitation speed: 150 rpm; contact time: 72 h; pH: 4; C_0: 5–400 mg·L^{-1}).

Table 1 shows the fitting values of the Langmuir, Freundlich, and Sips models; the experimental data were fitted with Langmuir equations with better accuracy. Sips equation is a common adaptation of the Langmuir–Freundlich models; values for 1/ns close to zero are generally associated with heterogeneous sorbents, while values closer to 1.0 correspond to sorbents with relatively homogeneous binding sites [36]. Sorption capacity of ChiFer(III) is almost four times higher than raw chitosan particles, it is consistent with the values obtained in the pH study: the introduction of iron into the composite increases the removal of Nd (the maximum sorption capacities obtained are 3.5 mg·g^{-1} and 13.8 mg·g^{-1} for chitosan and ChiFer(III) respectively); taking into account that iron content in the composite is 0.33 g per gram of dried ChiFer(III), the contribution of iron is calculated as 30.9 mg Nd/g Fe, which is very close to the value reported by Tu et al. [4] using magnetic iron oxide (Fe$_3$O$_4$) particles as sorbent (Table 2).

Table 1. Langmuir, Freundlich, and Sips constants of ChiFer(III) and chitosan particles.

Experimental			Langmuir			Freundlich			Sips			
Sorbent	q_{exp} (mg·g^{-1})	q_{max} (mg·g^{-1})	b (L·mg^{-1})	r^2	K_F (mg$^{1-1/n}$ g^{-1}·L$^{1/n}$)	n	r^2	q_{max} (mg·g^{-1})	Ks (L·mg^{-1})	ns	r^2
Chitosan	3.77	3.54	1.91	0.971	2.78	20.44	0.957	3.51	1.75	0.80	0.966
ChiFer(III)	11.51	13.76	0.03	0.973	1.65	2.63	0.922	13.80	0.03	1.00	0.969

Moreover, the configuration of the material under the form of spherical beads contributes to improving the handling of the sorbent for a fixed packed (column) system; it avoids the typical operational difficulties found for non-regular particles (as chitosan): (i) high head-loss and clogging effect; (ii) diffusional problems; (iii) reduction of the mass transfer performance due to a non-uniform arrangement of the bed in the column. The findings can be compared with those obtained by Wang et al. [24]; the authors evaluated calcium alginate under beads configuration and a novel hybrid gel (ALG-PGA: prepared from calcium alginate (ALG) and γ-poly glutamic acid (PGA)); the results demonstrated that the introduction of PGA molecules into the alginate matrix, significantly enhanced the sorption capacity of the final sorbent. Thus, polymeric structures may improve the management of active materials and according to the functional groups may contribute to increasing metal uptake.

Table 2 shows a comparison of the maximum sorption capacities of several sorbents found in the literature; ChiFer(III) has a sorption uptake of the same order of magnitude. Although SiO_2/CMCH material [42] has a higher performance, the simplicity of the manufacturing process and the hydro-dynamic ability for columns make ChiFer(III) beads a promising sorbent for neodymium removal from aqueous solutions.

Table 2. Comparison of sorption capacities for several sorbents.

Sorbent	pH Range	qmax ($mg \cdot g^{-1}$)	Authors
C. colliculosa yeast	1.5	10.0	Vlachou et al. [52]
K. marxianus yeast	1.5	12.0	Vlachou et al. [52]
D. hansenii yeast	1.5	10.0	Vlachou et al. [52]
ChiFer(III) beads	4.0–5.0	13.8	This work
Cysteine-functionalized chitosan magnetic particles	6.0	15.3–17.1	Galhoum et al. [25]
Magnetic iron oxide (Fe_3O_4)	8.2	24.8	Tu et al. [4]
Ion imprinted polymer	7.0–7.5	35.2	Guo et al. [53]
SiO_2/Carboxymethyl chitosan (CMCH)	6.9	53.0	Wang et al. [42]

3.4. Influence of Contact Time

Figure 6 shows the kinetic profiles for sorption of neodymium using freeze-dried beads (FD), air-dried beads (AD), and raw chitosan particles; in the literature, it was reported that the drying procedure may impact on the kinetic profiles [43]. Different techniques have been used for maintaining the original structures of hydrogels after drying. Borisova et al. [44] suggested that drying under supercritical CO_2 conditions may improve the storage and lead to conserving the porosity of the manufactured materials (as a consequence, accessibility into active sites is not strongly modified); however it could be cost-expensive (in terms of energy) for drying large amounts of materials (at industrial scale production). In turn, freeze-drying and air-drying techniques (which are easy to operate at large-scale production) were compared in order to evaluate their contribution to the kinetic profiles.

ChiFer(III) beads and chitosan particles have a similar trend, and regardless of the type of drying, three pseudo steps can be considered: (i) an initial fast step that takes around 20 min for chitosan particles and 30–45 min for FD and AD beads (it represents 12%, 14%, and 31% of the total metal sorption using AD, chitosan and FD beads respectively); (ii) a second step which takes 3–6 h and accounts for an increase of 2%–13%; (iii) a third step which accounts for a lesser metal uptake (<5%). The difference in metal sorption (of each step) can be explained by the fact that the concentration gradient decreases progressively with the time: at longer contact time, the mass transfer is slower, since the driving force decreases.

It is noteworthy that kinetic uptakes are controlled by several mechanisms, including: (i) bulk diffusion; (ii) external diffusion (also called film diffusion); (iii) intraparticle diffusion; and (iv) reaction rate. Benettayeb et al. [54] pointed out that by maintaining a sufficient agitation speed and avoiding the settling of the sorbent, the contribution to the bulk diffusion could be neglected: this consideration was especially taken into account for performing the kinetic experiments and the system was kept continuously agitated (150 rpm).

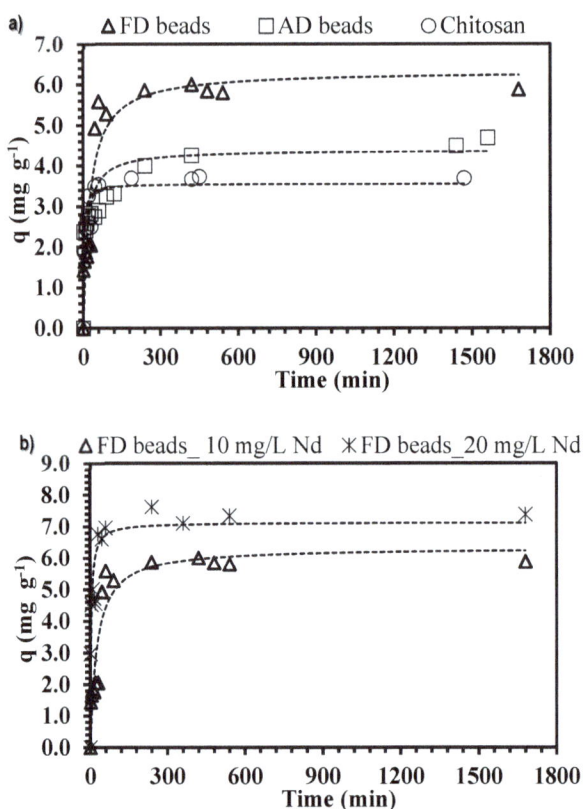

Figure 6. Effect of contact time on ChiFer(III) material performance. (**a**) Effect of drying method. (**b**) Effect of initial metal concentration. (Dashed line: Pseudo-second order rate equation (PSORE); T: 20 °C; sorbent dosage, SD: 0.5 $g \cdot L^{-1}$; agitation speed: 150 rpm; contact time: 72 h; pH: 4; C_0: 10–20 $mg \cdot L^{-1}$).

As expected (Figure 6a), the time for achieving the equilibrium is shortened for chitosan due to the small particles size (0.5 mm < Sp < 1 mm) which results in more available surface area, impacting on the resistance to the external film diffusion and a greater number of active sites are quickly available. Moreover, the time required for FD is less than for AD beads (5 h for FD and 7 h for AD); it can be explained by the open structure of the sorbent, produced as a result of the freeze-drying technique, which improves the diffusion and the accessibility of neodymium into the active sites. Table 3 reports the comparison of the experimental sorption capacities at equilibrium with calculated values for both the pseudo-first order and the pseudo-second order rate equation models (PFORE and PSORE); the correlation coefficients confirm a better fitting with PSORE.

Additionally, the simplified equation of Weber and Morris, [37] (W&M) was tested for evaluating the contribution of the resistance in the intraparticle diffusion. The plots on Supplementary Materials Section (Figure S5) have three slopes (corresponding to the external surface diffusion, intraparticle diffusion, and reaction rate steps), which do not pass through the origin; this is an indication that intraparticle diffusion is not the unique mechanism that controls neodymium removal. Thus, K_p values were obtained from the slope of the linear portions of the second phase of the kinetics profiles, these do not show a significant trend (Table 3), but it is evident that the K_p value for FD beads is higher than for AD beads at the same operating conditions (this confirms the easy accessibility and a better diffusion is performed into the pores of FD material). Nevertheless, for chitosan particles, the initial slope (attributed to the external diffusion) is considerably steeper than the second step (attributed to the intraparticle diffusion); this verifies that intraparticle diffusion is not only the controlling step of the process.

Table 3. Kinetic parameters of sorbent materials.

Experimental		Pseudo-First Order Rate Equation (PFORE)			Pseudo-Second Order Rate Equation (PSORE)			Weber and Morris Equation (W&M)
Sorbent	q_{exp} $(mg \cdot g^{-1})$	K_1 (min^{-1})	q_1 $(mg \cdot g^{-1})$	r_2	K_2 $(g \cdot mg^{-1} \cdot min^{-1})$	q_2 $(mg \cdot g^{-1})$	r^2	K_p $(mg \cdot g^{-1} \cdot min^{-1/2})$
Air-dried, AD ($C_0 = 10$ mg·g^{-1})	4.50	5.95×10^{-2}	3.96	0.455	1.69×10^{-2}	4.39	0.619	0.14
Freeze-dried, FD ($C_0 = 10$ mg·g^{-1})	5.89	2.78×10^{-2}	5.93	0.908	6.16×10^{-3}	6.34	0.887	1.24
Chitosan ($C_0 = 10$ mg·g^{-1})	3.74	0.177	3.44	0.757	0.11	3.56	0.851	2.81×10^{-2}
FD ($C_0 = 20$ mg·g^{-1})	7.39	0.129	6.88	0.778	3.9×10^{-2}	7.13	0.885	0.64

Furthermore, increasing the initial concentration of neodymium (Figure 6b), increased the uptake in the initial and the second sections of the kinetic profiles: the contact time for achieving the equilibrium is reduced (4.5 h are required for reaching complete saturation plateau). It means that an increase of the metal concentration involves an increase of the concentration gradient between the solution and the internal/external surface of the ChiFer(III) material; as a consequence, the driving force increases which also involves a faster migration of ions and a higher sorption rate [45].

Moreover, Rorrer et al. [55] pointed out that a pore blockage mechanism may occur at low metal concentrations; the sorbed species flux is low, resulting in the accumulation of the sorbate species at the entrance to the porous network, which could finally block the external cavity of the pore. This mechanism was not completely evidenced in this study.

3.5. Dynamic System

In sorption processes, an important task for determining the feasibility of a promising sorbent is the capability of being used in a continuous system. Columns experiments are very useful to approach the real applications in the treatment of industrial effluents; although most of the results, currently present in the literature are reported on the batch system; the relevance of columns must be highlighted since the assessment of the monitoring parameters (such as flow rate, inlet metal concentration, pH, internal diameter of the column, and bed-depth of the packed sorbent) leads to improve the design and enhances the scaling-up of the sorption technique as a function of the sewage characteristics.

The simplicity of using a fixed amount of packed sorbent for treating large volumes of polluted waters (instead of a batch configuration) involves a better engineering conception of the material; the operating time could be easier optimized and the unnecessary treatment steps could be avoided (such as filtration of the loaded sorbent). According to the experimental data, the breakthrough curves were plotted for neodymium removal from dilute aqueous solutions, the sorption capacities were calculated until the breaking (q_{BP}) and exhaustion (q_{exp}) points.

Figure 7 shows the typical S-shape of the curves, by using two different flow rates (0.01 and 0.02 L·h^{-1}); two common steps can be noted: (i) a first step where the neodymium species are totally sorbed; and (ii) a second step characterized by the progressive appearance of the metal, due to the continuous competition for occupying the active sites into the sorbent [56].

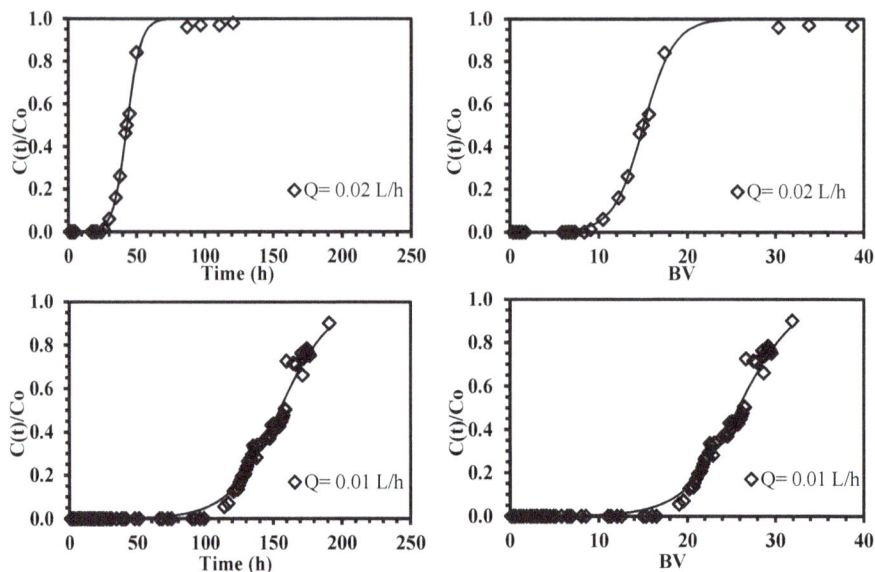

Figure 7. Continuous sorption of neodymium using ChiFer(III) material as fixed-packed sorbent. (**Left side**: Breakthrough curves as a function of time; **right side**: Breakthrough curves as a function of bed-volume; solid line: Thomas model; T: 20 °C; internal diameter of the column, Ø: 1.8 cm; bed depth: 23 cm; flow rate: 0.01–0.02 L·h^{-1}; pH: 4; C_0: 10.2 mg·Nd(III)·L^{-1}).

The breakthrough curves are similar in shape, but in terms of breakthrough time are substantially different; a decrease in the superficial velocity involves an increase of the effectiveness of the process; approximately 5% of the feed concentration is leached from the column after nine bed-volumes using a flow rate of 0.02 L·h^{-1}; this result is improved when the flow rate is halved (i.e., 0.01 L·h^{-1}), the sorption efficiency is increased from 25% to 79% and 19 bed-volumes are required for achieving the breaking point (Table 4). These differences are directly related to the residence time for mass transfer; when the higher superficial velocity is used, the external film diffusion is reduced, thus faster breaking and saturation times are produced. These findings are in agreement with the results obtained by Muhamad et al. [57] and Demey et al. [45] on heavy metal removal on using the biosorption process.

Table 4. Sorption parameters for continuous system using ChiFer(III) material as sorbent.

	Experimental					Thomas Parameters		
Sorbate	Q (L·h^{-1})	q_{exp} (mg·g^{-1})	Sorption Efficiency (%)	q_{BP} (mg·g^{-1})	Bed-Volume (BV$_{BP}$)	q_T (mg·g^{-1})	K_T (L·h^{-1}·mg^{-1})	r^2
Nd	0.02	4.02	24.97	2.12	9.06	3.46	1.98×10^{-2}	0.984
Nd	0.01	6.40	79.17	4.79	19.03	6.55	5.27×10^{-3}	0.981

Additionally, the sorption capacity obtained in the exhaustion point (q_{exp}) is two times lower of that obtained in the batch system (i.e., 13.76 mg·g^{-1} in batch system and 6.40 mg·g^{-1} in column system); this fact was attributed by Kleinübing et al. [58] to the resistance to the film diffusion which was particularly active in dilute concentrations (below 0.5 mmol·L^{-1}), on the continuous removal of copper and nickel from aqueous solutions with marine alga species; the mass transfer could be improved by reducing the flow rate.

ChiFer(III) material was evaluated in this work through three consecutive sorption/desorption cycles (in the batch system using 0.1 mmol·L^{-1} neodymium solutions, Figure S6); the performance of the beads drastically decreased from the third sorption cycle (i.e., it drops from 99.1% in the first cycle to 30.3% in the third cycle). It is noteworthy that ChiFer(III) was previously tested for boron recovery from seawater in several sorption–desorption cycles; the stability of the material and its performance was not affected under extreme saline conditions [23]. Thus, further studies will be performed in order to enhance the reusability of the sorbent with real effluents (containing boron and neodymium), which will be the scope of a future work. In order to verify the selectivity sequence of the sorbed metals, an equimolar solution of boron and neodymium (0.1 mmol·L^{-1}) was passed into the column; this influent was prepared to approach a real application of metal recovery from dilute waste waters of permanent magnets (Nd-Fe-B).

The breakthrough curves from the binary mixture are plotted in Figure 8; three relevant features are highlighted: (i) the first step of the sorption curve is more prolonged for neodymium than for boron, which is evidence of the higher selectivity toward Nd ions; (ii) a significant reduction of the bed-volumes on neodymium appearance was obtained (in comparison with single metal solution, the bed-volume values were reduced from 19 to 3.9); the process becomes less efficient, this is a consequence of the strong competition of species for active sites into the sorbent material; iii) boron in presence of neodymium is eluted from the column in a shorter time (5% of the feed boron concentration is leached after 0.5 bed-volumes), thus, boron overshoot is produced: the outlet concentration is higher than inlet concentration (i.e., $C/C_0 > 1$). The overshoot in a column system is based on the proper nature of the sorption itself, the sorbent has a limited sorption capacity and the species with lower affinity are pushed off and displaced by the species of higher affinity [59]; this phenomenon is in agreement with Sag and Kutsal, [60].

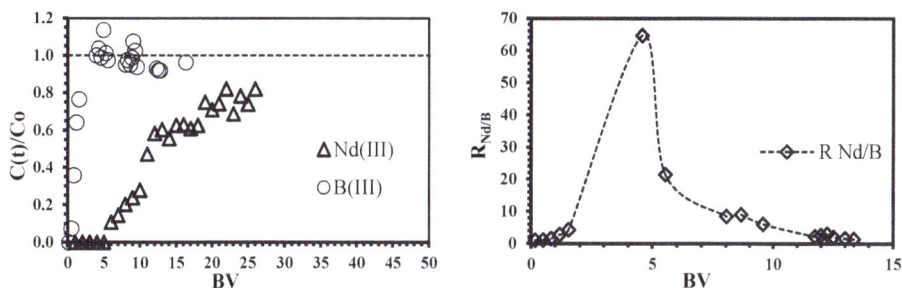

Figure 8. Simultaneous sorption of neodymium and boron using ChiFer(III) material as fixed-packed sorbent. (**Left side**: Breakthrough curves as a function of Bed-volume (BV), dashed line: overshooting guide line; **right side**: Separation coefficient $R_{Nd/B}$ as a function of BV; dashed line: experimental trend; T: 20 °C; internal diameter of the column, Ø: 1.8 cm; bed depth: 23 cm; flow rate: 0.01 L·h^{-1}; pH: 4; C_0: 0.1 mmol·L^{-1}).

Although the sorption of boron species (as co-ions) is almost negligible, due to the higher selectivity to Nd ions, the removal uptake of neodymium is inhibited (the sorption performance decreases from 79% in single solution to 49% in binary system, Tables 4 and S1). Some authors have suggested that the non-binding co-ions contribute to balancing the negative charges on the biomass surface (i.e., the electro-neutrality principle); this could inhibit electrostatically the binding of the main metal even though the co-ion is not bound to the sorbent [61]. The sorption mechanism onto iron surface was reported by Smith and Ghiassi, [62] and Grossl et al. [63] by using iron particles and goethite, respectively, as sorbents for chromate removal; the modeling results suggested the formation of monodentate inner-sphere complexes onto iron particles and bidentate inner-sphere complexes onto goethite. However, the interaction between the iron(III) hydroxide and boron species occurs similarly to an esterification reaction: both boric acid and borate ions can react with the hydroxide compounds to form the boric and borate esters [64]; this mechanism can only take place if the distance between adjacent OH^- groups in the $Fe(OH)_3$ structure are similar to those of OH^- groups of boron species [23,64].

The separation factor for equimolar solutions ($R_{Nd/B}$) was plotted as a function of bed-volume (Figure 8). The values of $R_{Nd/B} >>1$ confirm the higher neodymium selectivity of the sorbent; as expected, the point of maximum separation degree corresponds to 4.6 bed-volumes, which is also the point of the maximum overshooting of boron; consequently, the affinity of ChiFer(III) follows Nd $>>$B. The regeneration of the column was carried out with dilute HCl solution (demineralized water at pH 3.5); Figure 9 shows that only 0.6 L of eluent is enough for metal recovery: 20% of the total sorbed Nd and 30% of boron (it represents the remaining boron which was not previously overshot). The possibility of recovering the sorbed metals from the loaded ChiFer(III), makes this sorbent interesting for future separation of neodymium from real effluents.

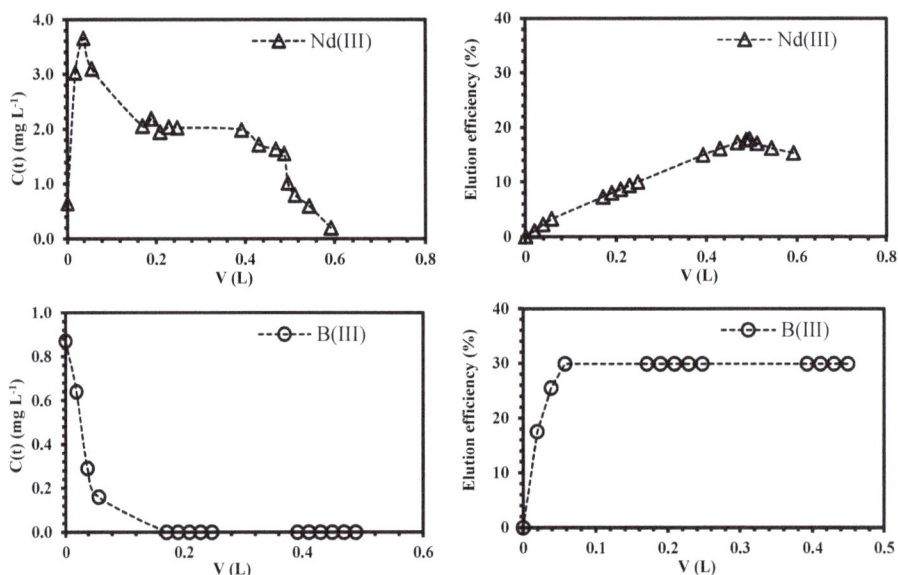

Figure 9. Simultaneous desorption of neodymium and boron from loaded ChiFer(III) material as fixed-packed sorbent. (**Left side**: Recovery metal concentration as a function of the volume (L), dashed line: desorption trend; **right side**: Elution efficiency as a function of the volume (L); dashed line: efficiency trend; T: 20 °C; internal diameter of the column, Ø: 1.8 cm; bed depth: 23 cm; flow rate: 0.01 L·h^{-1}; Eluent: demineralized water at pH: 3.5).

4. Conclusions

ChiFer(III) material is a promising and efficient sorbent for neodymium recovery from effluent waters. The incorporation of iron(III) hydroxide into the chitosan matrix, improves the sorption efficiency of the resulting material. The maximum sorption capacities obtained were 3.5 mg·g^{-1} and 13.8 mg·g^{-1} by using chitosan and ChiFer(III), respectively. The Sips model and PSORE fitted accurately the sorption isotherms and the kinetic experimental data, respectively. Column experiments demonstrated the feasibility of recycling the sorbents; the regeneration of the packed-bed can be performed with only 0.6 L of eluent (demineralized water at pH 3.5). The evaluations with simulated effluent of boron and neodymium demonstrated the higher selectivity of the sorbent toward Nd ions, which is useful for future evaluations with real wastewater effluents; the faster overshooting of boron

Polymers **2018**, *10*, 204

in the columns enhances the separation of neodymium. The continuous sorption data can be described by the Thomas equation.

Supplementary Materials: The supplementary materials are available online at http://www.mdpi.com/2073-4360/10/2/204/s1.

Acknowledgments: This work was supported by the Spanish Ministry of Economy and Competitiveness, MINECO (Projects CTM20014-52770-R and CTM2017-83581-R). The authors would like to thank M. Dominguez and T. Trifonov (from Universitat Politecnica de Catalunya) for their technical assistance in this work. Byron Lapo would like to thank to C. Quezada, A. Borja, and J. Perez from Universidad Técnica de Machala for their special interest in this research topic.

Author Contributions: Hary Demey, Byron Lapo, Montserrat Ruiz, Agustin Fortuny, Muriel Marchand and Ana M. Sastre conceived and designed the experiments; Hary Demey and Byron Lapo performed the tests and collected the experimental data; the manuscript was written by Hary Demey and Byron Lapo. The revision of the results interpretations were performed by Ana M. Sastre, Montserrat Ruiz, Agustin Fortuny, and Muriel Marchand.

Conflicts of Interest: The authors declare no conflict of interest.

References

1. Cossu, R.; Williams, I.D. Urban mining: Concepts, terminology, challenges. *Waste Manag.* **2015**, *45*, 1–3. [CrossRef] [PubMed]
2. Oddo, G. Die molekularstruktur der radioaktiven atom. *Z. Aanorg. Allg. Chem.* **1914**, *87*, 253–268. [CrossRef]
3. Harkins, W.D. The evolution of the elements and the stability of complex atoms. I. A new periodic system which shows a relation between the abundance of the elements and the structure of the nuclei of atoms. *J. Am. Chem. Soc.* **1917**, *39*, 856–879. [CrossRef]
4. Tu, Y.J.; Lo, S.C.; You, C.F. Selective and fast recovery of neodymium from seawater by magnetic iron oxide Fe₃O₄. *Chem. Eng. J.* **2015**, *262*, 966–972. [CrossRef]
5. Rademaker, J.H.; Kleijn, R.; Yang, Y.X. Recycling as a strategy against rare earth element criticality: A systemic evaluation of the potential yield of NdFeB magnet recycling. *Environ. Sci. Technol.* **2013**, *47*, 10129–10136. [CrossRef] [PubMed]
6. U.S. Department of Energy. *Critical Materials Strategy, Advanced Research Projects Agency-Energy*; U.S. Department of Energy: Washington, DC, USA, 2011.
7. Mirbagheri, S.A.; Hosseini, S.N. Pilot plan investigation on petrochemical wastewater treatment for the removal of copper and chromium with the objective of reuse. *Desalination* **2005**, *171*, 85–93. [CrossRef]
8. Özverdi, A.; Erdem, M. Cu²⁺, Cd²⁺ and Pb²⁺ adsorption from aqueous solutions by pyrite and synthetic iron sulphide. *J. Hazard. Mater.* **2006**, *137*, 626–632. [CrossRef] [PubMed]
9. Aji, B.A.; Yavuz, Y.; Koparal, A.S. Electrocoagulation of heavy metals containing model wastewater using monopolar iron electrodes. *Sep. Purif. Technol.* **2012**, *86*, 248–254. [CrossRef]
10. Chen, G. Electrochemical technologies in wastewater treatment. *Sep. Purif. Technol.* **2004**, *38*, 11–41. [CrossRef]
11. Blöcher, C.; Dorda, J.; Mavrov, V.; Chmiel, H.; Lazaridis, N.K.; Matis, K.A. Hybrid flotation-membrane filtration process for the removal of heavy metal ions from wastewater. *Water Res.* **2003**, *37*, 4018–4046. [CrossRef]
12. Samper, E.; Rodriguez, M.; De la Rubia, M.A.; Prats, D. Removal of metal ions at low concentration by micellar-enhanced ultrafiltration (MEUF) using sodium dodecyl sulfate (SDS) and linear alkylbenzene sulfonate (LAS). *Sep. Purif. Technol.* **2009**, *65*, 337–342. [CrossRef]
13. Obon, E.; Fortuny, A.; Coll, M.T.; Sastre, A.M. Mathematical modeling of neodymium, terbium and dysprosium solvent extraction from chloride media using methyl-tri(octyl/decyl)ammonium oleate ionic liquid as extractant. *Hydrometallurgy* **2017**, *173*, 84–90. [CrossRef]
14. Edebali, S.; Pehlivan, E. Evaluation of chelate and cation exchange resins to remove copper ions. *Powder Technol.* **2016**, *301*, 520–525. [CrossRef]
15. Kang, S.Y.; Lee, J.U.; Moon, S.H.; Kim, K.W. Competitive adsorption characteristics of Co²⁺, Ni²⁺, and Cr³⁺ by IRN-77 cation exchange resin in synthesized wastewater. *Chemosphere* **2004**, *56*, 141–147. [CrossRef] [PubMed]

16. Elwakeel, K.Z.; Guibal, E. Selective removal of Hg(II) from aqueous solution by functionalized magnetic-macromolecular hybrid material. *Chem. Eng. J.* **2015**, *281*, 345–359. [CrossRef]

17. Mahfouz, M.G.; Galhoum, A.A.; Gomaa, N.A.; Abdel-Rehem, S.S.; Atia, A.A.; Vincent, T.; Guibal, E. Uranium extraction using magnetic nano-based particles of diethylenetriamine-functionalized chitosan: Equilibrium and kinetic studies. *Chem. Eng. J.* **2015**, *262*, 198–209. [CrossRef]

18. Demey, H.; Tria, S.A.; Soleri, R.; Guiseppi-Elie, A.; Bazin, I. Sorption of his-tagged Protein G and Protein G onto chitosan/divalent metal ion sorbent used for detection of microcystin-LR. *Environ. Sci. Pollut. Res. Int.* **2017**, *24*, 15–24. [CrossRef] [PubMed]

19. Zhuo, N.; Lan, Y.; Yang, W.; Yang, Z.; Li, X.; Zhou, X.; Liu, Y.; Shen, J.; Zhang, X. Adsorption of three selected pharmaceuticals and personal care products (PPCPs) onto MIL-101(Cr)/natural polymer composite beads. *Sep. Purif. Technol.* **2017**, *177*, 272–280. [CrossRef]

20. Pavon, S.; Kutucu, M.; Coll, M.T.; Fortuny, A.; Sastre, A.M. Comparison of Cyanex 272 and Cyanex 572 on the separation of Neodymium from a Nd/Tb/Dy mixture by pertraction. *J. Chem. Technol. Biotechnol.* **2017**. [CrossRef]

21. Yurekli, Y. Removal of heavy metals in wastewater by using zeolite nano-particles impregnated polysulfone membranes. *J. Hazard. Mater.* **2016**, *309*, 53–64. [CrossRef] [PubMed]

22. Oladipo, A.A.; Gazi, M. Targeted boron removal from highly-saline and boron-spiked seawater using magnetic nanobeads: Chemometric optimisation and modelling studies. *Chem. Eng. Res. Des.* **2017**, *121*, 329–338. [CrossRef]

23. Demey, H.; Vincent, T.; Ruiz, M.; Nogueras, M.; Sastre, A.M.; Guibal, E. Boron recovery from seawater with a new low-cost adsorbent material. *Chem. Eng. J.* **2014**, *254*, 463–471. [CrossRef]

24. Wang, F.; Zhao, J.; Wei, X.; Huo, F.; Li, W.; Huo, Q.; Liu, H. Adsorption of rare earths (III) by calcium alginate–poly glutamic acid hybrid gels. *J. Chem. Technol. Biotechnol.* **2014**, *89*, 969–977. [CrossRef]

25. Galhoum, A.A.; Mafhouz, M.G.; Abdel-Rehem, S.T.; Gomaa, N.A.; Atia, A.A.; Vincent, T.; Guibal, E. Cysteine-functionalized chitosan magnetic nano-based particles for the recovery of light and heavy rare earth metals: Uptake kinetics and sorption isotherms. *Nanomaterials (Basel)* **2015**, *5*, 154–179. [CrossRef] [PubMed]

26. Zhao, F.; Repo, E.; Meng, Y.; Wang, X.; Yin, D.; Sillanpää, M. An EDTA-β-cyclodextrin material for the adsorption of rare earth elements and its applications in preconcentration of rare earth elements in seawater. *J. Colloid Interface Sci.* **2016**, *465*, 215–224. [CrossRef] [PubMed]

27. Zhao, F.; Repo, E.; Song, Y.; Yin, D.; Hammouda, S.B.; Chen, L.; Kalliola, S.; Tang, J.; Tam, K.C.; Sillanpää, M. Polyethylenimine-cross-linked cellulose nanocrystals for highly efficient recovery of rare earth elements from water and mechanism study. *Green Chem.* **2017**, *19*, 4816–4828. [CrossRef]

28. Wang, G.H.; Liu, J.S.; Wang, X.G.; Xie, Z.Y.; Deng, N.S. Adsorption of uranium (VI) from aqueous solutions onto cross-linked chitosan. *J. Hazard. Mater.* **2009**, *168*, 1053–1058. [CrossRef] [PubMed]

29. Hosoba, M.; Oshita, K.; Katarina, R.K.; Takayanagi, T.; Oshima, M.; Motomizu, S. Synthesis of novel chitosan resin possessing histidine moiety and its application to the determination of trace silver by ICP-AES coupled with triplet automated-pretreatment system. *Anal. Chim. Acta* **2009**, *639*, 51–56. [CrossRef] [PubMed]

30. Thomas, H.C. Heterogeneous ion exchange in a flowing system. *J. Am. Chem. Soc.* **1944**, *66*, 1664–1666. [CrossRef]

31. Ruiz, M.; Sastre, A.M.; Zikan, M.C.; Guibal, E. Palladium sorption on glutaraldehyde-crosslinked chitosan in fixed-bed systems. *J. Appl. Polym. Sci.* **2001**, *81*, 153–165. [CrossRef]

32. Guibal, E.; Larkin, A.; Vincent, T.; Tobin, J.M. Chitosan sorbents for platinum sorption from dilute solutions. *Ind. Eng. Chem. Res.* **1999**, *38*, 401–412. [CrossRef]

33. Langmuir, I. The adsorption of gases on plane surfaces of glass, mica and platinum. *J. Am. Chem. Soc.* **1918**, *40*, 1361–1403. [CrossRef]

34. Freundlich, H.M.F. Über die adsorption in lösungen. *Z. Phys. Chem.* **1906**, *57A*, 385–470. [CrossRef]

35. El-Korashy, S.A.; Elwakeel, K.Z.; El-Hafeiz, A.A. Fabrication of bentonite/thiourea-formaldehyde composite material for Pb(II), Mn(VII) and Cr(VI) sorption: A combined basic study and industrial application. *J. Clean. Prod.* **2016**, *137*, 40–50. [CrossRef]

36. Zhao, F.; Repo, E.; Yin, D.; Chen, L.; Kalliola, S.; Tang, J.; Iakovleva, E.; Tam, K.C.; Sillanpää, M. One-pot synthesis of trifunctional chitosan-EDTA-β-cyclodextrin polymer for simultaneous removal of metals and organic micropollutants. *Sci. Rep.* **2017**, *7*, 1–14. [CrossRef] [PubMed]

37. Weber, W.J.; Morris, J.C. Kinetics of adsorption on carbon from solutions. *J. Sanit. Eng. Div. ASCE* **1963**, *89*, 31–60.

38. Lagergreen, S. Zur theorie der sogenannten adsorption gelöster stoffe. Kungliga Svenska Vetenskapsakademiens. *Handlingar* **1898**, *24*, 1–39.

39. Shi, X.; Li, Q.; Wang, T.; Lackner, K.S. Kinetic analysis of an anion exchange absorbent for CO_2 capture from ambient air. *PLoS ONE* **2017**, *12*, 1–12. [CrossRef] [PubMed]

40. Ho, Y.S.; McKay, G. Sorption of dye from aqueous solution by peat. *Chem. Eng. J.* **1998**, *70*, 115–124. [CrossRef]

41. Chu, K.H. Fixed bed sorption: Setting the record straight on the Bohart-Adams and Thomas models. *J. Hazard. Mater.* **2010**, *177*, 1006–1012. [CrossRef] [PubMed]

42. Wang, F.; Zhao, J.; Zhou, H.; Li, W.; Sui, N.; Lui, H. O-carboxymethyl chitosan entrapped by silica: Preparation and adsorption behavior toward neodymium (III) ions. *J. Chem. Technol. Biotechnol.* **2013**, *88*, 317–325. [CrossRef]

43. Ruiz, M.; Sastre, A.M.; Guibal, E. Pd and Pt recovery using chitosan gel beads: I. Influence of drying process on diffusion properties. *Sep. Sci. Technol.* **2002**, *37*, 2143–2166. [CrossRef]

44. Borisova, A.; De Bruyn, M.; Budarin, V.L.; Shuttleworth, P.S.; Dodson, J.R.; Segatto, M.L.; Clark, J.H. A sustainable freeze-drying route to porous polysaccharide with tailored hierarchical meso-and macroporosity. *Macromol. Rapid Commun.* **2015**, *36*, 774–779. [CrossRef] [PubMed]

45. Demey, H.; Vincent, T.; Guibal, E. A novel algal-based sorbent for heavy metal removal. *Chem. Eng. J.* **2018**, *332*, 582–595. [CrossRef]

46. Gazi, M.; Shahmohammadi, S. Removal of trace boron from aqueous solutions using iminobis-(propylene glycol) modified chitosan beads. *React. Funct. Polym.* **2012**, *72*, 680–686. [CrossRef]

47. Demey, H.; Vincent, T.; Ruiz, M.; Sastre, A.M.; Guibal, E. Development of a new chitosan/Ni(OH)$_2$-based sorbent for boron removal. *Chem. Eng. J.* **2014**, *244*, 576–586. [CrossRef]

48. Shinde, T.J.; Gadkari, A.B.; Vasambekar, P.N. Influence of Nd^{+3} substitution on structural, electrical and magnetic properties of nanocrystalline nickel ferrites. *J. Alloy Compd.* **2012**, *513*, 80–95. [CrossRef]

49. Sakai, Y.; Hayano, K.; Yoshioka, H.; Yoshioka, H. A novel method of dissolving chitosan in water for industrial applications. *Polym. J.* **2001**, *33*, 640–642. [CrossRef]

50. Sorlier, P.; Viton, C.; Domard, A. Relation between solution properties and degree of acetylation of chitosan: Role of aging. *Biomacromolecules* **2002**, *3*, 1336–1342. [CrossRef] [PubMed]

51. El-Sonbati, A.Z.; El-Deen, I.M.; El-Bindary, M.A. Adsorption of hazardous azorhodanine dye from an aqueous solution using rice straw fly ash. *J. Dispers. Sci. Technol.* **2016**, *37*, 715–722. [CrossRef]

52. Vlachou, A.; Symeopoulos, B.D.; Koutinas, A.A. A comparative study of neodymium sorption by yeast cells. *Radiochim. Acta* **2009**, *97*, 437–441. [CrossRef]

53. Guo, J.; Cai, J.; Su, Q. Ion imprinted polymer particles of neodymium: Synthesis, characterization and selective recognition. *J. Rare Earth* **2009**, *27*, 22–27. [CrossRef]

54. Benettayeb, A.; Guibal, E.; Morsli, A.; Kessas, R. Chemical modification of alginate for enhanced sorption of Cd(II), Cu(II) and Pb(II). *Chem. Eng. J.* **2017**, *316*, 704–714. [CrossRef]

55. Rorrer, G.L.; Hsien, T.Y.; Way, J.D. Synthesis of porous-magnetic chitosan beads for removal of cadmium ions from waste water. *Ind. Eng. Chem. Res.* **1993**, *32*, 2170–2178. [CrossRef]

56. Chu, K.H. Improved fixed bed models for metal biosorption. *Chem. Eng. J.* **2004**, *97*, 233–239. [CrossRef]

57. Muhamad, H.; Doan, H.; Lohi, A. Batch and continuous fixed-bed column biosorption of Cd^{2+} and Cu^{2+}. *Chem. Eng. J.* **2010**, *158*, 369–377. [CrossRef]

58. Kleinübing, S.J.; Da Silva, E.A.; Da Silva, M.G.C.; Guibal, E. Equilibrium of Cu(II) and Ni(II) biosorption by marine alga *Sargassum filipendula* in a dynamic system: Competitiveness and selectivity. *Bioresour. Technol.* **2011**, *7*, 4610–4617. [CrossRef] [PubMed]

59. Escudero, C.; Poch, J.; Villaescusa, I. Modelling of breakthrough curves of single and binary mixtures of Cu(II), Cd(II), Ni(II) and Pb(II) sorption onto grape stalks waste. *Chem. Eng. J.* **2013**, *217*, 129–138. [CrossRef]

60. Sag, Y.; Kutsal, T. Recent trends in the biosorption of heavy metals: A review. *Biotechnol. Bioprocess Eng.* **2001**, *6*, 376–385. [CrossRef]

61. Shiewer, S.; Volesky, B. Ionic strength and electrostatic effects in biosorption of divalent metal ions and protons. *Environ. Sci. Technol.* **1997**, *31*, 2478–2485. [CrossRef]

62. Smith, E.; Ghiassi, K. Chromate removal by an iron sorbent: Mechanism and modeling. *Water Environ. Res.* **2006**, *78*, 84–93. [CrossRef] [PubMed]

63. Grossl, P.R.; Eick, M.J.; Sparks, D.L.; Goldberg, S.; Ainsworth, C.C. Arsenate and chromate retention mechanisms on goethite. 2. Kinetic evaluation using a pressure-jump relaxation technique. *Environ. Sci. Technol.* **1997**, *31*, 321–326. [CrossRef]

64. Ruiz, M.; Tobalina, C.; Demey-Cedeño, H.; Barron-Zambrano, J.A.; Sastre, A.M. Sorption of boron on calcium alginate gel beads. *React. Funct. Polym.* **2013**, *73*, 653–657. [CrossRef]

![polymers logo] *polymers*

MDPI

Article

Bioactive Sr(II)/Chitosan/Poly(ε-caprolactone) Scaffolds for Craniofacial Tissue Regeneration. In Vitro and In Vivo Behavior

Itzia Rodríguez-Méndez [1,†], Mar Fernández-Gutiérrez [2,3,†], Amairany Rodríguez-Navarrete [4], Raúl Rosales-Ibáñez [4], Lorena Benito-Garzón [5] , Blanca Vázquez-Lasa [2,3,*] and Julio San Román [2,3]

1   Faculty of Chemistry, Autonomous University of San Luis Potosi, San Luis Potosi 6, Salvador Nava Martínez, 78210 San Luis, S.L.P., Mexico; itziamdz@gmail.com
2   Institute of Polymer Science and Technology, ICTP-CSIC, C/Juan de la Cierva 3, 28006 Madrid, Spain; ictf339@ictp.csic.es (M.F.-G.); jsroman@ictp.csic.es (J.S.R.)
3   CIBER, Carlos III Health Institute, C/Monforte de Lemos 3-5, Pabellón 11, 28029 Madrid, Spain
4   Faculty of Higher Studies, National Autonomous University of Mexico, Av. Chalma s/n Col. La Pastora, Cuautepec Barrio Bajo. Delegación Gustavo A. Madero, Ciudad de México 07160, Mexico; cd.anyrn@outlook.com (A.R.-N.); dr.raul.rosales@gmail.com (R.R.-I.)
5   Faculty of Medicine, University of Salamanca, C/Alfonso X el Sabio, s/n, 37007 Salamanca, Spain; lorenabenito@usal.es
*   Correspondence: bvazquez@ictp.csic.es; Tel.: +34-915618806
†   These authors contributed equally to this work.

Received: 31 January 2018; Accepted: 2 March 2018; Published: 7 March 2018

Abstract: In craniofacial tissue regeneration, the current gold standard treatment is autologous bone grafting, however, it presents some disadvantages. Although new alternatives have emerged there is still an urgent demand of biodegradable scaffolds to act as extracellular matrix in the regeneration process. A potentially useful element in bone regeneration is strontium. It is known to promote stimulation of osteoblasts while inhibiting osteoclasts resorption, leading to neoformed bone. The present paper reports the preparation and characterization of strontium (Sr) containing hybrid scaffolds formed by a matrix of ionically cross-linked chitosan and microparticles of poly(ε-caprolactone) (PCL). These scaffolds of relatively facile fabrication were seeded with osteoblast-like cells (MG-63) and human bone marrow mesenchymal stem cells (hBMSCs) for application in craniofacial tissue regeneration. Membrane scaffolds were prepared using chitosan:PCL ratios of 1:2 and 1:1 and 5 wt % Sr salts. Characterization was performed addressing physico-chemical properties, swelling behavior, in vitro biological performance and in vivo biocompatibility. Overall, the composition, microstructure and swelling degree (≈245%) of scaffolds combine with the adequate dimensional stability, lack of toxicity, osteogenic activity in MG-63 cells and hBMSCs, along with the in vivo biocompatibility in rats allow considering this system as a promising biomaterial for the treatment of craniofacial tissue regeneration.

Keywords: chitosan; PCL; strontium; scaffolds; craniofacial engineering

1. Background

Reconstruction of large bone defects still continue as a major challenge for orthopedists, and craniomaxillofacial surgeons. The craniofacial hard tissues work as a functional unit and provide structural support, protection, sensation and allow movement. Defect and dysfunction of bone can result in devastating deficits of bone in the craniofacial skeleton [1]. Total or partial loss of bone has many psychological and behavioral problems associated with facial deformities [2]. The repair of complex craniofacial bone defects is challenging [3] and a successful result mainly lies in the choice of reconstructive method [4].

Bone tissue engineering approaches have been developed as an alternative to conventional use of autologous bone grafts, allografts or demineralized bone matrix from a donor tissue into the patient. Bone substitutes are formed by a biomaterial scaffold that acts as mimetic extracellular matrix (ECM) to induce new functional bone regeneration. The scaffolds usually loaded with osteoconductive/osteoinductive components and stem cells [5,6] are intrinsically biocompatible and some of them have reached clinical use with minimal adverse immunological reports [5]. Osteoinductive components such as bioactive glasses [7,8], phosphate-based glasses [9] or hydroxyapatite (HAp) [10] have been investigated. In addition, strontium (Sr), zinc (Zn), magnesium (Mg) or copper (Cu) have been used to dope or modified biomaterials [7]. Particularly, Sr(II) is known to play an important role in promoting bone formation and osteoblasts stimulation while inhibiting osteoclasts resorption [11]. However, in clinical practice the medication of Sr(II) salts has been restricted after the secondary problems associated with the systemic administration. Thus, the application of systems based on a local delivery of the Sr(II) derivatives can be considered as an adequate strategy in order to take advantage of the excellent properties of Sr(II) avoiding the secondary problems. Consequently, different biomaterials containing Sr(II) ions have been prepared in recent years, some of them based on ceramics [7–9,12,13] or composites of synthetic polymers [14–16].

It is clear that the latest trends in the preparation of constructs for bone tissue engineering, especially craniofacial repair, are directed towards the use of biomaterials scaffolds that accommodate stem cells [17]. Abundant literature has been reported using mesenchymal stem cells (MSCs) [18,19] for this application [20–25] as well as using human dental pulp stem cells (hDPSCs) [26] and adipose tissue derived mesenchymal stem cells [19].

Bone substitutes for craniofacial bone repair can be made of natural and/or synthetic polymers [27,28], calcium phosphate ceramics [29], metals [30–32] and composites [33,34]. Different biomaterials have been employed to mimic ECM in cleft palate reconstruction [35,36]. Resorbable bioactive systems based on synthetic polylactic acid (PLA), poly(lactic-*co*-glycolic acid) (PLGA), poly(ε-caprolactone) (PCL) or natural collagen have been employed as barrier membranes for guided bone regeneration (GBR) in oral and maxillofacial reconstruction [37]. A revision focused on therapy methods, growth factors and scaffolds in alveolar cleft defects has been published by Khojasteh et al. [38].

Natural polymers offer the advantage of good biocompatibility and are bioactive as they can interact with the host tissues [39]. Among natural polymers, chitin [40–43] and chitosan are excellent candidates [44,45]. Recently Anitha et al. published a review on their applications including a discussion about the chitinous scaffolds obtained from marine sponges [46]. Chitosan, the deacetylated form of chitin, offers some advantages [45,47–49] which extend its capabilities in the field [46,50]. Its disadvantages are the weak mechanical properties and high rate of degradability. Thus, it is usually cross-linked and/or combined with other natural/synthetic polymers (e.g., blends) or used in composites. Chitosan has ability to promote proliferation and mineral matrix deposition by osteoblasts in culture [51] and it allows osteoconduction [52]. Enhanced osteoconductive properties and osteoinductive behavior can be achieved using composite scaffolds with ceramics [53] and incorporating growth factors; all this provides osteogenic response [54,55]. Thus, the use of chitosan in orthopedic/periodontal applications [46,49,56] and craniofacial bone defects repair [37,57] has increased over the years as well as the development of hybrid systems [58–61]. As mentioned above, in order to increase bioactivity many chitosan composites have been developed [44,62]. These composites enhance the osteogenic potential of the calcium compounds at the time that the polymer matrix inhibits migration of calcium compounds [63]. A variety of chitosan composites have been tested in vitro or in vivo for bone and craniofacial regeneration [44,64,65]. Vaca-Cornejo et al. evaluated the effects of chitosan in combination with HAp to promote alveolar bone growth in patients with periodontitis. After twelve months of the therapeutic strategy the chitosan/HAp implant reduced the pocket depth of the supporting tissue, the grading of tooth mobility and promoted alveolar bone growth; the patients conserved the dental organ, favoring a better quality of life [66]. Composite membranes formed by chitosan/alginate polymers and octacalcium phosphate/bioactive glasses were suitable for adhesion and growth of human bone marrow mesenchymal stem cells (hBMSCs) [67]. In 2017, Zhou et al. evaluated whitlockite (WH)/chitosan composite membranes and HAp/chitosan scaffolds, and they concluded that WH/chitosan scaffold can significantly promote bone regeneration in calvarial defects [68]. Recently, Lu et al. published that high-activity chitosan/nano HAp (nHAp)/zoledronic acid scaffolds had a multifunction of tumor therapy, bone repair, and antibacterial [69]. Even though numerous strategies that are currently used to regenerate bone depend on employing biocompatible materials exhibiting a scaffold structure. In this sense, Guzmán et al. have immobilized calcium phosphate salts and/or bone morphogenetic protein 2 (BMP-2) into chitosan scaffolds. Interestingly, they found that this multicomponent scaffold exhibited a superior efficacy in bone regeneration than the scaffolds containing only one of the components, either calcium phosphate salts or rhBMP2, separately in maxillary sinus augmentation procedure [70].

Numerous composite systems containing chitosan have found application for guided tissue regeneration (GTR) in periodontal tissue engineering. Some examples are: HAp–chitin–chitosan composite formulated as a self-hardening paste [71], membranes composed of electrospun chitosan fibers [72], two-layer nanofibrous membranes made of polyglycerol sebacate (PGS)/PCL/β-tricalcium phosphate (β-TCP) and PCL/PGS/chitosan that provide flexibility, osteoconductivity and barrier properties [73]. In vivo experiments with hybrid composite nanofibers composed of fish collagen/chitosan/bioactive glasses (BG) demonstrated bone regeneration in a furcation defect of dogs [74]. Tamburaci and Tihminlioglu mentioned that the incorporation of diatomite to chitosan polymer matrix significantly enhanced the osteoblast-like cell proliferation on membrane surface and can be used as an ideal candidate for GTR [75].

In addition, the preparation of chitosan based scaffolds doped with strontium has been addressed for bone tissue engineering and craniomaxillofacial repair and many of them are based on Sr-doped ceramics. Chitosan/strontium HAp (SrHAp) nanohybrid scaffolds with interconnected macropores and SrHAp nanocrystals produced favorable adhesion, spreading and proliferation of hBMSCs [10]. In addition, the Sr(II) ions released from the nanohybrid scaffolds enhanced alkaline phosphatase (ALP) activity and ECM mineralization [10]. Three-dimensional Ag-loaded SrHAp/chitosan porous scaffold also provided good support for the adhesion, spreading and proliferation of hBMSCs [76] showing that the Sr element increased the ALP activity, ECM mineralization, and the expression levels of osteogenic-related genes BMP-2 and collagen-I. Masaeli et al. studied the performance of a SrHAp additive in calcium phosphate cement. In vitro biological characteristics revealed that incorporation of 3 wt % SrHAp could cause ALP activity increase, which may be due to the presence of strontium ions [77]. Recently, our research group has developed semi-interpenetrating polymer networks (semi-IPNs) of biohybrid scaffolds composed of chitosan/polyethylene glycol dimethacrylate/β-TCP scaffolds loaded with a biocompatible strontium salt (i.e., strontium folate (SrFO) [78]). The scaffolds were seeded with stem cells obtained from hDPSCs to study the regeneration of bone using a critical sized defect model of calvaria in rats. The in vitro and in vivo results demonstrated excellent cytocompatibility with resorption of scaffolds in a period of 4–6 weeks and a total regeneration of the defect, with a more rapid and dense bone formation in the group with SrFO compared with unloaded scaffold [79].

The aim of this work focuses on the preparation of Sr(II) hybrid bioactive scaffolds applicable for regeneration of craniofacial defects. The designed scaffolds intend to cover the actual niche in clinical practice. The rational is to produce a biomaterial that is able of regenerating good quality of bone in the short or medium terms using biomaterials that are clinically employed in biomedical devices, i.e., chitosan [80] and PCL [81], to favor their translation to the commercial and clinical fields. In addition, the scaffolds contain Sr(II) as an osteogenic compound. Up to our knowledge, scaffolds of this composition are not reported yet in literature.

Thus, the paper describes the fabrication of Sr(II) impregnated chitosan/PCL hybrid scaffolds by a relatively simple two-steps method, aiming the Sr(II) availability for regeneration processes. Their morphology, structural characterization, physicochemical properties as well as swelling behavior are analyzed. Likewise, in vitro cytotoxicity and biological performance studying their osteogenic response are evaluated using osteoblast-like cells (MG-63) and hBMSCs. Finally in vivo biocompatibility experiments are carried out applying a subcutaneous pocket rat model.

2. Experiment

2.1. Materials

Chitosan (Ch) with degree of acetylation DA = 15% and intrinsic viscosity = 457 mL/G (25 °C, 0.1 M AcOH + 0.2 M NaCl) was gently provided by IDEBIO S.L. (Salamanca, Spain). Pharmaceutical grade chitosan with DA = 10% and M_w = 300 kDa purchased from Altakitin (Aveiro, Portugal) was used for biological and in vivo experiments. Poly(ε-caprolactone) (PCL, M_w = 14 kDa, Sigma-Aldrich, Madrid, Spain), sodium tripolyphosphate (TPP, 85% Sigma-Aldrich, Madrid, Spain), strontium fluoride (SrF$_2$, Sigma-Aldrich, Madrid, Spain), 1,2-dichlorometane (DCM, Sigma-Aldrich, Madrid, Spain) and phosphate buffered saline solution (PBS) (pH = 7, Scharlau, Barcelona, Spain) were used as received.

2.2. Preparation of Scaffolds

2D membrane scaffolds with Ch:PCL ratios (*wt/wt*) of 1:2 and 1:1 were obtained by a casting/solvent evaporation technique. Briefly, in the first step chitosan (1 wt %) was dissolved in an aqueous solution of glacial acetic acid (0.25 wt %); separately, the PCL was dissolved in DCM and added to the chitosan solution under stirring. The final mixture was poured onto a Teflon mold, and dried at room temperature until constant weight. Then, dried membranes were cross-linked by dipping in a TPP (20 wt % respect to chitosan) aqueous solution for 24 h at room temperature. Then, membranes were removed from the solution, washed with 50 mM NaCl solution and distilled water after neutral pH, and dried until constant weight. In the second step, cross-linked membranes were treated by a drop-by-drop deposition of the SrF$_2$ (5 wt % respect to chitosan) aqueous solution until complete wetting and left at room temperature for evaporation of solvent. Afterwards, treated membranes were washed with distilled water and dried at room temperature. The codes and compositions of the membrane scaffolds are shown in Table 1.

Table 1. Names and compositions of blank and Sr(II) containing membrane scaffolds.

Name	Code	Ch/PCL (*wt/wt*)	SrF$_2$ (wt % Respect to Ch)
Blank membranes	Ch/2PCL	1:2	-
	Ch/PCL	1:1	-
Sr(II) membranes	Sr/Ch/2PCL	1:2	5
	Sr/Ch/PCL	1:1	5

Ch: chitosan; PCL: poly(ε-caprolacone).

2.3. Characterization Techniques

Structural characterization was performed by attenuated total internal reflectance Fourier transform infrared (ATR-FTIR) spectroscopy with a Spectrum One apparatus (Perkin-Elmer, Madrid, Spain) spectrometer equipped with an ATR accessory.

Atomic composition of membranes was determined using a FE-SEM (Field emission scanning electron microscope, Tokyo, Japan) Hitachi SU-8000 with an energy dispersive X-rays (EDS) analyzer Bruker XFlash model Detector 5030 using a voltage of 8 keV.

Morphology of membranes was examined by scanning electron microscopy (SEM, Eindhoven, Hollad) using a Philips XL 30 microscope at an accelerating voltage of 25 kV. Thermal properties were analyzed by thermogravimetry (TGA) in a thermogravimetric analyzer TGA Q500 (TA Instruments, Cerdanyola del Vallés, Spain) apparatus, with a heating rate of 10 °C/min in a range of 40–500 °C and under nitrogen atmosphere (10 mL/min).

2.4. In Vitro Swelling Study

Swelling experiments were performed in PBS buffer (pH = 7) at 37 °C. Each sample was immersed in 5 mL of the medium and left to attain equilibrium under static conditions. The medium was replaced every 2 days. The percentage of the water uptake (*WU*) was calculated by Equation (1), where W_t is the weight of the sample at time t and W_d is the initial dry weight. Swelling measurements were performed at 1, 2, 7, 15, 30 and 45 days after immersion.

$$\% \ WU = [(W_t - W_d)/W_d] \times 100 \tag{1}$$

In all the experiments, a minimum of four replicates of each composition were measured and results averaged. Results are given as mean ± standard deviation (sd).

Additionally, to evaluate changes on the surface topology samples soaked for 30 days were washed with distilled water and dried for SEM analysis.

2.5. In Vitro Biological Assays

2.5.1. Cell Cultures

MG-63 osteoblast-like cell line (ECACC, Sigma, Madrid, Spain) and human bone marrow mesenchymal stem cells, hBMSCs (Innoprot, Vizcaya, Spain), were used to study the biological performance of membranes scaffolds. The culture medium for MG-63 line was Dulbecco's modified Eagle's medium enriched with 4500 mg/mL glucose (DMEM) (Sigma, Madrid, Spain) supplemented with 10% fetal bovine serum (FBS), 200 mM L-glutamine, 100 units/mL penicillin and 100 μg/mL streptomycin, and modified with HEPES (complete medium). In the case of hBMSCs the culture medium was basal medium supplemented with 5% of FBS, 5 mL of mesenchymal stem cell growth supplement (MSCGS), 100 units/mL penicillin and 100 μg/mL streptomycin (Innoprot, Vizcaya, Spain). Thermanox® (TMX) discs (Nunc) were used as negative control. Tested sample membranes (1.5 cm diameter) were sterilized with a UV lamp (HNS Osram, 263 nm, 3.6 UVC/W) at a power of 11 W for 4 h.

2.5.2. Biological Assays

MTT test [82] was used for indirect cytotoxicity. Tested samples were set up in 5 mL of FBS-free supplemented DMEM, placed on a shaker at 37 °C and extracts were obtained at 1, 2, 7, 14 and 21 days under sterile conditions. Cells were seeded at a density of 9×10^4 cells/mL in complete medium in a sterile 96-well culture plate and incubated to confluence. After 24 h incubation the medium was replaced with the corresponding extract and incubated at 37 °C in humidified air with 5% CO_2 for 24 h. A solution of MTT (0.5 mg/mL) was prepared in warm FBS-free supplemented DMEM and the plates were incubated at 37 °C for 3–4 h. Excess medium and MTT were removed and DMSO was added to all wells in order to dissolve the MTT taken up by the cells. This was mixed for 10 min and the absorbance was measured with a Biotek Synergy HT detector using a test wavelength of 570 nm and a reference wavelength of 630 nm. Relative cell viability was calculated from Equation (2):

$$\text{Relative cell viability (\%)} = 100 \times (OD_S - OD_B)/(OD_C - OD_B) \tag{2}$$

where OD_S, OD_B and OD_C are the optical density for the sample (S), blank (B) and control (C), respectively. Results are given as mean ± standard deviation (sd) ($n = 5$). Analysis of variance (ANOVA) of the results was performed comparing samples with TMX (* $p < 0.05$).

Quantitative analysis for cell adhesion and proliferation on membrane scaffolds was carried out by means of the Alamar Blue (AB) test [83]. Cells were seeded at a density of 4×10^4 cell/mL for 24 h over the specimens in a 24-well culture plate. At determinate times (1, 4, 7, 14 and 21 days), 1 mL of AB dye (10% AB solution in phenol red free DMEM medium) was added to each specimen. After 4 h of incubation 100 μL ($n = 4$) of culture medium for each test sample was transferred to a 96-well plate, and the fluorescence emission was measured at 590 nm in a Biotek Synergy HT. The specimens were washed twice with PBS to remove rest of the reagent, and 1 mL of culture medium was added to monitor the cells over the materials. Results are given as mean ± sd. ANOVA of the results of tested materials was performed comparing the corresponding Sr(II) and blank groups at the same time (* $p < 0.05$).

Total DNA was measured using the PicoGreen dsDNA quantitation kit (P-7589, Molecular Probes, Fisher Scientific, Madrid, Spain). The recently introduced fluorescent dye, PicoGreen, has several advantages over other methods because it is sensitive and specific for double-stranded DNA (dsDNA) [84].

Biochemical detection of ALP activity was used as an indicator of osteoblast phenotype [85]. The ALP activity was evaluated in confluent cells cultured in the presence of tested sample. ALP catalyzes the hydrolysis of *p*-nitrophenyl phosphate (pNPP) to *p*-nitrophenol. It has a strong absorbance at 405 nm. The rate of the increased absorbance at 405 nm is proportional to the enzyme activity. Determination of the ALP/DNA ratio is indicative of the amount of ALP activity per cell. The variations caused by the different shape of the test samples can be eliminated using this approach [86]. Both were measured from cell lysate. Results are given as mean ± sd. ANOVA of the results of tested materials was performed comparing the corresponding Sr(II) and blank groups at the same time (* $p < 0.05$).

Cell morphology was examined by SEM. To that end, samples were placed in a 24-well tissue-culture plate. Cells (4 × 10^4 cells per well) were added and allowed to attach at 37 °C. Samples were washed three times with distilled sterile water and fixed with 2.5% glutaraldehyde for 2 h at room temperature. The dried samples were mounted on aluminum stumps and sputter-coated with gold/palladium mix before examination under a SEM apparatus (Philips XL 30) at an accelerating voltage of 15 kV.

2.6. In Vivo Biocompatibility

2.6.1. Animal Experimentation

All animal studies were performed according to the national guidelines and conducted in accordance with Spanish law (RD 53/2013) and international standards on animal welfare as defined by European Directive (2010/63/EU). In addition, surgery protocols were approved by the Ethical Committee (Project Identification Ethical Committee: 211, 17 November 2017) of the University of Salamanca, Salamanca, Spain. The animals were housed in cages with pelleted food and water in a temperature-controlled room with a 12 h artificial day/night cycle at the Animal Experimentation Unit (A.E.U.) of the University of Salamanca. They were acclimatized far at least 2 weeks prior surgery.

2.6.2. Subcutaneous Implantation in Rats

The biocompatibility of the membranes was assessed in the subcutaneous of rats. Twenty-one albino Wistar male rats, body weight 250–300 g, were purchased from a certificated stockbreeder (Charles River, Barcelona, Spain). The rats were placed under general anesthesia by inhalation of 1.5% isofluorane (Forane®). Pre- and post-operative analgesia was provided by subcutaneous injection of buprenorphine (0.01–0.05 mg/kg). The back of each rat was depilated on using an electric shaver and disinfected with povidone iodine 10% solution (Betadine®). A sterile field was placed on the back of the animal. Incisions were made through the skin on each side to the midline, along the vertebral column, to made unconnected subcutaneous pouches for each sample (three incisions in total) with 2 cm distance from each other. Three unconnected subcutaneous pockets were created by means of blunt dissection. Each rat received 3 membranes (1 × 1 cm²), each one in a separate subcutaneous pocket: control membrane of bovine collagen (RCM6, ACE Surgical Supply Inc., Brockton, MA, USA) and selected studied membranes: Ch/PCL and Sr/Ch/PLC. After membrane implantation skin was sutured using non-absorbable 4/0 silk suture (Aragó, Barcelona, Spain). At each time period, 1, 2 and 4 weeks of implantation, animals (*n* = 7) were euthanized by lethal injection of 5% sodium pentobarbital

(Dolethal®) and implants retrieved for histological evaluation. After dissection, samples isolating each membrane were fixed in 4% neutral formaldehyde.

2.6.3. Histological Analysis

Once fixed, samples were placed in cassettes and dehydrated in ascending series of 70%, 80%, 90% and 100% ethanol solutions. Then, they were placed into ethanol/toluene and pure toluene before being immersed in liquid paraffin at 60 °C. Afterwards, samples were embedded in paraffin blocks at −20 °C. Blocks were cut by using a standard rotatory microtome (Micron HM310, Walldorf, Germany). Thin histology sections (5 μm) were made perpendicular to the plane of the skin and stained with hematoxylin and eosin (H-E).

Sections were microscopically blinded viewed to determine the histological reaction to the membranes, specifically the presence and degree of inflammatory cell response. Sections were also examined to observe fibrous tissue and vascularization.

3. Results

3.1. Preparation and Characterization of Membrane Scaffolds

Biohybrid Sr(II) containing Ch/PCL membranes were fabricated using a two-step way methodology as it is described in the experimental section. Chemical composition of the blank samples was analyzed by ATR-FTIR spectroscopy. ATR-FTIR spectra (Figure S1) presented the characteristic bands belonging to pure precursor polymers but with some differences. In the hybrid scaffolds the band between 3500 and 3200 cm^{-1} (associated υ O–H and υ N–H) broadened compared to pure chitosan. In the region between 1750–1500 cm^{-1}, the band at 1722 cm^{-1} for Ch/2PCL and 1723 cm^{-1} for Ch/PCL (υ C=O in ester groups belonging to PCL) shifted with respect to pure PCL (1721 cm^{-1}); the bands at 1656 cm^{-1} for Ch/2PCL and 1646 cm^{-1} for Ch/PCL (υ C=O in amide groups, amide I of chitosan) shifted compared to pure polymer (1654 cm^{-1}), mainly for the Ch/2PCL sample; the bands at 1587 cm^{-1} for Ch/2PCL and 1586 cm^{-1} for Ch/PCL (δ N–H in NH_3^+ groups) also appeared shifted respect to those of pure chitosan (1578 cm^{-1}). On the other hand, bands in the region between 1300–900 cm^{-1} were assigned to υ_{as} PO_2 groups in TPP ions [87], υ C–O in the pyranose ring of chitosan [88] and υ_s and υ_{as} C–O in PCL [89].

The atomic composition of dried Sr(II) samples was examined by EDS analysis. For both sample compositions EDS spectra exhibited the peaks of C and O pertaining to both PCL and chitosan polymers, the peak of N belonging exclusively to the polysaccharide and the peak of Sr centered in 1.8 keV, indicating the presence of the bioactive Sr(II). In addition, a peak centered at 2 keV (P) belonging to TPP polyanions was appreciated [90]. Figure 1 shows the EDS results for the main elements of Sr(II) samples. Although this analysis must be considered semi-quantitative, a higher amount of Sr was measured in the sample with higher content of PCL.

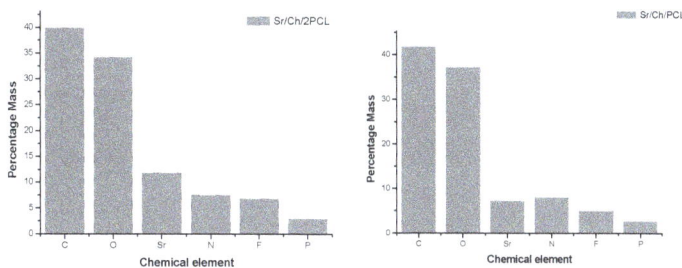

Figure 1. EDS chemical element percent mass of Sr/Ch/2PCL and Sr/Ch/PCL dried samples. Ch: chitosan; PCL: poly(ε-caprolactone).

Surface morphology of membranes with and without Sr(II) was analyzed by SEM and images are displayed in Figure 2. It is clear that all membranes presented phase separation morphology consisted of PCL microparticles dispersed in a continuous matrix of cross-linked chitosan. As far as Sr(II) membranes is concerned, the Sr salt was preferably located on the hydrophilic polysaccharide matrix. It seems that the hydrophobic character of PCL in water does not favor the diffusion of the Sr salt solution into the microparticles. For Ch:PCL 1:2 ratio, the higher concentration of PCL microparticles makes the chitosan phase more concentrated in the Sr salt than for the 1:1 ratio.

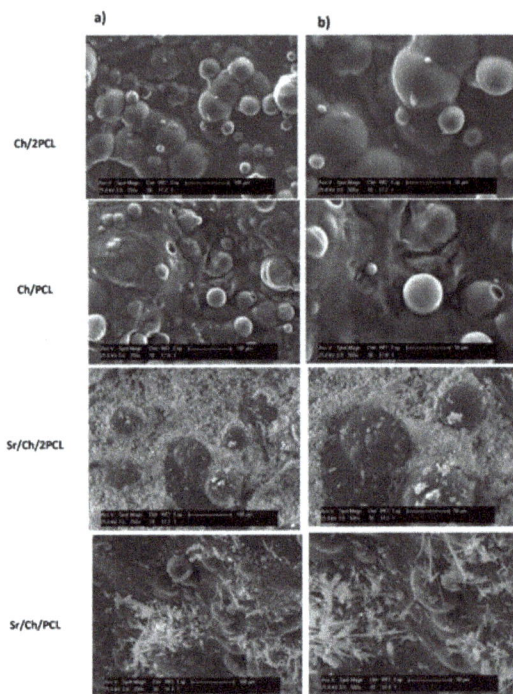

Figure 2. SEM images of dried blank and Sr(II) samples. (**a**) 250×; (**b**) 500×. Ch: chitosan; PCL: poly(ε-caprolactone).

3.2. Thermal Properties

Thermograms and Derivative Thermogravimetric analysis (DTG) of membranes obtained under nitrogen atmosphere are shown in Figure 3. Thermogravimetric results of both pure PCL and chitosan are shown in Figure S2.

Thermal degradation of PCL occurred in two steps of which DTG curve showed a T_{max1} at 307 °C and a T_{max2} at 419 °C. The first degradation step generates H_2O, CO_2, and 5-hexenoic acid as evolved products and the second one leads to the formation of ε-caprolactone (cyclic monomer) as a result of an unzipping depolymerization process [91]. TGA of chitosan showed a main degradation stage with maximum rate at 287 °C ascribed to deacetylation of the main chain and cleavage of their glycosidic linkages [88]. Blank and Sr(II) membranes thermally degraded in two steps, with T_{max1} and T_{max2} in DTG curves in the ranges 250–350 °C and 350–450 °C. The first step was ascribed to degradation of the chitosan matrix and part of PCL whereas the second one was only due to chain scission of the synthetic polymer. Table S1 shows the thermogravimetric results of all samples. There was little difference between blank and Sr(II) samples. In general, T_{max1} decreased compared to the average of those of both pure components. The drop was more market for the Ch/2PCL sample and it was closer to T_{max1} of the PCL.

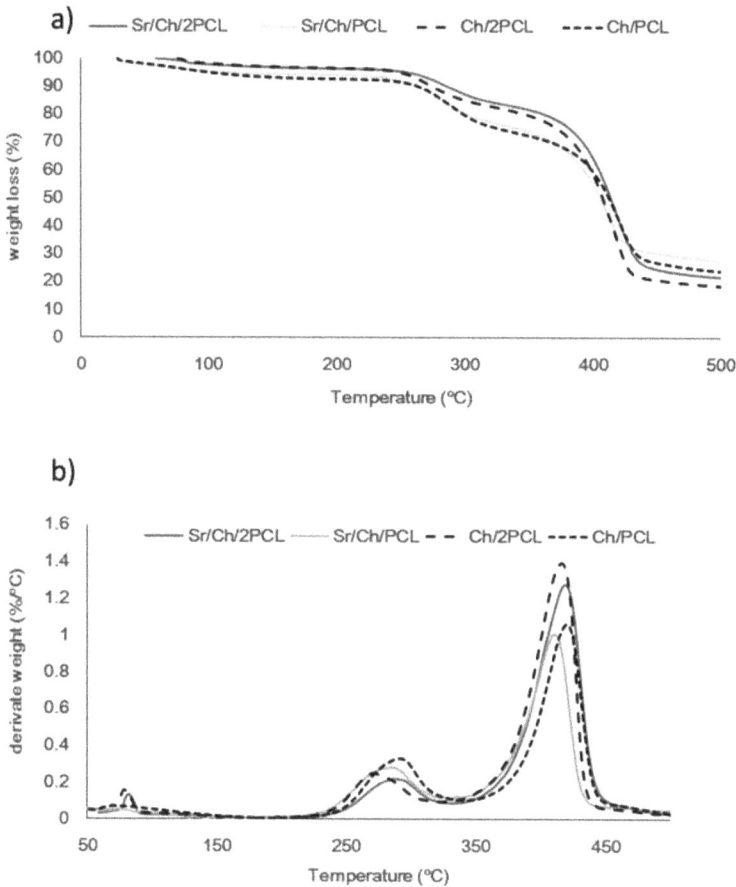

Figure 3. TGA (Termogravimetric analysis) (**a**) and DTG (Derivative Thermogravimetric analysis) curves (**b**) of blank and Sr(II) membranes under nitrogen atmosphere. Ch: chitosan; PCL: poly(ε-caprolactone).

3.3. Swelling Behaviour

In vitro swelling of all samples was analyzed in PBS buffer at 37 °C. Results are represented in Figure 4. All samples exhibited complete swelling, very rapid within the first 24 h, and from then on, hydration progressively and slowly increased up to reach a maximum value. Ch/2PCL sample absorbed 167% of water at 1 day whereas Ch/PCL sample 190%, and maximum WU values were 243% and 328% respectively, with decreasing PCL content. After the maximum, WU slightly decreased to values of 200% and 300% respectively; this phenomenon can be attributed to some degradation of matrix by breakdown of some ionic cross-linking. Sr bioactive membrane scaffolds absorbed 130% (Sr/Ch/2PCL) and 121% (Sr/Ch/PCL) of water at one day of soaking and absorption progressed until, maximum WU values were around 245% for both samples at 45 days. Comparing blank and Sr(II) samples, maximum WU were higher for the blanks in all the studied period, what can be attributed to delivery of Sr in the latter samples. Additionally, changes in morphology surface under wet conditions for Sr(II) samples at 30 days were analyzed by SEM (Figure S3). Few differences were observed between samples of varying composition. Generally, Sr(II) crystals considerably decreased after 30 days immersion, and signs of chitosan matrix erosion were incipient but PCL microparticles remained unaltered at this time.

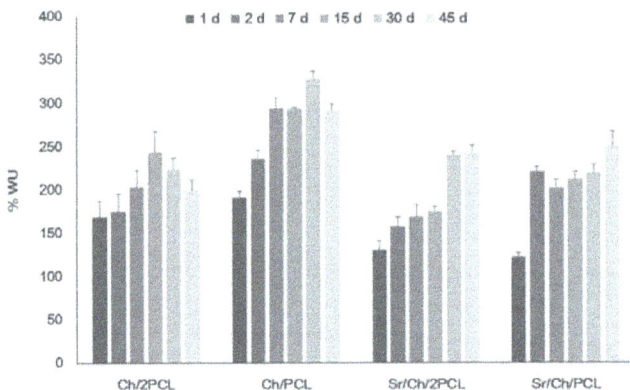

Figure 4. Variation of water uptake of membranes after immersion in PBS buffer at 37 °C. Ch: chitosan; PCL: poly(ε-caprolactone).

3.4. In Vitro Biological Behaviour

Cytotoxicity of samples was evaluated against MG-63 cell and hBMSCs with the MTT assay that measures the succinate mitochondrial dehydrogenase enzyme activity [82]. The results are shown in Figure S4; cell viability values in presence of lixiviates taken at 1, 2, 7, 14 and 21 days of all samples ranged around 100% reflecting absence of in vitro cytotoxicity according to standard specifications [92].

3.4.1. Osteoblasts-Like Cells

SEM examination of blank and Sr(II) samples directly seeded with MG-63 cells was performed to study adhesion and cellular morphology. Figure 5 shows the SEM images of the osteoblast-like cells colonization on samples at different times. Cells showing adhesion and spread morphology are signalized by white arrows on the micrographs. Cells showed much better adhesion and spreading on the systems containing Sr(II) compared with those detected on blank samples. Interestingly, SEM examination revealed that qualitatively, cell growth and extracellular matrix formation was much higher on the Sr/Ch/PCL sample in the studied period but especially at six days. However, it is worth mentioning that blank samples also provided adhesion of cells with good spreading and adaptation to the surface although in a much lower extent.

To quantify cell proliferation AB assay [93] was carried out. The results of this test are shown in Figure 6. In the Sr(II) membranes the increase of fluorescence from 1 to 21 days indicated a higher number of viable cells over time. In particular, the Sr/Ch/2PCL formulation showed a significantly higher cell growth than its blank at 21 days, and the Sr/Ch/PCL sample showed same result at both 14 and 21 days. Blank membranes, on the other hand, behaved differently, and cell viability remained low in all studied times. Therefore, in the light of SEM and AB findings, the Sr(II) membrane with lower PCL content was selected for further cellular studies.

Osteogenic response of Sr/Ch/PCL sample was evaluated and ALP levels were normalized for DNA. Results are plotted in Figure 7. A significant increase in ALP activity was observed for the Sr/Ch/PCL materials compared with the blank at 14 days culture period.

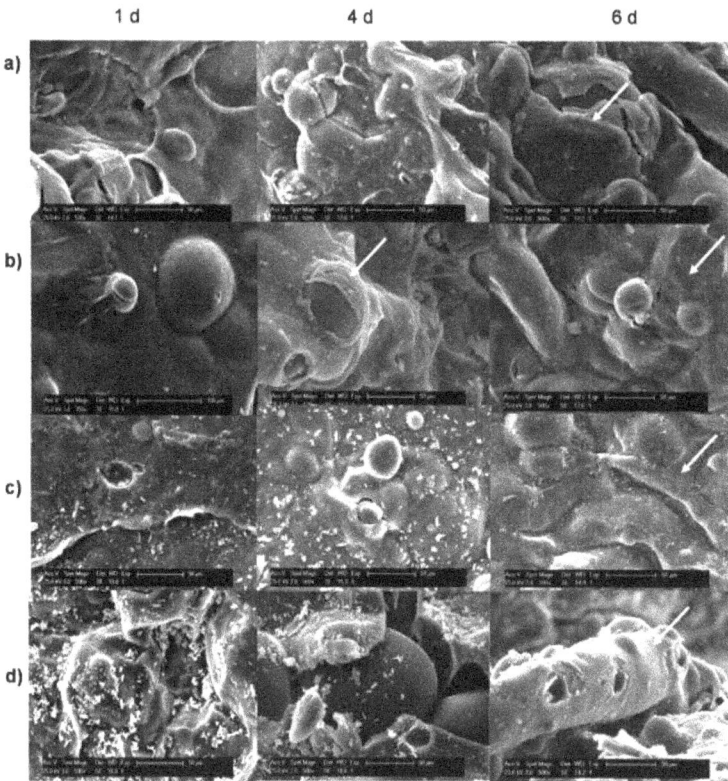

Figure 5. SEM images of MG-63 cells colonization on Sr(II) and blank membrane scaffolds at different post-seeding times. (**a**) Ch/2PCL; (**b**) Ch/PCL; (**c**) Sr/Ch/2PCL and (**d**) Sr/Ch/ PCL. Ch: chitosan; PCL: poly(ε-caprolactone).

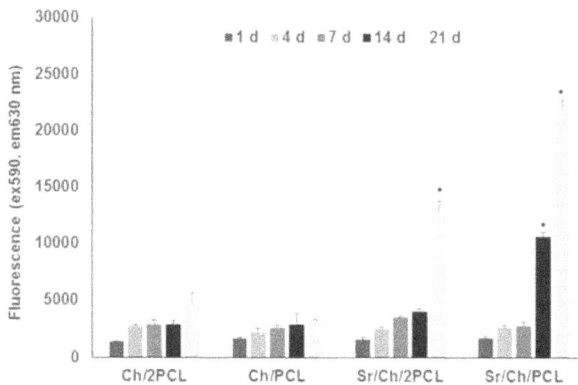

Figure 6. Alamar Blue results for blank and Sr(II) membranes in MG-63 cells over a period of 21 days. Results are given as mean $\pm$ sd ($n = 5$). Asterisks (*) indicate a significant difference comparing the corresponding Sr(II) and blank groups at the same time (* $p < 0.05$). Ch: chitosan; PCL: poly(ε-caprolactone).

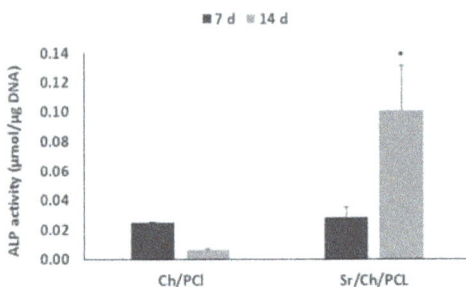

Figure 7. ALP/DNA activity in MG-63 cells cultured directly on test materials over a period of 14 days. Results are the mean ± sd (*n* = 5). Asterisks (*) indicate a significant difference comparing the two groups at the same time (* $p < 0.05$). Ch: chitosan; PCL: poly(ε-caprolactone).

3.4.2. Human Bone Marrow Mesenchymal Stem Cells

Blank and Sr/Ch/PCL membranes were seeded with hBMSCs and the morphology of cells grown was studied by SEM, after two and seven days (Figure 8). Cells adhesion and spread morphology are signalized by white arrows on the micrographs. In the case of membranes with Sr(II), good adhesion and proliferation at both studied times can be observed. We can also see that cells grew on the blank membranes but only at 7 days after seeding and in a lower extent.

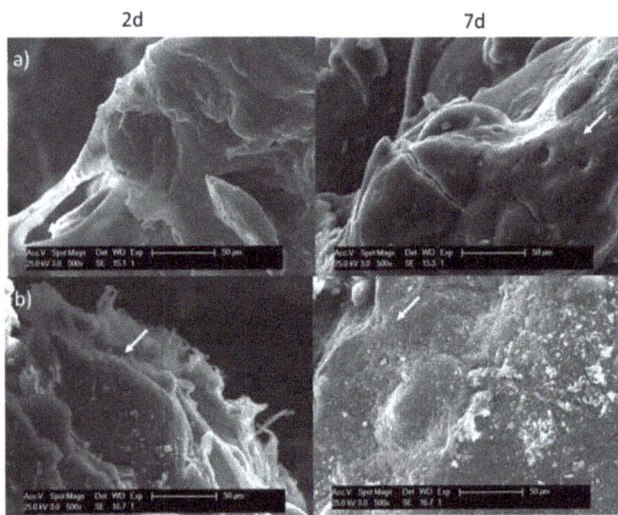

Figure 8. SEM images of hBMSCs colonization on blank and Sr(II) scaffolds at different times post-seeding. (**a**) Ch/PCL and (**b**) Sr/Ch/PCL. Ch: chitosan; PCL: poly(ε-caprolactone).

AB assay results (Figure 9a) showed a significantly lower cell proliferation for the Sr(II) membranes compared with the blank at 14 days although cell viability recuperated giving no significant differences at 21 days. Examination of DNA content in the period between seven and 14 days (Figure 9b) indicated that DNA was significantly higher in Sr(II) membranes compared with the blank at 14 days and ALP content normalized for DNA (Figure 9c) showed a significantly higher ALP activity compared with the blank over the 14 days culture period.

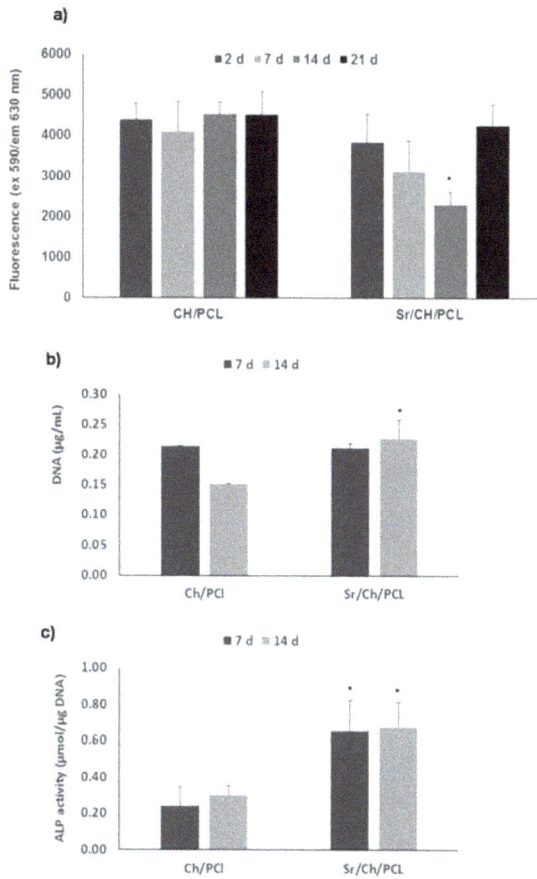

Figure 9. (**a**) Cell proliferation results obtained in AB assay of hBMSCs culture directly on test materials over a period of 21 days; (**b**) DNA (µg/mL) in cell lysate over a period of 14 days; (**c**) ALP/DNA activity over a period of 14 days. Results are the mean ± sd ($n = 4$). Asterisks (*) indicate a significant difference comparing blank and Sr(II) samples at the same time (* $p < 0.05$). Ch: chitosan; PCL: poly(ε-caprolactone).

3.5. In Vivo Biocompatibility

In vivo biocompatibility was studied in a subcutaneous model in rats in the system Sr/Ch/PCL and its blank using collagen membranes as control. Animals were sequentially sacrificed at each time point in post-operative days (7, 14 and 28 days). No behavioral changes or visible signs of physical impairment indicating systemic or neurological toxicity were observed between post-operative examination and the time of sacrifice. Macroscopic examination showed good wound closures without signs of inflammation. After 28 days all membranes were visible.

Detail microscopic histology images of the control (collagen), Ch/PCL and Sr/Ch/PLC membranes at specific time points are shown in Figures 10 and 11. The inflammatory responses of all membranes decreased with the time after implantation. At early experimentation time, typical inflammation repair process was developed composed by an infiltrate of inflammatory cells surrounded the membranes. However at 28 days of study, due to the resolution of the inflammatory (repair) process, less cellularity response was evident, where only few focal inflammatory reactions were evident.

The studied membrane specimens showed no degeneration of the structure but they were fragmented and showed cracks. However, control collagen membranes appeared unstructured forming like a mesh.

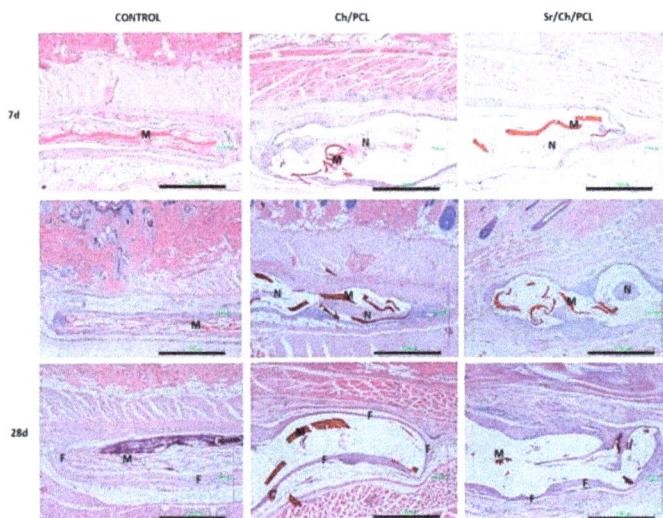

Figure 10. Micrographs of rat subcutaneous tissue responses to Control, Ch/PCL and Sr/Ch/PLC membranes after different implantation times (7, 14 and 28 days). M: membrane, N: necrotic tissue, F: fibrous tissue, C: calcification (H-E, 4×, scale bar 1 mm). Ch: chitosan; PCL: poly(ε-caprolactone).

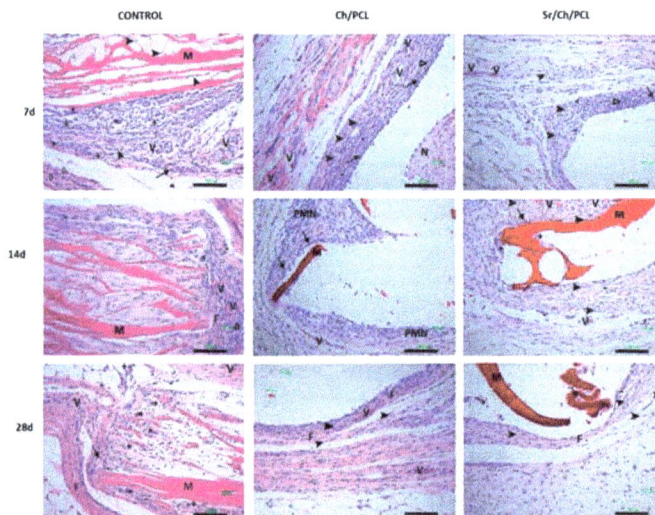

Figure 11. Micrographs of rat subcutaneous tissue responses to Control, Ch/PCL and Sr/Ch/PLC membranes after different implantation times (7, 14 and 28 days). M: membrane; N: necrotic tissue; F: fibrous tissue; V: blood vessels; →: macrophages; *: multinucleated cells; +: mast cells; ➤: lymphocytes; ◊: eosinophils; ○: plasmatic cells; −: monocytes; Δ: leucocytes; PMN: polymorph nuclear cells. (H-E, 20×, scale bar 100 μm). Ch: chitosan; PCL: poly(ε-caprolactone).

3.5.1. Control Group

Histology images of control collagen membranes are shown in Figures 10 and 11. After 7 days, slight acute inflammation was evident, characterized by the infiltration of mononuclear macrophages, mast cells, and eosinophils. It could be noticed the beginning of multinucleated cells formation. In the collagen membrane, there was little cell colonization, composed by polymorph nuclear cells and lymphocytes, particularly after short implantation time. The membrane was disorganized, forming a mesh-like structure. Focal development of small blood vessels was observed.

At 14 days, control collagen membranes appeared fragmented. Collagen membranes evoked a mild inflammatory reaction characterized by the infiltration of plasmatic cells, monocytes, leucocytes and eosinophils. Additionally, formation of multinucleated cells and local fibrous tissue were evident. An evidence of membrane degradation increased. Moderate focal blood vessels were developed.

Finally, at 28 days, control group evoked a mild local inflammatory reaction characterized by minimal infiltration of monocytes, macrophages, mast cells and scarce multinucleated cells. Fibroblasts presented activate synthesis of collagen fibers and formation of local fibrous tissue. Calcification process was evident in the collagen membranes remnants due to the augmented appetence to the basophilic stain (dark purple stain). An increased degradation and delamination of membrane material was observed. It could be observed less collagen fibers which were thinner and separated/spread during time.

3.5.2. Ch/PLC

Histology images of Ch/PLC membranes (Figures 10 and 11) after 7 days of subcutaneous implantation showed a typical acute inflammatory infiltrate with occasional areas of fibrinoid necrosis as well as some foci of foreign body reactions. At the membrane periphery the cellular infiltrated was characterized by macrophages and leucocytes. The surrounding tissue showed a mild inflammatory reaction composed of monocytes, macrophages, lymphocytes and few leucocytes. Moderate neovascularization was observed surrounded the membranes. Fibroblasts were presented but no fibrosis was evident.

At 14 days, the necrotic areas were reduced and vascularized, with infiltration composed mainly by macrophages and polymorphonuclears cells. Macrophages were active and phagocytosis processes were evident. A mild inflammation infiltrate was mainly composed of macrophages and leucocytes with progressive fibrosis.

After 28 days of implantation, mild focal inflammatory reaction with little vascularization was evident. The infiltrate was predominantly composed by macrophages and lymphocytes. Histological observations proved highly organized collagen fibers, demonstrating fibroblasts activity and focal fibrosis formation around membranes. Despite fibrous tissue proliferation no capsule formation occurred.

3.5.3. Sr/Ch/PLC

The histological response induced by Sr/Ch/PLC membranes is displayed in Figures 10 and 11. After 7 days of implantation, Sr/Ch/PLC evoked a mild local inflammatory reaction characterized by the infiltration of monocytes, macrophages and lymphocytes. Occasional small areas of necrosis were evident with an infiltrate of polymorphonuclear cells. At the membrane periphery the minor cellular infiltrated was characterized by leucocytes. No significant neovascularization was developed. Fibroblasts were present but no fibrosis was formed.

At 14 days, the necrotic areas almost disappeared and were vascularized allowing the cellular infiltration of macrophages and polymorphonuclear cells. The surrounding tissues showed signs of a slight focal inflammation at a very advanced stage of resolution, exhibiting good blood supply. The infiltrate was predominantly composed of monocytes, macrophages and lymphocytes. When small membrane fragments were present, formation of multinucleated cells occurred surrounding them. Minimal fibrous tissue proliferation was detected.

Finally, after 28 days, slight focal inflammatory reaction at an advanced stage of resolution with little vascularization was evident. The inflammatory infiltrate was characterized by the presence of monocytes, macrophages and lymphocytes. Also multinucleated cells were seen in association with small membrane fragments, if were present. Minimal fibrous tissue proliferation was detected composed by few collagen fibers parallel organized. Focal fibrosis formation occurred around membrane so no capsule formation happened.

4. Discussion

The main goal of this work deals with the preparation of bioactive hybrid scaffolds in respond to the clinical demand of bioactive systems intended for craniofacial defects regeneration. The scaffolds studied are membranes of a chitosan and PCL hybrid system, finally doped with Sr(II). The biodegradable polysaccharide with ionizable groups in its chemical structure and highly hydrophilic character will endow swelling and biological properties [71] whereas the PCL, a semi-crystalline and biodegradable polymer with a hydrophobic character, will contribute to the biomechanical stability [81]. Strontium component, on the other hand, will confer osteogenic response. Thus, the Sr(II) membranes were fabricated using a two-steps procedure. In the first step, membranes were obtained by casting a mixture of polymers solutions to be later ionically cross-linked by an immersion process in a TPP solution [87,94,95]. It is worth saying that the mixture of the solution of PCL in DCM into the aqueous solution of chitosan gives rise to the formation of PCL microparticles dispersed in a continuous matrix of chitosan after evaporation of solvents. In the second step, the cross-linked membranes were impregnated with a Sr(II) salt solution what conducted to incorporation of Sr(II) in scaffold areas close to or on the surface. The preparative methodology intends to get interactions of Sr(II) ions with the hybrid system while maintaining strontium availability to the biological medium. Systems based on chitosan and PCL without strontium element are well documented in literature for general tissue engineering. Actually, some of them have recently developed. They can consist of blends [96,97], bilayer systems [98] and functionalized coatings [99]. Other systems have been prepared following nano [100], micro and macro approaches [101]. As far as our knowledge is concerned, systems containing Sr(II) composed of chitosan/PCL have not been reported yet. However, there are systems containing Sr(II) that are separately based on these polymers. They were fabricated following different approaches. Composite scaffolds of chitosan/polymethacrylates loaded with SrFO were obtained by free radical polymerization of macromonomers, giving rise to porous scaffolds in which SrFO was homogeneously distributed along the polymeric matrix [79]. Freeze-drying fabrication of Ch/SrHAp composite scaffolds were reported by Xu et al. [76] and Lei et al. [10]. Respect to PCL approaches, Ren et al. developed a PCL/Sr-substituted 45S5 Bioglass® (SrBG) composite scaffold produced by melt electrospinning [102]. Other PCL/SrBG composite scaffolds were fabricated using an additive manufacturing technique. The bioactive particles were distributed in the PCL bulk across the scaffold (micro-CT examination) but SEM images revealed visible SrBG particles on their surface [16]. PCL and strontium porous composite scaffolds were fabricated using a simple one-step process to simultaneously foam PCL containing a strontium- and calcium carbonate. Integration of the inorganic educts into the scaffolds was observed by EDS-spectroscopy [15].

Morphology of membranes prepared in this work showed segregation of both components with microdomains of PCL microparticles distributed in a continuous matrix of the natural polymer as it has been commented above. This phenomenon can derive from both the structural characteristics of polymers and the methodology applied. On the one hand, the different hydrophilicicity of the synthetic and natural polymers contributes to the formation of a non-compatible system. However, the methodology applied is based on a mixture of both aqueous and organic solutions; this fact makes that during the fabrication process, a microemulsion of discrete droplets of the organic phase in a continuous aqueous phase is formed leading to segregation. Phase separation was reported for Ch/PCL blends obtained by casting; in this case the polysaccharide (10–30%) segregated in nano or microaggregates depending on the content and it appeared inside the PCL spherulites [103]. In other blends, no phase separation was formed in membranes with 25–75% PCL when using homogeneous solutions and dilute concentrations [96]. Returning to morphology of our Sr(II) membranes, the SEM images revealed agglomerates of the strontium crystals mainly on the surface of the continuous matrix; these agglomerates presented a morphology that resembles the "cauliflower" observed in the apatite-like layer usually formed upon immersion of sample in physiological body fluid (SBF). The exposition of the salt crystals on the surface will favor the availability of the Sr(II) ions to interact with the biological medium.

Structural characterization of Ch/2PCL and Ch/PCL samples reveals differences in some FTIR absorption bands compared to those of pure components. The broadness of band between 3500 and 3200 cm^{-1} indicates that during the cross-linking process some hydrogen bonds in the chitosan structure were destroyed and other new between the chitosan and TPP were formed [87]. In the region between 1750–1500 cm^{-1}, the band due to υ C=O in ester groups shifts with respect to pure PCL as reported by García Cruz et al. for biodegradable porous Ch/PCL semi-interpenetrating polymer networks (semi-IPNs) [104]; the bands due to υ C=O (amide I) and to δ N–H in NH_3^+ groups shift compared to pure chitosan what can be attributed to ionic interactions between the protonated amino groups and the negatively charged phosphate groups, as it was reported for other chitosan membranes cross-linked with TPP [87,95]. But this shift has also been ascribed to hydrogen bond interactions between the natural polymer and PCL [104]. Finally, in the region between 1300–900 cm^{-1}, the appearance of bands due to υ_{as} PO_2 groups in TPP ions indicates the presence of the cross-linking agent [87]. Likewise, EDS spectra of Sr(II) membranes show the peaks of Sr and P coming from TPP. TGA evaluation of samples supports the formation of hybrid composite materials showing different thermal behavior than that of a simple mixture of pure components.

Water uptake is important for tissue engineering but particularly when the scaffold will perform in the oral cavity [105]. The studied membranes demonstrated their ability to absorb and retain a noticeable amount of water after immersion in PBS. However, some differences were observed between blank and Sr(II) samples. In the blanks the Ch:PCL ratio plays the main role in the sense that the higher the PCL content the lower the maximum *WU*, suggesting that water preferably absorbs into the hydrophilic polysaccharide. However, both compositions of Sr(II) membranes swell following a similar pattern and attain maxima *WU* values ($\approx$245%) suggesting that the Sr(II) located on the surface is playing the dominant role. Morphology of Sr(II) membranes after 30 days soaking analyzed by SEM shows signs of erosion of the chitosan matrix as a consequence of water intrusion, scarce Sr(II) crystals and unaltered PCL microparticles attributed to the long-term degradability of the synthetic polymer [106]. The observed microstructure indicates that Sr(II) scaffolds have dimensional stability for the studied period and that most of the Sr(II) crystals have interchanged with the medium within this time.

The biocompatibility of biomaterials is closely related to cell-materials interactions and, in particular to cell adhesion to their surface. Attachment, adhesion and spreading of cells are the first step of these interactions and the quality of these processes will influence the cells capability to proliferate and to differentiate itself on contact with the scaffold [107]. In particular, cell adhesion and proliferation is highly important in biomaterials designed for tissue engineering purposes. Cell adhesion onto the studied membranes was examined by SEM using MG-63 and hBMSCs cell lines. Micrographs showed that both types of cells adhered, proliferated and formed ECM best onto the Sr(II) membranes compared to the blanks, what, additionally indicated that there was no direct toxic effects and cellular metabolism was normal. Additionally, the Sr/Ch/PCL composition distinguished with a higher cell adhesion and proliferation of MG-63 cells compared to the Sr/Ch/2PCL system, in the period from 1 to 6 days. Accordingly, several authors have been demonstrated that the presence of strontium in ceramics showed enhancing effects on osteoblasts cell growth and also ALP activity [108].

In further studies, the Sr/Ch/PCL system also demonstrated its capacity to adhere and grow hBMSCs, and cell population was much more abundant compared with the blank sample, especially at 7 days post seeding what can be attributed to the presence of Sr(II) ions. Cao et al. developed a modified chitosan/PCL electrospun fibrous scaffold charged with bone morphogenetic protein-2 (BMP-2) and found that MSCs attached readily with increasing spreading [109]. Similar observations were reported by composite chitosan scaffolds based on bactericide SrHAp/chitosan systems synthetized by Lei et al. [10] and composites of nSrHAp/chitosan developed by Xu et al. [76] produced favorable adhesion, spreading and proliferation of hBMSCs and enhanced osteoinductivity.

Quantitative analysis on cell proliferation after direct seeding of cells is usually assessed by AB assay [83]. Evaluation of our Sr(II) membranes using osteoblast-like cells and hBMSCs showed satisfactory cell viability after 21 days. Particularly, the Sr/Ch/PCL membrane highlighted by showing significant increase of MG-63 cells viability compared to the blank at 7 and 21 days. These findings seem to suggest again that both presence of the Sr(II) ions and the hydrophilic/hydrophobic balance of the sample can play an important role in the cell proliferation processes. Osteogenic response of our Sr/Ch/PCL membranes was investigated by DNA and ALP quantification. ALP activity is one of the characteristic parameters of osteoblast cells differentiation and a signal for the subsequent production of the proteins leading to mineralization [86,110]. In our studied Sr(II) membranes ALP content normalized for DNA was significantly higher compared to the blank in MG-63 cell culture at 14 days and in hBMSCs culture at 7 and 14 days. These results show the osteogenic capacity of the hybrid membranes that can be tentatively attributed to the presence of Sr(II) ions, although according to Seol et al., chitosan sponges can enhance ALP [51]. Generally, osteoconductive properties are attributed to the presence of inorganic components such as HAp, BG, or calcium phosphate salts. This behavior was also found by Kong et al. for chitosan/nHAp composite scaffolds using the MC 3T3-E1 preosteoblast cell line derived from newborn mouse calvaria [62] and by Shalumon et al. for nHAp and nBG incorporating PCL/chitosan composite scaffolds tested with human periodontal ligament fibroblast cells (hPLFs) and osteoblast like cells (MG-63 cell line) [111]. Both researchers found that the ALP activity of cells on the composite scaffolds increased compared with those of only chitosan, showing a higher differentiation level. Respect to the chitosan composite scaffolds with Sr(II), Su et al. have demonstrated that the strontium phosphate can affect the ALP activity in the osteogenic process when they test hydrogels formed with chitosan and strontium in presence of MSCs. The increase in the osteogenic expression, ALP activity, and calcium deposition indicates the effect of strontium in enhancing bone remodeling and bone structure stabilization [112]. Martín-del-Campo et al. demonstrated that the presence of Sr(II) ions stimulate hDPSCs proliferation, matrix mineralization and ALP activity [79]. More recently, Lei et al. observed that the release of Sr(II) ions from SrHAp/chitosan scaffolds enhanced ALP activity of hBMSCs and ECM mineralization [10].

In vivo biocompatibility was studied by implantation of Ch/PLC and Sr/Ch/PLC membranes using collagen (commercial collagen membrane material) as control. All systems where implanted in the backs of rats which were sacrificed at specific time points (7, 14 and 28 days) to observe the histological response using light microscopy. In a previous work where biocompatibility of cross-linked chitosan hydrogels was studied [113], non-degraded chitosan hydrogels were stained with eosin (pink) whereas degradable chitosan hydrogels were stained with the basophilic hematoxylin (blue). Azab et al. [113] proposed that the change in the staining patter was due to the shift in charge of the degraded chitosan from positive to negative. In our work, histological images (Figures 10 and 11) showed chitosan membranes stained with eosin (pink). At physiological conditions chitosan is positively charged, which leads to its staining by negative charged eosin. In the in vivo subcutaneous biocompatible study in rats, all studied and control membranes exhibited inflammatory and tissue responses. This acute inflammatory response was somewhat expected in line with the typical reported host reactions following biomaterial implantation [114]. This kind of reaction has been reported even for nontoxic biomaterials such as silk, collagen or PLA [115]. The inflammatory reaction observed after subcutaneous implantation of the materials represented a typical acute response upon injury.

Regarding membrane degradation, in Bavariya et al. work [116], significant degradation of chitosan membranes was not evident in histological sections until 16–20 weeks post implantation as compared with significant degradation of collagen membrane at 12 weeks. In our work, collagen membranes appeared more disorganized along time, in contrast with the studied membranes, which were stable for all the experimental time. In some cases, histology images showed membrane fragments possible due to the characteristic chitosan erosion. Clinicians have suggested ideal membrane degradation time 4–6 months for large defects to provide sufficient time for bone regeneration [117,118].

Tissue inflammatory response to the two types of studied membranes was mild compared to the inflammatory response caused by the absorbable collagen control membrane. Collagen membranes showed signs of acute and chronic inflammation in the rat, mostly due to the presence of multinucleated giant cells surrounded fragments and rest of the membrane. The presence of membrane fragments evoked a more cellular reaction in comparison with stable studied membranes. However, after subcutaneous implantation of membranes of the three groups, no physiological signs of severe inflammation were observed. The studied materials induced a mild inflammatory infiltrate mainly composed by mononuclear cells, i.e., monocytes, macrophages, lymphocytes and some fibroblasts.

The extension of the observed inflammatory infiltrate was slightly higher in Ch/PLC group than in Sr/Ch/PLC one demonstrating that the resolution of the inflammatory process was faster for Sr/Ch/PLC membranes. The inflammatory response triggered by Ch/PLC membranes was often accompanied by abundant monocytes and macrophages population, and more extensive necrosis when compare with Sr/Ch/PLC membrane. The granulation tissue showed progressive increased replacement by fibrous connective tissue being less developed at Sr/Ch/PLC membranes than in Ch/PLC. The reaction to Sr/Ch/PLC membranes evoked less development of fibrosis, with a thinner fibrous tissue compared with Ch/PCL. The focal fibrous tissue was formed by the parallel deposit of few collagen fibers.

Based on the histological observation all types of membranes can therefore be considered biocompatible, highlighting the most favorable response to Sr/Ch/PCL system. Additionally, the studied membranes Ch/PLC and Sr/Ch/PCL maintain their structural integrity for 28 days period highly recommended for bone regeneration purposes, opposed to collagen membranes which disintegrated and appeared forming a mesh-like structure.

5. Conclusions

Sr(II) containing scaffolds consisting of a continuous matrix of TPP cross-linked chitosan and PCL microparticles were fabricated in a facile two-step casting/evaporation method using Ch:PCL ratios of 1:2 and 1:1. Both compositions presented good physico-chemical properties, swelling content around 250% and presented dimensional stability for at least one month. They did not show in vitro cytotoxicity against neither MG-63 cell line nor hBMSCs. The system Sr/Ch/PCL showed qualitatively and quantitatively higher cell proliferation, offering good support for adherence and proliferation of cells. Specifically, cellular studies on the Sr/Ch/PCL system using hBMSCs demonstrated a noticeable enhancement of DNA content and ALP activity as well as good and adequate niche for adherence and proliferation of cells. In vivo experiments in rats manifested good biocompatibility for all studied membranes standing out the results obtained for the Sr/Ch/PCL system. All these results allow considering this Sr(II) hybrid system as a promising biomaterial for application of bioactive scaffolds in bone tissue engineering such as the treatment of craniofacial regeneration as well as on the active regeneration of other bone tissues, considering that the developed formulation acts as a well-defined route of local delivery of Sr(II) ions. In addition, the in vivo response of the implanted systems did not apparently show signs of toxicity avoiding the negative secondary effects described after administration of strontium by systemic route.

Supplementary Materials: The following are available online at www.mdpi.com/2073-4360/10/3/279/s1, Figure S1: ATR-FTIR spectra of pure Ch and PCL and Ch/2PCL and Ch/PCL membranes, Figure S2: TGA (Termogravimetric analysis) (a) and DTG (Derivative Thermogravimetric analysis). (down) of pure Ch and PCL,

Figure S3: SEM images of Sr(II) membranes after 30 d immersion in PBS buffer at 37 °C. a) 200x; b) 500x; c) 1000x, Figure S4: In vitro cytotoxicity results of blank and Sr(II) membranes using MG-63 cells (a) and hBMSCs (b). Results are given as mean ± sd (*n* = 5), Table S1: DTG (Derivative Thermogravimetric analysis) results of pure PCL (poly(ε-caprolactone), chitosan (Ch), blank and Sr(II) membranes.

Acknowledgments: Authors thank R. Ramírez and R. de Roba of ICTP-CSIC, Spain, for assistance in the biological culture assays, and María Rosa Sánchez González of University of Salamanca, Spain, for assistance in the histological preparations.

Author Contributions: Itzia Rodríguez-Méndez performed the chemical experiments; Mar Fernández-Gutiérrez performed the in vitro experiments; Lorena Benito Garzón performed in vivo experiments; all authors contributed to the analysis, discussion and redaction of the paper.

Conflicts of Interest: The authors declare no conflict of interest.

Funding: Authors thank CIBER-BBN, Spain, the Spanish Ministry of Economy and Competitivity (project MAT2017-84277-R) and Programa de Apoyo a Proyectos de Investigación e Innovación Tecnológica (PAPIIT) UNAM, Mexico, (project IA20947) for supporting this work.

References

1. Maobin, Y.; Hongming, Z.; Riddhi, G. Advances of mesenchymal stem cells derived from bone marrow and dental tissue in craniofacial tissue engineering. *Curr. Stem Cell Res. Ther.* **2014**, *9*, 150–161.

2. Amini, A.R.; Laurencin, C.T.; Nukavarapu, S.P. Bone tissue engineering: Recent advances and challenges. *Crit. Rev. Biomed. Eng.* **2012**, *40*, 363–408. [CrossRef] [PubMed]

3. Mossey, P. Addressing the global challenges of craniofacial anomalies. In *Report of a WHO Meeting on International Collaborative Research on Craniofacial Anomalies*; WHO: Geneva, Switzerland, 2004.

4. Tevlin, R.; McArdle, A.; Atashroo, D.; Walmsley, G.G.; Senarath-Yapa, K.; Zielins, E.R.; Paik, K.J.; Longaker, M.T.; Wan, D.C. Biomaterials for craniofacial bone engineering. *J. Dent. Res.* **2014**, *93*, 1187–1195. [CrossRef] [PubMed]

5. Black, C.R.M.; Goriainov, V.; Gibbs, D.; Kanczler, J.; Tare, R.S.; Oreffo, R.O.C. Bone tissue engineering. *Curr. Mol. Biol. Rep.* **2015**, *1*, 132–140. [CrossRef] [PubMed]

6. Yousefi, A.M.; James, P.F.; Akbarzadeh, R.; Subramanian, A.; Flavin, C.; Oudadesse, H. Prospect of stem cells in bone tissue engineering: A review. *Stem Cells Int.* **2016**, *2016*, 1–13. [CrossRef] [PubMed]

7. Wang, X.; Li, X.; Ito, A.; Sogo, Y. Synthesis and characterization of hierarchically macroporous and mesoporous CaO–MO–SiO$_2$–P$_2$O$_5$ (M = Mg, Mn, Sr) bioactive glass scaffolds. *Acta Biomater.* **2011**, *7*, 3638–3644. [CrossRef] [PubMed]

8. Owens, G.J.; Singh, R.K.; Foroutan, F.; Alqaysi, M.; Han, C.M.; Mahapatra, C.; Kim, H.W.; Knowles, J.C. Sol-gel based materials for biomedical applications. *Prog. Mater Sci.* **2016**, *77*, 1–79. [CrossRef]

9. Lakhkar, N.J.; Lee, I.H.; Kim, H.W.; Salih, V.; Wall, I.B.; Knowles, J.C. Bone formation controlled by biologically relevant inorganic ions: Role and controlled delivery from phosphate-based glasses. *Adv. Drug Deliv. Rev.* **2013**, *65*, 405–420. [CrossRef] [PubMed]

10. Lei, Y.; Xu, Z.; Ke, Q.; Yin, W.; Chen, Y.; Zhang, C.; Guo, Y. Strontium hydroxyapatite/chitosan nanohybrid scaffolds with enhanced osteoinductivity for bone tissue engineering. *Mater. Sci. Eng. C* **2017**, *72*, 134–142. [CrossRef] [PubMed]

11. Marie, P.J.; Ammann, P.; Boivin, G.; Rey, C. Mechanisms of action and therapeutic potential of strontium in bone. *Calcif. Tissue Int.* **2001**, *69*, 121–129. [CrossRef] [PubMed]

12. Zhang, J.; Zhao, S.; Zhu, Y.; Huang, Y.; Zhu, M.; Tao, C.; Zhang, C. Three-dimensional printing of strontium-containing mesoporous bioactive glass scaffolds for bone regeneration. *Acta Biomater.* **2014**, *10*, 2269–2281. [CrossRef] [PubMed]

13. Al Qaysi, M.; Walters, N.J.; Foroutan, F.; Owens, G.J.; Kim, H.W.; Shah, R.; Knowles, J.C. Strontium- and calcium-containing, titanium-stabilised phosphate-based glasses with prolonged degradation for orthopaedic tissue engineering. *J. Biomater. Appl.* **2015**, *30*, 300–310. [CrossRef] [PubMed]

14. Zhang, Q.; Chen, X.; Geng, S.; Wei, L.; Miron, R.J.; Zhao, Y.; Zhang, Y. Nanogel-based scaffolds fabricated for bone regeneration with mesoporous bioactive glass and strontium: In vitro and in vivo characterization. *J. Biomed. Mater. Res. Part A* **2017**, *105*, 1175–1183. [CrossRef] [PubMed]

15. Zehbe, R.; Zehbe, K. Strontium doped poly-ε-caprolactone composite scaffolds made by reactive foaming. *Mater. Sci. Eng. C* **2016**, *67*, 259–266. [CrossRef] [PubMed]

16. Poh, P.S.P.; Hutmacher, D.W.; Holzapfel, B.M.; Solanki, A.K.; Stevens, M.M.; Woodruff, M.A. In vitro and in vivo bone formation potential of surface calcium phosphate-coated polycaprolactone and polycaprolactone/bioactive glass composite scaffolds. *Acta Biomater.* **2016**, *30*, 319–333. [CrossRef] [PubMed]

17. Kruijt Spanjer, E.C.; Bittermann, G.K.P.; van Hooijdonk, I.E.M.; Rosenberg, A.J.W.P.; Gawlitta, D. Taking the endochondral route to craniomaxillofacial bone regeneration: A logical approach? *J. Cranio Maxillofac. Surg.* **2017**, *45*, 1099–1106. [CrossRef] [PubMed]

18. Bianco, P.; Robey, P.G. Stem cells in tissue engineering. *Nature* **2001**, *414*, 118–121. [CrossRef] [PubMed]

19. Liceras-Liceras, E.; Garzón, I.; España-López, A.; Oliveira, A.C.X.; García-Gómez, M.; Martín-Piedra, M.Á.; Roda, O.; Alba-Tercedor, J.; Alaminos, M.; Fernández-Valadés, R. Generation of a bioengineered autologous bone substitute for palate repair: An in vivo study in laboratory animals. *J. Tissue Eng. Regen. Med.* **2017**, *11*, 1907–1914. [CrossRef] [PubMed]

20. Maraldi, T.; Riccio, M.; Pisciotta, A.; Zavatti, M.; Carnevale, G.; Beretti, F.; La Sala, G.B.; Motta, A.; De Pol, A. Human amniotic fluid-derived and dental pulp-derived stem cells seeded into collagen scaffold repair critical-size bone defects promoting vascularization. *Stem Cell Res. Ther.* **2013**, *4*, 53–65. [CrossRef] [PubMed]

21. Petridis, X.; Diamanti, E.; Trigas, G.C.; Kalyvas, D.; Kitraki, E. Bone regeneration in critical-size calvarial defects using human dental pulp cells in an extracellular matrix-based scaffold. *J. Cranio Maxillofac. Surg.* **2015**, *43*, 483–490. [CrossRef] [PubMed]

22. Giuliani, A.; Manescu, A.; Langer, M.; Rustichelli, F.; Desiderio, V.; Paino, F.; De Rosa, A.; Laino, L.; D'Aquino, R.; Tirino, V.; et al. Three years after transplants in human mandibles, histological and in-line holotomography revealed that stem cells regenerated a compact rather than a spongy bone: Biological and clinical implications. *Stem Cells Trans. Med.* **2013**, *2*, 316–324. [CrossRef] [PubMed]

23. Mankani, M.H.; Kuznetsov, S.A.; Wolfe, R.M.; Marshall, G.W.; Robey, P.G. In vivo bone formation by human bone marrow stromal cells: Reconstruction of the mouse calvarium and mandible. *Stem Cells* **2006**, *24*, 2140–2149. [CrossRef] [PubMed]

24. Miranda, S.C.C.C.; Silva, G.A.B.; Mendes, R.M.; Abreu, F.A.M.; Caliari, M.V.; Alves, J.B.; Goes, A.M. Mesenchymal stem cells associated with porous chitosan-gelatin scaffold: A potential strategy for alveolar bone regeneration. *J. Biomed. Mater. Res. Part A* **2012**, *100 A*, 2775–2786. [CrossRef] [PubMed]

25. Miura, M.; Miura, Y.; Sonoyama, W.; Yamaza, T.; Gronthos, S.; Shi, S. Bone marrow-derived mesenchymal stem cells for regenerative medicine in craniofacial region. *Oral Dis.* **2006**, *12*, 514–522. [CrossRef] [PubMed]

26. Chamieh, F.; Collignon, A.M.; Coyac, B.R.; Lesieur, J.; Ribes, S.; Sadoine, J.; Llorens, A.; Nicoletti, A.; Letourneur, D.; Colombier, M.L.; et al. Accelerated craniofacial bone regeneration through dense collagen gel scaffolds seeded with dental pulp stem cells. *Sci. Rep.* **2016**, *6*, 1–11. [CrossRef] [PubMed]

27. Sell, S.A.; Wolfe, P.S.; Garg, K.; McCool, J.M.; Rodriguez, I.A.; Bowlin, G.L. The use of natural polymers in tissue engineering: A focus on electrospun extracellular matrix analogues. *Polymers* **2010**, *2*, 522–553. [CrossRef]

28. Madhavan Nampoothiri, K.; Nair, N.R.; John, R.P. An overview of the recent developments in polylactide (PLA) research. *Bioresour. Technol.* **2010**, *101*, 8493–8501. [CrossRef] [PubMed]

29. Coathup, M.J.; Hing, K.A.; Samizadeh, S.; Chan, O.; Fang, Y.S.; Campion, C.; Buckland, T.; Blunn, G.W. Effect of increased strut porosity of calcium phosphate bone graft substitute biomaterials on osteoinduction. *J. Biomed. Mater. Res. Part A* **2012**, *100 A*, 1550–1555. [CrossRef] [PubMed]

30. Matassi, F.; Botti, A.; Sirleo, L.; Carulli, C.; Innocenti, M. Porous metal for orthopedics implants. *Clin. Cases Miner. Bone Metab.* **2013**, *10*, 111–115.

31. Sagomonyants, K.B.; Hakim-Zargar, M.; Jhaveri, A.; Aronow, M.S.; Gronowicz, G. Porous tantalum stimulates the proliferation and osteogenesis of osteoblasts from elderly female patients. *J. Orthop. Res.* **2011**, *29*, 609–616. [CrossRef] [PubMed]

32. Skoog, S.A.; Kumar, G.; Goering, P.L.; Williams, B.; Stiglich, J.; Narayan, R.J. Biological response of human bone marrow-derived mesenchymal stem cells to commercial tantalum coatings with microscale and nanoscale surface topographies. *JOM* **2016**, *68*, 1672–1678. [CrossRef]

33. Singh, D.; Tripathi, A.; Zo, S.; Singh, D.; Han, S.S. Synthesis of composite gelatin-hyaluronic acid-alginate porous scaffold and evaluation for in vitro stem cell growth and in vivo tissue integration. *Colloids Surf. B Biointerfaces* **2014**, *116*, 502–509. [CrossRef] [PubMed]

34. Kim, S.S.; Sun Park, M.; Jeon, O.; Yong Choi, C.; Kim, B.S. Poly(lactide-co-glycolide)/hydroxyapatite composite scaffolds for bone tissue engineering. *Biomaterials* **2006**, *27*, 1399–1409. [CrossRef] [PubMed]

35. Lakhani, R.S. New biomaterials versus traditional techniques: Advances in cleft palate reconstruction. *Curr. Opin. Otolaryngol. Head Neck Surg.* **2016**, *24*, 330–335. [CrossRef] [PubMed]

36. Moreau, J.L.; Caccamese, J.F.; Coletti, D.P.; Sauk, J.J.; Fisher, J.P. Tissue engineering solutions for cleft palates. *J. Oral Maxillofac. Surg.* **2007**, *65*, 2503–2511. [CrossRef] [PubMed]

37. Karfeld-Sulzer, L.S.; Weber, F.E. Biomaterial development for oral and maxillofacial bone regeneration. *J. Korean Assoc. Oral Maxillofac. Surg.* **2012**, *38*, 264–270. [CrossRef]

38. Khojasteh, A.; Kheiri, L.; Motamedian, S.R.; Nadjmi, N. Regenerative medicine in the treatment of alveolar cleft defect: A systematic review of the literature. *J. Cranio Maxillofac. Surg.* **2015**, *43*, 1608–1613. [CrossRef] [PubMed]

39. García-Gareta, E.; Coathup, M.J.; Blunn, G.W. Osteoinduction of bone grafting materials for bone repair and regeneration. *Bone* **2015**, *81*, 112–121. [CrossRef] [PubMed]

40. Wan, A.C.A.; Khor, E.; Hastings, G.W. Preparation of a chitin-apatite composite by in situ precipitation onto porous chitin scaffolds. *J. Biomed. Mater. Res.* **1998**, *41*, 541–548. [CrossRef]

41. Chow, K.S.; Khor, E. Novel fabrication of open-pore chitin matrixes. *Biomacromolecules* **2000**, *1*, 61–67. [CrossRef] [PubMed]

42. Chow, K.S.; Khor, E.; Wan, A.C.A. Porous chitin matrices for tissue engineering: Fabrication and in vitro cytotoxic assessment. *J. Polym. Res.* **2001**, *8*, 27–35. [CrossRef]

43. Jayakumar, R.; Chennazhi, K.P.; Srinivasan, S.; Nair, S.V.; Furuike, T.; Tamura, H. Chitin scaffolds in tissue engineering. *Int. J. Mol. Sci.* **2011**, *12*, 1876–1887. [CrossRef] [PubMed]

44. Khor, E.; Lim, L.Y. Implantable applications of chitin and chitosan. *Biomaterials* **2003**, *24*, 2339–2349. [CrossRef]

45. Islam, S.; Bhuiyan, M.A.R.; Islam, M.N. Chitin and chitosan: Structure, properties and applications in biomedical engineering. *J. Polym. Environ.* **2017**, *25*, 854–866. [CrossRef]

46. Anitha, A.; Sowmya, S.; Kumar, P.T.S.; Deepthi, S.; Chennazhi, K.P.; Ehrlich, H.; Tsurkan, M.; Jayakumar, R. Chitin and chitosan in selected biomedical applications. *Prog. Polym. Sci.* **2014**, *39*, 1644–1667. [CrossRef]

47. Tsai, W.B.; Chen, Y.R.; Li, W.T.; Lai, J.Y.; Liu, H.L. RGD-conjugated UV-crosslinked chitosan scaffolds inoculated with mesenchymal stem cells for bone tissue engineering. *Carbohydr. Polym.* **2012**, *89*, 379–387. [CrossRef] [PubMed]

48. Palao-Suay, R.; Gómez-Mascaraque, L.G.; Aguilar, M.R.; Vázquez-Lasa, B.; Román, J.S. Self-assembing polymer systems for advanced treatment of cancer and inflammation. *Prog. Polym. Sci.* **2016**, *53*, 2017–2248. [CrossRef]

49. Kim, I.Y.; Seo, S.J.; Moon, H.S.; Yoo, M.K.; Park, I.Y.; Kim, B.C.; Cho, C.S. Chitosan and its derivatives for tissue engineering applications. *Biotechnol. Adv.* **2008**, *26*, 1–21. [CrossRef] [PubMed]

50. Hayashi, T. Biodegradable polymers for biomedical uses. *Prog. Polym. Sci.* **1994**, *19*, 663–702. [CrossRef]

51. Seol, Y.J.; Lee, J.Y.; Park, Y.J.; Lee, Y.M.; Ku, Y.; Rhyu, I.C.; Lee, S.J.; Han, S.B.; Chung, C.P. Chitosan sponges as tissue engineering scaffolds for bone formation. *Biotechnol. Lett.* **2004**, *26*, 1037–1041. [CrossRef] [PubMed]

52. Seeherman, H.; Li, R.; Wozney, J. A review of preclinical program development for evaluating injectable carriers for osteogenic factors. *J. Bone Jt. Surg. Ser. A* **2003**, *85*, 96–108. [CrossRef]

53. Srinivasan, S.; Jayasree, R.; Chennazhi, K.P.; Nair, S.V.; Jayakumar, R. Biocompatible alginate/nano bioactive glass ceramic composite scaffolds for periodontal tissue regeneration. *Carbohydr. Polym.* **2012**, *87*, 274–283. [CrossRef]

54. Haidar, Z.S.; Hamdy, R.C.; Tabrizian, M. Delivery of recombinant bone morphogenetic proteins for bone regeneration and repair. Part A: Current challenges in bmp delivery. *Biotechnol. Lett.* **2009**, *31*, 1817–1824. [CrossRef] [PubMed]

55. Haidar, Z.S.; Tabrizian, M.; Hamdy, R.C. A hybrid rhop-1 delivery system enhances new bone regeneration and consolidation in a rabbit model of distraction osteogenesis. *Growth Factors* **2010**, *28*, 44–55. [CrossRef] [PubMed]

56. Dash, M.; Chiellini, F.; Ottenbrite, R.M.; Chiellini, E. Chitosan—A versatile semi-synthetic polymer in biomedical applications. *Prog. Polym. Sci.* **2011**, *36*, 981–1014. [CrossRef]

57. Oktay, E.O.; Demiralp, B.; Demiralp, B.; Senel, S.; Cevdet Akman, A.; Eratalay, K.; Akincibay, H. Effects of platelet-rich plasma and chitosan combination on bone regeneration in experimental rabbit cranial defects. *J. Oral Implantol.* **2010**, *36*, 175–184. [CrossRef] [PubMed]

58. Duruel, T.; Çakmak, A.S.; Akman, A.; Nohutcu, R.M.; Gümüşderelioğlu, M. Sequential IGF-1 and BMP-6 releasing chitosan/alginate/PLGA hybrid scaffolds for periodontal regeneration. *Int. J. Biol. Macromol.* **2017**, *104*, 232–241. [CrossRef] [PubMed]

59. Haidar, Z.S.; Azari, F.; Hamdy, R.C.; Tabrizian, M. Modulated release of OP-1 and enhanced preosteoblast differentiation using a core-shell nanoparticulate system. *J. Biomed. Mater. Res. Part A* **2009**, *91*, 919–928. [CrossRef] [PubMed]

60. Joo, V.; Ramasamy, T.; Haidar, Z.S. A novel self-assembled liposome-based polymeric hydrogel for cranio-maxillofacial applications: Preliminary findings. *Polymers* **2011**, *3*, 967–974. [CrossRef]

61. Li, Z.; Ramay, H.R.; Hauch, K.D.; Xiao, D.; Zhang, M. Chitosan-alginate hybrid scaffolds for bone tissue engineering. *Biomaterials* **2005**, *26*, 3919–3928. [CrossRef] [PubMed]

62. Kong, L.; Gao, Y.; Lu, G.; Gong, Y.; Zhao, N.; Zhang, X. A study on the bioactivity of chitosan/nano-hydroxyapatite composite scaffolds for bone tissue engineering. *Eur. Polym. J.* **2006**, *42*, 3171–3179. [CrossRef]

63. Ito, M. In vitro properties of a chitosan-bonded hydroxyapatite bone-filling paste. *Biomaterials* **1991**, *12*, 41–45. [CrossRef]

64. Peter, M.; Binulal, N.S.; Nair, S.V.; Selvamurugan, N.; Tamura, H.; Jayakumar, R. Novel biodegradable chitosan-gelatin/nano-bioactive glass ceramic composite scaffolds for alveolar bone tissue engineering. *Chem. Eng. J.* **2010**, *158*, 353–361. [CrossRef]

65. Iqbal, H.; Ali, M.; Zeeshan, R.; Mutahir, Z.; Iqbal, F.; Nawaz, M.A.H.; Shahzadi, L.; Chaudhry, A.A.; Yar, M.; Luan, S.; et al. Chitosan/hydroxyapatite (HA)/hydroxypropylmethyl cellulose (HPMC) spongy scaffolds-synthesis and evaluation as potential alveolar bone substitutes. *Colloids Surf. B Biointerfaces* **2017**, *160*, 553–563. [CrossRef] [PubMed]

66. Vaca-Cornejo, F.; Reyes, H.; Jiménez, S.; Velázquez, R.; Jiménez, J. Pilot study using a chitosan-hydroxyapatite implant for guided alveolar bone growth in patients with chronic periodontitis. *J. Funct. Biomater.* **2017**, *8*, 29. [CrossRef] [PubMed]

67. Xu, S.; Chen, X.; Yang, X.; Zhang, L.; Yang, G.; Shao, H.; He, Y.; Gou, Z. Preparation and in vitro biological evaluation of octacalcium phosphate/bioactive glass-chitosan/alginate composite membranes potential for bone guided regeneration. *J. Nanosci. Nanotechnol.* **2016**, *16*, 5577–5585. [CrossRef] [PubMed]

68. Zhou, D.; Qi, C.; Chen, Y.X.; Zhu, Y.J.; Sun, T.W.; Chen, F.; Zhang, C.Q. Comparative study of porous hydroxyapatite/chitosan and whitlockite/chitosan scaffolds for bone regeneration in calvarial defects. *Int. J. Nanomed.* **2017**, *12*, 2673–2687. [CrossRef] [PubMed]

69. Lu, Y.; Li, M.; Li, L.; Wei, S.; Hu, X.; Wang, X.; Shan, G.; Zhang, Y.; Xia, H.; Yin, Q. High-activity chitosan/nano hydroxyapatite/zoledronic acid scaffolds for simultaneous tumor inhibition, bone repair and infection eradication. *Mater. Sci. Eng. C Mater. Biol. Appl.* **2018**, *81*, 225–233. [CrossRef] [PubMed]

70. Guzmań, R.; Nardecchia, S.; Gutíerrez, M.C.; Ferrer, M.L.; Ramos, V.; Del Monte, F.; Abarrategi, A.; López-Lacomba, J.L. Chitosan scaffolds containing calcium phosphate salts and rhBMP-2: In vitro and in vivo testing for bone tissue regeneration. *PLoS ONE* **2014**, *9*, e87149. [CrossRef] [PubMed]

71. Rinaudo, M. Chitin and chitosan: Properties and applications. *Prog. Polym. Sci.* **2006**, *31*, 603–632. [CrossRef]

72. Qasim, S.B.; Najeeb, S.; Delaine-Smith, R.M.; Rawlinson, A.; Ur Rehman, I. Potential of electrospun chitosan fibers as a surface layer in functionally graded GTR membrane for periodontal regeneration. *Dent. Mater.* **2017**, *33*, 71–83. [CrossRef] [PubMed]

73. Masoudi Rad, M.; Nouri Khorasani, S.; Ghasemi-Mobarakeh, L.; Prabhakaran, M.P.; Foroughi, M.R.; Kharaziha, M.; Saadatkish, N.; Ramakrishna, S. Fabrication and characterization of two-layered nanofibrous membrane for guided bone and tissue regeneration application. *Mater. Sci. Eng. C* **2017**, *80*, 75–87. [CrossRef] [PubMed]

74. Zhou, T.; Liu, X.; Sui, B.; Liu, C.; Mo, X.; Sun, J. Development of fish collagen/bioactive glass/chitosan composite nanofibers as a GTR/GBR membrane for inducing periodontal tissue regeneration. *Biomed. Mater.* **2017**, *12*, 055004. [CrossRef] [PubMed]

75. Tamburaci, S.; Tihminlioglu, F. Diatomite reinforced chitosan composite membrane as potential scaffold for guided bone regeneration. *Mater. Sci. Eng. C* **2017**, *80*, 222–231. [CrossRef] [PubMed]

76. Xu, Z.L.; Lei, Y.; Yin, W.J.; Chen, Y.X.; Ke, Q.F.; Guo, Y.P.; Zhang, C.Q. Enhanced antibacterial activity and osteoinductivity of Ag-loaded strontium hydroxyapatite/chitosan porous scaffolds for bone tissue engineering. *J. Mater. Chem. B* **2016**, *4*, 7919–7928. [CrossRef]

77. Masaeli, R.; Kashi, T.S.J.; Yao, W.; Khoshroo, K.; Tahriri, M.; Tayebi, L. Preparation of strontium-containing calcium phosphate cements for maxillofacial bone regeneration. *Dent. Mater.* **2016**, *32* (Suppl. 1), e49. [CrossRef]

78. Rojo, L.; Radley-Searle, S.; Fernandez-Gutierrez, M.; Rodriguez-Lorenzo, L.M.; Abradelo, C.; Deb, S.; San Roman, J. The synthesis and characterisation of strontium and calcium folates with potential osteogenic activity. *J. Mater. Chem. B* **2015**, *3*, 2708–2713. [CrossRef]

79. Martin-Del-Campo, M.; Rosales-Ibañez, R.; Alvarado, K.; Sampedro, J.G.; Garcia-Sepulveda, C.A.; Deb, S.; San Román, J.; Rojo, L. Strontium folate loaded biohybrid scaffolds seeded with dental pulp stem cells induce: In vivo bone regeneration in critical sized defects. *Biomater. Sci.* **2016**, *4*, 1596–1604. [CrossRef] [PubMed]

80. Shive, M.S.; Stanish, W.D.; McCormack, R.; Forriol, F.; Mohtadi, N.; Pelet, S.; Desnoyers, J.; Méthot, S.; Vehik, K.; Restrepo, A. Bst-cargel® treatment maintains cartilage repair superiority over microfracture at 5 years in a multicenter randomized controlled trial. *Cartilage* **2015**, *6*, 62–72. [CrossRef] [PubMed]

81. Woodruff, M.A.; Hutmacher, D.W. The return of a forgotten polymer-polycaprolactone in the 21st century. *Prog. Polym. Sci.* **2010**, *35*, 1217–1256. [CrossRef]

82. Mosmann, T. Rapid colorimetric assay for cellular growth and survival: Application to proliferation and cytotoxicity assays. *J. Immunol. Methods* **1983**, *65*, 55–63. [CrossRef]

83. Nakayama, G.R.; Caton, M.C.; Nova, M.P.; Parandoosh, Z. Assessment of the alamar blue assay for cellular growth and viability in vitro. *J. Immunol. Methods* **1997**, *204*, 205–208. [CrossRef]

84. Singer, V.L.; Jones, L.J.; Yue, S.T.; Haugland, R.P. Characterization of picogreen reagent and development of a fluorescence- based solution assay for double-stranded DNA quantitation. *Anal. Biochem.* **1997**, *249*, 228–238. [CrossRef] [PubMed]

85. Magnusson, P.; Larsson, L.; Magnusson, M.; Davie, M.W.J.; Sharp, C.A. Isoforms of bone alkaline phosphatase: Characterization and origin in human trabecular and cortical bone. *J. Bone Miner. Res.* **1999**, *14*, 1926–1933. [CrossRef] [PubMed]

86. Méndez, J.A.; Aguilar, M.R.; Abraham, G.A.; Vázquez, B.; Dalby, M.; Di Silvio, L.; San Román, J. New acrylic bone cements conjugated to vitamin E: Curing parameters, properties, and biocompatibility. *J. Biomed. Mater. Res.* **2002**, *62*, 299–307. [CrossRef] [PubMed]

87. Gierszewska, M.; Ostrowska-Czubenko, J. Chitosan-based membranes with different ionic crosslinking density for pharmaceutical and industrial applications. *Carbohydr. Polym.* **2016**, *153*, 501–511. [CrossRef] [PubMed]

88. Fernández-Quiroz, D.; González-Gómez, Á.; Lizardi-Mendoza, J.; Vázquez-Lasa, B.; Goycoolea, F.M.; San Román, J.; Argüelles-Monal, W.M. Effect of the molecular architecture on the thermosensitive properties of chitosan-*g*-poly(*N*-vinylcaprolactam). *Carbohydr. Polym.* **2015**, *134*, 92–101. [CrossRef] [PubMed]

89. Elzein, T.; Nasser-Eddine, M.; Delaite, C.; Bistac, S.; Dumas, P. FTIR study of polycaprolactone chain organization at interfaces. *J. Colloids Interface Sci.* **2004**, *273*, 381–387. [CrossRef] [PubMed]

90. Fernández-Gutiérrez, M.; Bossio, O.; Gómez-Mascaraque, L.G.; Vázquez-Lasa, B.; Román, J.S. Bioactive chitosan nanoparticles loaded with retinyl palmitate: A simple route using ionotropic gelation. *Macromol. Chem. Phys.* **2015**, *216*, 1321–1332. [CrossRef]

91. Persenaire, O.; Alexandre, M.; Degée, P.; Dubois, P. Mechanisms and kinetics of thermal degradation of poly(ε-caprolactone). *Biomacromolecules* **2001**, *2*, 288–294. [CrossRef] [PubMed]

92. International Organization for Standardization. *Biological Evaluation of Medical Devices—Part 5: Tests for in Vitro Cytotoxicity*, 3rd ed.; ISO: Geneva, Switzerland, 2009.

93. Ansar Ahmed, S.; Gogal, R.M., Jr.; Walsh, J.E. A new rapid and simple non-radioactive assay to monitor and determine the proliferation of lymphocytes: An alternative to [3H] thymidine incorporation assay. *J. Immunol. Methods* **1994**, *170*, 211–224. [CrossRef]

94. Sacco, P.; Borgogna, M.; Travan, A.; Marsich, E.; Paoletti, S.; Asaro, F.; Grassi, M.; Donati, I. Polysaccharide-based networks from homogeneous chitosan-tripolyphosphate hydrogels: Synthesis and characterization. *Biomacromolecules* **2014**, *15*, 3396–3405. [CrossRef] [PubMed]

95. Lima, H.A.; Lia, F.M.V.; Ramdayal, S. Preparation and characterization of chitosan-insulin-tripolyphosphate membrane for controlled drug release: Effect of cross linking agent. *J. Biomater. Nanobiotechnol.* **2014**, *5*, 211–219. [CrossRef]

96. Sarasam, A.; Madihally, S.V. Characterization of chitosan-polycaprolactone blends for tissue engineering applications. *Biomaterials* **2005**, *26*, 5500–5508. [CrossRef] [PubMed]

97. Sarasam, A.; Krishnaswamy, R.K.; Madihally, S.V. Blending chitosan with polycaprolactone: Effects on physicochemical and antibacterial properties. *Biomacromolecules* **2006**, *7*, 1131–1138. [CrossRef] [PubMed]

98. Sundaram, M.N.; Sowmya, S.; Deepthi, S.; Bumgardener, J.D.; Jayakumar, R. Bilayered construct for simultaneous regeneration of alveolar bone and periodontal ligament. *J. Biomed. Mater. Res. Part B Appl. Biomater.* **2016**, *104*, 761–770. [CrossRef] [PubMed]

99. Cui, Z.; Lin, L.; Si, J.; Luo, Y.; Wang, Q.; Lin, Y.; Wang, X.; Chen, W. Fabrication and characterization of chitosan/ogp coated porous poly(ε-caprolactone) scaffold for bone tissue engineering. *J. Biomater. Sci. Polym. Ed.* **2017**, *28*, 826–845. [CrossRef] [PubMed]

100. Ghaee, A.; Nourmohammadi, J.; Danesh, P. Novel chitosan-sulfonated chitosan-polycaprolactone-calcium phosphate nanocomposite scaffold. *Carbohydr. Polym.* **2017**, *157*, 695–703. [CrossRef] [PubMed]

101. Ozkan, O.; Turkoglu Sasmazel, H. Hybrid polymeric scaffolds prepared by micro and macro approaches. *Int. J. Polym. Mater. Polym. Biomater.* **2017**, *66*, 853–860. [CrossRef]

102. Ren, J.; Blackwood, K.A.; Doustgani, A.; Poh, P.P.; Steck, R.; Stevens, M.M.; Woodruff, M.A. Melt-electrospun polycaprolactone strontium-substituted bioactive glass scaffolds for bone regeneration. *J. Biomed. Mater. Res. Part A* **2014**, *102*, 3140–3153. [CrossRef] [PubMed]

103. García Cruz, D.M.; Coutinho, D.F.; Martinez, E.C.; Mano, J.F.; Ribelles, J.L.G.; Sánchez, M.S. Blending polysaccharides with biodegradable polymers. Ii. Structure and biological response of chitosan/polycaprolactone blends. *J. Biomed. Mater. Res.* **2008**, *87*, 544–554. [CrossRef] [PubMed]

104. García Cruz, D.M.; Coutinho, D.F.; Mano, J.F.; Gómez Ribelles, J.L.; Salmerón Sánchez, M. Physical interactions in macroporous scaffolds based on poly(ε-caprolactone)/chitosan semi-interpenetrating polymer networks. *Polymer* **2009**, *50*, 2058–2064. [CrossRef]

105. Yin, N.; Zhang, Z. Bone regeneration in the hard palate after cleft palate surgery. *Plast. Reconstr. Surg.* **2005**, *115*, 1239–1244. [CrossRef] [PubMed]

106. Sun, H.; Mei, L.; Song, C.; Cui, X.; Wang, P. The in vivo degradation, absorption and excretion of PCL-based implant. *Biomaterials* **2006**, *27*, 1735–1740. [CrossRef] [PubMed]

107. Anselme, K. Osteoblast adhesion on biomaterials. *Biomaterials* **2000**, *21*, 667–681. [CrossRef]

108. Qiu, K.; Zhao, X.J.; Wan, C.X.; Zhao, C.S.; Chen, Y.W. Effect of strontium ions on the growth of ROS17/2.8 cells on porous calcium polyphosphate scaffolds. *Biomaterials* **2006**, *27*, 1277–1286. [CrossRef] [PubMed]

109. Cao, L.; Yu, Y.; Wang, J.; Werkmeister, J.A.; McLean, K.M.; Liu, C. 2-N, 6-O-sulfated chitosan-assisted BMP-2 immobilization of PCL scaffolds for enhanced osteoinduction. *Mater. Sci. Eng. C* **2017**, *74*, 298–306. [CrossRef] [PubMed]

110. Stein, G.; Lian, J. *Molecular Mechanisms Mediating Proliferation/Differentiation Interrelationships During Progressive Development of the Osteoblast Phenotype*; Node, M., Ed.; Academic Press: London, UK, 1993; pp. 424–442.

111. Shalumon, K.T.; Sowmya, S.; Sathish, D.; Chennazhi, K.P.; Nair, S.V.; Jayakumar, R. Effect of incorporation of nanoscale bioactive glass and hydroxyapatite in PCL/chitosan nanofibers for bone and periodontal tissue engineering. *J. Biomed. Nanotechnol.* **2013**, *9*, 430–440. [CrossRef] [PubMed]

112. Su, W.T.; Chou, W.L.; Chou, C.M. Osteoblastic differentiation of stem cells from human exfoliated deciduous teeth induced by thermosensitive hydrogels with strontium phosphate. *Mater. Sci. Eng. C* **2015**, *52*, 46–53. [CrossRef] [PubMed]

113. Azab, A.K.; Doviner, V.; Orkin, B.; Kleinstern, J.; Srebnik, M.; Nissan, A.; Rubinstein, A. Biocompatibility evaluation of crosslinked chitosan hydrogels after subcutaneous and intraperitoneal implantation in the rat. *J. Biomed. Mater. Res. Part A* **2007**, *83*, 414–422. [CrossRef] [PubMed]

114. Anderson, J.M.; Rodriguez, A.; Chang, D.T. Foreign body reaction to biomaterials. *Semin. Immunol.* **2008**, *20*, 86–100. [CrossRef] [PubMed]

115. Meinel, L.; Hofmann, S.; Karageorgiou, V.; Kirker-Head, C.; McCool, J.; Gronowicz, G.; Zichner, L.; Langer, R.; Vunjak-Novakovic, G.; Kaplan, D.L. The inflammatory responses to silk films in vitro and in vivo. *Biomaterials* **2005**, *26*, 147–155. [CrossRef] [PubMed]

116. Bavariya, A.J.; Andrew Norowski, P., Jr.; Mark Anderson, K.; Adatrow, P.C.; Garcia-Godoy, F.; Stein, S.H.; Bumgardner, J.D. Evaluation of biocompatibility and degradation of chitosan nanofiber membrane crosslinked with genipin. *J. Biomed. Mater. Res.* **2014**, *102*, 1084–1092. [CrossRef] [PubMed]

117. Simion, M.; Scarano, A.; Gionso, L.; Piattelli, A. Guided bone regeneration using resorbable and nonresorbable membranes: A comparative histologic study in humans. *Int. J. Oral Maxillofac. Implants* **1996**, *11*, 735–742. [PubMed]
118. Hämmerle, C.H.F.; Jung, R.E. Bone augmentation by means of barrier membranes. *Periodontology* **2003**, *33*, 36–53. [CrossRef]

polymers

MDPI

Article

Recyclable Heterogeneous Chitosan Supported Copper Catalyst for Silyl Conjugate Addition to α,β-Unsaturated Acceptors in Water

Lei Zhu [1,*] , Bojie Li [1], Shan Wang [1], Wei Wang [1], Liansheng Wang [1], Liang Ding [2,*] and Caiqin Qin [1,*]

[1] School of Chemistry and Materials Science, Hubei Engineering University, Hubei Collaborative Innovation Center of Conversion and Utilization for Biomass Resources, Xiaogan 432000, China; libojie0614@hotmail.com (B.L.); smallcoral@live.cn (S.W.); 15327403696@163.com (W.W.); wangls@hbeu.edu.cn (L.W.)

[2] Department of Polymer and Composite Material, School of Materials Engineering, Yancheng Institute of Technology, Yancheng 224051, China

* Correspondence: zhuleio@mail.ustc.edu.cn or Lei.zhu@hbeu.edu.cn (L.Z.); dl1984911@ycit.edu.cn (L.D.); qincq@hbeu.edu.cn (C.Q.); Tel.: +86-712-2345464 (L.Z.)

Received: 7 February 2018; Accepted: 26 March 2018; Published: 1 April 2018

Abstract: The first example of an environmentally-benign chitosan supported copper catalyzed conjugate silylation of α,β-unsaturated acceptors was accomplished in water under mild conditions. This protocol provides an efficient pathway to achieve an important class of β-silyl carbonyl compounds and the desired products were obtained in good to excellent yields. Gram-scale synthesis and easy transformation of obtained β-silyl products were also been demonstrated. Remarkably, this chitosan supported copper catalyst can be easily recycled and reused six times without any significant decrease of catalytic activity. The advantages of this newly developed method include operational simplicity, good functional group tolerance, scale-up ability, ready availability, and easy recyclability of catalyst.

Keywords: chitosan supported copper; heterogeneous catalyst; organosilicon compound; easily recyclable

1. Introduction

The development of methods for the synthesis of organosilicon compounds has attracted intense attention because of their myriad uses in organic synthesis [1,2] as well as their potential applications in materials science [3] and biological chemistry [4,5]. Traditionally, most β-silyl carbonyl compounds are prepared via conjugate addition process by using stoichiometric amounts of silyl metal reagents [6–8]. Besides, the methods that utilized a catalytic amount of metal in the presence of silylzincates [9–11] or silylcuprates [12] were also developed for further improvement of efficiency. In recent years, the utilization of Suginome's silylboron reagent [13], dimethylphenylsilyl pinacolatoboronate (1), provided an alternative and effective strategy to directly install the dimethylphenylsilyl moiety to α,β-unsaturated acceptors. Although, Rh(I) [14–17], Cu(I) [18–20], and metal-free NHC [21] catalysts were successfully applied for activating the Si–B bond of Suginome's reagent, Cu(II)-catalysis is still in high demand because its protocol is much more convenient with lower cost. Until now, only a few methods based on Cu(II) catalysis have been disclosed for this transformation. For example, our group [22] and Santos [23] independently reported that $Cu_2(OH)_2CO_3$ and $CuSO_4$ respectively could catalyze the silylation of α,β-unsaturated compounds in the presence of a base. Kobayashi and co-workers [24] carried out this reaction in an enantioselective manner by using a chiral complex obtained from Cu(acac)$_2$ and a chiral 2,2'-bipyridine ligand. In consideration of limited examples of Cu(II) catalysis and environmental impact, it is necessary to develop a green, efficient, and much milder strategy using Cu(II) catalysis to obtain functionalized organosilicon compounds bearing a carbonyl moiety. Furthermore, the recycling and reuse of Cu(II) catalyst still remains a challenge due to the deactivation or decomposition of non-immobilized Cu(II) salts.

Transition metal catalysts immobilized on a heterogeneous support played an important role in discovering unique reactivities and selectivities different from homogeneous systems [25,26]. The observed heterogeneous catalyst displays several advantages such as easy isolation, operational simplicity, environmental compatibility and remarkable reusability. Various supports—including magnetic materials [27], zeolites [28,29], silica [30], and polymers [31]—have been adopted for the immobilization of transition metals. Recently, chitosan (CS) has received extensive interest due to its non-toxicity, biodegradability, and reasonable ability of chelation [32,33]. It was proven that chitosan supported metal complexes were efficient to catalyze C–C [34–39], C–N [40], C–O [41], and C–S [42] bonds forming reactions. Meanwhile, we reported that chitosan supported copper is capable of catalyzing the transfer of pinacolatoboron moiety to α,β-unsaturated carbonyls, leading to the formation of C–B bond [43]. However, to the best of our knowledge, very few examples describing the C–Si formation in chitosan supported heterogeneous system were reported previously and the topic still represents a challenge.

With our continuous efforts in exploring applications of chitosan supported metal catalysts [22,43], we were interested in developing a green and mild protocol for synthesis of useful organosilicon compounds. Hence, we herein report a simple and easily available chitosan supported Cu(II) material as a highly reactive and recyclable catalyst for the desired β-silyl conjugate additions of α,β-unsaturated acceptors.

2. Materials and Methods

2.1. Materials

Chitosan (degree of acetylation = 5%; M_W: 10,000–50,000, determined by GPC) was purchased from Aladdin (Shanghai, China), dimethylphenylsilyl pinacolatoboronate (CAS: 185990-03-8) was purchased from Energy Chemical (Shanghai, China) and Cu(II) salts were purchased from J&K (Beijing, China). All α,β-unsaturated acceptors, bases were obtained commercially from Energy Chemical (Shanghai, China) and used without further purification. Chitosan supported copper catalysts CS@CuSO$_4$, CS@Cu(OAc)$_2$, and CS@Cu(acac)$_2$ were prepared according to the procedures reported [42].

2.2. Analytical Methods

Nuclear magnetic resonance (NMR) spectra were recorded on a Bruker Avance III 600 MHz spectrometer (Karlsruhe, Germany), operating at 600 for ^{1}H and 150 MHz for ^{13}C NMR in CDCl$_3$ unless otherwise noted. CDCl$_3$ is served as the internal standard (δ = 7.26 ppm) for ^{1}H NMR and (δ = 77.0 ppm) for ^{13}C NMR. Data for ^{1}H-NMR is reported as follows: chemical shift (ppm, scale), multiplicity (s = singlet, d = doublet, t = triplet, q = quartet, m = multiplet and/or multiplet resonances, br = broad), coupling constant (Hz), and integration. Data for ^{13}C NMR is reported in terms of chemical shift (ppm, scale), multiplicity, and coupling constant (Hz). Flash column chromatographic purification of products was accomplished using forced-flow chromatography on silica gel (200–300 mesh). The weight percentage and metal leaching of copper were determined by inductively coupled plasma-optical emission spectroscopy (ICP-OES) (PerkinElmer, Waltham, MA, USA) analysis. The copper loading of CS@CuSO$_4$, CS@Cu(OAc)$_2$, and CS@Cu(acac)$_2$ were found to be 1.85, 1.42, and 1.50 mmol/g, respectively.

2.3. General Procedure for Preparation of Chitosan Supported Copper Catalyst

Chitosan (5 g) was suspended in 100 mL of water. Copper salts (1 g) was added to this suspension and the mixture was stirred for 3 h. The catalyst was separated using a centrifuge (5000 rpm, 10 min), and dried under vacuum at 50 °C.

2.4. General Procedure for the Sample Preparation for ICP Analysis to Determine Catalyst Loading

Chitosan supported copper catalyst (~20 mg) was placed in a clean test tube and heated with H_2SO_4 (1 mL) at 200 °C. After 30 min, several drops of concentrated HNO_3 were added carefully and the tube was shaken occasionally. HNO_3 was continuously added until a clear solution was obtained and excess amount of HNO_3 was allowed to evaporate under heating. After the solution was cooled to room temperature, 1 mL of aqua regia was added carefully. Effervescence of gas was observed and the solution become clearer. The solution was then transferred to a volumetric flask and made up to 50 mL with water which was submitted for ICP analysis.

2.5. General Procedure for the Sample Preparation for ICP Analysis to Determine Metal Leaching

After the reaction was finished, the reaction mixture was filtered. The filtrate obtained was concentrated and diluted with 10 mL of THF. Then 50% v/v of the crude THF solution (5 mL) was then passed through a membrane filter (0.25 or 0.45 μm) into a clean test tube. After evaporation of solvent, the solid obtained in the test tube was heated to 200 °C and 1.0 mL of concentrated H_2SO_4 was added. Following similar procedure described above, concentrated HNO_3 were added at regular intervals until the resulting solution was clear. After the solution was cooled to room temperature, 1 mL of aqua regia was added carefully. Effervescence of gas was observed and the solution become clearer. The solution was then transferred to a volumetric flask and made up to 50 mL with water which was submitted for ICP analysis.

2.6. General Procedure for CS@Cu-Catalyzed Silylation of α,β-Unsaturated Acceptors in Water

CS@Cu(acac)$_2$ (10.0 mg, 5 mol % Cu loading) and 4-picoline (3.3 mg, 6 mol %) were mixed in water (2 mL). The mixture was stirred for 1 h at room temperature, followed by successive addition of substrate (**2**) (0.3 mmol) and PhMe$_2$Si-B(pin) (**1**) (94.4 mg, 0.36 mmol). After stirring for 12 h at room temperature, the reaction mixture was filtered and the filtrate was extracted with EtOAc (20 mL × 3). Combined organic layers were dried over anhydrous Na$_2$SO$_4$. After being concentrated under reduced pressure, the crude mixture was purified by flash column chromatography (200–300 mesh silica gel, petroleum ether/EtOAc = 10:1~20:1) to afford the desired product **3**.

2.7. General Procedure for Gram-Scale Synthesis of 3a

CS@Cu(acac)$_2$ (150 mg, 5 mol % Cu loading) and 4-picoline (49.5 mg, 6 mol %) were mixed in water (30 mL). The mixture was stirred for 1 h at room temperature, followed by successive addition of chalcone (**2a**) (0.94 g, 4.5 mmol) and PhMe$_2$Si-B(pin) (**1**) (1.42 g, 5.4 mmol). After stirring for 12 h at room temperature, the reaction mixture was filtered and the filtrate was extracted with EtOAc (30 mL × 3). Combined organic layers were dried over anhydrous Na$_2$SO$_4$. After being concentrated under reduced pressure, the crude mixture was purified by flash column chromatography (200–300 mesh silica gel, petroleum ether/EtOAc = 10:1) to afford the desired product **3a**.

2.8. Recycling and Reuse of CS@Cu Catalyst

In order to demonstrate the recyclability of CS@Cu(acac)$_2$ catalyst, the silyl conjugate addition was repeated six times with the same sample. The initial amount of catalyst was 50.0 mg (5 mol % Cu loading). Reactions were carried out for 12 h. After the reaction, catalyst was filtered off, washed by EtOAc and water, and then dried for 4 h at 60 °C before next run. The recovery rate of the catalyst at the time of reusing is about 92%.

3. Results and Discussion

For the initial investigation, we tested the silyl conjugate addition of $PhMe_2SiB(pin)$ (**1**) to chalcone (**2a**) catalyzed by an array of chitosan supported copper catalysts in water (Table 1, entries 1–3). To our delight, $CS@CuSO_4$, $CS@Cu(OAc)_2$, and $CS@Cu(acac)_2$ were all able to catalyze this transformation to obtain desired β-silyl product in 30%, 22%, and 41% yield respectively. Encouraged by this observation, we turned to screen different bases, including inorganic bases and a series of pyridine derivatives to further improve the conversion (Table 1, entries 4–9). It was found that the reactivity obviously increased in a more basic aqueous solution when sodium carbonate or potassium carbonate were used as base (Table 1, entries 4 and 5). However, 2,2'-bipyridine (**B1**) and its analog (**B2**) resulted in no effect to the reaction system, probably due to their poor dispersability in water (Table 1, entries 6 and 7). Furthermore, we had previously disclosed that 2,2'-bipyridine was effective to activate Si-B bond in the presence of a non-ionic surfactant [22]. Gratifyingly, significant improvements were achieved when pyridine (**B3**) and 4-picoline (**B4**) were used as additives (Table 1, entries 8 and 9). 4-Picoline is favorable for this reaction because it may perform the roles of the ligand for copper and Brønsted base for activating water molecules [23]. In the absence of catalyst, the reaction did not proceed at all (Table 1, entry 10). Subsequently, several aprotic solvents were examined (Table 1, entries 11–14), upon using dichloromethane, tetrahydrofuran, diethylether and toluene as solvents, product **3a** was not observed due to the lack of proton source. Protic solvents such as methanol and ethanol only resulted in very low conversion because of their poor ability to provide proton (Table 1, entries 15 and 16). Notably, it was feasible to reduce the catalyst loading to half without compromising on yield (Table 1, entry 17). In addition, the reaction atmosphere did not affect the reactivity as illustrated (Table 1, entry 18). Therefore, the optimized reaction conditions were determined to run the reaction at room temperature in H_2O with 6 mol % 4-picoline, using $CS@Cu(acac)_2$ as catalyst (Table 1, entry 9).

Table 1. Optimization of reaction conditions [a].

Entry	Catalyst	Base	Solvent	Yield (%) [b]
1	$CS@CuSO_4$	none	H_2O	30
2	$CS@Cu(OAc)_2$	none	H_2O	22
3	$CS@Cu(acac)_2$	none	H_2O	41
4	$CS@Cu(acac)_2$	Na_2CO_3	H_2O	50
5	$CS@Cu(acac)_2$	K_2CO_3	H_2O	65
6	$CS@Cu(acac)_2$	B1	H_2O	40
7	$CS@Cu(acac)_2$	B2	H_2O	38
8	$CS@Cu(acac)_2$	B3	H_2O	86
9	$CS@Cu(acac)_2$	B4	H_2O	92
10	-	B4	H_2O	NR
11	$CS@Cu(acac)_2$	B4	DCM	NR
12	$CS@Cu(acac)_2$	B4	THF	NR
13	$CS@Cu(acac)_2$	B4	Et_2O	NR
14	$CS@Cu(acac)_2$	B4	Toluene	NR
15	$CS@Cu(acac)_2$	B4	MeOH	25
16	$CS@Cu(acac)_2$	B4	EtOH	8
17 [c]	$CS@Cu(acac)_2$	B4	H_2O	89
18 [d]	$CS@Cu(acac)_2$	B4	H_2O	91

[a] Reaction conditions: substrate **2** (0.3 mmol), Me_2PhSi-B(pin) **1** (1.2 equiv.), catalyst (5 mol % Cu loading), base (6 mol %), H_2O (2 mL), room temperature, air, 12 h; [b] Isolated yield of product; [c] 2.5 mol % Cu loading was used; [d] Performed under Ar atomosphere.

With the optimized conditions in hand, the substrate scope of α,β-unsaturated acceptors was surveyed and the results were summarized in Figure 1. Chalcone derivatives bearing a chloro-substituent at *ortho*- or *para*-position both proceeded smoothly to give corresponding products in good yields (**3b,3c**) (**3b–3q** are in Supplementary Materials). Electron-donating group as well as the electron-withdrawing one was tolerated in this kind of silylation process (**3d**). Next, di-substituted chalcones were also found to be suitable substrates besides mono-substituted derivatives (**3e,3f**). Silylation of α,β-unsaturated ketones having aromatic moieties on one side while aliphatic ones on the other, all performed well under the optimized conditions (**3g,3h**), even with sterically congested substrate (**3i**). It is worthy to note that great improvements were achieved for methyl- and ethyl-esters by using our newly developed strategy, compared to the previous report (**3j,3k**) [23]. Not only acyclic α,β-unsaturated carbonyls, but also cyclic ketones were applicable and exhibited comparable reactivities (**3l,3m**). Remarkably, lactones which are common motifs of many natural products could also be reacted with **1** to yield β-silyl products in 75% and 79%, respectively (**3n,3o**). Notably, 1,4-addition products were obtained exclusively without the formation of 1,6-addition byproducts when α,β,γ,δ-unsaturated dienones were used as substrates (**3p,3q**). The allylsilane products **3p** and **3q** could be further employed as a useful synthon in the Hosomi–Sakurai reaction [44]. Therefore, it was demonstrated that a wide range of α,β-unsaturated acceptors could be silylated effectively which were catalyzed by chitosan supported copper catalyst with 4-picoline as base.

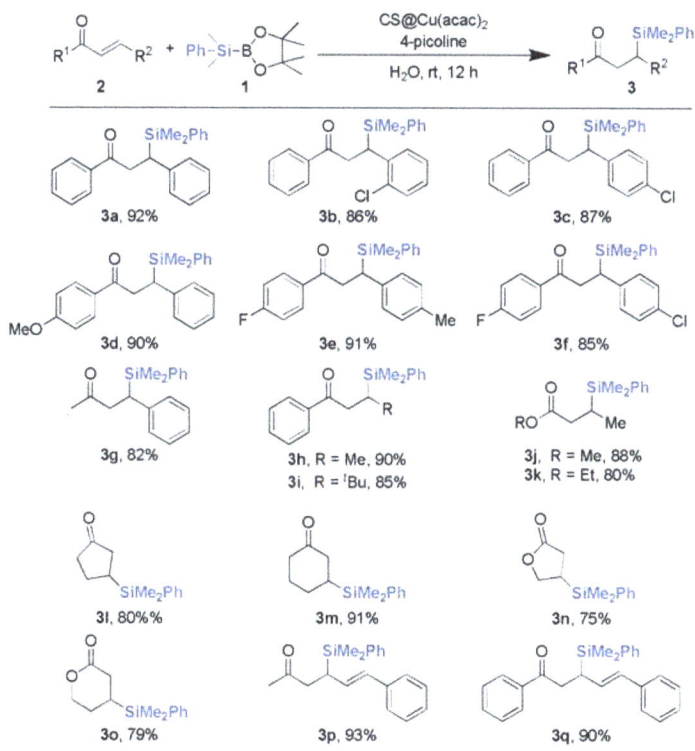

Figure 1. Substrate scope of α,β-unsaturated acceptors [a,b]. [a] Reaction conditions: substrate **2** (0.3 mmol), Me$_2$PhSi-B(pin) **1** (1.2 equiv.), CS@Cu(acac)$_2$ (5 mol % Cu loading), 4-picoline (6 mol %), H$_2$O (2 mL), room temperature, air, 12 h; [b] Isolated yields were listed.

For the purpose of practical application, we further carried out this reaction in a gram scale. As shown in Scheme 1a, β-dimethylphenylsilyl substituted chalcone (**3a**) (in Supplementary Materials) was successfully synthesized in 89% yield under the standard conditions. Next, when this reaction was performed in deuterium oxide instead of water, about 50% deuterium was observed in α-methylene, as revealed by ^{1}H NMR spectrum (Scheme 1b). This result indicated that water performs not only as a solvent but also as an important proton source for the protonation step of the whole catalytic cycle. To be mentioned, using water in this silyl conjugate addition made our method green and environmental friendly, because water is non-toxic, cheap, and non-flammable unlike organic solvents. In order to confirm the utility of obtained β-silyl products, further conversion of **3a** by Fleming–Tamao oxidation [45] easily produced corresponding β-hydroxyl compound **4a** (Scheme 1c). This process supplied a simple and efficient protocol to introduce hydroxyl group into complicated structure of natural products and biological active compounds.

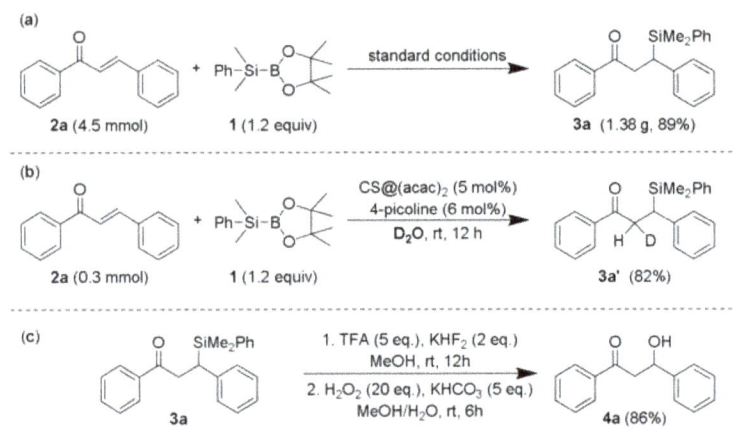

Scheme 1. (**a**) Gram scale synthesis β-silyl compound; (**b**) reaction in deuterium oxide; (**c**) further conversion of β-silyl products to β-hydroxyl compounds.

Recovery and recycling of heterogeneous catalyst are usually crucial considerations for transition metal-catalyzed reactions, from the aspects of economy and sustainability. Thus, the recyclability of chitosan supported copper catalyst was evaluated in the reaction of chalcone (**2a**) with **1** as shown in Figure 2. In each cycle, recovered CS@Cu(acac)$_2$ was treated with another new substrate for the next run under the optimized conditions. Remarkably, the catalyst still remained catalytically active after being reused six times. To date, there has only been one report about the recycling of catalyst for just two runs in similar silylation of cyclopentenone, along with apparent decline of reactivity [24]. In addition, we halted the reaction after six hours when about 60% conversion of starting material was received. The catalyst was then removed by filtration at the reaction temperature and the residue reacted for the rest time. It was found that n additional product was formed in the absence of catalyst. Finally, ICP analysis of the filtered aqueous solution after reaction showed no detectable leaching of copper. Both the filtration experiment and metal leaching test strongly suggested chitosan supported copper catalysis is heterogeneous in nature.

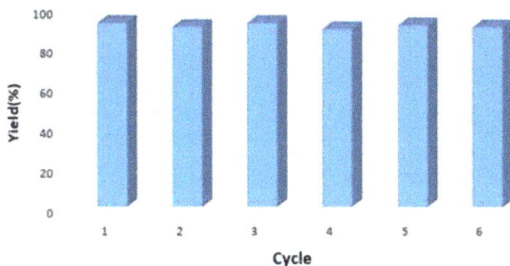

Figure 2. Recycling of chitosan supported copper catalyst.

4. Conclusions

In summary, we have prepared a heterogeneous chitosan supported copper catalyst which is highly active to catalyze the silyl conjugate addition of α,β-unsaturated acceptors with Suginome's silylboron reagent. The catalytic reaction was performed in good to excellent yields in water under mild conditions. Various substituted acyclic and cyclic α,β-unsaturated ketones, esters proceeded well under the optimized conditions. Besides, α,β,γ,δ-unsaturated acyclic dienones could also be silylated regioselectively to give corresponding 1,4-addition products. Water possibly plays a predominant role in accelerating the protonation step to achieve high yields. Gram scale synthesis and conversion to β-hydroxy compound made this method more versatile. Remarkably, the catalyst could be easily recovered and recycled six cycles without any significant decrease of reactivity. As a development towards the application of chitosan, our method stands a chance over the previously reported methods in terms of substrate scope, operational simplicity, and recyclability of catalyst.

Supplementary Materials: The following are available online at http://www.mdpi.com/2073-4360/10/4/385/s1, Characterization data, and spectra for the ^{1}H and ^{13}C NMR of compounds **3a–3q**.

Acknowledgments: The authors acknowledge the financial support from the National Natural Science Foundation of China (Nos. 21774029, 21774107, 31371750), the Natural Science Foundation of Hubei Province of China (No. 2016CFB104) and Hubei Engineering University.

Author Contributions: Lei Zhu and Caiqin Qin conceived and designed the experiments; Bojie Li, Shan Wang, and Wei Wang performed the experiments; Liang Ding analyzed the structure data and prepared the draft manuscript; Liansheng Wang contributed reagents/materials/analysis tools; Lei Zhu and Caiqing Qin reviewed and modified the manuscript.

Conflicts of Interest: The authors declare no conflict of interest.

References

1. Colvin, E.W. *Silicon in Organic Synthesis*, 1st ed.; Butterworths: London, UK, 1981; ISBN 0-408-10619-0.
2. Weber, W.P. *Silicon Reagents for Organic Synthesis*, 1st ed.; Springer: Berlin, Germany, 1983; ISBN 0-387-11675-3.
3. Allen, R.B.; Kochs, P.; Chandra, G. *Organosilicon Materials*, 1st ed.; Springer: Berlin, Germany, 1997; ISBN 978-3-540-68331-5.
4. Mutahi, M.W.; Nittoli, T.; Guo, L.; Sieburth, S.M. Silicon-based metalloprotease inhibitors: Synthesis and evaluation of silanol and silanediol peptide analogues as inhibitors of angiotensin-converting enzyme. *J. Am. Chem. Soc.* **2002**, *124*, 7363–7375. [CrossRef] [PubMed]
5. Sieburth, S.M.; Chen, C.A. Silanediol protease inhibitors: From conception to validation. *Eur. J. Org. Chem.* **2006**, *2006*, 311–322. [CrossRef]
6. Still, W.C. Conjugate addition of trimethysilyllithium, a preparation of 3-silyl ketones. *J. Org. Chem.* **1976**, *41*, 3063–3064. [CrossRef]

7. Fleming, I.; Lee, D. Conjugate addition of silyl groups to β-unsubstituted enones, & Si-to-OH conversion: A synthesis of (±)-lavandulol. *Tetrahedron Lett.* **1996**, *37*, 6929–6930. [CrossRef]
8. Dambacher, J.; Bergdahl, M. Empolying the simple monosilylcopper reagent, Li[PhMe₂SiCuI], in 1,4-addition reactions. *Chem. Commun.* **2003**, 144–145. [CrossRef]
9. Lipshutz, B.H.; Sclafani, J.A.; Takanami, T. Silyl cuprate couplings: Less silicon, accelerated yet catalytic in copper. *J. Am. Chem. Soc.* **1998**, *120*, 4021–4022. [CrossRef]
10. Oestreich, M.; Weiner, B. Copper-catalyzed conjugate addition of a bis(triorganosilyl) zinc and a methyl(triorganosilyl) magnesium. *Synlett* **2004**, 2139–2142. [CrossRef]
11. Auer, G.; Weiner, B.; Oestreich, M. Copper-free and copper-promoted conjugate addition reactions of bis(triorganosilyl) zincs and tris(triorganosilyl) zincates. *Synthesis* **2006**, 2113–2116. [CrossRef]
12. Weickgenannt, A.; Oestreich, M. Silicon- and tin-based cuprates: Now catalytic in copper! *Chem. Eur. J.* **2010**, *16*, 402–412. [CrossRef] [PubMed]
13. Suginome, M.; Matsuda, T.; Ito, Y. Convenient preparation of silylboranes. *Organometallics* **2000**, *19*, 4647–4649. [CrossRef]
14. Walter, C.; Auer, G.; Oestreich, M. Rhodium-catalyzed enantioselective conjugate silyl transfer: 1,4-addition of silyl boronic esters to cyclic enones and lactones. *Angew. Chem. Int. Ed.* **2006**, *45*, 5675–5677. [CrossRef] [PubMed]
15. Walter, C.; Oestreich, M. Catalytic asymmetric C-Si bond formation to acyclic α,β-unsaturated acceptors by Rhᴵ-catalyzed conjugate silyl transfer using a Si–B linkage. *Angew. Chem. Int. Ed.* **2008**, *47*, 3818–3820. [CrossRef] [PubMed]
16. Walter, C.; Fröhlich, R.; Oestreich, M. Rhodium(I)-catalyzed enantioselective 1,4-addition of nucleophilic silicon. *Tetrahedron* **2009**, *65*, 5513–5520. [CrossRef]
17. Hartmann, E.; Oestreich, M. Asymmetric conjugate silyl transfer in iterative catalytic sequences: Synthesis of the C7–C16 fragment of (+)-neopeltolide. *Angew. Chem. Int. Ed.* **2010**, *49*, 6195–6198. [CrossRef] [PubMed]
18. Lee, K.; Hoveyda, A.H. Enantioselective conjugate silyl additions to cyclic and acyclic unsaturated carbonyls catalyzed by Cu complexes of chiral N-heterocyclic carbenes. *J. Am. Chem. Soc.* **2010**, *132*, 2898–2900. [CrossRef] [PubMed]
19. Welle, A.; Petrignet, J.; Tinant, B.; Wouters, J.; Riant, O. Copper-catalyzed domino silylative aldol reaction leading to stereocontrolled chiral quaternary carbons. *Chem. Eur. J.* **2010**, *16*, 10980–10983. [CrossRef] [PubMed]
20. Ibrahem, I.; Santoro, S.; Himo, F.; Córdova, A. Enantioselective conjugate silyl additions to α,β-unsaturated aldehydes catalyzed by combination of transition metal and chiral amine catalysts. *Adv. Synth. Catal.* **2011**, *353*, 245–252. [CrossRef]
21. O'Brien, J.M.; Hoveyda, A.H. Metal-free catalytic C-Si bond formation in an aqueous medium. Enantioselective NHC-catalyzed silyl conjugate additions to cyclic and acyclic α,β-unsaturated carbonyls. *J. Am. Chem. Soc.* **2011**, *133*, 7712–7715. [CrossRef] [PubMed]
22. Wang, W.; Li, B.; Xiao, Z.; Yan, F.; Wei, P.; Wang, L.; Zhu, L. Basic copper carbonate-catalyzed silyl conjugate additions to α,β-unsaturated carbonyls in water. *J. Chin. Chem. Soc.* **2018**, *65*, 81–86. [CrossRef]
23. Calderone, J.A.; Santos, W.L. Copper(II)-catalyzed silyl conjugate addition to α,β-unsaturated conjugated compounds: Brønsted base-assisted activation of Si-B bond in water. *Org. Lett.* **2012**, *14*, 2090–2093. [CrossRef] [PubMed]
24. Kitanosono, T.; Zhu, L.; Liu, C.; Xu, P.; Kobayashi, S. An insoluble copper(II) acetylacetonate-chiral bipyridine complex that catalyzes asymmetric silyl conjugate addition in water. *J. Am. Chem. Soc.* **2015**, *137*, 15422–15425. [CrossRef] [PubMed]
25. Mizuno, N.; Misono, M. Heterogeneous catalysis. *Chem. Rev.* **1998**, *98*, 199–218. [CrossRef] [PubMed]
26. Yin, L.; Liebscher, J. Carbon-carbon coupling reactions catalyzed by heterogeneous palladium catalysts. *Chem. Rev.* **2007**, *107*, 133–173. [CrossRef] [PubMed]
27. Schultz, A.M.; Salvador, P.A.; Rohrer, G.S. Enhanced photochemical activity of α-Fe₂O₃ films supported on SrTiO₃ substrates under visible light illumination. *Chem. Commun.* **2012**, *48*, 2012–2014. [CrossRef] [PubMed]
28. Dey, R.; Screedhar, B.; Ranu, B.C. Molecular sieves-supported palladium(II) catalyst: Suzuki coupling of chloroarenes and an easy access to useful intermediates for the synthesis of irbesartan, iosartan and boscalid. *Tetrahedron* **2010**, *66*, 2301–2305. [CrossRef]

29. Zhu, Y.; Hua, Z.; Zhou, X.; Song, Y.; Gong, Y.; Zhou, J.; Zhao, J.; Shi, J. CTAB-templated mesoporous TS-1 zeolites as active catalysts in a desulfurization process: The decreased hydrophobicity is more favourable in thiophene oxidation. *RSC Adv.* **2013**, *3*, 4193–4198. [CrossRef]

30. Opanasenko, M.; Štěpnička, P.; Čejka, J. Heterogeneous Pd catalysts supported on silica matrices. *RSC Adv.* **2014**, *4*, 65137–65162. [CrossRef]

31. Zhang, J.; Han, D.; Zhang, H.; Chaker, M.; Zhao, Y.; Ma, D. In Situ recyclable gold nanoparticles using CO_2-switchable polymers for catalytic reduction of 4-nitrophenol. *Chem. Commun.* **2012**, *48*, 11510–11512. [CrossRef] [PubMed]

32. Kumar, M.N.V.R.; Muzzarelli, R.A.A.; Muzzarelli, C.; Sashiwa, H.; Domb, A.J. Chitosan chemistry and pharmaceutical perspectives. *Chem. Rev.* **2004**, *104*, 6017–6084. [CrossRef] [PubMed]

33. Kadib, A.E. Chitosan as a sustainable organocatalyst: A concise overview. *ChemSusChem* **2015**, *8*, 217–244. [CrossRef] [PubMed]

34. Hardy, J.J.E.; Hubert, S.; Macquarrie, D.J.; Wilson, A.J. Chitosan-based heterogeneous catalysts for Suzuki and Heck reactions. *Green Chem.* **2004**, *6*, 53–56. [CrossRef]

35. Kadib, A.E.; Molvinger, K.; Bousmina, M.; Brunel, D. Improving catalytic activity by synergic effect between base and acid pairs in hierarchically porous chitosan@titania nanoreactors. *Org. Lett.* **2010**, *12*, 948–951. [CrossRef] [PubMed]

36. Primo, A.; Quignard, F. Chitosan as efficient porous support for dispersion of highly active gold nanoparticles: Design of hybrid catalyst for carbon-carbon bond formation. *Chem. Commun.* **2010**, *46*, 5593–5595. [CrossRef] [PubMed]

37. Makhubela, B.C.E.; Jardine, A.; Smith, G.S. Rh(I) complexes supported on a biopolymer as recyclable and selective hydroformylation catalysts. *Green Chem.* **2012**, *14*, 338–347. [CrossRef]

38. Sk, M.P.; Jana, C.K.; Chattopadhyay, A. A gold-carbon nanoparticle composite as an efficient catalyst for homocoupling reaction. *Chem. Commun.* **2013**, *49*, 8235–8237. [CrossRef]

39. Shen, C.; Xu, J.; Ying, B.; Zhang, P. Heterogeneous chitosan@copper(II)-catalyzed remote trifluoromethylation of aminoquinolines with Langlois reagent by radical cross-coupling. *ChemCatChem* **2016**, *8*, 3560–3564. [CrossRef]

40. Baig, R.B.N.; Varma, R.S. Copper on chitosan: A recyclable heterogeneous catalyst for azide-alkyne cycloaddition reactions in water. *Green Chem.* **2013**, *15*, 1839–1843. [CrossRef]

41. Ying, B.; Xu, J.; Zhu, X.; Shen, C.; Zhang, P. Catalyst-controlled selectivity in the synthesis of C2- and C3-sulfonate esters from quinoline N-oxides and arylsulfonyl chlorides. *ChemCatChem* **2016**, *8*, 2604–2608. [CrossRef]

42. Shen, C.; Xu, J.; Yu, W.; Zhang, P. A highly active and easily recoverable chitosan@copper catalyst for the C-S coupling and its application in the synthesis of zolimidine. *Green Chem.* **2014**, *16*, 3007–3012. [CrossRef]

43. Xu, P.; Li, B.; Wang, L.; Qin, C.; Zhu, L. A green and recyclable chitosan supported catalyst for the borylation of α,β-unsaturated acceptors in water. *Catal. Commun.* **2016**, *86*, 23–26. [CrossRef]

44. Hosomi, A.; Sakurai, H.J. Chemistry of organosilicon compounds. 99. Conjugate addition of allylsilanes to .alpha,beta-enones. A New method of stereoselective introduction of the angular allyl group in fused cyclic alpha, beta-enones. *J. Am. Chem. Soc.* **1977**, *99*, 1673–1675. [CrossRef]

45. Fleming, I.; Barbero, A.; Walter, D. Stereochemical control in organic synthesis using silicon-containing compounds. *Chem. Rev.* **1997**, *97*, 2063–2192. [CrossRef] [PubMed]

MDPI

St. Alban-Anlage 66

4052 Basel

Switzerland

Tel. +41 61 683 77 34

Fax +41 61 302 89 18

www.mdpi.com

Polymers Editorial Office

E-mail: polymers@mdpi.com

www.mdpi.com/journal/polymers

www.ingramcontent.com/pod-product-compliance
Lightning Source LLC
Chambersburg PA
CBHW051706210326
41597CB00032B/5388